THERMAL ANALYSIS OF POLYMERS

THERMAL ANALYSIS OF POLYMERS
Fundamentals and Applications

EDITED BY

JOSEPH D. MENCZEL
Alcon Laboratories
Fort Worth TX

R. BRUCE PRIME
San Jose, CA

A JOHN WILEY & SONS, INC., PUBLICATION

About the cover: Image of an optoelectronics device in the middle circle on the cover reproduced with permission from CyOptics, Inc., Breinigsville, PA.

Copyright © 2009 by John Wiley & Sons, Inc. All rights reserved

Published by John Wiley & Sons, Inc., Hoboken, New Jersey
Published simultaneously in Canada

No part of this publication may be reproduced, stored in a retrieval system, or transmitted in any form or by any means, electronic, mechanical, photocopying, recording, scanning, or otherwise, except as permitted under Section 107 or 108 of the 1976 United States Copyright Act, without either the prior written permission of the Publisher, or authorization through payment of the appropriate per-copy fee to the Copyright Clearance Center, Inc., 222 Rosewood Drive, Danvers, MA 01923, (978) 750-8400, fax (978) 750-4470, or on the web at www.copyright.com. Requests to the Publisher for permission should be addressed to the Permissions Department, John Wiley & Sons, Inc., 111 River Street, Hoboken, NJ 07030, (201) 748-6011, fax (201) 748-6008, or online at http://www.wiley.com/go/permission.

Limit of Liability/Disclaimer of Warranty: While the publisher and author have used their best efforts in preparing this book, they make no representations or warranties with respect to the accuracy or completeness of the contents of this book and specifically disclaim any implied warranties of merchantability or fitness for a particular purpose. No warranty may be created or extended by sales representatives or written sales materials. The advice and strategies contained herein may not be suitable for your situation. You should consult with a professional where appropriate. Neither the publisher nor author shall be liable for any loss of profit or any other commercial damages, including but not limited to special, incidental, consequential, or other damages.

For general information on our other products and services or for technical support, please contact our Customer Care Department within the United States at (800) 762-2974, outside the United States at (317) 572-3993 or fax (317) 572-4002.

Wiley also publishes its books in a variety of electronic formats. Some content that appears in print may not be available in electronic formats. For more information about Wiley products, visit our web site at www.wiley.com.

Library of Congress Cataloging-in-Publication Data:

Thermal analysis of polymers: fundamentals and applications / edited by Joseph D. Menczel, R. Bruce Prime.
   p. cm
Includes bibliographical references and index.
ISBN 978-0-471-76917-0 (cloth)
1. Polymers–Analysis. 2. Thermal analysis. I. Menczel, Joseph D. II. Prime, R. Bruce.
QD139.P6.T445 2008
547'.7046—dc22

2008024101

Printed in the United States of America

10 9 8 7 6 5 4 3 2

CONTENTS

PREFACE ix

1 INTRODUCTION 1
Joseph D. Menczel, R. Bruce Prime and Patrick K. Gallagher

2 DIFFERENTIAL SCANNING CALORIMETRY (DSC) 7
Joseph D. Menczel, Lawrence Judovits, R. Bruce Prime, Harvey E. Bair, Mike Reading, and Steven Swier

    2.1. Introduction / 7
    2.2. Elements of Thermodynamics in DSC / 9
    2.3. The Basics of Differential Scanning Calorimetry / 18
    2.4. Purity Determination of Low-Molecular-Mass Compounds by DSC / 37
    2.5. Calibration of Differential Scanning Calorimeters / 41
    2.6. Measurement of Heat Capacity / 52
    2.7. Phase Transitions in Amorphous and Crystalline Polymers / 58
    2.8. Fibers / 115
    2.9. Films / 123
    2.10. Thermosets / 130
    2.11. Differential Photocalorimetry (DPC) / 154
    2.12. Fast-Scan DSC / 162
    2.13. Modulated Temperature Differential Scanning Calorimetry (MTDSC) / 168
    2.14. How to Perform DSC Measurements / 208
    2.15. Instrumentation / 217
        Appendix / 225
        Abbreviations / 225
        References / 229

3 THERMOGRAVIMETRIC ANALYSIS (TGA)   241
R. Bruce Prime, Harvey E. Bair, Sergey Vyazovkin, Patrick K. Gallagher, and Alan Riga

- 3.1. Introduction / 241
- 3.2. Background Principles and Measurement Modes / 242
- 3.3. Calibration and Reference Materials / 251
- 3.4. Measurements and Analyses / 256
- 3.5. Kinetics / 277
- 3.6. Selected Applications / 295
- 3.7. Instrumentation / 308
  - Appendix / 311
  - Abbreviations / 312
  - References / 314

4 THERMOMECHANICAL ANALYSIS (TMA) AND THERMODILATOMETRY (TD)   319
Harvey E. Bair, Ali E. Akinay, Joseph D. Menczel, R. Bruce Prime, and Michael Jaffe

- 4.1. Introduction / 319
- 4.2. Principles and Theory / 320
- 4.3. Instrumental / 326
- 4.4. Calibration / 332
- 4.5. How to Perform a TMA Experiment / 335
- 4.6. Key Applications / 340
- 4.7. Selected Industrial Applications (with Details of Experimental Conditions) / 363
  - Appendix / 378
  - Abbreviations / 380
  - References / 381

5 DYNAMIC MECHANICAL ANALYSIS (DMA)   387
Richard P. Chartoff, Joseph D. Menczel, and Steven H. Dillman

- 5.1. Introduction / 387
- 5.2. Characterization of Viscoelastic Behavior / 394
- 5.3. The Relationship between Time, Temperature, and Frequency / 401
- 5.4. Applications of Dynamic Mechanical Analysis / 410
- 5.5. Examples of DMA Characterization for Thermoplastics / 424
- 5.6. Characteristics of Fibers and Thin Films / 432

5.7. DMA Characterization of Crosslinked Polymers / 438
5.8. Practical Aspects of Conducting DMA Experiments / 456
5.9. Commercial DMA Instrumentation / 477
Appendix / 488
Abbreviations / 489
References / 491

6 DIELECTRIC ANALYSIS (DEA) 497

Aglaia Vassilikou-Dova and Ioannis M. Kalogeras

6.1. Introduction / 497
6.2. Theory and Background of Dielectric Analysis / 502
6.3. Dielectric Techniques / 520
6.4. Performing Dielectric Experiments / 528
6.5. Typical Measurements on Poly(Methyl Methacrylate) (PMMA) / 538
6.6. Dielectric Analysis of Thermoplastics / 553
6.7. Dielectric Analysis of Thermosets / 576
6.8. Instrumentation / 592
Appendix / 599
Abbreviations / 599
References / 603

7 MICRO- AND NANOSCALE LOCAL THERMAL ANALYSIS 615

Valeriy V Gorbunov, David Grandy, Mike Reading, and Vladimir V. Tsukruk

7.1. Introduction / 615
7.2. The Atomic Force Microscope / 616
7.3. Scanning Thermal Microscopy / 618
7.4. Thermal Probe Design and Spatial Resolution / 620
7.5. Measuring Thermal Conductivity and Thermal Force-Distance Curves / 624
7.6. Local Thermal Analysis / 628
7.7. Performing a Micro/Nanoscale Thermal Analysis Experiment / 633
7.8. Examples of Micro/Nanoscale Thermal Analysis Applications / 637
7.9. Overview of Local Thermal Analysis / 644
Abbreviations / 647
References / 648

INDEX 651

PREFACE

This book is about thermal analysis as applied to polymers. It is organized by thermal analysis techniques and thus contains chapters on the core techniques of differential scanning calorimetry (DSC), thermogravimetric analysis (TGA), thermomechanical analysis (TMA), and dynamic mechanical analysis (DMA). Although it can be argued that dielectric analysis (DEA) is more frequency than temperature oriented, we decided to include it because we believe it is an integral part of the thermal analysis of polymers. And we felt that it was necessary to include micro/nano-TA (μ/n-TA) because we believe that with the ever increasing ability to probe the macromolecular size scale, this field will become increasingly more important in the characterization and development of new materials. Each chapter describes the basic principles of the respective techniques, calibration, how to perform an experiment, applications to polymeric materials, instrumentation, and its own list of symbols and acronyms & abbreviations. Several examples are given where thermal analysis was instrumental in solving industrial problems.

In undertaking this project we wanted to write a book that described the underlying principles of the various thermal analysis techniques in a way that could be easily understood by those new to the field but sufficiently comprehensive to be of value to the experienced thermal analyst looking to refresh his or her skills. We also wanted to describe the practical aspects of thermal analysis, for example, how to make proper measurements and how best to analyze and interpret the data. We wrote this book with a broad audience in mind, including all levels of thermal analysts, their supervisors, and those that teach thermal analysis. Our purpose was to create a learning tool for the practioner of thermal analysis.

We were very fortunate to be able to assemble an international team of distinguished scientists to contribute to this book. These are truly the experts in the field and in some cases the people who invented the techniques. They are scientists and educators with the uncommon ability to explain complex principles in a manner that is thorough but still easy to comprehend. Note that all chapters have multiple authors, illustrating the collaborative nature of this undertaking. We took our jobs as editors seriously by becoming intimately involved in every chapter, and we express our appreciation to each and every contributor not only for their outstanding contributions but also for their understanding and patience with the editors.

We would like to recognize two people who have been role models for us: Professors Bernhard Wunderlich and Edith Turi. Both have been significant influences on our professional careers. As it is our hope that this book will benefit thermal analysis education, it is important to note that both Professors Wunderlich and Turi dedicated much of their professional lives to promoting and furthering education in thermal analysis. Professor Wunderlich was advisor to one of us (RBP) and post-doctoral advisor to the other (JDM) at Rensselaer Polytechnic Institute, giving us a fundamental grounding in the principles of thermal analysis and instilling a lifelong love for the subject. The roots of our understanding of the basics of thermal analysis stem from that time, and they can be noted in his novel teaching efforts and the founding of the ATHAS (Advanced Thermal Analysis System) Research Group. These efforts consisted first of audio tapes, allowing independent study, and then as technology developed, computer-based courses (novel for the time). Professor Turi taught thermal analysis to thousands of scientists and engineers during her renowned short courses at the Polytechnic Institute of New York (Brooklyn Poly) and for the American Chemical Society in addition to several national and international venues. Several of the contributors to this book cut their teeth as instructors in these short courses and/or as contributors to her classic book *Thermal Characterization of Polymeric Materials* (1981 and 1997).

Many people have contributed to the making of this book, and we thank them all. Special recognition goes to Larry Judovits, who not only led the collaboration on the modulated temperature DSC section but also critically reviewed much of the book. And to Harvey Bair, who contributed several personal examples of the ability of thermal analysis techniques to solve real industrial problems. We want to acknowledge those who read chapters or parts of chapters and offered many helpful comments, including Professor Sue Ann Bidstrup-Allen and Richard Siemens. We express appreciation to Professor Henning Winter for helpful discussions on measurement of the gel point. A huge thank you to our editor at Wiley, Dr. Arza Seidel, who always had the right answer to our many questions and steered us through the maze of transforming a vision into reality. One of us (R.B.P.) would like to acknowledge my long-term collaboration with Professor James Seferis from whom I have learned so much and, last but not least, my wife Donna for generously contributing her graphic arts skills and for her patience and encouragement. The other one of us (JDM) would like to express his gratitude to Judit Simon, Editor-in-Chief of the *Journal of Thermal Analysis and Calorimetry*, who supported him so much when he entered the field of thermal analysis.

JOSEPH D. MENCZEL and R. BRUCE PRIME
June 2008

CHAPTER 1

INTRODUCTION

JOSEPH D. MENCZEL
Alcon Laboratories, Fort Worth, TX

R. BRUCE PRIME
IBM (Retired)/Consultant, San Jose, CA

PATRICK K. GALLAGHER
The Ohio State University (Emeritus), Columbus, OH

Thermal analysis (TA) comprises a family of measuring techniques that share a common feature; they measure a material's response to being heated or cooled (or, in some cases, held isothermally). The goal is to establish a connection between temperature and specific physical properties of materials. The most popular techniques are those that are the subject of this book, namely differential scanning calorimetry (DSC), thermogravimetric analysis (TGA), thermomechanical analysis (TMA), dynamic mechanical analysis (DMA), dielectric analysis (DEA), and micro/nano-thermal analysis (μ/n-TA).

This book deals almost exclusively with studying polymers, by far the widest application of thermal analysis. In this area, TA is used not only for measuring the actual physical properties of materials but also for clarifying their thermal and mechanical histories, for characterizing and designing processes used in their manufacture, and for estimating their lifetimes in various environments. For these reasons, thermal analysis instruments are routinely used in laboratories of the plastics industry and other industries where polymers and plastics are being manufactured or developed. Thus, thermal analysis is one of the most important research and quality control methods in the development and manufacture of polymeric materials as well as in industries that incorporate these materials into their products.

Not withstanding its importance, educational programs in thermal analysis at universities and colleges are almost nonexistent; certainly they are not sys-

Thermal Analysis of Polymers: Fundamentals and Applications, Edited by Joseph D. Menczel and R. Bruce Prime
Copyright © 2009 by John Wiley & Sons, Inc.

tematic. Thermal analysis training in the United States is for the most part limited to short courses, such as the short course at the annual meeting of the North American Thermal Analysis Society (NATAS) and earlier the short course at the Eastern Analytical Symposium. Our goal was to write a book that could be used as a text or reference to accompany thermal analysis courses and that would enable both beginners and experienced practitioners to do some self-education. This book is for experimenters at all levels that addresses both the fundamentals of the thermal analysis techniques as well as the practical issues associated with the running of experiments and interpretation of the results. Several examples are given where thermal analysis played a key role in solving a practical problem, and they are presented in a manner that will allow readers to apply the lessons to their own problems.

This book is organized by measuring techniques rather than by material classification. These techniques all follow the change of specific physical properties as the temperature and possibly atmosphere are controlled. Table 1.1 indicates the classification of the more common techniques by the physical property measured.

Most thermal analysis studies today are conducted with commercial instruments. Manufactures have striven to provide complete "systems" capable of a wide range of analyses and frequently sharing modular components. Naturally, this is a market-driven phenomenon, and the current driving forces are speed, miniaturization, and automation. The goals of a modern industrial quality control facility, a state-of-the-art research institution, or those of a teaching laboratory are quite different. This difference leads to a broad spectrum of available instrumentation in terms of ultimate capabilities, simplicity, and cost.

Commercial thermal analysis instrumentation is relatively new, a product of the last four decades or so. Mass production of TA instruments started in the early 1950s. From then to the 1970s, several major TA instruments were marketed, and some of them are still manufactured even today. This was the

TABLE 1.1. The Most Important General Methods and Techniques of Thermal Analysis

General Method	Acronym	Property Measured
Differential Scanning Calorimetry	DSC	ΔT, differential power input
Differential Thermal Analysis	DTA	ΔT
Thermogravimetry or Thermogravimetric Analysis	TG or TGA	Mass
Thermomechanical Analysis, Thermodilatometry	TMA, TD	Length or volume
Dynamic Mechancal Analysis	DMA	Viscoelastic properties
Dielectric Analysis	DEA	Dielectric properties
Micro/Nano-Thermal Analysis	μ/n-TA	Penetration, ΔT

time when the two types of DSC (power compensation and heat flux) appeared, and the principle of the measurements is still the same today.

A gigantic leap forward in the evolution of scientific instrumentation and data analysis took place with the advent of the digital computer. The replacement of such things as chart recorders and analog computers has dramatically improved our ability to control, measure, and evaluate experiments precisely. This process in TA introduced revolutionary changes. No longer was there a need to rerun the measurement if the sensitivity was not properly adjusted, the accuracy of the measurements dramatically increased, the time necessary for data evaluation decreased significantly, and many measurements became automated. The use of autosamplers doubled or tripled the throughput of the instruments, allowing them to run day and night. To complement the autosamplers, cooling units can be turned on and off at a pre-programmed time in the absence of the operator.

At the same time, problems developed because of certain software issues. When software is automatically capable of performing calculations, the operator often tends to be lazy and fails to learn the theoretical basis of the measurement and the calculation. Although it is certainly not the intention of any instrument manufacturer or software purveyor to deceive or mislead the user, slavish reliance on software without adequate comprehension can be dangerous. As an example, the quest for nice-looking plots can lead to excessive smoothing or dampening of the results, with the possible consequence of missing meaningful, even critical, subtle events. Another negative aspect of blind software use is the question of significant figures. Modern scientific literature is replete with insignificant figures. The ability of a computer to generate an unending series of digits is no indication or justification of their relevance. And although the manufacturers often provide the option to change the number of displayed digits, with the rush in modern laboratories, operators often fail to make this change. But eventually, it is incumbent on investigators to evaluate the results of their analyses, with regard to both the significance of the numbers and their conclusions.

The number of instrument manufacturers decreased somewhat during the last decade or so, but there are still a significant number on the market. Today it seems that most of these corporations will survive. Although some are highly specialized, almost all thermal analysis instrument companies produce one or more DSCs. As mentioned, popularity has its disadvantages: DSC and TGA are the two most popular TA techniques, and these are the ones that many operators routinely use, often without the necessary theoretical knowledge. This lack of understanding creates such an absurd situation that the essence of DSC measurements is reduced to recording a melting peak, whereas in TGA, all they look for is the start of the mass loss. Similarly, TMA is often reduced to looking for the shift in the slope of the dimension versus temperature curves to measure T_g. Interestingly, experimenters who use the relaxation techniques (DMA and DEA), and μ/n-TA, tend to rely more on theory and do fewer simple repetitive measurements.

Temperature in thermal analysis is the most important parameter. The strict definition of TA stipulates a programmed (i.e., time- or property-dependent) temperature. From the standpoint of instrumentation and methodology, however, isothermal measurements are included. Some applications concerning kinetics, as an example, involve a series of isothermal measurements at different temperatures or measuring an isothermal induction time to reaction. Other isothermal techniques may involve time to ignition and changes in the measured property with a changing atmosphere or force applied to the sample. Temperature is conveniently and most often measured by thermocouples, either individually or coupled in series as a thermopile to increase the sensitivity and/or to integrate the measurement over a greater volume. Some instruments use platinum resistance thermometers. Optical pyrometry has been applied in rare instances. These latter two methods are the methods of choice, depending on the specific temperature, as set forth in the definition of the International Temperature Scale. Regardless of the particular sensor used, calibration of the temperature is generally dependent on the specific technique and will be discussed with each class of instrumentation. Careful consideration must always be given when equating sensor temperature with actual sample temperature. Depending on the enthalpy of the processes or reactions occurring, thought must also be given to equating a "bulk" sample temperature with that at the interface where the actual reaction may be taking place. Such considerations are particularly important for meaningful kinetic analyses.

Thermal analysis can be used in a variety of combinations. The most common combinations all share the same sample as well as thermal environment. A key distinction is made between true simultaneous methods like TGA/DTA and TGA/DSC, in which there is no time delay between the measurements, and near-simultaneous measurements like TGA/MS and TGA/FTIR, in which the time delay is small between the mass loss and the respective gas detector. Such combined techniques not only represent a saving in time, but also they help to alleviate or minimize uncertainties in the comparison of such results. And TGA/MS and TGA/IR can be instrumental in identifying complex processes that involve mass loss. Although combined instruments are not described in this book in detail, some examples are given in the TGA chapter.

Unfortunately, the proper selection of the run parameters is often ignored in thermal analysis measurements, even though it is a critical part. Of all the run parameters, the sample mass, the ramp rate, and the purge gas are the most important. The sample size and its physical shape play a significant role in the results. The proper sample size and the heating rate are interconnected because in several techniques faster rates require smaller samples and substantially improved conditions for rapid thermal transport between the sample and its controlled environment. Therefore, compromises are necessary between sample size and heating and cooling rates. And thermal conductivity and flow of the atmosphere thus become significant factors. Transport conditions may change substantially with temperature as the nature of the thermal path and the relative roles of conduction, convection, and radiation are altered.

Traditionally, simple combinations of linear heating or cooling rates and isothermal segments have been employed. Modern methods, however, frequently impose cyclic temperature programs coupled with Fourier analyses to achieve particular advantages and added information. These approaches are referred to as modulated techniques, and temperature is the most commonly modulated parameter. Note that in DMA, stress or strain is the modulated parameter and that in DEA, the electric field is modulated, but in modulated temperature DSC and modulated temperature TMA, it is the temperature that is modulated.

Often interaction of the sample with its atmosphere is important, for example, in oxidation/reduction reactions or catalytic processes. Reversible processes will be influenced by product accumulations, and hence, the ability of the flowing atmosphere to purge these volatile products becomes important. A common example of such a reversible process is dehydration or solvent removal. Clearly the degree of exposure of the sample to its atmosphere then becomes a factor. Deliberate modifications of the sample holder or compartment are made to control these interactions. Maximum exposure can be attained with the sample in a thin bed with the atmosphere flowing over or even percolating through it. Minimum exposure can be achieved using sealed sample containers or ones with a small orifice to alleviate the changes of pressure resulting from temperature changes and/or reactions with the gas phase. Diffusion of species into and out of the sample also needs to be considered. For example, in oxidative processes, diffusion of oxygen into the sample becomes important, and in mass loss processes, the volatile products need to diffuse to the surface where evaporation occurs. In these cases, sample size and shape, e.g., surface-to-volume ratios, may influence the results.

The above considerations all dictate that the sample size and form can be a significant factor in the effort to achieve the desired analysis and its reliability. The ability to impose a rapidly changing thermal environment on the sample may be necessary to simulate the true process conditions properly or simply to obtain the results more quickly. As mentioned, this necessity dictates a small sample in order to follow the temperature program, but this in turn demands a representative sample. Thus, it may be difficult or impossible to achieve for a very small sample of some materials. Composites, blends, and naturally occurring materials may lack the necessary homogeneity. Reproducibility of the measurements or even other analytical data is needed to assure that the sample is indeed representative.

Even though the inhomogeneity of the sample at fast heating rates can be compensated for with a smaller sample size, there are time-dependent phenomena in most thermal analysis techniques. One can change the heating rate and ensure an acceptable temperature gradient in the sample, and different physical processes may and will take place at different heating rates as demonstrated in Chapters II and III (DSC and TGA). Thus, the selection of a proper ramp rate is important, and just by changing the sample size, one cannot compensate for all the variabilities in the sample.

A word about mass is appropriate. The International Union of Pure and Applied Chemistry (IUPAC) and the International Confederation of Thermal Analysis and Calorimetry (ICTAC) have determined that the property measured by TGA should be referred to as "mass" and not as "weight." Although some figures are reproduced with their original ordinates, which may be weight or weight percent, we adhere to this terminology and refer to sample mass, mass percent, and mass loss. It is still correct to refer to the weighing of samples and to standard reference weights.

CHAPTER 2

DIFFERENTIAL SCANNING CALORIMETRY (DSC)

JOSEPH D. MENCZEL
Alcon Laboratories, Fort Worth, Texas

LAWRENCE JUDOVITS
Arkema, King of Prussia, Pennsylvania

R. BRUCE PRIME
IBM (Retired)/Consultant, San Jose, California

HARVEY E. BAIR
Bell Laboratories (Retired)/Consultant, Newton, New Jersey

MIKE READING
University of East Anglia, Norwich, United Kingdom

STEVEN SWIER
Dow Corning Corporation, Midland, Michigan

2.1. INTRODUCTION

Differential scanning calorimetry (DSC) is the most popular thermal analysis technique, the "workhorse" of thermal analysis. This is a relatively new technique; its name has existed since 1963, when Perkin-Elmer marketed their DSC-1, the first DSC. The term DSC simply implies that during a linear temperature ramp, quantitative calorimetric information can be obtained on the sample. According to the ASTM standard E473, DSC is a technique in which the heat flow rate difference into a substance and a reference is measured as a function of temperature, while the sample is subjected to a controlled temperature program. As will be seen from this chapter, the expression "DSC"

Thermal Analysis of Polymers: Fundamentals and Applications, Edited by Joseph D. Menczel and R. Bruce Prime
Copyright © 2009 by John Wiley & Sons, Inc.

refers to two similar but somewhat different thermal analysis techniques. It is a common feature of these techniques that the various characteristic temperatures, the heat capacity, the melting and crystallization temperatures, and the heat of fusion, as well as the various thermal parameters of chemical reactions, can be determined at constant heating or cooling rates. It is important to note that the acronym DSC has two meanings: (1) an abbreviation of the technique (i.e., differential scanning calorimetry) and (2) the measuring device (differential scanning calorimeter).

Since the 1960s the application of DSC grew considerably, and today the number of publications that report DSC must amount to more than 100,000 annually.

One of these techniques that brought into science the name *DSC*, called today *power compensation DSC*, was created by Gray and O'Neil at the Perkin-Elmer Corporation in 1963. The other technique grew out of differential thermal analysis (DTA), and is called *heat flux DSC*. Differential thermal analysis itself originates from the works of Le Chatelier (1887), Roberts-Austen (1899), and Kurnakov (1904) (see Wunderlich, 1990). It needs to be emphasized that both of these techniques give similar results, but of course, they both have their advantages and disadvantages.

The major applications of the DSC technique are in the polymer and pharmaceutical fields, but inorganic and organic chemistry have also benefited significantly from the existence of DSC. Among the applications of DSC we need to mention the easy and fast determination of the glass transition temperature, the heat capacity jump at the glass transition, melting and crystallization temperatures, heat of fusion, heat of reactions, very fast purity determination, fast heat capacity measurements, characterization of thermosets, and measurements of liquid crystal transitions. Kinetic evaluation of chemical reactions, such as cure, thermal and thermooxidative degradation is often possible. Also, the kinetics of polymer crystallization can be evaluated. Lately, among the newest users of DSC we can list the food industry and biotechnology. Sometimes, specific DSC instruments are developed for these consumers.

DSC is extremely useful when only a limited amount of sample is available, since only milligram quantities are needed for the measurements. As time goes by, newer and newer techniques are introduced within DSC itself, like pressure DSC, fast-scan DSC, and more recently modulated temperature DSC (MTDSC). Also, with development of powerful mechanical cooling accessories, low-temperature measurements are common these days. DSC helps to follow processing conditions, since it is relatively easy to fingerprint the thermal and mechanical history of polymers. Although computerization, enormously accelerated the development of DSC, this has its own negatives; many operators tend to use the software without first understanding the basic principles of the measurements. Nevertheless, newer and more powerful software products have been were marketed that increased the productivity of the thermal analyst by significantly reducing the time for calculation of experimental results, and sometimes, interpretation of the data. It is unfortunate that few of these software applications are available to the research personnel for modification and thus for use in special conditions.

Today DSC is a routine technique; a DSC instrument can be found virtually in every chemical characterization laboratory, since the instruments are relatively inexpensive. Unfortunately, this has its drawback. It is a popular misconception that if you recorded a DSC peak, you did your job. In this chapter we would like to prove, from one side, that this is not true, but from the other side that despite this, DSC is still a simple and easily applied technique. Our goal is to present a simple but consistent picture of the present state of this measuring technique.

2.2. ELEMENTS OF THERMODYNAMICS IN DSC

Thermodynamics studies two forms of energy transfer: heat and work. *Heat* can be defined as transfer of energy caused by the difference in temperatures of two systems. Heat is transferred spontaneously from hot to cold systems. It is an extensive thermodynamic quantity, meaning that its value is proportional to the mass of the system. The SI (Système International de Unités) unit of the heat is the joule (J). The earlier unit of "calorie" is not in use any more.

The goal of thermodynamics is to establish basic functions of state, the most important of which (for differential scanning calorimetry) are U, internal energy; H, enthalpy; p, pressure; V, volume; S, entropy; and C_p, heat capacity at constant pressure.

In thermodynamics, the description of reversible processes is the one that is most widely used. This is called *equilibrium thermodynamics*, because it deals with equilibrium systems. *Nonequilibrium thermodynamics*, which deals with irreversible processes, and thus has time as an additional variable to the basic parameters of state, exists, but is rarely used by chemists. It was Onsager and Prigogine [see, e.g. Onsager (1931a,b), Prigogine (1945, 1954, 1967), and Prigogine and Mayer (1955)] who did the most for development of this branch of science. We will not spend time on describing nonequilibrium thermodynamics in this book, but simply mention that the whole nonequilibrium system can be subdivided into small subsystems being in equilibrium, and the whole system is described as a sum of these subsystems. The interested reader can find more information in the book by de Groot and Mazur (1962).

As mentioned above, equilibrium thermodynamics deals with reversible processes, and is based on the following four laws of thermodynamics, which are empiric laws rather than theoretically deduced laws:

1. The zeroth law of thermodynamics introduces the concept of temperature.
2. The *first law of thermodynamics*, also known as the *principle of energy conservation*, states that the change of internal energy of a thermody-

namic system equals the difference of the heat added to the system and all the work done by the system. The internal energy of a thermodynamic system is a function of state; its value depends only on the state of the system, not on the path, that is, how the system arrived at this state. Therefore, in a cyclic process $\Delta U = 0$. Thus, energy cannot be created or destroyed. The first law can be expressed as

$$\Delta U = Q + W \qquad (2.1)$$

where ΔU is the change in internal energy, Q is the heat, and W is the work.

3. There are several formulations of the *second law of thermodynamics*. The so-called Clausius statement says that in spontaneous processes heat cannot flow from a lower-temperature body to a higher-temperature body. The Thomson (Lord Kelvin) statement says that heat cannot be completely converted into work.
4. The *third law of thermodynamics* (Nernst law) states that the entropy of perfectly crystalline materials at 0 K is zero.

So, the zeroth law says that, there is a game (heat-to-work conversion game), and that you've got to play the game. The first law says you can't win; at best, you can only break even. But according to the second law, you can break even only at 0 K. And the third law says, you can never reach 0 K (Moore 1972; Wunderlich 2005).

The most important functions of state used in DSC are described in the following subsections.

2.2.1. Temperature

Temperature is the most important quantity in differential scanning calorimetry. With DSC, in essence temperature is the only *measured* quantity. Everything else is calculated from the changes of temperature, from the difference between the sample and reference temperatures. We can define temperature as a primary thermodynamic parameter of a system, which is a measure of the average kinetic energy of the atoms or molecules of the system. In everyday language, we use the words "hot," "warm," and "cold" to characterize the temperature of materials and bodies.

Temperature can be defined for *equilibrium systems* only, in which the velocity of the particles are described by the *Boltzmann distribution*. The temperature controls the flow of heat between two thermodynamic systems.

There are two laws of thermodynamics that help define "temperature" as a parameter of the system.

The *zeroth law of thermodynamics* states that if systems A and B are separately in thermal equilibrium with system C, then they are in thermal equilibrium with each other as well. Since all these systems are in thermal equilibrium with each other, some thermodynamic parameter must exist that has the same value in all of them. This parameter is called *temperature*. In other words, in the state of equilibrium, all thermodynamic systems have an intensive variable of state called the *temperature*.

The second law of thermodynamics helps define temperature mathematically as

$$T \equiv \left(\frac{\partial U}{\partial S}\right)_V \qquad (2.2)$$

in other words, temperature is the rate of increase of internal energy of the system with increasing entropy.

There are three temperature scales in use:

1. In the English-speaking countries, especially the United States, the Fahrenheit scale is still in everyday use. In the Fahrenheit scale, the melting point of ice is 32 °F, while the boiling point of water is 212 °F.
2. The most popular temperature scale for everyday use is the Celsius scale, in which the melting point of ice is taken as 0 °C, while the boiling point of water is 100 °C.
3. In the thermodynamic temperature scale (previously called "absolute temperature scale"), the base point, the triple point of water, is 273.16 K, while the boiling point of water is 373.15 K.

Temperatures throughout the Universe vary widely. To show this, here are several important temperature values: (1) the average temperature of the Universe is ~−270 °C; (2) the temperature in the core of the Sun is ~12 million °C; and (3) finally, several temperatures are important in thermal analysis: the melting point of indium, 156.60 °C, the melting point of tin, 231.93 °C; the melting point of lead, 327.47 °C; and the melting point of mercury is −38.8 °C.

Temperature is measured by thermometers. The first temperature-measuring device was a special type of liquid thermometer, the thermoscope (discovered by Galilei in the sixteenth century). These days there are gas thermometers, liquid thermometers, infrared thermometers, liquid crystal (LC) thermometers (cholesteric), thermocouples, and resistance thermometers (these latter ones are the most important in thermal analysis).

2.2.2. Heat

Heat is a form of energy, which in spontaneous processes flows from a higher-temperature body to a lower-temperature body (the second law of thermodynamics). Therefore, *heat flow* can be defined as a process in which two thermodynamic systems exchange energy. The flow of heat continues until the temperature of the two systems or bodies becomes equal. This state is called *thermal equilibrium*.

For infinitesimal processes Eq. (2.1) can be rewritten as

$$dU = \delta Q + \delta W \qquad (2.3)$$

(the quantities of δQ and δW are not differential, because Q and W are not functions of state in general; the heat (Q), becomes a function of state only for reversible processes).

In the case when only volume work takes place, one obtains

$$dU = \delta Q + \delta W = \delta Q - p\,dV \qquad (2.4)$$

so

$$\Delta U = Q + W = Q - \int_{V_1}^{V_2} p\,dV \qquad (2.5)$$

and

$$\delta Q = dU + p\,dV = \left(\frac{\partial U}{\partial V}\right)_T dV + \left(\frac{\partial U}{\partial T}\right)_V dT + p\,dV \qquad (2.6)$$

which can be rewritten as

$$\delta Q = \left(\frac{\partial U}{\partial T}\right)_V dT + \left[\left(\frac{\partial U}{\partial V}\right)_T + p\right] dV \qquad (2.7)$$

This equation states that if a process occurs at constant volume, and no other work is taken into account, then

$$dU = \delta Q \qquad (2.8)$$

Thus, at these conditions the change in the internal energy equals the amount of heat added to the system or extracted from the system, and heat (Q) becomes a function of state.

2.2.2.1. Heat Flow We need to mention here the flow of heat, which is especially important in calorimetry. There are three major forms of heat flow: conduction, convection, and thermal radiation:

1. In the case of *conduction*, heat travels from the hotter part of a body to a cooler part, or if heat conduction takes place between two bodies, these

bodies have to come into physical contact with each other. Heat conduction takes place by diffusion, as the atoms or molecules give over part of their vibrational energy to their neighbors.
2. In *heat convection*, the heat is transferred from a solid surface to a flowing material (like a gas or a liquid), and vice versa.
3. In *thermal radiation*, energy is radiated from the surfaces of the systems in the form of electromagnetic energy. For this type of heat transfer, no medium is needed; it usually takes place between solid surfaces. The intensity and the frequency of the radiation depend only on the surface temperature of the body (they both increase with increasing temperature).

2.2.2.2. Latent Heat The *latent* ("hidden") *heat* is the amount of heat absorbed or emitted by a material during a phase transition (it is called "latent" because the temperature of the material does not change during the phase transition despite the absorption or release of heat). This expression is used less frequently now; the current term is the *heat of transition*.

2.2.3. Enthalpy

Equation (2.8) indicates that in thermodynamics, the internal energy is used as a function of state to characterize the system at constant volume, and also when no work is being performed on or by the system. But the majority of real processes, especially for polymers, take place at constant pressure, because solids and liquids (the only physical states for polymers) are virtually incompressible. For such processes (i.e., those taking place at constant pressure), Gibbs introduced a new function of state, enthalpy H

$$H \equiv U + pV \tag{2.9}$$

where p is pressure and V is volume.

Thus, enthalpy as a function of state is similar to internal energy, but it contains a correction for the volume work. However, the change of volume with temperature for solids and liquids is small; therefore the difference between enthalpy and internal energy is also small. Enthalpy is especially useful for processes taking place at constant pressure. For such processes

$$dH = dU + p\,dV \tag{2.10}$$

and

$$\Delta H = Q \tag{2.11}$$

Therefore, the enthalpy increase of the system in equilibrium processes is identical to the heat added to the system.

In practice, enthalpy differences are generally used. The total absolute enthalpy of the sample cannot be directly measured, but it can be calculated if the heat capacity in the whole temperature range (from absolute zero) is known.

In DSC, the enthalpy change is calculated from the temperature difference between the sample and the reference. In endothermic processes (processes with energy absorption, such as melting or evaporation), the enthalpy of the system increases, while in exothermic processes (condensation, crystallization) the enthalpy (and the internal energy) of the system decreases.

Similar to the SI unit for heat, the SI unit for enthalpy is J (joule). Calories are no longer used.

2.2.4. Entropy

Entropy is probably the most important function of state. Every aspect of our life is governed by entropy; it is the function of state characterizing the disorder of the system.

Entropy was introduced by Clausius in 1865 (Clausius 1865):

$$\Delta S = S_B - S_A \geq \int_A^B \frac{\delta Q}{T} \qquad (2.12)$$

so

$$dS \geq \frac{\delta Q}{T} \qquad (2.13)$$

where the greater than (>) part of the sign refers to irreversible processes, while the equal to part of the sign refers to reversible processes. Therefore, in a reversible process that proceeds at constant temperature from state A to state B, we have

$$S_B - S_A = \Delta S = \frac{1}{T} \int_A^B \delta Q = \frac{Q}{T} \qquad (2.14)$$

For cyclic processes, according to the second law of thermodynamics

$$\oint \frac{\delta Q}{T} \leq 0 \qquad (2.15)$$

where the "equal" part of the ≤ sign refers to the reversible processes, and the "less than" part refers to irreversible processes.

From Eq. (2.14) it follows that in an isolated system irreversible processes can proceed spontaneously if the entropy of the system increases:

$$\Delta S > 0 \qquad (2.16)$$

ELEMENTS OF THERMODYNAMICS IN DSC 15

2.2.5. Helmholtz and Gibbs Free Energy

Two more functions of state play an extremely important role in thermodynamics: the *Helmholtz free energy* (F) and the *Gibbs free energy*, or, as it is often called, the *free enthalpy* (G). These functions indicate what part of the internal energy or enthalpy can be converted into work at constant temperature and volume or pressure, respectively:

$$F \equiv U - TS \tag{2.17}$$

$$G \equiv U - TS + pV = H - TS \tag{2.18}$$

As can be seen, the usable portion of the internal energy and enthalpy decreases with increasing temperature and increasing entropy.

2.2.6. Heat Capacity

An extremely important function of state in differential scanning calorimetry is the heat capacity at constant pressure (C_p) and constant volume (C_v), because in the absence of chemical reactions or phase transitions, the amplitude of the DSC curve is proportional to the heat capacity of the sample at constant pressure.

Heat capacity indicates how much heat is needed to increase the sample temperature by 1 °C. The heat capacity of a unit mass of a material is called *specific heat capacity*. The SI units for heat capacity are J/(K·mol) or J/(K·kg). There are two major heat capacities:

Heat capacity at constant volume:

$$C_V \equiv \left(\frac{\delta Q}{\partial T}\right)_V = \left(\frac{\partial U}{\partial T}\right)_V \tag{2.19}$$

Heat capacity at constant pressure:

$$C_p \equiv \left(\frac{\delta Q}{\partial T}\right)_p = \left(\frac{\partial H}{\partial T}\right)_p \tag{2.20}$$

Differential scanning calorimetry always determines C_p, because it is impossible to keep the samples at constant volume when temperature changes. When necessary, C_v can be calculated from C_p using one of the following relationships:

$$C_p - C_v = \left[\left(\frac{\partial U}{\partial V}\right)_T + p\right]\left(\frac{\partial V}{\partial T}\right)_p \tag{2.21}$$

Equation (2.21) can be modified to the following equation

$$C_v = C_p - V\gamma^2 T \beta_T \qquad (2.22)$$

where V is the volume, γ is the coefficient of volumetric thermal expansion, and β_T is the isothermal compressibility (the reciprocal of bulk modulus) (Wunderlich 1997a).

2.2.7. Phase Transitions

When a thermodynamic system changes from one phase to another as a result of changing temperature and/or pressure, we call this a phase transition. Ehrenfest (1933) was the first physicist who classified the thermodynamic *phase transitions*. In his scheme, a transition is called a *first-order* when a first partial derivative of the free energy with a thermodynamic variable (e.g., temperature, pressure) exhibits discontinuity. These first derivatives are the volume, entropy, and enthalpy. Thus, melting, evaporation, sublimation, crystal-to-crystal transitions, crystallization, condensation, and deposition (also called desublimation) are first-order transitions. Similarly, a transition is called a second-order transition when the just-mentioned first derivatives are continuous, but a second partial derivative of the free energy exhibits discontinuity. Since C_p/T is one of these second derivatives, a break (jump) in the $C_p = f(T)$ DSC curve indicates a second-order transition. Examples of true second order phase transitions are the magnetic transition at the Curie point, the superfluid transition of liquid helium, and various sub-T_g transitions in glassy or crystalline polymers, like the δ transition in bisphenol A polycarbonate (Heijboer 1968; Sacher 1974). In differential scanning calorimetry of polymers, the recordable first-order phase transitions are melting of crystalline polymers, crystallization, and crystal-to-crystal transitions. Evaporation and sublimation are nonexistent transitions for polymers, because, owing the high molecular mass of polymers, they cannot be transferred intact into the gaseous phase without undergoing decomposition.

2.2.8. Melting Point and Heat of Fusion

The *melting point* (T_m) is the temperature at which a crystalline solid changes to an isotropic liquid. From a DSC curve the melting point of a low-molecular-mass, high-purity substance can be determined as the point of intersection of the leading edge of the melting peak with the extrapolated baseline (see Section 2.6 of this chapter). This determination of the melting point is not suitable for low-molecular-mass substances of low purity and semicrystalline polymers. In both cases the melting range is somewhat broad, but for semicrystalline polymers it is often extremely broad. In such cases, the melting point is determined as the last, highest-temperature point of the melting endotherm, because this is the temperature at which the most perfect crystallites melt. Also, the melting point determined by the method described above can be correlated with the melting point determined by polarization optical microscopy. In addition to the melting point of semicrystalline polymers, often the peak temperature of melting (T_{mp}) is also reported; in the case of polymers, this temperature corresponds to the maximum rate of the melting process.

The *heat of fusion* (ΔH_f) is the amount of heat that has to be supplied to 1 g of a substance to change it from a crystalline solid to an isotropic liquid.

The equilibrium melting point (T_m°) of a crystalline polymer is the lowest temperature at which macroscopic equilibrium crystals completely melt (Prime and Wunderlich 1969; Prime et al. 1969).

The heat of fusion of 100% crystalline polymer, or as it is sometimes called, the *equilibrium heat of fusion* (ΔH_f°), is the heat of fusion of the equilibrium polymeric crystals at the equilibrium melting point (the heat of fusion of 100% crystalline polymer depends somewhat on the melting temperature; that is why ΔH_f° is given at T_m°).

2.2.9. Crystallization Temperature

The crystallization temperature [often called the *freezing point* (T_c)] is the temperature at which an isotropic liquid becomes a crystalline solid during cooling. As a result of supercooling, the freezing point is almost always lower than the melting point. For low-molecular-mass, pure substances the freezing point is determined as the point of intersection of the leading edge of the crystallization exotherm with the extrapolated baseline. For semicrystalline polymers, the crystallization temperature is the highest temperature of the crystallization exotherm (designated as T_{c0}). When reporting DSC data, in addition to the freezing point, often the peak temperature of crystallization (indicating the highest rate of crystallization) is also reported (T_{cp}). Since usually both the melting and crystallization of semicrystalline polymers are far from equilibrium, the heating and cooling rates should be given when reporting data.

2.2.10. Glass Transition Temperature

The *glass transition temperature* (T_g) is the temperature beyond which the long-range translational motion of the polymer chain segments is active. At this temperature (on heating) the glassy state changes into the rubbery or melt state. Below T_g, the translational motion of the segments is frozen, only the vibrational motion is active. Formally, the glass transition resembles a thermodynamic second-order transition because at the glass transition temperature there is a heat capacity jump. But this heat capacity increase does not take place at a definite temperature as would be required by equilibrium thermodynamics, but rather in a temperature range. Therefore the glass transition is a kinetic transition. When the glass transition temperature is determined by a relaxational technique (DMA, DEA), it is often called the *temperature of the α relaxation* or the α *dispersion* (or β *relaxation* or β *dispersion* if a crystalline relaxation exists at higher temperatures).

2.3. THE BASICS OF DIFFERENTIAL SCANNING CALORIMETRY

As previously mentioned in 2.1, ASTM standard E473 defines *differential scanning calorimetry* (DSC) as a technique in which the heat flow rate difference into a substance and a reference is measured as a function of temperature while the substance and reference are subjected to a controlled temperature program. It should be noted that the same abbreviation, DSC, is used to denote the technique (differential scanning calorimetry) and the instrument performing the measurements (differential scanning calorimeter).

As Wunderlich (1990) mentioned, no heat flow meter exists that could directly measure the heat flowing into or out of the sample, so other, indirect techniques must be used to measure the heat. Differential scanning calorimetry is one of these techniques; it uses the temperature difference developed between the sample, and a reference for calculation of the heat flow. An exotherm indicates heat flowing out of the sample, while an endotherm indicates heat flowing in.

Two types of DSC instruments exist: heat flux and power compensation. Historically, heat flux DSC evolved from differential thermal analysis (DTA). The basic design of a DTA consists of a furnace adjoined to separate sample and reference holders—A programmer heats the furnace containing the sample and the reference holders at a linear heating rate. The signals from the DTA sensors, usually thermocouples, are then fed to an amplifier. Unlike DSC, in DTA the sample and reference holders hold the sample and the reference material directly, without any additional packing (this additional packing is the sample and reference pans in DSC); the sample holder is loaded with the sample, while the reference holder is filled with an inert reference material such as aluminum oxide. Since the sample and the reference are heated from the outside, the DTA response is now susceptible to heat transport effects through the sample because of the usually large amount of the sample (up to several grams in older-type DTAs). Such factors could be the amount, packing, or thermal conductivity of the sample. These problems are reduced for DSC when the sample is separated from direct contact with the sensor and encapsulated in a pan constructed of a high-thermal-conductivity material. This is typically high-purity-aluminum, although other metals, such as copper, gold, or platinum can also be used. A normally empty sample pan is used as a reference. Newer DTAs now include sensors that are separated from the sample and lie outside the container; however, the sample is still directly packed into the holder. An example of this is the 1600 °C DTA attachment to the TA Instruments 2920 module. In this instrument the sample is put into platinum sample containers (TA Instruments 1993).

2.3.1. Temperature Gradient, Thermal Lag, and Thermal Resistance

Since the sample is heated from one specific source (usually from outside), potentially significant temperature gradients exist within the sample. It is an important task in thermal analysis to create conditions in which the temperature gradients within the sample can be minimized. The *temperature gradient* is the unequal distribution of temperature within the sample. The temperature gradient in the sample depends on the heating rate, the sample size, and the thermal diffusivity of the sample and the sample holder. Thermal diffusivity (m²/s) is determined as the ratio of thermal conductivity λ [W/(m·K)] and the volumetric heat capacity [(J/(kg·K)]

$$k = \frac{\lambda}{\rho C_p} \tag{2.23}$$

where ρC_p (the product of the density and the specific heat capacity) is the volumetric heat capacity.

This means that sample holders with high thermal diffusivity are desirable in DSC or DTA, because they conduct heat rapidly. Since the design of the DSC cell is set and has been optimized by the manufacturer, one can minimize the temperature gradient only by selecting appropriate sample size and heating rate. The temperature gradient within a sample can be calculated from the heating rate and the sample thickness (Wu et al. 1988); it increases with increasing heating rate and sample thickness.

The temperature gradient is not to be confused with thermal lag, which is another physical property that should also be minimized in DSC experiments. *Thermal lag* is the difference between the average sample temperature and the sensor temperature and is caused by so-called thermal resistance, which characterizes the ability of the material to hinder the flow of heat. Thermal lag is smaller in DSC than in DTA because of smaller sample size (milligrams in DSCs), but more types of thermal resistance develop in DSC than in DTA. These effects are caused by introduction of the sample and reference pans into the DSC sample and reference holders. Thus, in DTA thermal resistance develops between the sample holder (in some instruments called the sample pod) and the sample (analogously, between the reference holder and the reference material), and within the sample and the reference materials. On the other hand, in DSC thermal resistance will develop between the sample holder and the bottom of the sample pan and the bottom of the sample pan and the sample (these are called *external thermal resistances*), and within the sample itself (this is called internal thermal resistance). These thermal resistances should be taken into account since they determine the thermal lag. Let us suppose that the cell is symmetric with regard to the sample and reference pods or holders, the instrumental thermal resistances are identical for the sample and reference holders, the contact between the pans and the pods are intimate, no crosstalk exists between the sample and reference sensors (i.e., the electrical signals from the sensors do not influence each other), and the temperature distribution in the sample (and reference) is uniform (i.e., there is no temperature gradient in the sample). In such a case in steady state, the sum of all the thermal lags described can be expressed by the following equation

$$\dot{Q} = -\lambda A \frac{\Delta T_{sbl}}{\Delta X} = \frac{1}{R} \Delta T_{sbl} \qquad (2.24)$$

where $\dot{Q}$ is the heat flow rate, R is the thermal resistance (including both internal and external contributions), λ is the thermal conductivity, A is the contact area, ΔT_{sbl} is the temperature difference between the sample and the block (which is the heat sink), and ΔX is the linear heat conduction pathway (Hemminger and Höhne 1984). The thermal resistance can be calculated from the slope of the leading edge of the melting peak of a pure low molecular mass substance such as indium. The directional heat pathway formed by the temperature difference is referred to as the *heat leak*. Here we mention that in differential scanning calorimetry one comes to steady state when in nonisothermal mode (i.e., during heating or cooling), the $\Delta T (= T_s - T_r)$ signal reaches a constant value. In steady state the ΔT signal may change slightly because of the slight increase of the heat capacity of the sample with temperature.

If the cell is not symmetric or the thermal resistances are not identical, then one will not measure equal contributions from the sample and the reference sensors, which in turn will manifest itself in a nonlinear baseline. This can if the cell is not machined for exact symmetry or even if sample and reference pans of unequal masses are used, but software can be used to compensate for these imbalances. Imbalances can result in a curved baseline. Other influences like crosstalk between the sensors will result in unequal contributions from the sample and reference sensors. The operator can control pan contact to some extent. The pan bottoms should be crimped flat without any deformities and sit steady on the sensor. Although this is less of a problem for the reference pan, the sample pan may become deformed if a bulky material is encapsulated, such as irregularly shaped hard pieces with sharp edges. How to best encapsulate a bulky sample depends on the situation, but one can use sample pans pressed out of thick sheet, pulverize the sample into a powder, press the sample into a film, or use a specially designed crimper (like the Tzero™ crimper of TA Instruments). The last factor considered here is temperature uniformity throughout the sample. As mentioned above, a good uniform temperature distribution is dependent on sample size, heating rate, and packing. When preparing the sample, one needs to use a mass that does not result in a large temperature gradient for the heating (or cooling) rate used and ensure that there is good contact with the pan. Wunderlich (1990) calculated the maximum sample size for various heating rates when the maximum temperature gradient in the sample did not exceed ±0.5 °C for a disk-shape sample with a radius of 2.5 mm. His calculations showed that at a heating rate of

10 °C/min the maximum sample mass is 20 mg, at 1000 °C/min it is 2 mg, and at 100,000 °C/min it is 200 μg.

Finally, the thermal lag should not be confused with the lag time, also called *time to steady state*, although the thermal lag does influence the time necessary to reach steady state (the lag time is the time necessary to reach steady state after a DSC run has begun). The thermal lag becomes a factor once steady state has been achieved, while the initial startup of the DSC run characterizes the instrument response time or the time to steady state. This can be illustrated if one proceeds from an isothermal hold (where sample temperature, reference temperature, and block temperature are all identical, i.e., $T_s = T_r = T_{bl}$; see 1, below) to a heating experiment that results in an initial exponential rise of the heat flow until the steady-state condition is achieved. This initial rise represents a nonlinear response, and this part of the DSC curve does not contain information that could be used to evaluate transitions. For this reason the starting temperature should be far away from the thermal event of interest.

2.3.2. Heat Flux DSC

Heat flux DSC usually consists of a cell containing reference and sample holders separated by a bridge that acts as a heat leak surrounded by a block that is a constant-temperature body (see Fig. 2.1). The block is the housing that contains the heater, sensors, and the holders. The holders are raised platforms on which the sample and reference pans are placed. The heat leak permits a fast transfer of heat allowing a reasonable time to steady state. The differential behavior of the sample and reference is used to determine the

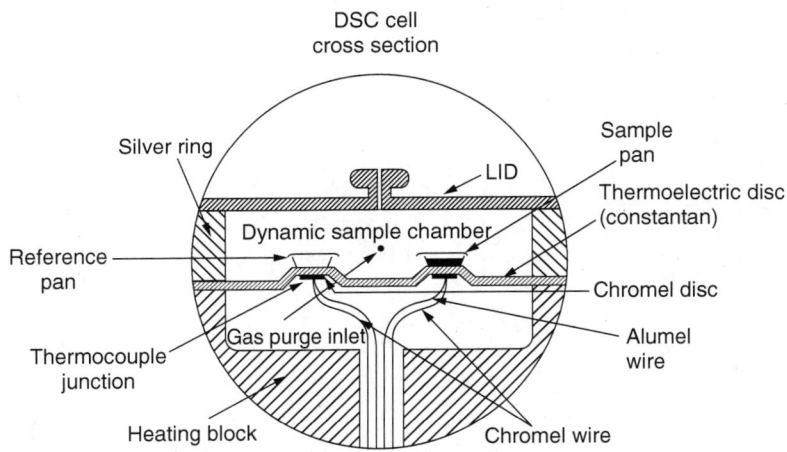

Figure 2.1. Cross section of a Du Pont (now TA Instruments) 910, 2910, and 2920 DSC heat flux cell (Blaine, Du Pont Instruments bulletin; courtesy of TA Instruments).

thermal properties of the sample. A temperature sensor is located at the base of each platform. Associated with the cell are a furnace and a furnace sensor. The furnace is designed to supply heating at a linear rate. However, not only the heating rate must be linear, but also the cooling rate during cooling experiments. This can be accomplished by cooling the block or housing of the instrument to a low temperature, where the heater fights against a cold block, or a coolant can be nebulized into the block. Finally, some inert gas, called the *purge gas*, flows through the cell.

The operation of the heat flux DSC is based on a thermal equivalent of Ohm's law. Ohm's law states that current equals the voltage divided by the resistance, so for the thermal analog one obtains

$$\dot{Q} = \frac{\Delta T}{R} \tag{2.25}$$

where $\dot{Q}$ a is the heat flow rate, ΔT is the temperature difference between the sample and the reference sensors, and R is the thermal resistance of the heat leak disk. This equation can also be derived from Newton's law of cooling when the heat flow rate is substituted for the slope of the cooling curve (Wunderlich 1990). Newton's law of cooling can be given by the following equation

$$-\frac{dT}{dt} = K(T - T_{\text{sur}}) \tag{2.26}$$

where the slope of the cooling curve, barring no transitions, is equal to the difference in temperature times a constant, K. The slope in this equation is denoted by a negative sign to indicate that it is a cooling rate, T is the temperature at any time t, and T_{sur} is the temperature of the heat sink (or surroundings).

2.3.3. TA Instruments Modules

2.3.3.1. Modules 910, 2910, and 2920 The TA Instruments Q series DSCs evolved from their 910, 2910, and 2920 modules. The DSC 910, 2910, and 2920 cells use a thermoelectric heat leak made of constantan (a copper/nickel alloy) as noted in Fig. 2.2. The sample and reference pans sit on raised platforms or pods with the constantan disk at their base. The temperature sensors are disk-shaped chromel/constantan "area" thermocouples and chromel/alumel thermocouples. The thermocouple disk sensors sit on the underside of each platform. The ΔT output from the sample and reference thermocouples is fed into an amplifier to increase their signal strength. The heating block is made of silver for good thermal conductivity and also provides some reflectivity for any emissive heat.

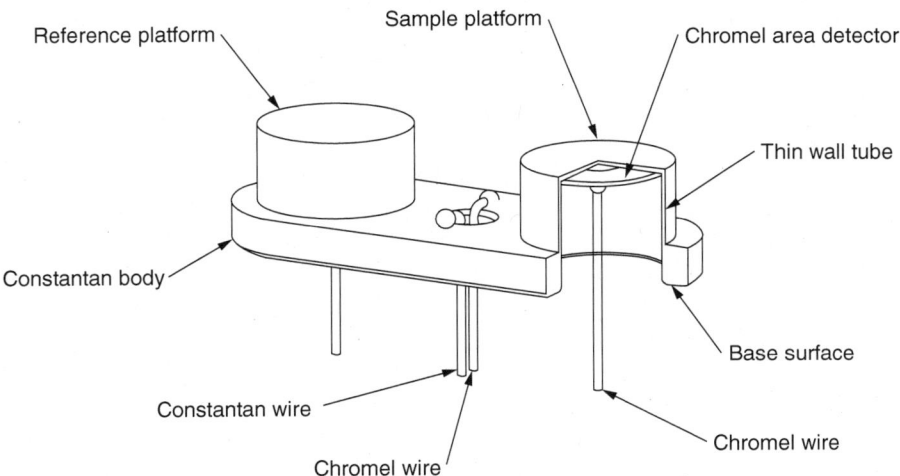

Figure 2.2. DSC sensor assembly for the TA Instrument Q10, Q20, Q100, Q200, Q1000, and Q2000 modules. Note the three thermocouple heat flow sensor design as compared to the two thermocouple heat flow sensor design as seen in Fig. 2.1. [From Danley (2003a); reprinted with permission of Elsevier and TA Instruments.]

Before the DSC experiment is started, the two calorimeters (i.e., the sample and reference pods, since they are separate calorimeters with one heater) are in equilibrium, they are at the same temperature: $T_{bl} = T_s = T_r$, where T_{bl} is the block temperature, T_s is the sample temperature and T_r is the reference temperature. When the operator starts the heating experiment, the block will be heated at a linear rate; therefore the sample and the reference calorimeters will also be heated. They will lag behind the block temperature, but to a different extent since the heat capacity of the sample calorimeter is higher because of the additional mass of the sample as compared with an empty pan for the reference. The sample temperature will lag behind T_r. Assuming the pan masses are identical, the $T_{bl} - T_s$ and $T_{bl} - T_r$ temperature differences will be proportional to the heat capacity of the sample and reference calorimeters, respectively. The temperatures T_{bl}, T_s, and T_r are measured by thermocouples.

In most cases the $\Delta T = T_s - T_r$ is displayed as a function of block temperature T_{bl}, or temperature of the Tzero sensor for the Q series DSCs. Thus, when heating starts, the DSC signal ($\Delta T = T_s - T_r$) will shift from zero (at the starting isothermal) to a steady state value of $T_s - T_r$. This shift is proportional to the heating rate and the sample heat capacity. Essentially, this phenomenon is used to measure the heat capacity of the sample (see Section 2.6, on heat capacity).

2.3.3.2. Q Series Modules The TA Instruments cell for the Q series DSC utilizes three thermocouples (see Fig. 2.2) and the associated Tzero technology

(Danley 2003, 2004). In addition to the sample and reference sensors, an additional center thermocouple, denoted T_0 (Tzero), is utilized for the heat flow measurements. Similar to the 910, 2910, and 2920 modules, there are two raised platforms for the sample and the reference on a constantan disk, which acts as a heat leak. The sample and reference disk thermocouples are attached to the underside of each platform. Two ΔT measurements are made. The first is taken between the chromel wires that are attached to the chromel disk area detectors. In addition, ΔT_0 is measured between chromel wires attached to the sample chromel disk and the T_0 sensor. A chromel wire is looped between the sample chromel disk and the T_0 sensor, which measures the sample temperature at the raised pod. The T_0 sensor temperature is measured at the junction of the constantan and chromel wires attached at the center of the heat leak base.

The operation of the Tzero (T_0) technology, as for the TA Instruments 910, 2910, and 2920 DSC modules, is also based on a thermal equivalent of Ohm's law. With the assumptions that the cell offers thermal resistance analogous to an electrical resistance and that the heat capacity of the platform pods needs to be accounted for, a heat balance equation can be constructed for each sensor pod:

$$\dot{Q}_s = \frac{T_0 - T_s}{R_s} - \frac{C_s dT_s}{dt} \tag{2.27}$$

$$\dot{Q}_r = \frac{T_0 - T_r}{R_r} - \frac{C_r dT_r}{dt} \tag{2.28}$$

In these equations, $\dot{Q}$ is the heat flow rate with the subscript s or r denoting the sample or reference pod, respectively; R is the thermal resistance, C is the heat capacity of the sensor pod (the thermal capacitance), T_s is the sample temperature, T_r is the reference temperature, T_0 is the temperature of the center sensor (which is called the *Tzero thermocouple*), and dT/dt is the heating rate.

Rearranging these equations, where $\dot{Q} = \dot{Q}_s - \dot{Q}_r$, results in the following four-term heat flow expression:

$$\dot{Q} = -\frac{\Delta T}{R_r} + \Delta T_0 \left[\frac{1}{R_s} - \frac{1}{R_r}\right] + [C_r - C_s]\frac{dT_s}{dt} - C_r\frac{d\Delta T}{dt} \tag{2.29}$$

Here $-\Delta T/R_r$ is the principal heat flow rate (which is roughly equivalent to that of the TAI 2920 model, which uses $\Delta T/R$, where R is the thermal resistance of the constantan heat leak disk), $[C_r - C_s]dT_s/dt$ is the thermal capacitance imbalance, and finally, $-C_r(d\Delta T/dt)$ is the heating rate imbalance (TA Instruments 1993).

It should be noted that ΔT_0 is the difference $T_s - T_0$ and therefore is different from ΔT, which is the temperature difference between the sample and the reference. The values to the four-term heat flow equation can be determined in the calibration routine using the calorimeter time constants. Calorimeter time constants (τ) can be obtained through the rearrangement of the preceding equations, where now

$$\tau_s = C_s R_s = \frac{\Delta T_0}{dT_s/dt} \tag{2.30}$$

$$\tau_r = C_r R_r = \frac{\Delta T_0 - \Delta T}{(dT_s/dt) - (d\Delta T/dt)} \tag{2.31}$$

The time constants are determined by running a *baseline* in an empty cell (i.e., without the sample pan and the reference pan), while the thermal capacitance is calculated using a sapphire run; two similar mass sapphire disks are placed on the sample and reference platforms, and heated at the same rate as the above mentioned baseline run. The thermal resistance is defined as the time constant divided by the thermal capacitance. Three modes are possible for the Q1000 and Q2000 DSCs:

1. The $T1$ mode, which uses only the first term in Eq. (2.6) and thus makes the mathematical treatment similar to that for the older 910, 2910, and 2920 models.
2. The $T4$ mode, which uses all four terms in Eq. (2.6), including a correction for the different heating rates of the sample and reference calorimeters, due to their different heat capacities.
3. The $T4P$ mode, which also includes a correction for the differences in mass of the sample and reference pans, and a correction for thermal resistance of the pan material (aluminum, copper, etc.) in addition to the terms listed for the $T4$. One should correct for the difference in pan masses to account for their effects on the heat capacities of the sample and reference calorimeters. Only the Q1000 and now the Q2000 are capable of operating in the $T4P$ mode.

The additional "P" correction is done using a model based on the pan type and consists of summations of the thermal resistances between the sample and the sensors. This value is then substituted back into Eq. (2.6) to correct for smearing of the heat flow due to the additional thermal resistances. Also allowed in the $T4P$ mode is the correction for the difference between the sample and reference pan masses to account for slight differences in the pan heat capacities if the pan masses were not matched.

2.3.4. Mettler Toledo DSC

In most instruments, if the DSC sample holder is damaged, then the entire cell needs to be replaced. However, for the Mettler Toledo units the sensors are replaceable. Two sensor types are available, the FRS5 and the HSS7. Instead of using one thermocouple, Mettler Toledo uses a grouping of thermocouples called a *thermopile* (see Fig. 2.3 for an example). Thermocouples in a thermopile may be connected in parallel or in series; the series arrangement is generally used when the output signal of the thermocouple needs to be intensified. For the Mettler Toledo units the thermocouples are connected in series using a symmetric zigzag pattern. The difference between the sample and reference thermopiles is proportional to the difference between the heat flows to the sample and reference.

For the Mettler Toledo thermopiles, the thermocouples form a star-shaped symmetric pattern around a measuring point. Because of the symmetric arrangement of the thermocouples, the thermopiles will cause imbalances to cancel that would otherwise result in distortions in heat flow. Each thermocouple in a thermopile forms a junction under the pan and away from the pan so that each has an associated ΔT signal at both measuring points. Each thermopile signal is the summation of the ΔT signals from each thermocouple junction. A final ΔT signal is obtained from the difference between the sample and reference thermopiles. The heat flow rate ($\dot{Q}$) can now be expressed as

$$\dot{Q} = \sum_{i=1}^{n} \dot{Q}_i = \frac{1}{R}\left(\sum_{i=1}^{n} \Delta T_{si} - \sum_{i=1}^{n} \Delta T_{ri}\right) \tag{2.32}$$

where R is the thermal resistance of the heat leak but the summation of the differential temperatures are substituted for the temperature of the sample and reference (Riesen 1998). A thick-film screen process is used where a conductive paste is laid on an electrically nonconductive substrate to form thermocouple

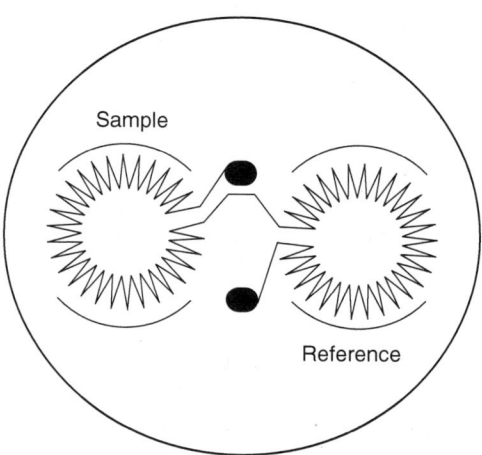

Figure 2.3. Schematic of a thermopile used in a Mettler Toledo DSC (courtesy of Mettler Toledo).

wires. The paste is then heat-treated by firing thereby forming and permanently affixing the wire to the substrate (Kehl and van der Plaats 1991).

The FRS5 sensor, which is usually recommended for polymer analysis by the manufacturer, has an arrangement of 56 thermocouples (28 for each sensor). The HSS7 sensor, which is recommended for the detection of low energy transitions, has a layered arrangement of 120 thermocouples (i.e., three layers of 20 for each sensor). The reported temperature (x-axis display) is monitored at the furnace block by a platinum resistance thermometer. Two different upper-limit temperatures are available (500 and 700 °C), depending on the associated furnace power amplifier. The higher-power amplifier also allows fast heating rates up to 300 °C/min. The silver furnace block is located below the sample chamber but above the flat heater element, which is located above a cooling flange. The cooling flange is associated with a cold finger that allows the attachment of a number of cooling units. The assembly is held together by compression springs, which allow for thermal expansion and contraction (Mettler Toledo 2005). The TOPEM software of Mettler Toledo is capable of making correction for the difference between the sample pan and reference pan masses for heat capacity determination.

2.3.5. Power Compensation DSC

The first power compensation instrument, the Perkin-Elmer DSC1, was marketed in 1963. This was the first instrument termed a "differential scanning calorimeter" (O'Neill 1964). Presently only PerkinElmer Life and Analytical Sciences (also referred to as either Perkin-Elmer or Perkin Elmer) is marketing power compensation DSCs. Unlike the heat flux DSC, which is based on one furnace only, the power compensation DSC has two separate identical holders (sample and reference holders), each with its own heater and sensor (Fig. 2.6). The material of the holders is a platinum/rhodium alloy. These two calorimeters are placed in a common block of constant temperature. Depending on the method of cooling, the block can be at various temperatures (e.g., –100 °C with a mechanical cooling accessory, –196 °C when liquid nitrogen is the coolant, etc.). The sample holder contains the sample in a sample pan (made of high-purity aluminum, copper, gold, or platinum), and the reference container contains an empty sample pan as a reference. Both holders are covered with a platinum lid. Platinum resistance thermometers (Pt sensors) measure the temperature. These thermometers are built into the base of the calorimeter holder. Next to the thermometers, the two individual heaters are also built into the bottom of the holders (see Fig. 2.4). The heaters and the sensors are identical; in the early 1980s when using an "AutoZero Accessory" to straighten the baseline, the role of the sensor and the heater could even be inverted with the help of a special cable. As in any other type of DSC, an inert purge gas of constant and predetermined rate flows through the cells.

In this DSC, a programmer sends a signal to the average amplifier that heats the cells at a constant linear heating rate. When talking about a "heating rate"

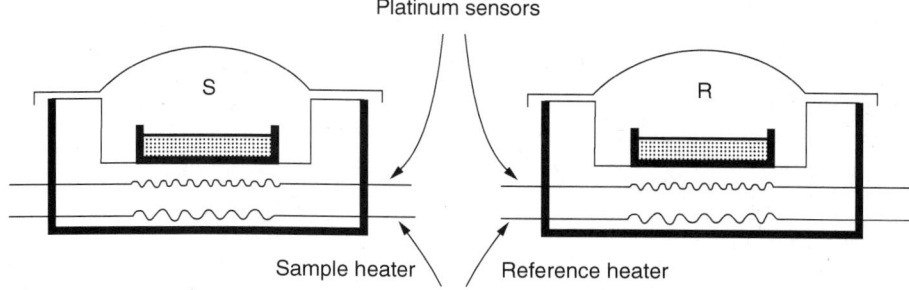

Figure 2.4. Typical power compensation sample holder with twin furnaces and sensors [from Wunderlich (1990); reprinted with permission of Elsevier, Perkin-Elmer and B. Wunderlich].

in a power compensation DSC, one refers to the program temperature of the sample holder and the reference holder. Mathematically this is expressed as

$$T_{av} = \frac{T_s + T_r}{2} \tag{2.33}$$

The signal that is returned from the sensors of the sample and reference holders is compared to the programmer's signal, and the average amplifier uses this difference signal to ensure that the average temperature during the heating is as close to the programmed temperature as possible. The two holders are matched at the factory on the basis of their mass, heat capacity, and electrical properties. When one of the holders has higher heat capacity (usually the sample holder, due to the contained mass between the sample), it will be heated slower; thus it will lag behind the other holder. This temperature difference signal is fed into a differential amplifier, which in turn sends power into the lagging calorimeter to correct for the imbalance between the sample holder and reference holder temperatures. Thus, this differential power attempts to equalize the temperature of the sample and reference holders. The circuit without the correction of the differential amplifier is called an *open loop* and a *closed loop* when the differential amplifier steps in to correct the imbalance of temperatures (see Fig. 2.5). At the same time, the supplied differential power is displayed as a function of the program temperature, and this display is called the *DSC curve*. The schematics of the power compensation DSC is shown in Fig. 2.6.

If the two calorimeters were absolutely identical, a straight, horizontal "flat baseline" would be recorded as the DSC curve of the empty calorimeters containing only two empty sample pans of identical masses. In practice this is not the case—two absolutely identical calorimeters cannot be found; therefore, since the heating of the calorimeters is accomplished with two separate individual heaters, usually there is some curvature in the baseline. Adjusting the "ΔT Balance" and the "Slope" knobs of the older types of power compen-

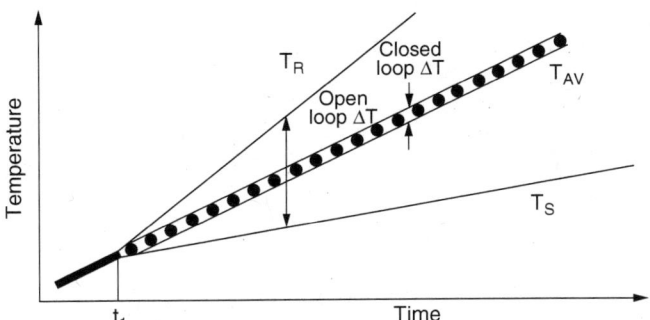

Figure 2.5. Perkin-Elmer power compensation DSC: open loop similar to the ΔT of a heat flux DSC and closed loop after adjusting from feedback to keep both DSC cells at approximately the same temperature [from Wunderlich (1990); reprinted with permission of B. Wunderlich and Elsevier].

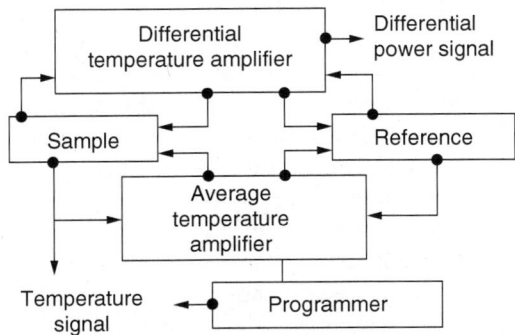

Figure 2.6. Simplified schematic of the operation of a power-compensating DSC [from Wunderlich (1990); reprinted with permission of B. Wunderlich and Elsevier].

sation DSCs (e.g., DSC7) can substantially decrease this curvature and the possible sloping of the baseline. If any residual curvature is left, the baseline should be subtracted. This means that

- The instrumental baseline (viz., the DSC curve with two empty sample pans as close in mass as possible; i.e., difference in pan weights should be <0.01%) should be run and saved.
- The sample should be encapsulated in a sample pan of the same mass as the two previously mentioned sample pans (used for recording the baseline); then the sample run must be carried out at identical conditions as the instrumental baseline, and finally, the baseline should be subtracted from the sample run.

It is important that all the run parameters should be identical during the sample and baseline runs. Under these conditions, if condensation on the swingaway enclosure cover (in the DSC2, DSC4, and DSC7), and also on the DSC block surface, is avoided, the baseline is usually quite reproducible, so the subtraction gives DSC curves similar to those of the heat flux DSCs.

It has been shown that the measured temperature difference between the sample holder and the reference holder (ΔT) is proportional to the power necessary to establish an approximately equal temperature of the two holders (Perkin-Elmer 1970). The heat flow rate into the sample and the reference calorimeters can be expressed by the equations

$$\frac{dQ_r}{dt} = \lambda(T_r^{meas} - T_r) = W_{av} \tag{2.34}$$

$$\frac{dQ_s}{dt} = \lambda(T_s^{meas} - T_s) = W_{av} + W_d \tag{2.35}$$

where dQ/dt is the heat flow rate ($= \dot{Q}$), T_r^{meas} is the temperature of the reference holder measured at the platinum thermometer, and T_r is the true reference temperature (T_s^{meas} and T_s are defined analogously), λ is the thermal conductivity, and the subscripts s and r refer to the sample and the reference, respectively; W_{av} is the power from the average amplifier, and W_d is the power from the differential amplifier in an attempt to equalize the temperatures of the sample and reference holders. To simplify this mathematical treatment, we will consider the total W_d going to only the sample calorimeter. In practice both the sample and reference calorimeters are adjusted for a quicker instrumental response with $\frac{1}{2}W_d$ added to the sample holder and $\frac{1}{2}W_d$ subtracted form the reference holder. The total W_d can be mathematically described as equal to the amplifier gain signal X from the feedback loop multiplied by the difference of the measured temperatures between the reference and sample holder, so

$$W_d = X(T_r^{meas} - T_s^{meas}) = X \Delta T^{meas} \tag{2.36}$$

Thus, the mentioned open loop is similar to the heat flux DSCs ΔT, while the closed loop accounts for the power feedback from the heater. Therefore

$$\lambda(T_s^{meas} - T_s) = \lambda(T_r^{meas} - T_r) + W_d \tag{2.37}$$

and

$$W_d = -\lambda \Delta T^{meas} + \lambda \Delta T \tag{2.38}$$

where ΔT expresses the actual temperature difference between the sample and the reference. Since

$$W = \lambda \Delta T \tag{2.39}$$

where W is the true differential heat flow needed to achieve isothermal conditions between the sample and the reference, this equation leads to the following:

$$\Delta T^{\text{meas}} = \frac{W - W_d}{\lambda} \tag{2.40}$$

Combining this with Eq. (2.36), the following equation is obtained if $\lambda \ll X$ (Wunderlich 1990):

$$W = W_d\left(\frac{\lambda}{X} + 1\right) \approx W_d \tag{2.41}$$

This latter equation describes the heat flow needed to achieve identical sample and reference temperatures. It indicates that the heat flow from the differential power (W_d) is approximately equal to the actual heat flow W if the amplifier gain X is made very large, so that $\lambda/X \ll 1$.

Summarizing, in power compensation DSCs, the sample and the reference holders are heated with an average amplifier, so that to obtain a constant average heating rate. Since the sample holder has higher heat capacity than the reference holder (i.e., when the difference between them is the heat capacity of the sample), during the heating its temperature will lag behind that of the reference holder, and another power source, the differential amplifier, will provide additional heat to the lagging sample holder in order to equalize its temperature with the reference holder. In the DSC trace, the power from the differential amplifier (W_d) is displayed as a function of program temperature. In the absence of a chemical reaction or a thermal transition, W_d is approximately proportional to the heat capacity of the sample.

In addition to the description of the DSCs of the three best-known commercial manufacturers in the previous paragraphs, Netzsch, Setaram, Seiko Instruments, and Shimadzu are often encountered on the US market. These are all heat flux DSCs; Perkin-Elmer is the only power compensation DSC manufacturer.

After describing the principle of differential scanning calorimetry measurements, we briefly describe the most important concepts below.

2.3.6. Heating Rate

The *heating rate*, the most important parameter of the DSC runs, expresses how fast or slow the sample is heated. The unit is degrees Celsius per minute (°C/min). The value of the heating rate is important because considerable time can be saved if the DSC runs are made at fast (high) heating rates. However, this is not always possible, because the heating rate influences the shape of the DSC curves and also the results calculated from them. At fast heating rates the temperature resolution dramatically decreases because of the large temperature gradients within the sample and hence a dramatic increase in the thermal lag. Likewise, too large a sample mass will lead to broadened transitions with decreased resolution and increased thermal lag due to considerable temperature gradients in the sample. Thus, for high resolution, slow heating rates (and small sample masses) are desirable, but the instrument sensitivity decreases, and longer times are required to perform the DSC runs. On the other hand, the sensitivity can be increased by increasing the sample mass and the heating rate at the expense of resolution. This is one way to increase sensitivity in heat capacity measurements.

Also, as will be described in Section 2.7, different melting results are obtained at slow and fast heatings as a result of time-dependent melting phenomena. In addition, changing the heating rate will also change the shape of the glass transition. Therefore, the heating rate should be selected according to the information that the operator seeks. The most frequently used heating rate is 10 °C/min; for purity determinations it is 1 °C/min. Most commercial instruments are capable of running experiments with heating rates of ranging from 0.1 °C/min~100 °C/min, some instruments, like the Perkin-Elmer Hyper® DSC are capable of heating at rates of ≤500 °C/min. It should be remembered that the instrument should be calibrated at the heating rate at which the actual runs are made.

Specific recommendations and guidance on how to carry out DSC measurements are given in sections on specific measurements.

2.3.7. Cooling Rate

The cooling rate is another important parameter during the DSC runs. Especially in polymers, changing the cooling rate may drastically change the results. First, the crystallization temperature and the glass transition temperature measured on cooling will change when the cooling rate is changed. The crystallinity of the polymer can be strongly influenced by the cooling rate; at fast coolings many samples can be quenched (i.e., avoiding crystallization and transferred directly into the glassy state). For determination of accurate temperature values during cooling, the instruments need to be calibrated with cooling standards (Menczel and Leslie 1990, 1993; Menczel 1997). The cooling experiments, in addition to determining the characteristic temperatures of crystallization and glass transition, are important for introduction of reproducible thermal history into the samples when comparison with other samples is necessary (see discussions of phase transitions in amorphous and semicrystalline polymers in Section 2.7). The available cooling rate depends on the instrument type, temperature range of the experiment, and the cooling medium. Most frequently liquid nitrogen, cold nitrogen vapor, or a mechanical cooling accessory is used to enhance the cooling capabilities of the instruments.

2.3.8. Sample Mass

Before running the DSC, one should weigh the sample to an accuracy of at least ±0.2%, for example, ±0.02 mg for a 10-mg sample or ±0.002 mg for a 1-mg sample. For these purposes electronic microbalances are used. The manufacturers may recommend certain balances to the buyer.

Also, selecting the correct sample mass is critical in a DSC experiment. As with the heating and cooling rates, some optimum value should be selected with the sample mass as well. Too large a sample mass leads to an undesirable increase in the thermal lag and temperature gradient within the sample. With extremely large sample masses, the linear proportionality between the sample mass and the measured heat effect may be lost. On the other hand, too small a sample mass may cause undesirable decrease in the sensitivity. For most DSC experiments the sample mass should be between 3 and 10 mg. Larger sample masses (up to 20–40 mg) can be used for specific heat capacity measurements; lower masses (as small as possible, but definitely less than 2–3 mg) are necessary in purity determinations in order to stay close to equilibrium conditions during low-molecular-mass melting experiments. Small sample masses are desired in MTDSC experiments as well.

2.3.9. Purge Gas

The DSC cells should be continuously purged with a high-purity gas, usually dry. This is usually some inert gas, although oxygen, air, or other reactive gases also can be used if the purpose of the experiment is to study the behavior of the sample in such an atmosphere. There are several reasons for purging the cell with an inert gas:

- To prevent condensation of water on the cell and on the sample at sub-ambient temperatures
- To carry away contaminants released by certain samples at higher temperatures
- To ensure that a constant atmosphere exists around the sample in the sample holder
- To prevent formation of gas turbulences as the temperature is being increased
- To prevent deposition of degradation products inside various parts of the instrument
- To increase heat transfer to the sample
- To prevent oxidation in high-temperature measurements

Whenever a high thermal conductivity purge gas is needed, or DSC analysis is performed at temperatures lower than −196 °C (the boiling point of nitrogen), helium is preferred. But in practice almost always nitrogen is used because of the high cost of helium [the thermal conductivity values of He and N_2 at $27\,°C$ are 156.7 and $25.8 \times 10^{-3}\,W/(m\cdot K)$, respectively (Lide 1998–1999)]. Table 3.2 (in Chapter 3), lists thermal conductivities of several common purge gases.

The purge gas should be dry, unless special experiments are carried out in humid atmosphere such as studying thermal degradation in a humid environment. The flow rate of the purge gas is very important, and it should be set as suggested by the manufacturer. In the actual DSC experiments, the same purge gas and the same flow rate must be used as in the calibration procedure.

2.3.10. Encapsulating the Sample

The sample should be encapsulated in some kind of a DSC pan. The DSC pans are made of high-purity metal (Al, Pt, Au, Ag, Cu, or stainless steel). DSC pans for various purposes are available:

- *Standard DSC Pans.* When using these pans, the sample is placed in a disk shaped metal crucible, a flat lid is then placed on top of the sample, and a special device ("crimper") pushes down the lid on the sample, simultaneously folding the edges of the pan on the lid (see Figs. 2.7 and 2.8). These are the most frequently used DSC pans for polymer measurements; the

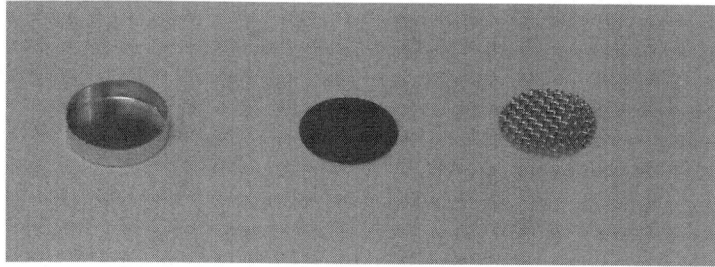

Figure 2.7. A DSC standard pan, a standard lid, and a lid punched out of aluminum mesh (the pan and the standard lid are from Perkin-Elmer).

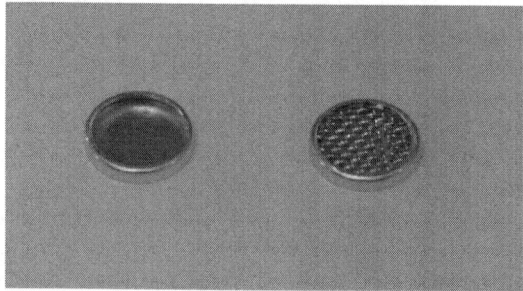

Figure 2.8. A standard DSC aluminum pan crimped with a standard lid and a lid punched out of aluminum mesh.

pressure inside the pan remains constant at any temperature. The purge gas has access to the sample, although this access is not completely free because of the tight folding of the pan edge on the lid. When it is important that the purge gas be in close contact with the sample, the lids must be punched out of fine Al (or gold, silver, platinum, or copper) mesh.

- *Hermetically Sealed Pans.* The sample is put in the pan, which is then closed with a special lid, and a crimper cold-welds the bottom and top parts of the pan (see Fig. 2.9). These pans are usually able to withstand 0.2 MPa pressure (2 atm); at higher pressure these pans rupture, which can be clearly seen in the DSC curve—when the pan ruptures, it suddenly moves, and this will appear as a spike. Special, low-volume (15 μL), hermetically sealed aluminum sample pans are available from Perkin-Elmer with capability of withstanding 30 atm of internal pressure without rupturing.
- *High-Pressure DSC Capsules.* These are made of thick stainless steel. The sample is put into the bottom part of the capsule, a gold-plated copper disk is then placed on it as a sealing agent, and the top of the capsule is screwed onto the bottom part (Fig. 2.10). Such DSC capsules

Figure 2.9. Bottom and top of a hermetically sealed DSC pan (left and center pieces), and a crimped pan (Perkin-Elmer pans).

Figure 2.10. Bottom and top parts of a high-pressure DSC capsule, and gold-plated copper plate (top row). Also shown is sealed capsule (bottom row) (Perkin-Elmer high-pressure capsules).

can withstand up to 20 MPa pressure (~200 atm). The disadvantage of these capsules is their heavy weight; since no two capsules have the same mass, the baseline with these capsules is often heavily sloped. Therefore, it is recommended to run a baseline with the empty capsule (and of course, an empty reference capsule). The bottom and top parts of the empty sample capsule should not be screwed together; rather, the copper ring and the top should be simply placed on the bottom part. Then the same pan should be used for the sample run, and the baseline should be subtracted.

Other special capsules, such as high-volume DSC capsules, or pans made of other materials, are also available.

2.3.11. Autosamplers

Many DSCs can be purchased with an autosampler. Some of these devices will load up to 64 devices the DSC after the previous run has been completed. Considerable time can be saved with these devices, because the instruments can operate overnight and on weekends.

2.3.12. Baseline

First, we need to mention that the word *baseline* in differential scanning calorimetry has three basic meanings: (1) the *instrumental baseline*, which is the DSC curve recorded with two empty pans; (2) the extrapolated part of the DSC curve during a transition over which the peak should be integrated; and (3) the *pre-* and *postmelting baselines* (see Section 2.7).

Since the cell of power compensation DSCs is composed of two calorimeters, each with its own heater, the baseline of these calorimeters is not as flat as the baseline of heat flux DSCs. This is no reason for alarm; when running power compensation DSCs, the instrumental baseline should be run from time to time, and the baseline should be subtracted from the sample run. Of course, all the parameters during the baseline and sample runs should be identical. But the baseline is sensitive to the environment; therefore care should be taken to keep the environmental temperature constant.

2.3.13. Different Types of Differential Scanning Calorimeters

As the reader may know by now, two types of DSCs are available: heat flux and power compensation DSCs. Generally, both types can give identical, or almost identical, results; and both have their advantages and disadvantages. The advantage of the heat flux DSCs is the flat and perhaps more reproducible baseline. Power compensation DSCs perform better when isothermal measurements are needed. They can heat to the desired temperature faster, avoiding unwanted thermal events during the heating and establishing steady state more quickly because of their smaller furnace mass.

2.4. PURITY DETERMINATION OF LOW-MOLECULAR-MASS COMPOUNDS BY DSC

The purity determination of low-molecular-mass organic crystalline compounds by DSC (the so-called calorimetric purity method) is a popular technique because it is simple and quick. While this technique is not specific regarding the contaminations in low-molecular-mass crystalline substances, it can give purity estimations with reasonable accuracy in about 1–2 h. Also, there is no need for a highly skilled scientist; within a couple of hours, an operator can be trained to carry out the measurements.

This method cannot be used for determination of the impurity concentrations in macromolecular systems, because, as previously mentioned, the melting of polymeric crystals is a process far from equilibrium. Therefore, the purity of crystalline polymers cannot be evaluated on the basis of melting point lowering. But even in polymer chemistry and physics, there is often a need to deal with low-molecular-mass crystalline compounds, such as certain monomers, initiators, and additives. In those cases, the method described below provides a valuable way for quick and simple determination of purity.

Gray (1966a,b) was the first to recognize that the melting point lowering of a crystalline substance due to the presence of certain impurities can be used for estimation of its purity. However, this method is valid only if the impurity does not form a solid solution with the primary crystal, but is soluble in the melt of the major component. Therefore, only low-molecular-mass eutectic systems can be evaluated with this method. Also, this method is applicable in the purity range of ~95–100 mol%, preferably 98–100 mol%. If the impurity is soluble or partially soluble in the major component of the crystalline phase, the method described here does not work.

The melting point lowering of a low-molecular-mass material due to impurities can be described by the so-called van't Hoff equation (Gray 1966a,b; Wunderlich 1990; Brennan et al. 1984; Davis and Porter 1969):

$$T_m^\circ - T_m = \frac{RT_m^{\circ 2} x_{imp}}{\Delta H_f} \qquad (2.42)$$

where R is the universal gas constant [8.314 J/(K·mol)], T_m° is the melting point of the 100% pure sample, T_m is the melting point of the sample containing the impurity, x_{imp} is the mole fraction of the impurity, and ΔH_f is the heat of fusion of the major component (J/mol).

The final equation for the purity determination by DSC as derived by Gray (1966a,b) is

$$T_m^\circ - T_s = \frac{RT_m^{\circ 2} x_{imp}^2}{\Delta H_f} \frac{1}{F} \qquad (2.43)$$

where T_s is the sample temperature, and F is the fraction of the material (major component) melted to temperature T_s. The fraction F can be determined by integrating the melting peak from the beginning of melting to temperature T_s, and then dividing it by the total area under the peak:

$$F = \frac{\Delta H_{tF}}{\Delta H_f} \qquad (2.44)$$

Thus, T_s as a function of $1/F$ must give a straight line. This line is called the *van't Hoff plot*. The mole fraction of the impurity can be calculated from the slope of this line, while extrapolation to $F = 1$ will determine the melting point of the 100% pure material.

It needs to be mentioned that two melting peaks should exist in eutectic systems for most impurity concentrations: a small melting peak at lower temperatures corresponding to the melting of the major component + the impurity, and an intense melting peak of the major component at higher temperatures. Because of the small concentration of the impurity and deviation from equilibrium, and also the fact that the melting of the impurity is not considered in Eq. (2.1), the T_s-versus-$1/F$ $[T_s-(1/F)]$ plot is not linear, it is curved upward (see Fig. 2.12) (Davis and Porter 1969; Gray and Fyans 1973). The deviation from linearity is somewhat proportional to the impurity concentration. Therefore, in order to handle this discrepancy, the purity software adds a small constant number to x (i.e., $x = x + k$), and recalculates the purity with these new data. Then the software calculates the standard deviation of the points used in Eq. (2.43), and it continues to add k to x until the smallest standard deviation is obtained when fitting a straight line to the $T_s-(1/F)$ plot. The results obtained are reasonably close to those determined by other methods.

The important parameters during a DSC purity determination are the following:

- *Sample Size.* This should be as small as possible, in order to keep the melting of the sample close to equilibrium conditions, up to 5 mg.
- *Heating Rate.* This should be very slow for the same reason (close to equilibrium conditions, usually 0.5 °C/min or 1 °C/min).
- *Pan Type.* The larger the contact area between the DSC pan and cell, the more reliable the data. So, whenever sublimation of the sample is not a factor, use of standard DSC pans rather than hermetically sealed pans is recommended.
- *Pan Crimping.* It is critical to decrease both the external and internal thermal resistance (see Section 2.3) during the purity runs. For this reason it is important to ensure the maximum area of contact between the bottom of the DSC pan and the cell, and also between the sample and the bottom of the DSC pan. Therefore, it is not recommended to run chunky samples. Also, when running powdery samples, the particle size of the powder should be as small as possible. Whenever possible, the sample

should be melted and crystallized so that it covers the bottom of the DSC pan with a continuous thin film, and purity should be determined on the second heating.
- *DSC Calibration.* Calibration of the DSC must be carried out at a heating rate identical to the heating rate of the purity run. Also, as can be seen in Fig. 2.14 of Section 2.5, the angle between the leading edge of the calibration standard depends on the mass of the standard, so the sample mass of the purity run should be close to the mass of the standard (the temperature of any point of an endothermic peak is determined by dropping a straight line to the baseline, not perpendicularly, but at the same angle as the indium leading edge with the temperature axis).

These days, the purity determination by DSC is a relatively simple job; most, if not all, commercial instruments are provided with the software necessary to perform the purity determination. All the operator needs to do is calibrate the instrument well, have a clean sample holder with a good baseline, and predetermine the melting point of the sample in question using a faster heating rate (e.g., 10°C/min). Then, from this faster heating rate run, the operator can determine the starting and end temperatures of the purity run. The actual purity determination by DSC should be carried out at a slow heating rate (most often 1°C/min), so that melting will occur at close to equilibrium conditions. Finally, the molecular mass of the sample needs to be entered, and the melting peak of the sample needs to be integrated in the purity software, and the value of the molar purity is displayed. Partial areas between 10% and 60% are generally used for the purity determination. The operator has an option to change these limits in special cases (e.g., if the melting gives a multiple endotherm, and only the first partially overlaid peak can be used for the calculations).

A good example of an application of the van't Hoff calorimetric purity method is the determination of purity of 4,4'-thiobis(3-methyl-6-*tert*-butylphenol), a commercial antioxidant, with the trade name Santonox R and a molecular mass of 358 g/mol. This type of phenolic antioxidant is used to prolong the life of polyolefins exposed to oxygen. Two comparative melting scans of different lots of Santonox R revealed a startling difference in melting behavior as shown in Fig. 2.11 (Bair 1997).

The antioxidant with the lower and broader melting curve (solid line in Fig. 2.11) is clearly less pure than second sample of Santonox R (broken line).

A plot of temperature versus $1/F$ for the higher-melting-point antioxidant produced a curve that is concave upward (filled circles, Fig. 2.12). The van't Hoff plot was made linear by increasing F by 1.2%, and the purity computed from the slope of the line was 99.7 mol%. Purity analysis of the lower-melting-point antioxidant sample yielded a purity of 95.9 mol%.

Fortunately, the large amount of impurity in the broader melting lot of hindered phenol does not have a deleterious effect on the additive's oxidation induction time (OIT), as compared to the OIT of the higher-purity antioxidant.

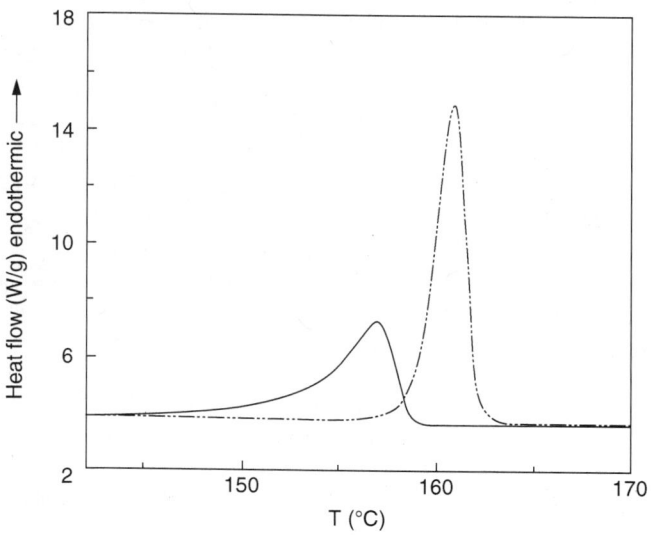

Figure 2.11. Melting curves of two lots of a commercial phenolic antioxidant at a heating rate of 12 °C/min [from Bair (1997); reproduced with permission of Elsevier].

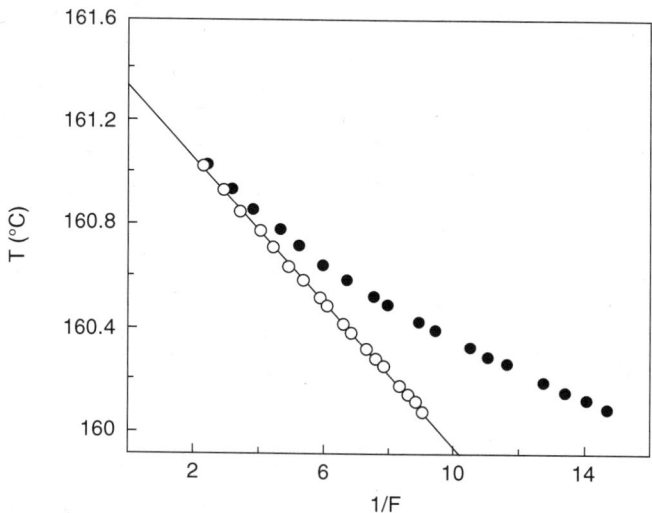

Figure 2.12. Sample temperature plotted against the reciprocal of the melted fraction of the higher melting lot of Santonox R [from Bair (1997); reproduced with permission from Elsevier].

Hence, both additives should provide similar protection in retarding the degradation of a polyolefin polymer such as polyethylene if employed in equal concentrations. Recently this purity method has proved effective in determining how long a dopant molecule can be held in the melt during a continuous fiber extrusion process before degradation results in undesirable changes in the material's optical properties. These two practical examples demonstrate the utility of the DSC purity technique in indirectly monitoring the use of additives in polymers in two very different industrial applications.

2.5. CALIBRATION OF DIFFERENTIAL SCANNING CALORIMETERS

The calibration of DSCs is one of the most important jobs a thermal analyst needs to perform. Thermal analysis instruments need to be calibrated because the indicated temperatures, heats, and heat capacities do not reflect the real values. Thus, certain procedures are necessary to enable the instrumental software to recalculate these values indicated by the instrument, to real temperature, heat, and heat capacity values. This procedure, called *calibration*, consists of measuring thermal properties of standard materials whose thermal properties are well known. All the results of subsequent actual measurements depend on the validity of the calibration; therefore all calibrations have to be carried out carefully.

Before carrying out the actual DSC experiments, both the abscissa and the ordinate have to be calibrated. In a DSC curve, the abscissa displays the temperature (or time for isothermal experiments), while the ordinate displays the heat flow signal (dQ/dt) or heat capacity signal (C_p).

When and how often should a DSC be calibrated? The calibration should be checked regularly. Although there is no rule of thumb as to how often the calibration should be checked, it should be done at least once a week for an instrument in continuous operation.

2.5.1. Temperature Calibration of DSCs

2.5.1.1. Temperature Calibration on Heating For DSC experiments, the temperature should be calibrated on heating and also on cooling if cooling experiments are also conducted (see next subsection). For isothermal experiments, determination of the true value of temperature is more difficult; it should be done by calibrating the DSC at various heating and cooling rates and interpolating the $T_{real} = F(T_{ind})$ dependence to zero heating rate (where T_{real} is the real melting point of the standard and T_{ind} is the indicated temperature or the melting point of the standard of the uncalibrated instrument).

Since the vast majority of calibration experiments are done on heating, we will first describe the calibration for heating runs (*heating calibration*). Traditionally, the true temperature in heating calibration experiments is taken from melting points of high-purity water and very high-purity metal standards. The energy (heat of fusion) calibration is usually carried out simultaneously with the temperature calibration.

To enable analysis of the melting peak of water and metal standards, the characteristic temperatures of an endothermic peak of a low-molecular-mass crystalline compound will be described (see Figs. 2.13a and 2.13b). The

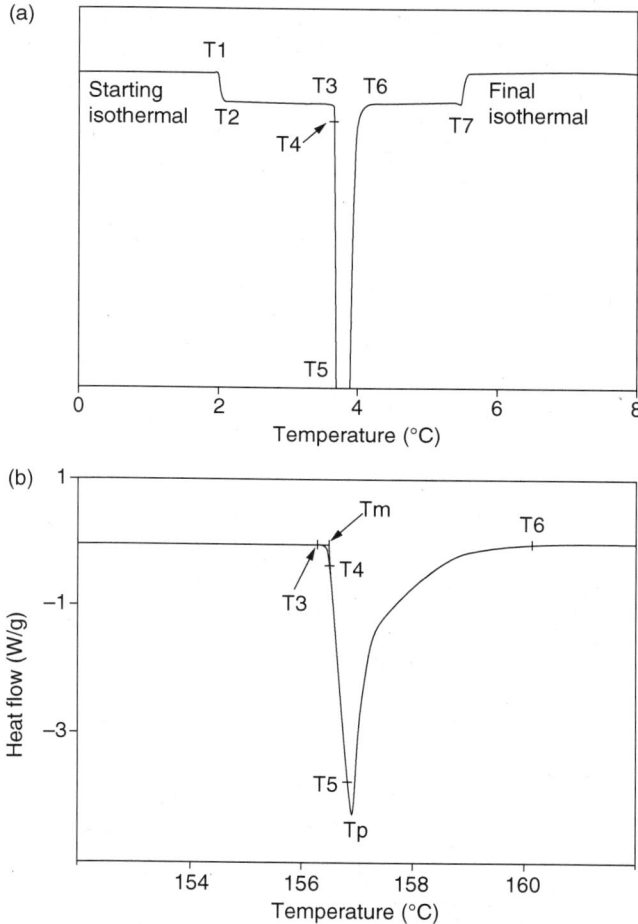

Figure 2.13. Analysis of a melting peak of a low-molecular-mass crystalline substance (in the present case, indium): (a) total DSC curve, including beginning and ending isothermals (as a function of time); (b) heating part of DSC curve. In both diagrams T_1 = temperature at which heating starts, T_2 = temperature where steady-state conditions have been achieved, T_3 = temperature where melting starts due to impurities, T_m = melting point; between T_4 and T_5 the sample temperature is constant during the melting process, T_p = peak temperature (melting process ends at T_p), and from T_p the sample temperature starts catching up with the reference temperature; by T_6 we have steady state again; at T_7 the heating stops and the final isothermal is recorded. Endotherm is down in both diagrams.

experiment is started with holding the sample isothermally at temperature T_1. Then, after a certain isothermal period (e.g., 2 min), the heating starts and a transient signal can be observed, the magnitude of which is proportional to the heat capacity of the sample and the heating rate as described in Section 2.6. After the transient signal ends at temperature T_2, steady state is established, and a horizontal "baseline" is recorded between temperatures T_2 and T_3. At T_3, the melting process of the sample starts, and this is indicated by a shift of the DSC signal in the endothermic direction. At the beginning, deviation from the baseline has an accelerating character (up to temperature T_4), then the curve becomes linear in the temperature range from T_4 to T_5. This part of the curve is called the "leading edge." The continuous signal rising in the T_3–T_4 temperature range is caused by minor amounts of impurities.

It should be remembered that the abscissa of a DSC trace displays the block temperature or, in the case of the TAI Q series DSCs, the temperature of the Tzero sensor (heat flux DSCs), or the program temperature (power compensation DSCs). Therefore, although the block temperature or the program temperature increases linearly in the temperature range T_4–T_5, the sample temperature is constant as required by the laws of thermodynamics for equilibrium processes; once the melting process has started, the temperature of the sample remains constant (at T_m) until the sample is completely melted. Therefore, the point of intersection of the leading edge with the extrapolated baseline is the melting point of a low-molecular-mass crystalline substance. After the melting ends (at temperature T_p), the sample temperature begins to increase, catches up with the reference temperature, and the steady-state conditions are achieved again at temperature T_6.

Thus, for a temperature calibration experiment on heating, the most important temperature is T_m (the melting point). It should be mentioned that the determination of the melting point described above (i.e., a well-defined, sharp melting point) is valid only for substances of high purity (>99.9%). For materials with purity less than 99.9%, the determination of the melting point is different, and it is described in Section 2.7.

The most frequently used calibration standards together with their melting points and heats of fusion, as well as some manufacturers of these standards, are listed below. We do not wish to advertise any particular company that sells high-purity water or metals; rather, the listed source is simply one of the possible sources, and not necessarily the only one:

- High-purity water WX0004-1 [suitable for high-performance liquid chromatography (HPLC), spectrophotometry, gas chromatography, gradient analysis], from EMD Chemicals, Gibbstown, NJ (www.emdchemicals.com), $T_m = 0.0\,°C$, $\Delta H_f = 333.6\,J/g$
- Indium ingot, Puratronic®, 99.9999+% purity, from Alfa Aesar, Ward Hill, MA (www.alfa.com), catalog item 10618; $T_m = 156.60\,°C$, $\Delta H_f = 28.42\,J/g$
- Tin rod, 6 mm diameter, Puratronic, 99.9985%, from Alfa Aesar, Ward Hill, MA catalog item 11472; $T_m = 231.93\,°C$, $\Delta H_f = 59.21\,J/g$

- Lead rod 5 mm diameter, Puratronic, 99.9999%, from Alfa Aesar, Ward Hill, MA, catalog item 12450; T_m = 327.47 °C, ΔH_f = 23.02 J/g
- Zinc rod, 10 mm diameter, 99.9999% purity, from Alfa Aesar, Ward Hill, MA, catalog item 12718; T_m = 419.53 °C, ΔH_f = 111.97 J/g

The temperature scale of most modern commercial instruments is linear, but this should be checked at least once at the installation of the newly purchased instrument with at least three standards. Then, running melting experiments with two standards is usually sufficient to perform the temperature calibration on heating.

Sometimes inorganic compounds are also used in the calibration experiments, but these are mainly high-melting-point standards. Since 420 °C is high enough for the vast majority of the polymer DSC experiments, we do not list any standard having a melting point higher than that of zinc.

The following precautions have to be taken when using the standards listed above:

1. The metal standards must be in the physical shape of ingots or thick rods. When preparing a sample for calibration, a thin layer is removed from the surface of the ingot or rod, and discarded since the surface may be covered with an oxide layer. Then, fresh "inside" (bulk) parts of the metal ingot (that was previously not in contact with air) are cut for encapsulating as a calibrating standard. This part is weighed and crimped in standard aluminum DSC pans, and used for calibration.
2. Since the metal standard piece just crimped and the bottom of the DSC pan are not in good thermal contact, it is recommended first to melt the standard 10–15 °C above the melting point, and crystallize it by cooling 50–60 °C below the melting point. This procedure ensures that a thin, continuous metal film is formed in the crimped DSC pan. In other words, part of the external thermal resistance (characterizing the thermal contact of the metal standard with the bottom of the DSC pan) is decreased to the minimum possible value. Then a second heating is performed on the just-treated sample. The DSC run recorded on the second heating should be used for calibration calculations.
3. The metal standards cannot be used for a prolonged period of time (especially Sn and Pb) because their melting points tend to decrease with time. A reasonable period for using a specific metal standard sample is one day only.
4. It is not recommended to use metal powders as melting point standards. The specific surface area of such samples is huge, which can affect the melting point. In addition, a considerable amount of metal oxide can be incorporated into the final metal film for the second heating.
5. Water and indium are considered the best melting point standards. The leading edges of the melting curves of Sn and Zn are not always straight,

and Pb is easily oxidized. Nevertheless, if a Pb standard is used the same day as it has been prepared, it is considered the second best melting point metal standard after indium.

6. All metal standards should be encapsulated in standard aluminum DSC pans. On the other hand, water should be encapsulated in a hermetically sealed aluminum pan.

7. Several laboratories use gallium (Ga) as a melting point standard, because it is the only available metal standard at around room temperature (its melting point is 29.76 °C). The use of this standard is not recommended, because Ga is toxic, and, in addition, can easily alloy with the material of the aluminum sample pan.

8. The room (or environment) temperature in the lab should not change considerably between the calibration and the actual DSC measurements on samples. The sample holders of commercial DSCs are relatively well insulated, so a couple of degrees' change in the room temperature should not cause a significant shift in the calibration.

9. The mass of the standard should be relatively small, under no circumstances exceeding 20 mg; with larger masses, the thermal lag can become excessive. Using a larger mass standard will cause only a small (or no) difference in the melting point, but the width of the melting curve will increase, the slope of the leading edge will be less steep, and this will introduce errors when determining thermal resistivity (this tendency is illustrated in Fig. 2.14). Also, with larger sample masses the linearity of

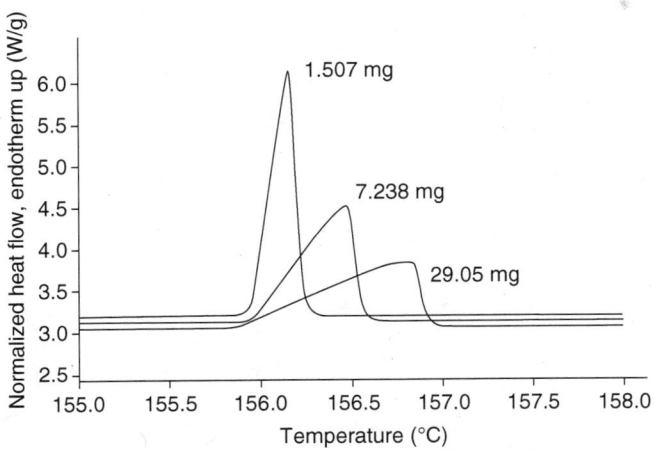

Figure 2.14. Melting curves of high-purity indium samples at various sample masses. The Perkin-Elmer DSC7 instrument used for these measurements was calibrated for a heating rate of 10 °C/min, the runs were made at 1 °C/min. Using Eq. (2.50), the displayed melting point of indium should be 155.83 °C; this is the value obtained in these experiments. (Menczel, unpublished results.)

the ordinate calibration is gradually lost. The area under the melting peak is linear with increasing mass of the sample only up to a certain limit. This limit is determined not only by the sample mass, but also by the heat that the instrument has to supply to the sample (i.e., the heat of fusion).

10. The importance of the coolant of the sample holder should be emphasized. If a mechanical cooling accessory is used, it should be turned on at least 2 h prior to the actual measurements in order to achieve thermal equilibrium.
11. The temperature calibration depends on the purge gas and its flow rate; all the measurements should be done at conditions identical to the calibration runs. See Table 3.2 (in Chapter 3), which lists the thermal conductivities of the most common purge gases.
12. The sample holder should be cleaned periodically, as suggested by the manufacturer.
13. All efforts should be made to avoid unnecessary heat losses from the cell during the calibration and the actual measurements. For instance, in the Perkin-Elmer DSCs the cell covers tend to tilt slightly if not replaced carefully, which may significantly alter the calibration and render it irreproducible.
14. The heating rate in the actual DSC runs of the samples should be identical to the heating rate of the calibration.

2.5.1.2. Temperature Calibration on Cooling

As mentioned already, the temperature calibration of DSCs is usually done on heating, because only the melting point has reproducible values for all the suitable calibrating standards. Cooling calibration could not be carried out for a long time, because the crystallization temperature of crystalline substances does not have a reproducible value, since supercooling commonly takes place. The crystallization temperature of the standard depends on factors such as the purity of the substance, the presence of nucleating agents, the composition, and the surface of the sample pan. A separate temperature calibration for cooling experiments is important mostly for polymer crystallization measurements, especially at high cooling rates. Such a cooling calibration procedure has been developed by Menczel and Leslie (Menczel and Leslie 1990, 1993; Menczel 1994, 1997) by using liquid crystals as calibrating standards. This procedure was applied to two DSCs (Perkin-Elmer DSC7 and TA Instruments 2920), and it was indicated that the cooling calibration was necessary for both DSCs for accurate determination of the temperature in the cooling mode. Some manufacturers may indicate that their instrument does not need a cooling calibration. Since the procedure described here has not been applied to other DSCs yet, only experiments can tell whether the cooling calibration is needed for a particular DSC or not. Therefore, we recommend that the reader carry out a cooling calibration procedure at selected cooling rates to determine whether this type of calibration is necessary.

The basic idea behind the cooling calibration is that the nematic (or cholesteric)-to-isotropic transition does not have a hysteresis on cooling, that is, that the transition on cooling occurs at the same temperature as on heating if the purity of the standard is 99.8–99.9% or higher. The transition temperature is called the *clearing point* (T_{cl}). The transition occurs at the same temperature on heating as on cooling because the nuclei of the nematic (or cholesteric) phase are always present in the isotropic liquid owing to density fluctuations (see Fig. 2.15). Three major classes of liquid crystals are known:

- *Nematic Liquid Crystals.* The molecules do not have positional order, but preferentially align themselves in one direction, the so-called optic axis or director.
- *Cholesteric Liquid Crystals.* These crystals are often called *twisted nematic* or *chiral nematic*, because only chiral molecules possess this special type of liquid crystalline phase. These liquid crystals are similar to the nematic liquid crystals, but along the director, each molecule is twisted as compared to the previous molecule by a certain angle. Similar to (i.e., in much the same way as do) nematic liquid crystals, cholesteric liquid crystals become isotropic liquids at the clearing point.
- *Smectic Liquid Crystals.* These liquid crystals have two-dimensional order; they are the most perfect liquid crystals. The rodlike molecules form layers, and often there is some mobility of the molecules inside the layers, but these layers can also move relative to each other. Depending on the arrangement of the molecules in the smectic plates, several different smectic phases are known, and sometimes transitions can be observed between these smectic phases (smectic A, B, C, etc.). Most smectic liquid crystals become nematic at a certain temperature before changing into an isotropic liquid at the clearing point.

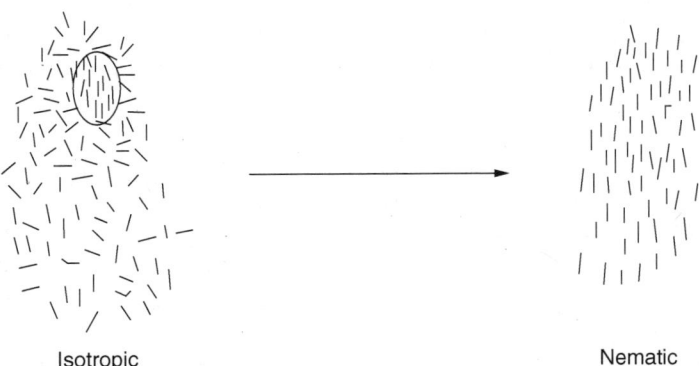

Isotropic   Nematic

Figure 2.15. Schematics of the transition of a nematic liquid crystal from the isotropic melt to the nematic liquid crystalline phase; the encircled area in the isotropic phase indicates a nucleus of the nematic phase formed due to density fluctuations.

Most liquid crystals are mechanically fluid, but there is some residual orientation (anisotropy) order of the long, rodlike molecules (Fig. 2.15). When these liquid crystals are heated to sufficiently high temperatures, they will pass through the clearing point, where they loose their anisotropy and become an isotropic melt.

Thus, the transitions in liquid crystals can be described schematically as

$$Cr \to (Sm_1 \to Sm_2 \to Sm_i \ldots) \to N(Ch) \to I \qquad (2.45)$$

where Cr denotes a crystalline substance, Sm is a smectic phase (parentheses indicate the fact that not all liquid crystals possess this phase), N is the nematic phase, Ch is the cholesteric phase (which can be present instead of the nematic phase at certain molecular geometries), and I is the isotropic phase.

The liquid crystalline standards are organic substances. Therefore, for a reliable calibration these materials should not degrade in the isotropic phase, since even minor amounts of an impurity formed as a result of thermal degradation will cause a decrease in the transition temperature; thus the transition temperature will differ on cooling and heating. The stability of the standard can be checked by slow heating–cooling. If the liquid crystal is stable, the transition temperature should be the same in the subsequent heating runs and the subsequent cooling runs (if there is a decrease of more than 0.1–0.2 °C in two subsequent cycles, the standard is degrading).

Menczel and Leslie (1990, 1993) indicated that the temperature gradient in the sample even at 80 °C/min does not exceed 0.1 °C if the mass of the standard is not higher than 4 mg. Also, they determined that, in addition to the nematic (cholesteric) → isotropic transition, other liquid crystal → liquid crystal transitions (e.g., transitions between various smectic phases and a smectic → nematic transition) do not have hysteresis, and therefore that all these transitions can be used for calibrating a DSC on cooling.

Thus, the following procedure is recommended for calibration of a DSC on cooling:

1. Calibrate the instrument at the heating rate of interest using high-purity metal standards and/or water. The instrument should be calibrated to a accuracy of ±0.1 °C.
2. Select and obtain a liquid crystalline standard with a purity of at least 99.8%. If the purity is lower than 99.8%, purify the standard to that level (the purity of the standard can be determined by applying the van't Hoff equation to the crystal → first liquid crystal transition). It is advantageous to have a standard with several liquid crystal phases, and to use all the transition temperatures for calibrating the instrument on cooling.
3. Perform the heating and cooling runs for the standard at the desired cooling rate, and determine the transition temperatures from both the heating and cooling runs. The results should be evaluated using the following equation

$$T_{\text{real}} = T_{\text{ind}} + \Delta T \qquad (2.46)$$

where the real temperature T_{real} is the transition temperature determined from the heating run and the indicated temperature T_{ind} is the transition temperature determined from the cooling run.

From here, a ΔT correction factor can be calculated that should be applied to all the temperatures measured in cooling runs. When this correction factor is calculated from several LC → LC and LC → I (where LC = liquid crystal) transition temperatures, it must be averaged to obtain an effective correction factor.

Menczel and Leslie (1990, 1993) and Menczel (1994, 1997) found three liquid crystal standards suitable for cooling calibration of DSCs:

1. N-(4-n-octyloxy-2-hydroxybenzal)-4′-n-butylaniline
2. (+)-4-n-hexyloxyphenyl-4′-(2″-methylbutyl)-biphenyl-4-carboxylate (with a commercial name of CE-3)
3. (+)-(4-(2′-methylbutyl)phenyl-4′-n-octylbiphenyl-4-carboxylate (with a commercial name of CE-8) (see also Fig. 2.16)

The use of these standards was advantegous, because they had several liquid crystalline transitions well separated from each other. Unfortunately, these standards are no longer available. But ASTM issued a standard procedure for cooling calibration (E2069-06), and they now recommend the following three liquid crystal standards:

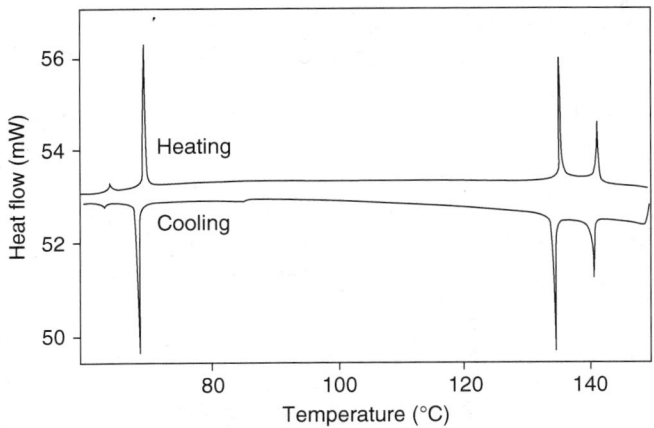

Figure 2.16. Heating and cooling DSC curves of an E. Merck liquid crystal standard CE-8 [(+)-(4-(2′-methylbutyl)phenyl-4′-n-octylbiphenyl-4-carboxylate] recorded on a Perkin-Elmer DSC7. Heating rate is 5 °C/min, cooling rate is 5 °C/min. In order of increasing temperature, the transitions are S_J^* to S_I^*, S_I^* to S_C^*, S_A to Ch, and Ch to isotropic. [From Menczel and Leslie (1993), reproduced with permission of Springer-Verlag.]

- M24, which is 4-cyano-4′-octyloxybiphenyl having a smectic A → nematic transition at 67.1 °C and a nematic → isotropic transition at 79.8 °C [ASTM 2069-06; see Menczel (to be published)]
- HP-53, which is 4-(4-pentyl-cyclohexyl)-benzoic acid-4-propylphenyl ester having a smectic A → nematic transition at 92.9 °C and a nematic → isotropic transition at 120.4 °C [ASTM 2069-06; Menczel (to be published)]
- BCH-52, which is 4′-ethyl-4-(4-propyl-cyclohexyl)-biphenyl having a smectic B → nematic transition at 146.4 °C [ASTM 2069-06; Menczel (to be published)]

2.5.2. Enthalpy Calibration of DSCs

The DSC runs carried out with the purpose of temperature calibration with metal standards are used at the same time to perform the enthalpy (heat of fusion) calibration. The basis for this calibration, using the so-called baseline method (see Section 2.3), is that the area under the endothermic peak reflecting the melting is related to the heat of fusion by the following equation:

$$\Delta H_f = \frac{K_{DSC} A}{mq} \qquad (2.47)$$

where ΔH_f is the heat of fusion (J/g), A is the area under the peak, m is the mass of the sample (mg), q is the heating rate (°C/min), and K_{DSC} is a proportionality coefficient, usually called the *instrument constant*.

For determination of A, the peak must be integrated. The integration on modern DSC instruments is done by commercial software. For the integration, the start and the end of the melting must be determined (Fig. 2.13b).

2.5.3. Heat Capacity Calibration of DSCs

Measurement of the heat capacity is described in Section 2.6, and heat capacity calibration of DSCs is also described in that section.

2.5.4. Miscellaneous Calibrations

2.5.4.1. Calibration of Thermal Lag In the Mettler Toledo DSC to compensate for thermal lags, an adjustment, denoted by Mettler Toledo as "tau lag," is used. The DSC tau lag performs two compensations: (1) it matches the program temperature to the reference temperature (as noted in Fig. 2.17), and (2) it adjusts for the heating rate as noted in the following equation:

$$\Delta T_{lag} = q\, \tau_{lag} \qquad (2.48)$$

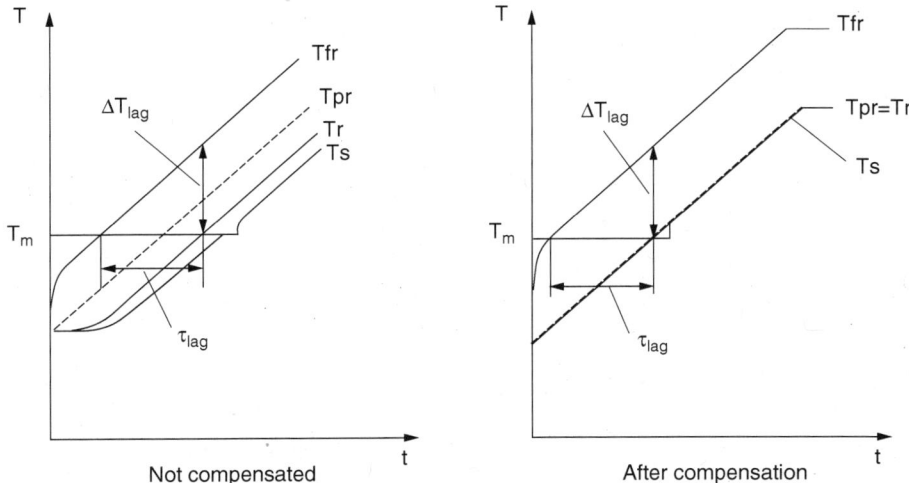

Figure 2.17. Temperature–time profile of the Mettler DSC before and after compensation for the thermal lag. In the figure, T_{fr} = furnace temperature, T_r = reference temperature, T_s = sample temperature, T_{pr} = program temperature, T_m = melting point, ΔT_{lag} = extent that reference temperature lags behind furnace temperature, and τ_{lag} = tau lag compensation of the furnace temperature. The tau lag compensation synchronizes the program temperature with the reference and sample temperatures. [From Mettler Toledo (2004); courtesy of Mettler Toledo.]

where q is the heating rate, ΔT_{lag} is the difference between the furnace and reference temperatures, and τ_{lag} is the adjustment factor. According to Mettler Toledo, with proper calibration the Mettler unit is calibrated for all heating and cooling rates as well as the isothermal condition (see Fig. 2.18). However, the specific value of the tau lag adjustment needs to be calculated from multiple ramp rates and temperatures, i.e., the tau lag is determined by interpolation between the ramp rates and temperatures used higher and lower than the one for which it is calculated. A similar calibration exists for Perkin-Elmer's power compensation DSC (see text below).

Similar to the Mettler Toledo DSC, an advantage of the Perkin-Elmer power compensation DSC is the simplicity of its temperature calibration for any heating rate (Perkin-Elmer 1976). This was true for the sample holder of the Perkin-Elmer DSC-2, and it is true for the sample holder of the Diamond Pyris DSC, since the structure of the sample holder did not change. In general, the following equation describes the temperature calibration of a power compensation DSC on heating

$$T_{real} = T_{disp} - C_1 q + C_2 \qquad (2.49)$$

where T_{real} is the actual temperature during a heating scan or in an isothermal measurement, T_{disp} is the temperature displayed by the DSC instrument. C_1 and C_2 are constants; C_1 is called the *thermal lag constant* (its value depends

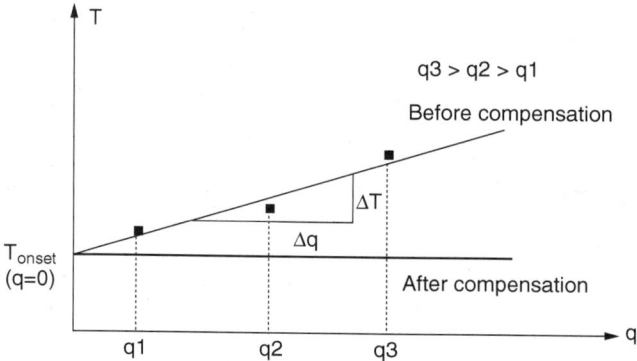

Figure 2.18. Melting points or extrapolated onset calibration temperatures for tau lag adjustment factor after compensation as a function of heating rate q. The extrapolated onset temperatures (T_{onset}) are determined from calibration standards; note that there is no change with heating rate after the tau lag compensation has been applied. [From Mettler Toledo (2004); courtesy of Mettler Toledo.]

on the material of the sample pan), and C_2 is called the *offset* (its value depends on the heating rate used to calibrate the given DSC instrument). If a Perkin-Elmer DSC has been calibrated at a heating rate of 10°C/min, then

$$T_{\text{real}} = T_{\text{disp}} - 0.085\, q + 0.85 \qquad (2.50)$$

because 0.085 is a frequent value for aluminum pans. Nevertheless, when installed, every sample holder should be calibrated at several different heating rates, and the thermal lag constant should be determined [this number ranges from 0.065 to ~0.085; see Menczel and Leslie (1993)]. After this, for heating runs a power compensation DSC should be calibrated at 10°C/min, and all the temperatures from runs with other heating rates can be calculated using Eq. (2.49).

2.5.4.2. Other Calibrations In addition to the calibrations described above, special calibration procedures may exist in various commercial instruments, such as the Tzero calibration in the Q series DSCs of TA Instruments. The reader should consult the instruction manuals for understanding and carrying out these calibrations.

2.6. MEASUREMENT OF HEAT CAPACITY

As described in Section 2.2, the heat capacity indicates how much heat is needed to raise the temperature of 1 g of the sample by 1°C. In equilibrium thermodynamics heat capacity is a function of state. Two major heat capacity variants are in use:

C_v, which is the heat capacity at constant volume
C_p, which is the heat capacity at constant pressure

Specific heat capacity refers to 1 g or 1 mol of the sample. Its dimension is J/(g·K), or J/(mol·K), which indicates how much heat is needed to raise the temperature of 1 g or 1 mol of the material by 1 °C. Mathematical definitions of the constant volume and constant pressure heat capacities are given by Eqs. (2.19) and (2.20).

As the reader probably knows by now, polymers can exist only in solid or liquid states—and since it is extremely difficult (often impossible) to keep solid and liquid samples at a constant volume when their temperature changes, it is virtually impossible to directly measure C_v. Thus, differential scanning calorimetry measures C_p of polymers. The DSC signal is closely proportional to the heat capacity of the sample.

Wunderlich and his research group [see, e.g., Gaur and Wunderlich (1980, 1981) and Gaur et al. (1983)] collected a large amount of experimental data on heat capacity of various polymers, and these data were entered into the ATHAS databank (see http://athas.prz.rzeszow.pl/). They also performed theoretical heat capacity calculations for a number of polymers and compared them to experimental data.

The major advantage of DSC is the speed with which the heat capacity can be measured. With careful measurements the accuracy of these measurements can reach or exceed ±1%.

As indicated above, C_v of polymers cannot be directly measured, but it can be calculated from C_p using one of the following relationships:

$$C_p = \frac{dQ}{dT} = C_v + \left[\left(\frac{\partial U}{\partial V}\right)_T + p\right]\left(\frac{\partial V}{\partial T}\right)_p \tag{2.51}$$

or

$$C_p = C_v + T\left(\frac{\partial p}{\partial T}\right)_V \left(\frac{\partial V}{\partial T}\right)_p \tag{2.52}$$

and

$$C_p = C_v + \frac{TV\gamma^2}{\beta_T} \tag{2.53}$$

where $\gamma \equiv (1/V)(\partial V/\partial T)_p$ is the volumetric coefficient of thermal expansion and $\beta_T \equiv -(1/V)(\partial V/\partial p)_T$ is the isothermal compressibility.

Pan et al. (1989) obtained a universal constant for polymers in a relationship for calculating C_v of polymers based on a modified Nernst-Lindemann expression for macromolecules, namely

$$C_p - C_v = \frac{3RA_0 C_p T}{T_m^\circ} \tag{2.54}$$

where T_m° is the equilibrium melting point of the polymer in question, and the constant A_0 changes little from polymer to polymer, and has an average value of 3.9×10^{-3} (K·mol)/J. (see Wunderlich 1990, pp. 245–246).

A general rule is that the heat capacity in the absence of chemical processes increases with temperature; therefore the heat capacity of the liquid is always higher than the heat capacity of the glass or the crystal. Also, for the majority of polymers, the heat capacity of the glass is close to the heat capacity of the crystals.

2.6.1. Measurement of Heat Capacity with Traditional DSC

With traditional DSC, the C_p determination is as follows. Figure 2.19a indicates three DSC runs simultaneously: a blank run (carried out with two empty pans and often called *baseline*), a calibrant run, and a sample run. In all three runs, the DSC cell is held isothermally for a short time at temperature T_1, then heating commences to produce a constant heating rate. After heating starts, the DSC signal shifts in an exponential manner, and this continues until steady-state conditions are achieved. At temperature T_2 the heating stops and the cell temperature is held isothermally at this temperature, where there is another exponential signal shift (from the steady-state condition to the isothermal baseline) (Wunderlich 1990). Since the area encompassed by the sample and the blank run curves (the shaded area in Fig. 2.19a) is proportional to the heat input into the sample, an average heat capacity of the sample for the temperature range T_1–T_2 can be calculated from the following relationship:

$$\bar{C}_p = \frac{Q}{T_2 - T_1} \tag{2.55}$$

With such heat capacity determination it would take a long time to determine temperature dependence of the heat capacity. However, the curves from T_1 to steady state and steady state to T_2 have similar shape; therefore the $h_{\text{s-bl}}$ (for sample and baseline) amplitude differences are proportional to the heat capacity at every temperature and there is no need to determine the $S_{\text{s-bl}}$ area (shaded area in Fig. 2.19a), but simply to measure the $h_{\text{s-bl}}$ amplitude differences. So, if the instrument is calibrated with a standard for which the temperature dependence of the heat capacity is well known, the heat capacity of the sample can be measured at any temperature. For this, the $h_{\text{sap-bl}}$ amplitude differences must be used (see Fig. 2.19a, curves "Sapphire run" and "Baseline"). The standard is usually sapphire (crystalline Al_2O_3), which is readily available from the instrument companies. For low-temperature heat capacity

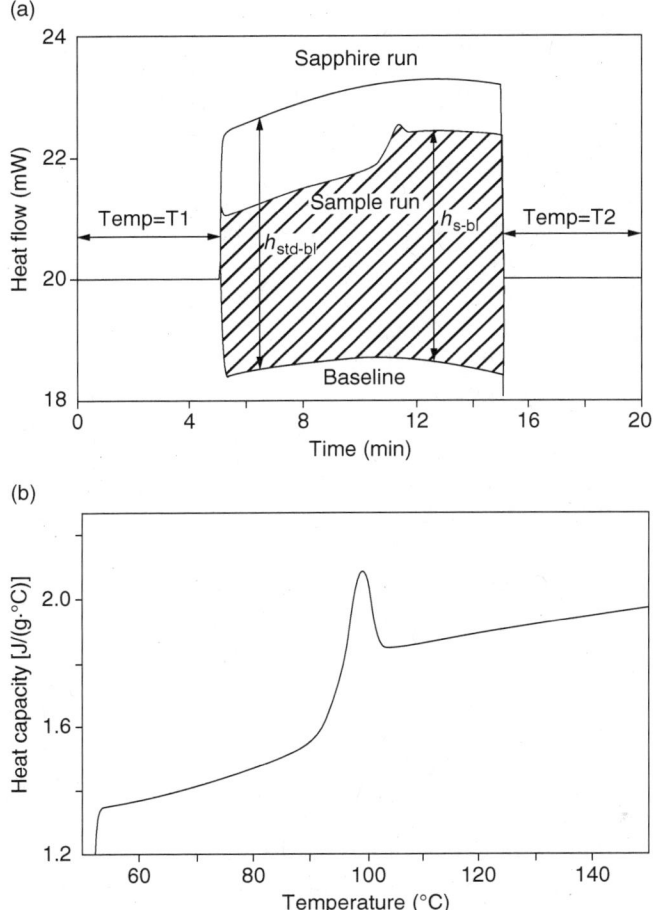

Figure 2.19. (a) Combination of the three DSC runs for determination of the heat capacity. The measurements were made on a Perkin-Elmer DSC7. The shaded area is the area proportional to the heat input necessary to raise the sample temperature from T_1 to T_2. Endotherm is up. (b) Heat capacity calculated from the three runs shown in Fig. 2.19a. Perkin-Elmer DSC7, 10°C/min heating rate. (Menczel, unpublished results.)

measurements, benzoic acid is often used as a standard, because its heat capacity is more linear with temperature in the low-temperature region than is sapphire.

It should be mentioned that not all instrument companies have standard software for heat capacity measurements. In those cases the heat capacity can be calculated using the procedure described below.

If possible, all the pan masses (i.e., sample pan, reference pan, and the empty pan in the sample position for the baseline run) should be identical in heat capacity measurements. Bares and Wunderlich (1973) found that better results can be obtained if the sample pan masses are closely matched than if the pan masses are different. It should be mentioned that the software of the TA Instruments Q1000 and Q2000 is capable for correction of different pan masses, but no systematic study has been done on how such a correction affects the accuracy of C_p determination.

The masses of the sample pans, the sample, and the standard should be determined to an accuracy of at least ±0.1% in order to ensure a ±1% accuracy of the heat capacity determined. In all of the three runs the temperature of the DSC cell has to be brought to T_1 (the starting temperature of the runs), at which it should be held for at least 10 min to achieve temperature equilibration. Then the actual runs start. There is an isothermal period at temperature T_1 for a specific time (usually 2–5 min), and then heating begins at a constant rate and continues to temperature T_2, where there is another isothermal period. All the run parameters in all three runs must be identical. The heat capacity can be calculated only at those temperatures where steady-state conditions exist.

After the calibration and the blank runs are completed, the instrument constant K_{DSC} can be calculated using the following equation:

$$K_{DSC} = C_{p,sap}(T) \frac{m_{sap} q}{h_{sap\text{-}bl}(T)} \qquad (2.56)$$

where $C_{p,sap}(T)$ is the heat capacity of the calibrating standard (usually sapphire) at temperature T, m_{sap} is the mass of the calibrating standard, q is the heating rate, and $h_{sap\text{-}bl}(T)$ is the steady-state amplitude difference of the calibrating standard and the blank run at temperature T.

Then, the sample needs to be run, and the heat capacity can be calculated from the following equation

$$C_{p,s}(T) = K_{DSC} \frac{h_{s\text{-}bl}(T)}{m_s q} \qquad (2.57)$$

where $C_{p,s}$ is the heat capacity of the sample, $h_{s\text{-}bl}$ is the steady-state amplitude difference between the sample curve and the blank run curve, and m_s is the mass of the sample.

The accuracy of heat capacity determination with a traditional DSC can be better than ±1% when the measurements are done carefully (Bares and Wunderlich 1973).

Since the instrumental baseline depends on the environment temperature, for good reproducible results all the runs should be done at an ambient temperature constant to ±1 °C. Naturally, all the runs (sample, baseline, and standard) should be performed with identical parameters (identical starting isothermal in both temperature and time, identical heating rate, and identical final isothermal in both temperature and time). Of course, the purge gas and the purge gas flow rate should also be identical. The temperature range of the determination should not exceed 100 °C.

The DSC runs for actual C_p determination are done as follows:

1. Select the temperature range in which you need the heat capacity data. Keep in mind that the first approximately 20 °C interval will be lost; this is the temperature interval needed for achieving steady-state conditions. The sample should be equilibrated at T_1 for at least 10 min. Then the isothermal holding time at T_1 is usually 2 min.
2. When at temperature T_2, if the sample is in the molten state, care must be taken to avoid thermal degradation.
3. The isothermal baseline at T_1 and T_2 must be horizontal. If some curvature is observed in the baselines, erroneous heat losses may be taking place (e.g., the sample holder cover may be tilted, or some chemical or physical processes may be occurring).
4. The suggested heating rate is 10–20 °C/min. The sample size should be 10–20 mg.
5. The sample must be conditioned before running. Many polymers (e.g., nylons and polyesters) contain water, and the sample needs to be dried but carefully, so that no degradation takes place during the drying process.
6. Care must be taken to ensure that the DSC sample cell is dry. For example, in a Perkin-Elmer DSC it is important to prevent ice condensation on the swingaway enclosure cover. Therefore, the purge gas must be dry, and dry conditions must be ensured in the dry boxes.
7. When necessary, thermal lag corrections can be made (thermal lag is the difference between the displayed temperature and the actual temperature of the sample; see Section 2.3). In common C_p determinations this is not a problem, because at a heating rate of 20 °C/min and a sample weight of 20 mg, the thermal lag is less than 1 °C.

Performing heat capacity measurements on cooling can often be advantageous, especially when the temperature dependence of the melt heat capacity is needed. To determine the temperature dependence of the melt heat capacity in traditional heating experiments is not easy, because the available temperature range in the melt is often limited with the onset of thermal degradation. As described in Section 2.7, supercooling is a common phenomenon in polymer crystallization; therefore the temperature range available for heat capacity determination of the melt during cooling can be expanded when compared to heating runs. Naturally, the other two measurements (standard and blank runs) will also have to be carried out on cooling, and the temperature calibration of the DSC instrument also should be done on cooling (see Section 2.5).

2.6.2. Measurement of Heat Capacity with Modulated Temperature DSC in the Quasi-isothermal Mode

MTDSC can measure C_p quasi-isothermally (Jin et al. 1993; Boller et al. 1994; Thomas 2006; Reading and Hourston 2006) to a high degree of accuracy (Ishikiriyama and Wunderlich 1997), but the measurements take more time because the temperature dependence of C_p is determined point by point. In MTDSC the amplitude of the modulated heat flow, not its average value, is used for determination of reversing heat capacity (referred to as complex heat capacity for the 2920 DSC), therefore eliminating the need for a stable baseline as in traditional DSC.

Thus, in MTDSC the reversing heat capacity is determined using the following formula:

$$C_{p,\text{rev}} = \frac{\text{heat flow amplitude}}{\text{heating rate amplitude}} K(C_{p,\text{rev}}) \qquad (2.58)$$

where $K(C_{p,\text{rev}})$ is the calibration constant for the reversing heat capacity, which if not adjusted would be the traditional DSC calibration constant.

In these heat capacity determinations it is important to use modulation slow enough to ensure good heat transfer between the sample and the sensor. If the time during the modulation is insufficient, the amplitude of the heat flow will be decreased and the reversing heat capacity value will be reduced. Temperature modulation of 0.5–1.0 °C and a modulation period of 100–120 s are suggested.

Often hermetically sealed pans must be used for various purposes. In such cases, the modulation period must be increased somewhat, because the contact area between the pan and the sensor for hermetically sealed pans is smaller than that for standard DSC pans.

2.7. PHASE TRANSITIONS IN AMORPHOUS AND CRYSTALLINE POLYMERS

In this section we present an overview of the transitions of amorphous and crystalline polymers that can be observed by differential scanning calorimetry. These are the glass transition, crystallization from the melt during cooling (the so-called melt crystallization), cold crystallization (taking place when an amorphous polymer crystallizes during heating from the glassy state), and melting. There are two other types of transitions that a thermal analyst can sometimes encounter: the liquid crystal → isotropic melt transition of sidechain liquid crystalline polymers, and crystal → crystal transitions. However, these transitions are not common, therefore they are out of the scope of this book.

2.7.1. The Glass Transition

When a polymer melt is being cooled, it often partially crystallizes. But this "melt crystallization" sometimes cannot take place because of a lack of regularity of the repeating units along the polymer chain (as in the case of atactic polymers) or because the translational motion of the segments of the polymer chains freezes owing to the sudden increase of the viscosity of the melt. In these cases, at a certain temperature (or rather, in a certain temperature range) the glassy phase is formed, in which the molecular conformation of the melt is preserved. The glass is a metastable state, but at temperatures much lower than the glass transition temperature, it can be preserved for an indefinite time.

The temperature at which the rubbery or melt state changes into the glassy state is called the *glass transition temperature* (T_g). From the microscopic perspective, T_g is the temperature at which the long range micro-Brownian motion of the chain segments freezes on cooling. As will be seen later, the transition temperature is not the same in cooling as in heating experiments. It would be better to measure T_g on cooling, but for historical reasons, the majority of the measurements are done on heating.

In heat flux DSCs on heating, the glass transition is observed as a jump in the baseline (i.e., heat flow signal) pointing downward, while in the Perkin-Elmer power compensation DSCs the glass transition is a jump in the baseline pointing upward, although lately a number of manufacturers have provided an option to select the direction of the endoterm–exotherm.

It is well known that the "glass" is a noncrystalline, mechanically solid, rigid material. In the molten state, one of the major types of motion is largely translational, while in the glassy state motion of the segments is mainly vibrational. Thus, the polymer is in the melt or rubbery state above T_g, while below T_g the material is in the glassy state, where it is rigid. This definition of the glass transition temperature is valid for static measurements (like traditional DSC). In relaxation methods, such as DMA or DEA, and also MTDSC (see Chapters 5 and 6, and Section 2.13 of this chapter), the glass transition temperature is reached when the timescale of the experiment becomes comparable to the relaxation time of the segments.

Sometimes it is still debated whether the glass transition is a purely kinetic transition or a second-order thermodynamic transition (van Krevelen 2003). On one hand, it is true that the crystallization process for a number of (atactic) polymers would not take place even at infinite time, and this transition possesses the characteristics of a second-order thermodynamic transition (at least formally, in the Ehrenfest sense; see definition of the phase transition in Section 2.2). But the absence of crystallization does not prove that the glass transition is a thermodynamic second-order transition, and it is also true that the glass transition does not occur as a definite sharp transition as would be required by equilibrium thermodynamics. Therefore, the glass transition must be considered a kinetic transition.

The glass transition temperature is extremely important from a practical perspective; for many polymers it determines the highest use temperature, while at the same time it defines the lowest possible processing temperature. DSC is a suitable technique for measuring the characteristics of the glass transition, because the specific-heat capacity at the glass transition temperature exhibits a generally abrupt change, and as the reader by now probably knows, the DSC signal (heat flow) is proportional to the specific-heat capacity of the sample.

The glass transition takes place in amorphous polymers or in the amorphous fraction of semicrystalline polymers. Thus, in a semicrystalline polymer (which contains both amorphous and crystalline regions), two transitions can be identified: the melting and the glass transition. The melting reflects the transition of the crystalline portions, while the glass transition comes from the change in the amorphous regions of the polymer. The value of T_g depends on the mobility of the polymer chains; polymers possessing more mobile chains have a lower glass transition temperature, and restriction of the rotational motion of the polymer chain segments leads to an increase in T_g.

For semicrystalline polymers the ratio T_m/T_g varies between 1.5 and 2.0, with several exceptions like PTFE (Wunderlich 1960, 1990). This empirical rule also holds for many small-molecular-mass substances, such as ethanol, water, and SiO_2 (Wunderlich 1990).

In publications describing DSC experiments, the glass transition temperature is usually reported as the temperature at the half-height of the heat capacity increase ($\frac{1}{2}\Delta C_p$, also called the "temperature of half-unfreezing"; see Fig. 2.20). T_g can also be taken as the inflection point, which is slightly different and corresponds to the peak in the derivative of the heat flow or heat capacity versus temperature. Most software programs of commercial instruments do the determination of T_g almost automatically. There are two popular methods

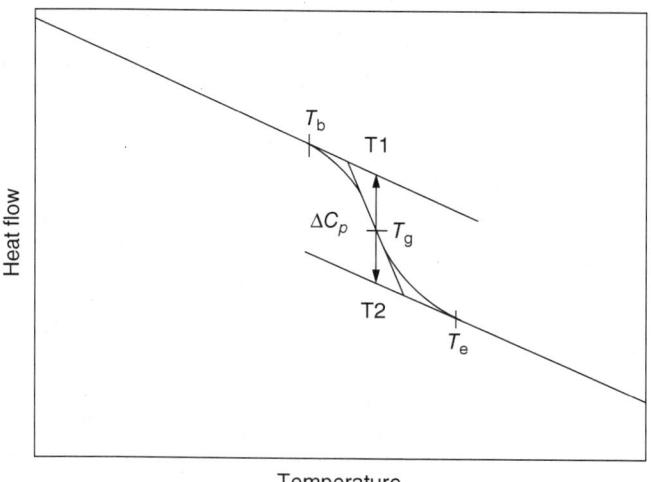

Figure 2.20. Determination of glass transition temperature in a heating experiment from an idealized DSC curve (endotherm down). In this DSC curve no hysteresis peak is present, which is a very rare case. The height of the double arrow is proportional to the heat capacity increase at the glass transition.

for calculation of T_g from the DSC curves. In the first one, the operator can vary the position of six points—two should be put down on the linear portion of the DSC curve in the glassy state, two should be selected on the linear portion of the DSC curve in the melt, and two should be selected in the transition region itself. In the other method, the operator selects two points: one on the linear portion of the DSC curve before the glass transition, and one after the glass transition. Finally, the software gives two straight lines, and the operator should adjust them if necessary so they become a continuation of the linear portions of the DSC curve below and above T_g. Then, in both cases, calculation of T_g and ΔC_p is done and displayed by the software. For amorphous polymers, where the glass transition is reasonably narrow, and can be easily seen, the precision is good, about ±0.5 °C.

There is no heat of transition at the glass transition, since only first-order transitions have a heat of transition. In addition to the transition temperature, the glass transition is characterized by the heat capacity jump or heat capacity increase at the glass transition (ΔC_p). ΔC_p is a characteristic constant number for a given amorphous polymer. However, ΔC_p depends on the crystallinity and the so-called rigid amorphous fraction for semicrystalline polymers (see later in this section). There is an empirical rule for amorphous polymers, often called *Wunderlich's rule*; the specific heat capacity increase (ΔC_p) at the glass transition is usually 11 J/(K·mol) for one mobile unit of the polymer mainchain in the case of relatively small mobile units. For larger units (such as the phenylene rings), ΔC_p may be double or triple this value [i.e., 22 or 33 J/(K·mol)] (Wunderlich 1960, 1990).

The melt → glass and the glass → melt (or glass → rubber for crosslinked polymers) transitions do not take place at a definite temperature, but in a generally wide temperature range, which is at least 5 °C for amorphous polymers, but sometimes as broad as 50–60 °C or more for semicrystalline or crosslinked polymers. For polymer blends it can be even wider. This is an indication that the glass transition is not an equilibrium transition, and the temperature and the width of the transition depend on the heating and/or cooling rate(s).

Therefore, in order to compare the glass transition behaviors of various polymers, a specific temperature, called the *glass transition temperature* (T_g) had to be determined; one selected temperature needed to be defined as T_g from the transition region. Originally, before the widespread use of DSC, the glass transition temperature was determined primarily by dilatometry, that is, from the dependence of volume on temperature. The volume expansion of the glass and the melt are practically linear over a short temperature range below and above T_g, so that in dilatometric measurements the glass transition temperature was defined as the point of intersection of the temperature dependences of volume of the glass and the melt expansion curves. However, in DSC measurements this had to be changed, because here the recorded physical quantity is not the thermodynamic analog of volume, which is enthalpy (H), but the derivative of enthalpy, which is heat capacity (C_p).

For DSC measurements it is more convenient to define the glass transition temperature as shown in Figs. 2.20 and 2.21. Figure 2.21 indicates that on *cooling* five temperatures are necessary to define the glass transition [see, e.g., Wunderlich (2006)]: T_b, which is the beginning of the deviation of the DSC curve from linearity; T_1, the extrapolated onset temperature of the glass transition; T_g, the glass transition temperature, which is the temperature at half-height of the heat capacity decrease; T_2, the extrapolated end temperature of the glass transition, and T_e, the end temperature of the glass transition, where the heat capacity dependence becomes linear again. Unfortunately, when T_g is defined as just described, it rarely coincides with the dilatometric glass transition temperature. The consequences of this discrepancy are disturbing, when comparing T_g values determined by these two methods. The T_g determination from a DSC heating curve is even more controversial, because in most cases an endothermic hysteresis peak is superimposed on the high-temperature side of the glass transition (see Fig. 2.22a). Such a peak appears when the heating rate is faster than the cooling rate when the glass was formed, or when the glass was annealed below T_g. Also, sometimes an exothermic peak is superimposed on the DSC curve at the low-temperature side of the glass transition (Fig. 2.22b). This type of peak will occur when the heating rate is considerably slower than the cooling rate used for preparing the glass, that is, when the sample was quenched and is heated up at moderate heating rates. The thermal analyst should be careful in such measurements. Many operators do not measure the DSC curves in sufficiently wide temperature ranges; in such cases the broad exothermic hysteresis peak may be undetected. One important difference between the T_g values determined by dilatometry and DSC is that the glass transition temperature determined on heating from dilatometric

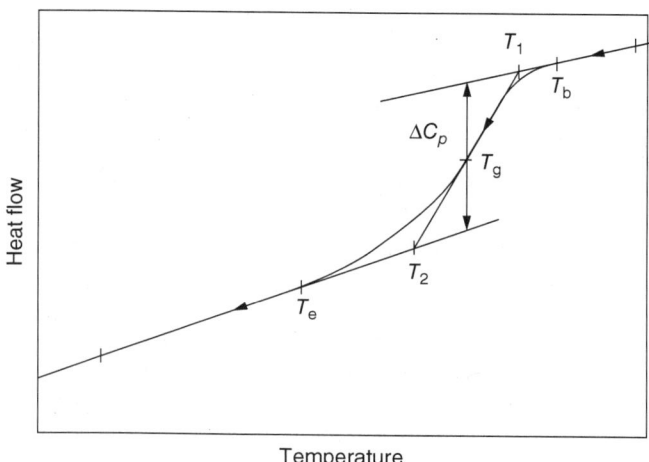

Figure 2.21. Characterization of glass transition during recording of a cooling DSC curve (endotherm is down).

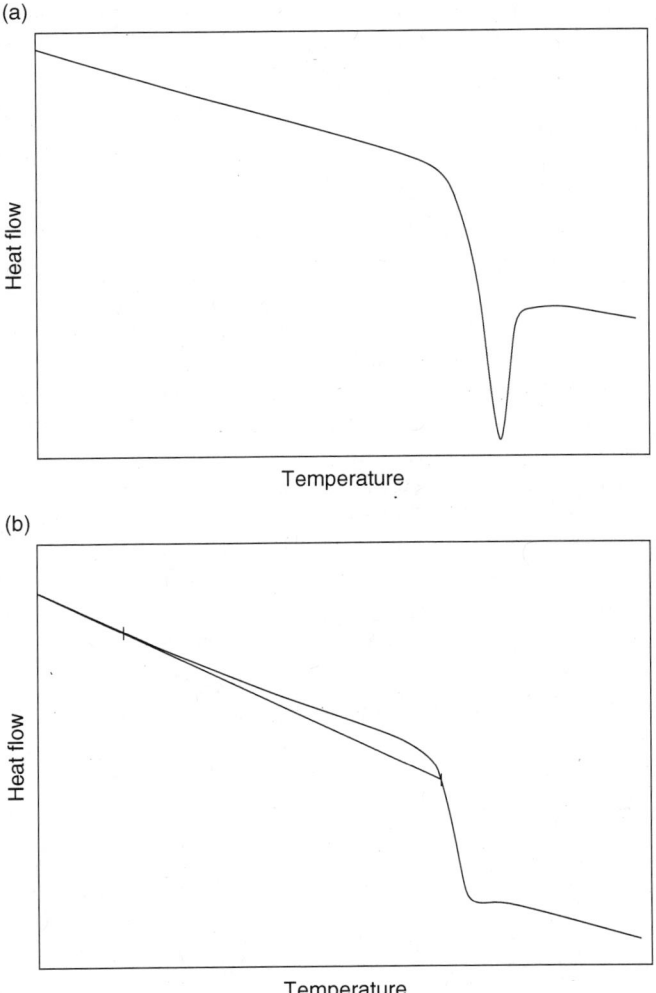

Figure 2.22. DSC curves showing glass transitions of amorphous polymers with (a) an endothermic hysteresis peak on high-temperature side of glass transition (endotherm down) and (b) an exothermic hysteresis peak on low-temperature side of glass transition (endotherm down). The heat capacity of the glass is extrapolated to higher temperatures to make the broad exothermic peak more visible.

measurements decreases as the rate of previous cooling decreases, but the opposite applies when T_g is determined from DSC measurements. However, there is no reason for alarm. This phenomenon is well explained by Richardson (1993a), who observed that when T_g is calculated from DSC measurements as the point of intersection of the temperature dependences of the glassy and melt *enthalpies* (after integration of the DSC curve), results similar to dilatometry are obtained. It should be emphasized that there is absolutely no controversy between the different definitions of the glass transition temperature from dilatometry and DSC; all that is happening is that two differently defined points of a temperature region have different dependences on the cooling rate used for preparing the glass.

As already mentioned, for characterization of a polymer, it is best to measure the glass transition temperature on cooling at a constant cooling rate. Of course, in these cases the instrument should be calibrated on cooling. The measurement of T_g is preferred on heating when the effects of processing and thermal history need to be known. Several DSC curves showing the glass transition on cooling for polystyrene at various cooling rates are displayed in Fig. 2.23 (see also Fig. 2.24). When T_g is measured on cooling, there is a decrease in the specific heat capacity at the glass transition temperature, and T_g and ΔC_p can be determined from the cooling curves. In this case, T_g depends only on the cooling rate, and the temperature range of the glass transition can be characterized by four (sometimes six) numbers (see Fig. 2.21): the cooling rate; T_g (i.e., taken as the so-called midpoint value); and the extrapolated start and end temperatures of the glass transition, T_1 and T_2. Sometimes T_b and T_e are also added, but these numbers are difficult to determine precisely because of the subjectivity of the determination.

In practice, almost everyone measures T_g on heating. There are historical reasons for this. In the early days of DSC measurements, the experiments at constant cooling rates were not simple, because few instruments had cooling

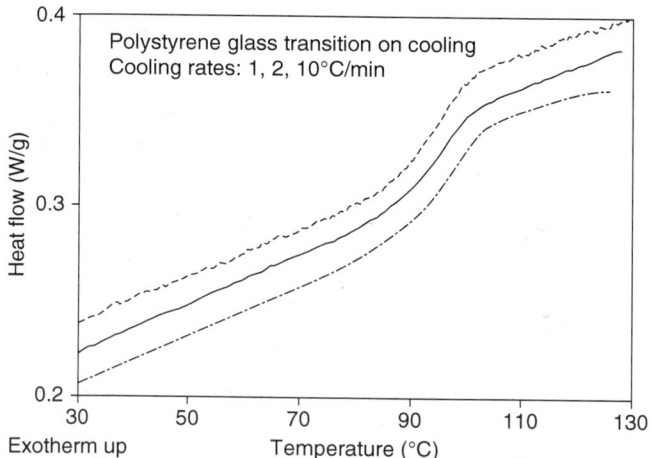

Figure 2.23. DSC curves indicating glass transition of atactic polystyrene on cooling. Cooling rate for the curves from top to bottom: 1, 2, and 10 (°C/min). When comparing the curves, it can be clearly seen that T_e (the end temperature of the glass transition; compare to Fig. 2.21) is considerably lower for the 10°C/min cooling than for the other two cooling rates. (Menczel, unpublished results.)

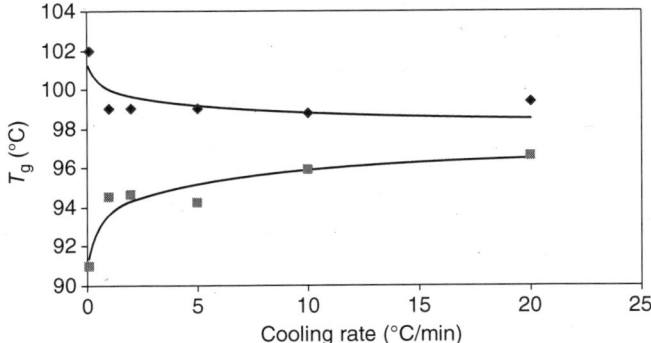

Figure 2.24. Glass transition temperature of polystyrene ($M_w = 19{,}300$, $M_n = 18{,}100$) measured on cooling at various rates and on reheating at a rate of 10°C/min. Squares—cooling; diamonds—reheating (Menczel, unpublished results).

units, and temperature control was often lost during cooling. Also, the instrumental baseline in cooling experiments was rarely linear. Therefore, the glass transition measurements were, and even today mostly still are, carried out on heating. The disadvantages of the T_g determination on heating are that (1) more numbers are needed to characterize the glass transition, since in addition to the four numbers recorded on heating (T_1, T_2, T_g, and heating rate), the thermal history of the glass has to be given (e.g., it had been cooled at 2°C/min, perhaps under pressure, etc.); and (2) almost always, hysteresis peaks appear on the high, and sometimes on the low-temperature side of the glass transition (see Figs. 2.22a and 2.22b).

However, there are some advantages to measuring the glass transition on heating; for instance, the DSC curve recorded on heating (see example plot in Fig. 2.25) contains information about the thermal history, including processing and physical aging of the sample.

The glass transition temperature depends strongly on molecular mass up to mass ~10,000. Note that the lower limit of the molecular mass of a polymer is generally considered to be about 10,000, so that the T_g of a polymer shows little to no dependence on the molecular mass. Such a dependence is shown in Fig. 2.26 for polystyrene of narrow molecular mass distribution (T_g was measured by DSC at $\frac{1}{2}\Delta C_p$).

The use of a polymer is, of course, determined by the application. Polymers can be used in the glassy state (examples of this are PMMA or Plexiglas, polystyrene, and epoxy), or in the rubbery or elastomeric state, where they are soft (examples include polyisoprene or polybutadiene). The limit of usability for both groups is determined by the glass transition temperature.

The definitions and characteristics of the glass transition described above refer to a static measurement technique: traditional DSC. Another complication may arise when the glass transition temperature is measured by a

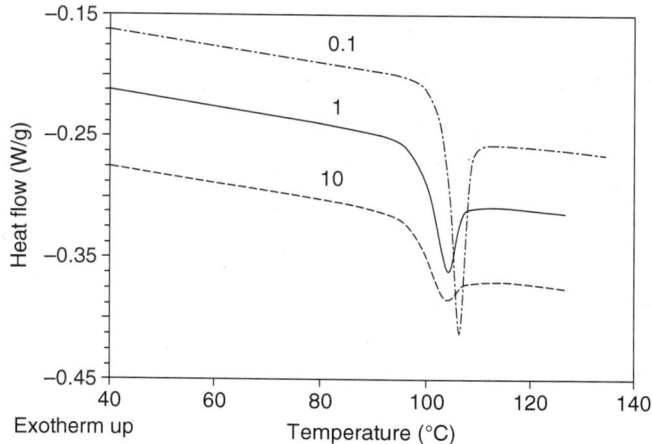

Figure 2.25. DSC curves indicating glass transition of a narrow-molecular-mass polystyrene on heating (M_w = 19,300, M_n = 18,100) of samples previously cooled at 0.1, 1, and 10 (°C/min). Heating rate: 10°C/min; endotherm is down. (Menczel, unpublished results.)

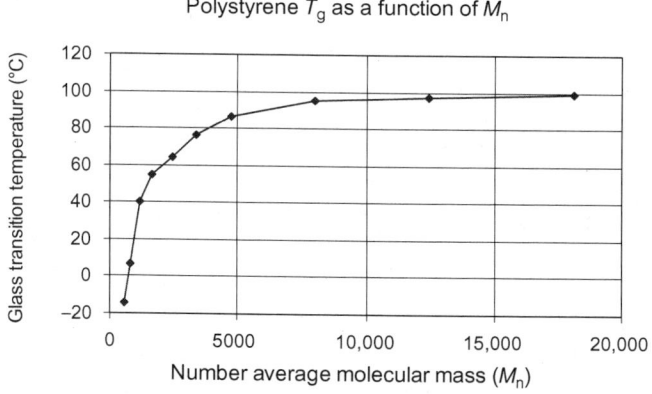

Figure 2.26. Dependence of the glass transition temperature of polystyrenes of narrow molecular mass distribution on the number average molecular mass (M_w/M_n for all samples ranged between 1.01 and 1.12). Heating experiments, with heating rate of 10°C/min. (Menczel, unpublished results.)

relaxation technique, such as DMA or DEA or modulated temperature DSC. In such cases, T_g occurs when the experimental timescale becomes comparable to the molecular relaxation time. Using these techniques, T_g depends on the frequency of the measurement as well.

It was mentioned before that endothermic or exothermic hysteresis peaks are common in the DSC curves of the glass transition. How can the presence of these peaks be explained, remembering that the glass transition is not a first-order transition?

Figure 2.20 shows an "ideal" DSC curve of a glass transition recorded on heating. However, such a DSC curve without some hysteresis peak can rarely be realized. When the melt is cooled more slowly than the subsequent heating rate to "glassify" the sample, an endothermic peak appears on the high-temperature side of the heat capacity increase. The magnitude and the temperature position of this peak increase with decreasing cooling rate as shown in Fig. 2.25 (polystyrene, 10, 1, 0.1 °C/min). A similar hysteresis peak appears when the sample is annealed. The case for semicrystalline polymers is more complex, as discussed in the next subsection. The reason behind the appearance of the endothermic hysteresis peak can be seen in Fig. 2.27. Let us suppose that the melt is being cooled at a cooling rate CR, then reheated at a heating rate HR, and |CR| < |HR|. Since the heating rate is higher than the previous cooling rate, the glass transition, which is a kinetic transition, will shift to higher temperatures on heating, as indicated by the upward arrows in Fig. 2.27. Also

$$\int_0^{T_f} C_p dT = H_{0-T_f} \tag{2.59}$$

must not depend on the value of the cooling or heating rate according to the first law of thermodynamics; this integral determines the area under the $C_p = f(T)$ curves starting from absolute zero temperature, so it reflects the total enthalpy of the sample, which should have a fixed value according to thermodynamics. Therefore, the two shaded areas in Fig. 2.27 must be equal. Thus, the appearance of the hysteresis peak is due to the glass transition shifting to higher temperatures at higher heating rates. It is noteworthy that an endothermic hysteresis peak appears even in the case of equal cooling and heating rates because the glass is being annealed during the heatup. The shape and the

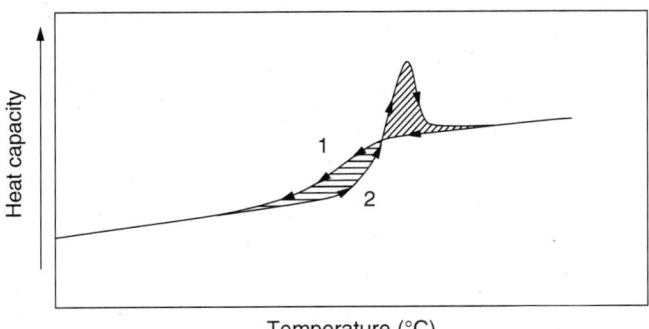

Figure 2.27. Cooling (1) and reheating (2) a glassy sample through glass transition. The cooling rate is less than heating rate on subsequent heating (endotherm is up).

TABLE 2.1. Glass Transition Temperatures for Common Polymers

Polymer	T_g (°C)
Polyethylene (PE)[a]	−36
Isotactic polypropylene (PP)[a]	−14
Poly(ethylene terephthalate) (PET)[b]	79
Poly(butylene terephthalate) (PBT)[a] (semicrystalline)	40–50
Poly(ethylene-2,6-naphthalene dicarboxylate)(PEN)[b]	115–125
Nylon 6[a]	47
Nylon 66[a]	48
Nylon 12[a]	50
POM[a]	−80
PEO[a]	−70
PAN[a]	80–95
PPS[b]	92.5
PVA[a]	70–85
PEEK[a]	146–157
PS[c]	100
PC[c]	145
PTFE[a]	−73
PVF2[a]	−40
PVC[c]	81
Atactic PMMA[c]	105

[a]Semicrystalline polymers.
[b]Glass transition temperature T_g of a quenched polymer that is amorphous but able to crystallize at slow cooling rates.
[c]Amorphous polymers.

position of the heat capacity increase at the glass transition depend on the thermal (and mechanical) history of the polymer, so DSC in this case not only characterizes the material but also presents a kind of "fingerprint" of what happened to the sample before the DSC run. Thus, for characterization of a polymer it is more advantageous to measure the glass transition on cooling, as shown in Fig. 2.21. A cooling run will show only the heat capacity change at the glass transition and will be free of any hysteresis phenomena or previous history.

Table 2.1 lists the glass transition temperatures of several common polymers.

2.7.1.1. Glass Transition in Semicrystalline Polymers

In addition to amorphous polymers, the glass transition can also take place in the amorphous regions of semicrystalline polymers. Although there is no rule, in most cases the glass transition temperature in semicrystalline polymers is somewhat higher than that for the same material when totally amorphous. This can be seen from the data in Table 2.2. The T_g values of semicrystalline materials reported in Table 2.2 were measured on samples cooled at 1 °C/min from above the equilibrium melting point.

TABLE 2.2. Glass Transition Parameters of Poly(ethylene Terephthalate) (PET), Poly(ethylene 2,6-Naphthalenedicarboxylate) (PEN), and Poly(phenylene Sulfide) (PPS)

Polymer	Amorphous (Quenched) Polymer		Semicrystalline Polymer[a]	
	T_g (°C)	ΔC_p [J/(g·°C)]	T_g (°C)	ΔC_p [J/(g·°C)]
PET	77.5	0.400	79.0	0.181
PEN	124	0.361	127	0.168
PPS	92.5	0.253	99.0	0.107

[a] Experimental conditions: 10 °C/min heating after 1 °C/min cooling from the melt.

Sources: Menczel, unpublished results. PET sample from Scientific Polymer Products (Cat. # 138); PEN sample from Aldrich (435317-100G), extrusion-grade PPS sample from Hoechst Celanese.

It is often difficult to determine the glass transition temperature of semicrystalline polymers from DSC curves; the glass transition can be very broad and smeared out, and long term experience is needed to see the glass transition on such DSC curves. In these cases, the derivative signal ($d\dot{Q}/dT$ vs. T or dC_p/dT vs. T) (where $\dot{Q}$ is the heat flow) can help, because the heat capacity increase will be replaced by a broad peak. However, calculation of the ΔC_p at T_g is still often difficult. Care must be taken when reporting T_g from these DSC curves, because the peak temperature on the $(d\dot{Q}/dT)$–T curve corresponds to the inflection point and not the half-height heat capacity increase of the C_p–T (or $\dot{Q}$–T) curves. The precision of the T_g determination in these DSC curves is much worse than for amorphous polymers; in the best case it is ±1 °C.

According to the two-phase model of polymers, semicrystalline polymers are composed of crystalline and amorphous regions. In this model it is accepted that the extensive properties of these materials can be calculated by multiplying the relative amounts of these two phases by the extensive properties of the individual phases. DSC measurements indicated that the glass transition of the amorphous portion in semicrystalline polymers is considerably influenced by the presence of the crystalline phase. Figures 2.28a and 2.28b show the DSC curves of poly(ethylene-2,6-naphthalene dicarboxylate) (PEN) and poly(ethylene terephthalate) (PET) in the glass transition region for samples with various crystallinities (i.e., cooled from the melt at various cooling rates). Thus, these samples have different crystallinities. As can be seen, the heat capacity jump at the glass transition decreases as the cooling rate decreases. The reason for the decrease of ΔC_p is that the crystallinity of the polymer increases with decreasing cooling rate (at lower cooling rates more time is available for the sample to crystallize), so the amount of the amorphous fraction simultaneously decreases.

As mentioned before, a polymer's glass transition can be characterized by the glass transition temperature (T_g) and the heat capacity jump at the glass

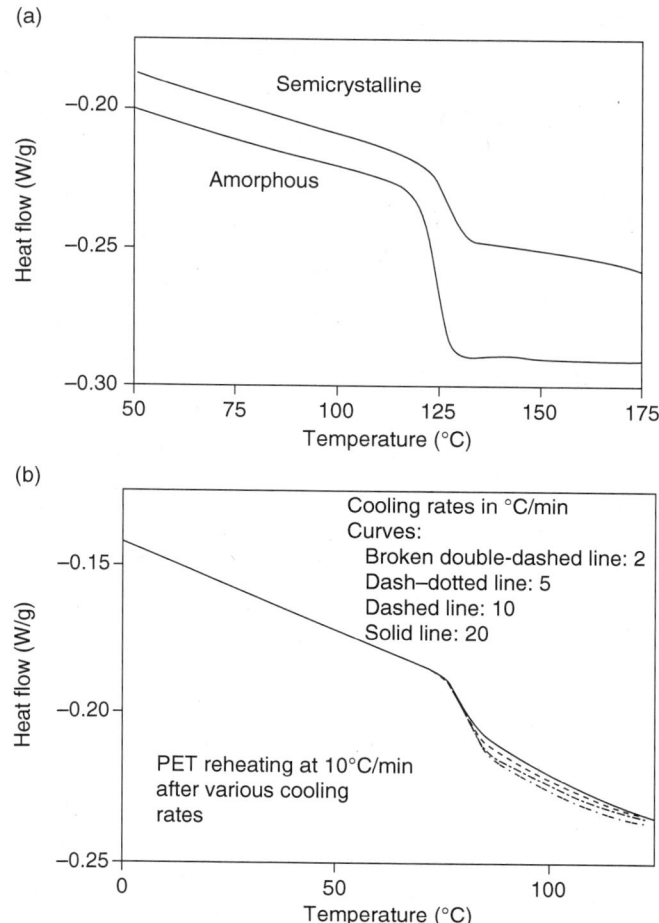

Figure 2.28. (a) DSC heating curves showing the glass transition of amorphous and semicrystalline PEN at a heating rate of 10°C/min. The amorphous sample was prepared by quenching it with liquid nitrogen from above the melting point. The semicrystalline sample was prepared by cooling the sample at a rate of 5°C/min from the melt. [PEN sample from Aldrich (435317-100G) (Menczel, unpublished results).] (b) DSC heating curves of poly(ethyelene terephthalate) in the glass transition region: the heat capacity jump at the glass transition decreases with decreasing cooling rate because of increasing crystallinity of the sample. The samples were prepared by cooling at various rates from the melt. These cooling rates are indicated in the figure. (Endotherm is down in both diagrams.) [PET sample from Scientific Polymer Products (Cat. # 138) (Menczel, unpublished results).]

transition (ΔC_p). Therefore, when a semicrystalline polymeric sample is heated in a DSC, the amorphous fraction can be determined from the ΔC_p of the sample (provided that the ΔC_p of the pure amorphous polymer is known), and the crystalline fraction (or as it is often called, "crystallinity") can be determined from the heat of fusion (provided that the heat of fusion of the 100% crystalline polymer is known). Therefore, if the mentioned two-phase model is valid, then

$$\frac{\Delta C_{p,\text{sc}}}{\Delta C_{p,\text{am}}} + \alpha = 1 \qquad (2.60)$$

where $\Delta C_{p,\text{sc}}$ is the heat capacity increase at the glass transition of the semicrystalline sample [J/(K·mol)], $\Delta C_{p,\text{am}}$ is the heat capacity increase at the glass transition for the totally amorphous polymer, and α is the crystallinity calculated from the heats of fusion of the sample and of the 100% crystalline polymer (α may roughly be calculated by dividing the heat of fusion of the sample by the heat of fusion of the 100% crystalline polymer, although, for higher accuracy, the temperature dependence of the heat of fusion must be factored in, as is done by the crystallinity software of the Perkin-Elmer DSC7 or Pyris DSC).

As it turns out, for the vast majority of semicrystalline polymers Eq. (2.60) does not hold, the $\Delta C_{p,\text{sc}}$ is usually much smaller than it should be, and the sum of the amorphous fraction and the crystallinity is considerably less than 1, often close to 0.7 (Menczel and Wunderlich 1980, 1986; Menczel and Jaffe 2006, 2007). Part of the amorphous fraction behaves as it does in amorphous polymers, but part of it, which shows no jump in the heat capacity, exhibits unfreezing over a wide range of temperatures. This latter part is called "the rigid amorphous fraction," which roughly represents the interface between the crystals and the so-called traditional (or mobile) amorphous phase. A peculiar characteristic of the glass transition of the traditional amorphous phase in semicrystalline polymers is the absence of the hysteresis peak; contrary to amorphous polymers, no endothermic hysteresis peak appears on the high-temperature side of the glass transition when the sample is cooled slowly and reheated at a higher rate (see the glass transition inset in Fig. 2.29). The existence of the rigid amorphous phase indicates that the two-phase model of semicrystalline polymers is merely a rough approximation for describing their structure.

Since the ΔC_p of the sample is smaller than it should be on the basis of the two-phase model, it is reasonable to assume that only part of the amorphous fraction exhibits unfreezing at T_g (this is the "mobile" or "traditional" amorphous fraction), the other part unfreezes in a wider temperature range (this is the "rigid" amorphous fraction).

If the two-phase model of semicrystalline polymers were valid, the following relationship would hold beyond the glass transition temperature:

$$(1-\alpha)C_{p,\text{am}}(T) + \alpha C_{p,\text{c}}(T) = C_{p,\text{s}}(T) \qquad (2.61)$$

where α is the crystallinity; $C_{p,\text{am}}$ is the heat capacity of the amorphous polymer at temperature T, $C_{p,\text{c}}(T)$ is the heat capacity of the polymer crystal at

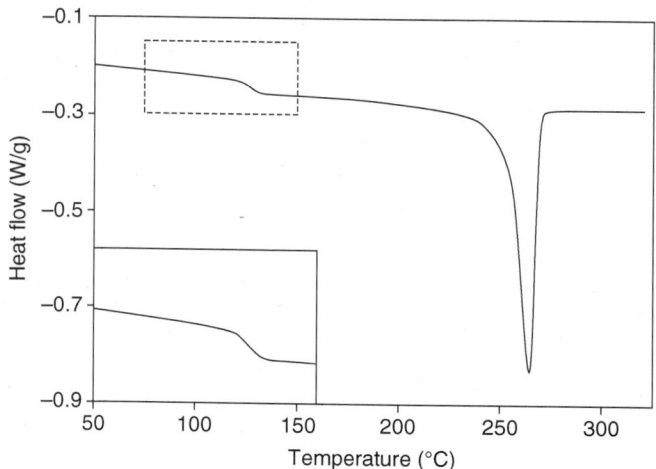

Figure 2.29. The DSC heating curve of a poly(ethylene 2,6-naphthalenedicarboxylate) sample at a heating rate of 10 °C/min previously cooled at 1 °C/min from the melt. There is no hysteresis peak at the glass transition; see that part of the curve in the inset. Endotherm is down. [PEN sample from Aldrich (435317-100G) (Menczel, unpublished results).]

temperature T, and $C_{p,s}(T)$ is the heat capacity of the sample at temperature T. This equation simply says that the amorphous and crystalline heat capacities are additive. But the beyond the transition of the traditional amorphous phase (traditional T_g), the DSC curve [which is proportional to the heat capacity (right hand side of the equation)] is different from the sum of the crystalline and traditional amorphous heat capacities. Therefore, it is reasonable to suggest that part of the amorphous phase dose not unfreeze at the traditional T_g, and when the left hand side (LHS) of Eq. (2.61) is plotted against temperature, it will intercept the DSC curve at a temperature where the unfreezing of the rigid amorphous phase is completed. This case is shown in Fig. 2.30 for a PET sample cooled at 0.3 °C/min, and reheated at 10 °C/min.

The absence of the hysteresis peak for the samples cooled slowly and reheated at a faster rate indicates that the glass transition in semicrystalline polymers should not be cooling or heating rate dependent. The cooling curves recorded for an amorphous polymer (PET) and a semicrystalline polymer (Nylon 6) are shown in Figs. 2.31a and 2.31b, respectively. As expected, T_g depends on the cooling rate for the amorphous polymer, while no cooling rate dependence is observed for the semicrystalline polymer. It can also be expected that the curves recorded during cooling and heating are superimposable as well. However, these curves recorded for PET (Fig. 2.32) cooling (CR = 1 °C/min) and heating (HR = 10 °C/min) indicate that the glass transition as a whole is different when recorded on cooling and heating. It seems that the midpoint (where the two curves intercept) is unchanged and can be rate-independent. However, the course of the glass transition varies on cooling and on heating;

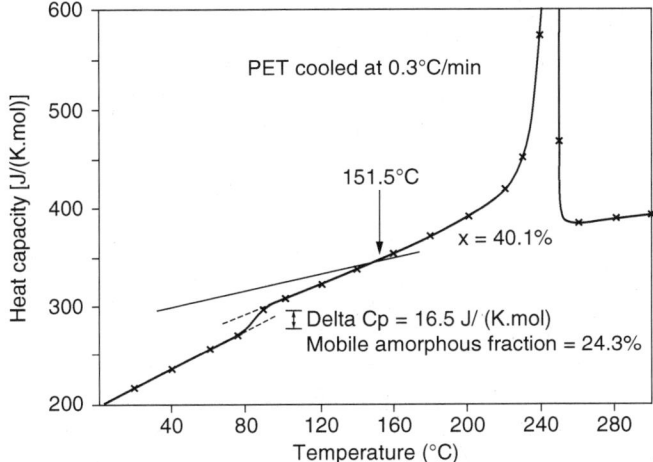

Figure 2.30. Determination of endpoint of unfreezing of rigid amorphous fraction: the straight line represents the heat capacity of the crystalline fraction plus the heat capacity of the amorphous fraction above T_g [from Menczel and Wunderlich (1986)].

the transition on heating is narrower, but the two areas between the two curves are approximately equal. When this is compared with Fig. 2.27, it must be understood why the hysteresis peak is missing for semicrystalline polymers (Menczel and Jaffe 2006, 2007).

2.7.1.2. Physical Aging and Annealing of Glasses

Amorphous polymers and the amorphous portions of semicrystalline polymers are not in thermodynamic equilibrium below their glass transition temperatures. In glassy amorphous polymers and the glassy regions of semicrystalline polymers the material solidifies without developing some crystalline structure; specifically the polymer chains retain the molecular conformation of the melt. Therefore, the volume, enthalpy, and entropy of the glass are higher than they would be at equilibrium. Therefore the state of the glass is metastable, and these metastable glasses undergo slow processes toward equilibrium with decrease of enthalpy and specific volume. Simultaneously, their mechanical properties also change. This gradual approach to equilibrium is referred to as *physical aging* (when it takes place at the use temperature of the polymer), because no irreversible chemical changes occur during this process. The same phenomenon at other, usually higher temperatures (but below the glass transition temperature) is called *annealing*, so physical aging is essentially annealing at the ambient temperature of the polymer. Physical aging affects all the temperature-dependent properties that change more-or-less abruptly at T_g (Struik 1978, 1990; Tant and Wilkes 1981).

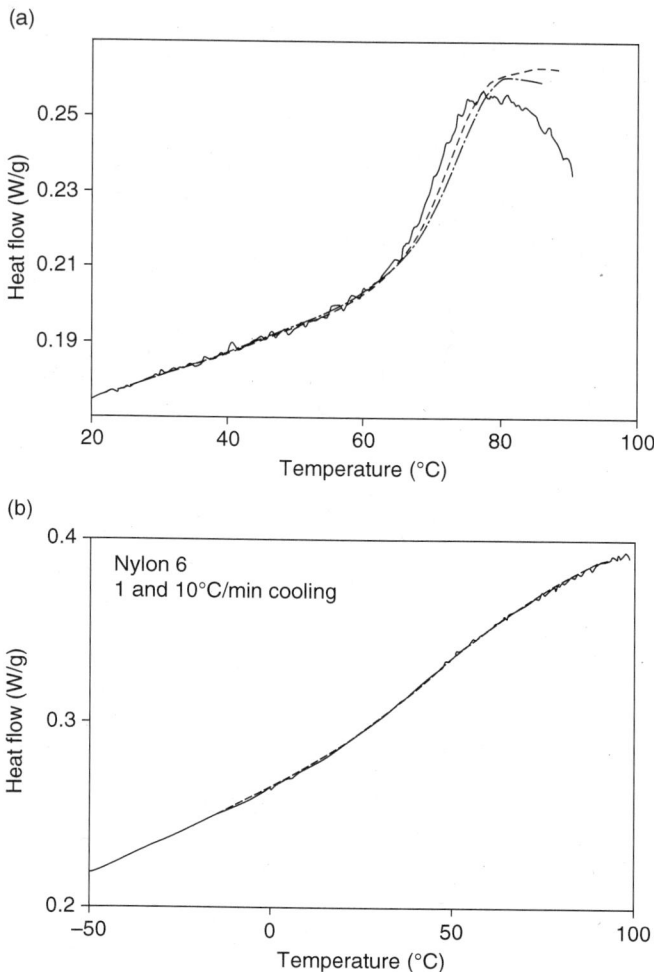

Figure 2.31. (a) DSC curves showing glass transition of amorphous poly(ethylene terephthalate) at various cooling rates: T_g decreases with decreasing cooling rate. Cooling rates: noisy curve 1 C/°min, dashed-line curve 5 °C/min, dotted-line curve 20 °C/min. Endotherm is down. (b) The glass transition of Nylon 6 on cooling at two different cooling rates (1 and 10 °C/min). The curves are superimposable, indicating that the glass transition on cooling is not rate-dependent [compare with plot (a) for amorphous polymers]. (Endotherm is down). [Both (a) and (b) from Menczel and Jaffe (2006, 2007); reprinted with permission of Springer-Verlag and the North American Thermal Analysis Society.]

For a long time it was believed that annealing changes the DSC curves of the glass transition in a manner similar to low cooling rates; longer and longer time annealing times at temperatures below the glass transition temperature will have an effect similar to that for lower and lower cooling rates; specifically, the intensity of the hysteresis peak increases with increasing annealing

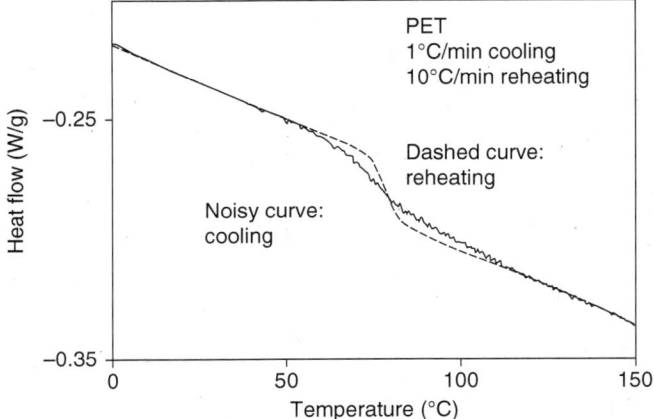

Figure 2.32. The glass transition of (semicrystalline) PET recorded on cooling (CR = 1 °C/min) and on reheating (HR = 10 °C/min); since the two areas between the two curves are equal, there is no hysteresis peak (Endotherm is down) [from Menczel and Jaffe (2006, 2007); reprinted with permission of Springer-Verlag and the North American Thermal Analysis Society].

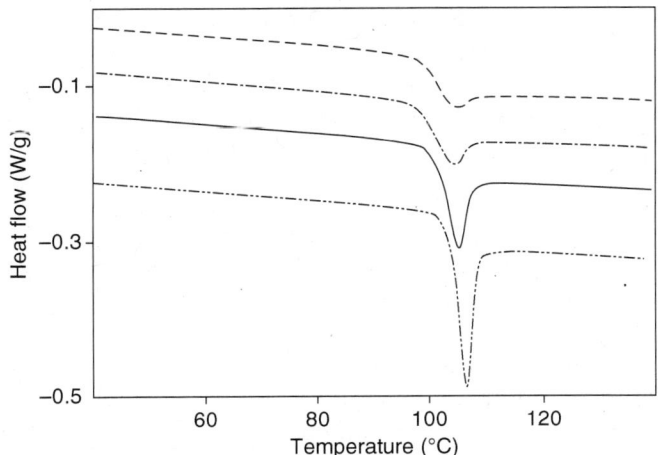

Figure 2.33. DSC curves showing glass transition of an amorphous polymer (polystyrene, M_w = 19,300, M_n = 18,100) annealed for various periods of time; annealing times at 80 °C from top to bottom: 0 h, 1 h, 5 h, 20 h (Endotherm is down) (Menczel, unpublished results).

temperature (up to a certain temperature, where the sample begins to soften), and at constant temperature it increases with time of annealing. While this is true for amorphous polymers (see Fig. 2.33), it is not true for semicrystalline polymers. It was shown that for semicrystalline polymers slow cooling does not introduce a hysteresis peak. On contrary, a hysteresis peak appears as a

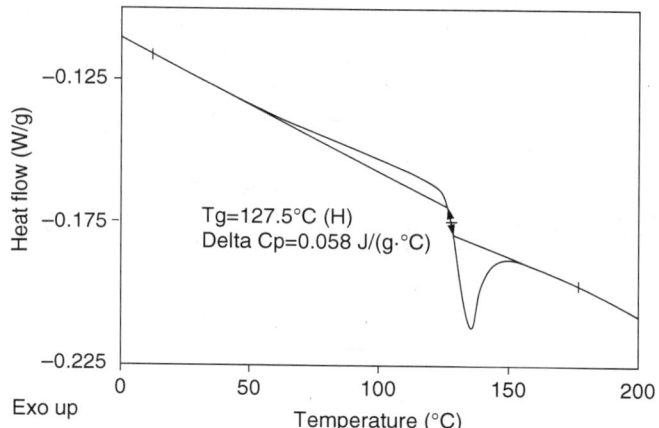

Figure 2.34. The glass transition of poly(ethylene 2,6-naphthalenedicarboxylate) cooled from the melt at a rate of 1 °C/min, then annealed for 31 days at 102 °C; heating rate is 10 °C/min, Endotherm is down. (Menczel and Jaffe 2006, 2007) [Reprinted with permission of Springer-Verlag and the North American Thermal Analysis Society].

result of sub-T_g annealing for these same semicrystalline polymers (Fig. 2.34). However, the picture now is more complicated than in the case of amorphous polymers; a broad exothermic hysteresis peak precedes the glass transition. While the origin of these peaks is not clear yet, it can be suggested that the reason for the endothermic hysteresis peak is not lowering of the enthalpy of the polymer due to annealing, but change in the kinetics of the unfreezing.

2.7.1.3. Effect of Pressure Pressure is one of the physical factors that often influences the glass transition, since polymeric products are frequently prepared by injection molding. Here, the skin–core effect needs to be mentioned; that is, a thin surface layer of an injection-molded polymeric product solidified at a temperature different from that of the bulk. If the sample solidified at an elevated pressure, endothermic peaks would appear at the low-temperature side of the glass transition curve. These peaks shift to lower temperatures with increasing pressure. The DSC curves get rather complex, but little quantitative information can be extracted besides the fact that the sample was cooled under high pressure.

2.7.1.4. Effect of Orientation When an oriented polymeric melt sample is glassified, and heated in the DSC, generally it is difficult to seen the glass transition; the heat capacity jump becomes shallow and blurred, and, similar to semicrystalline polymers, a lot of experience is needed to evaluate the glass transition. Unless the heat capacity increase at the glass transition needs to be determined, an inexperienced operator is advised to use dynamic mechanical analysis or dielectric analysis to determine T_g.

2.7.1.5. Effect of Crosslinking The glass transition temperature increases and ΔC_p decreases with increasing crosslink density of the polymer, since the segmental mobility decreases. The glass transition of crosslinked polymers is discussed in further detail in Section 2.10 of this chapter.

2.7.1.6. Effect of Low-Molecular-Mass Additives The presence of any foreign material may affect the course of the glass transition and the T_g. Although many fillers are inert (i.e., they do not interact with the polymer segments), during cooling they can be a source of creating stresses, since their coefficient of thermal expansion is different from that the matrix polymer.

Often the goal of using low-molecular-mass additives is to lower the glass transition temperature of the given polymer. This can be accomplished by *plasticizers*. These are low-molecular-mass compounds whose primary job is to decrease the interaction between the neighboring polymer chains. A plasticizer increases the free volume in the polymer; thus, the translational mobility of the segments is retained to temperatures lower than in the pure polymer; therefore T_g will decrease. The best-known case of a practical application of the effect of plasticization is soft poly(vinyl chloride) (PVC). Hard PVC (i.e., PVC without any plasticizer) has a T_g at 89 °C (Bair and Warren 1981), and adding 20% dioctyl phthalate pushes down the glass transition temperature to 30 °C (Sichina 2000). Soft tubing like Tygon versus rigid PVC pipe are examples, both of which are quite common and available at retail stores such as Home Depot and Lowe's.

2.7.1.7. The Glass Transition in Polymer Blends and Copolymers Historically, polymers have resulted from at least three stages of development: (1) many new polymers were formed from a large supply of monomers, (2) various chemical techniques were utilized to copolymerize different monomers in order to gain various chemical and physical properties than were unavailable from the monomers alone, and (3) the plastics industry has ventured into blending two or more polymers to yield new plastics with properties superior to those of the blend's individual polymeric components. Here we describe the glass transition behavior of polymer blends and copolymers, but cannot present an exhaustive review. For that purpose the reader is directed to several excellent reviews (Noshay and McGrath 1977; Paul and Barlow 1982; MacKnight and Karasz 1989; Bates and Fredrickson 1990; Folkes and Hope 1993; Hale and Bair 1997).

2.7.1.8. Miscibility in Polyblends Polymers are miscible if they form a single phase. Usually the components in a polymer blend are not miscible unless there is an attractive interaction between groups on two or more polymer chains. The most widely used technique to determine miscibility uses the measurement of the glass transition. The observation of a single glass transition with T_g somewhere between the T_g values of the individual component polymers is an indication of a miscible system. Conversely, a nonmiscible blend will show two glass transitions with T_g values corresponding to those of the individual components. In some nonmiscible, binary polyblends (partial miscibility), the lower T_g in the blend is shifted toward the higher T_g when compared to the pure component, while the higher T_g is lowered with respect to its pure component's T_g. This latter case involves some mixing between the two phases (Hale and Bair 1997).

Several equations have been proposed for calculation of the glass transition temperature of miscible polymer blends. The earliest of these was the Fox equation (Fox 1956):

$$\frac{1}{T_g} = \frac{m_1}{T_{g1}} + \frac{m_2}{T_{g2}} \qquad (2.62)$$

where m_1 and m_2 are the mass fractions of the components, T_{g1} and T_{g2} are the glass transition temperatures of the pure components, and T_g is the glass transition temperature of the miscible blend. However, for most blends positive or negative deviations are observed from the Fox equation, because this equation does not take into account the interaction between the components. Another equation, which does factor in this interaction, is the Gordon–Taylor equation (Gordon and Taylor 1952)

$$T_g = \frac{m_1 T_{g1} + K m_2 T_{g2}}{m_1 + K m_2} \qquad (2.63)$$

where K is a dimensionless binary constant described by the equation

$$K = \frac{\Delta\gamma_2 V_2}{\Delta\gamma_1 V_1} \qquad (2.64)$$

where $\Delta\gamma_1$ and $\Delta\gamma_2$ is the change of the volumetric expansion coefficient at the glass transition temperatures of components 1 and 2, respectively, and V_1 and V_2 are their specific volumes. This equation originally was derived for random copolymers. There are several more similar equations in use (Couchman 1980; Couchman and Karasz 1978; Kwei 1984; Lu and Weiss 1992, 1993).

The use of the simple glass transition criterion to determine polymer–polymer miscibility is shown in Figs. 2.35 and 2.36. The first figure for polymer blends depicts DSC cooling scans at 15 °C/min of polyisoprene (PI) and polybutadiene (PBD) and of their blends. Note that each blend shows a single glass transition where T_g is intermediate between those of the PI and PBD homopolymers. Also, this miscible system shows significant broadening of the blend's glass transition compared to that of each homopolymer. Under the scanning conditions employed, the width of the glass transition ranges from 10 °C for the pure polymers to 39 °C for the 60/40 PBD/PI blend (where *width* is defined

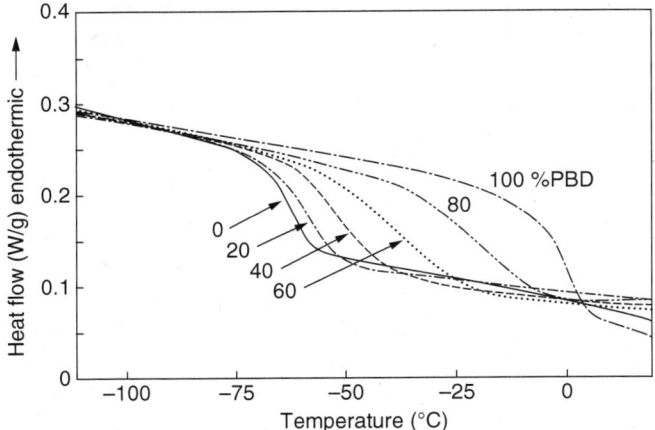

Figure 2.35. DSC curves of polybutadiene–polyisoprene blends recorded at a cooling rate of 15 °C/min. The blends had different polybutadiene (PBD) contents; the PBD mass percentages are indicated at each curve. A single glass transition was recorded for every sample, with T_g depending on the compoisition. The single glass transition indicates miscibility. [From Hale and Bair (1997); reprinted with permission of Elsevier.]

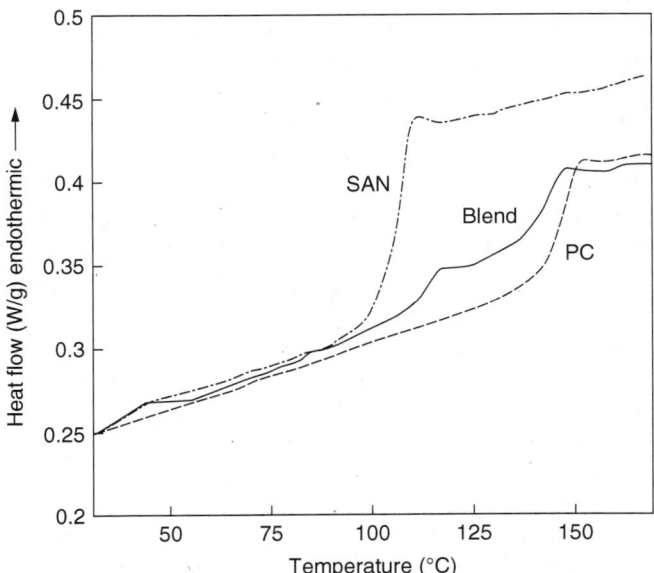

Figure 2.36. DSC heating curves of styrene–acrylonitrile copolymer (SAN), polycarbonate (PC), and unknown polymer blend recorded at 15 °C/min. Endotherm is up [reprinted from Hale and Bair (1997) with permission of Elsevier].

as the difference between the extrapolated end temperature and extrapolated onset temperature similarly as in Fig. 2.35). This broadening of the glass transition is due to small-scale compositional fluctuations in the blend.

Figure 2.36 shows the DSC curve of the blends of a random copolymer of styrene and acrylonitrile (SAN) with polycarbonate (PC). T_g's of the SAN and PC components are 116 and 147 °C, respectively. The blend shows two glass transitions related to the glass transitions of the component polymers; however, SAN's T_g is shifted up to 111 °C, while PC's T_g is lowered to 137 °C. These shifts in the blend's glass transition temperatures are due to limited phase mixing between PC and SAN (Bair et al. 1980, 1981).

If portions of a polyblend are grafted together as is often the case in an impact modified thermoplastic such as the terpolymer, ABS, the blend's rubber content will be underestimated by DSC measurement if the mass of SAN grafted to PBD exceeds a 1:1 ratio (Bair et al. 1981; Bair 2007). When the graft ratio is higher than this level, ΔC_p at T_g for PBD will be suppressed and consequently, the rubber concentration will be significantly less than the actual amount present. For example, when two commercial ABS resins having identical rubber levels, namely 17 mass percent but with SAN:PBD graft ratios of 0.7:1 and 2.4:1 were scanned in a DSC their calculated PBD masses were 15 and 9%, respectively based on ΔC_p values. This 40% reduction in DSC determined rubber content was accompanied by a 47% decrease in the polyblend's impact strength, IS. The ABS with the lower graft content has an IS of 6.0 while the higher graft ABS resin's IS is 47% lower or equal to 3.2 ft-lb/in notch (Bair et al. 1981). It is important to note that the level of rubber found in these two polyblends by DSC measurements roughly scales with the terpolymer's impact strength at room temperature. In other words, the DSC estimate of rubber in an ABS resin gives a direct measure of the amount of PBD that is free to modify the deformational behavior of the terpolymer and, hence, is an important tool in analyzing the behavior of impact modified polyblends.

2.7.1.9. Measurement of Glass Transition Temperature with Modulated Temperature DSC Additional information can be obtained on the glass transition when characterized by modulated temperature DSC as described in detail in Section 2.13 of this chapter. The DSC heat capacity jump signal in this case appears in the reversing heat flow (or reversing heat capacity) signal, while the endothermic or exothermic hysteresis peak is displayed on the nonreversing signal. Naturally, as in the case of relaxation thermal analysis techniques (DMA and DEA), the measured glass transition temperature depends on the frequency of the measurement, in this case on the modulation period. The T_g taken from the reversing signal is almost always higher than the value taken from the total heat flow signal (which is identical to the traditional DSC signal). The advantage of recording the glass transition with MTDSC is that the heat capacity jump is separated from the hysteresis phenomena; the disadvantage is that the measurement time is longer, since it is not recommended for MTDSC runs at heating rates higher than 10 °C/min, and usually slower, such as 2–5 °C/min.

2.7.2. Crystallization

Crystallization, one of the two first-order transitions encountered in the thermal analysis of polymers, is a process in which a material from the amorphous state is transformed into the crystalline state from either solution or the melt. Crystallization of macromolecules is different from the crystallization of low-molecular-mass materials. First, similar to the melting process, it takes place at conditions far from equilibrium. When compared to low-molecular-mass substances, the crystallization process of polymers is much slower because of the lower mobility of the polymer chain segments; therefore in nonisothermal conditions this process takes place over much wider temperature ranges. Crystallization of low-molecular-mass materials is mentioned here very briefly, and only for the purpose of comparison with macromolecules.

The thermodynamic basis for the first-order phase transitions is summarized in Fig. 2.37. Without getting into details, we simply mention that the most stable phase has the smallest Gibbs free energy, and that the free energy decreases with increasing temperature.

In Fig. 2.37, the Gibbs free energy curves for the equilibrium crystals and equilibrium melt are shown by solid lines. The rate of decrease of the free energy–temperature curve for the crystal differs from that of the melt; therefore the two curves intercept, and the point of interception is the equilibrium melting point ($T_m°$). Below the melting point the crystals are stable, because their free energy is lower than the free energy of the melt. Above the melting point, the melt is more stable, because its free energy is smaller than the free energy of the crystals. However, in most cases the state of polymeric crystals is far from equilibrium. In Fig. 2.37 the free energy of nonequilibrium (metastable) polymer crystals is denoted by a dashed curve, which runs above the free-energy curve of the equilibrium crystals. This curve intercepts the

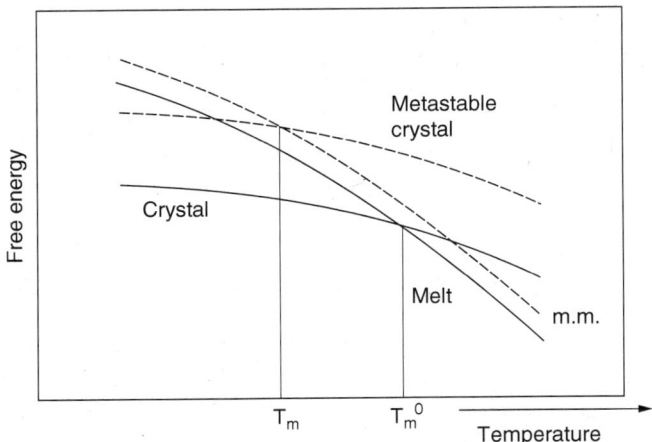

Figure 2.37. The Gibbs free energy as a function of temperature for the crystalline and melt phases (mm = metastable melt) [reprinted from Wunderlich (1990) with permission of Elsevier and B. Wunderlich].

equilibrium melt curve at a lower temperature than the equilibrium melting point, showing that the melting point of metastable crystals is lower than $T_m°$. In the absence of reorganization and superheating (see later in this section) the melting at this point is called *zero-entropy-production melting*; at this point there is no entropy production because the metastable crystals are transformed into a melt of equal metastability (Wunderlich 1997a). This is not an equilibrium process, and Fig. 2.37 indicates that the zero-entropy-production melting point is much lower than the equilibrium melting point. Often the melt of polymers is oriented, for example, in the production of films and fibers. The melt orientation leads to an entropy decrease, so the Gibbs free-energy curve of the oriented melt will run above the curve of the equilibrium melt (dashed curve denoted by mm = metastable melt in Fig. 2.37). Figure 2.37 illustrates that in this case a melting point higher than the zero-entropy-production melting point will be recorded.

2.7.2.1. Crystallization of Low-Molecular-Mass Substances

When the melt of a low-molecular-mass substance is cooled, then, at a certain temperature, where the free enthalpy of the crystals becomes smaller than the free enthalpy of the melt, and there is a sufficient number of nuclei in the melt after passing through the energy barrier of nucleus formation, crystallization will take place, which indicates formation of a long-range three-dimensional order of the motifs or recurring structures of the material. Since crystallization is a first-order transition, there is an exothermic latent heat involved; thus, an exothermic DSC peak is recorded at the transition. However, crystallization does not occur at the melting point; in most cases considerable supercooling is involved, so $T_c < T_m$. Sometimes the rate of crystallization is so fast that the heat released at the beginning of the process simply cannot be dissipated or conducted away by the cell, so it raises the sample temperature back toward the melting point. At the temperature at which the continuing heat release gets into equilibrium with the cooling capacity of the sample holder, the mentioned temperature rise ceases. Such a crystallization curve is shown in Fig. 2.38a for crystallization of indium at a cooling rate of 10°C/min. This DSC curve was recorded on a TA Instrument Q200 DSC in $T4$ mode, which displays ΔT versus the *sample temperature* (see Section 2.3). It can be clearly seen that the sample starts crystallizing at a temperature of T_{c0} (154.9°C) and is heated back to 155.66°C (the melting point of indium is 156.60°C). In older-type TA Instruments DSCs or in the $T1$ mode in Q200 and Q2000 DSCs, the crystallization curve of low-molecular-mass substances leans back toward the melting point. This phenomenon cannot be seen with the power compensation (Perkin-Elmer) DSCs, where the output of the differential amplifier is displayed as a function of the program temperature (see Fig. 2.38b).

2.7.2.2. Crystallization of Polymers

Crystallization from solution will not be considered here because it is slow and the heat effect per unit mass of the solution is usually too small for DSC experiments. More detailed description

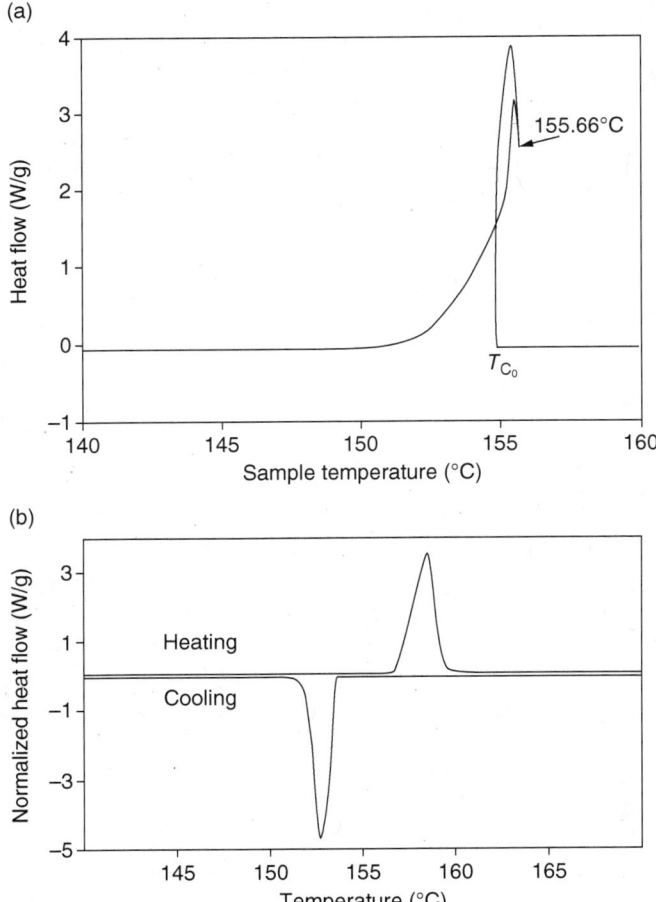

Figure 2.38. (a) Crystallization of indium at a cooling rate of 10 °C/min: ΔT as a function of sample temperature recorded on a TA Instruments Q200 DSC (T4 mode); exotherm is up. Crystallization starts at T_{C_0}, then the sample temperature is raised back toward the melting point, where the crystallization is completed. Then the sample temperature is lowered again. (b) Melting and crystallization of indium as recorded by a Perkin-Elmer DSC7 power compensation DSC at a heating and cooling rates of 10 °C/min. Supercooling can be clearly seen since the crystallization temperature is lower than the melting point. Exotherm is down.

of polymer crystallization can be found in two books (Wunderlich 1976; Mandelkern 1964).

Thus, in this section we describe *melt crystallization*, in which the crystal formation takes place from the polymer melt. Melt crystallization of polymers can be subdivided into (1) *isothermal crystallization*, a crystallization process that takes place at a definite, constant temperature; and (2) *nonisothermal melt crystallization*, in which the polymer sample is melted and crystallization takes place during cooling at a constant rate. A special case of melt crystallization is what is called cold crystallization. This term describes the crystallization process of crystallizable polymers that had been quenched into the amorphous glassy state by extremely fast cooling of the polymer melt. Cold crystallization takes place above T_g and can be both isothermal and nonisothermal, the latter taking place during heating.

Similar to low-molecular-mass materials, polymer crystals are unstable beyond the melting point, and the melt is stable, because at these temperatures the free energy of the melt is smaller than the free energy of the crystals. The curves of the free energy of the crystals and the melt intercept at the melting point, and crystallization takes place when the temperature is decreased (see Fig. 2.37).

Crystallization itself is a two-step process, consisting of crystal nucleation and crystal growth.

2.7.2.3. Nucleation

At the beginning of the crystallization process there is a barrier in the free energy that needs to be overcome for nuclei to start growing. Density fluctuations help to overcome this barrier, and to proceed to a stage of stable, ordered regions of critical size. These are called *primary nuclei*, and the process of their formation is called *primary nucleation*. Thus, *nucleation* describes the formation of embryonic crystallites of critical dimensions that are stable at a given temperature and can initiate the growth of crystals. In this context "growth" means an increase of the crystallite size. When discussing nucleation, most chemists and physicists imply primary nucleation. The primary nuclei are the ones that initiate crystal growth in a polymer melt containing no crystals, provided that they have reached the critical size and are stable. In polymer physics, three other types of nucleation are important for the *primary nucleation* itself, or for the growth of crystals. The first one is molecular nucleation, in which the initial part of a polymer molecule is incorporated into a crystal. This may occur in the primary nucleation itself, or take place as part of a secondary or tertiary nucleation (see below). This type of nucleation is especially important for formation of fringed micelles (see later in this section). *Secondary nucleation* refers to growth of a new layer of a polymer crystal on a smooth crystal surface, while *tertiary nucleation* characterizes attachment of macromolecular segments to edges of a growing crystal.

The nucleation process can also be classified on the basis of the mechanism and origin of nucleus formation. When the nuclei are formed in a chemically homogeneous space containing no foreign surfaces, the process is called homogeneous nucleation. However, the presence of foreign surfaces [nucleating agents or other solid (or sometimes even gaseous) surfaces like the bottom of the DSC pan] may lead to a decrease of the size of critical nuclei, and in this case the nucleation is called heterogeneous. Thus, the terms "homogeneous" and "heterogeneous" indicate the origin of the nucleation process. "Thermal" and "athermal" nucleation refer to the time dependency of the nucleation process. Thermal nucleation is used to designate the nucleation process that occurs throughout the crystallization, while athermal nucleation refers to a nucleation process in which all crystal started to grow at the same time. Homogeneous nucleation is always thermal, but heterogeneous nucleation can be thermal or athermal; in most cases it is athermal. Polarization optical microscopy is often a good technique to decide whether the nucleation is thermal or athermal. If the lines of impingement of the spherulites are straight, the nucleation must be athermal, because in this case all the spherulites started to grow at the same time. If the lines of impingements are curved, the spherulites started to grow at different times, so the nucleation is thermal.

Finally, a very special type of the nucleation during crystallization of polymers is "self-nucleation," indicating that an originally crystalline (semicrystalline) polymer sample is melted, and for various reasons, nuclei of the melted crystals are preserved in the melt at temperatures beyond the melting point. When this sample is cooled, the preserved nuclei help to initiate the crystallization process. An example of this process is when highly oriented polymeric samples are melted, and "row nuclei" may remain in the melt (Varga et al. 1981). Figure 2.39 shows isothermal and nonisothermal crystallization curves of biaxially oriented polypropylene. Optical microscopy revealed that after the melting of biaxially oriented samples, row nuclei were preserved in the melt, and higher temperature and more time were necessary to destroy these oriented regions. As a result of cyclic crystallization and melting, or holding the sample at sufficiently high temperature, these row nuclei continuously diminished. In the first crystallization two types of nucleation took place simultaneously (self-nucleation with row nuclei, and usual heterogeneous nucleation after the row nuclei were exhausted). In the subsequent crystallization cycles, the number of row nuclei gradually decreased. For rapid destruction of the row nuclei, the temperature of the melt should be raised well beyond the equlibrium melting point [for polypropylene this is likely to be 220°C or about 50°C above the non-equilibrium T_m (Samuels 1975a,b)].

2.7.2.4. Crystal Growth As mentioned above, crystal growth is a process in which the dimensions of stable nuclei, and later the dimensions of polymeric crystallites, increase. The most important physical quantity used to characterize crystal growth is the linear growth rate, which is constant in time, but changes with temperature. This is the rate of crystal size increase in one dimension, and can be determined from polarization optical microscopy experiments.

There are several types of polymeric crystallites. Of these, the following need to be mentioned:

- *Extended-Chain (Equilibrium) Crystals.* In these crystals, as the term implies, the polymer chains have extended conformation (i.e., there is no chain folding). Extended-chain crystals have been prepared for a limited number of polymers. Wunderlich and coworkers were the first to

86 DIFFERENTIAL SCANNING CALORIMETRY (DSC)

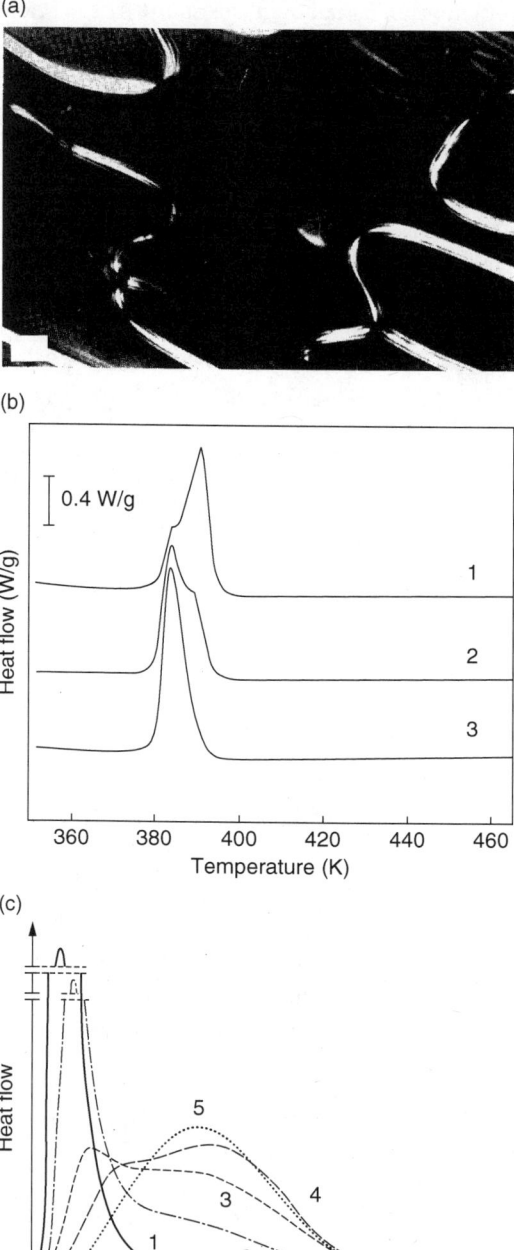

crystallize polyethylene under high pressure from the melt, and obtained extended chain macroscopic crystals (Geil et al. 1964; Davidson and Wunderlich 1969; Prime and Wunderlich 1969; Prime et al. 1969). He and his research group characterized the properties of these crystals. Later it was found that at high pressure, polyethylene has a hexagonal crystal form, and the constituting macromolecular chains have high mobility along the c axis of the crystals. This mobility serves as an explanation for the formation of the extended-chain crystals.

- *Folded-Chain Lamellae.* These are thin, 5–50-nm-thick platelet crystals in which the direction of the polymer mainchains is perpendicular to the plane of the lamellae, and the chains in most cases fold back into the crystal, referred to as *adjacent re-entry*. Development of folded-chain crystals is guided by the so-called chain-folding principle. According to this principle, "a sufficiently regular, flexible linear macromolecule crystallized from the mobile random state will always crystallize first in a chain folded macroconformation" (Wunderlich 1976, p. 350). Therefore, an optimum condition of preparing folded-chain crystals of a linear polymer is crystallization from solution at ambient conditions or elevated temperatures, because high segmental mobility is needed for chain folding. But chain folding is more difficult in melt crystallization owing to steric problems and mobility issues.

- *Fringed Micelles.* These are the crystals formed during crystallization from the melt, and they are the most important in terms of melt crystallization measured with DSC. The elements of all three major conformations, the folded-chain lamellae and the extended-chain crystals, and the random coil configuration of the amorphous phase are present in the fringed micelles. In fringed micelles the polymer chains pass through several crystallites and the amorphous portions surrounding these crystallites.

Figure 2.39. Crystallization of biaxially oriented polypropylene (samples taken from polypropylene bottle walls): (a) polarization optical micrograph at 180°C (melting point, i.e., end of melting of sample, is 170°C); (b) nonisothermal crystallization curves recorded in cyclic heating–cooling experiments: in the heating DSC experiments the samples were heated to 180°C at 10°C/min, held at this temperature for 5 min, and then cooled at 10°C/min. This temperature program was then repeated in a cyclic manner on the same sample; 1 indicates first cooling; 2, second cooling; 3, third cooling. Exotherm is up. (c) Isothermal crystallization curves recorded at similar conditions as the nonisothermal crystallization curves; the numbers, similarly to nonisothermal crystallization, indicate the sequence number of the crystallization (i.e., 1, first crystallization; 2, second crystallization; etc.). Double crystallization in both nonisothermal and isothermal crystallization indicates the presence of two simultaneous nucleations. Exotherm is up. After several melting–crystallization cycles, the crystallization curves become single, indicating that the ordered regions (row nuclei) in the polymer melt have been destroyed. [From Varga et al. (1981), reprinted with permission of Springer-Verlag.]

- *Spherulites.* As the term implies, these are spherically shaped crystalline bodies (sometimes called "supermolecular structures") are aggregates smaller fibrillar, lamellar, or needlelike crystallites. The spherulites are well known to most polymer chemists from polarization optical microscopy observations. It should be mentioned that, although spherulites are birefringent, they are not 100% crystals, they do contain considerable concentrations of amorphous regions.

In DSC experiments only the heat effects of crystallization can be recorded. This means that only the crystal growth, not the nucleation, can be measured. However, indirect conclusions about the primary nucleation can be made by analyzing DSC data from isothermal crystallization as described later in this section.

2.7.2.4.1. Isothermal Melt Crystallization. The most frequent quantitative evaluation of crystallization is done by applying the Avrami equation to isothermal crystallization (Avrami 1939, 1940, 1941). This is an idealized equation that describes the growth of crystals as a function of time. In a DSC crystallization experiment, the sample is melted, and heated beyond the equilibrium melting point to completely erase any thermal history. This step is necessary when the goal is characterization of the polymer itself, and not that of thermal history (if the sample is not heated beyond the equilibrium melting point, ordered regions may remain in the polymer melt, and self-nucleation may take place). After melting the sample and waiting for ~2–10 min for temperature equilibration and destruction of any remaining ordered regions, the temperature is quickly lowered to the temperature of isothermal crystallization (T_c), where it is held until the process is completed. A good indication of the completion of the crystallization process is a stable horizontal baseline. Of course, this type of evaluation is suitable when *one* definite type of nucleation occurs and when *one* crystal form is developing. In the case of mixed crystallization, the crystallization curve consists of two peaks, and for quantitative evaluation it will be necessary to deconvolute the peaks. An example is the simultaneous growth of α- and γ-crystal forms of polypropylene. But the overall crystallization cannot be simply represented as the sum of two crystallization subprocesses, because the presence of the crystal form growing in the first step can and often does influence the second crystallization subprocess.

The Avrami equation for low-molecular-mass substances has the following form

$$\alpha(t) = 1 - \exp[-K(t-\tau_0)^n] \qquad (2.65)$$

where $\alpha(t)$ is the crystallinity during the isothermal crystallization at time t, K is the rate constant of crystallization, t is the time (min), τ_0 is the induction time of crystallization indicating how much time is needed for nuclei to appear, and n is the Avrami exponent. τ_0 is often zero, especially for heterogeneous nucleation, but it may have non-zero values, especially for homogeneous nucleation. Since macromolecules during their melt crystallization never reach 100% crystallinity, this equation was modified (normalized) for polymer crystallization to the following form

$$\frac{\alpha(t)}{\alpha(\infty)} = 1 - \exp\left[-K(t-\tau_0)^n\right] \qquad (2.66)$$

where $\alpha(\infty)$ is the crystallinity at the completion of the crystallization process. After rearrangement, this equation can be rewritten as

$$\ln\left[-\ln\left(1-\frac{\alpha(t)}{\alpha(\infty)}\right)\right] = \ln K + n\ln(t-\tau_0) \qquad (2.67)$$

that is plotting the LHS of Eq. (2.67) versus $\ln t$ should give a straight line that intercepts with the y axis at K and whose slope determines the Avrami exponent.

There is some confusion regarding the induction time of crystallization. While it is definitely present in the case of homogeneous nucleation, its existence is not that obvious in heterogeneous nucleation. While here it is often regarded as 0, there are some experimental indications for its presence even for heterogeneous nucleation (Menczel and Varga 1983).

The exponent n (Avrami exponent) contains some information about the type of nucleation. Table 2.3 lists the physical meaning of the Avrami exponent in terms of one-dimensional, two-dimensionsal, and three-dimension crystal growth.

As can be seen from Table 2.3, in the case of athermal nucleation the Avrami exponent equals the dimension of the geometry of the growing crystal entities: 1 for fibrillar crystals, 2 for lamellar crystals, and 3 for spherulites. In the case of thermal nucleation, the Avrami exponent equals the geometry of the growing entities plus 1 (however, this is not true for diffusion-controlled processes).

The cases above described are only the simplest cases of isothermal crystallization. In cases when the growing crystals branch, the Avrami exponent can be as high as 5 or 6 (Booth and Hay 1972; Wunderlich 1976). There are other

TABLE 2.3. Avrami Exponent for Various Nucleation Types

Dimension of Crystal Growth	Avrami Exponent in	
	Athermal nucleation	Thermal nucleation
1 (fibrillar)	1	2
2 (lamellar)	2	3
3 (spherulitic)	3	4

problems with the Avrami evaluation. The Avrami exponent often has a fractional value, and there have been some attempts to interpret this as development of distorted lamellae or spherulites. Also, toward the end of the crystallization process, the Avrami exponent often changes, and there may be several reasons for this. One of these could be the onset of secondary crystallization, which is slow crystallization of the amorphous portions of the sample after the main ("primary") crystallization is over. Another effect that can cause a change in the Avrami exponent close to the end of the primary crystallization, is crystal perfection, when the just-developed crystals tend to improve. Also, for diffusion controlled processes, the Avrami exponent even theoretically has non-integer values for three-dimensional growth.

Figure 2.40 shows an isothermal crystallization curve of a high-density polyethylene sample recorded by a Perkin-Elmer DSC7 (see Fig. 2.41 for Avrami plots). In the study of isothermal crystallization, the power compensation DSC is the preferred instrument among the presently available commercial DSCs, because the temperature difference between the sample and the reference cells is negligible, as described in Section 2.3.

The major difficulty of the DSC recording of isothermal crystallization is the beginning part of the peak. As described in Section 2.6, switching the DSC from the cooling mode to the isothermal mode (see Fig. 2.19a of Section 2.6) causes a severe signal transient, which is proportional to the heat capacity of the sample and the heating or cooling rate. Thus, when the sample temperature is dropped to T_c, the DSC signal suddenly shifts, and it shifts back when the isothermal crystallization temperature is reached. In most cases the crystallization process starts before the second signal shift is over. The consequence is that the beginning part of the peak is often missing. Therefore, DSC can be reliably applied to isothermal crystallization only in a limited temperature range, where the crystallization does not start too soon after reaching the crystallization temperature.

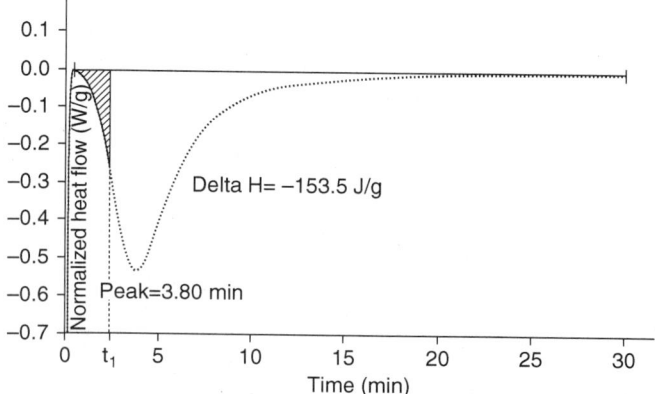

Figure 2.40. An isothermal crystallization curve of high-density polyethylene (courtesy of Perkin-Elmer). The area calculation is shown for the Avrami evaluation. Exotherm is down.

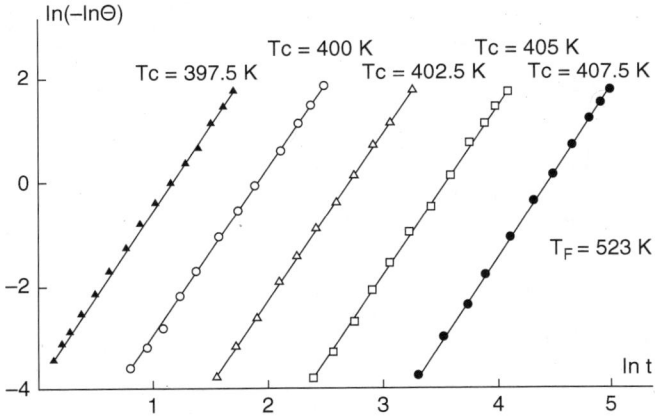

Figure 2.41. Avrami plots of the crystallization isotherms for biaxially oriented polypropylene after erasing the thermal history of the samples by heating them to 250 °C, which is much higher than the equilibrium melting point of polypropylene ($T_m^\circ = 220\,°C$).

2.7.2.4.2. Nonisothermal Melt Crystallization. In routine DSC crystallization measurements of polymers, nonisothermal crystallization is typically used. This method is simple and gives practical results. At the same time, the information obtained is limited to the characteristic temperatures of crystallization. The sample is heated beyond the equilibrium melting point of the polymer, held there for several minutes for temperature equilibration, and also to destroy all nuclei, and then it is cooled at a specific constant rate to room temperature or below to record the crystallization exotherm.

Figure 2.42 shows the nonisothermal crystallization curve of poly(butylene terephthalate) at a cooling rate of 10 °C/min. The peak is considerably broader than the crystallization peak of low-molecular-mass substances. When performing such experiments, usually three characteristic temperatures are reported: the starting temperature of crystallization (T_{c0}), the extrapolated onset temperature of crystallization ($T_{c,\text{onset}}$), and the peak temperature of crystallization (T_{cp}). The peak temperature of crystallization is important from a practical perspective, because it indicates the maximum rate of crystallization at a given cooling rate.

The determination of heat of crystallization is not always simple, because the crystallization process in nonisothermal crystallization often lasts to the glass transition temperature. Also, since the absolute values of the melt and crystal heat capacities and their temperature dependences are different, application of the sigmoidal baseline is needed for generally accurate determination of the heat of crystallization (for the sigmoidal baseline, see Section 2.7.3). However, because of the frequent absence of a linear baseline above T_g, but after completion of the crystallization process, the sigmoidal baseline cannot

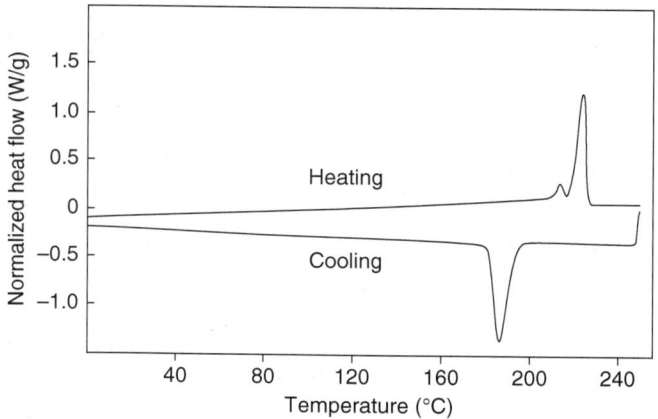

Figure 2.42. Melting and nonisothermal crystallization curve of poly(butylene terephthalate) at a heating and cooling rates of 10 °C/min (sample from Scientific Polymeric Products Polymer Kit). Considerable supercooling: $T_m = 226.5\,°C$, $T_{c0} = 200.0\,°C$. Endotherm is up. [Perkin-Elmer DSC (Menczel, unpublished results).]

be applied; only the linear baseline option is a viable option and can lead to errors as high as 20%. Therefore, determination of the heat of crystallization (which should be identical to the heat of fusion) in melting experiments is recommended.

There have been attempts to carry out the Avrami evaluation of nonisothermal crystallization of polymers by the so-called Ozawa method (Ozawa 1971; Collins and Menczel 1992). This is a very complex method, because the temperature change affects both nucleation and the crystal growth. But this method can give some idea about the change of the crystallization mechanism as temperature changes during crystallization: considerable curvature can be seen in the Avrami plots also, depending on the cooling rate. It was suggested that the crystallization mechanism depends on the cooling rate as well.

2.7.2.4.3. Isothermal Cold Crystallization. When a polymer sample is quenched, and its temperature is raised, isothermal cold crystallization can also be carried out. This is not a very popular technique, but it is useful when the rate constant (K) of crystallization from Eqs. (2.65)–(2.67) needs to be determined in a wide temperature range.

2.7.2.4.4. Nonisothermal Cold Crystallization. As mentioned before, the term *cold crystallization* means crystallization of a polymer from the amorphous state when it is heated from the glassy state. First, the molten polymer has to be quenched (i.e., cooled at an extremely high rate) in order to freeze it to an amorphous glass before crystallization can take place. Some (not all) polymers can be easily quenched to an amorphous state. A typical example is

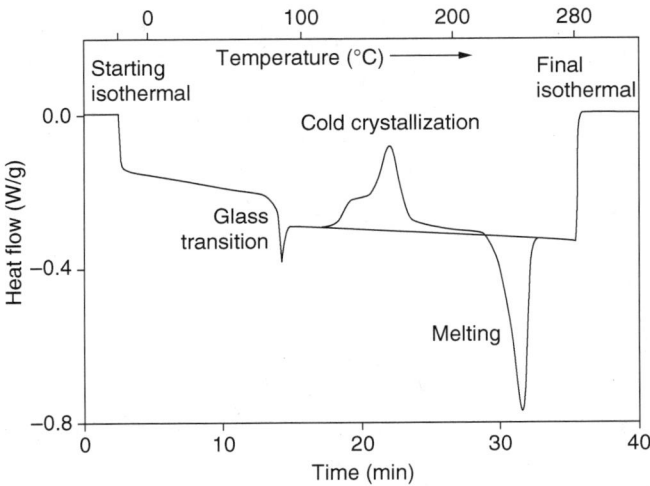

Figure 2.43. A full DSC curve of quenched (amorphous) poly(ethylene terephthalate) recorded at a heating rate of 10 °C/min; the glass transition (at ~80 °C) is followed by cold crystallization and melting of crystals formed in nonisothermal cold crystallization. The baseline during the transition was drawn to show that the crystallization process lasts to the beginning of melting. Endotherm is down. (Menczel, unpublished results.)

poly(ethylene terephthalate) (PET). When such an amorphous polymer is heated, it shows peculiar behavior. The heating curve of quenched PET is shown in Fig. 2.43. This sample was prepared by melting it at a temperature beyond the equilibrium melting point (which is 280 °C for PET) and then quenching it by immersing the DSC pan containing the sample into liquid nitrogen. Such an amorphous glassy sample cannot crystallize below the glass transition temperature, because the segmental mobility is frozen. During heating this sample goes through the glass transition at about 80 °C. Once it is above T_g, the chain segments acquire some translational mobility, and the material has enough time to crystallize, since the heating is relatively slow (in this case it is 10 °C/min). This cold crystallization process takes place in the temperature range 120–224 °C. It is important to note that, although the crystallization definitely slows down at ~163 °C, it does not stop. The cold crystallization, as can be seen from the linear baseline drawn in Fig. 2.43, continues to ~224 °C, that is, to the beginning of the melting of crystals formed during the cold crystallization.

Cold crystallization is not characteristic of all polymers. Many crystalline polymers crystallize so fast that they always crystallize on cooling, at least at the usually applicable cooling rates or usual quenching conditions. The truth is that quenching with liquid nitrogen is not very efficient. When the DSC pan containing the molten polymer is suddenly immersed in liquid nitrogen, a nitrogen bubble is formed around the pan that somewhat insulates the sample from the −196 °C (which is the temperature of liquid N_2 at its boiling point). Therefore the cooling is not fast enough. Other, more efficient cooling (quenching) methods are available: using other cooling media (e.g., Freon) or some other cooling methods [something similar as in Quick Freeze Deep Etch (QFDE)] in which cooling rates of several thousand °C/s can be achieved, but most of these cooling methods produce small amounts of sample. At such conditions, many polymers can be quenched, and although thermal analysis of these amorphous polymers is feasible, such experiments have not been done yet.

2.7.3. Melting

Melting is one of the two major first-order phase transitions of crystalline polymers. It is the transition from the crystalline state to the melt. Melting is a transition with an increase of disorder of the system, so it is accompanied by an entropy increase.

Since melting is a first-order transition, the entropy increases abruptly at the melting point. The entropy jump at melting can be estimated by Richard's rule, which states that the entropy increase at melting is 7–14 J/(mol·K) for substances having monoatomic melt. This rule is analogous to Trouton's rule (Trouton 1883) or Hildebrand's rule (Hildebrand 1915, 1918) giving 88 J/(mol·K) for the molar entropy of vaporization of various liquids at their boiling points. All three of these rules deal with positional entropy, because they are valid for order–disorder transitions of spherical motifs. However, for linear macromolecules two other entropy contributions need to be considered: the orientational and the conformational entropies. For polymers these are more important than the positional entropy, which for linear macromolecules is negligible because of the extreme length of the individual polymer chains. The conformational entropy increases with increasing length of the macromolecular chains.

In addition to the entropy change, the change of two other thermodynamic parameters is important for melting: the volume change and the change in specific-heat capacity (therefore, enthalpy). There is no rule to describe the volume change on melting. For most materials volume increases at melting, but there are exceptions. Among these, water is well known for its volume decreases on melting. On the other hand, the specific heat capacity and enthalpy always increase on melting.

Melting is an endothermic event, and shows up in the DSC curve as an endothermic peak. One important task in DSC measurements is determination of the melting point and heat of fusion of both low-molecular-mass and macromolecular crystals. In addition to melting of polymers, we briefly describe here the melting of low-molecular-mass substances. Every thermal analyst must be familiar with this if for no other purpose than for calibration of the instruments with metal standards and for measuring melting properties of low-molecular-mass substances used in the plastics industry.

2.7.3.1. Melting of Low-Molecular-Mass Substances

The melting of low-molecular-mass substances is almost always a process close to equilibrium, because even imperfect low-molecular-mass crystallites perfect themselves during heating. Therefore, in the vicinity of the melting point close to equilibrium crystals are available, and the melting of such substances is called *reversible melting*.

The melting point for pure samples (purity >99.9%) is determined as the extrapolated onset of the endotherm (see Fig. 2.13b in Section 2.5, on calibration of DSCs). For determination of the heat of fusion, let us consider the picture presented in Fig. 2.44. Before the melting experiment starts, the temperature of the sample and the reference is the same as the block temperature: $T_s = T_r = T_{bl}$. When the operator starts the experiment, the temperature of the sample and the reference begins to lag behind the furnace temperature, and also, the sample temperature starts lagging behind the reference temperature, so there will be an increase in the $\Delta T = T_r - T_s$ signal. After a certain period t_{ssi} (which depends on the heating rate), the steady-state conditions are reached and a horizontal "baseline" follows, indicating that the ΔT value is constant in this region (ΔT_i). This horizontal line continues until the sample temperature T_s reaches the melting point of the sample (at time t_i). At this moment the temperature increase of the sample stops, because according to equilibrium thermodynamics, the temperature cannot increase until the whole sample is melted. From this point an endothermic peak starts, and the DSC curve displays a steady linear increase in the $\Delta T = T_r - T_s$ signal. This part of the endothermic peak is called the "leading edge." Thus, in the $t_i - t_f$ region the sample temperature remains constant (and in the power compensation DSCs, the reference temperature will be higher than the program temperature, i.e., the heating rate of the reference will exceed the average heating rate). As soon

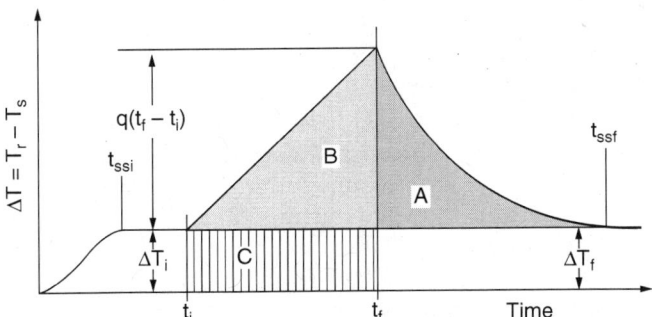

Figure 2.44. Schematic representation of the baseline method. Area A is the energy accounted for in the return to steady state, while areas B and C represent the energy necessary for melting the sample. ΔT = difference of reference temperature T_r minus the lagging sample temperature T_s. The subscripts i and f indicate initial and final, while q represents the ramp rate and t is time. Endotherm is up. [From Wunderlich (1990); reprinted with permission of B. Wunderlich and Elsevier].

as the melting process ends (at time t_f), the sample temperature starts catching up to the reference temperature, and a steady-state condition is achieved at temperature T_{ssf}. At this temperature, the temperature difference between the reference and the sample is ΔT_f (see Fig. 2.44). From this description it is clear that the extrapolated onset temperature of the endotherm is the melting point. Usually, there is a small curved section around t_i that is attributed to minor amounts of impurities.

For determination of the heat of fusion, refer to Fig. 2.44 again. Three characteristic areas can be singled out: the horizontally shaded area B, the vertically shaded area C, and the crosshatched area A. Since the melting terminates at the peak (at time t_f), the heat of fusion must be proportional to the sum of areas B and C. However, experimental determination of area C is difficult, because the zero line would also need to be recorded (which involves recording the starting and final isothermals). However, it can be proved that area C is equal to area A if $\Delta T_i = \Delta T_f$. Therefore, the heat of fusion can be obtained with good approximation if one integrates the melting peak (i.e., determines the sum of areas B and A), and for this the instrumental zero line does not have to be determined. This method is called the *baseline method*. A more detailed description of the baseline method is given by Wunderlich (1990, 1997a).

This picture is somewhat different for melting of substances whose purity is less than ~99.9%. If the operator sees that the leading edge is not straight, the melting point determination is different from that described above. In such cases, the composition of the melt and the crystal at any moment are not identical during the melting process, and therefore the temperature of the sample is not constant during the melting. For such samples the last, highest temperature point of the melting endotherm is chosen as the melting point. To determine this, one needs to determine the onset slope of melting of high-purity indium (or another calibrating substance), and drop a straight line to the baseline at the angle of the indium melting peak leading edge. The intersection of this straight line with the baseline will be the melting point, as shown in Fig. 2.45. The temperature difference between this melting point and the one that would be obtained if the straight line had been dropped at a 90° angle is not great (in Fig. 2.45 $\Delta T = 0.8\,°C$ for a sample of 97% purity 1-phenylcyclohexanol). However, this difference is very important in purity determinations (see Section 2.4). In this case, the determination of the heat of fusion is the same as that described above, so the baseline method can be used, but not always with a linear baseline. As a consequence of the broadened melting peak, the heat capacity difference before and after the melting may become considerable, so, as shown in Fig. 2.47b, often the sigmoidal baseline option is needed (see Section 2.7.3.4).

2.7.3.2. Melting of Crystalline Macromolecules

2.7.3.2.1. Equilibrium Melting of Polymers. The state of polymers in most cases is far from equilibrium. Therefore the melting points measured in every-

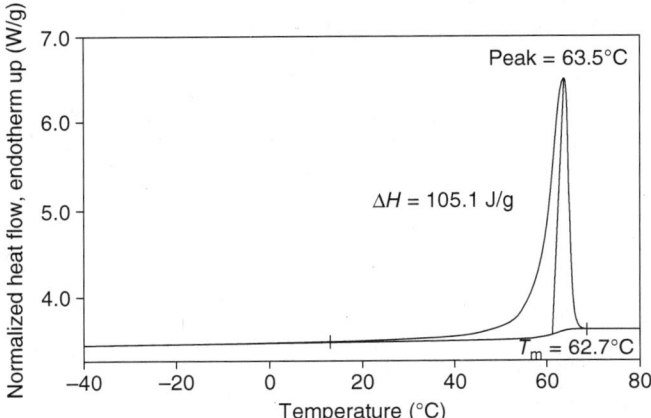

Figure 2.45. Determination of melting point of a low-molecular-mass substance with purity less than 99.9% (1-phenylcyclohexanol, 97% purity); a straight line is dropped from the peak maximum to the baseline at an angle of the indium leading edge (which is not 90°), and the point of intersection of this line with the baseline determines the melting point (Menczel, unpublished results).

day experiments are almost always nonequilibrium ones. It is much more difficult to determine the equilibrium melting points of polymers. The equilibrium melting point (T_m°) can be defined as the melting point of macroscopic-size equilibrium (i.e., extended-chain) polymeric crystals. T_m° can be determined either by direct measurements on equilibrium crystals (if available) or by various extrapolations. The equilibrium heat of fusion (i.e., the heat of fusion of 100% equilibrium crystals) also can be determined by direct measurements on equilibrium crystals or, again, by various extrapolations.

2.7.3.2.2. The Equilibrium Melting Point. The equilibrium melting point of a polymer is the melting point of large, macroscopic equilibrium crystals (these are the extended-chain crystals) without superheating. It is designated as T_m°. Direct experimental determination of the equilibrium melting point for most polymers is not possible, simply because macroscopic-size extended-chain crystals could not be prepared. In these cases the equilibrium melting point is estimated by various extrapolations. The following are the most important equilibrium melting point determinations:

1. *Direct Measurement of T_m° when Equilibrium Samples are Available.* Extended-chain crystals have been prepared. For a limited number of polymers. Polyethylene was the first (Geil et al. 1964; Prime and Wunderlich 1969; Prime et al. 1969); in addition, extended-chain crystals have been prepared from polyoxymethylene (Jaffe and Wunderlich 1967; Iguchi 1976) poly(tetrafluoroethylene), poly(ethylene terephthalate)

(Siegmann and Harget 1973; Harget 1987) (although T_m° for PET has not been determined experimentally). Polarization optical microscopy is the best method for measurement of T_m°. In these measurements the sample is heated extremely slowly, and not continuously, but at small temperature intervals; the operator waits at every temperature, because superheating is a common phenomenon during the melting of extended-chain crystals. Applying this method, 141.4°C has been determined as T_m° for polyethylene (Arakawa and Wunderlich 1965). The equilibrium heat of fusion has been obtained as 293 J/g (Wunderlich and Cormier 1967; Wunderlich and Czornyj 1977).

2. *The Hoffman–Weeks Method.* Lauritzen and Hoffman (Lauritzen and Hoffman 1960; Hoffman and Weeks 1962) observed a linear relationship between the melting points and the crystallization temperatures of a polymer crystallized isothermally at various temperatures. Hoffman and Weeks (1962) then realized that the lamellar thickness and the crystal perfection increases with increasing crystallization temperature, and they proved that the $T_m = f(T_c)$ curves must intercept the $T_m = T_c$ curve at the equilibrium melting point (T_m°). The Hoffman–Weeks method is widely used in the literature because of its simplicity, but the authors seldom focus on the basic premises of this method, namely, that neither considerable reorganization nor superheating should take place during the melting. That is to say, the zero-entropy-production melting points must be used for the extrapolation. Failure to meet this condition will lead to erroneous results. Figure 2.46 shows determination of the equilibrium melting point for poly(2-methylpentamethylene terephthalamide).

2.7.3.2.3. Other Extrapolations. Several other extrapolations exist to estimate the equilibrium melting point of linear macromolecules. Some of these are relatively complicated, but the most important are the small crystal melting point extrapolation (in which the Thomson–Gibbs equation is used to extrapolate the melting points of thin lamellae to infinite thickness; see later in this section) and the small molecule extrapolation (in which the melting point of low-molecular-mass homologs is extrapolated to infinite molecular mass) (Wunderlich 1980). Wunderlich (1990, p. 193) collected melting point data from the literature for the melting point of polyethylene of different lamellar thicknesses, and plotted them as a function of the reciprocal lamellar thickness. A straight line was obtained, from which the melting point for infinite lamellar thickness (i.e., the equilibrium melting point) was determined as $T_m^\circ = 414.2\,K\,(= 141.0°C)$, in good agreement with the experimentally measured equilibrium melting point of 141.4°C (Arakawa and Wunderlich 1965). When building such a plot, only zero-entropy-production melting point data can be used.

2.7.3.3. The Equilibrium Heat of Fusion Another equilibrium quantity is the heat of fusion for 100% crystalline polymer. The value of this parameter must be known for determining the crystallinity of polymers by DSC.

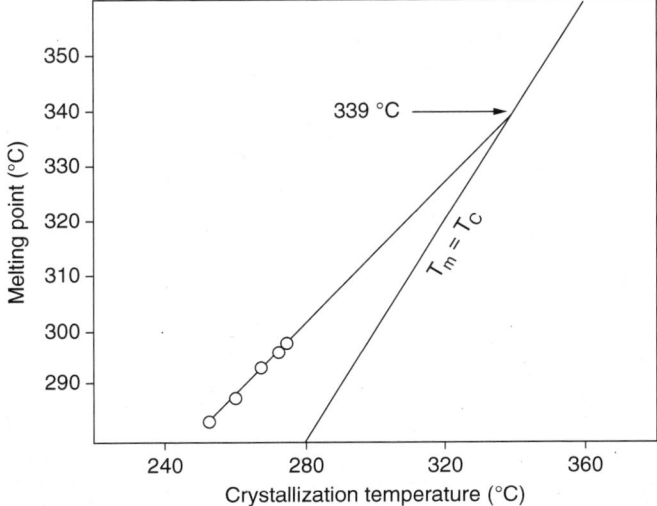

Figure 2.46. Determination of equilibrium melting point for poly(2-methylpentamethylene terephthalamide) by the Hoffman–Weeks method. Heating rate = 10 °C/min. Melting point data obtained on samples isothermally crystallized at 252.5 °C, 260 °C, 267.5 °C and 275 °C were used to determine the equilibrium melting point. [From Menczel et al. (1996); reprinted with permission of Springer-Verlag.]

It is not easy to determine the heat of fusion of 100% crystalline polymers (ΔH_f°). Of course, the easiest estimation is if extended-chain macroscopic size crystals are available, because these are virtually 100% crystalline. Polyethylene was the first polymer for which ΔH_f° has been determined, with a value of 293 J/g. Sometimes the crystallinity of individual samples is determined by wide-angle X-ray diffraction (WAXD), and the heat of fusion of such a sample is measured by DSC, and ΔH_f° is calculated from these two data points. The disadvantage of this method is that it can be used only for samples containing large crystals: X-ray diffraction tends to be insensitive to crystals below a certain size, so the value of ΔH_f° determined in this manner is not necessarily accurate.

The heat of fusion of a polymer sample can be used to determine the mass fraction crystallinity (x). As a first approximation, the following relationship can be used:

$$x = \frac{\Delta H_f}{\Delta H_f^\circ} \tag{2.68}$$

where ΔH_f is the experimentally determined heat of fusion and ΔH_f° is the equilibrium heat of fusion. However, the heat of fusion of polymer crystals does have some temperature dependence as the Thomson–Gibbs equation indicates (see later in this section). Since the melting of polymers is extremely broad, this temperature dependence needs to be factored in for exact determination of the crystallinity. The dependence that is used is as follows (Wunderlich 1997a)

$$\frac{d\Delta H_f}{dT} = C_{pa} - C_{pcr} \qquad (2.69)$$

where C_{pcr} is the crystalline heat capacity and C_{pa} is the amorphous heat capacity of the polymer in question. Such a determination of crystallinity is performed by the "Crystallinity Calculation" software of Perkin-Elmer.

2.7.3.4. Nonequilibrium Melting of Polymers In a DSC heating experiment the sample temperature lags behind the reference temperature in an increasing fashion as the transition proceeds, so that ΔT ($= T_s - T_r$) increases, and an endothermic peak is displayed in the DSC during the melting.

However, the melting of semicrystalline polymers is a very broad process because of the broad distribution of the crystallite sizes and the imperfection of these crystallites. This is a nonequilibrium process, so, in contrast to the equilibrium melting of pure low-molecular-mass substances (where the sample temperature does not change during melting), the sample temperature of a crystalline polymeric sample continuously increases during melting. The width of the melting peak is often 50 °C or more; thus it was necessary to define the melting point by convention. The melting point was selected as the highest temperature of the melting endotherm, that is, the temperature of the melting of the most perfect crystallites (235.1 °C in Fig. 2.47b). The reason for this convention was to get the same melting point by DSC and polarization optical microscopy (in polarization optical microscopy, the melting point is the temperature at which birefringence disappears). Nevertheless, in many publications the melting is characterized by the peak temperature of melting (T_{mp}). No doubt, this is an easy and simple method, but the peak temperature of melting simply indicates the temperature at which the melting proceeds with the maximum rate, and not the highest temperature at which crystals melt.

The heat of fusion or heat associated with any other endothermic event can be determined by integrating the peak with the appropriate baseline method. For this, as for low-molecular-mass substances, a baseline should be drawn connecting parts of the DSC curve before and after the melting (the part of the DSC curve before the melting peak is called the "premelting baseline," while that after the melting peak is called the "postmelting baseline"). Since the melting of polymers is a broad transition, significant heat capacity differences may exist before and after the melting. Therefore, in most cases the straight baseline is not adequate, and the sigmoidal baseline should be used. The use of the straight and sigmoidal baselines is illustrated in Figs 2.47a and 2.47b. First, the operator should determine the starting and ending points of melting. It is known that the heat capacity of the crystals and the melt changes

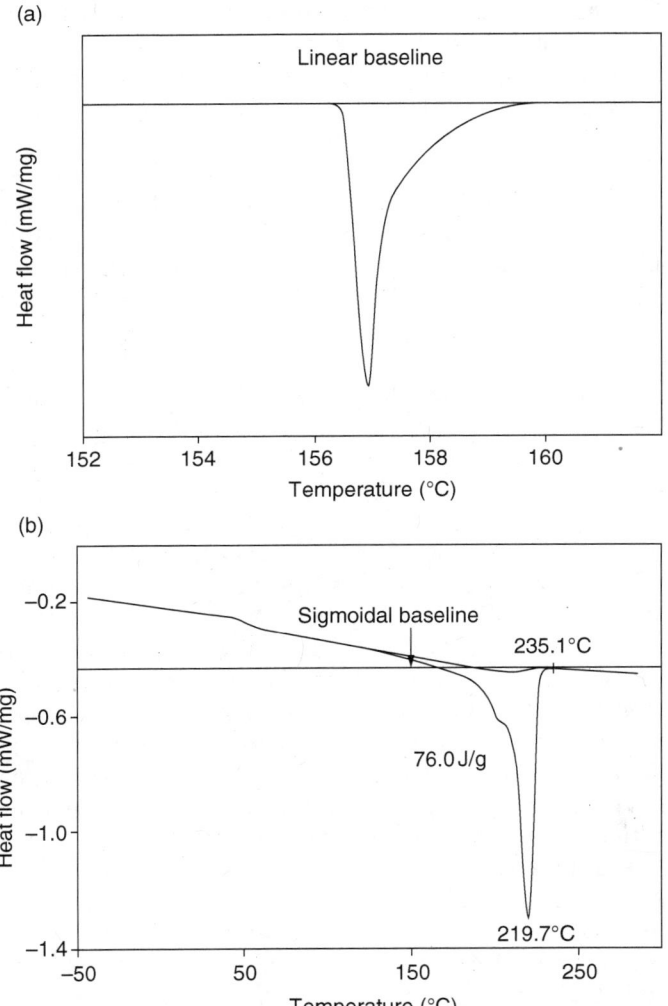

Figure 2.47. (a) Use of a straight (linear) baseline for determination of the heat of fusion (indium). The postmelting baseline is a linear continuation of the premelting baseline (i.e., the temperature dependence of the heat capacity of the crystals and the melt are identical). Since the heat flows in both pre- and postmelting baselines have identical values, these heat capacities are also almost the same (due to the narrow temperature range). (b) Drawing of the sigmoidal baseline. Nylon 6 crystallized at a rate of cooling of 1 °C/min from melt. Heating rate = 10 °C/min. Endotherm is down in both diagrams. (Menczel, unpublished results.)

linearly with temperature, and the apparent heat capacity during the melting is nonlinear. So a simple ruler is sufficient to determine the temperature range of the melting. One needs to determine where the linear change of the premelting baseline ends, and this temperature will be the temperature at which the melting starts. The end of melting can be determined in a similar manner from the postmelting baseline. Then, it should be determined what baseline should be used. If (again, with the use of a ruler) the postmelting baseline is the exact continuation of the premelting baseline (meaning that the temperature dependences of the heat capacities of the solid and the melt are identical, a rare case for polymers), a linear (straight) baseline can be used as shown in Fig. 2.47a: the starting and the ending points of the melting should be connected with a straight line, and the area under the peak should be determined. However, if this is not the case, the sigmoidal baseline must be used. All instrumental companies provide software for the sigmoidal baseline. For its determination one needs to mark four points on the DSC curve: two on the premelting baseline and two on the postmelting baseline (all four points on the linear portions of the DSC curve!), and the software will do the rest.

The physical reason for the necessity of the sigmoidal baseline is that the specific heat capacity of the sample needs to be determined during the melting at every point, and when connecting these heat capacity values, a sigmoidal shape is obtained. Thus, in the instrument software a straight baseline is initially drawn as the first approximation, and the partial areas of the peak are determined at every point. Using these partial areas, the software determines some approximate specific heat capacities during the melting for every point using the extrapolated specific heat capacities of the solid and the melt. Then, this iteration continues until the difference in the heat of fusion of two successive iterations falls below a certain percentage. For accurate determination of the heat of fusion it is important to use the sigmoidal baseline option. If the linear baseline is used, the error in the determination of the heat of fusion can be as high as 20–25%.

For reporting the values of measurements, one needs to determine the melting point (this is the first point on the postmelting baseline, i.e., 235.1 °C in Fig. 2.47b), and the heat of fusion obtained from integration of the melting peak.

So, with what experimental parameters should a melting curve of a semicrystalline polymer be determined? Obviously, there are no rules, but it should be remembered that at low heating rates reorganization is likely to take place (see text below). If this is the case, the melting point will decrease with increasing heating rate. On the other hand, if the melting point increases with increasing heating rate, superheating is present. If one wants to properly characterize a polymer, a heating rate should be selected at which neither reorganization nor superheating takes place. This is the so-called zero-entropy-production melting point (see the discussion in Section 2.7.3.5, below). The experiment should be started well below the temperature at which the melting process starts. Even for common polymers such as low-density polyethylene (LDPE), this may be at subambient temperatures. The sample should be heated up so that a sufficient length of the postmelting baseline is recorded in order to ensure accurate extrapolation of that postmelting baseline to lower temperatures by the software when determining the sigmoidal baseline. When reporting the melting data for crystalline polymers, the following characteristic data should be reported (see Fig. 2.48):

- The starting point of melting (T_{st}). Often this is not easy to determine, because this is a gradual process, and several polymers (e.g., LDPE) start melting at temperatures near or below room temperature. Also, the reproducibility of T_{st} of the melting is poor as it is a very subjective value, and considerable differences may arise because of slight changes in the instrumental baseline.
- The peak temperature of melting (T_{mp}). Unfortunately, in many publications, this is reported as the melting point. It should be remembered that this temperature indicates the maximum rate of melting.
- The melting point (T_m), the highest temperature point of the melting endotherm (this not the same as the extrapolated ending point of the melting, see Fig. 2.48). To determine the melting point, the sensitivity on the DSC trace must be increased considerably so that the determination procedure will be less subjective.

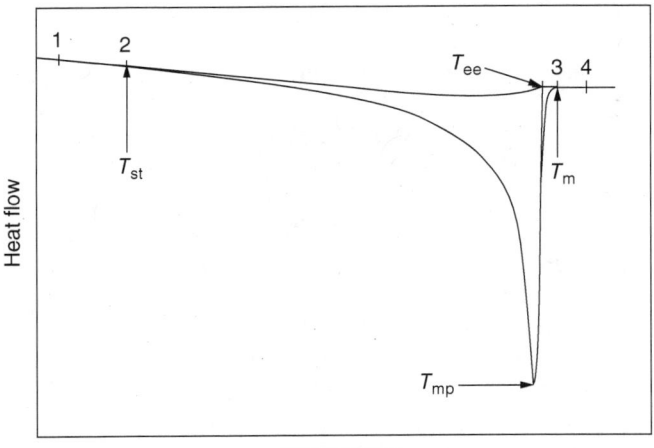

Figure 2.48. Characteristic temperatures of a polymer melting, and determination of the sigmoidal baseline: numbers 1–4 indicate spots where the cursor should be put down to determine the sigmoidal baseline. Unless the four points can be put on the same line, the sigmoidal baseline option should be used. If the four points are on the same line, you can still use the sigmoidal baseline option, but then the baseline will be straight. (T_{st} = starting temperature of melting, T_{mp} = peak temperature of melting, T_{ee} = extrapolated end temperature of melting, T_m = melting point.) Endotherm is down.

- The extrapolated ending-point of melting. This is the intersection of the closing edge of the melting peak and the postmelting baseline (Fig. 2.48).
- The heat of fusion (ΔH_f), in joules per gram (J/g) calculated with the sigmoidal baseline.

In publications sometimes the "extrapolated onset temperature" is reported for polymers. The reader should be careful with such data, as they have no physical meaning for polymers. Also, such an "extrapolated onset temperature" cannot be objectively determined because the leading edge of polymer melting peaks does not have a linear portion; rather, it is continuously curved.

Finally, several words on significant figures. Many DSC instrument manufactures display the determined temperature values with three digits after the decimal point (although more recent designs allow the operator to change this). Obviously, this is beyond the precision or accuracy of the measurement. The temperature of a characteristic point on a DSC curve cannot be determined with a precision of better than ±0.02 °C (obviously, the accuracy is even less, at best ±0.1 °C), and this would be the optimum case, for example, for the melting point of a pure low-molecular-mass substance (e.g., indium). For determination of the peak temperature of melting or the melting point and the glass transition temperature of a polymer, the accuracy will not be better than ±0.5 or ±1 °C. For these reasons we recommend reporting results to at most one significant figure beyond the decimal point. Similarly, in the case of heat of fusion, the precision that can be achieved does not exceed ±1–3%.

2.7.3.5. Time-Dependent Processes during Polymer Melting

Since the melting curves recorded at different heating rates cannot be superimposed on each other, this shows that the melting of crystalline polymers is not an equilibrium process. These melting curves differ from each other because of the presence of time-dependent processes (rearrangement, recrystallization, etc.) taking place during the heating.

In most cases the polymeric crystallites are small, and the lamellae in these crystallites are very thin (their thicknesses can be as small as ~10nm). The specific surface area of these crystallites is so large that the contribution of the surface free energy cannot be ignored. Thus, the melting process of these crystallites is much more complicated than the melting of macroscopic-size equilibrium crystals.

Since the melting of polymers is a nonequilibrium process, three major time-dependent phenomena must be considered:

1. *Reorganization of Poor Crystals into More Perfect Crystals during Melting.* This is crystal perfection during melting. The original crystals of the sample continuously improve during the melting without an apparent crystal → melt → crystal transition. One characteristic of this process is that

the melting point (i.e., the final melting point) of the sample decreases with increasing heating rate, because with increasing heating rate less and less time is available for the reorganization process. Changing the timescale (i.e., heating rate) also changes the shape of the melting curve. Therefore, in order to see the melting curve of the original crystallites, the heating rate in these experiments must be high, sometimes so high that the presently available commercial instruments are unable to provide sufficiently rapid heating. Of course, when performing melting experiments at high heating rates, the DSC instrument also needs to be calibrated at these high heating rates. MTDSC can help in the identification of reorganization, because the nonreversing signal may contain much information about this process.

2. *Recrystallization during Melting.* In this case, at least two melting peaks are observed, with an exotherm between them. Since this is a time-dependent process, changing the heating rate will alter the shape of the melting curve. The ratio of the amplitudes of the two peaks can be used to decide whether the multiple melting reflects recrystallization. Let us designate the height (amplitude) of the lower-temperature melting peak with h_L, and the amplitude of the higher temperature melting peak with h_H. As Fig. 2.49 indicates, if recrystallization is occurring, the h_H/h_L ratio will decrease with increasing heating rate, because less time is available for the recrystallization process. Often it is difficult to distinguish between reorganization and recrystallization processes, and MTDSC experiments are needed.

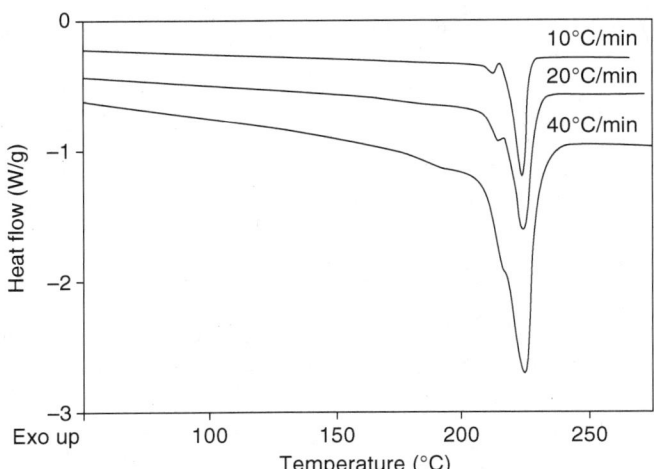

Figure 2.49. Melting curves of poly(butylene terephthalate) at various heating rates showing recrystallization during melting. Measurements made on Hoechst Celanese PBT 1600A. Sample preparation: 10°C/min cooling from 250°C, then the samples were heated for analysis at the rates indicated at the curves. (Endotherm is down) (Menczel, unpublished results.)

3. *Superheating.* There is a phenomenon in thermal analysis of polymers that causes an elevation of the melting point with increasing heating rate, which is just the opposite of the melting point decrease due to reorganization. This is "superheating," a phenomenon that may occur because at high heating rates the heat is supplied to the sample faster than the melting of the crystals can take place. Therefore, at high heat supply rates the interior of the crystal often heats above the equilibrium melting point of the polymer. Since melting is a surface phenomenon, the interior of the crystals at high heat supply rates can be higher than the equilibrium melting point. In melting experiments reflecting superheating, the start of the melting is seldom affected by the heating rate. This phenomenon is typical for extended-chain crystals [see, e.g., Hellmuth and Wunderlich (1965)]. There is another kind of superheating with the presence of the strained amorphous portions usually encountered with drawn fibers or highly oriented polymer films when the length of the sample is fixed mechanically in order to maintain the strain in the amorphous portions. Here the tie molecules between the crystals play a major role; they immobilize the crystals, and if these tie chains are removed by etching the amorphous phase, superheating disappears. This phenomenon is described in more detail in Section 2.8.

2.7.3.6. Zero-Entropy-Production Melting

So-called zero-entropy-production melting plays an important role in differential scanning calorimetry of polymers. If during the heating of metastable crystals conditions can be achieved (usually at faster heating rates), at which no reorganization takes place during the heating, the crystals will melt into a supercooled melt of the polymer (T_{mz}), which is also metastable. In general, the entropy production during melting without reorganization can be expressed by the following relationship (Wunderlich 1990)

$$d_i S = \frac{\Delta g_f dm_c}{T} - \frac{2\sigma dm_c}{T\rho l} \qquad (2.70)$$

where Δg_f is the specific free enthalpy on melting of a large crystal, dm_c is the rate of melting, σ is the surface free energy of the lamellae perpendicular to the chain direction, ρ is the lamellar density, and l is the lamellar thickness. From this equation it follows that $d_i S = 0$ (i.e., there is no entropy production) if

$$\Delta g_f = \frac{2\sigma}{\rho l} \qquad (2.71)$$

that is, at these conditions the melting of polymer crystals will proceed without entropy production. We will not discuss this case here any more, but refer the interested reader to Wunderlich (1990). It is important that under conditions of zero-entropy production the lamellar crystals of the polymers will melt without reorganization, and the melting point obtained will reflect the true melting point of the lamellar crystals, that is, uncomplicated by any reorganization or superheating. We also mention that this is the melting point that should be used in the Hoffman–Weeks method to determine the equilibrium melting point. The zero-entropy-production melting point can be measured at relatively high heating rates (so as to avoid reorganization). The zero-entropy-production melting point can characterize the metastability of the crystallites in the polymer sample. Briefly, the zero-entropy-production melting point is the melting point of the crystallites without reorganization and superheating.

A peculiarity of the polymer melting process is that special crystals, folded-chain lamellae, are involved in the melting process. These are very small crystallites (the thickness of the lamellae can be as small as ~10 nm); therefore they have very large specific surface area, and the contribution of the surface free energy cannot be neglected. Mathematically an equation called the *Thomson–Gibbs equation* describes the melting of these crystals

$$T_m^\circ - T_m = \frac{2\sigma_e T_m^\circ}{\Delta H_f l} \qquad (2.72)$$

where σ_e is the free energy per unit area of the chain folds (J/cm^2), ΔH_f is the heat of fusion per unit volume (J/cm^3), T_m° is the equilibrium melting point of the given polymer, T_m is the experimentally measured melting point, and l is the lamellar thickness. Therefore, this equation can be used to determine the equilibrium melting point of polymers by measuring the zero-entropy-production melting points of samples containing lamellae of different thicknesses, provided that reorganization can be suppressed.

2.7.3.7. Multiple-Peak Melting of Polymers

Most polymers are only partially crystalline. The temperature range of their melting is often extremely broad, and their melting points are significantly lower than the equilibrium melting point, and most polymers are only partially crystalline. In addition, it is quite common to observe multiple melting peaks during melting of polymers, for which several sources may be responsible:

1. Recrystallization during melting as described above may occur.

2. Several crystal forms, as with PVDF or PP, may be present simultaneously. In this case both crystal forms will melt, giving two endotherms in the heating curve.

3. The sample may also exhibit crystal → crystal transitions during heating. Figure 2.50 shows the DSC heating curve of quenched Nylon M5T {poly(2-methylpentamethylene terephthalamide), [CH$_2$–CH(CH$_3$)–CH$_2$–CH$_2$–CH$_2$–

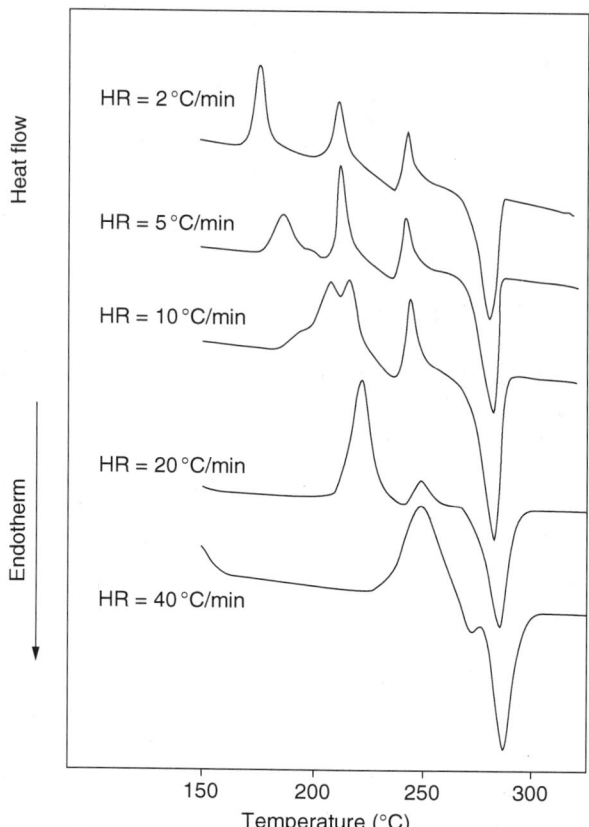

Figure 2.50. Melting curves of quenched Nylon M5T (or as-spun Nylon M5T fiber) beyond the glass transition temperature. The crystal-to-crystal transitions are clearly visible. At 40 °C/min the crystal form A→crystal form B transition does not take place (probably due to kinetic reasons), and crystal form A changes directly to crystal form C. [From Menczel et al. (1996). Reprinted with permission of Springer-Verlag.]

HNOC–C_6H_4–CONH–$]_n$}. Similar to the case of quenched PET (see Fig. 2.43), the glass transition is followed by an exothermic peak. This peak corresponds to an amorphous melt → crystal transition. However, several crystal forms exist for this polymer, and crystal → crystal transitions take place as the temperature increases. These crystal forms have not been identified, but were clearly seen by wide-angle X-ray diffraction (Menczel et al. 1996). So if the three existing crystal forms are designated with A, B, and C, then the lowest-temperature exotherm (the cold crystallization exotherm) corresponds to the melt → crystal form A transition, followed by melting of crystal form A. After the melting of crystal form A is completed, the sample crystallizes into crystal form B. Then crystal form B melts, and the melt crystallizes into crystal form C, and finally crystal form C will melt. Some of these transitions can be avoided

PHASE TRANSITIONS IN AMORPHOUS AND CRYSTALLINE POLYMERS   **109**

at high heating rates (e.g., 40 °C/min). This conclusion has been supported by DMA measurements as shown in Fig. 5.37 (of Chapter 5).

4. Annealing of crystalline polymers means heat treatment accompanied by changes in crystal perfection and size that are revealed in the melting properties. When a polymeric sample is annealed at a certain temperature much below the melting point, a small melting peak appears 5–10 °C above the heat treatment temperature (Fig. 2.51) in addition to the main melting

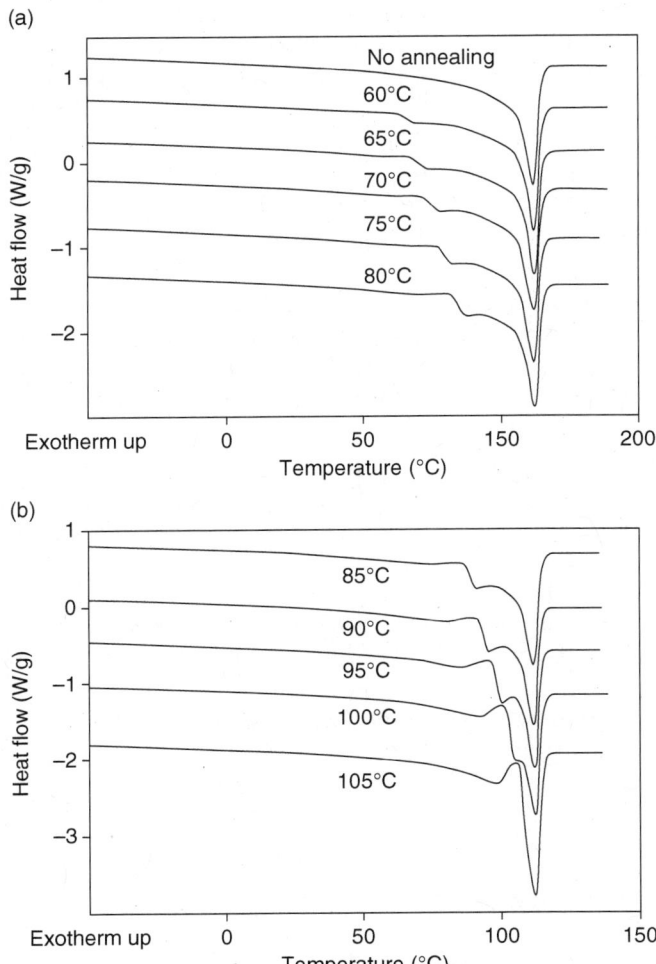

Figure 2.51. Melting curves of low-density PE after various annealing treatments: the sample was cooled from above the equilibrium melting point at 10 °C/min; then it was heated to the annealing temperature for 1.5 h, cooled to −80 °C, and reheated for analysis at a rate of 10 °C/min. The annealing temperature is indicated at each curve. (Menczel, unpublished results.)

peak. The reason for this small melting peak is that at a given annealing temperature (T_{ann}) all the crystallites with melting points of less than T_{ann} will melt, and recrystallize having perfection corresponding to crystallization at temperature T_{ann}. The newly developed crystallites will have a melting point somewhat higher than T_{ann}. However, if the annealing is carried out at temperatures close to the melting point, the whole melting endotherm shifts to higher temperatures, the crystallinity (i.e., the heat of fusion) of the sample increases considerably, and in most cases a relatively narrow single (or sometimes double) melting endotherm is obtained (Fig. 2.51). The intensity of the annealing peaks is usually heating-rate-dependent. When the intensity grows fast with the heating rate, this indicates that considerable reorganization takes place at slow heating. The curve shown in Fig. 2.51 (melting curve of annealed LDPE) often exhibits a third, low-temperature melting peak. The reason for this peak is that the crystallinity of LDPE is strongly temperature-dependent, so after the annealing at a particular temperature, enough crystallizable material is left in the sample that can form only very imperfect crystallites during cooling to low temperatures (this is due to branching of the polyethylene chains).

5. Stepwise isothermal crystallization (Varga et al. 1979) (see Fig. 2.62 in Section 2.9), especially of branched polymers (e.g., low-density polyethylene) leads to a melting curve where each melting peak corresponds to an isothermal crystallization step. It seems that this kind of multiple melting peak heating curve can be obtained when the melting peak is very broad, that is, when the crystallinity of the polymer is strongly temperature-dependent.

6. It also needs to be mentioned that the same polymer may exhibit multiple melting endotherms of various sources at various times. Above it was shown that Nylon M5T shows multiple melting due to crystal → crystal transitions. However, isothermally crystallized Nylon M5T may also show a different, double endothermic melting peak where the presence of the double endotherm is due to recrystallization during melting. Thus, the source of multiple melting needs to be clarified in every case. The heating rate dependence of the melting point and in general, the melting curve itself is a good source of information to explain why the melting curve contains several endotherms, but to fully comprehend the behavior, sometimes other analytical techniques may have to be applied.

Finally, Table 2.4 indicates the equilibrium melting point and heat of fusion of several common polymers.

2.7.3.8. Melting of Copolymers
Study of the melting of random copolymers is justified for such practically important copolymers as polyacrylonitrile (PAN), polyacetals, propylene copolymers, ethylene copolymers, and ethylene vinyl acetate copolymers. The majority of PAN polymers are copolymers of acrylonitrile with halogenated monomers (with the purpose of imparting

TABLE 2.4. Equilibrium Melting Points and Heats of Fusion of Some Common Polymers

Polymer	T_m (°C)	ΔH_f° (J/g)
Polyethylene	141.4	293
Polypropylene	220	165
Poly(ethylene terephthalate)	280	140
Poly(butylene terephthalate)	245	140
Poly(ethylene naphthalate)	337	190
Nylon 6, α	270	230
Nylon 66	280	300
Poly-ε-caprolactone	64	142
Polyoxymethyelene	198.5	326
Poly(ethylene oxide)	68.9	197
Poly(phenylene sulfide)	348.5	80
PEEK	382–395	130
Poly(vinylidene fluoride), α	210	104.5
PTFE	332	132

flame-retardant properties to the PAN fiber). A polyacetal or polyoxymethylene copolymer is the Celanese Corporation's Celcon® copolymer, which has an advantage over the polyoxymethylene homopolymer in that the comonomer units stop the "unzipping" reaction during the degradation of the polymer, thus, considerably increasing its thermal stability. A large variety of ethylene-propylene copolymers are used with 93–99% propylene unit content and 1–7% ethylene unit contents (see, e.g., the Profax® series of Hercules), and the ethylene–vinyl acetate copolymers (Evatane® of Arkema and Elvax® of DuPont) have the advantage of resilience and flexibility. Last but not least, the famous ethylene copolymers [low-density polyethylene (LDPE) and linear low-density polyethylene (LLDPE)] are an essential part of our everyday life. LDPE is basically a linear polymer with a large number of short branches, and LLDPE is one of the newest plastic materials formed by copolymerization of ethylene with longer-chain olefins; it has reduced crystallinity, lower density, lower tensile strength, higher resilience, than has high-density polyethylene. Finally, the Vectra liquid crystalline polymer is also a copolymer (of p-oxybenzoate units with 2,6-naphthalenedicarboxylate units). These are mainly random copolymers, but lately block copolymers have also experienced wide use. Here we will limit the description to melting of random copolymers.

All random copolymers have a broader melting range, reduced crystallinity, and melting point compared to the basic homopolymer. Figure 2.52

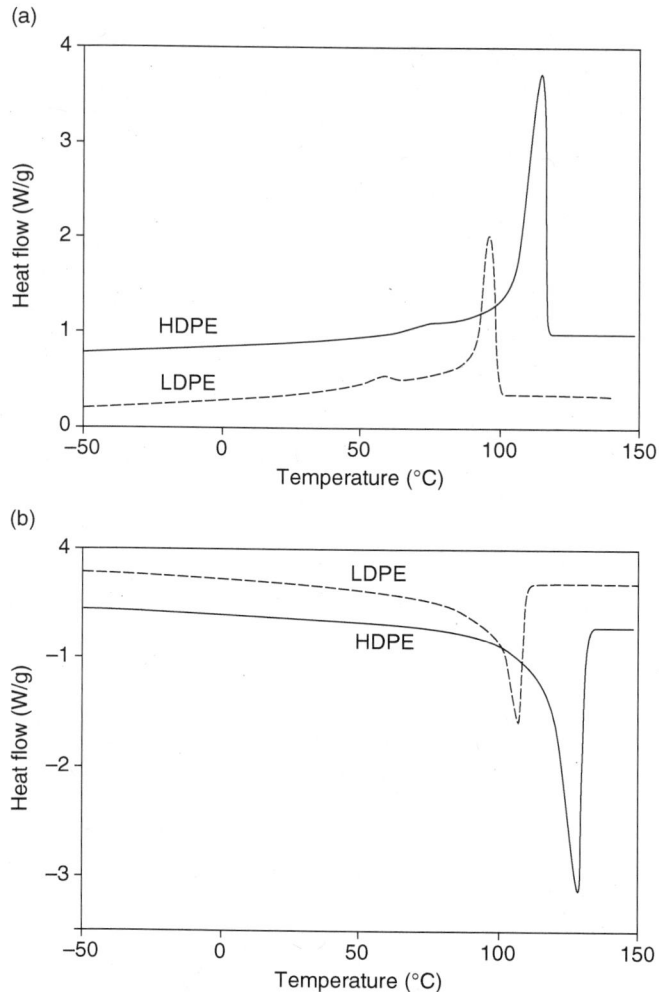

Figure 2.52. (a) Crystallization curves of low- and high-density polyethylene; cooling rate is 10 °C/min. Exotherm is up. (b) melting curves after 10 °C/min cooling, heating rate = 10 °C/min. Endotherm is down. (Menczel, unpublished results).

shows the crystallization and melting curves of high- and low-density polyethylenes (all samples from Scientific Polymer Products kit). The samples were crystallized by cooling from a temperature beyond the equilibrium melting point (141.4 °C) of polyethylene. Figure 2.53 shows the crystallization and melting curve of several ethylene/vinyl acetate copolymers with various vinyl acetate contents. It is clear that both the melting point and the crystallinity (heat of fusion) decrease with increasing comonomer content. It is also clear that the crystallization temperature of these copolymers, as is the case

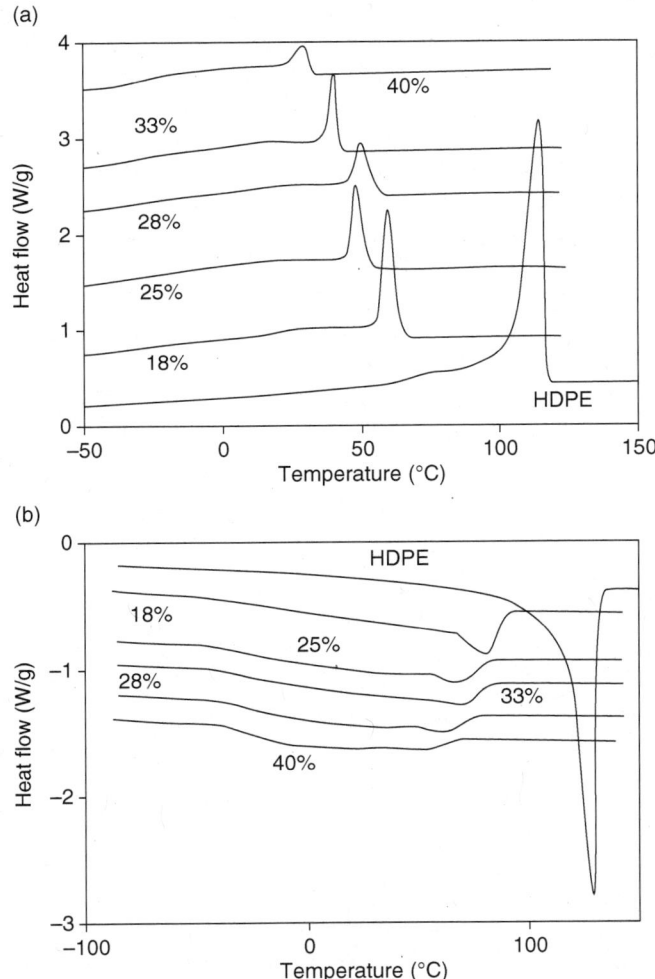

Figure 2.53. (a) Crystallization curves of high-density polyethylene and ethylene–vinyl acetate random copolymers with various vinyl acetate content, cooling rate = 10 °C/min. The vinyl acetate content is indicated at each curve. Exotherm is up. (b) melting curves of the same samples recorded at 10 °C/min heating rates. Endotherm is down. Samples from Scientific Polymer Products, Inc. (Menczel, unpublished results.)

with the melting point, is lower than the crystallization temperature of the homopolymer PE.

As just demonstrated, random copolymers have a considerably lower melting point and heat of fusion (i.e., crystallinity) than do the parent homopolymers. The reason behind the decrease of the melting point and the crystallinity of random copolymers compared to the homopolymer is that the repeating units of the comonomer may not be able to get incorporated into the crystal structure of the homopolymer. If this cannot be done, the repeating units of the second component have to be moved to the surface of the crystallites. Thus, the relative amount of these comonomer units will decide whether the crystallinity simply decreases or disappears, and the extent of the decrease of the crystallite size. In the case of alternating copolymers the question of whether there will be any crystallinity in the sample, will be determined by the relative size of the two comonomer units. Here we need to mention that most polymers that we simply consider homopolymers are essentially alternating copolymers, since various repeating units alternate in their main chains, such as nylons, PET, and polycarbonate. Strictly speaking, there are a limited number of true homopolymers, like polyethylene and poly(tetrafluoro ethylene).

The basic equation that can be used to describe the melting of random copolymers was derived by Flory (1951, 1953). Originally, this equation showed how much the melting point of a polymer decreases in a solvent:

$$\frac{1}{T_m} - \frac{1}{T_m^\circ} = \frac{R}{\Delta H_f} \cdot \frac{V_2}{V_1}(v_1 - \chi_1 v_1^2) \tag{2.73}$$

Here, T_m is the melting point of the polymer in the solvent, T_m° is the equilibrium melting point of the polymer, ΔH_f° is the equilibrium heat of fusion (J/mol) of the repeating unit, V_2 is the molar volume of the polymer, V_1 is the molar volume of the solvent; v_1 is the volume fraction of the solvent, χ_1 is the polymer–solvent interaction parameter, and R is the universal gas constant. This equation can be simplified to the following form:

$$\frac{1}{T_m} - \frac{1}{T_m^\circ} = -\frac{R}{\Delta H_f}\ln(N_1) \approx -\frac{R}{\Delta H_f}x\ln(1-N_2) \approx \frac{R}{\Delta H_f}N_2 \tag{2.74}$$

According to Flory, the effect of the comonomer units on the melting of a polymer is similar to the effect of a diluent, so in the last equation N_2 simply means the mole fraction of noncrystallizable units, while ΔH_f is the molar heat of fusion of the crystallizable component. Therefore, the slope of a plot of $1/T_m - 1/T_m^\circ$ versus the mole fraction of noncrystallizable units will be $R/\Delta H_f$. When ΔH_f is determined from such plots, it is somewhat smaller than the expected value. The reason for this discrepancy is that the lamellar thickness is ignored by this equation; moreover, this equation does not account for the presence of the repeating units of the second component in the crystals. In addition, no blockiness is allowed by this equation; it describes only ideally random copolymers.

The Thomson–Gibbs equation can also be used to describe the melting of copolymers

$$T_m = T_m^\circ\left(1 - \frac{2\sigma}{n\Delta H_F^\circ}\right) \tag{2.75}$$

where T_m° is the equilibrium melting point of the crystallizable homopolymer, ΔH_F° is the heat of fusion of equilibrium crystals of the crystallizable homopolymer, and σ is the surface free energy of the lamellae perpendicular to the molecular chain direction (J/cm^2).

2.8. FIBERS

Fibers, and often films, are oriented polymers. Because of the widespread use of fibers in all areas of human life (e.g., textiles, carpets, ropes, twine, cordage, furniture, curtains; in the healthcare area, with synthetic arterial replacement and sutures; fibers for tire reinforcement, protective fabrics, e.g., in ballistic protection), their thermal analysis is very important. A detailed description of the thermal analysis of fibers is given by Jaffe et al. (1997). Fiber and film preparation is also briefly described in the Section 5.5 of this book.

As in most polymeric materials, the structure of the macromolecules in fibers is far from equilibrium, and thermal analysis is an excellent tool to determine the thermal and mechanical history of fibers (and films). This technique is called structural "fingerprinting." A good example of fingerprinting is described in Section 2.9 [see Fig. 2.62 (Varga et al. 1979), where the melting curve of a polyethylene film clearly reflects the thermal (and mechanical) history of the sample]. The goal of thermal analysis measurements can be various, but monitoring the phase transitions (melting, crystallization, and glass transition) is certainly of major importance. With thermal analysis, it is easy to measure the end-use properties, study the effects of various additives on fiber properties, and determine the process–structure–property relationship of the fiber.

After spinning, synthetic fibers are almost always stretched or drawn in order to introduce or increase crystallinity and orientation. The intensity of the drawing is characterized by the *draw ratio*, which is the ratio of the initial cross-sectional area of the fiber before drawing to its final cross-sectional area after drawing. The orientation of the fiber increases with the draw ratio.

One crucial problem in DSC measurements of fibers is the sample preparation due to the long length of the fiber samples. Another problem is that most commercial fibers shrink during heating because they are highly drawn and thus highly oriented. As a consequence, there are two principally different ways to measure the thermal properties of fibers: (1) *constrained-state* or *fixed-length measurement*, in which the glass transition and melting are measured while the length of the fiber is kept constant throughout the DSC measurement; and (2) "free-to-shrink" or *unconstrained measurement*, where the fiber is chopped up into small pieces and therefore the ends of the fiber pieces are free to shrink during the heating.

The reason for the two types of DSC fiber measurements is that the melting and glass transition show considerable dependence on the experimental conditions, i.e., if the length of the fiber is held constant during the heating or is allowed to shrink. In the constrained measurements the constraint can be achieved by fixing the ends of a monofilament sample around a thin circular steel plate in which two V-shape cuts are made at opposite ends (see Fig. 2.54). The fiber is wound around the cut groves of the steel plate, and its ends are fixed by either gluing or tying them together. Then this steel plate with the "loaded" fiber sample is put into a standard DSC pan, a lid is placed on it, and the sample is crimped (naturally, the fiber weight has to be determined).

Obviously, the constrained-state and free-to-shrink measurements may give vastly different results depending on the orientation of the fiber. The melting point of drawn fibers measured in a constrained state is always higher than in the free-to-shrink measurements. It can be intuitively understood that in a constrained measurement, the entropy increase is smaller than in the free-to-shrink measurement, since in the constrained measurement the macromolecules are stretched, and they will remain partially stretched in the melt. They will lose their orientation in the melt, but that will require some time. Then, supposing that the heat of fusion is the same in both types of measurement (which is not always true, but can be accepted as a first approximation)

$$T_m = \frac{\Delta H_f}{\Delta S_f} \tag{2.76}$$

(where T_m is the experimental melting point, ΔH_f is the heat of fusion, and ΔS_f is the entropy of fusion), the melting point will increase in the constrained-state measurements because of the decrease in the conformational entropy of fusion.

Another difference between the two types of measurement is that the crystalline portion exhibits reorganization during heating in both the free-to-shrink and constrained heatings, but the crystallites in the constrained fiber

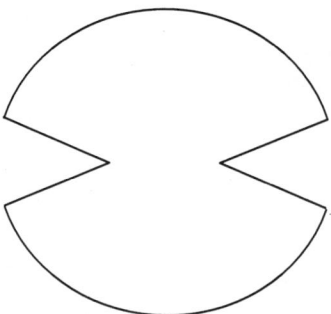

Figure 2.54. Shape of the steel plate with the two V-shaped cuts for recording DSC curves of fibers at fixed length. This steel plate can be prepared by cutting out two V-shaped pieces from a circular steel piece.

can superheat as well, even at modest heating rates [q > 0.5 °C/min, see Miyagi and Wunderlich (1972)]. Naturally, when reporting the results of measurements, it should be specified whether the measurement was carried out in the constrained or free-to-shrink state. For drawn fibers, especially at high draw ratios, the difference in the melting points can be considerable when determined by these two techniques. On the other hand, there is no difference in the glass transition or melting behavior of the as-spun fibers measured in the free-to-shrink and constrained states, since these fibers are rarely oriented.

With fibers, the heating rate dependence often contains useful information. If the melting point decreases with increasing heating rate in free-to-shrink measurements, this is an indication of reorganization (crystal perfection) during heating (see Section 2.7). If the melting point increases with increasing heating rate (in constrained measurements), this indicates superheating (Miyagi and Wunderlich 1972).

The most important specific commercial fibers are described below.

2.8.1. Gel-Spun Polyethylene Fibers

One of these commercial fibers is gel-spun polyethylene from ultra-high-molecular-mass polymer (e.g., Spectra, Honeywell) with extremely high modulus [for PE, $T_g = -36\,°C$, $T_m^° = 141.4\,°C$, $\Delta H_f^° = 293$ J/g (Gaur and Wunderlich 1980; Wunderlich and Czornyj 1977)]. The idea promoted by Penning (1967) is used in preparation of such fibers; in dilute solution (in decalin in the case of polyethylene), the polymer chains disentangle from each other. When this solution is spun into a coagulation bath, solid gel is formed. These low-orientation fibers can then be drawn to extremely high draw ratios (DR > 30), giving fibers with very high tensile strength. This fiber preparation technique is used for other polymers as well [poly(vinyl alcohol), polyacrylonitrile, polyesters, and nylons]. Figure 2.55 shows the melting curves of gel-spun, ultra-high-molecular-mass polyethylene fibers hot-drawn to various draw ratios. The melting curves were recorded in both free-to-shrink and constrained states. As the draw ratio increased, the crystallinity of the fibers (calculated from the heat of fusion) also increased; the crystallinity was ~52.5% for the as-spun fiber, and 93.5% for DR = 80. Considerable superheating was observed in the constrained-state measurements: a melting peak that increased with draw ratio was recorded beyond 150 °C [i.e., well beyond the equilibrium melting point of polyethylene, which is 141.4 °C (!); see Wunderlich and Czornyj (1977)]. The considerable difference in the heat of fusion measured between constrained and unconstrained states (see Fig. 2.55c) was explained by formation of partially oriented melt in the constrained measurements. In the constrained measurements three melting peaks were observed. In order of increasing temperature, these were attributed to melting of folded-chain lamellae, an orthorhombic → hexagonal crystal form change, and finally, superheated melting of the hexagonal phase. To prove this point, Murthy et al. (1990) found a high-temperature melting peak at ~160 °C. Simultaneous DSC-X-ray

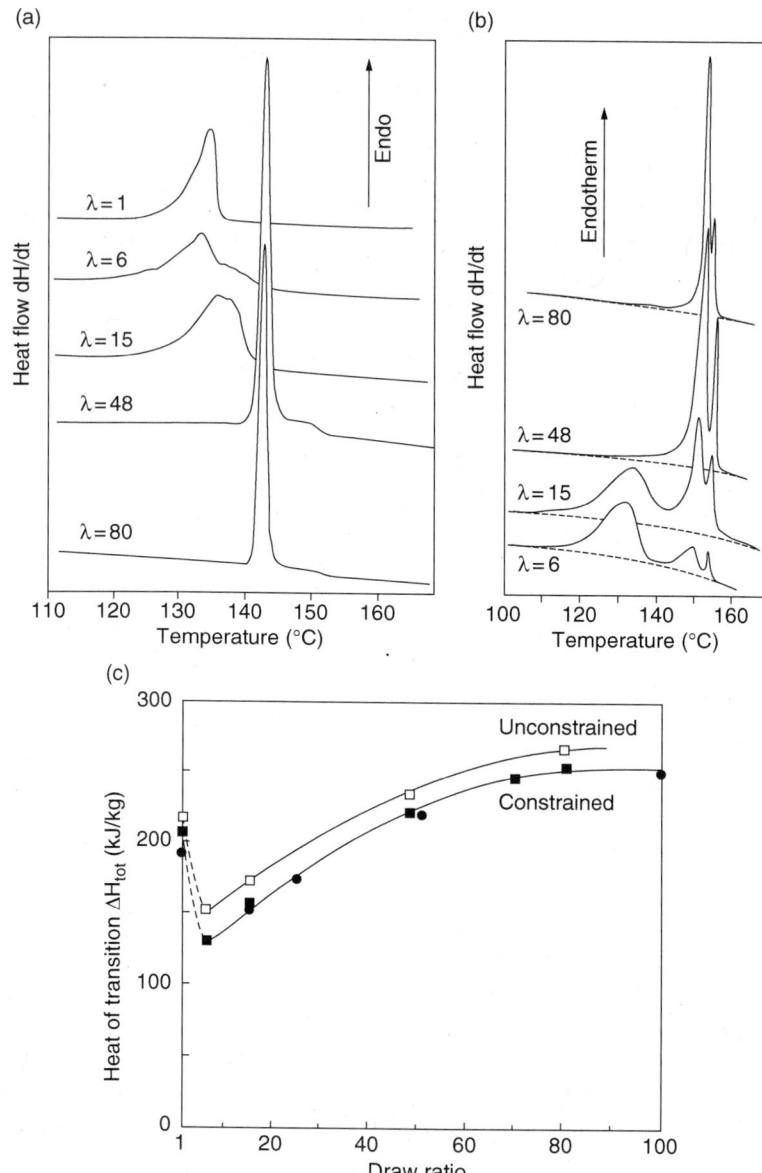

Figure 2.55. (a,b) DSC melting curves of hot-drawn gel-spun high-density polyethylene fibers (measurements in free-to-shrink state in a Perkin-Elmer DSC2 at a heating rate of 5 °C/min; (c) heat of fusion of ultra-high-molecular-mass PE fiber at various draw ratios measured in constrained and free-to-shrink states. The difference in heat of fusion measured in the two measuring modes was explained by formation of partially oriented melt in the constrained measurements. [From Smook and Pennings (1984); reprinted with permission of Springer-Verlag.]

diffraction measurements assigned this peak to the melting of the hexagonal phase developed in orthorhombic → hexagonal phase transformation at somewhat lower temperatures.

2.8.2. Isotactic Polypropylene Fibers

Fibers cannot be made from atactic polypropylene. For PP, $T_g = -14\,°C$ (Gaur and Wunderlich 1981), $T_m^° = 220\,°C$ (Samuels 1975a,b), $\Delta H_f^° = 165$ J/g (Wunderlich 1980). The melting curves of PP fibers, drawn to various draw ratios, and then heat-set, are shown in Fig. 2.56 (Samuels 1975a). Two melting peaks can be observed for the drawn fibers. The position of the lower-temperature melting peak depends on the orientation of the amorphous regions of the fiber, but the origin of this peak is unknown. The position of the higher-temperature melting peak also depended on the orientation of the amorphous regions, but not on the orientation of the crystallites in the fiber. When the position of this peak was extrapolated to full orientation of the amorphous region, a temperature of 220 °C was obtained, which is often accepted as the equilibrium melting point of PP.

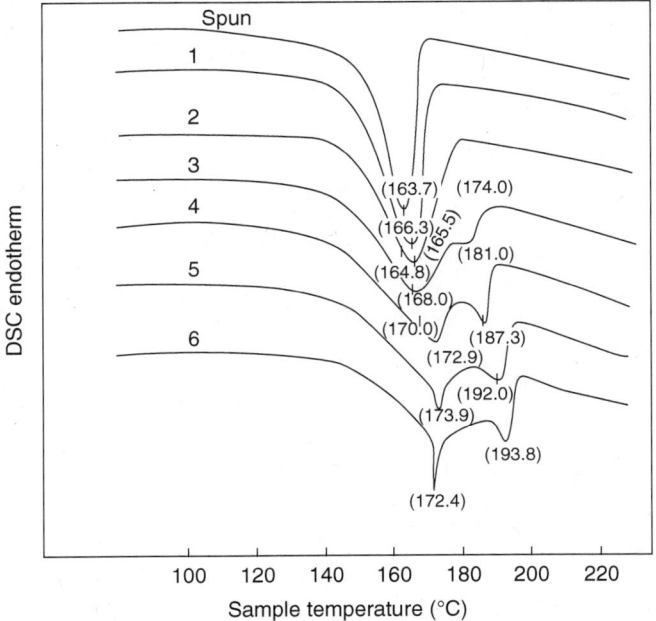

Figure 2.56. DSC heating curves of as-spun and drawn polypropylene fibers. The drawing was done at 90 °C to the following draw ratios: curve 1—DR = 1.39; curve 2—DR = 1.84; curve 3—DR = 2.29; curve 4—DR = 2.70; curve 5—DR = 3.21; curve 6—DR = 3.58. The measurements were carried out in a Perkin-Elmer DSC2, heating rate = 20 °C/min. Endotherm is down. [Reprinted from Samuels (1975a) with permission of John Wiley and Sons.]

2.8.3. Polyesters

Poly(ethylene terephthalate) and poly(ethylene naphthalate) are the major representatives of this group of synthetic fibers.

Poly(ethylene terephthalate) (PET) [$T_g = 79\,°C$ (Menczel and Wunderlich 1980), $T_m° = 280\,°C$ (Ikeda 1967), $\Delta H_f° = 140$ J/g (Wunderlich 1980)] is the largest-volume synthetic fiber; its major uses are in textile and tire cord applications. Miyagi and Wunderlich (1972) were the first to study DSC behavior of this fiber. They found that the fiber exhibits reorganization during heating in the free-to-shrink state. They also found strong superheating in constrained-state DSC measurements.

A similar fiber is poly(ethylene naphthalate) or poly(ethylene-2,6-naphthalene dicarboxylate). The use of this fiber may increase considerably in tire cord applications, since its glass transition temperature and melting point are higher than those of PET ($T_g = 120\,°C$, $T_m° = 337\,°C$, $\Delta H_f° = 103$ J/g).

Figure 2.57 shows the DSC heating curves of as-spun and drawn poly(ethylene naphthalate) fibers (Saw et al. 1997). When heated, the as-spun fiber exhibits a glass transition at 125 °C, followed by cold crystallization and finally melting. The heat of cold crystallization and the heat of fusion are nearly identical, indicating that the as-spun fiber is almost completely amorphous. The drawn fiber is highly crystalline, and, as expected, T_m is higher in constrained measurements than in free-to-shrink measurements. In the constrained-state measurements, the melting point of this fiber increases with increasing draw ratio up to DR = 4.3, and then stays constant. The crystallinity of these fibers can

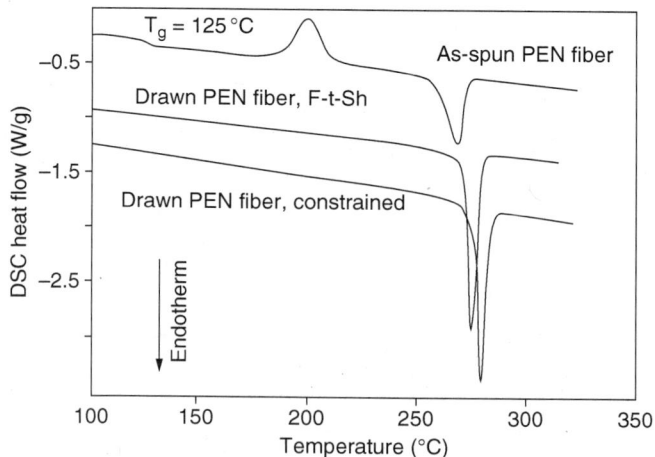

Figure 2.57. Melting curves of as-spun and drawn PEN fiber (draw ratio = 5.2); as can be seen, the as-spun fiber is amorphous, and the crystallinity of the drawn fiber is 50% on the basis of the heat of fusion [From Saw et al. (1997); reprinted with permission of Society of Plastics Engineers].

be calculated from the heat of fusion as described in Section 2.7, and it increases from 45% to 50% as the draw ratio increases from 2 to 5.4.

2.8.4. Polyacrylonitrile Fibers

These are important fibers, used primarily as staple fibers in knitted textiles. The glass transition temperature of this polymer is still debated (it is somewhere between 85°C and 140°C). Similarly, the heat of fusion of 100% crystalline polymer is not known exactly, and $T_m^°$ is somewhere between 317°C and 335°C (beyond the degradation temperature of the polymer). In addition, this polymer is strongly plasticized by water, and since most acrylic fibers are copolymers, their thermal analysis is complicated and not at all clear. For this reason no more time will be spent discussing these fibers. The interested reader is referred to a more detailed description by Jaffe et al. (1997).

2.8.5. Nylons (Polyamides)

These are extremely important fibers, and their applications are very versatile. They are used as reinforcement fibers in airplane and truck tires and fibers in seatbelts, fishing lines, upholstery fabric, textiles, and more. The moisture regain of nylons is high, potentially it reaching 8% of the mass of the fiber (Papir et al. 1972). A classical work published by Todoki and Kawaguchi (1977a,b) describes the melting behavior of Nylon 6 fibers in different measuring modes (free-to-shrink and constrained). These fibers were measured as-received, then the amorphous regions of the fibers were crosslinked through the amide groups of nylon. The crosslinking was carried out with acetylene initiated by γ irradiation. As expected, in all measuring modes the constrained measurements gave the highest melting points. However, when comparing the melting points obtained on the crosslinked and as-received fibers, the melting point of the as-received fibers was 65–80°C higher than the melting point of the crosslinked fibers, indicating that considerable annealing (crystal perfection) takes place during the heating of as-received fibers. In the acetylene-treated fibers this crystal perfection is prevented by the crosslinks because the amorphous chain segments become rigid and cannot enter the crystals, which prevents reorganization during the melting process. As a result, in the crosslinked fibers essentially the zero-entropy-production melting point is measured (Figs 2.58a and 2.58b).

2.8.6. Liquid Crystalline Polymeric Fibers

Two major groups are known in this class of fibers. The first one is Kevlar® [DuPont, chemically poly(phenylene terephthalamide) (PPT)], which is prepared by modified wet spinning from a lyotropic liquid crystalline solution, and coagulated in water. The second one, which forms a thermotropic nematic melt, is the material with the trade name of Vectra® (Celanese, for injection-

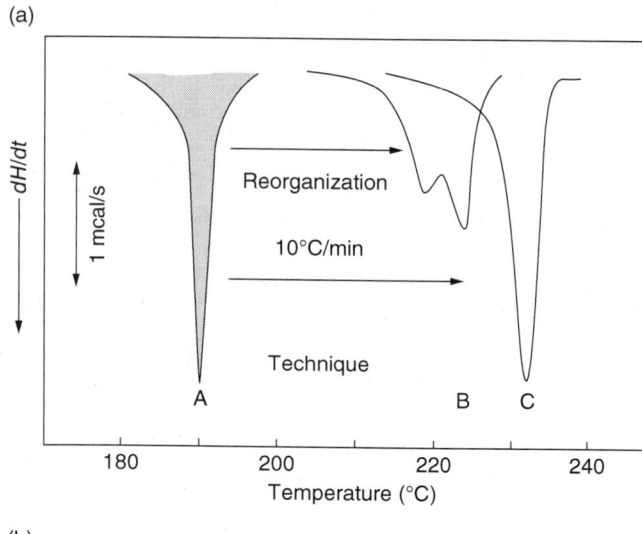

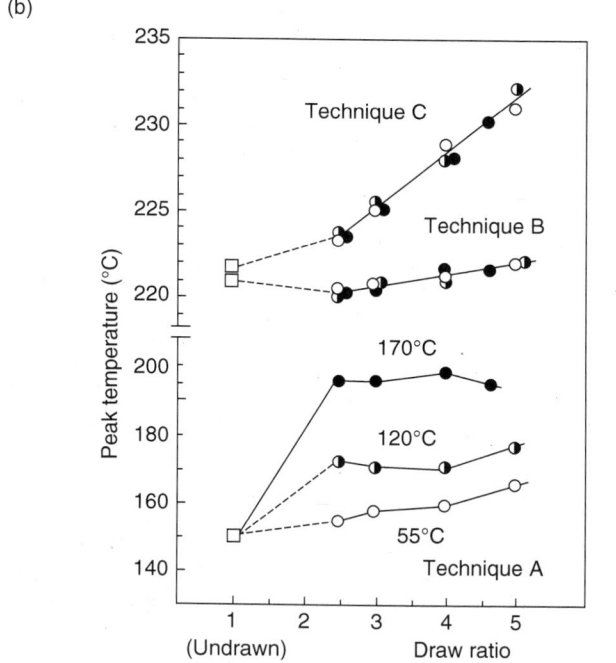

Figure 2.58. (a) Melting peak of nylon obtained with different measuring techniques: (curve A) the fibers subjected to γ-radiation in the presence of acetylene, which cross-links the amorphous portions; (curve B) traditional DSC melting experiment on as-received fibers (free-to-shrink measurement); (curve C) melting curve of drawn nylon fiber (DR = 5×) in constrained measurement. (b) Peak temperature of melting of nylon fibers by various techniques: (A) for the fibers cross-linked at 55 °C, 120 °C and 170 °C; (B) traditional (free-to-shrink) DSC measurements; (C) constrained DSC measurements. Heating rate was 10 °C/min. [from Todoki and Kawaguchi, 1997b. Reprinted with permission of John Wiley and Sons]

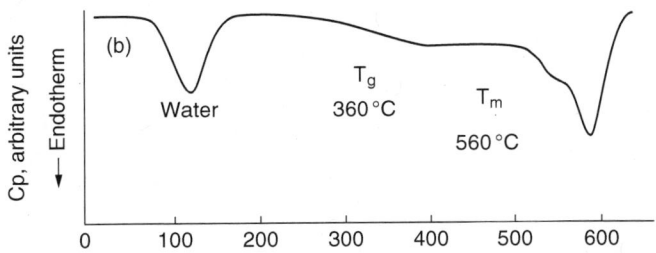

Figure 2.59. DSC heating curve of Kevlar fiber (Jaffe and Jones 1985). Heating rate was 10°C/min. (Reprinted with permission of Taylor&Francis)

molded products) or Vectran® (Celanese, for fibers and films). Because of the alignment of the polymer chains preferentially in the direction of the fiber axis, these fibers possess superior mechanical properties.

Not much DSC has been done on Kevlar (Penn and Larsen 1979; Jaffe and Jones 1985); however, an endothermic peak indicating evaporation of absorbed water was seen with a peak temperature close to 100°C followed by the glass transition at ~360°C, and another endothermic peak assigned to melting and decomposition of PPT beyond 500°C (see Fig. 2.59). The peak area can be used to approximate the amount of absorbed water. In such calculations it is assumed that the heat of vaporization of water from the polymer is identical to heat of vaporization from pure water: 2256 J/g (which may not always be true).

The DSC curve of Vectran fiber, which is a random copolymer of p-hydroxybenzoic acid (HBA), and 2-hydroxy-6-naphthoic acid (HNA), shows a very broad glass transition at T_g ~100°C, and an extremely broad melting peak between 200°C and 280°C consisting of three peaks (Menczel et al. 1997a). Variable-heating-rate experiments indicated crystal perfection during heating (these are the second and the third peaks; see Fig. 2.65 in Section 2.9, on films). The relative intensity of the lowest temperature peak did not change with changing heating rate at the heating rates applied in these measurements. On the basis of modulated DSC and X-ray measurements, it was suggested that this process may also indicate crystallinity increase, but much faster heating rates would be necessary to prove this.

2.9. FILMS

Polymeric films are used primarily as substrates for a variety of media (e.g., recording tapes), and as packaging materials. Similar to fibers, the thermal analysis of films may give information about orientation and thermal and mechanical history of the sample. A detailed description of thermal analysis of polymeric films can be found in Menczel et al. (1997b). Fiber and film preparation is also briefly described in Section 5.5 (of Chapter 5).

From a methodological perspective, DSC of films is more important than DSC of fibers. For good thermal contact, it is often desirable to run films in DSC, since in the DSC pans a film will have the maximum contact with the bottom of the pan, thus minimizing the external thermal resistance.

Similar to fibers, films are often oriented. The simplest case of orientation is called *uniaxial orientation*, when the macromolecular segments are oriented in one preferred orientation. But a film may also be oriented in two directions (called *biaxial* orientation) having no fiber analog. Isotropic, unoriented film would correspond to the as-spun fiber (both having no preferred molecular orientation). Thus, DSC experiments carried out on unoriented films give results similar to those obtained from experiments performed on chips, powders, or as-spun fibers, but with minimum external thermal resistance.

Performing constrained-state DSC measurements on oriented films is more complex than similar measurements on drawn fibers, especially when the film has biaxial orientation; it is more difficult to immobilize the total film sample during the DSC measurement, than simply tying together the ends of a single fiber after winding it up on a steel plate. Therefore, various approaches have been suggested to solve this problem. The two most important of these are as follows:

1. For film samples with uniaxial orientation, a thin stripe can be cut and handled for sample preparation in the same way as a fiber, by gluing its ends together after winding it up on a steel plate. However, this type of sample preparation is more complicated for films than that for fibers; the film often may break during the sample preparation, especially if it is not very thin. Also, sufficiently long film pieces are not always available for cutting out a thin stripe.
2. Clough (1970a,b, 1971) used polyethylene film samples lightly crosslinked with electron beams, and then, using specially made DSC pans shown in Fig. 2.60, carried out isothermal crystallization in constrained state at different temperatures and different extensions.

Clough is one of the very few authors who have performed constrained-state film measurements. Most DSC film experiments reported in the literature are prepared unconstrained (free-to-shrink) because of the technical difficulties of sample preparation just mentioned.

The most important commercial films are described in the following subsections.

2.9.1. Polyethylene Films

Polyethylene (all three: low-density, high-density, and linear low-density) is used mainly for food packaging, trashbags, food carryout bags, and so on and has a T_g of $-36\,°C$ (Gaur and Wunderlich 1980) or $-128\,°C$ (Beatty and Karasz 1979) and an equilibrium melting point of $141.4\,°C$ as directly measured on

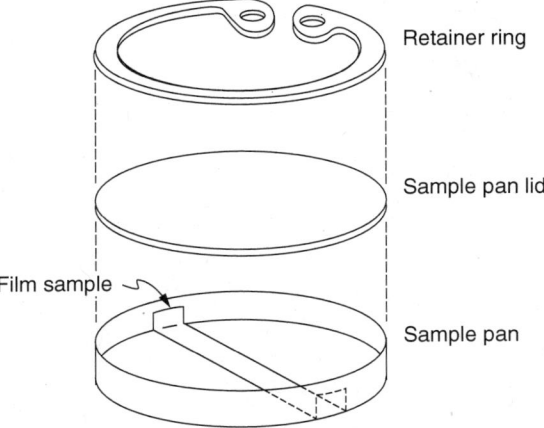

Figure 2.60. The sample pan developed by Clough (1970a) for measuring melting of stretched polymeric films in constrained state (Clough 1970a). (Reprinted with permission of Taylor&Francis)

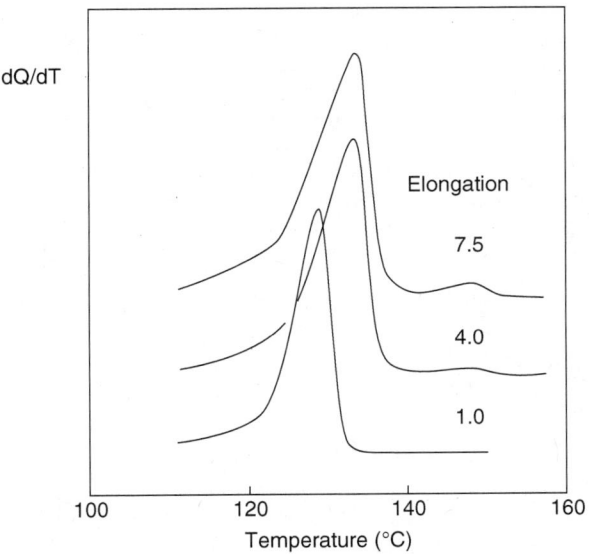

Figure 2.61. Melting curves of polyethylene films crosslinked with electron beams; after the irradiation, these samples were crystallized at various elongations at 121 °C. Endotherm is up. (Clough 1970a). (Reprinted with permission of Taylor&Francis)

specially prepared extended-chain crystals (Wunderlich and Czornyj 1977). On highly drawn film samples, several authors observed a small melting peak at 150–152 °C. This peak was attributed to superheated melting of extended-chain PE crystals as shown in Fig. 2.61 (Clough 1970a). Similar conclusions were reached by Illers (1970).

Similar to that for PE fibers, the gel-processing technology also exists for PE films (Sakai and Miyasaka, 1988, 1990; Sakai et al. 1993), but biaxial orientation can be created on films as well. For most of these samples, two melting peaks were observed: the lower-temperature peak corresponding to melting of folded-chain lamellae, and the higher-temperature peak, which indicated the melting of extended-chain crystals. On gel films drawn at a ratio of 300×, a crystallinity of 96.2% could be achieved (Sawatari and Matsuo 1985).

Varga et al. (1981) carried out isothermal stepwise crystallization of low-density polyethylene. These experiments could be done because the crystallinity of LDPE is strongly temperature-dependent. In these experiments they crystallized polyethylene isothermally at temperature T_1, then after the crystallization process was completed, they lowered the temperature to T_2. At T_2 the maximum possible crystallinity of LDPE was considerably higher than at T_1, so additional portion of the sample crystallized. Then the temperature was lowered to T_3, and this process continued to room temperature. On reheating, a multiple-peak-melting curve was recorded, and the number of peaks equaled the number of isothermal temperature steps (see Fig. 2.62, curve labeled 0.0%). When this film was drawn, the melting curve did not change until the elongation reached ~200%, and at higher elongations the individual melting peaks started to become blurred.

2.9.2. Poly(ethylene terephthalate) Films

Poly(ethylene Terephthalate) [$T_g = 79\,°C$, $T_m^\circ = 280\,°C$, $\Delta H_f^\circ = 140\,J/g$, Menczel and Wunderlich, 1981; Gaur et al. 1983, Illers, 1970) primarily used for magnetic media and for packaging. Illers (1980) measured the heat of fusion of various PET film samples as a function of their specific volumes at room temperature. This dependence gave a straight line with $\Delta H_f^\circ(PET) = 166\,J/g$ (Figure 2.63). However, when data from other authors are taken into account, 140 J/g seems a more reasonable value (Wunderlich, 1980).

2.9.3. Nylon 6 Film

Nylon 6 [$T_g = 47\,°C$ (Illers 1977), $T_m^\circ(\alpha\text{-crystal form}) = 270\,°C$, $\Delta H_f^\circ = 230\,J/g$ (Wunderlich 1980)] film is mainly used as a packaging material in the healthcare and food industries. This film has excellent oxygen barrier properties, and can pick up water up to 8% (Papir et al. 1972).

Figure 2.64 shows the melting curves of undrawn and various drawn Nylon 6 films (Chuah and Porter 1986). The undrawn film has two melting peaks: a lower-temperature peak assigned to melting of lamellar crystals, and another peak reflecting the melting of mixed fibrillar–lamellar crystals. As the draw ratio increases, the relative intensity of the lower-temperature peak decreases, reflecting decrease in concentration of the lamellar crystals, as expected. Simultaneously, the relative intensity of the higher-temperature peak increases,

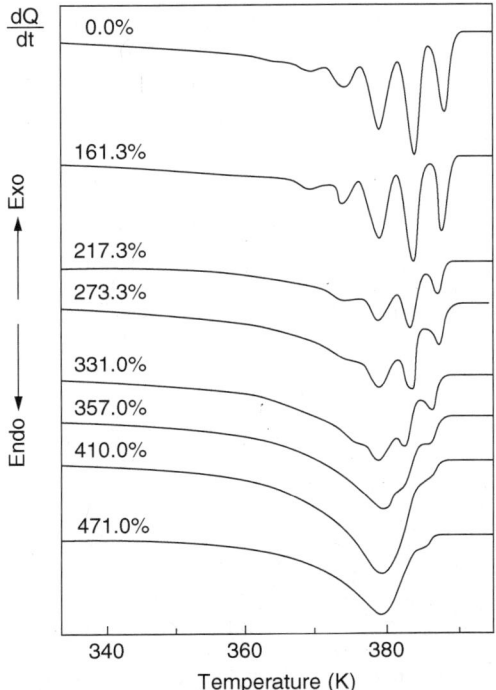

Figure 2.62. Melting curves of low-density polyethylene thin film crystallized in a stepwise manner (sample 0.0%). Then the film was drawn at room temperature to various draw ratios. The elongation is indicated at each curve. It is remarkable that the crystallites responsible for the multiple melting behavior are mechanically stable up to ~200% elongation, and almost 500% elongation is needed to erase the thermal history mechanically. Perkin-Elmer DSC2, heating rate 10°C/min. [From Varga et al. (1979); reprinted with permission of Springer-Verlag.]

which was also expected, since drawing should facilitate development of fibrillar crystals and disappearance of the lamellar crystals. However, the crystallinity of the films decreased with increasing draw ratio, which was quite unexpected and could not be explained.

2.9.4. Poly(vinylidene fluoride) Films

Poly(vinylidene fluoride) [$T_g = -40$°C (Gaur et al. 1983), T_m° (α-crystal form) = 210°C, $\Delta H_f^\circ = 104.5$ J/g Nakagawa and Ishida (1973a,b)] is used primarily for its piezoelectric and pyroelectric properties (in military applications as a large-area hydrophone array for submarines, motion detectors, traffic counters, etc.).

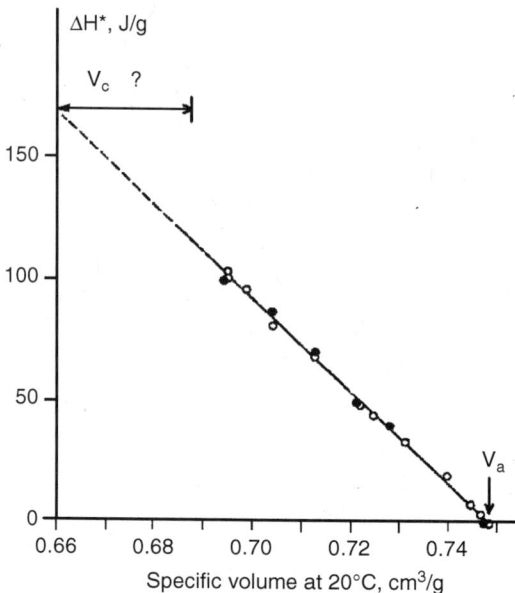

Figure 2.63. The heat of fusion of PET film samples as a function of their specific volume at 20 °C (open circles—measurements on a DuPont 990 DSC; filled circles—measurements on a Perkin-Elmer DSC1B) [From Illers (1980); reprinted with permission of Springer-Verlag].

2.9.5. Liquid Crystalline Polymer Films

Similar to fibers from liquid crystalline polymers (LCPs), films from LCPs are also manufactured. The major product is the film with the trade name of Vectran, which has the same composition as the Vectran fiber, namely, a copolymer of *p*-hydroxybenzoic acid and 2-hydroxy-6-naphthoic acid. The primary consumer of this film is the electronic industry. The DSC curve of Vectran film (see examples in Fig. 2.65) shows a very broad glass transition at T_g ~100 °C, and an extremely broad melting peak between 200 °C and 280 °C consisting of three peaks Menczel et al. (1997a). Heating rate experiments indicated crystal perfection during heating, when the intensities of the second and the third peaks are compared. However, the relative intensity of the lowest-temperature peak did not change with varying heating rate at the HR ranges applied in these measurements. On the basis of modulated DSC and X-ray measurements, it was suggested that this process may also indicate crystal perfection process, but much higher heating rates would be necessary to achieve change in the relative intensity of this peak.

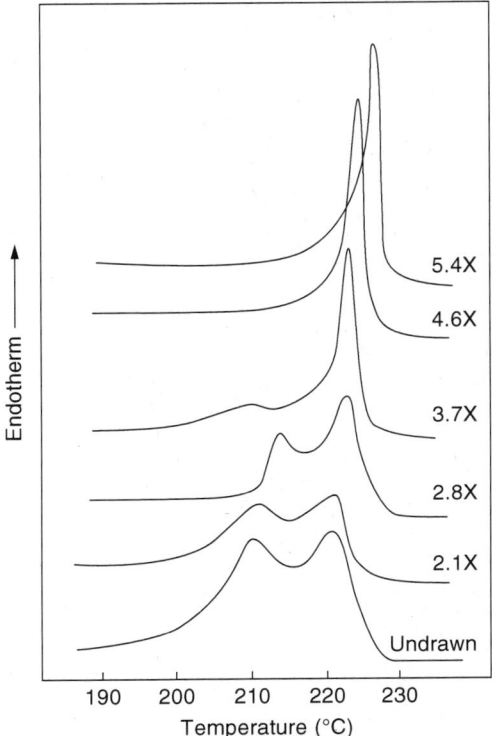

Figure 2.64. Free-to-shrink DSC melting curves of various Nylon 6 films: undrawn, and drawn to various elongations [From Chuah and Porter (1986); reprinted with permission of Elsevier Ltd.]

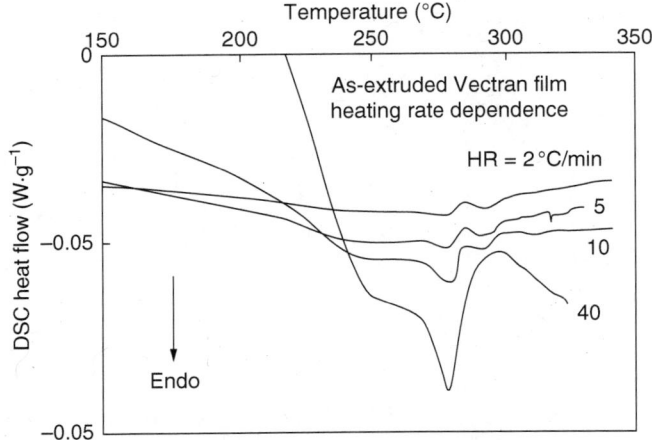

Figure 2.65. The DSC heating curves of as-extruded Vectran film recorded at various heating rates [from Menczel et al. (1997a); reprinted with permission of Springer-Verlag].

2.9.6. Poly(tetrafluoroethylene) (PTFE)

This is an extremely important polymer, since chemically it is more inert than platinum. Most chemists know that PTFE is very popular as a pressure-sensitive tape; thus it is often used for insulation of pressure regulators of gas cylinders, since under pressure it has a transition into a mesophase. Also, for the electronic industry it is important that its dielectric constant is very stable at various frequencies in a wide temperature range. The thermal analysis of PTFE is extremely rich in transitions; in addition to four subambient relaxations and a glass transition at $-73\,°C$ (Lau et al. 1984a,b), it has a crystal $\rightarrow$ crystal transition at $19\,°C$, and a triclinic $\rightarrow$ mesophase (haxagonal) transition at $30\,°C$. Finally, it melts at temperatures higher than $300\,°C$ [$T_m° = 332\,°C$, $\Delta H_f° = 132$ J/g (Lau et al. 1984a,b; Starkweather et al. 1982)].

Figure 2.66 shows a DSC trace of an as received PTFE tape and also the heating curve of this tape sample after a $10\,°C/\text{min}$ cooling (second heating). Three transitions can be seen: the crystal $\rightarrow$ crystal transition at $20\,°C$, the triclinic $\rightarrow$ hexagonal transition with a peak temperature of $29\,°C$, and the melting. The as-received tape (first heating) has a melting point at $342\,°C$, while the sample crystallized at $10\,°C/\text{min}$ melts at $328\,°C$. Since $342\,°C$ is higher than the equilibrium melting point ($T_m° = 332\,°C$), superheating takes place.

2.10. THERMOSETS

Thermosets are network-forming polymers. Unlike thermoplastic polymers, whose processing requires only physical changes such as melting, thermosets are distinguished by the bulk chemical reactions that are involved in their use.

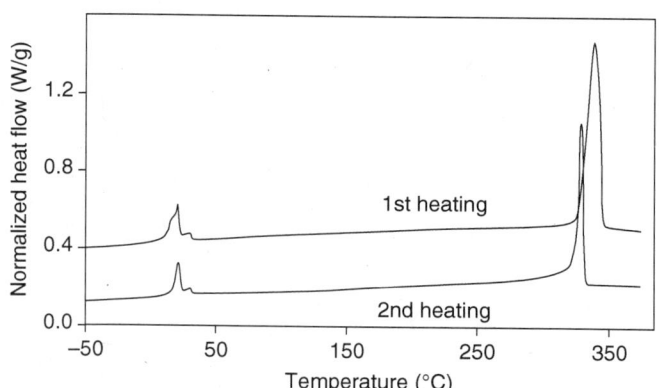

Figure 2.66. Heating curves of PTFE (Teflon®) tape at $10\,°C/\text{min}$: as-received tape and heating after $10\,°C/\text{min}$ cooling. The two crystal→crystal transitions are clearly visible in the low-temperature region, while the main melting peak is $300–350\,°C$. Perkin-Elmer DSC7. Endotherm is up. (Menczel, unpublished results.)

As a result of these reactions the materials eventually crosslink and become *set*; that is, they lose the ability to flow or to be dissolved. Cure most often is thermally activated; hence the term *thermoset*, but network-forming materials whose cure is light- or radiation-activated are also considered to be thermosets (see Section 2.11, on differential photocalorimetry). Some thermosetting materials, such as certain adhesives, crosslink by a dual-cure mechanism, that is, by either heat or light activation. In contrast to the T_g values for crosslinked elastomers or rubbers, the glass transition temperature of thermosets is generally above room temperature.

The focus of this section is to provide a description of the behavior of thermosets measurable by DSC. This consists primarily of cure and properties developed during cure, including cure kinetics. It also includes measurement of T_g and conversion to establish the important T_g–conversion relationship.

Uncured thermosets are mixtures of small reactive molecules, often a resin and a hardener, or a mixture of resins and hardeners. They may contain additives such as catalysts to promote or accelerate cure. Many thermosets are used in filled or reinforced form to reduce cost, modify physical properties, act as a binder for particles or fibers, or reduce shrinkage during cure. While epoxy is probably the best-known thermoset, there are many other network-forming polymers such as phenolic, unsaturated polyester, polyurethane, dicyanate, bismaleimide, and acrylate. The studies by Gotro and Prime (2004), Pascault et al. (2002), Prime (1997), and Van Mele et al. (2004) can be listed as references describing DSC of thermosets. A brief general "Introduction to Thermosets" may be found at www.primethermosets.com. A good review of the thermal analysis of elastomer systems that crosslink may be found in Sircar (1997).

Thermosets are distinct from thermoplastic polymers in one major respect—their processing includes the chemical reactions of cure. As illustrated in Fig. 2.67, cure begins with the growth and branching of chains. As the reaction proceeds, the increase in molecular mass accelerates, and eventually several chains become linked together into a network of infinite molecular mass. The abrupt and irreversible transformation from a viscous liquid to an elastic gel or rubber is called the *gel point*. Cure is illustrated schematically in Fig. 2.67 for a material with coreactive monomers such as an epoxy–diamine system. Gelation is the incipient formation of a crosslinked network, and it is the most distinguishing characteristic of a thermoset. A thermoset loses its ability to flow and can no longer be processed above the gel point, and therefore gelation defines the upper limit of the work life, the time after mixing of reactants that a thermoset remains sufficiently liquid that it can be worked. As an example, for a "5-minute epoxy," which can be found in any hardware store, the "5 minutes" (5 min) refers to the gel point. After the two parts are mixed, the user must form an adhesive joint within 5 min before the material becomes rubbery and is incapable of flow, and then keep the repaired part fixtured until cure is sufficiently complete, typically for several hours. A distinction may be drawn between the phenomenon of molecular gelation and its

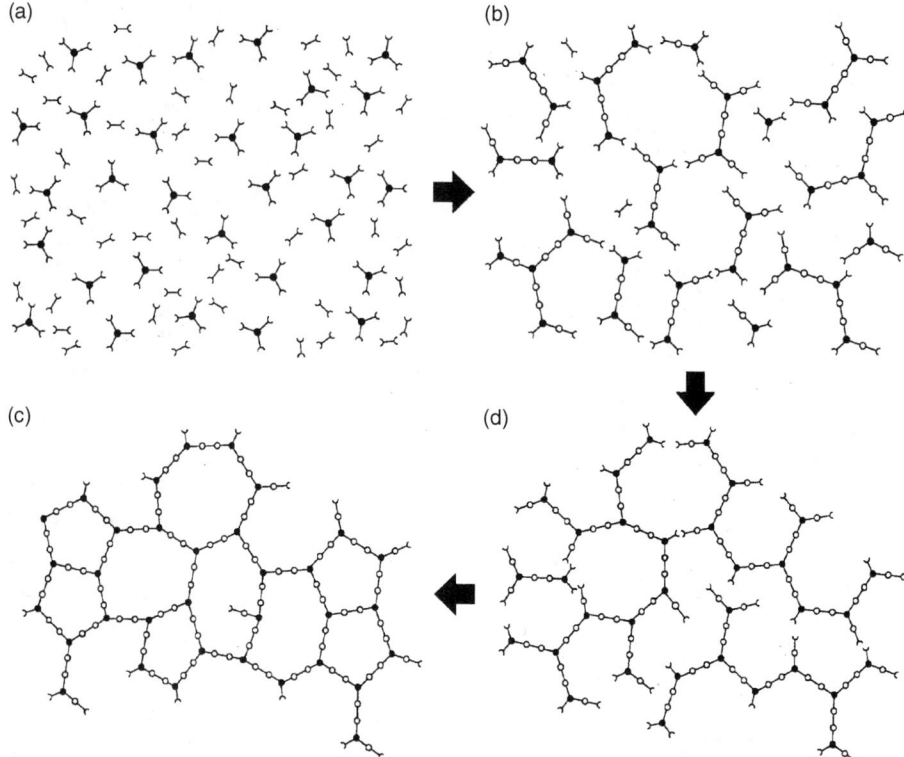

Figure 2.67. Schematic, two-dimensional representation of thermoset cure. For simplicity difunctional and trifunctional coreactants are depicted. Cure starts with A-stage monomers (a), proceeds via simultaneous linear growth and branching to a B-stage material below the gel point (b), continues with formation of a gelled but incompletely crosslinked network (c), and ends with the fully cured, C-stage thermoset (d). [From Prime (1997), with permission of Elsevier.]

consequence, macroscopic gelation. The degree of conversion at the gel point α_{gel} is constant for a given thermosetting system, independent of cure temperature; that is, gelation is isoconversional. (See Table 2.5 for glossary of terms.) For this reason the time to gel versus reciprocal absolute temperature can be used to measure the activation energy of cure. Gelation does not usually inhibit cure (e.g., the reaction rate remains unchanged on passing through gelation), and it cannot be detected by techniques sensitive only to the chemical reaction, such as DSC. Beyond the gel point, the reaction proceeds toward the formation of an infinite network with substantial increase in crosslink density, glass transition temperature, and ultimate physical properties.

Vitrification, a phenomenon completely different from gelation, may or may not occur during cure depending on the cure temperature relative to $T_{g\infty}$, the T_g for the fully cured network. T_g of the network increases continuously with

TABLE 2.5. Glossary of Characteristic Cure Parameters

α	Chemical conversion (e.g., of epoxide groups), degree of cure
α_{gel}	α at gel point
t_{gel}	Time to gelation, gel time
t_{vit}	Time to vitrification
T_{cure}	Cure temperature, a process parameter
T_g	Glass transition temperature, a material property
$T_{g,0}$	T_g for uncured thermoset with degree of conversion $\alpha = 0$
$_{gel}T_g$	T_g for thermoset with degree of conversion α_{gel}
$T_{g,\infty}$	T_g for fully cured thermoset with degree of conversion $\alpha = 1$

increasing cure, that is, increasing crosslink density, to $T_{g\infty}$. Vitrification is glass formation due to T_g increasing from below T_{cure} to above T_{cure} as a result of the cure reaction, and is defined as the point where $T_g = T_{cure}$ (Gillham 1986). Vitrification can occur anywhere during the reaction to form either an ungelled glass or a gelled glass. It can be avoided by curing at or above $T_{g\infty}$. In the glassy state, the rate of cure will usually undergo a significant decrease and fall below the chemically controlled reaction rate as the reaction becomes controlled by the diffusion of reactants. It is common for complete vitrification to result in a decrease in the rate of reaction by two or three orders of magnitude. Unlike gelation, vitrification is reversible by heating, and chemical control of cure may be reestablished by heating to devitrify the partially cured thermoset. Vitrification may be detected by a stepwise decrease in heat capacity by modulated temperature DSC (MTDSC) as described in Section 2.13. It may also be measured by dynamic mechanical analysis (DMA; see Chapter 5) as a frequency-dependent transition resulting in a glassy modulus typically >1 GPa and by dielectic analysis (DEA; see Chapter 6).

Generally, the shift from chemical control to diffusion control of the reaction may be observed by a decay of the reaction rate, which is often seen when T_g reaches 10–15 °C above T_{cure} (see Figs. 2.73 and 2.74). Although not common, it is possible for diffusion to control the cure kinetics prior to vitrification; it is also possible for sluggish reactions to remain under chemical control well into the glassy state (Gotro and Prime 2004; Pascault et al. 2002; Prime 1997; Van Mele et al. 2004).

Three characteristic temperatures are listed in Table 2.5: T_{g0}, the glass transition temperature of the completely unreacted thermoset; $_{gel}T_g$, the temperature at which gelation and vitrification coincide; and $T_{g\infty}$, the glass transition temperature of the fully cured network. At temperatures below T_{g0}, reaction takes place in the glassy state and is therefore slow to occur. To minimize reaction during storage, unreacted systems should be stored well below T_{g0}; 20–50 °C below T_{g0}, deep into the glassy state, is recommended. Between T_{g0} and $_{gel}T_g$, the liquid resin will react without gelation until its continuously rising glass transition temperature becomes coincidental with the cure temperature, at which stage vitrification begins and the reaction may become diffusion-controlled. Note that $_{gel}T_g$ is the temperature at which gelation and vitrification occur simultaneously.

At temperatures between $_{gel}T_g$ and $T_{g\infty}$, the viscous liquid changes to a viscoelastic fluid, then to a rubber, and finally to a glass. Gelation precedes vitrification, and a crosslinked rubbery network forms and grows until its glass transition temperature coincides with the cure temperature, where the reaction may become diffusion-controlled. At temperatures above $T_{g\infty}$, the network remains in the rubbery state after gelation unless other reactions occur, such as thermal degradation or oxidative crosslinking.

The handling and processing of thermosets are closely dependent on gelation and vitrification. For example, thermosets are often identified at three stages of cure: *A stage*, referring to an unreacted resin; *B stage*, referring to a partially reacted and usually vitrified system, *below the gel point*, which, on heating to devitrify, can flow and be processed and the cure completed; and *C stage* to the completely cured network. Prepreg (woven or unidirectional fibers previously impregnated with resin and partially reacted) is a common example of a B-stage thermoset.

2.10.1. Measurements

In this subsection we describe how to make the basic DSC measurements to characterize cure. These measurements provide the necessary input for kinetic analyses and to construct T_g–conversion relationships. Measurements include the heat of reaction; the extent and rate of conversion; T_g, including vitrification; characteristic cure parameters; and enthalpy relaxation associated with physical aging that often occurs with the glass transition of partially cured thermosets (see Section 2.7). Cure may be accomplished thermally or by irradiation such as ultraviolet light, leading to measurements by differential photocalorimetry (DPC; see the following section). Although gelation is not directly detectable by DSC, when DSC is combined with techniques such as dynamic mechanical analysis (DMA; Chapter 5), it can yield valuable information such as the extent of cure at the gel point (α_{gel}) and the isothermal time to reach α_{gel} as a measure of the gel time. Barton (1985), Richardson (1993a,b), and Prime (1981, 1997) reviewed the application of DSC to the cure of thermosets. Van Mele et al. (2004) reviewed the application of modulated temperature DSC (MTDSC) to thermosets, which is described in Section 2.13. MTDSC is especially useful for its ability to detect vitrification during cure and to separate the glass transition region when it overlaps with the cure exotherm of a partially cured thermoset. Kinetic methods and models are described later in this section.

2.10.1.1. Sample Preparation Careful sample preparation is the first step to obtaining good data on thermosets. Sample preparation must account for the fact that these materials are reactive, and care must be taken to avoid premature or unwanted reaction. When studying one-part systems, such as UV or dual-cure adhesives, premixed and frozen adhesives or prepregs, strict adherence to the manufacturer's instructions for sample freshness as well as shipping and storage temperatures cannot be stressed too strongly. For two-part systems care must be exercised in sample preparation and storage to prevent significant reaction prior to the measurement. A good and usually achievable criterion is to allow less than 1% of reaction prior to analysis. A technique that has worked well for slow to moderately reactive systems at room temperature, such as many epoxy–aromatic diamine systems, is to mix solutions of ingredients at or below room temperature, freeze-dry or remove the solvent under vacuum, and use immediately or store at temperatures below T_{g0} (Acitelli et al. 1971; Hale et al. 1991). A temperature of 30 °C or more below T_{g0} is recommended, but even under these conditions some reaction will occur over long periods, and samples should be periodically discarded. A technique that has worked well for slowly reacting systems even at moderate temperatures, such as the system shown in Figs. 2.69–2.71 and high-performance epoxies, such as tetrafunctional epoxies such as tetraglycidyl-methylene dianaline (TGMDA) reacted with high-temperature amines such as methylenedianiline (MDA) or diaminodiphenylsulfone (DDS), is to mix the ingredients for a short time at elevated temperature and then immediately cool them (Riccardi and Williams 1986; Wisanrakkit and Gillham 1990a,b). This technique will avoid the presence of solvent and can result in negligible initial conversion. For fast-reacting systems, such as "5-" or "30-min" epoxy or polyurethane pultrusion systems, the same <1% initial reaction criterion still applies. (Here we mention that pultrusion is a continuous molding process where fiber reinforcements are pulled through a thermosetting resin bath and then cured resulting in a composite material. The reinforcements are often fiberglass or can be more exotic materials such as carbon or aramid fibers). It has been demonstrated that a fast-reacting polyurethane [see Prime et al. (2005); described later in this section; see Figs. 2.74 and 2.75] can be mixed and DSC measurements commenced in less than 2 min and that the abovementioned criterion is satisfied. For 5- and even 30-min epoxy, where these times are approximately equivalent to the time to reach the gel point, reaction is so fast that this criterion cannot be met. Recommended sample size is 5–20 mg, depending on the amount of filler and intensity of the exotherm. Inhomogeneous materials, like fiber-reinforced prepreg, may require subsequent measure of fiber content to obtain reproducible data (Noel et al. 1988). The heat capacity of the reference cell should approximate that of the empty sample cell. Liquid sample pans, where the top can be cold-welded after the sample is introduced to form a hermetic seal, are recommended. They are usually adequate to contain the small amount of volatiles, which may be volatile reactants or residual solvent or reaction products associated with cure. Some workers prefer the use of high-pressure capsules made from stainless steel that withstand much higher pressures (Freeberg and Alleman 1966) or pressure DSC (Levy et al. 1970), especially if volatile loss is significant. If there is concern that the containment of volatiles may alter the desired reaction, as can happen with phenolic systems, then other forms of analysis such as TGA or DMA may be more appropriate than DSC.

2.10.1.2. Heat of Reaction ΔH_{rxn} A common property of all thermosetting systems is the liberation of heat accompanying cure:

$$\text{Reactants} \rightarrow \text{products} + \Delta H_{rxn} \qquad (2.77)$$

where ΔH_{rxn} is the exothermic heat of reaction, expressed as heat per mole of reacting groups (kJ/mol) or per mass of material (J/g). The heat of reaction is a fundamental property of thermosets that is required for the measurement of conversion as well as for all kinetic measurements. Great care should be taken to obtain as accurate a measurement as possible, as this will affect the outcome of kinetic analyses.

In some cases it will be possible to calculate a value for ΔH_{rxn} of the thermosetting system. One example is an epoxy–amine system where the composition is known. A value of $\Delta H_{rxn} = 107 \pm 4$ kJ/mol of epoxide (Prime 1997) can be used to calculate the heat of reaction for such a system. Values for other thermosetting systems, including epoxy–phenol, dicyanate ester, and carbon–carbon double bonds, are also available (Prime 1997). But in the vast majority of cases ΔH_{rxn} will need to be measured experimentally.

The heat of reaction ΔH_{rxn} is defined as the total heat liberated when the thermosetting material is taken from 0% to 100% conversion. For this reason it is very important to start with completely unreacted material, as discussed in the previous paragraphs, and to finish with all reactive groups consumed. Precision and accuracy of ±3–5% may be expected. However, complete reaction cannot always be accomplished, such as for stoichiometrically balanced systems (e.g., equivalent amounts of amine and epoxy) or for systems with high functionality (e.g., tetrafunctional amine reacted with tetrafunctional epoxy). Toward the end of reaction in such systems large-scale motion of reactive groups in the network can become so restricted as to render the groups unable to achieve the necessary proximity for reaction. The heat measured for cure of these systems is referred to as ΔH_{ult}, the ultimate heat of reaction that can be achieved that may be significantly less than ΔH_{rxn}. If ΔH_{rxn} is known or can be calculated for such systems, it should be used to measure conversion and other kinetic parameters. Otherwise ΔH_{ult} may be used, in which case only relative conversion is measured and the kinetic analyses may be affected.

The shape of the exotherm contains important information. A symmetric exotherm, such as those in Figs. 2.69, 2.70, and 2.77, suggests a relatively simple reaction without the complexity of multiple reactions. In Figs. 2.69 and 2.70, for example, the reaction of epoxy with primary amine is indistinguishable from its reaction with secondary amine, resulting in the symmetric peak. If the secondary amine–epoxy reaction is significantly slower, the exotherm will be skewed to the high-temperature side and in extreme cases a second peak may develop.

It is recommended that the value for ΔH_{rxn} be determined from scanning experiments to avoid heat that can go unrecorded at the beginning of an isothermal experiment during the ramp to temperature, and unrecorded at the end because of vitrification or the reaction falling below the dectectability limit of the DSC. It is recommended that the value be taken as the average of three or more measurements at heating rates within 2–10 °C/min. The heating rate should be slow enough to avoid high temperatures that can introduce unwanted competing reactions or degradation. It is further recommended that measurements be made at different heating rates to verify that ΔH_{rxn} is independent of heating rate. A dependence on heating rate indicates a more complex reaction and possibly multiple reactions. If multiple reactions differ significantly in reaction rate, multiple exothermic peaks may be observed. Data at different heating rates can also be used to estimate the activation energy for cure and as input to model-free kinetics analyses (described later in this section and in Section 3.5 of Chapter 3).

2.10.1.3. Conversion or Degree of Cure, Reaction Rate, and T_g

The basic assumption underlying the application of DSC to thermoset cure is that the measured heat flow dH/dt is proportional to the rate of conversion or rate of cure ($d\alpha/dt$), as represented by Eq. (2.78). Sometimes the symbol x is used instead of α to denote conversion or degree of cure. Note that dH/dt is the ordinate or y axis of the DSC measurement and is measured in joules per second (J/s) or watts (W). It is also assumed that, on integration, the total heat detected during the DSC measurement is identical to the heat evolved by the cure reaction. In cases such as the epoxy–amine reaction, it is further assumed that the heat of reaction for epoxy with primary amine is the same as that for the reaction of epoxy with secondary amine, and that the activation energies are the same. For systems with a symmetric DSC exotherm, these parameters are usually within a few percentage points of each other (Gotro and Prime 2004; Pascault et al. 2002; Prime 1981, 1997; Van Mele et al. 2004). There is usually a substitution effect where the secondary amine is somewhat less reactive than the primary amine, but the effect on the overall kinetics measured by DSC is seldom significant. In the small number of systems where steric hindrance is significant, the substitution effect is much larger, and the DSC exotherm exhibits a high-temperature shoulder or even two distinct exotherms, and an overall reaction cannot be assumed. Under these conditions multiple heating rate methods may be appropriate. The assumptions mentioned above have proved to be valid provided that other thermal events such as evaporation or degradation are not occurring at the same time to interfere with measurement of the cure reaction (Gotro and Prime 2004; Pascault et al. 2002; Prime 1981, 1997; Van Mele et al. 2004). One example of an interfering event is the endothermic evaporation of water of condensation from some phenolic and amino resins, which occurs simultaneously with and as a result of the cure reaction:

$$\frac{d\alpha}{dt} = \frac{dH/dt}{\Delta H_{rxn}} \qquad (2.78)$$

Isothermal kinetic measurements fall into two categories: *method 1*, in which the rate and extent of reaction at constant temperature are continuously monitored in the DSC; and *method 2*, in which a partially cured sample is heated in the DSC to measure the residual heat of reaction. An advantage of method 1 is the simultaneous measurement of conversion and rate of conversion, which are necessary for some kinetic analyses. It should be noted that vitrification will occur during method 1 measurements if T_{cure} is less than $T_{g\infty}$. Method 2 has the advantage of simultaneous measurement of T_g and conversion, from which the T_g–conversion relationship can be established. Both thermal and UV cure reactions can be measured by these methods.

Isothermal Method 1. This method capitalizes on the ability of DSC to simultaneously monitor both the conversion and the rate of conversion over the entire course of the cure reaction. This allows direct use of derivative forms of the rate equation, such as Eq. (2.86), which are necessary for kinetic analysis of autocatalytic reactions such as epoxy–amine. Experimentally this method is well suited to autocatalytic reactions that do not reach maximum rate until later in the reaction after the instrument has achieved thermal equilibrium. Even so, at high temperatures a significant portion of the reaction can take place before the calorimeter equilibrates and go unrecorded. Widmann (1975) and Barton (1983) have proposed a means to correct for such unrecorded heat by rerunning the experiment on the reacted sample, under the same conditions, to obtain an estimate of the true baseline and the unrecorded heat that should be added to the measured heat, as illustrated in Fig. 2.68. Note that this system appears to follow nth-order kinetics where the maximum reaction rate occurs at $t = 0$. For the sample shown, Widmann reports that 5% of ΔH_{rxn} goes unrecorded at 150 °C and 20% at 170 °C. In practice, method 1 is limited to a range of temperatures below which the reaction rate is too slow to give statistically reliable data and above which unrecorded heat of reaction becomes excessive or competing reactions or degradation occur. Note that when $T_{cure} < T_{g\infty}$, vitrification will occur accompanied by a slowing of the reaction.

There are two method 1 experimental techniques. In one the calorimeter is preheated to the desired reaction temperature before the unreacted sample is placed in the calorimeter cell. This technique may be preferred for higher-sample-holder-mass heat flux instruments. In the other, the sample is placed in the calorimeter cell at a temperature at which no significant reaction will take place over a short time period, and the temperature is ramped as rapidly as possible to the reaction temperature. This technique is recommended for

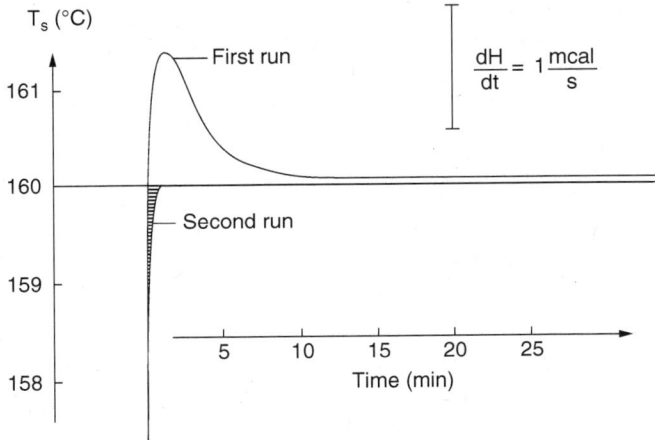

Figure 2.68. Illustration of DSC isothermal method 1. Hatched area between first and second runs used to measure unrecorded heat. Epoxy–anhydride. Exotherm is up. [From Widmann (1975); reprinted with permission of Heydon, London.]

lower-sample-holder-mass power compensation instruments. With both techniques the rate of heat generation is followed with time. The degree of cure or conversion is given by

$$\alpha_t = \frac{\Delta H_t}{\Delta H_{rxn}} \qquad (2.79)$$

where α_t is the conversion and ΔH_t the heat generated up to time t. The instantaneous rate of conversion at time t is proportional to the heat flow at time t $[(dH/dt)_t]$, according to Eq. (2.78).

Isothermal Method 2. This method is necessary to obtain cure data at low temperatures where the rate of heat evolution is too small for method 1 to be reliable. It is also recommended for nth-order reactions where the maximum rate of cure occurs at $t = 0$, and to obtain simultaneous T_g and conversion data to construct T_g–conversion plots. The conversion–time data can be fit to integrated forms of the rate equation, such as Eqs. (2.83)–(2.85). Several samples are cured isothermally, for example, in an oven, in the calorimeter or at ambient temperature, for various times until no additional curing can be detected. The samples are subsequently scanned in the DSC at a fixed heating rate, from which T_g and the residual heat of cure (ΔH_{res}), the heat evolved during completion of the reaction, are measured, as illustrated in Fig. 2.69.

The degree of cure or conversion at time t for method 2 (α_t) is calculated from

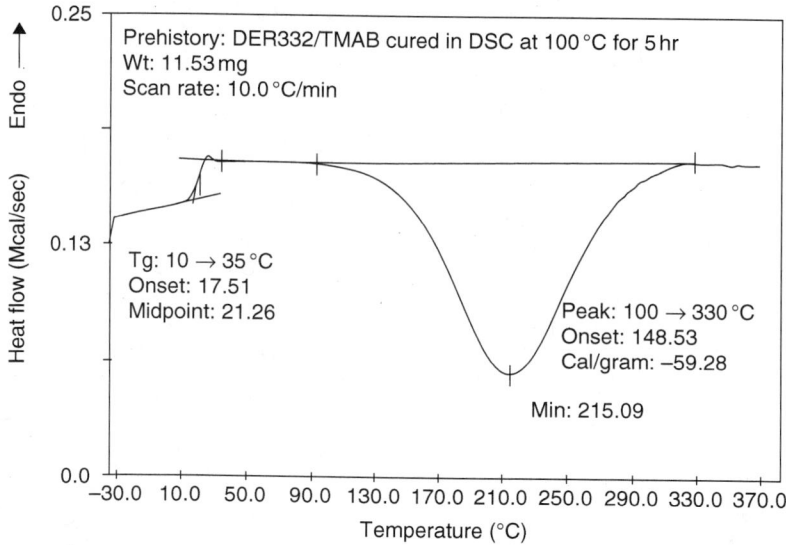

Figure 2.69. Typical DSC scan of a partially cured thermoset. The glass transition appears as an endothermic shift in the heat flow over a temperature interval. The residual heat ΔH_{res} appears as an exothermic peak following the glass transition. Epoxy–amine. Exotherm is up. [From Wisanrakkit and Gillham (1990a); reprinted with permission of John Wiley & Sons, Inc.]

$$\alpha_t = \frac{\Delta H_{rxn} - \Delta H_{res}}{\Delta H_{rxn}} \equiv \frac{\Delta H_t}{\Delta H_{rxn}} \qquad (2.80)$$

where ΔH_{rxn} and ΔH_t are as previously defined.

Wisanrakkit and Gillham (1991) present an excellent example of DSC isothermal method 2 for an epoxy–amine system (see Fig. 2.70). Figure 2.69 shows DSC scans for the uncured and partially cured thermoset, illustrating the measurement of both ΔH_{rxn} and ΔH_{res}. Figure 2.71 shows conversion–ln(time) curves for the same epoxy for cure temperatures of 100–180 °C. Conversion was calculated from Eq. (2.80). Note that the curves are parallel during the first part of cure, and at the lower cure temperatures, less than 100% conversion is achieved.

2.10.1.4. Multiple Heating Rate Methods

Multiple heating rate methods are isoconversional kinetic methods that are sometimes referred to as *model-free kinetics*. They allow estimation of activation energy from relationships between the heating rate and the temperatures to reach constant conversion at those heating rates. Figure 2.77 shows an example of an epoxy system cured at three different heating rates. These methods are especially useful for systems whose activation energy is independent of temperature but may change with

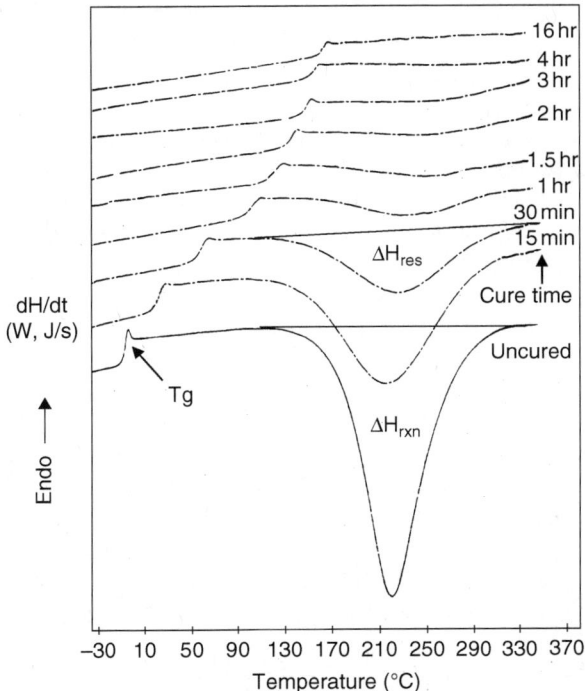

Figure 2.70. A series of DSC temperature scans at 10 °C/min of an epoxy–amine cured isothermally at 160 °C for different times, showing T_g increasing and the residual exotherm decreasing with increasing cure times Endotherm is up. [from Wisanrakkit and Gillham (1990a); reprinted with permission of John Wiley & Sons, Inc.].

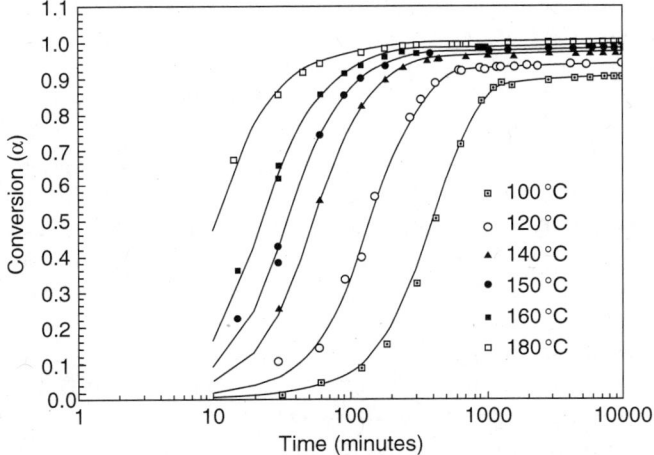

Figure 2.71. Conversion versus time data for the same epoxy–amine system shown in Fig. 2.70 [from Wisanrakkit and Gillham (1990a); reprinted with permission of John Wiley & Sons, Inc.].

conversion. A minimum of three heating rates is required, but four to five are recommended. Care should be taken to choose heating rates that are fast enough to avoid vitrification during cure but slow enough to avoid high temperatures that may promote unwanted side reactions or degradation. Most instrument manufacturers offer ASTM E698 software for determining the activation energy. This method applies the principles of Ozawa (1965, 1970) and Flynn and Wall (1966a,b). It measures the activation energy for different conversion levels, including the exothermic peak from the variation in isoconversional temperature with heating rate. Mettler Toledo and Perkin-Elmer offer model-free kinetics (MFK) software, based on the work of Vyazovkin (1997, 2001), that utilizes conversion–temperature data from the entire reaction exotherm. This method allows regeneration of the DSC curve and predictions of cure behavior from the mathematical modeling.

These methods are useful for measuring changes in reaction mechanism, via changes in activation energy, with conversion (Flynn 1978a,b; Barton 1974a,b, 1985). They are able to yield kinetic information on thermosetting systems not easily studied by other methods, for example, those with multiple reactions, due, for example, to off-stoichiometric formulations or a mixture of curing agents. The DSC ASTM E698 methodology (see Appendix) can treat systems with unresolvable baselines, or residual solvent, providing the peak exotherm temperature is unaffected.

2.10.2. T_g–Conversion Relationship and Vitrification

Figure 2.72 graphically illustrates the relationship between the glass transition temperature and conversion measured from the residual exotherm of the DSC curves of the epoxy–amine system shown in Fig. 2.70. Several workers have shown that for most thermoset systems there is indeed a unique relationship between the chemical conversion of a thermoset and its glass transition temperature, independent of the cure temperature and thermal history [see, e.g., Pascault and Williams (1990), Hale et al. (1991), Venditti and Gillham (1997), and, for a general overview, Prime (1997)]. It is good practice to establish the T_g–conversion relationship as part of a cure study. Conversion can be difficult or impossible to measure directly, for example, for systems where mass loss accompanies cure. In these cases this relationship may be invoked in order to use T_g as a measure of cure. As Fig. 2.72 suggests, T_g may be preferred as a measure of cure for the final 5–10% of cure where it is usually more sensitive than the residual heat of reaction.

The empirical DiBenedetto equation was developed in the late 1960s to mathematically relate T_g and conversion (Nielson 1969; DiBenedetto 1987). Excellent theoretical treatises on the T_g–conversion relationship can be found in Pascault and Williams (1990), Hale et al. (1991), and Venditti and Gillham (1997). Venditti and Gillham (1997) developed an equation based on thermodynamic considerations put forth by Couchman and Karasz (1978) to predict T_g versus mole fraction of constituents of a linear copolymer:

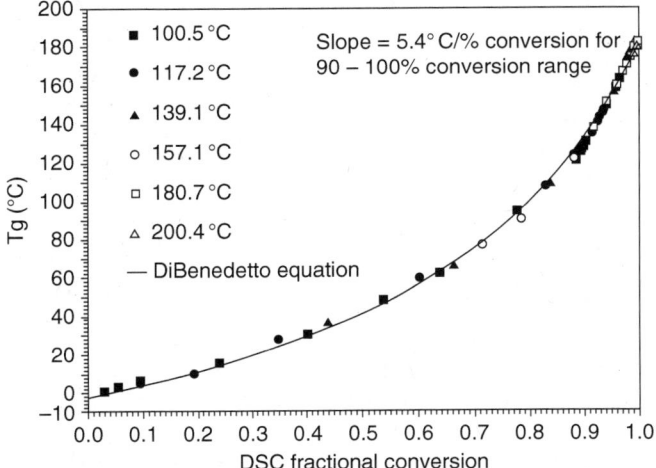

Figure 2.72. Relationship between T_g and fractional conversion for the same epoxy–amine system shown in Figs. 2.69–2.71. Different symbols represent material cured at different temperatures. The solid curve represents the best-fit calculation from the modified DiBenedetto equation (Nielson 1969; DiBenedetto 1987). [From Wisanrakkit and Gillham (1990a); reprinted with permission of John Wiley & Sons, Inc.].

$$\ln(T_g) = \frac{(1-\alpha)\ln(T_{g0})(\Delta C_{p\infty}/\Delta C_{p0})\alpha \ln(T_{g\infty})}{(1-\alpha) + (\Delta C_{p\infty}/\Delta C_{p0})\alpha} \tag{2.81}$$

Derivation of Eq. (2.81) includes the assumptions that the system at conversion α is a regular mixture of unreacted end segments with concentration $(1-\alpha)$ and $T_g = T_{g0}$, and reacted end segments with concentration (α) and $T_g = T_{g\infty}$. It is also assumed that ΔC_p is independent of temperature. The parameters T_{g0}, $T_{g\infty}$, ΔC_{p0}, and $\Delta C_{p\infty}$ can be measured from one or two DSC experiments (see Fig. 2.110 later in this chapter).

Figure 2.72 illustrates the T_g–conversion relationship for an epoxy–amine system cured at several temperatures. All the data, which can be seen to collapse to a single curve, are fit to the empirical DiBenedetto equation. As can be seen from the increasing curvature of the data in Fig. 2.72, T_g is a more sensitive measure of cure than is conversion, as measured by ΔH_{res}, in the latter stages of cure, which are often the most critical. From the slope of the curve for the final 10% of cure (5.4 °C/% conversion) and assuming an ability to measure T_g to ±2 °C, it would be possible to control degree of cure to ~0.5% from measurement of T_g.

This is a very important relationship for several reasons. For example, for thermosetting systems in which mass loss is associated with cure or for which there is no simple or unambiguous measure of cure, T_g is often the only practical means to monitor the progress or extent of cure.

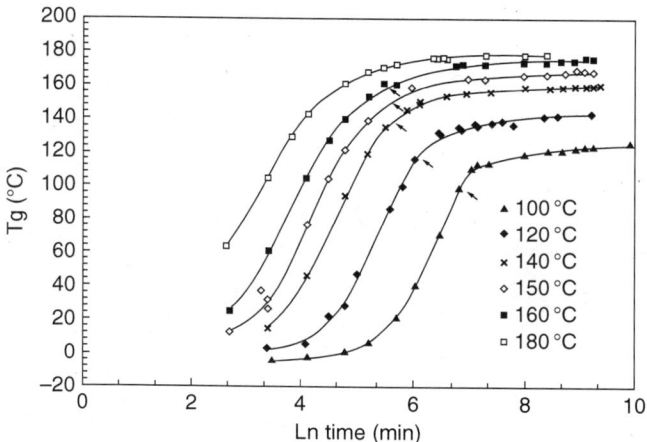

Figure 2.73. Glass transiton temperature T_g versus ln(time) for data at different cure temperatures. Isothermal vitrification ($T_g = T_{cure}$) at each cure temperature is designated by an arrow. Same epoxy–amine system as in Figs. 2.69–2.72. [From Wisanrakkit and Gillham (1990a); reprinted with permission of John Wiley & Sons, Inc.].

The similarity of the T_g–time data in Fig. 2.73 with the conversion–time data of Fig. 2.71 is a consequence of the T_g–conversion relationship and illustrates the ability to monitor cure through measurement of T_g. Figure 2.73 also directly illustrates vitrification, defined as T_g increasing to T_{cure} as a result of cure, and designated at each cure temperature by an arrow. Note that the progress of cure is significantly impeded shortly after vitrification, which marks the shift from chemical control to diffusion control of the reaction, as described at the beginning of this section.

2.10.3. Cure Kinetics

This topic is important in two respects: characterization of the cure behavior of a material and the design and optimization of cure processes. Cure kinetics generally falls into two categories: (1) aging or storage kinetics after reactants have been mixed but prior to their use (reaction of prepregs, premixed and frozen adhesives, and powder coatings during storage fall into this category)— the objective in this case is to minimize chemical reaction between the components; and (2) (of primary significance) the cure or completion of cure of the thermoset after the prepreg has been formed into a structure, the resin has infused into a preform in a vacuum-assisted resin transfer molding (VARTM) process, or the adhesive joint has been formed. Included here are the characterization, design, and optimization of cure processes, where issues such as the optimum time to apply pressure in an autoclave process, the time–temperature to reach full cure, and the increase in temperature required to reduce the cycle time by 50% are examples. Development of a kinetic equation is often needed for software that models the thermal cure process. When combined with rheological models, the flow behavior or chemorheology of the reacting system can be projected (May 1983).

Cure kinetics is the mathematical relationship between time, temperature, and conversion. Three approaches to the measurement of cure kinetics are considered. The first is traditional isothermal kinetic analysis, which involves chemical models such as nth-order and autocatalytic. In this approach DSC conversion–time data, or conversion–rate of conversion–time data, taken at several temperatures are fit to the chemical models to determine reaction orders and measure rate constants. Analysis of data taken at several temperatures provides measurement of the activation energy E from the Arrhenius equation [Eq. (2.87)]. The shift from chemical control to diffusion control of the reaction, which often occurs at vitrification, can be modeled by traditional isothermal kinetic methodologies (see later in this section). The second approach is time–temperature superposition or TTS kinetics, which includes the assumption that cure can be described by a single or overall activation energy, a reasonable assumption for many thermosetting systems. TTS kinetics creates master cure curves by superimposing conversion–time, or T_g–time curves through the T_g–conversion relationship, utilizing the Arrhenius equation. It is the only method that allows kinetic analysis of T_g–time data. TTS kinetics does not require a chemical or mathematical model (i.e., it is model-free), but the master curves may be fit to appropriate models. It may be applied to the characterization of cure as well as to the monitoring and control of systems that cure according to complex time–temperature profiles. Another approach to cure kinetics is multiple heating rate kinetics, which includes ASTM kinetics and what is often referred to as *model-free kinetics*. ASTM E698 kinetics (see Appendix A), based on the methodologies developed by Ozawa (1965, 1970) and Flynn and Wall (1966a,b), performs an isoconversional analysis on data from DSC experiments at three or more heating rates. This method is capable of measuring activation energy E as a function of conversion. However, it is often used to estimate E from the shift in peak exotherm temperature with heating rate [see Prime (1997) for a detailed discussion]. Note that model-free kinetics (MFK) also performs an isoconversional analysis on DSC data taken at three or more heating rates. It can also perform an isoconversional analysis on data from isothermal DSC experiments at three or more temperatures. An advanced version can analyze data from complex time–temperature profiles. In this methodology E is treated as a variable. MFK is therefore useful for characterizing and modeling a wide variety of systems, including those whose the activation energy is not constant with conversion.

An experimenter will have several options when embarking on a cure kinetics study. The ultimate approach will depend on many factors, including availability of software, objectives of the study, and characteristics of the particular system being studied. Described below are recommended methodologies.

Single-heating-rate methods and those that assume at the outset a specific reaction model (e.g., first-order) are not recommended.

2.10.3.1. Traditional Isothermal Kinetics Measurements of conversion–rate of conversion–time data by isothermal method 1 and conversion–time data by isothermal method 2 were described earlier in this section. Isothermal method 1 measurements have the advantage of simultaneously measuring conversion (α) and rate of conversion ($d\alpha/dt$), which allows use of derivative forms of the rate equation such as Eqs. (2.82) and (2.86). Both conversion and rate of conversion data are necessary to model autocatalytic kinetics [e.g., using Eq. (2.86)]. Isothermal method 2 yields both T_g and conversion at a series of times and temperatures (see Fig. 2.70 as an example). However, the measurements are time-consuming, and since the reaction rate is not measured, it must be calculated mathematically, or integrated forms of the rate equation such as Eqs. (2.83)–(2.85) must be used. To perform these analyses generally requires use of a spreadsheet.

Equation (2.7) shows the general nth-order rate equation

$$\frac{d\alpha}{dt} = k(1-\alpha)^n \qquad (2.82)$$

where k is the rate constant and n is the reaction order. An example of nth-order kinetics is the second-order reaction of an isocyanate with an alcohol to produce a polyurethane, as illustrated in Scheme 2.1.

Integrated forms of Eq. (2.82) for first-, second- and general nth-order reactions are given in (2.83)–(2.85), respectively. Since there are only two unknowns, simple linear regression analysis may be used to determine values for k and n. For a first-order reaction ($n = 1$), one obtains

$$-\ln(1-\alpha) = kt \qquad (2.83)$$

for a second-order reaction ($n = 2$), one obtains

$$\frac{1}{1-\alpha} = 1 + kt \qquad (2.84)$$

and for an nth-order reaction ($n \neq 1$), one obtains

$$-\ln(1-\alpha) = \frac{1}{n-1}\ln[1+(n-1)kt] \qquad (2.85)$$

$$\text{R-N=C=O} + \text{HO-R'} \longrightarrow \text{R-NH-}\underset{\underset{\text{O}}{\|}}{\text{C}}\text{-O-R'}$$

(1)

Scheme 2.1.

Equation (2.11) shows a common autocatalytic rate equation proposed by Sourour and Kamal (1976)

$$\frac{d\alpha}{dt} = (k_1 + k_2 \alpha^m)(1-\alpha)^n \tag{2.86}$$

where k_1 is the catalyzed rate constant, attributed to catalytic species initially present, such as impurities and/or fiber or particle surfaces; k_2 is the autocatalyzed rate constant, due to catalytic species produced by the reaction; and m and n are reaction orders. An example is the reaction of an epoxy resin with a diamine to produce an epoxy thermoset, as illustrated in Schemes 2.2 and 2.3 for reaction of epoxide with primary amine and epoxide with secondary amine, respectively.

The chemistry for a stoichiometrically balanced reaction suggests that $m = 1$ and $n = 2$ in Eq. (2.86). For real systems, values are often close to these values but not identical. In the epoxy–amine reaction the alcohol, which may be present initially in small concentrations but is also a product of the reaction, catalyzes further reaction, resulting in autocatalysis. Since there are four unknowns (k_1, k_2, m, and n) nonlinear regression analysis must be employed, although k_1 can be evaluated independently as the extrapolated reaction rate at $\alpha = 0$. Autocatalytic kinetics are usually evaluated by the derivative form of the autocatalytic rate equation [Eq. (2.86)] with data collected by isothermal method 1 measurements. Activation energy E and preexponential factor A are measured from the Arrhenius equation

$$k = A \exp\left(\frac{-E}{RT}\right) \tag{2.87}$$

where k is the appropriate rate constant, R is the universal gas constant (8.314 J/K·mol), and T is absolute temperature. Note that two activation energies (E_1 and E_2) are measured for autocatalytic cure, corresponding to the two rate constants k_1 and k_2 in Eq. (2.86).

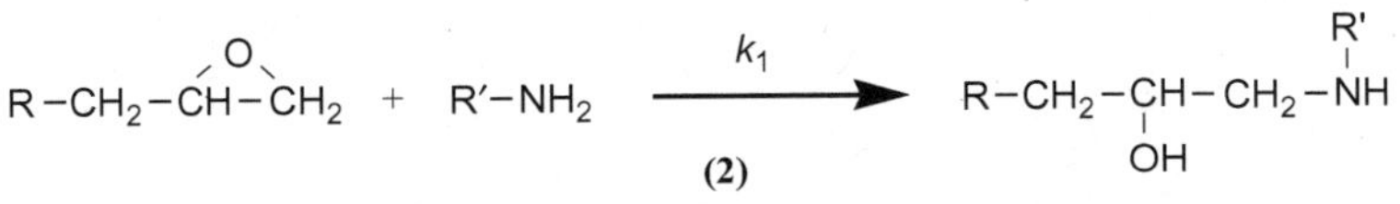

$$R-CH_2-\overset{O}{\overset{\diagup \diagdown}{CH-CH_2}} + R'-NH_2 \quad \xrightarrow{k_1} \quad R-CH_2-\underset{OH}{\overset{R'}{\underset{|}{CH}}}-CH_2-\overset{R'}{\underset{|}{NH}}$$

(2)

Scheme 2.2.

$$R-CH_2-\overset{O}{\overset{\diagup \diagdown}{CH-CH_2}} + R-CH_2-\underset{OH}{\overset{R'}{\underset{|}{CH}}}-CH_2-NH \quad \xrightarrow{k_2} \quad R-CH_2-\underset{OH}{\overset{R'}{\underset{|}{CH}}}-CH_2-N-CH_2-\underset{OH}{\overset{R'}{\underset{|}{CH}}}-CH_2-R$$

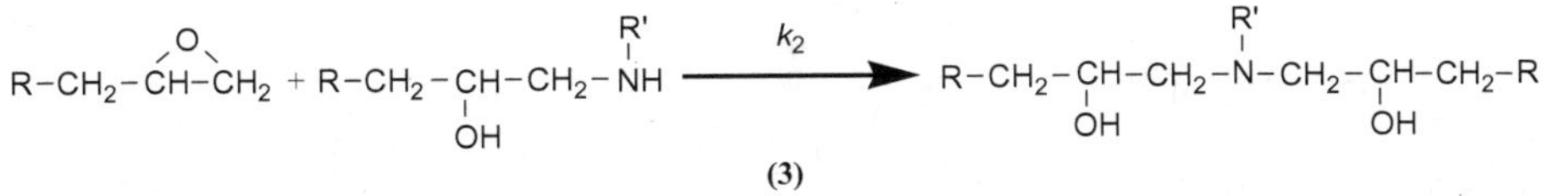

(3)

Scheme 2.3.

2.10.3.2. Time-Temperature Superposition (TTS) Kinetics

For applications of TTS kinetics we consider cure reactions that can reasonably be described by a single or overall activation energy, independent of conversion and temperature. A large number of cure reactions fall into this category, and they typically exhibit a single, symmetric DSC exotherm. This includes epoxy–amine cure, when $k_1 \ll k_2$ as in Eq. (2.86). While the primary amine–epoxy reaction and the secondary amine–epoxy reaction may have different activation energies, they are usually within a few percentage points of each other, and an average or overall activation energy gives a good description of the cure (Wisanrakkit et al. 1990, see Fig. 2.74 for an example). In cases where the activation energies are significantly different, the DSC exotherm will be skewed or there will be two distinct exotherms, in which case a more complex analysis of cure, such as multiple heating rate methods, must be undertaken [see Prime (1997); also Section 3.5]. Activation energies for cure are typically between 40 and 125 kJ/mol. More detailed discussions of kinetics, not only related to cure but also to storage life before cure and aging after cure, can be found in Gotro and Prime (2004), Van Mele et al. (2004), Pascault et al. (2002), and Prime (1981, 1997).

The assumption of a single or overall activation energy means that the only effect of temperature is to speed up or slow down the reaction. As illustrated in Fig. 2.71, when E is constant, conversion–time curves (or T_g–time curves through the T_g–conversion relationship; see Fig. 2.72) will be parallel on a

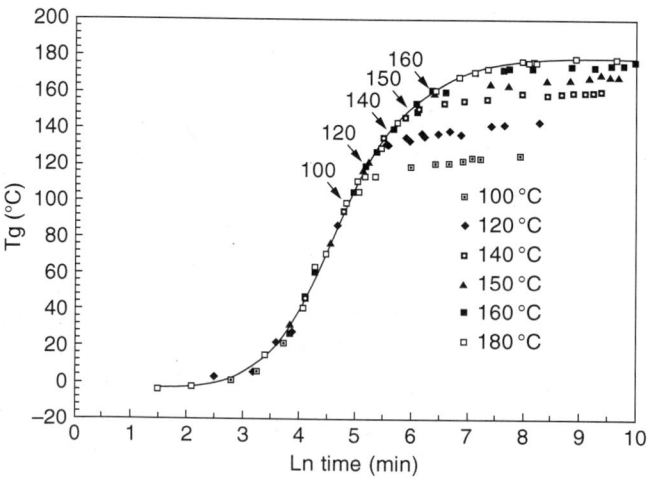

Figure 2.74. Superposition of the T_g versus ln(time) data to form a master curve at 140 °C by shifting each curve in Fig. 2.73 by a constant factor [ln a_T = ln($t_{140°C}$) − ln(t_T)] [see Eq. (2.88)] along the ln(time) axis. Isothermal vitrification points at different cure temperatures are marked by arrows. Note that vitrification points at all cure temperatures lie on the master curve. [From Wisanrakkit and Gillham (1990a); reprinted with permission of John Wiley & Sons, Inc.].

ln(time) plot, allowing construction of master cure or aging curves by shifting the data along the x axis (Neag and Prime 1991; Prime 1997). T_g–conversion–time data were measured from isothermal method 2 experiments (Wisanrakkit and Gillham 1991). The shift factor a_T is the ratio of times t_2/t_1 to reach a fixed level of conversion or fixed T_g at temperatures T_1 and T_2, respectively. Equation (2.88) relates the shift factor to activation energy through the Arrhenius equation. Master curves are useful for succinctly summarizing all of the kinetic data taken at different temperatures and for predicting behavior at times and temperatures that may be of interest but differ from those that were measured. Note the parallel nature of the curves prior to vitrification in Figs. 2.71 and 2.73. Figure 2.74 shows the master curve from shifting the data of Fig. 2.72 along the ln(time) axis so that its beginning section ($T_g < 90\,°C$) coincides with the curve for $T_{cure} = 140\,°C$. Note that this is equivalent to shifting the data at all temperatures to the reference temperature of $140\,°C$ by means of Eq. (2.88) using the measured overall activation energy of 63.5 kJ/mol (Wisanrakkit and Gillham 1991). Figure 2.74 clearly shows the reaction under chemical control (master curve, solid line). Only the chemically controlled reaction, which follows kinetics described by the Arrhenius equation, is subject to time–temperature superposition. Vitrification points, marked with an arrow and the respective cure temperature, can be clearly observed where individual curves deviate from the master curve as the reaction slows and becomes controlled by diffusion of reactants in the glassy state:

$$a_T = \frac{t_2}{t_1} = \exp\left(\frac{E(T_1 - T_2)}{RT_1 T_2}\right) \tag{2.88}$$

Note that in most thermoset cure processes, temperatures are high enough relative to $T_{g\infty}$ to avoid vitrification. A good example is a fast-curing, two-component polyurethane used in pultrusion processes (Prime et al. 2005). Parts are made by mixing the components inline and rapidly processing and curing. In this study conversion–time data were collected from isothermal method 2 measurements. Note that this system was too reactive to use isothermal method 1, where a significant amount of reaction would have gone unrecorded at all temperatures. Isothermal method 2 temperatures and times were chosen to minimize reaction during heating to temperature (arguing for low temperatures) and to avoid vitrification (arguing for high temperatures). The master cure curve shown in Fig. 2.75 was obtained by shifting these data to times at a reference temperature of $80\,°C$ by means of Eq. (2.88), using the activation energy for cure measured from the peak maxima from multiple heating rate DSC measurements [see Prime (1997); see also Section 2.10.3.3, below]. To obtain additional data at high conversion but without the complication of vitrification, heat–hold–cool DSC measurements were made that simulated process cure profiles. Equivalent cure times at the same reference temperature of $80\,°C$ were calculated from the profiles by shifting individual points to times at $80\,°C$ by means of Eq. (2.88) and summing those points

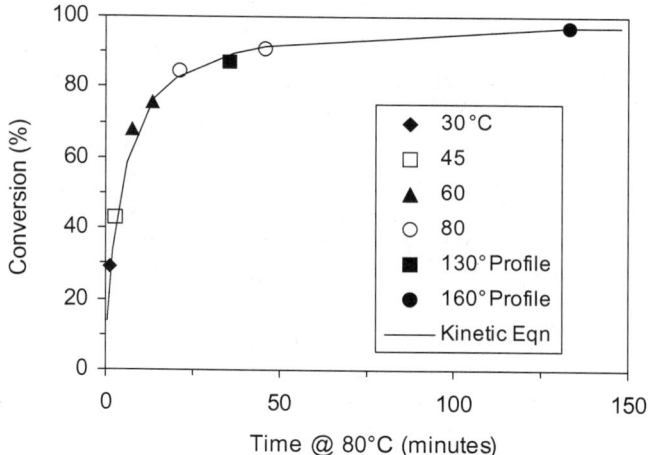

Figure 2.75. Master cure curve for polyurethane pultrusion resin system. Data points from isothermal DSC and DSC simulated cure profiles were shifted to times at 80°C by means of Eq. (2.88) using the activation energy measured from multiple-heating-rate DSC. The solid line is calculated from the second-order kinetic equation found to fit the master curve. [From Prime et al. (2005); reprinted with permission of Elsevier.]

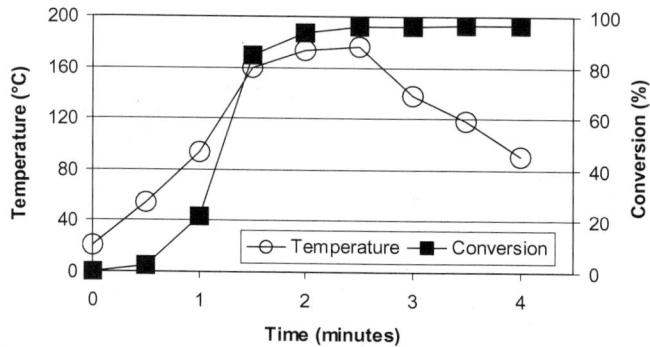

Figure 2.76. Time–temperature profile for pultrusion process and development of conversion along the process [from Prime et al. (2005); reprinted with permission of Elsevier].

(Neag and Prime 1991; Prime 1997). The master curve was found to fit the second-order kinetics equation [Eq. (2.84)], and a complete mathematical description of cure was provided. From this kinetic equation the development of conversion along the profile shown in Fig. 2.76 was determined. A conversion of 97% for the polyurethane resin cured according to this profile was estimated. The kinetic equation for cure also allowed the computation of the master curve shown in Fig. 2.75 as "Kinetic Eqn" (see Fig. 2.75 inset).

2.10.3.3. Multiple Heating Rate Kinetics Multiple heating rate methods are described in some detail in Section 3.5. For DSC they include ASTM E698 (see Appendix A) and what is often referred to as *model-free kinetics* (Vyazovkin 1997, 2001). These methods are isoconversional; that is, they measure the temperature to reach a fixed conversion versus heating rate to calculate activation energy. They are often referred to as *model-free kinetics* because this calculation does not require a chemical reaction model, such as autocatalytic or *n*th-order. All thermal analysis vendors provide ASTM E698 kinetics software. Mettler Toledo and Perkin-Elmer provide model-free kinetics (MFK) software. Figure 2.77 illustrates multiple heating rate data for a typical epoxy thermoset.

The ASTM method contains a simple relationship such as that in Eq. (2.89) between the activation energy E, the heating rate q, and isoconversional temperatures T_i. It is based on the work of Ozawa (1965, 1970), and Flynn and Wall (1966a,b). Activation energy is calculated from the slope of a plot of the natural logarithm of the heating rates (°C/min) versus the reciprocal of the corresponding isoconversional temperatures [in degrees Kelvin (K)]. A more accurate value of E may be obtained by recalculating the constant 1.052 from tables in Doyle (1961). This correction is incorporated in the software provided by the manufacturers. The ASTM method allows activation energy to be calculated by the method of Kissinger (1957) in place of Eq. (2.89). Performing kinetic analyses beyond the estimation of activation energy is not recommended, since they involve the assumption of a first-order reaction, which is almost always incorrect for thermoset cure:

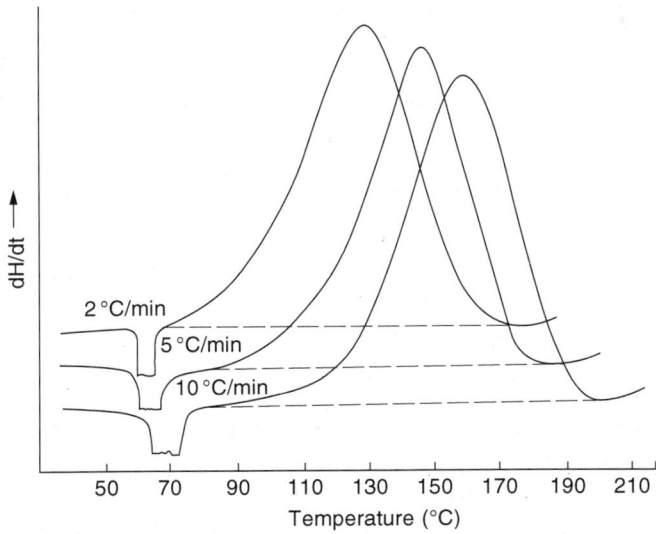

Figure 2.77. Effect of heating rate on the DSC cure exotherms of an epoxy powder paint (epichlorohydrin–bisphenol A type with amine hardener). Exotherm is up. [from Swarin and Wims (1976); reprinted with permission of Springer].

$$E \cong \frac{-R}{1.052} \frac{\Delta \ln q}{\Delta T_i^{-1}} \tag{2.89}$$

This method is often used to estimate activation energy from the variation in peak exotherm temperature with heating rate, from data such as those shown in Fig. 2.77. Note that the peak temperature at the fastest heating rate is well below 200 °C, which should preclude unwanted side reactions or degradation. Several authors have observed that the extent of reaction at the peak exotherm is constant and independent of heating rate for several thermosetting systems (Prime 1997), supporting the assumption that this is an isoconversional temperature. This application is useful for estimating activation energy for reactions where E is constant[1] and for studying thermosetting systems with unresolvable baselines or residual solvent, provided the peak exotherm is unaffected. This method was used by Prime et al. (2005) (see also Figs. 2.75 and 2.76 and related discussion) to estimate the activation energy for cure of a thermosetting polyurethane system.

Utilizing Eq. (2.89), the ASTM method may also be used to analyze the entire reaction exotherm for activation energy versus conversion, from isoconversional temperature–heating rate data at series of conversions such as 10%, 20%, 30%, ..., 80%, 90%. This method is useful to ascertain whether activation energy is constant with conversion and for characterizing thermoset systems where the activation energy is not constant (e.g., due to a change in reaction mechanism with conversion). However, this does requires that a baseline be drawn between a clear beginning and a clear end of the reaction (see Fig. 2.77 as a good example). In general, the fastest heating rate should be chosen so that the reaction is complete below 200 °C, as is the case in Fig. 2.77, to avoid unwanted side reactions or degradation. It is also recommended that the slowest heating rate be chosen so as to avoid vitrification during the DSC experiment, discussed later in this section. Note that, unlike the epoxy in Fig. 2.77, thermosets with a nonconstant activation energy for cure will often exhibit an asymmetric exotherm or multiple exotherms (see footnote 1 above). Barton (1985) used a similar approach to characterize cure of an epoxy with multiple reaction exotherms at four different heating rates. An activation energy of 70 kJ/mol was measured between 10% and 50% conversion, which increased to 128 kJ/mol at 70–90% conversion.

We recommend using the ASTM E698 method to estimate activation energy from the entire reaction exotherm, including the peak exotherm temperature, to estimate the average activation energy and evaluate the constancy of E with conversion. Predictions can be made using only the first-order kinetic model, which thermosets rarely follow. This method is not recommended to predict cure or design cure processes. The model-free kinetics (MFK) methodology, described next, overcomes these shortcomings.

---

[1] If the DSC peak is symmetric and the exotherms are uniformly spaced with heating rate, as in Fig. 2.77, it is reasonable to expect E to be constant to within ±10%. An asymmetric peak, a peak with a shoulder, or two exothermic peaks indicate that E will most likely not be constant.

Vyazovkin (1997, 2001) developed an enhanced isoconversional method that allows evaluation of an effective activation energy (E_α) as a function of the extent of reaction (α). This methodology, often referred to as *model-free kinetics* (MFK), is described in Section 3.5. The MFK software allows calculations such as conversion–time plots at selected temperatures that can be compared with actual measured data. It also allows the calculation of DSC curves that can be compared with actual measured DSC curves to help validate the analyses. The same curves analyzed by ASTM kinetics may be evaluated by MFK kinetics, and the same guidance is given. MFK kinetics is very comprehensive in that it is applicable to the simplest as well as to the most complex cure reactions, provided that a baseline can be drawn between a clear beginning and a clear end of the reaction. But it should be pointed out that the software is provided only by Mettler Toledo and Perkin-Elmer. For other users it is possible to measure E_α versus conversion by the ASTM method and generate a spreadsheet with the appropriate MFK equation [e.g., Eq. (3.31) in Chapter 3], to calculate conversion–time plots.

Model-free kinetics software employs numerical integration methods to measure activation energy versus conversion from cure exotherms at three or more heating rates, or from isothermal data at three or more temperatures. In both cases a minimum of four runs is recommended. Predictions like conversion–time plots and calculated DSC curves are made using Eq. (3.31). An advanced version of MFK software allows analysis of data from arbitrary heating programs, such as combined ramp and isothermal. A drawback of the commercial software is that a discrete mathematical relationship is not produced that can be exported and incorporated into cure models.

Model-free kinetics has been applied to thermoset cure with success (Vyazovkin and Sbirrazzuoli 1999; Sbirrazzuoli and Vyazovkin 2002; Zhang and Vyazovkin 2005). On occasion a decrease in E_α near the end of a reaction has been assigned to vitrification. The reader is reminded that vitrification is not an isoconversional phenomenon that can be characterized by model-free kinetics. It generally does not occur on heating except at slow heating rates, (e.g. <2 °C/min). For this reason the reader should use caution in the selection of the low end of their heating rates to avoid the possibility that vitrification will occur during cure and give erroneous kinetic results. Schawe (2002) demonstrated this in a study of the cure of an epoxy–amine system cured at 1, 2, 5 and 10 (°C/min). When the three fastest rates were used in an MFK analysis a constant activation energy of ~52 kJ/mol was observed. However, when the slowest rate was incorporated, E_α was observed to increase dramatically beyond 80% conversion, which was attributed to vitrification during cure at 1 °C/min.

2.10.3.4. Diffusion Control of the Cure Reaction

In the initial stages of thermoset cure the reaction is chemically controlled and continues under chemical control until some diffusional process becomes rate limiting. The onset of diffusion control occurs when the timescales of diffusion and the chemical reaction are of the same magnitude. During isothermal cure, the time scale for chemical reaction (the reciprocal of the normalized chemical rate constant) remains constant, whereas the time scale of diffusion increases as segmental mobility decreases and the concentration of reacting groups decreases. The same phenomena can occur during nonisothermal cure, where the onset of diffusion control is heating-rate-dependent and generally occurs only at slow heating rates, usually less than 2 °C/min. In many thermosetting systems, the most notable of which are epoxies, diffusion control begins soon after vitrification, as illustrated in Figs. 2.73 and 2.74. In Fig. 2.74 the shifts from chemical control to diffusion control are observed as deviations from the master curve for the chemically controlled reaction for cure temperatures below $T_{g\infty}$. An apparent absence of diffusion control has been observed for systems where the chemical reaction rate was inherently very slow, such as an uncatalyzed dicyanate ester system, and diffusion control has been observed to occur well before vitrification when the chemical reaction rate was very fast, such as a catalyzed cyanate ester system (Simon and Gillham 1993).

Chemical models that incorporate both chemically controlled and diffusion-controlled stages of cure have been described (Chern and Poehlein 1987; Matsuoka et al. 1989; Wisanrakkit and Gillham 1990a,b; Stutz et al. 1993; Deng and Martin 1994). See Prime (1997) for a general overview. The most common approach is to employ the Rabinowitch (1937) model for small-molecule reactions, where the timescale for the overall reaction can be modeled as the sum of the time for diffusion of reactants plus the time for chemical reaction

$$\frac{1}{k_a(\alpha, T)} = \frac{1}{k_T(T)} + \frac{1}{k_d(\alpha, T)} \qquad (2.90)$$

where k_a is the overall rate constant, k_T is the Arrhenius rate constant for the chemical reaction, and k_d is the diffusion rate constant ($1/k_d = \tau$, the average polymer segmental relaxation time). The overall rate constant is governed at one extreme by the Arrhenius rate constant, generally prior to vitrification, and at the other extreme by the diffusion rate constant, usually well after vitrification.

Van Mele et al. (1995) describe a mobility factor that can be measured by MTDSC and approximates the diffusion factor in Eq. (2.90); see Section 2.13 on modulated temperature DSC. Kenny and Trivisano (1991) and Kenny et al. (1992) present a phenomenological approach to the shift from chemical control to diffusion-control of the cure reaction. Schawe (2002) also presents a phenomenological approach that uses model-free kinetics to evaluate the chemically controlled kinetics and introduces a diffusion-controlled function.

2.11. DIFFERENTIAL PHOTOCALORIMETRY (DPC)

The polymerization and/or cross-linking of photoinitiated systems is an important process in many industries, including semiconductor, electronics, information storage, printing and dental. UV- and blue light-curable materials include inks, coatings and adhesives. Advantages of these materials over conventional heat-cured thermosets include short process times and cure at ambient conditions. Note that unless it is filtered out, e.g. by a water filter, the IR component of the UV lamp may heat the sample to above the isothermal cure temperature during DPC cure. By quantitatively measuring enthalpy or heat flow changes in a material during and after exposure to the light, differential photocalorimetry (DPC) extends the application of DSC to measuring and quantifying the complete reaction of photo-cross-linkable materials (see Prime (1997) for a more extensive review of this topic). While this section focuses on thermosetting systems to illustrate the technique, these same methods can be applied to elastomeric systems and those that react to form polymers that do not cross-link.

2.11.1. Background and Basic Principles

This technique is sometime referred to as photo DSC. In a DPC experiment the sample is irradiated with light of a specific wavelength or spectrum of wavelengths, and the DSC signal is recorded after or during the irradiation. DPC runs are almost always conducted in the isothermal mode. Photochemical reactions are usually too fast to consider a dynamic temperature experiment. With materials that cure by cationic photo-polymerization the chain growth and cross-linking reactions that follow irradiation are best studied by standard or modulated temperature DSC. Often a series of isothermal runs are made at several temperatures to obtain kinetic data. Thus, the actual DSC runs are conducted in a manner similar to those used with other reactive materials (e.g., Isothermal Method 1 described in the previous section on Thermosets). A typical example is shown in Fig. 2.78. The unreacted sample is first equilibrated at the desired isothermal temperature. When the shutter is opened the sample and reference are irradiated with light of the desired wavelength, causing a shift in the DSC baseline. [This baseline shift takes place because the sample is heated up more than the reference due to absorption of the UV-light, so T_s-T_r increases (see Section 2.3): an abrupt change in T_s-T_r will manifest itself as a jump in the baseline.] Hence, it is necessary to wait at least 1 minute to establish the non-irradiated baseline before opening the shutter. Also, after the run is completed and the shutter is closed, the operator should wait again for 1 minute in order to record the post-reaction non-irradiated baseline. In this manner it will be easy to observe the baseline shift caused by the irradiation. If the sample has reacted completely in the initial run, a second run will provide a measure of the effect of the illumination on the DSC baseline, and can be used to correct the first run by simply subtracting the second run from the first run. Note that in the first few seconds after the irradiation has begun a small amount of heat is liberated that is only detectable after making the rerun. This correction technique for tracking photochemical reactions is analogous to Widmann's (1975) and Barton's (1983) method proposed for conventional thermosets. The difference curve produced by subtracting run #2 from run #1 in Fig. 2.78 below yields a direct measure of conversion as a function of time by integrating the area

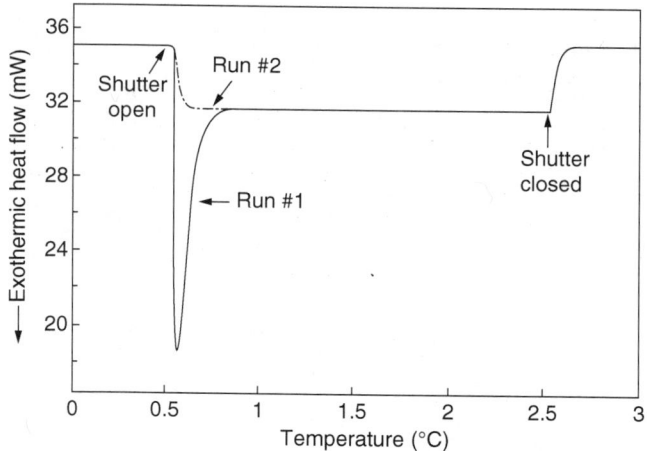

Figure 2.78. Typical example of a DPC run on a UV-curable resin covered with PVDF film (see discussion below) and irradiated for 2 min with a UV intensity of 53 mW/cm². Run 1 = uncured sample, and run 2 = same sample immediately following the first irradiation. Because the sample weight is unchanged and its UV absorption characteristics are similar before and after cure, run 2 may be used as the baseline and subtracted from run 1 to generate the true heat flow–time curve. A Perkin-Elmer DPC accessory with its water filter was used to minimize the IR component of the incoming radiation. [From Bair and Blyler (1985); reprinted with permission of the North American Thermal Analysis Society.]

under the difference curve for any selected time. The rate of conversion may be obtained from the ordinate of the difference curve. Refer to **Isothermal Method 1** in Section 2.10. For kinetic studies it may be necessary to reduce the intensity of the light impinging on the sample, e.g., to 10 mW/cm² or less as suggested by Fig. 80, so that the DSC can accurately follow the progress of the photochemical reaction in 'real time'.

In a kinetic study the conversion-time or conversion-rate of conversion-time data can be fit to kinetic equations. This fitting process will allow the experimenter to determine whether the reaction follows n^{th} order or autocatalytic kinetics and to obtain rate constants (k) and reaction orders (n or m). The activation energy may be obtained from an Arrhenius plot of $\ln k$ versus the reciprocal of absolute temperature (see discussion in **Traditional Isothermal Kinetics** in Section 2.10).

If the reaction is incomplete after the first irradiation, a residual heat of reaction will be detected on the second irradiation. A good practice is to repeat irradiation cycles until two cycles are obtained that are indistinguishable from each other. This same methodology can be used to ascertain the completeness of cure of samples cured in laboratory and production processes, although additional measurement of T_g as an indicator of conversion is recommended.

Commercial photocalorimetry accessories are now offered by several instrument manufacturers with a choice of light sources and optical configura-

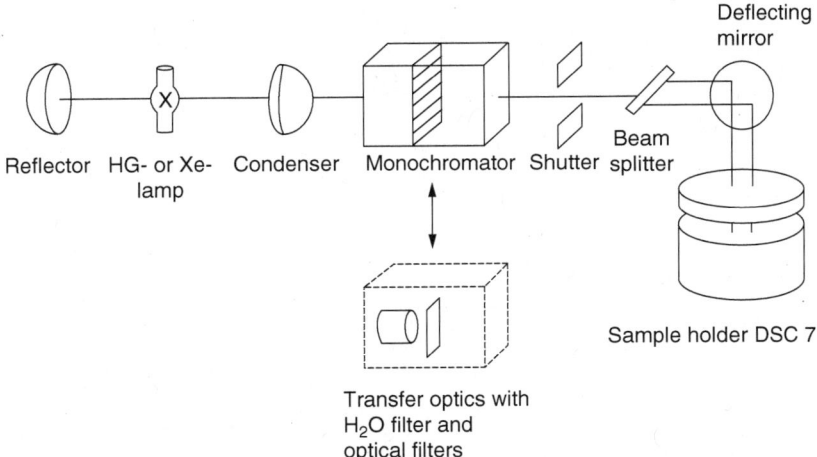

Figure 2.79. Schematic diagram of the Perkin-Elmer DPA7 double-beam photocalorimeter accessory used with a DSC7 [from Fischer et al. (1988a,b); courtesy of Perkin-Elmer].

tions that allow control of exposure time, lamp intensity and spectral distribution (see Fig. 2.79). Although the DPC experiment can be conducted in static air to simulate manufacturing processes, it is recommended that the cell be purged with a high purity inert gas like nitrogen, argon or helium, since oxygen is a well-known scavenger for free radicals formed in the UV or visible light irradiation. At the low intensities typical of differential photocalorimetry, oxygen inhibition can have a significant effect on the reaction. Copper tubing purge lines to the DSC are recommended to minimize oxygen in the purge gas; it is reported that oxygen levels can be reduced to less than 2 ppm (Kloosterboer and co-workers, 1984). Decker (1992) commented on some of the limitations of photocalorimetry, especially when comparison is made with practical systems that are usually cured under intense UV light in the presence of air. It is important to match the speed of the photo-initiated reaction to the response time of the DSC. This constraint requires the use of UV irradiation 2—3 orders of magnitude less intense than that used in manufacturing processes. Note that oxygen inhibition becomes more pronounced with low-intensity radiation. If the photon flux absorbed by the sample becomes too large, a saturation effect occurs and the DSC may lose the ability to distinguish differences such as photoinitiator concentration. Other limitations listed by Decker are lack of precise values for the heat of the photochemical reaction; the baseline shift often observed after UV curing; and poor wetting of aluminum pans that often leads to poor control of film thickness. A theoretical reaction enthalpy of 86.2 kJ/mol of double bonds for diacrylate monomers (Kurdikar and Peppas, 1994) and 86.6 kJ/mol of epoxy for cationically cured epoxies (Hale and co-workers, 1989) may be useful if the chemical composition of the system is known.

Another serious experimental difficulty that occurs in dealing with free radical and cationic polymerizations in a DPC experiment is the loss of volatiles. Most of these reactive systems consist of blends of monomers and oligomers and often these components are sufficiently volatile even at room temperature to cause the DSC baseline to drift during an isothermal run. One way to diminish the evolution of volatile species and thereby eliminate baseline drift, is to place a thin transparent, circular material such as a quartz disk or polyvinylidene fluoride (PVDF) film that matches the inner diameter of the sample pan on top of the liquid sample (Bair and Blyler, 1985). The transparent cover will cause the liquid drop to spread out and form into a thin film of uniform thickness. In addition, if the sample mass is controlled, e.g. to 0.70 ± 0.05 mg, films of nearly uniform thickness of ~20 µm will be produced when the transparent plastic film is placed on top of the liquid sample drop (Olsson et al., 2003). Sub-milligram size samples can normally be used in differential photocalorimetry since during cure most reactive systems liberate between 200 to 600 J/g of heat over a relatively short time frame, sufficient to be detected accurately by commercial DSC instruments. Lastly, it is critically important to produce thin, uniformly thick samples for DPC work because as light is absorbed from the top to the bottom of the sample, less light penetrates through the sample. Thus, the rate of the photochemical reaction on the topside of the sample is faster than that at the bottom of the sample. Only by using very thin samples will this effect be minimized.

In practice, one can develop a technique of placing a minute liquid drop on the tip of a syringe or wooden toothpick and touching this sample tool against the bottom center of the sample pan. If the sample pan's mass has been tared previously in an electronic microbalance one can readily ascertain the mass of the liquid drop by reweighing the sample pan. In this manner one can simply add or subtract mass from the sample using a "wet" or dry probe, respectively, until the sample mass falls within the accepted range. It is important to keep the mass close to this small, suggested value so that surface tension keeps the liquid from running out to the outer edge of the sample pan and up the walls of the pan when the transparent lid is added. Another benefit of sub-milligram size samples is that they reduce the total exothermic heat released during the curing reaction and hence, help the DSC maintain its isothermal state.

A major objective of differential photocalorimetry is to balance the light source so that light strikes the sample and reference cells of the DSC in equal amounts. Usually this is accomplished in either commercial or homemade set-ups by cutting holes in the top of the upper cover that are directly above the sample and reference cells. These openings are typically filled in with transparent quartz windows. Then, light can be directed through these quartz windows using a variety of UV sources (Bair and Blyler, 1985; Bair and co-workers, 1993).

An easy way to balance the intensity of the light beam is to use a carbon disk as both sample and reference and irradiate both cells simultaneously. The carbon disk is skipped in the TA Instruments Q-series DSC, where intensity is balanced by shining the light directly on the sample and reference pedestals. Some DPC accessories have built-in optics that enable one to adjust the relative intensity of light falling on either cell. In addition, the height of the light guide can be adjusted in several systems, thus changing the intensity at the level of the DSC pan. If the latter options are not available one can use neutral density filters to make intensity adjustments. Ideally the net output from the DSC should be zero milliwatts, the result being that no change is detected on the DSC ordinate when the light is switched from off to on. There may be an additional baseline shift during the measurements if the instrument is unable to maintain a constant temperature due to the released heat. Once the sample and reference beams are balanced one can use the same carbon disks to determine the intensity of light falling on the sample chamber. This procedure includes monitoring the DSC output while blocking the light striking the reference cell. When light hits the black disk in the sample cell it is converted to heat. This event is recorded in the DSC as a constant exothermic response as long as the light is on, when the shutter is closed the signal returns to its original value (this behavior is similar to that shown in Run #2 in Fig. 2.78 for a cured resin). Dividing the mW baseline shift by the cross-sectional area of the carbon disk gives a measure of light intensity in mW/cm^2. Alternatively, one can place a light meter just above the sample cell and directly measure the beam's intensity. With the use of a spectrometer the beam's intensity as a function of wavelength can be determined.

Although the technique of placing a thin plastic film over the sample diminishes the loss of volatiles in DPC experiments to an acceptable level during isothermal work at or near room-temperature, as isothermal reaction temperatures are increased typically to 60°C or above a hermetic capsule with a UV transparent lid is recommended to avert baseline drift (Bair and co-workers, 2003).

2.11.2. UV Cure of Free Radical Systems

Unfortunately, DPC instruments do not have sufficient response time to follow the very fast rate of free radical photochemical reactions encountered in rapid processing environments. As shown in Fig. 2.80 the time to 90% cure of a UV curable resin at a UV intensity of $400\,mW/cm^2$ is about one second at room temperature. This fast kinetic measurement was carried out by a technique known as real time infrared spectroscopy (RT-IR). By this method one follows quantitatively the development of cure by tracking the disappearance of unsaturation as a decrease in the area of an absorbance band that was initially associated with the uncured resin (Decker and Moussa, 1988). The data in Fig. 2.80 indicates that a power compensation DSC begins to give accurate conversion data when UV intensities are of the order of $10\,mW/cm^2$ or less for this resin. Importantly, the data also show that the free radical polymerization of this UV curable resin can be fitted to a linear log-log plot of time versus intensity over a range of intensities of nearly five orders of magnitude. It is

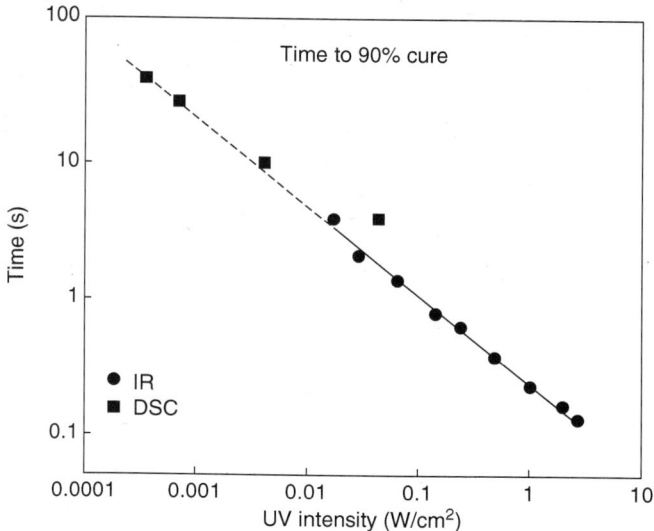

Figure 2.80. Time for a UV-curable resin to reach 90% conversion plotted against UV intensity as measured by RT-IR and DPC on a UV-curable resin at 23 °C. Standard Al DSC pans with PVDF lids were used with nitrogen purge gas. (A. L. Harris and H. E. Bair, unpublished work, 1994.)

also clear that reaction kinetic data collected by DSC at relatively low UV intensities can be used to estimate this resin's behavior under irradiation conditions close to those used in commercial applications (>100 mW/cm^2). Some improvement in DPC response time can be gained by substituting helium for nitrogen as the purge gas and replacing standard, 0.10 mm thick Al pans with thinner ones.

A major difference between the kinetics of step-reaction thermosets and photo-initiated free radical polymerization is their temperature dependence. As described in Section 2.10, increasing temperatures accelerate the cure of thermosets such as epoxies and urethanes. However, this situation does not hold for photo-initiated free radical systems. In such systems the rate of photo-initiation is relatively independent of temperature since the activation energy for the reaction comes from the absorption of light and not thermal activity. In the UV-initiated process the temperature dependence of the reaction rate is directly tied to activation energies for propagation and termination. Hence, if the termination process dominates over the propagation step, as is often the case at elevated temperatures, the overall reaction rate will slow with increasing temperature (Hale, 1992). In addition to the reaction rate, the cure temperature can also influence the mechanical properties of the system. It has been found that the elastic modulus of fully cured rubbery urethane acrylates is lower for samples cured at high temperatures than for samples of the same material reacted at lower temperatures (Levin and co-workers, 1995).

2.11.3. Cationic Photo-Polymerization

It was not until the invention of iodonium and sulfonium salts as photo-initiators by Crivello (1975) that cationic photo-polymerization became practical (see Crivello et al., 1977, 1990, 2000). Upon irradiation of these Crivello salts, acids are generated. Another significant difference between free radical and cationic polymerizations is the latter process is a "living" polymerization—once the acid species is formed, it remains active even after the irradiation is stopped. In contrast to this behavior free radicals "die" soon after irradiation is stopped. Also, unlike free radical polymerizations cationic reactions are not inhibited by oxygen. Quite often the dark reaction following irradiation can play an important role in enabling a cationic system to develop its full properties and this leads manufacturers of commercial cationic photopolymers to often recommend a thermal (dark) postcure after carrying out the photo-irradiation process.

In cationically cured systems, one usually finds that the reaction rate increases with increasing reaction temperature. In many cases elevated temperatures are needed to advance this type of reaction to near completion in a short time. Olsson and co-workers (2002, 2003) found that in the vicinity of room temperature in the DPC cationic polymerization of a cycloaliphatic epoxy only proceeds to a limited degree during the first few minutes of the irradiation period. For example in a 30 °C DPC measurement the cycloaliphatic epoxy showed what appeared to be a full symmetrical, exothermic peak after 60 seconds of exposure; however, integration of this peak yielded a heat of reaction of 46 J/g, equivalent to only 7% of the full heat of reaction expected for this epoxy resin. T_g of a sample measured fifteen minutes after irradiation at room temperature was only −20 °C, well below the reaction temperature. The result indicates that diffusion control due to vitrification is not the reason for this unusually low level of conversion after several minutes of curing at or near ambient. However, with a significant increase in time at room temperature the dark reaction causes the glass transition temperature and the modulus to increase significantly although the rate of heat being liberated is not sufficient to be tracked quantitatively. For example, in the case of this cycloaliphatic epoxy, there is a five-fold increase in storage modulus after aging in the dark for about 1¾ hours at room temperature. After about 4 hours this system vitrifies, indicating that T_g has increased to around room temperature or slightly above. As cure advances due to the dark reaction the same problems encountered with all partially cured thermosets arise. The glass transition, possibly with enthalpy relaxation, may overlap with the residual cure exotherm, making it difficult to accurately measure T_g and conversion. When traditional DSC fails modulated temperature DSC, described in Section 13 (see Fig. 22), can separate these processes and allow the analyses to be accomplished (Olsson, et al., 2002, 2003). Photo-polymerization can lead quickly to high levels of conversion if the irradiation is conducted at higher temperatures. In the case of the cycloaliphatic epoxy, irradiation curing for a few minutes at 100 °C in a hermetically sealed capsule released 560 J/g, which corresponds to a conversion of 84% as compared to the 7% found by reacting the material at 30 °C. This example shows the importance of understanding the possible role of polymerization temperature on the overall extent of conversion and how differential photocalorimetry experiments can be utilized to gain this kind of information.

In summary, differential photocalorimetry experiments similar to that illustrated in Fig. 2.78 are recommended to measure kinetics and compare different UV curable materials, and determine differences in composition such as photoinitiator type or level, or process variables such as irradiation wavelength or temperature. DPC experiments can contribute to the characterization and design of laboratory and manufacturing processes. As an example a series of samples may be cured in DSC pans as a function of process variables such as intensity and exposure time. It is important that lamp intensity at the sample be well calibrated. It is also advised that the increase in sample temperature due to irradiation be measured, as it can be considerable. Residual cure can be measured by a DPC cycle similar to that shown in Fig. 2.78 for free radical initiated processes. It is recommended to conduct multiple UV exposure cycles until the area under two successive cycles cannot be distinguished. Thermal scans to measure T_g are also recommended, as T_g may be a more sensitive measure of conversion, particularly in the latter stages of cure. For some systems, such as dual cure adhesives that cure by both thermal and UV processes or cationically cured epoxies following UV exposure, a thermal scan may yield both T_g and the residual heat of reaction. Lastly, it is often important to determine the modulus of a photo-reacted sample as a function of conversion and reaction temperature for this mechanical property may have a desired range in a specific engineering application.

2.12. FAST-SCAN DSC

Polymers are metastable materials whose properties, such as crystallinity, melting point, or glass transition temperature, can change during the course of an experiment. An example of this propensity for change is highlighted in past efforts to determine the equilibrium melting temperature (T_m°) of polyethylene (PE). Early attempts to measure T_m° of PE by plotting the observed melting points of crystals against known reciprocal lamellar thicknesses and extrapolating the data to infinite thickness failed, because the crystals partially melted and reorganized into thicker crystals while being scanned at a rate of 10 °C/min, and a multiple melting peak DSC–curve was obtained. Faster heating rates could reduce the amount of reorganization during the melting experiment, but not eliminate the multiple melting endotherms. In order to circumvent this experimental difficulty, crystals were irradiated to link together fold surfaces on adjacent crystals. The irradiation immobilized the amorphous regions, therefore slowed down crystal thickening and allowed T_m° to be determined from DSC melting data collected at rates as slow as 10 °C/min (Bair et al. 1968). Hence, it would be useful to heat a crystalline polymer through its melting point fast enough to avoid or minimize reorganization. Also, it is important to be able to mimic cooling conditions that a polymer may experience during manufacturing. Unfortunately, the fastest heating and cooling rates currently available in a commercial DSC are not sufficient to completely stop reorganization or enable one to cool thermoplastics at rates comparable to those typically found in injection-molding processes (>1000 °C/min). However, rates of several hundred degrees per minute are possible and sufficient, in some cases, to suppress crystallization during cooling, or to prevent cold crystallization on heating after devitrification of an amorphous polymer. As significant progress has occurred since the late 1990s, some of the techniques required to collect good thermal data at high rates are reviewed here (Pijpers et al. 2002).

Most conventional DSCs use rates falling in a range of 1–40 °C/min. Fast DSC rates with commercial heat flux instruments with furnaces that range in mass from 30 to 200 g rarely exceed 100 °C/min. Power compensation DSCs have small furnaces weighing only about one gram that allow them to scan at rates of ≤500 °C/min or nearly 4 times faster than the fastest commercial heat flux device. This fast-scan, power compensation method is referred to as HyperDSC™ by the Perkin-Elmer Corporation. More recent developments in thin-film (chip) calorimetry allow scan rates as high as 160,000 °C/min (Minakov et al. 2004; van Herwaarden 2005; Adamovsky et al. 2003; Adamovsky and Schick 2004). However, the instruments described in these studies are not currently available commercially and hence not considered in this discussion. At the 35th North American Thermal Analysis Society (NATAS) Conference in East Lansing, Michigan, TA Instruments announced creation of a new fast-scan heat flux DSC capable of heating rates of ≤2000 °C/min and cooling rates of ≤1000 °C/min operating with nanogram quantities of samples. This instrument is not commercialized yet, but has been distributed to a limited number of thermal analysts with the aim of giving feedback to TA Instruments to correct possible design flaws (TA Instruments 2007).

The effect of heating rate on the ordinate displacement of a DSC signal during melting as a function of heating rate is shown in Fig. 2.81 (Bair, 1994). In this case a thin flat, circular (dia. ~4-mm) indium sample weighing about 4 mg was heated at 0.15, 1.5, 15 and 150 °C/min in a standard aluminum pan whose wall thickness measured 0.10 mm. The temperature scale of the DSC was calibrated at 20 °C/min. Under these conditions the height of indium's melting peak is maximized at 150 °C/min, but unfortunately the sample's melting breath is about 25 °C at this heating rate. At the same time, at a rate of heating of 150 °C/min the metal standard's fusion peak is nearly as tall as found for the highest scan rate and its melting breath is only 2 °C. However, if one improves the flow of heat from the sensor to the indium sample by reducing the wall thickness of the aluminum pan and the mass of the sample (or its thickness), a significant increase in ordinate displacement can be gained at rates as fast as 300 °C/min or higher without a significant change in shape

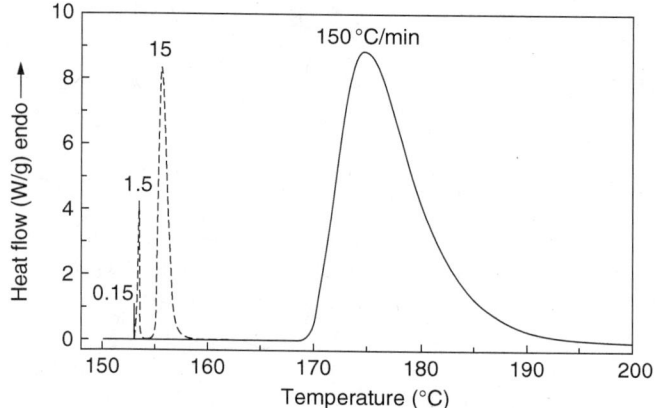

Figure 2.81. Effect of heating rate on a DSC ordinate during the melting of indium. Purge gas: nitrogen. Perkin-Elmer DSC7. The instrument was calibrated at a rate of heating of 20 °C/min [Bair 1994; reprinted with permission of ASTM International]

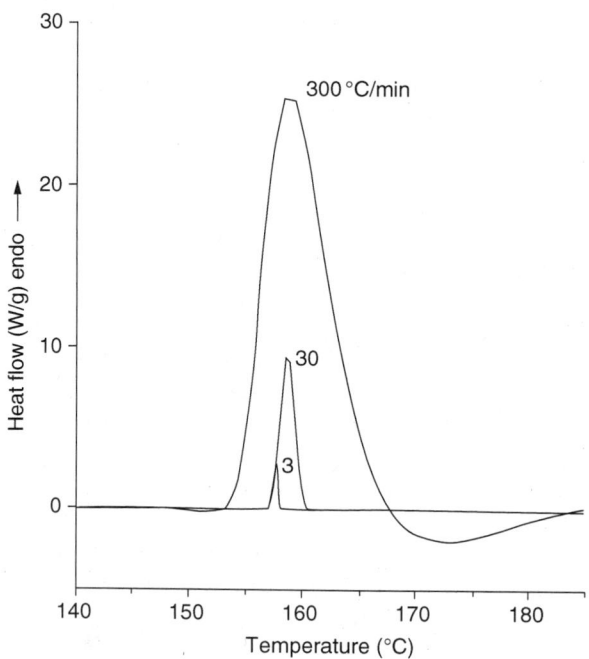

Figure 2.82. Effect of heating rate on indium melting peak height in nitrogen as the thickness of the aluminum foil used to encapsulate the sample was decreased from 0.10 to 0.06 to 0.01 mm with increasing heating rate. Perkin-Elmer DSC7 (Bair, unpublished data).

of indium's melting curve (see Fig. 2.82). In addition, it has been suggested that if the scanning rate is increased by a factor x, then the sample weight should be reduced by this same x factor (Pijpers et al. 2002, see also the end of Section 2.3.1 for the relationship of maximum sample mass and heating rate). Hence, a 10 mg sample scanned satisfactorily at 10 °C/min should have its mass reduced to 1 mg if scanned at 100 °C/min.

In Fig. 2.82 the ordinate maximum ranges from 2.8 to 9.5 to 25.6 W/g as the heating rate is increased from 3 to 300 °C/min, respectively. Note that the data collection rate was not rapid enough to yield a sharp peak at 30 and 300 °C/min. The weight of indium was 0.334, 1.046 and 9.041 mg for rates of 300, 30 and 3 °C/min, respectively. Indium was pounded and flattened to create as thin a sample as possible. In addition, the slope of the leading edge of each melting curve is similar, indicating good heat transfer is occurring in each melting run. Next, the melting of a polymer with its thermal conductivity three orders of magnitude less than indium will be examined not only for transition temperatures but also for curve shapes and quantities of heat.

The accepted thermal properties of semicrystalline and amorphous polycarbonate (PC) were reported in the 1960s (O'Reilly et al. 1963). This work was carried out in an adiabatic calorimeter that ran in an intermittent fashion at a rate of 0.2 °C/min or slower. Polycarbonate as polymerized was partially crystalline, and on first heating the melting process started at 180 °C and ended at about 243 °C. As usual with semicrystalline polymers, the glass transition was difficult to detect. The apparent heat of fusion equaled 23.0 J/g. On cooling below room temperature and reheating the sample, a glass transition was observed at 142 °C with ΔC_p equal to 0.25 J/(g·°C). During the second run the quenched polycarbonate showed no sign of crystallinity. A 1.164-mg sample of this originally powdered, semicrystalline polycarbonate and now aged for 40 years at room temperature was wrapped in a single sheet of 0.013-mm-thick aluminum foil and then placed on a smooth metal surface and tapped gently with a flat punch to smooth the foil's surface. This specially prepared sample was then placed in a Perkin-Elmer Pyris-1 DSC and heated from −70 °C to 265 °C at a rate of 200 °C/min (Fig. 2.83). On first heating a small glass transition occurred at 153 °C with ΔC_p equal to 0.11 J/(g·°C). Note that in the old, slow adiabatic run the glass transition was difficult to locate but in the HyperDSC scan it is relatively easy to observe. Above T_g, melting starts near 185 °C and terminates at about 235 °C with an apparent heat of fusion of 25.2 J/g. Note that the endotherm appears as two overlapping peaks. A subsequent run of the same aged sample at 20 °C/min (not shown in Fig. 2.83) exhibits the same two overlapping endotherms in about the same proportion; hence, the data indicate that reorganization is not occurring. When the quenched, amorphous polycarbonate was rerun at 200 °C/min, T_g was observed at 154 °C with ΔC_p equal to 0.25 J/(g·°C).

The quantitative data, curve shapes, and breadth of melting at 200 °C/min are in satisfactory agreement with the thermal properties established earlier by the slow adiabatic calorimetry (AC) measurements. The exception is the

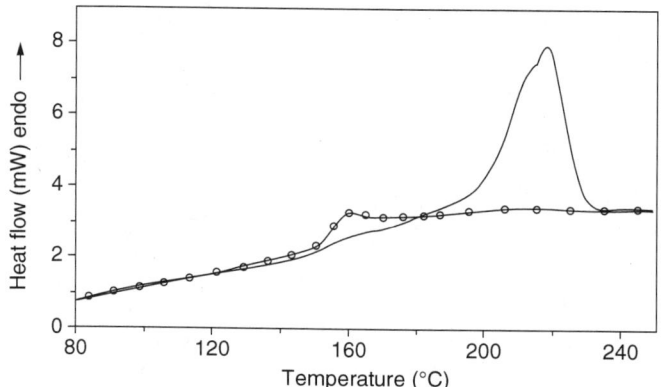

Figure 2.83. Pyris-1 DSC scan of Lexan® polycarbonate recorded from 80 to 265 °C at 200 °C/min heating with a nitrogen gas purge (first heating—solid line, second heating—circles; Bair, unpublished results).

glass transition temperature that is shifted 12 °C higher in the DSC scan when compared to the adiabatic calorimetry data. However, the DSC heating rate is about three orders of magnitude greater than in the AC run (with a 1000-fold rate increase, a T_g increase of this magnitude is expected).

The improvement in sensitivity with fast scans on a small polymeric sample is demonstrated with a DSC run on an amorphous, one-milligram polycarbonate sample shown in Fig. 2.84. In these experiments the purge gas was nitrogen, and the heating started at −80 °C. With an increase in heating rate from 20 °C/min to 400 °C/min, the glass transition temperature increases from 152 °C to 155 °C while the step increase in C_p at T_g scales with the rate increase, except at the slowest rate, where the potential for error is greatest. Note that each heating scan has a small enthalpy relaxation peak associated with T_g. In calculating the step increase in C_p, this endothermic peak must be eliminated by extrapolating the flat part of the curve above T_g backward until it intersects the scan near T_g. The step height increases were 0.04, 0.46, 0.86, and 1.69 W/g for heating rates 20, 100, 200, and 400 °C/min, respectively.

The data in Fig. 2.84 indicate that an amorphous polycarbonate sample as small as 50 μg could be heated at 400 °C/min and yield a glass transition that is as discernible as that shown in Fig. 2.84 for the 1-mg sample of polycarbonate scanned at 20 °C/min. These results indicate that it is now possible to determine thermal properties on polymeric samples that were previously deemed too small to analyze.

Pijpers et al. (2002) have shown the utility of using fast scan rates to prevent crystallization of poly(ethylene terephthalate) (PET). In Fig. 2.85 during cooling at 100 °C/min, no crystallization occurs, only vitrification. On reheating PET at 10 °C/min, devitrification is observed, followed by extensive cold crystallization near 150 °C. This is followed by melting at 230–260 °C. However, on the first heating of PET at 100 °C/min, cold crystallization is nearly eliminated and the subsequent melting is quite small.

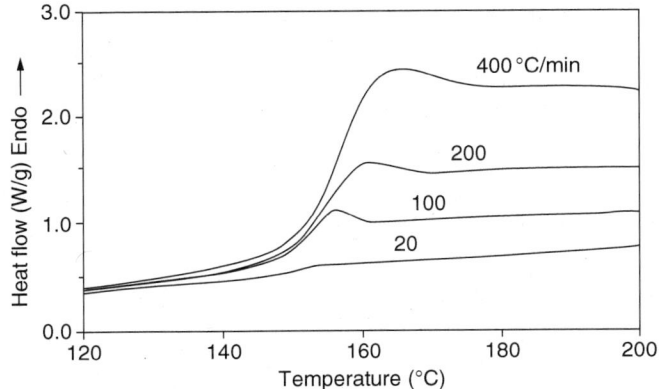

Figure 2.84. DSC scans of amorphous, 1 mg polycarbonate heated at 20, 100, 200, and 400 °C/min. Nitrogen purge gas, Perkin-Elmer Pyris-1 DSC (Bair, unpublished results).

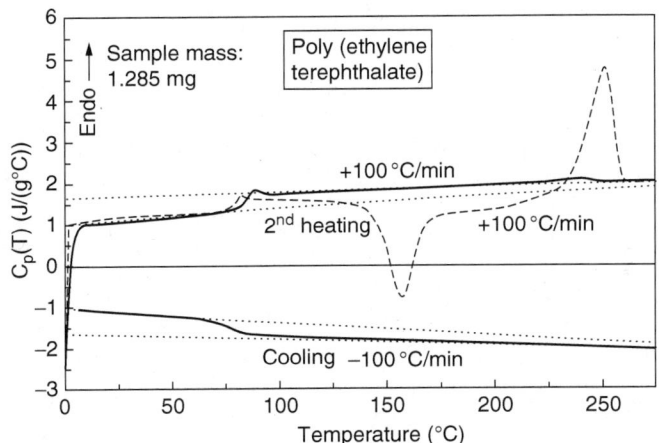

Figure 2.85. DSC cooling scan of PET at a rate of 100 °C/min followed by reheating at 100 °C/min, then cooling again at 100 °C/min and finally reheating at 10 °C/min. A modified Perkin-Elmer Pyris-1 DSC was purged with a purge gas mixture containing 10% helium and 90% neon. [Reprinted from Pijpers et al. (2002) with permission of the American Chemical Society.]

In conclusion, it can be said that if the sample size is reduced and a single sheet of thin aluminum foil is used to encapsulate the sample, it is possible to obtain good-quality data in a power compensation DSC at rates up to 500 °C/min.

High scan rates obviously enable measurement times to be reduced dramatically. With these fast DSC rates the thermal behavior of polymeric materials can be studied quantitatively as well as qualitatively. With fast scan rates comes greater sensitivity, which enables one to study microgram quantities of material. These capabilities hold promise for areas such as biochemistry and polymer physics as well as the fields of materials science and pharmaceutical science.

2.13. MODULATED TEMPERATURE DIFFERENTIAL SCANNING CALORIMETRY (MTDSC)

2.13.1. Background and Basic Principles

Modulated temperature differential scanning calorimetry (MTDSC) is a family of DSC techniques in which a temperature modulation is overlaid on a linear heating or cooling rate (also called the *underlying rate*) (see Figs. 2.86–2.88). Modulated temperature DSC (MTDSC) was introduced by Reading and coworkers (Seferis et al. 1992; Gill et al. 1993; Reading et al. 1993a,b; Reading 1993) and is generally accepted as a useful addition to conventional DSC. The family of MTDSC techniques includes modulated DSC (MDSC®) offered by TA Instruments, and two techniques not currently sold on the United States market: alternating DSC (ADSC) by Mettler Toledo, and oscillating DSC (ODSC) by Seiko Instruments. These techniques are based on a periodic modulation. In addition to ADSC, another type of MTDSC is offered by Mettler Toledo. This technique is called TOPEM®, which is based on a quasi-stochastic modulation. Finally, a third technique that could be considered an MTDSC technique is step scan DSC (SSDSC) from Perkin-Elmer, in which a pulsed temperature modulation is used over a very small incremental step

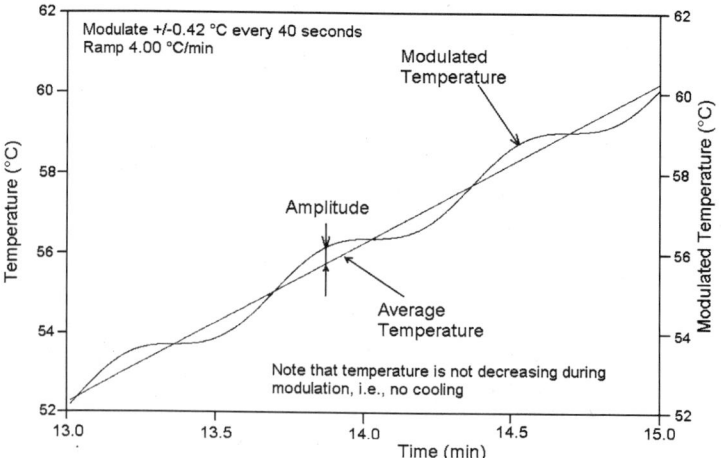

Figure 2.86. Modulation of the temperature around an average (or underlying) heating rate for a heat-only modulation [TA Instruments (2005); courtesy of TA Instruments].

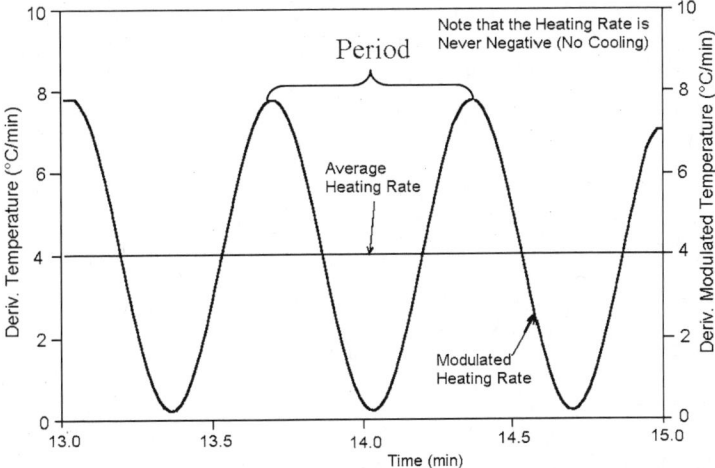

Figure 2.87. Determination of the average heating rate for a given modulated heating rate with its associated period in heat-only modulation mode [TA Instruments (2005); courtesy of TA Instruments].

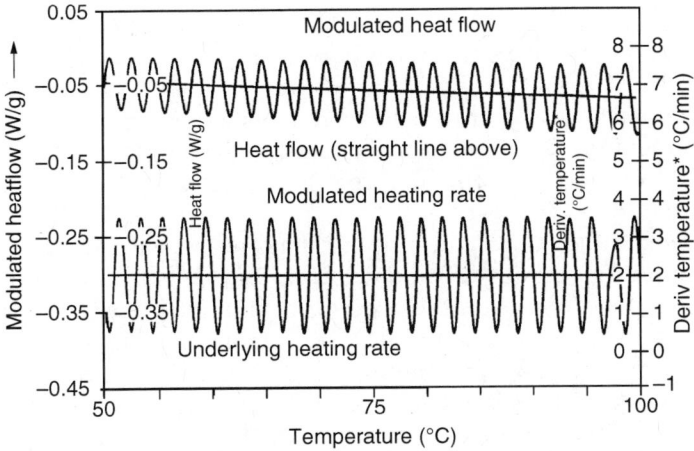

Figure 2.88. Graphical determination of the total heat flow from the modulated heat flow and the underlying heating rate from the modulated heating rate [Judovits (1997); reprinted with permission from the North American Thermal Analysis Society].

heating followed by an isothermal hold (for further discussion, see Section 2.13.12.2). MTDSC is also referred to as *temperature-modulated differential scanning calorimetry* (TMDSC), with or without the hyphen, although MTDSC is recognized as the correct nomenclature [see ASTM E473 (ASTM 2006b)].

As mentioned above, an MTDSC experiment usually involves the application of a sinusoidal perturbation to the linear heating program of a conventional DSC. This is combined with a mathematical procedure designed to deconvolute (i.e., disentangle) different types of contributions to the heat flow. The deconvolution procedure can most easily be understood in terms of a simple equation (Reading et al. 1992, 1994):

$$\frac{dQ}{dt} = C_p \frac{dT}{dt} + f(t,T) \tag{2.91}$$

where dQ/dt is the heat flow rate, C_p is the heat capacity, T is the temperature, t is the time, and $f(t,T)$ equals some function of time and temperature that governs the response associated with the physical or chemical transformation. The first term on the right-hand side (RHS) represents the heat flow associated with the sample's heat capacity. When a thermal event such as a chemical reaction is in progress, the enthalpy associated with the reaction will also contribute to the heat flow, and this is represented by the second term on the RHS of Eq. (2.91).

An MTDSC experiment can not only generate the total heat flow similar to the heat flow obtained in conventional DSC but also separate the total heat flow into its reversing and nonreversing components. The total heat flow is the sum of the thermal events and is generally equivalent to the heat flow seen in conventional DSC. The reversing heat flow is the heat capacity component (plus other terms in some cases; see text below) of the total heat flow [$C_p\, dT/dt$ as noted in Eq. (2.91)].

As explained above, the reversing heat flow derives from the heat capacity of the sample; in addition, some transitions contribute to the reversing heat flow. These are thermal events that respond directly to changes in the ramp rate, and the events are reversing at the time and temperature at which they are observed. In other words, transitions that are fast enough to be reversing on the timescale of the modulation will contribute to this signal. An example of this is the glass transition. Those events that do not respond to changes in the ramp rate will be observed in the nonreversing heat flow. Typical thermal events that are observed in the nonreversing signal are the enthalpy relaxation (hysteresis peak) at the glass transition, cold crystallization, evaporation, and chemical reactions including decomposition and cure. One should not confuse a thermodynamic reversible process with a reversing process of MTDSC. For example, vaporization is a reversible phase transition, but it appears in the nonreversing heat flow signal since the loss of mass results in a nonreversing event.

2.13.2. Advantages of MTDSC

A major advantage of MTDSC is its ability to resolve overlapping transitions that occur in many polymers, for example, if one of these transitions separates into the reversing signal and the other one, into the nonreversing signal. In particular, the glass transition may coincide with a number of processes, making it difficult to analyze. These processes may include an enthalpy relaxation or hysteresis peak, a curing exotherm, or even crystallization on heating (cold crystallization; see Section 2.7).

Another benefit of MTDSC is the increased sensitivity observed in the reversing heat flow because of decreased noise level. This is in part due to smoothing, better baseline linearity, and separation of some of the noise separated into the nonreversing signal.

The resolution is also increased by modulation. Two factors are involved here: (1) the separation allowed when one thermal effect is reversing and the other is nonreversing and (2) the separation of transitions due to the use of the constantly changing heating rate incorporated in the temperature modulation. An increased resolution of separating overlapping glass transitions may even be achieved by using a temperature modulation that has a cooling component (i.e., when the temperature oscillation has some cooling associated with it).

The most frequently used MTDSC technique is modulated DSC (MDSC) from TA Instruments. In MDSC a sinusoidal temperature modulation is superimposed on the underlying linear ramp resulting in cyclic heat flow. In Figs. 2.86 and 2.87, one can note the temperature oscillation around the average temperature without a cooling segment, specifically, only an increase and decrease in the ramp rate (TA Instruments 2005). Such a modulation permits determination of the total heat flow together with its reversing and nonreversing components (Reading et al. 1993a).

Unlike conventional DSC where the heat flow is directly related to the temperature difference between the sample and reference signals, the MDSC operational signals are calculated from the modulated signals. The total heat flow is calculated as the average value of the modulated heat flow signal.

The reversing heat flow is calculated from the reversing heat capacity, which is the heat capacity of the sample in the absence of any transitions or reactions. However, within a transition and a reaction we should are dealing with an apparent heat capacity. The heat capacity of a sample can be calculated from both the total heat flow and the amplitude of the modulated heat flow and temperature. In conventional DSC, the heat capacity can be calculated using the following equation:

$$C_p = K \times \frac{\text{heat flow (sample)} - \text{heat flow (empty pan)}}{\text{ramp rate}} \qquad (2.92)$$

where K ($C_{p,rev}$) is the calibration constant for the reversing heat capacity, which if not adjusted would be the DSC calibration constant. The DSC calibration constant itself can be applied to the heat flow before modulation.

$$C_{p,\text{rev}} = K \times \frac{\text{heat flow amplitude}}{\text{heating rate amplitude}} \qquad (2.93)$$

In other words, the heat flow difference is replaced by the heat flow amplitude, and the ramp rate is replaced by the heating rate amplitude. The amplitude of the heat flow signal is obtained by use of a Fourier transform. The reversing heat flow can now be obtained by multiplying the reversing heat capacity signal by (usually) the negative of the underlying (or average) ramp rate. Multiplying by −1 changes the heat flow direction, since TA Instruments usually records increasing heat flow in the downward direction (however, it should be noted that modern software allows the user to specify this, and so it is arbitrary). Multiplying by the heating rate, one obtains the heat flow:

$$\text{RHF} = -C_{p,\text{rev}} \times \text{UHR} \qquad (2.94)$$

where RHF is the reversing heat flow UHR is the underlying heating rate, and $C_{p,\text{rev}}$ is the reversing heat capacity (i.e., the heat capacity without any kinetic events or reactions).

The nonreversing heat flow can be calculated from the difference between the total heat flow and the reversing heat flow:

$$\text{NHF} = \text{THF} - \text{RHF} \qquad (2.95)$$

where NHF is the nonreversing heat flow and THF is the total heat flow. The total heat flow, reversing heat flow, and the nonreversing heat flow curves for a quenched PET sample are shown in Fig. 2.89.

The basis for the deconvolution procedure can be illustrated using a few simple equations. When used with a sinusoidal perturbation, the temperature program for an MTDSC experiment is given by

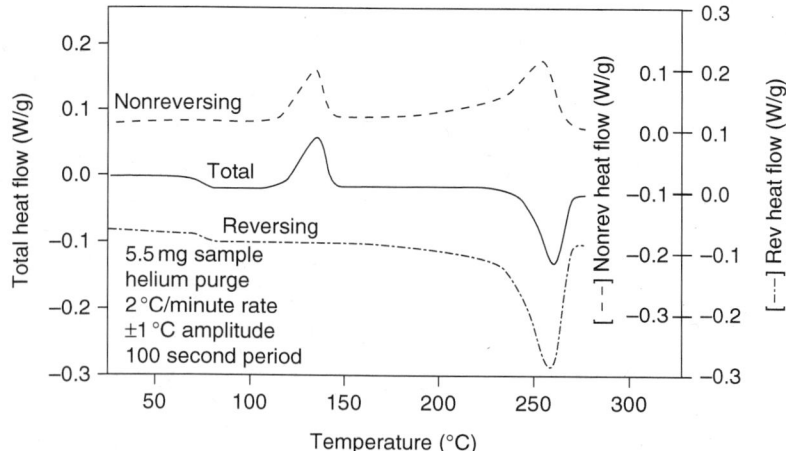

Figure 2.89. The total heat flow, reversing heat flow, and the nonreversing heat flow from an MTDSC heating of quenched poly(ethylene terephthalate) (courtesy of TA Instruments).

$$T = T_0 + qt + B\sin(\omega t) \qquad (2.96)$$

where T_0 is the starting temperature, q is the heating rate, B is the amplitude of the modulated temperature (or temperature oscillation), and ω is the frequency. This is in contrast to the temperature ramp equation for conventional DSC, which is given by $T = T_0 + qt$. The heat flow equation [Eq. (2.91)] can therefore be written as

$$\frac{dQ}{dt} = C_p(q + B\omega\cos(\omega t)) + f'(t,T) + C\sin(\omega t) \qquad (2.97)$$

where $f'(t,T)$ is the underlying kinetic function once the effect of the sine wave modulation has been subtracted, C is the amplitude of the kinetic response [from $f(t,T)$] to the sine wave modulation, and $(q + B\omega\cos(\omega t))$ equals the sinusoidal heating rate. It is important to note that this treatment assumes that the response of the sample is linear, and this will often be the case if the amplitude of the modulation is small. The heat flow signal will therefore contain a cyclic component that is dependent on the values of B, ω, and C.

The first step in the deconvolution procedure is to average the modulated signals over the period of the modulation (or integer multiples of this period). Where the response is linear, this removes the modulation and provides, to a good approximation, the results that would have been obtained had the modulation not been used:

$$\text{Total heat flow} = C_p q + f'(t,T) \qquad (2.98)$$

It can be seen from Eq. (2.97) that the heat flow signal comprises two sets of terms; the first, $C_p(q + B\omega\cos(\omega t))$, is dependent on the magnitude of the terms q, B, and ω, and so it is a measure of the sample's heat capacity. Considering only the modulation, we see that the heat capacity contributes a cosine wave (as opposed to the sine wave in the temperature). When there is no transition, the modulation in heat flow will, in principle, follow this cosine wave. Note that the phase difference between the modulation in heat flow and heating rate will, at least in this simple model, be zero when the convention is adopted that endothermic is up and it will be 180° when the convention is adopted that endothermic is down.

The second component in the heat flow, $f'(t,T) + C\sin(\omega t)$, is associated, in the simple example we are considering here, with an irreversible chemical reaction. We can consider this to be a kinetic or kinetically hindered process in contrast to the way heat can be very quickly stored in and removed from the sample's heat capacity. The quantity C arises because if the temperature is modulated, there will be a corresponding modulation in the heat flow associated with the modulated rate of reaction. This modulation in heat flow follows the sine wave in temperature and so is 90° out of phase with that associated with the heat capacity, which follows a cosine function. A correction for this phase lag can be applied where it is considered significant (see text below).

In generating the Fourier transform, one should keep in mind that only the first harmonic is considered for MDSC and a power series is not used. Errors may occur if there are substantial contributions from higher harmonics, which is what would occur if there were nonlinear behavior. Also, the way that the temperature is modulated may differ between instrument manufacturers. For example, the temperature modulation for the TAI 2900 DSC modules occurs at the block or furnace temperature (Slough 2006). Differences can also be found between modules for the same manufacturer. For example, the TA Instruments Q series bases its calculation of modulated signals on both the sample and reference heat flow signals (the temperature difference is calculated by subtracting the T_0 temperature from the sample and reference temperatures separately) while only one heat flow signal is used for the 2900 series (TA Instruments 2005). However, these differences do not substantially change the interpretation of the MTDSC signal.

Different from the other MTDSC techniques is the TOPEM technique, which is based on a quasistochastic temperature modulation superimposed on a conventional DSC temperature program as seen in Fig. 2.90 (Schubnell et al. 2005). This allows separate determination of the reversing and nonreversing heat flows and the quasistatic (zero-frequency) heat capacity, as well as a complex heat capacity (Schawe and Hutter 2005). Since the response at a desired frequency is determined after the run, multiple run parameters can be used to optimize the analysis of different transition regions during a run as noted in text below (see also Fig. 2.91).

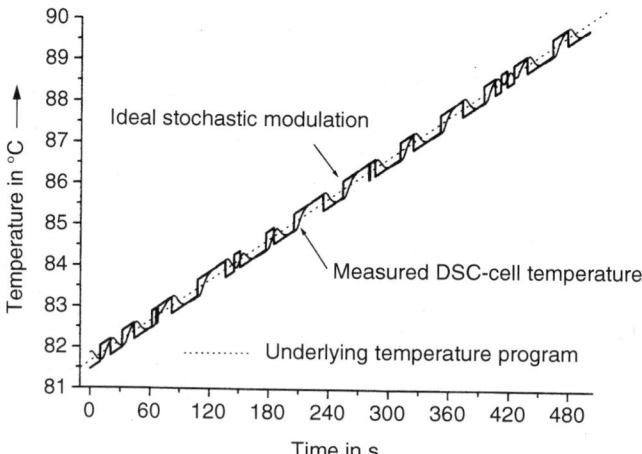

Figure 2.90. TOPEM (Mettler Toledo) temperature program and the DSC cell temperature [Schubnell et al. (2005); reprinted with permission of the North American Thermal Analysis Society and Mettler Toledo].

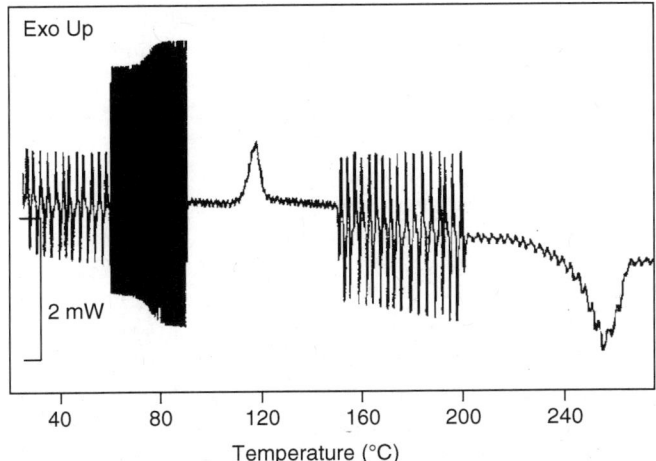

Figure 2.91. A TOPEM multiparameter run. The experimental conditions have been optimized to achieve the optimum run parameters for a transition within a single run [Sauerbrunn (2006); reprinted with permission from Mettler Toledo].

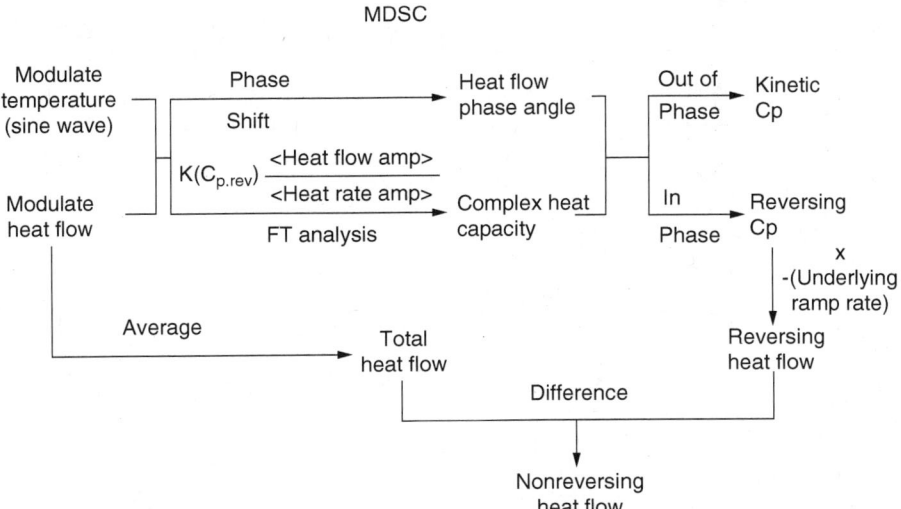

Figure 2.92. MDSC2920 flow diagram with application of phase correction on the basis of the TA Instruments (1997) reference [Fuller and Judovits (2002); reprinted with permission from ASTM International].

The MTDSC signals are treated differently by various manufacturers. A pictorial diagram representing the signals seen by MDSC based on the reference (TA Instruments 1997), and is given in Fig. 2.92 (Fuller and Judovits 2002).

As noted before, considerably more material has been published on the mathematical treatment of MDSC than on other modulated techniques, which is reflected in Fig. 2.92.

The MDSC diagram is more complicated than the description given so far since it introduces the phase lag correction. The heat flow phase angle arises from the modulation shift between the modulated and detected temperatures. Changes in the phase lag occur at transitions because there is a heat capacity and a kinetic component (see discussion above). Generally, however, this effect is small and can be neglected (Lacey et al. 1997). This is illustrated in Fig. 2.93 for the block and sample temperatures. Similarly, the heat flow or ΔT response has a lag, from which an in-phase signal and out-of-phase signal can be determined.

As it stands now, each manufacturer has its own MTDSC terminology. For the TAI 2900 series modules, the complex heat flow mentioned above would be exactly the same as the reversing heat flow when no phase correction is used. The use of the term "complex" instead of "reversing" may be more precise since it implies that the signal is a complex quantity containing in-phase and out-of-phase components. However, the term "reversing" appears to be more widely accepted for historical reasons, as it is descriptive of the process, and is the one used for the Q series modules (see Table 2.6). Similarly, we can have a reversing (or complex) heat capacity signal, which has the added

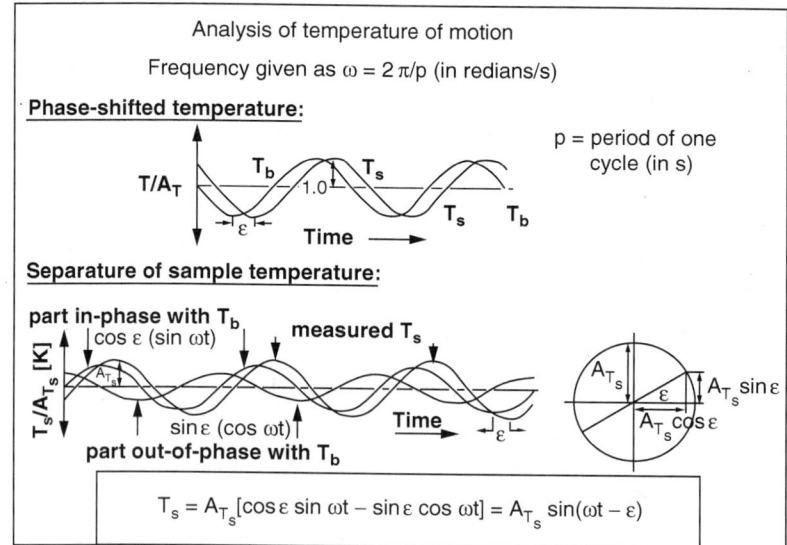

Figure 2.93. Phase-shifted sample temperature T_s relative to the modulated block temperature T_b; similarly, the heat flow (ΔT) has its phase angle, and this permits determination of the phase-corrected heat flow responses [from Wunderlich et al. (1994); reprinted with permission from Elsevier Ltd.].

TABLE 2.6. MDSC Phase-Corrected Signal Comparison for TAI Modules

2900 Series	Q Series
Modulated heat flow	Same
Modulated temperature	Same
Heat flow phase angle	Same
Complex heat capacity	Reversing heat capacity
Kinetic heat capacity	Out-of-phase heat capacity
Kinetic heat flow	Out-of-phase heat flow
Reversing heat capacity	In-phase heat capacity
Reversing heat flow	In-phase heat flow
Nonreversing heat flow	Kinetic heat flow
Nonreversing heat capacity	Kinetic heat capacity

value of corresponding to the heat capacity of the sample in the absence of thermal transitions. This signal can also be referred to as simply the "heat capacity signal" to point out the ability of MTDSC to measure the absolute value of this key thermodynamic property.

With the use of the phase correction, the TAI nomenclature again changes (see Section 2.13.3).

Use of the phase lag correction depends on the user's analysis needs. However, the effect of the correction is not significant unless it is applied to the melting transition.

2.13.3. Run Parameters for Thermoplastics

For an MTDSC experiment additional run parameters must be selected when compared with conventional DSC. For TAI modules, these parameters are

- Modulation amplitude
- Modulation period
- Underlying or average ramp rate.

When deciding what conditions for these parameters to select, one first needs to decide on what type of modulation is desired. MTDSC can be run in a number of modulation modes as noted by Wunderlich and illustrated in Fig. 2.94 (Wunderlich 1997b).

If one includes the cooling options, seven different modulation choices are available. These would be three heating modulations, three cooling modulations, and a modulation around a single temperature referred to as *quasi-isothermal*. The first heating option would be the *heat-only* mode. This is commonly used for initial scans, especially when the response of the sample to modulation may be unknown. The second option is the *heat-with-some cooling* mode, where some cooling in the modulation is allowed, although the

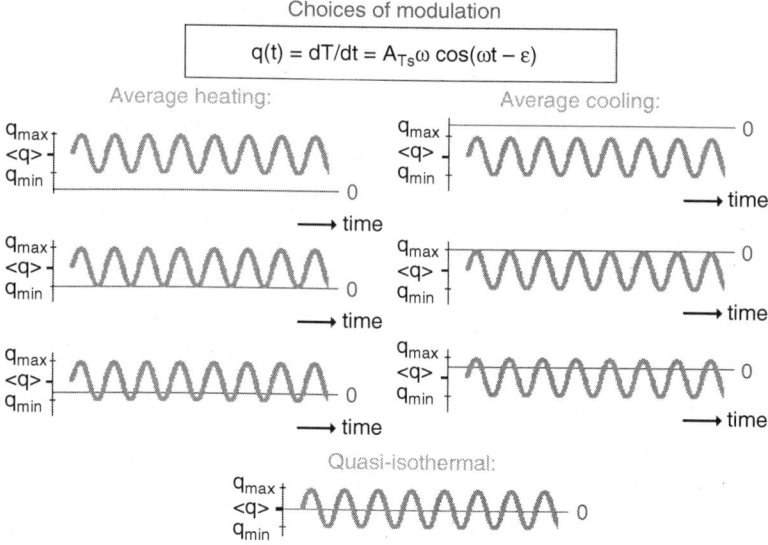

Figure 2.94. Different modulation selections are available for an MTDSC experiment. Seven choices are available: heat or cool only, heat with some cooling or conversely cooling with some heating, heat with an associated zero ramp or conversely cool with an associated zero ramp, and finally a modulation around a particular temperature, namely, quasi-isothermal [from Wunderlich (1997b); reprinted with permission from B. Wunderlich].

average heating rate is positive. This mode is used because of its enhanced resolution when glass transitions or overlapping transitions are analyzed. The modulation period is selected to yield maximum temperature values (vs. time) so that it exceeds the temperature maxima obtained in the heat-only modulation. Another selection would be to heat with a modulation that results with zero heating as its slowest rate. Since the modulation amplitude is the major factor that determines the resolution, this later leads to highest resolution without any cooling associated with it. To correctly match the amplitude and the modulation period for a particular ramp rate, TA Instruments provides a table (Table 2.7) with the parameters for the maximum heat-only mode already calculated. (This table is useful for the 2920 DSC, but it is not needed when operating a Q series instrument; there the software may select the parameters for a heat-only modulation.) Three similar modulation options exist for cooling ramps.

In general, TA Instruments recommends underlying heating rates of less than 5°C/min for MDSC experiments with modulation amplitudes between ±0.1°C and 2°C, and modulation periods between 40 and 100s (TA Instruments 2005). For quasiisothermal experiments, the nonreversing heat flow

TABLE 2.7. MDSC® Chart Describing Maximum Heat-Only Conditions[a]

Period (sec)	*Heating Rate (°C/min)*						
	0.1	0.2	0.5	1	2	5	10
10	0.003	0.005	0.013	0.027	0.053	0.133	0.265
20	0.005	0.011	0.027	0.053	0.106	0.265	0.531
30	0.008	0.016	0.040	0.080	0.159	0.398	0.796
40	0.011	0.021	0.053	0.106	0.212	0.531	1.062
50	0.013	0.027	0.066	0.133	0.265	0.663	1.327
60	0.016	0.032	0.080	0.159	0.318	0.796	1.592
70	0.019	0.037	0.093	0.186	0.372	0.929	1.858
80	0.021	0.042	0.106	0.212	0.425	1.062	2.123
90	0.024	0.048	0.119	0.239	0.478	1.194	2.389
100	0.027	0.053	0.133	0.265	0.531	1.327	2.654

$$T_{amp} = H_r * \left(\frac{P}{2\pi * 60} \right)$$

where: T_{amp} = maximum temperature amplitude for "heat only" (°C)
H_r = Average heating rate (°C/min)
P = period (seconds)
60 = converts seconds to minutes

[a] The values tabulated above indicate amplitude needed to maintain a heat-only ramp.
Source: TA Instruments, 1998; reprinted with permission from TA Instruments.

cannot be determined since the underlying heating rate is zero. In addition, long modulation periods and large amplitudes (such as ±0.5°C) are recommended for quasi-isothermal analyses.

2.13.4. Sample Mass

Modeling has shown that for polymeric materials, the phase lag depends on the thickness of the sample (Buchler and Seferis 2000). Thick samples were found to have a significant phase lag and difficulties reaching steady state. Therefore, large masses in MTDSC experiments should be avoided. However, the actual value of the mass depends on the modulation period, where longer periods tolerate larger masses. According to TA Instruments, the selection of the mass in an MDSC experiment depends on the type of analysis. Typical values range from 10 to 15 mg (TA Instruments 2005). In the case of barely visible transitions, it is suggested to increase the mass to 20 mg. On the other hand, for intense transitions the mass can be decreased to 5–10 mg. However, the use of the smallest possible mass is recommended for reducing any effects originating from temperature gradients.

2.13.5. MTDSC Parameter Selection for Thermosets

For nonisothermal cure experiments it is not necessary to perform the experiment in heat-only conditions. However, for curing experiments, it is advisable to use larger modulation amplitudes with low underlying heating rates, to improve the accuracy of the heat capacity measurement. The amplitude of the temperature modulation should not exceed a certain value in order to limit its effect on the cure kinetics. Typical modulation amplitudes range within 0.1–1 °C.

The range of modulation frequencies of practical importance is limited to about one decade, ranging from 0.01 Hz (100 s modulation period) to 0.1 Hz (10 s modulation period), as opposed to the conditions of dielectric spectroscopy or dynamic mechanical analysis, where applied frequencies can cover the range from 0.01 Hz to 10^5 Hz or higher. A period of 60 s is usually a good selection for most cure experiments.

2.13.6. Good MTDSC Runs

As noted before, thermal events that respond to changes in the ramp rate are reversing at the time and temperature at which they are observed. This means that the timescale of the thermal events is important. Processes that can respond faster than the timescale of the modulation will be time-independent and reversing (Scherrenberg et al. 1998). This also means that a change in the modulation amplitude should relate to a proportional change in the heat flow amplitude (Simon 2001). Factors that affect linearity (as described in Section 2.1) are the modulation amplitude and period. Maintaining steady state (see Section 2.3 for a definition of *steady state*) is also of importance since it is assumed in the calculation used to separate the total heat flow into its components. Finally, a factor that should be considered when quantifying the heat flow is *stationarity*, which refers to the sample not changing (at least significantly) during one modulation cycle. A detailed analysis of the errors is given by Lacey et al. (1997). The physical meaning of the MTDSC results may become questionable if stationarity is not maintained. This is rarely a problem unless analyzing the melting transition, where linear response and steady state are often lost. A number of equations are reported in the literature to evaluate stationarity and linearity of MTDSC runs (Lacey et al. 1997; Simon 2001).

Also possible is the simple use of an empirical evaluation involving of the examination of both the curve shape and identifying spurious peaks. This method allows for the use of faster ramp rates, but integration of the heat flow values should be done with caution. Several factors should be included with this approach:

1. It is essential that at least four or five complete modulation cycles can be completed within a given transition (TA Instruments 1998), although even more modulations are better. Lacey et al. (2006) suggest five modulations for a transition. Where a peak is noted, the width at half-height of the peak is best used, whereas for a glass transition, the inner breadth or the width of the glass transition region between the extrapolated onset and end is best.
2. The ramp rate and heat flow modulation should be considered. It is important to establish whether the peaks are spiky or well formed and uniform. Well-formed sine-wave shaped peaks are necessary for achieving a proper modulation with no distortion showing a uniform wave pattern. Figure 2.95 shows differences between modulations with no distortion and those that are highly distorted.
3. It is also important whether the heat flow cycle shows a complete modulation.

Figure 2.96 illustrates an example of a poor modulation through a melting transition, since the regular modulation of the heat flow is not maintained.

An example of verification of a proper modulation parameter can be found in the recrystallization process occurring during the melting of polyamide 12 shown in Fig. 2.97. The exotherm seen in the nonreversing heat flow is found to be real and not an artifact by noting that it also appears in the raw modulated heat flow (Judovits 1997). Just as the exotherm appears in the nonreversing heat flow during the melting process, likewise the modulated heat flow demonstrates an exothermic process compared to the baseline in the modulated heat flow signal before and after the melting.

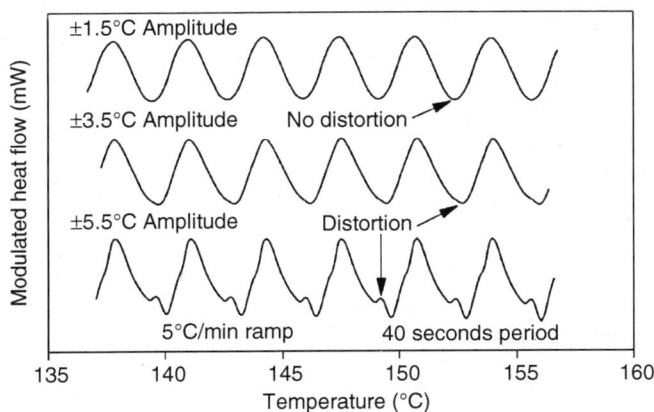

Figure 2.95. Selection of the proper modulation conditions through simple inspection of the modulated heat flow [TA Instruments (n.d.) Compendium; courtesy of TA Instruments].

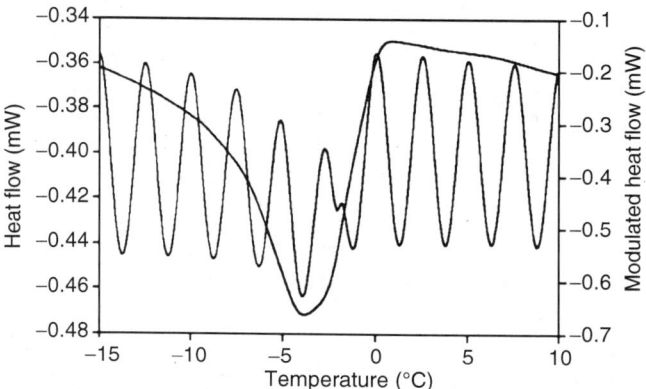

Figure 2.96. Distortion of the modulated heat flow due to an incomplete modulation [TA Instruments (2005); courtesy of TA Instruments].

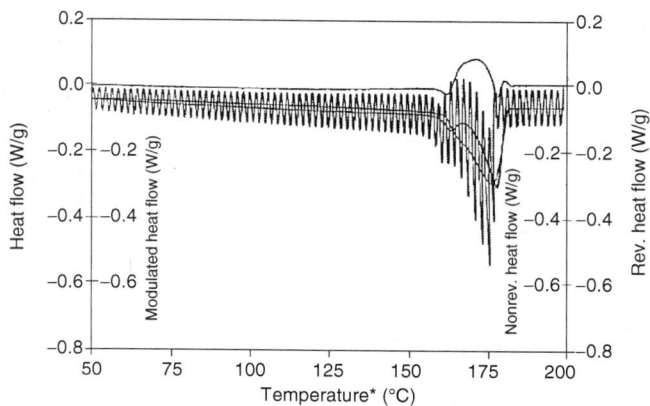

Figure 2.97. The crystallization exotherm during an MDSC heating experiment seen in the nonreversing heat flow (top curve) is confirmed by superimposing it on the modulated heat flow signal for polyamide 12 that had been annealed at 160 °C for one hour. The exothermic direction of this figure is upward [from Judovits (1997); reprinted with permission from the North American Thermal Analysis Society].

As mentioned above, one can also modulate around a particular temperature; this process is called *quasi-isothermal*. Highly accurate heat capacity values can be obtained using this mode. To evaluate whether appropriate modulation is used, Lissajous figures may be employed (Pyda and Wunderlich, 2000). Lissajous figures for MTDSC are plots of the modulated heat flow versus modulated temperature or modulated heating rate. Figure 2.98 shows an example of a Lissajous figure demonstrating its use to determine steady state. If proper modulation is obtained and steady state is achieved, the plots

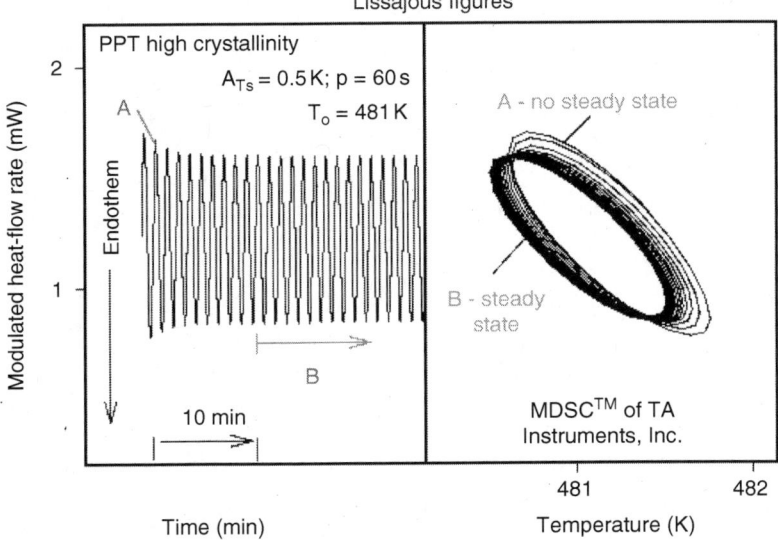

Figure 2.98. Construction of Lissajous figures from the modulated heat flow. The Lissajous figures can be used to evaluate whether steady state is achieved in an MTDSC experiment. Where the loops overlap, the system is in steady state. The above mentioned quasi-isothermal run was performed on a sample of poly(propylene terephthalate) (PPT). [From Pyda and Wunderlich (2000); reprinted with permission from John Wiley & Sons, Inc.]

should superimpose on each other in the absence of a transition. The time to steady state can be identified as well as errors such as drift that can affect the measurements. Ideally, the Lissajous plot should be an ellipse. When the heat flow lags behind the temperature, the slope of the long axis of the ellipse changes (Aubuchon et al. 1997). When the curves overlap, steady state has been achieved. Although commonly used to analyze quasi-isothermal runs, Lissajous figures can also be used to evaluate temperature ramps.

When attempting to use MTDSC, it should be kept in mind that it is a supplementary technique to traditional DSC and should be considered an extension of conventional DSC. It does not replace it. Before attempting an MTDSC experiment, conventional DSC measurements should first be done to evaluate the need for a modulated run. Care must be taken in a modulated experiment, since the use of modulation adds an additional complication, and any MTDSC experiment, like any DSC experiment, should be verified by other means whenever possible. There are several disadvantages associated with MTDSC. The choice of experimental parameters is difficult, and the use of incorrect parameters could lead to erroneous results. Additionally, interpretation of the heat flow components can be difficult. Finally, speed—MTDSC is usually run at a slower ramp than conventional DSC because of the need to obtain a sufficient number of modulation cycles over a transition region.

2.13.7. Glass Transition of Amorphous Polymers

Modulated temperature DSC offers a number of advantages when analyzing a glass transition. These advantages stem from its ability to separate multiple thermal events. As noted earlier, the glass transition can be separated into reversing and nonreversing components.

When this separation occurs, the glass transition temperature of the reversing heat flow is always higher than the glass transition temperature determined from the total heat flow [as seen in Figs. 2.99 and 2.100; a detailed explanation can be found in Lacey et al. (2006)].

The glass transition curve usually shows an enthalpy relaxation or hysteresis peak as well. The intensity of the hysteresis peak can be greatly increased if one anneals the glass just below the glass transition temperature or if the material was slowly cooled (see Section 2.7, on phase transitions in thermoplastic polymers). This poses two problems: (1) obtaining a comparative glass transition temperature and (2) evaluating the hysteresis behavior.

The DSC curve of the glass transition on heating recorded by conventional DSC depends on the thermal history and the run parameters during the heating; that is, one needs to reference the cooling rate or past thermal treatment and then the subsequent heating rate. However, with MTDSC, frequency effects must also be considered. It has been shown that if sufficient cycles are made through the glass transition and the frequency effects are taken into account, agreement is found for the measured and modeled reversing glass transition temperatures (Simon and McKenna 2000). The modeling was done by simulating the glass transition using the Tool–Narayanaswamy–Moynihan

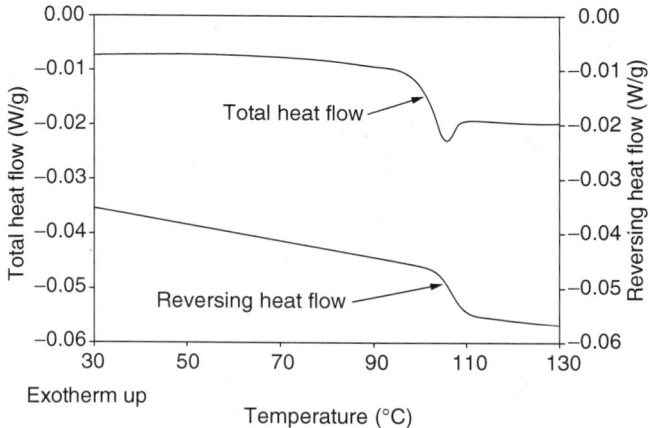

Figure 2.99. Glass transition of polystyrene on heating. Note that the glass transition of the reversing heat flow is higher than that of the total heat flow. The sample had been previously cooled at 2°C/min. Underlying heating rate is 2°C/min, modulation amplitude ±0.21°C, modulation period 40s. (Judovits, unpublished results.)

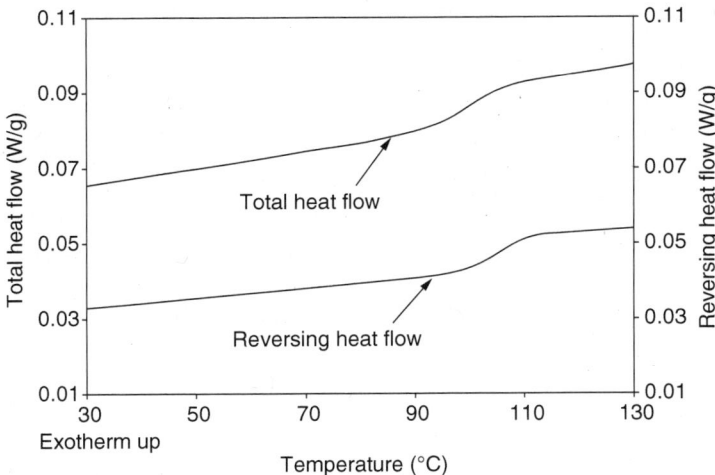

Figure 2.100. DSC curves of glass transition of polystyrene on cooling. As with the heating ramp, the glass transition for the reversing heat flow is higher than that on the total heat flow. Underlying cooling rate is 2°C/min, modulation amplitude ±0.21°C, modulation period 40s. (Judovits, unpublished results.)

model of structural recovery. Work on polystyrene showed that the glass transition from the reversing heat flow depended mainly on the modulation frequency and was weakly dependent on the sample thermal history (Boller et al. 1995). It should also be noted that the differences in glass transition temperatures between the reversing heat flows measured on cooling and heating as the ramp rate was reduced converged to that of the quasi-isothermal run (Wunderlich and Okazaki 1997).

Modulated temperature DSC also allows an easier analysis of the physical aging process since the enthalpy relaxation is separated into the nonreversing heat flow. Separation of reversing and nonreversing components for aged poly(methyl methacrylate) has been reported (Kubota et al. 2005). The reversing and nonreversing heat flows of the aged samples were subtracted using a nonaged sample as a control (see the nonsubtracted and subtracted nonreversing heat flows in Fig. 2.101). However, very large enthalpies of relaxation can lead to deviations in the linear relationship between the enthalpy seen in the nonreversing signal and the enthalpy loss on annealing, but this can be corrected for (Lacey et al. 2006).

Since MTDSC can resolve subtle transitions, it has been applied in studying miscibility of amorphous polymers. Riga and Sisk (1999) found that immiscible polymers can be distinguished if the difference in T_g values of the corresponding homopolymers is greater than 10°C. A more rigorous use has been to study the interphases in multicomponent polymer materials. This was accomplished by the use of the derivative of the reversing heat capacity (Hourston et al. 1997). These authors were able to differentiate glass transition temperatures

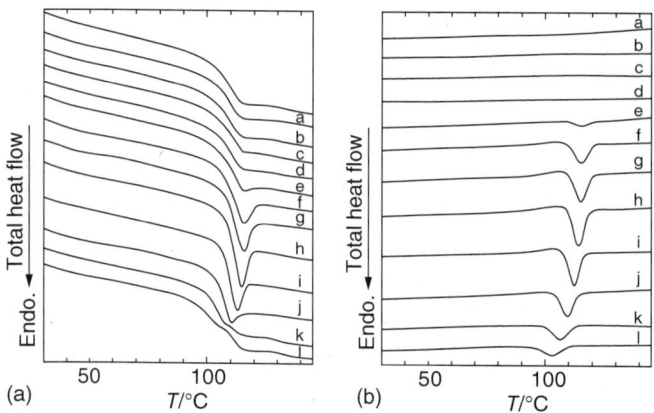

Figure 2.101. Nonreversing heat flow curves following isothermal aging of PMMA. Annealings for 6h at a = 140 °C, b = 130 °C, c = 120 °C, d = 115 °C, e = 110 °C, f = 105 °C, g = 103 °C, h = 100 °C, i = 95 °C, j = 90 °C, k = 85 °C. The two figures differ in that panel (a) shows the observed curves while (b) shows the curves after subtraction of the nonaged runs (top curve). [From Kubota et al. (2005); reprinted with permission from Elsevier Ltd.]

that were less than 15 °C apart. Furthermore, they were able to determine polymer–polymer miscibility with glass transitions 10 °C apart and resolve glass transitions for reduced concentrations of one of the components (Song et al. 1995). Song and coworkers investigated a number of blend systems using the derivative method (Song et al. 1999). This method has more recently been extended to polymer/modified clay nanocomposites (Hourston and Song 2005), (see Fig. 2.102).

Modulated temperature DSC has also been used to study plasticization of poly(ethylene terephthalate) by water. PET is somewhat hygroscopic and absorbs moisture, so the glass transition temperature will decrease. Quasi-isothermal analysis was used and the glass transition temperature was found to increase with time as the water evaporated (Toda et al. 1997d). A similar analysis was performed to investigate the effects of water on Nylon 6. A hermetically sealed pan was used and the plasticized glass transition separated into the reversing heat flow (TA Instruments 1994).

Stress effects in the glass transition region can also be studied using MTDSC. Premium vinyl formulations use alternative capstocks (surface layers) other than PVC over a PVC substrate layer. Stresses may develop between the two dissimilar layers. Stress relaxation may be noted after the glass transition for vinyl siding (Judovits and Crabb 2005). The signals due to these stresses can be separated into the nonreversing heat flow (Fig. 2.103) and are not seen in the reversing heat flow. Examination of the components indicates that the stress release originates from the PVC layer.

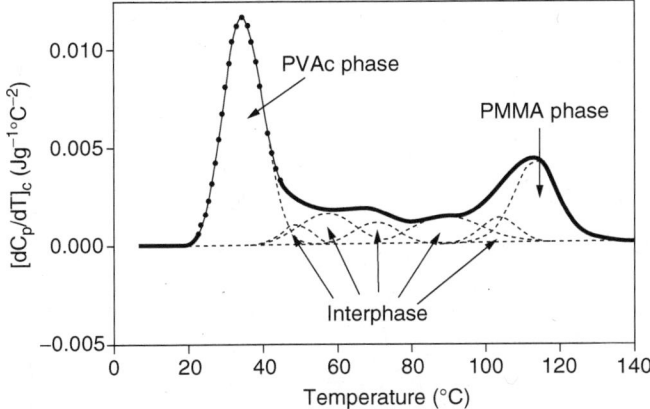

Figure 2.102. Use of the derivative reversing heat capacity to study interfacial effects; this figure shows the application of this method to poly(methylmethacrylate)–poly(vinyl acetate) (PMMA-PVAc), (50:50) latex film [from Song et al. (1999); reprinted with permission from Springer-Verlag].

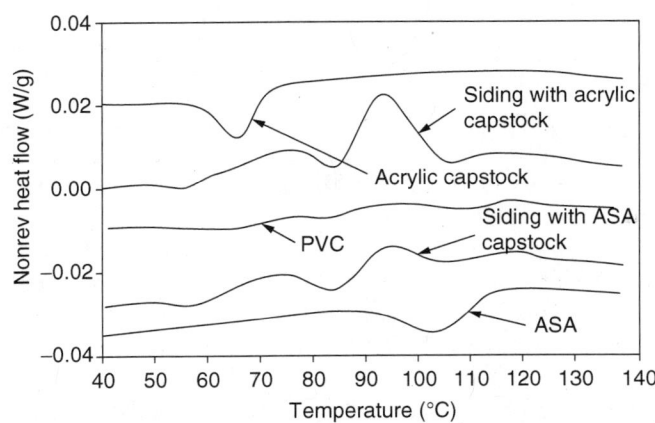

Figure 2.103. Nonreversing heat flow of vinyl siding and its components. One can note stress release in the siding, that is, additional exothermic peaks (in the upward direction in the figure) in the siding composite in comparison to the reference component runs. [From Judovits and Crabb (2005); reprinted with permission of the North American Thermal Analysis Society.]

2.13.8. Crystallization and Related Phenomena

Comparison of crystallization behavior measured in modulated (MTDSC) and nonmodulated (conventional DSC) conditions has been made for poly(p-phenylene sulfide) (Menczel 1999). A cool–heat mode was utilized. Although the total crystallinity was not found to differ between the two techniques, the

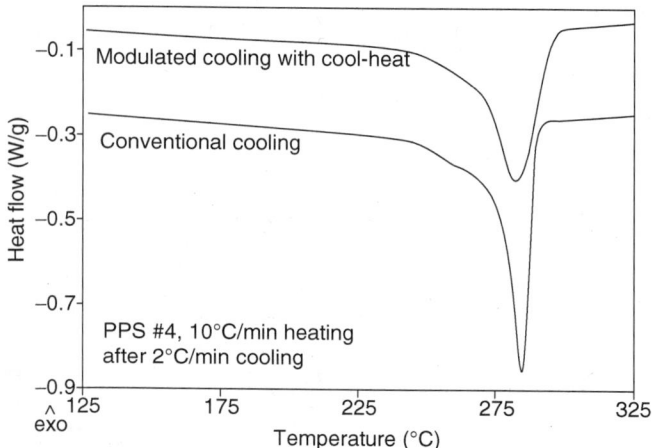

Figure 2.104. Conventional melting curves for poly(p-phenylene sulfide), PPS samples cooled by conventional and modulated cooling at the same underlying cooling rate [from Menczel (1999); reprinted with permission from Springer-Verlag].

crystal perfection was affected. On reheating by conventional DSC, the peak temperature was found to decrease while the end of the melting was found to increase for the samples cooled by modulation! This can be seen by comparison in Fig. 2.104. This would indicate that both an increase in forming some larger crystals and a general decrease in crystal perfection occurred because of the cool–reheat process. Similar effects have been noted for macromolecular segregation, where similar molecular masses cocrystallized with each other after annealing broadening the melting peak (Wunderlich 1976).

Crystallization processes have also been studied using a modified method of Toda (Toda et al. 1997a,b,c), where the temperature dependence of the crystal growth rate can be determined through modulated cooling. This modified method was used to study the crystallization kinetics of a polyester–imide (Chen et al. 2000). This method permitted the evaluation of transitions regarding their reversibility. Likewise, MTDSC was used in a step cooling to study the crystallization of poly(L-lactic acid) (Sanchez et al. 2005). The crystal growth was also modeled using the method proposed by Toda. Using quasi-isothermal crystallization, they were able to note changes in crystallization behavior versus temperature.

2.13.8.1. Cold Crystallization Cold crystallization takes place in samples which were prevented from crystallizing as a result of fast cooling. The crystallization of the amorphous material takes place when the material is heated above its glass transition temperature [see Wunderlich (1997a); see also Section 2.7, below]. Generally, this occurs for materials that were quenched through

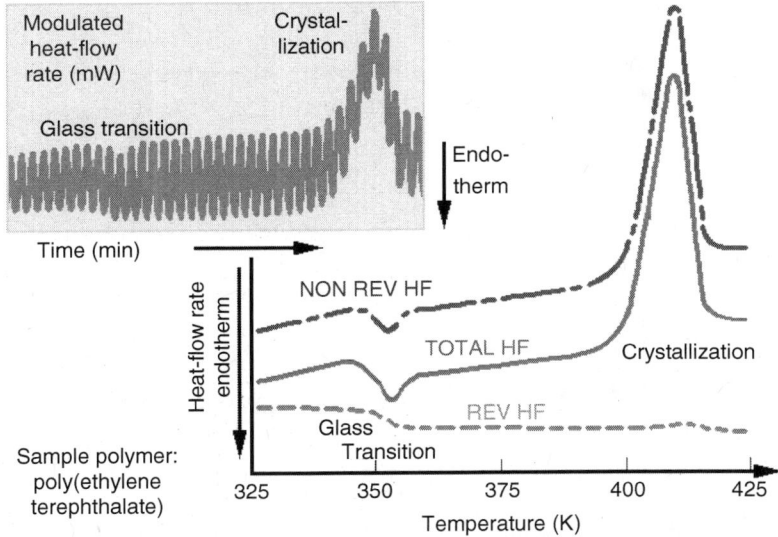

Figure 2.105. Heating of low crystallinity poly(ethylene terephthalate) with the cold crystallization noted in the raw modulated signal; the sample was heated at an underlying heating rate of 5 °C/min [from Wunderlich et al. (1994); reprinted with permission of Elsevier Ltd.].

their glass transitions. On reheating above the glass transition temperature, the material will crystallize rapidly at a low crystallization temperature.

Cold crystallization was one of the first examples of the ability of MTDSC to separate various processes (Reading et al. 1994). Cold crystallization was found, in the case considered, to be a nonreversing process; therefore the glass transition, which is a reversing process, could be studied separately as shown in Fig. 2.105.

2.13.8.2. Melting Modulated temperature DSC usually has difficulties in providing quantitatively interpretable measurements for narrow melting transitions noted for low-molecular-weight materials since the phenomenon is too fast for the applicable modulation parameters. During the melting, the large heat flow response over a short time period makes it difficult for the modulation to maintain itself, and one can note crosstalk between the cycles.

Analysis of the melting transition by MTDSC can be difficult, especially for quantification. Most MTDSC parameters result in a frequency dependence in separating the total heat flow for the melting transition. Differences in the effects of experimental conditions on the reversing and nonreversing heat flows in the melting region have been noted where the reversing and nonreversing signals will increase or decrease at the expense of each other [TA Instruments n.d. (*Compendium*)].

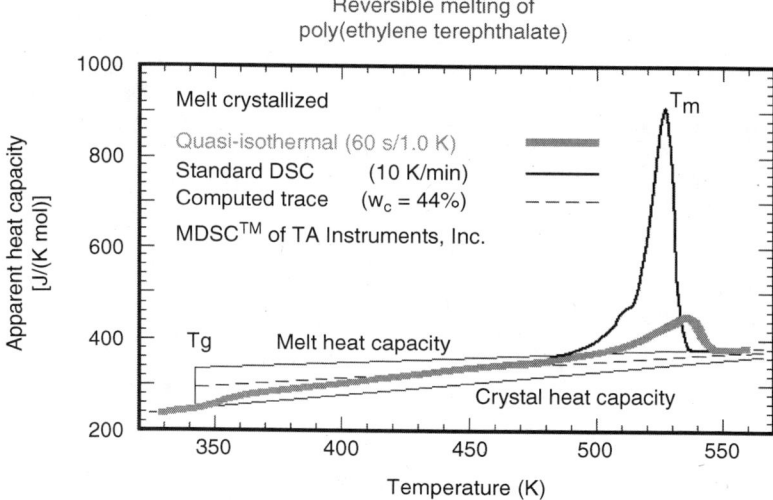

Figure 2.106. Reversible melting of poly(ethylene terephthalate) as recorded in a quasiisothermal step heating experiment. Comparisons can be made to a conventional DSC run performed at 10 °C/min. For the quasi-isothermal run, the conditions used were a period of 60 s, an amplitude of ±1 °C, and a sample mass of 5 mg. [From Wunderlich et al. (1998); reprinted with permission from Elsevier Ltd.]

The determination of initial crystallinity, that is, the crystallinity of the as-received sample, can sometimes be determined more reliably using MTDSC than conventional DSC. A discussion and detailed procedure for its determination is given by Reading et al. (2001). Generally, macromolecules melt irreversibly because of imperfection and small size of their crystallites. However, quasi-isothermal analysis has revealed a reversible component to the melting even at long step times as noted in Fig. 2.106. Although quasi-isothermal analysis offers the advantage of maintaining a small perturbation in the melting region, avoiding the issue of loss of steady state, one would not expect to note any melting after the material has been allowed to relax to an equilibrium condition.

Quasi-isothermal analysis of PET has found a relationship between the thermal history of the sample and the reversing component of the melting, as shown in Fig. 2.107 (Wunderlich et al. 1998); that is, the less perfect crystals developed during fast cooling have a higher height for the reversing melting peak than do the crystals that have been better crystallized. These crystallites have been linked to partially melted molecules that remain partially crystallized and therefore can melt without molecular nucleation when reheated.

2.13.8.3. Reorganization and Recrystallization during Melting

Reorganization is a process in which the initial metastable polymeric crystallites

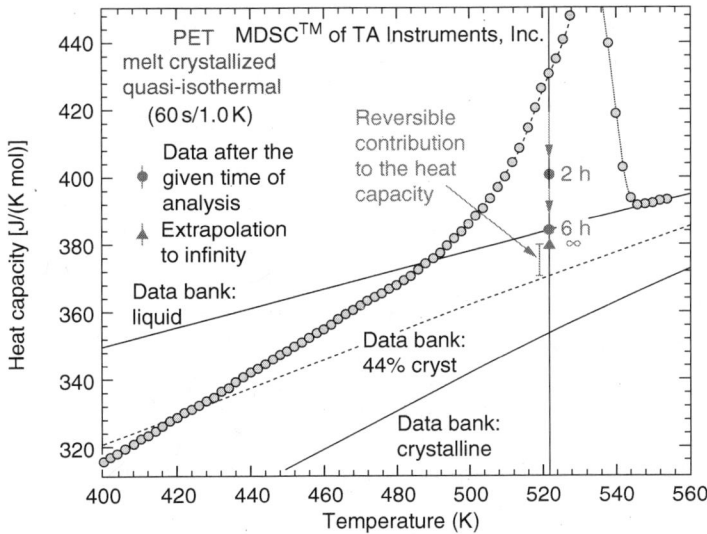

Figure 2.107. Contribution of reversible melting of PET when a quasi-isothermal run is extrapolated to infinite time; the time of measurement is indicated by the filled circles and is then extrapolated to infinite time [from Wunderlich et al. (1998); reprinted with permission from Elsevier Ltd.].

improve themselves during heating (Wunderlich 1990). *Recrystallization* is a similar process, but in this case a clear exotherm is seen between two endotherms during the melting process. A double endothermic peak was shown by Nakagawa and Ishida (1973b) for the melting of poly(vinylidene fluoride) (PVDF) using conventional DSC. As described in Section 2.7, the lower melting peak was noted to represent the original metastable crystals while the upper peak was assigned to the melting of the perfected crystals. Therefore, the lower temperature peak becomes more pronounced for faster heating rates while the upper temperature peak becomes more pronounced for slower heating rates. Modulated temperature DSC was also applied by Judovits et al. (1998) to study reorganization of PVDF, nylon 12 and poly(phenylene sulfide) during melting (Figure 2.108).

Use of MTDSC showed that the melting process is preceded by a growth process of the original metastable crystals, because a slight break is seen in the nonreversing signal curves at around 150 °C. This may indicate an exotherm due to starting mobility, since, as shown by the total heat flow curves, melting does proceed at this temperature. A second crystal perfection process is seen at 170 °C, where two endotherms are observed corresponding to the melting of a less perfect and perfected crystallites (Judovits et al. 1998).

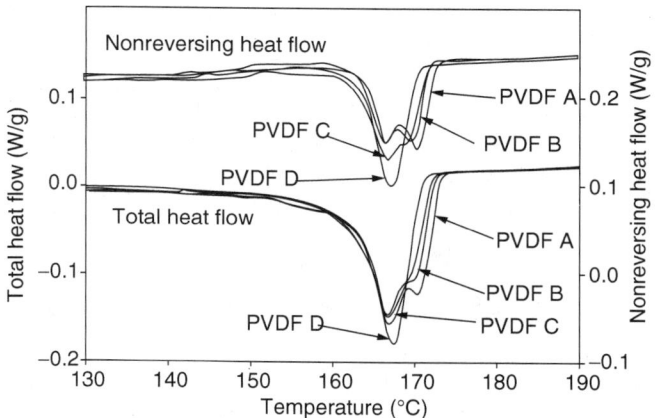

Figure 2.108. Reorganization of poyl(vinylidene fluoride) (PVDF) crystallites for samples of various molecular masses as seen in an MTDSC run. The molecular mass increases from A to D. Note the additional crystallization occurring in the nonreversing heat flow before the main melting peak. [From Judovits et al. (1998); reprinted with permission from Springer-Verlag.]

2.13.9. Heat Capacity

Heat capacity can be directly determined from a single-run MTDSC ramp. Furthermore, highly accurate heat capacity measurements can be made in the quasi-isothermal mode since more time can be devoted to reaching steady state. So, in addition to the standard heat capacity measurement made by running an empty pan, sapphire, and sample, two additional methods are offered by MTDSC. As with regular DSC heat capacity measurements, the operator must ensure that the instrument is well calibrated. This can be done by running sapphire under identical modulation conditions and determining a calibration factor. However, if a long period is used, a reasonable amplitude is selected, and the measurement is done on a thin specimen so that thermal gradients are negligible (see Section 2.3 for calculation of sample size), then the instrument calibration of the heat flow in the conventional mode is sufficient, and no special heat capacity calibration is needed. Figure 2.34 represents the effect of modulation period on the heat capacity calibration factor for the three Tzero modes of TA Instruments Q1000 or Q2000 (the Tzero modes are explained in Section 2.3) and the 2920 module DSC (see also Fig. 2.109).

The operator should consult the specific instruction manual for selecting the proper run parameters for the determination of heat capacity.

2.13.10. Curing and Reactive Systems

MTDSC has found considerable application in monitoring the degree of thermoset cure. It is important to monitor both an increasing glass transition tem-

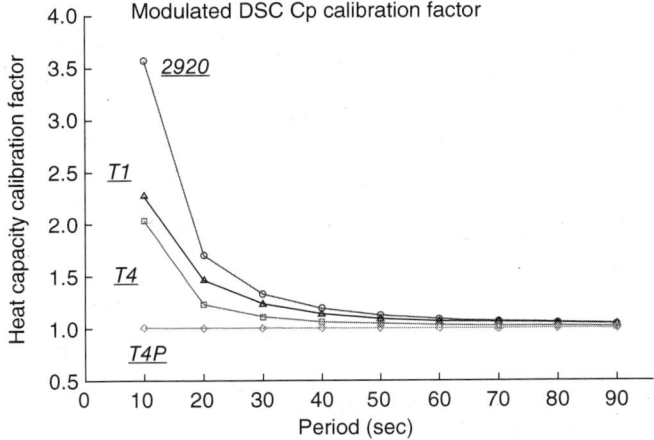

Figure 2.109. Effect of modulation period on the heat capacity calibration factor for TA instruments 2920 module and Q1000 module operated in different modes denoted $T1$, $T4$ and $T4p$ in the diagram [from TA Instruments (2005); courtesy of TA Instruments].

perature and the heat of the cure exotherm. In a ramp run, MTDSC can separate any relaxation effects appearing in the glass transition region without exposing the material to a previous thermal history, and additionally separate the cure exotherm if the glass transition overlaps with it. Use of the quasi-isothermal mode allows the operator to simultaneously monitor changes in heat capacity associated with curing versus time for a particular temperature (Cassettari et al. 1993; Van Assche et al. 1995).

The cure process for thermosets occurs with the formation of a crosslinked polymer network. Curing results in an increase of the glass transition temperature until vitrification occurs. At that point the kinetics can change from chemically controlled to diffusion-controlled.

The kinetic rate equation $[(dx/dt)_{kin}]$ can be corrected by the diffusion factor to give the observed rate of conversion $[(dx/dt)_{obs}]$:

$$\left(\frac{d\alpha}{dt}\right)_{obs} = \left(\frac{d\alpha}{dt}\right)_{kin} \cdot DF(x,T) \qquad (2.99)$$

where the diffusion factor DF accounts for the effect of the decrease in mobility on the rate of reaction and ranges from 1 to 0.

An estimate for the diffusion factor can be obtained through various models. Van Mele and coworkers described a mobility factor (MF)

$$MF(\alpha,T) = \frac{C_p(\alpha,T) - C_{pg}(\alpha,T)}{C_{pl}(\alpha,T) - C_{pg}(\alpha,T)} \qquad (2.100)$$

where α is the conversion, T is the temperature, C_p is the reversing heat capacity, C_{pg} is the glass heat capacity, and C_{pl} is the liquid heat capacity (Van Assche et al. 1995). The mobility factor changes from 1 to 0 as the liquid vitrifies, similar to how the diffusion factor changes when the reaction becomes diffusion-controlled. Note that MF can be obtained only from an MTDSC experiment since heat capacity information in quasi-isothermal conditions is needed. For most epoxy–amine systems and for typical modulation frequencies around 1/60 Hz, the stepwise decrease in C_p can be quantified to express the diffusion-controlled reaction (Van Mele et al. 2006).

In cases where diffusion control does not become dominant on vitrification, the mobility factor may not approximate the diffusion factor well (Meng and Simon 2005). Also note that the mobility factor is most useful for isothermal cure experiments, while larger deviations were found for nonisothermal experiments (Meng and Simon 2005). For thermosetting systems like isocyanates, unsaturated polyesters and inorganic polymer glasses with different reaction mechanisms and associated rate-controlling mobilities, the frequencies of correspondence deviate more significantly (Van Mele et al. 2006; Simon et al. 2005).

Modulated temperature DSC has been used to evaluate ambient cure conditions of thermoset coatings (Neff and Barsotti 1999). Many thermoset coatings are crosslinked under ambient conditions. This freezes in both residual solvents and thermal stresses. Heating above the glass transition to relieve these stresses will result in additional curing. Residual cure and detection of the glass transition can be disentangled using MTDSC. A similar use can be found in studying crosslinked foams (Siemens 2001).

2.13.11. Use of MDSC for Specific Thermosets

2.13.11.1. Nonisothermal Cure to Characterize a Thermoset Heating a thermosetting system twice through a wide temperature range provides characteristic cure parameters (see also Table 2.5 in Section 2.10) that form the basis for further in-depth analysis. A two-component 30-min Devcon glue was taken as an example of a readily available material (see Fig. 2.110). The reaction that occurs during mixing of this epoxy–amine system results in lost conversion before the actual experiment (~10–15%). As can be inferred from Fig. 2.110, the initial and final glass transitions can be obtained together with the reaction enthalpy by performing two consecutive heating experiments. Note that the enthalpy relaxation (or hysteresis) peaks separate into the non-reversing heat flow signal, leaving only a heat capacity jump in the reversing signal. Note that the reversing heat flow signal and heat capacity signal are proportional as indicated in Eq. (2.94).

The small second stepwise increase in the reversing signal of the first heating corresponds to the reaction heat capacity or the change in heat capacity when converting the epoxy–amine reactants into a fully cured product.

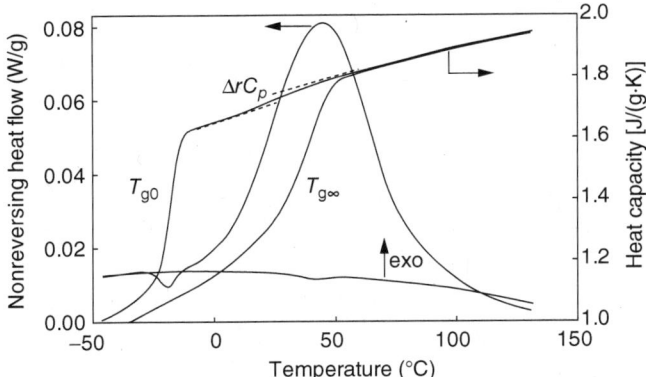

Figure 2.110. Nonisothermal experiment (1 °C/min underlying heating rate; 1 °C/60 s modulation, two consecutive heatings) on the 30-min Devcon glue, showing initial glass transition (at $T_{g,0}$), heat capacity increase at T_{g0} [$\Delta C_p(T_{g,0})$], reaction enthalpy (ΔH_T) and glass transition of fully cured thermoset (at $T_{g,\infty}$), heat capacity increase at $T_{g\infty}$ [$\Delta C_p(T_{g\infty})$], and reaction heat capacity ($\Delta r C_p$) (Swier, unpublished results).

2.13.11.2. Deconvolution of a Cure Exotherm with the Overlapping Glass Transition Region

Heating a partially cured thermoset in conventional DSC results in a complex heat flow signal where the glass transition, reaction heat, and, possibly, enthalpy relaxation overlap. The ability of MTDSC to separate fast (e.g., glass transition) from slow (e.g., reaction, enthalpy relaxation) processes is very beneficial in this respect. An isothermal cure experiment was stopped after different times, and the subsequent heating runs shown in Fig. 2.111 reflect different reaction stages. The glass transition is found in the reversing signal well separated from the peak, due to the chemical reaction and enthalpy relaxation found in the nonreversing heat flow. These experiments are useful to determine T_g–conversion relationships. Enthalpy relaxation peaks (an endothermic peak around 100 °C for curve 3 in Fig. 2.111) begin to emerge when the isothermal cure time exceeds the onset of vitrification (see also Fig. 2.112).

2.13.11.3. Isothermal Cure of Epoxy–Amine Systems: Benefit of Heat Capacity and Heat Flow Phase Angle Signals

The ability to separate overlapping phenomena proves to be especially useful in curing systems as reported by Van Assche et al. (1995, 1996, 1997). As an example, the simultaneous measurement of the nonreversing heat flow, heat capacity, and heat flow phase angle is shown during the quasi-isothermal cure of diglycidyl ether of bisphenol A (DGEBA) and methylenedianiline (MDA) at 70 °C, 80 °C, and 100 °C in Fig. 2.112. Stoichiometric mixtures were prepared and quenched in small amounts in liquid nitrogen to prevent reaction before the MTDSC experiment. Samples were introduced at low temperatures in the DSC cell

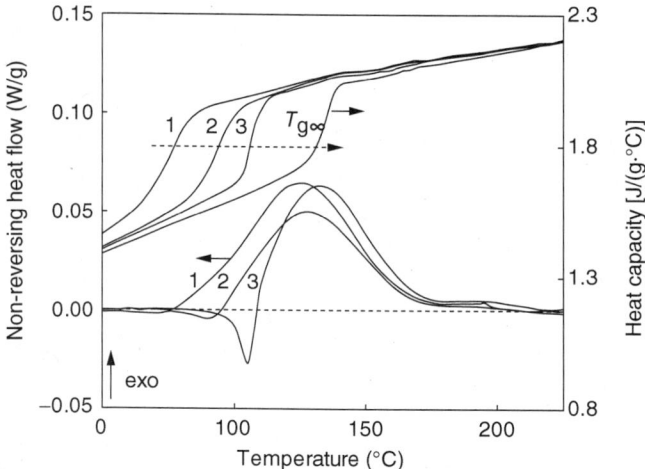

Figure 2.111. Separation of the reaction peak into nonreversing heat flow and glass transition in heat capacity signal (proportional to reversing heat flow signal) for DGEBA (diglycidyl ether of bisphenol A) + anhydride (methyl-tetrahydrophthalic anhydride, 1% 1-methyl imidazole); heating at 2.5°C/min (1°C underlying heating rate, 60s modulation period) after partial cure at 85°C for 165 min (1), 230 min (2), and 800 min (3) and second heating after full cure ($T_{g\infty}$) [data reproduced from Van Mele et al. (2006) with permission of Springer-Verlag].

(10–15°C) and quickly heated to the cure temperature of interest at a rate of 40°C/min.

The reaction heat appears in the nonreversing heat flow, which corresponds to the heat flow in a conventional DSC experiment. This signal can be linked to the heat evolved when amine groups react with epoxy rings, as explained earlier in this section. The reaction exotherm displays typical autocatalytic behavior; the buildup of hydroxyl groups that catalyze the epoxy–amine reaction is responsible for the initial heat flow increase, whereas the heat flow decrease in the later stages reflects the depletion of reactive groups or the completion of reaction under these conditions.

The initial increase in the heat capacity signal corresponds to the *reaction heat capacity* or the change in heat capacity from reactants to products (see arrow in Fig. 2.112). A thermodynamic analysis of the epoxy–aromatic amine reaction revealed that the primary amine–epoxy reaction contributes less to the increase in reaction heat capacity than does the secondary epoxy–amine reaction (Swier and van Mele 2003b). Information specific to the different steps in the reaction mechanism can therefore be deduced from the heat capacity signal, in contrast to the global conversion evolution obtained from the total heat flow signal.

The abrupt, stepwise decrease in C_p, first reported by Cassettari et al. (1993), is related to the vitrification transition of the thermoset (see right arrow in

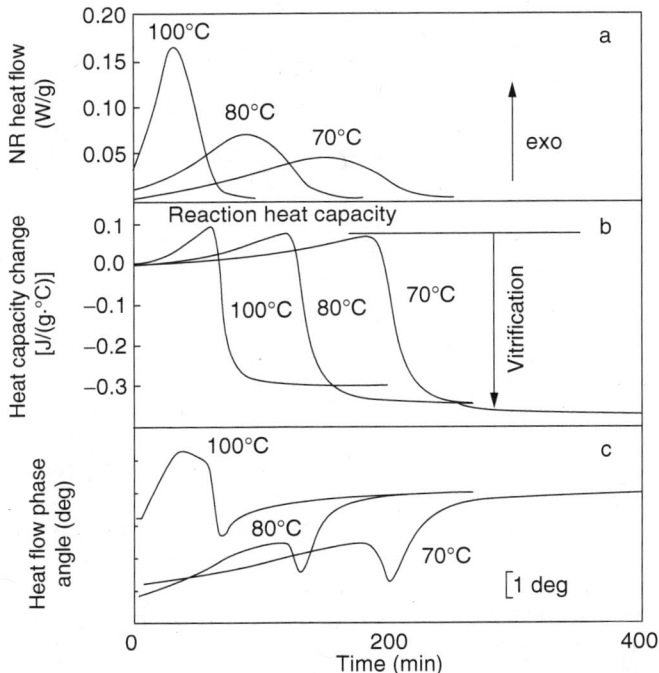

Figure 2.112. Nonreversing (NR) heat flow (a), heat capacity change (ΔC_p) (b), and heat flow phase (ϕ) (c); for the isothermal cure of stoichiometric diglycidyl ether of bisphenol A (DGEBA) and methylenedianiline (MDA) mixture at 70/80/100 °C; the increase in ΔC_p due to reaction and the stepwise decrease due to vitrification are indicated on the graph (1 °C/60 s) [data reproduced from Swier et al. (2004) with permission of John Wiley & Sons, Inc.].

Fig. 2.112b), or the transition from a rubbery or liquid state to a glassy state (Gillham and Enns 1994). The T_g–conversion relation, shown in Fig. 2.113, as compared to the cure temperature range in Fig. 2.112, predicts vitrification at some point in the reaction path. In other words, vitrification will occur during isothermal cure if the cure temperature is below the glass transition of the fully cured thermoset $T_{g\infty}$. This transition can be compared to the decrease in heat capacity observed when cooling a nonreactive material to a temperature below its T_g. Similar information can be gathered from the heat flow phase angle (see discussion around Fig. 2.92), which is especially suited to detect relaxation phenomena during cure (Van Assche et al. 1997). Parts of a time–temperature–transformation (TTT) diagram can be constructed on the basis of the conversion as a function of time from the heat flow signal and the vitrification time from the heat capacity (reversing) signal (gelation cannot be determined). This simultaneous determination is more accurate than using a separate rheology experiment to determine vitrification time.

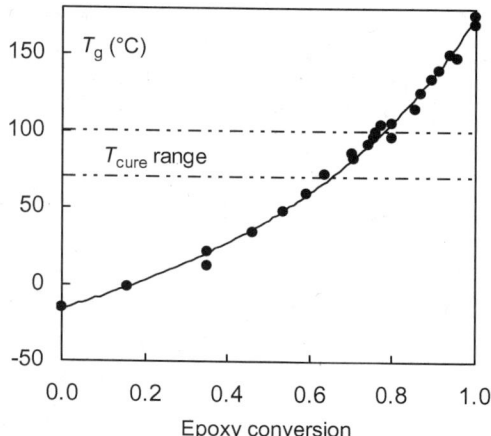

Figure 2.113. Glass transition temperature–conversion relation for stoichiometric DGEBA + MDA mixtures as obtained from MTDSC experiments; the cure temperature range (T_{cure}) is also shown (Swier, unpublished results).

2.13.11.4. Vitrification and Devitrification during Nonisothermal Cure of an Epoxy–Amine System

Vitrification will occur during nonisothermal cure if the glass transition increases faster than the cure temperature, which will be especially important at slow heating rates. This same phenomenon has been investigated by DMA, as described in Chapter 5. These conditions will result in diffusion-controlled reactions and can be beneficial for curing large samples, where the exothermic reaction can result in inferior mechanical properties if the rate of heat production is greater than the rate of heat dissipation (Gillham and Enns 1994). The selection of low heating rates or combined cure paths can be useful in reducing internal stresses and can also be used to control reaction rates of highly exothermic reactions. Breitigam et al. (1993) also noted that heating large parts slowly enough (e.g., 1.5°C/min) to ensure reaction in the vitrified state can replace lengthy isothermal cures, possibly reducing autoclave times by 75%.

Vitrification in nonisothermal conditions is shown for the DGEBA + MDA system in Fig. 2.114b. The region above the dotted line corresponds to a material in the glassy state ($T_g > T_{cure}$), while liquid or rubbery materials will exist below this line ($T_{cure} > T_g$) (van Assche et al. 1996). Initially, the material is in the glassy state and devitrification occurs around −16°C (T_{g0}), as can be seen in the heat capacity (reversing) signal [plot (c), 1st]. Reaction starts around at 50°C, where a slight increase in T_g is noticed. The steepest increase in T_g is seen at the maximum in the nonreversing heat flow, namely, at T_{cure} equal to 115°C. Since the increase in T_g is much faster than the increase in T_{cure}, T_g rises above T_{cure}, which induces vitrification. This can be seen in the heat capacity signal as a stepwise decrease superimposed on the increase in C_p due to reaction, starting at 130°C. Reaction-induced vitrification occurs partially as can

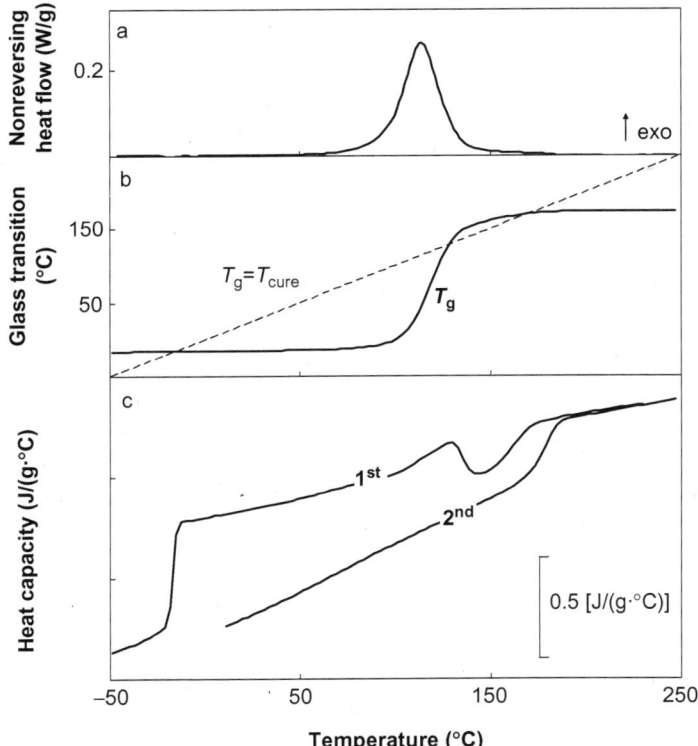

Figure 2.114. Nonisothermal cure at 1 °C/min of a stoichiometric DGEBA + MDA mixture: (a) nonreversing (NR) heat flow; (b) glass transition as a function of temperature calculated from the total heat flow signal; (c) heat capacity (reversing signal) for the first (1st) and second (2nd) heating (1 °C/60 s modulation period) [reprinted from Swier et al. (2004) with permission of John Wiley & Sons, Inc.].

be inferred from the second heating [plot (c), 2nd], where a lower C_p level is found in this temperature range.

2.13.11.5. Reaction-Induced Phase Separation

The demand for tailor-made epoxy composites and the need for application-specific design have initiated interest in multicomponent epoxy systems, for example, to achieve materials with improved impact resistance (Hodgkin et al. 1998; Williams et al. 1997). Typically, a homogeneous three-component reactive mixture, consisting of an epoxy resin, an amine hardener, and a polymeric modifier, phase-separates into a desired heterogeneous morphology, due to the unfavorable entropy contribution of the growing epoxy chain [reaction-induced phase separation (RIPS)]. This is shown schematically in Fig. 2.115, together with possible positions of the glass transition of the modifier ($T_{g,mod}$) with respect to the initial epoxy–amine T_g, the cure temperature (T_{cure}), and the final, fully

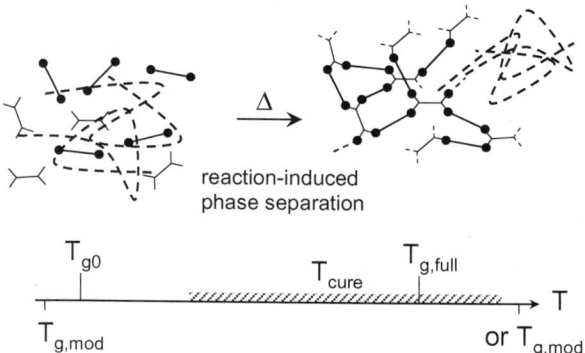

Figure 2.115. Reaction-induced phase separation (RIPS) occurs when a homogeneous mixture of epoxy (bifunctional "sticks"), amine (tetrafunctional "sticks"), and polymeric modifier (dashed line) is heated and the epoxy–amine network formation induces phase separation. The position of T_{cure} to the glass transition of the modifier $T_{g,mod}$ determines whether vitrification of a phase rich in this component will occur during cure.

cured epoxy–amine T_g ($T_{g\infty}$). While most modifiers result in morphologies in the micrometer range, more recent interest in nanostructured thermosets has initiated research in amphiphilic block copolymers as well (Mijovic et al. 2000; Hillmyer et al. 1997; Lipic et al. 1998).

2.13.11.5.1. Vitrification of a High-T_g Phase Formed during Reaction-Induced Phase Separation.

Cure of a linearly polymerizing epoxy–amine system (DGEBA+aniline) modified with 20 wt% poly(ether sulfone) (PES: T_g = 223 °C) is shown at three temperatures in Fig. 2.116. The nonreversing heat flow reflects the reaction rate, while the initial increase in the heat capacity signal contains complementary information on the epoxy–amine reaction. The decrease in the heat capacity signal accompanied by a relaxation peak in the heat flow phase angle marks the time at which the T_g of the PES-rich phase (T_g pure PES = 223 °C) rises above the cure temperature and the mobility of this phase is frozen in.

This effect is illustrated in Fig. 2.117, where glass transitions are measured at intermittent times during cure. Comparing the evolution in the heat capacity change (reversing signal) from time zero [real-time signal during the isothermal experiment at 100 °C) (a)] at different times with the resulting thermal properties obtained from intermittent heating experiments (b) indeed indicates that cure time "4" marks the time where the high-T_g phase crosses the cure temperature. After a long cure time "5," this phase is almost completely frozen. This will have important implications for the diffusion rates of the coexisting phases in this composite. Phase separation, deduced from the cloud point, occurs at cure time "2," where two overlapping glass transitions are also

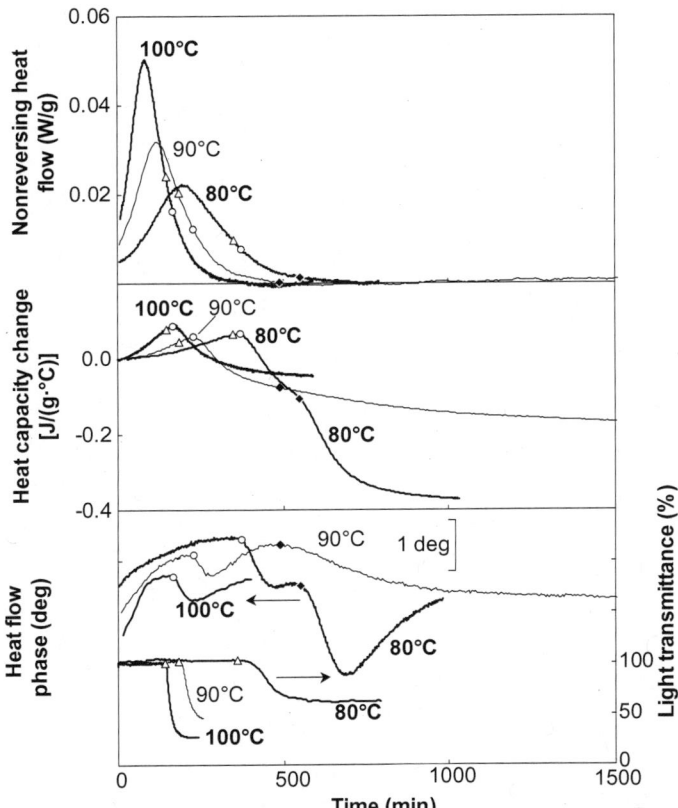

Figure 2.116. Nonreversing heat flow, change in heat capacity, and heat flow phase signal from MTDSC and percent of light transmittance from optical microscopy (OM) for the reactive blend DGEBA + aniline ($r = 1$)/20 wt% PES cured at 100/90/80 °C; the cloud point from OM; (Δ), onset of heat flow phase relaxation corresponding to vitrification of PES-rich phase ($\bigcirc$) and epoxy–amine-rich phase ($\blacklozenge$) are shown and also indicated in the heat flow and heat capacity signals [reprinted from Swier and Van Mele (2003a) with permission of Elsevier Ltd.].

detected (Fig. 2.117b). However, the decrease in the real-time heat capacity signal occurs later at cure time "4" since both phases are still mobile before that time. An indirect, postponed phase separation is therefore detected with MTDSC.

When the cure temperature is below the full cure glass transition of the epoxy–amine ($T_{g\infty} = 95\,°C$), vitrification of the epoxy–amine-rich phase also takes place. This clearly occurs in Fig. 2.116 for the reaction at 80 °C. The heat flow phase angle is important in this respect, showing a second relaxation peak. Partial vitrification is seen at 90 °C, and it is again most obvious in the heat flow phase or phase angle.

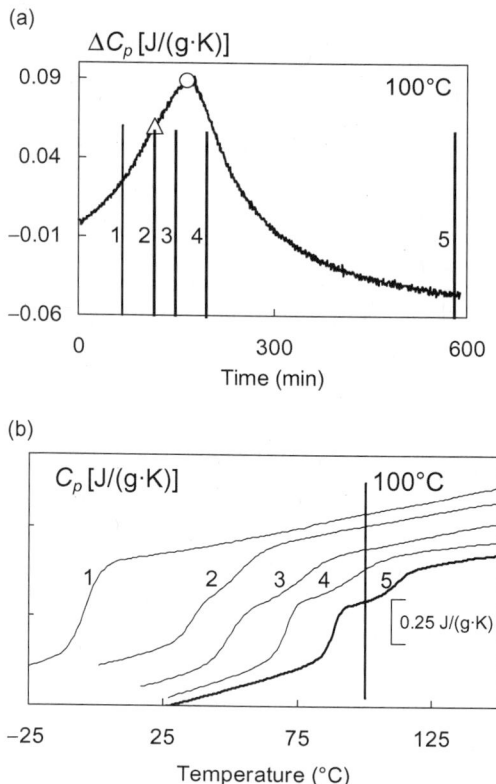

Figure 2.117. Cure of the reactive blend DGEBA + aniline ($r = 1$)/20 wt% PES at 100 °C: (a) change in heat capacity (reversing signal) during isothermal cure at 100 °C, with cloud point from OM (△), onset of heat flow phase relaxation (○); (b) glass transitions measured after different isothermal cure times at 100 °C (heating rate = 2.5 °C/min); numbers indicate cure times on (a) (Swier, unpublished results).

2.13.11.5.2. Excess Contribution in Heat Capacity (Reversing) Signal as an Indication for Phase Separation. The heat of phase separation can be detected by conventional DSC to determine the onset of *temperature*-induced phase separation in partially miscible polymer blends (Arnauts et al. 1994). However, the simultaneous occurrence of the much larger reaction enthalpy in RIPS dominates the heat flow signal. Miscibility gaps in nonreactive, low- and high-molecular-weight blends have been constructed using DSC, although poorly resolved signals are obtained that are difficult to interpret (Shen and Torkelson 1992; Ebert et al. 1986). MTDSC, and in particular the heat capacity (reversing) signal, proves to be useful for determining the cloud point temperature associated with phase separation in polymer blends showing LCST (lower critical solution temperature) behavior (Dreezen et al. 2001). This signal is an apparent heat capacity in this case since excess contributions arising from heat effects of demixing and remixing on the timescale of the modulation are retrieved. When this methodology is extended to RIPS, reaction and phase separation are separated into the nonreversing heat flow and the heat capacity (reversing) signal, respectively (Swier and Van Mele 2003a). Note that this method is suited mostly for analyzing thermosets modified with low-T_g polymers, in contrast to the technique outlined in the previous subsection a for high-T_g polymers.

In summary, the benefits of MTDSC in the characterization of reacting polymer systems are as follows:

1. Simultaneous measurement of the cure exotherm and glass transition in nonisothermal conditions, separated into the nonreversing heat flow and heat capacity (reversing) signal, respectively
2. Separation of relaxation effects associated with the glass transition into the nonreversing heat flow during nonisothermal experiments
3. Isothermal detection of the cure exotherm and heat capacity changes, reflecting effects such as the reaction heat capacity, vitrification, and reaction-induced phase separation
4. Development of mechanistic models where diffusion control is taken into account, using the mobility factor determined from the heat capacity

2.13.12. Other Methods

2.13.12.1 Thermal Conductivity *Thermal conductivity* is an intensive quantity and vector property that characterizes the ability of a material to conduct heat. ASTM defines thermal conductivity as the "time rate of heat flow under steady conditions, through unit temperature gradient in the direction perpendicular to the area" (ASTM E-1142). Thermal conductivity has the dimension W/(m·K). We find the use of thermal conductivity in our everyday lives, for example, in building materials. Therefore, thermal conductivity is an important physical quantity, although we are more familiar with an associated R value (thermal resistivity) (which is the inverse of the conductivity). Thermal diffusivity can be obtained if one knows the density and heat capacity of the material, that is

$$D = \frac{\lambda}{\rho C_p} \qquad (2.101)$$

where D is the thermal diffusivity, λ is the thermal conductivity, ρ is the density, and C_p is the specific-heat capacity.

A detailed experimental procedure for measuring thermal conductivity is given in the ASTM standard E1952 using MTDSC. Essentially, for determination of thermal conductivity, heat capacity measurements are made under two conditions: a good measurement and a poor one. In order to obtain heat capacity values under these two conditions, one can use two different circular specimens. The ASTM method recommends using a thin disk less than 0.5 mm thickness and a cylinder thicker than 3 mm. Good heat capacity results are obtained when experimental conditions are selected to obtain maximum thermal uniformity through the specimen. Therefore, one needs to use a thin specimen with the experimental parameters set to long oscillation periods. A complete encapsulation of the test specimen using sample pans of high conductivity such as aluminum is recommended. For poor heat capacity measurements, the thick specimen is placed over aluminum foil, wetted on both sides with silicone oil. Since the temperature oscillation is applied to only one side of the test specimen, a temperature gradient is created. The ratio between the two heat capacities is then used to determine the thermal conductivity. For right circular cylinders, the thermal conductivity can be determined from the following equation:

$$\lambda = \frac{8LC^2}{C_p m d^2 P} \qquad (2.102)$$

Here L is the sample length, d is the diameter, m is the mass, C_p is the specimen's specific-heat capacity, C is the measured apparent heat capacity of the thick sample, and P is the modulation period. Derivation of the thermal conductivity equations for MTDSC can be found in Blaine and Marcus (1998).

One should note that this method is not without its limitations. Notably, this covers only homogeneous, nonporous solid materials whose thermal conductivity is in the range of 0.10–1.0 W/(m·K) (ASTM E1952-2001). Included in this range are many polymeric, glass, and ceramic materials (Marcus and Blaine 1994). Examples of the use of this method can be found in the literature (Weese 2005; Lopes and Felisberti 2004). The thermal conductivity reproducibility standard deviation by this method given in the E1952 ASTM standard for polystyrene and poly(methyl methacrylate) is 27% and 10%, respectively.

2.13.12.2. Step Scan DSC (SSDSC) *Step scan DSC* (SSDSC) is a subset of modulated temperature DSC (MTDSC) where repeated ramps linked to a subsequent isothermal hold are used to separate reversing and nonreversing phenonena (Cassel et al. 1999). Different from other MTDSC techniques, this technique is used in either the heat or cool mode only, and not a mixture of both. As with other MTDSC techniques, SSDSC is highly dependent on the run parameter. The ramps are best performed at fast rates over short temperature intervals, although the hold times vary. However, in practice, slow to moderate ramp rates are generally utilized. The step scan technique can be used on any DSC, although if very fast heating rates are desired, this may limit the instrument selection. Power compensation DSC has been typically

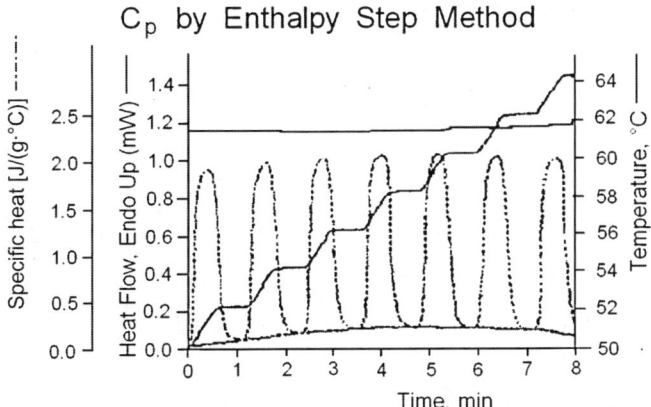

Figure 2.118. Determination of the reversing heat capacity from a SSDSC experiment [from Cassel et al. (1999); reprinted with permission from the North American Thermal Analysis Society (no experimental conditions were reported in that publication)].

employed for this technique because of the low thermal mass of their cells, allowing fast heating and cooling rates.

The SSDSC technique considers that the reversing phenomenon goes into the ramp segment while the nonreversing effects are separated in the isothermal step. If plotted in heat flow versus time, a peak can be noted for the ramp from one isothermal segment to the next, especially if a short step (e.g., 1 °C) is utilized. The reversing component can then be obtained by integrating the area of the peak created by the step and dividing by the step temperature difference. When converted to heat capacity this signal is denoted as the *thermodynamic C_p* (Fig. 2.118).

The nonreversing effects can be noted in the IsoK baseline (Ye 2006). The IsoK baseline is the heat flow curve at "zero ramping" or for the isothermal hold after the ramp. In Fig. 2.119 the signals for the step scan DSC are shown, the IsoK baseline can be noted in the raw data following the return to baseline after the ramp segment. Slight overshoots in the heat flow may occur as the calorimeter obtains steady state.

2.13.12.3. Glass Transition Step scan DSC has been used to monitor the glass transition during curing of an epoxy prepreg by separating it from the reaction exotherm (Bilyeu and Brostow 2002), as noted in Fig. 2.120. The sample in Fig. 2.120 is mixture of tetraglycidyl 4,4-diaminodiphenyl methane (TGDDM) and 4,4′-diaminodiphenylsulfone (DDS). The scans were heated at 10°C/min in 5°C steps with 30-s holds. Similar T_g values were obtained as a function of curing time and temperature when compared to standard MTDSC runs. However, one should be cognizant of obtaining a steady isotherm because of the curing reaction that occurs after heating through the glass transition.

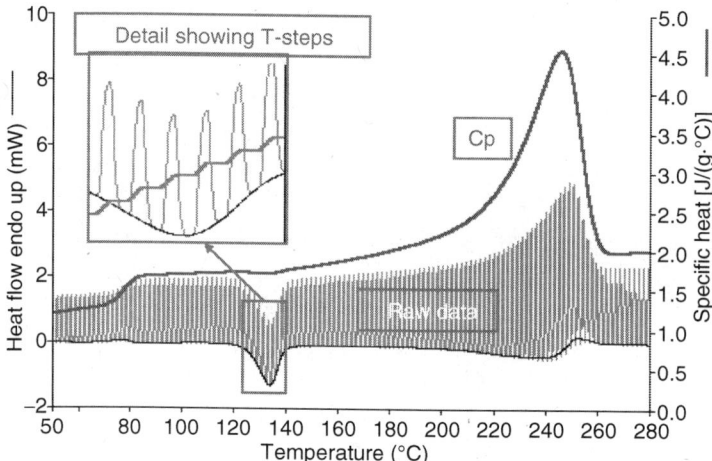

Figure 2.119. Separation of a scan of PET into the nonreversing component or IsoK baseline and the reversing or "thermodynamic heat capacity" from an SSDSC experiment from the raw data labeled above. Upper curve shows the reversing heat capacity component; bottom curve, is the IsoK baseline. An insert of the cold crystallization process demonstrates the temperature steps with the resultant modulated effect on the heat flow with its associated IsoK baseline. [From Ye (2006); courtesy of Perkin-Elmer.]

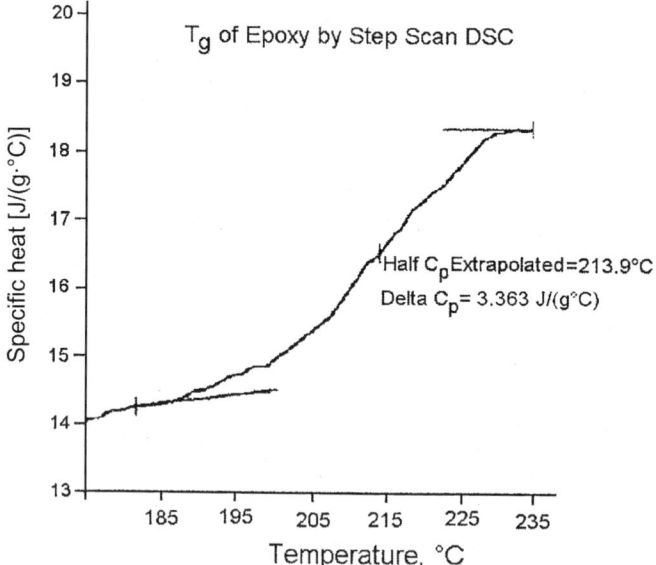

Figure 2.120. Separation of glass transition from the curing exotherm, into the thermodynamic or reversing heat capacity; the sample is a mixture of tetraglycidyl 4,4-diaminodiphenyl methane (TGDDM) and 4,4'- diaminodiphenylsulfone (DDS) [from Bilyeu and Brostow (2002); reprinted with permission from John Wiley & Sons, Inc.].

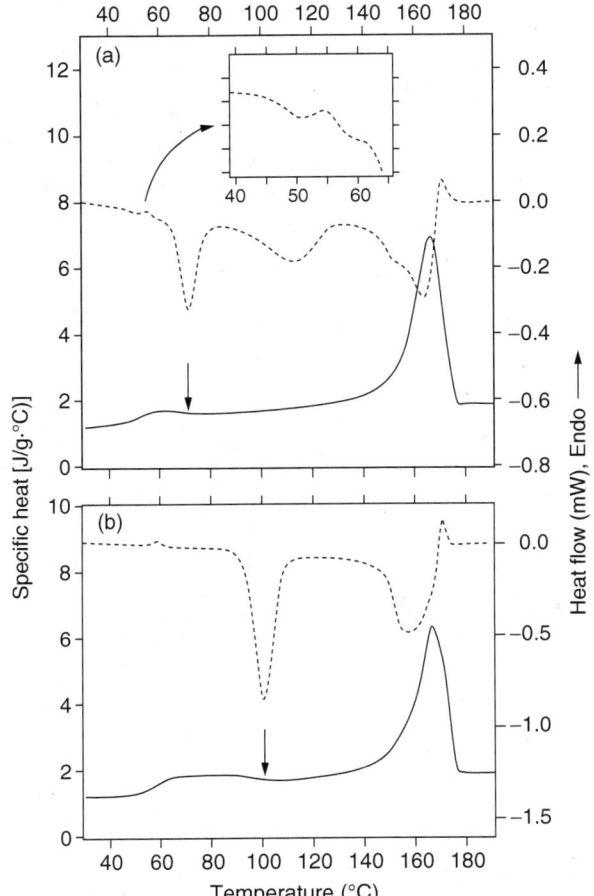

Figure 2.121. (a) As-prepared PLLA-FDA; (b) reference bulk sample. The reversing and nonreversing heat capacities are shown by the solid and broken curves, respectively. The arrows indicate cold crystallization. [From Sasaki et al. (2005); reprinted with permission from John Wiley & Sons, Inc.]

Step scan DSC has also shown the ability to separate cold crystallization from the glass transition as was noted for freeze-dried poly(L-lactide), PLLA (Sasaki et al. 2005). Separation of the cold crystallization can be noted in the Fig. 2.121 for PLLA sample A using a freeze-drying apparatus (PLLA-FDA). Heating rates of 5°C/min with 2°C steps were used.

2.13.12.4. Melting Step scan DSC has been used to study a number of melting processes. Papageorgiou et al. (2006) investigated the effects of annealing at different temperatures for different molecular masses of poly(1,3-propylene terephthalate) samples. Step ramp rates of 5°C/min were utilized

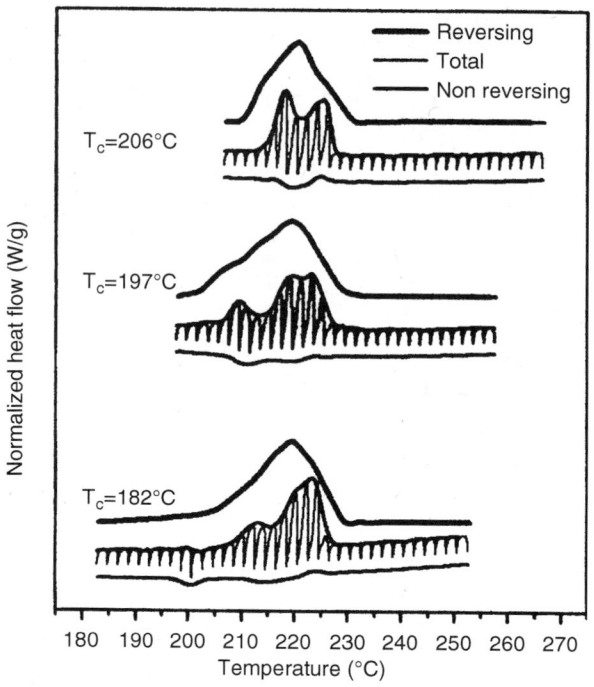

Figure 2.122. SSDSC traces for the low-molecular-weight poly(1,3-propylene terephthalate) [Papageorgiou et al. (2006); reprinted with permission from Elsevier Ltd.].

with an average heating rate of 2.5 °C/min in 2 °C steps. The authors were able to notice multiple melting peaks, as can be seen in Figs. 2.122 and 2.123, suggesting reorganization.

Multiple melting behavior of poly(3-hydroxybutyrate-co-hydroxyvalerate) was investigated using average ramp rates of 2 °C/min (Gunaraine and Shanks 2005a,b). The authors were able to note additional reorganization for the copolymers versus the poly(3-hydroxybutyrate) homopolymer by the exothermic activity noted in the IsoK baseline. SSDSC analysis of poly(ethylene oxide) showed that changing ramp rate, step size, and isothermal hold influenced the resulting curve (Pielichowski et al. 2004). For example, ramp rates of 0.25–2 °C/min were utilized for different runs of identical thermal history. With a faster heating rate, a decrease in the reversing component was noted (Fig. 2.124), which may be related to an incomplete separation of the responses.

2.14. HOW TO PERFORM DSC MEASUREMENTS

Before carrying out a DSC experiment, the operator needs to select several parameters. It should be emphasized that there are no optimum parameters

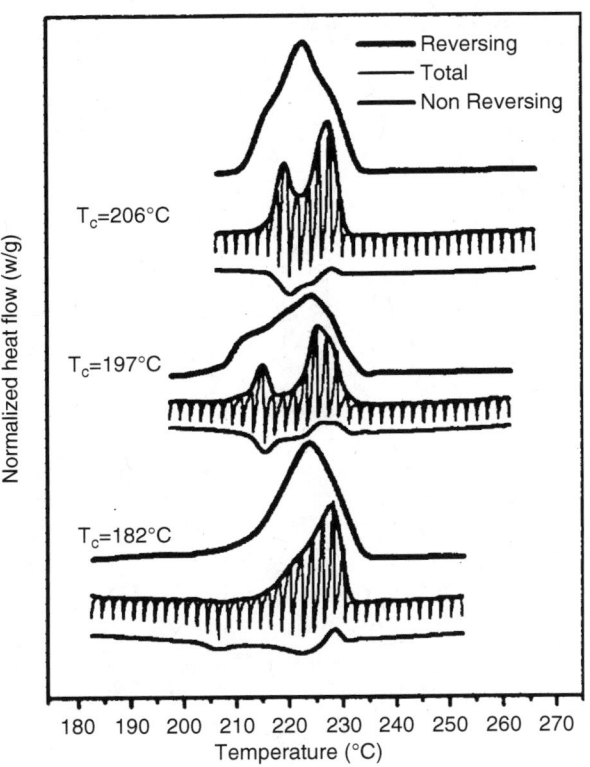

Figure 2.123. SSDSC traces for the high-molecular-weight poly(1,3-propylene terephthalate) [Papageorgiou et al. (2006); reprinted with permission from Elsevier Ltd.].

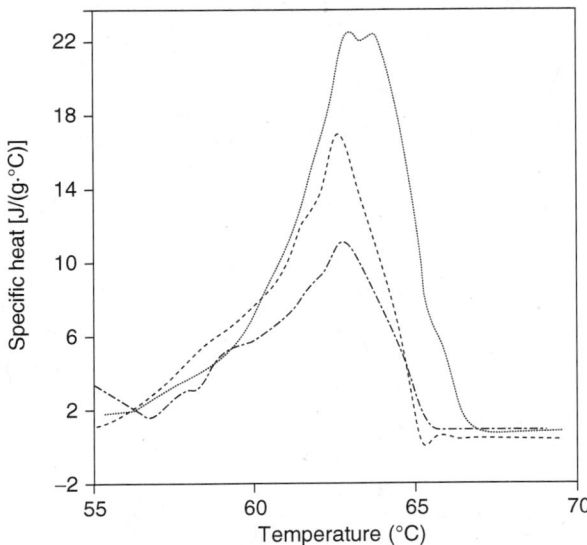

Figure 2.124. Reversing component of poly(ethylene oxide) at different ramp conditions: 0.5 °C/min for the dotted line, 1 °C/min for the dashed line, and 2 °C/min for the solid line [from Pielichowski et al. (2004); reprinted with permission from Elsevier Ltd.].

in general; rather, the run parameters depend on the information that is sought.

2.14.1. Sample Preparation

Before loading the sample into the DSC instrument, it has to be encapsulated into DSC pans. The pans used for polymers are usually made of high-purity aluminum. For high-temperature measurements of inorganic samples, often copper, gold, silver, or platinum (sometimes graphite and sapphire) pans are used, because aluminum melts at 660 °C, and alloys with the material of the sample holder. An overview of the most frequently used DSC pans is given in Section 2.3.

For "everyday" DSC measurements on polymers, the usual standard aluminum pans should be used (Figs. 2.7 and 2.8). These pans should never be used without a lid, because the major function of the lid is to push down the sample to the bottom of the pan, ensuring good thermal contact. The disadvantage of standard pans is that modern crimpers tightly fold down the pan edge on the lid, so when it is desirable to evaporate some solvent from the sample, the solvent may be slow to leave the pan. In these cases, it is better to punch out lids from aluminum mesh as shown in Figs. 2.7 and 2.8.

After crimping, the bottom of the sample pan should be examined. It should be flat. Usually there is no problem when fine polymer powder or polymer film has been crimped, but crimping grainy polymer (i.e., rough "powder" or pieces with an uneven shape) may and often does lead to an uneven surface of the bottom of the sample pan. Running such samples should be avoided, since the external thermal resistance will be considerably increased owing to poor contact between the sample pan and the DSC cell. In such cases, the grainy sample can be freezer-milled to obtain fine polymer powder. However, care should be taken when running such freezer milled samples, since many free radicals are generated by mechanical degradation during freezer milling, and often exothermic peaks can be observed above the melting point of the polymer. Such samples should not be subjected to first heating–cooling–second heating temperature programs. It should be mentioned that special Tzero pans are available from TA Instruments. Crimping these pans will ensure as flat a bottom of the pan as possible.

Hermetically sealed pans should be used for volatile (or liquid) samples. After the liquid sample is placed on the bottom of the DSC pan, it is covered with a lid, and the edges of the bottom and the lid are cold-welded together with a small instrument called a *crimper*. The process itself is called *crimping*. In addition to measurements on low-molecular-mass liquids, these sample pans should be used for liquid samples such as unreacted thermosets and when the sample contains volatile components such as residual solvent or low-mass additives whose evaporation one wishes to suppress (these pans are also used for polymers that can release degradation products). Evaporation of these materials is endothermic and may interfere with exothermic events such as thermoset cure, and they can lead to undesirable contamination of the sample holder. These DSC pans hermetically seal off the sample from the environment. Naturally, the pressure inside the pan increases with the increasing temperature during the DSC experiment, and these pans can withstand pressures from 0.2 MPa (~2 atm) to 3 MPa (30 atm) before rupturing. A definite disadvantage of these pans is that their area of contact with the cell is smaller than for the standard pans.

High-pressure DSC capsules are used in cases when the pressure in the DSC pan may reach 20 MPa (~200 atm) as a result of heating to high temperatures, and it is undesirable to release some solvent or gases of decomposition into the DSC cell. These pans are rarely used in measurements on polymeric samples.

The mass of the sample pan and the reference pan should be similar, especially for specific heat capacity measurements.

2.14.2. Sample Mass

The sample should be weighed on an electronic microbalance to at least ±0.1% accuracy, so a 10-mg sample should be weighed to ±0.01 mg. The optimum sample mass is different in different DSC experiments. The sample mass usually should be minimized so as to decrease the thermal lag and the temperature gradient within the sample (see Section 2.5 on calibration). But sometimes higher sample masses should be used in order to increase the sensitivity of the instrument. Very small (1–3 mg) sample mass must be used in purity determinations, since the applied van't Hoff equation is based on equilibrium thermodynamics, and it is desirable to be as close to equilibrium conditions as possible. Again, small (≤5 mg) sample masses must be used in MTDSC experiments, because the rate of the momentary temperature increase and decrease due to modulation can be considerable. At such high momentary heating or cooling rates, high-mass samples will not be able to follow the modulation. Of course, the sensitivity of the instruments will be smaller at small sample masses, but resolution will be improved because of small temperature gradients in the sample. On the contrary, in specific heat capacity measurements large sample masses (up to 20 mg) are desirable in order to increase the sensitivity. In such measurements, resolution is not an issue. In melting, nonisothermal and isothermal crystallization experiments the optimum sample weight is around 10 mg. Of course, it depends on the applied heating or cooling rate (see Section 2.10 for a discussion of appropriate sample sizes for thermosets).

2.14.3. Loading the Sample into the DSC

After the sample has been prepared, it is loaded into the DSC pan, usually at room temperature. If the experiment demands, the sample can be loaded into the cell at high temperature (although this is a rare case; isothermal cure in a heat flux DSC is an example). Technically it is more difficult to load the sample at low temperatures. An example of this is when a sample has been quenched to low temperatures, and its heating trace should be recorded from low temperatures without first warming to ambient. In these cases, if no precaution is taken, excessive condensation may damage the DSC sample holder. Therefore, in these experiments the purge gas (see Section 2.3 for purge gas selection criteria) has to be very dry (perhaps two or three dryers can be connected in series before the purge gas enters the cell), and the DSC cell should be well isolated from the environment, usually with a glovebox. It should be kept in mind that even with these precautions, the sample holder may be damaged when opened at negative temperatures.

2.14.4. Carrying out the Actual DSC Measurements

After all the preparations described above have been done, a proper temperature program should be selected or created. Depending on the purpose of the measurement, heating, cooling, or isothermal measurements can be applied. The primary purpose of everyday routine DSC experiments is to measure the melting point, heat of fusion, heat of reaction, crystallization temperature, and glass transition temperature of a given polymeric (or low-molecular-mass) sample, or to characterize the crystallization of a polymer or cure of a thermoset through measurements of isothermal and often multiple heating rate kinetics. The heat of fusion (which must be the same as the heat of crystallization) is usually determined in heating experiments, for historical reasons, such as better baseline, etc. In addition, it is technically easier to determine the heat of fusion than the heat of crystallization, because baselines of sufficient length are rarely available in crystallization experiments for using the sigmoidal baseline option at low temperatures due to supercooling. Similarly, the glass transition temperature and the heat capacity jump at the glass transition usually are determined in heating experiments, although it would be considerably more logical to determine these parameters during cooling (see Section 2.7).

Very often, in routine DSC measurements of polymers, first heating–cooling–second heating runs are made. The first heating and the cooling give information on the thermal and mechanical history (if the final temperature of the first heating does not exceed the equilibrium melting point of the polymer). At the same time the first heating erases the processing history if the final temperature of heating exceeds the equilibrium melting point, so that the second heating will characterize the polymer itself.

2.14.4.1. First Heating
The results calculated from the first heating provide a "fingerprint" of the sample, including its processing and thermal and mechanical history. It is well known that the crystallization of polymers in most cases takes place far from the equilibrium conditions. Cooling rate, pressure, orientation—all these will have an effect on the glass transition temperature, melting

point, and the shape of the DSC curve. Before the heating starts, the sample should be equilibrated, and this means a 5–10-min waiting period at the starting temperature (T_1). The heating should start well below the glass transition temperature. It should be remembered that in the first ~20 °C part of the DSC curve there are no steady-state conditions, so this part of the DSC curve cannot be used for thermal analysis purposes. If, for example, the glass transition is in this temperature region, the operator will not be able to properly evaluate it. Therefore, it is recommended that heating start ~50 °C below the glass transition temperature. Also, since the melting of polymers is extremely broad when compared to low-molecular-mass substances, and the smallest and most imperfect crystallites melt far below the melting point, heating should start at least 100 °C below the melting point if the melting transition is of interest. The heating rate should be carefully selected, since extended-chain crystals that may be present in the sample can superheat, and reorganization during melting can also take place (see the appropriate sections in this chapter).

2.14.4.2. Cooling and Second Heating

In the first heating the sample should be heated to a temperature (T_2) above the equilibrium melting point (T_m°) of the given polymer in order to destroy any ordered regions, like row nuclei in polypropylene, and also to record a sufficiently long postmelting baseline necessary for accurate integration with the sigmoidal baseline option. Some time (2–5 min) should be allowed at temperature T_2 so that the ordered regions in the sample can fall apart. It should be kept in mind that some polymers may thermally degrade in the vicinity of their equilibrium melting point. The selection of the cooling rate for the subsequent cooling experiment is important, and, of course, this depends on the information the operator is looking for. If the goal is to achieve high crystallinity, the cooling should be as slow as possible. If the cooling rate is high, the polymer sometimes can be quenched to an amorphous state. Also, at slow heating, the reorganization during the subsequent second heating may be considerable. The intensity of such reorganization will be determined by both the cooling rate during the cooling experiment and the heating rate applied during the second heating. Similar to heating rates, high cooling rates do improve the sensitivity of the recording, but the thermal lag may become excessive.

The DSC curve recorded during the second heating provides information on the structure of the polymer formed during the cooling, and can also be used when samples of the same polymer (e.g., polyethylene) from different sources are compared. Thus, this experiment provides information on the material itself without the effects of processing.

2.14.4.3. Amorphous Polymers

The first heating–cooling–second heating temperature program is also good for amorphous polymers; similar to that mentioned above, the first heating (shape of the glass transition) contains information about the processing and thermal history of the sample, while the second heating characterizes the given polymer when it is recorded after a definite cooling rate. In the first heating the sample should be heated to at least 30 °C above the glass transition temperature to ensure a sufficiently long linear baseline.

2.14.4.4. Importance of Cooling Rate Special emphasis should be placed on the cooling rate. The maximum realizable cooling rate depends on the block temperature of the instrument, and thus the method of cooling. In very old DSC instruments water cooling can sometimes be found, but not in modern DSCs. Lately, the use of mechanical cooling accessories and liquid nitrogen cooling have become popular. It should be remembered that the lower the base temperature of the block, the higher the applicable cooling rate can be. Many instruments do warn the operator if the loss of temperature control can be expected as a result of the high nominal cooling rate. Most instruments do have some means to show that the temperature control is lost. One good indication of loss of temperature control during cooling is a sudden break in the slope of the DSC curve.

2.14.4.5. Experimental Parameters When performing DSC runs, the following experimental parameters have to be carefully selected:

- *Heating Rate.* This should be as small as possible to minimize thermal lag and as large as possible to increase the signal at, for example, the glass transition. For melting and crystallization measurements, it depends on the information being sought. In polymeric melting experiments a faster heating rate helps to avoid reorganization during melting; in the case of slower heating, the melting curve seldom is the melting curve of the original crystallites in the sample, but contains information about the start of the melting of the original crystallites, crystal perfection during melting, and melting of the reorganized crystals (multiple melting phenomena). Heating rate for glass transition measurements is usually 10–20 °C/min; for purity determinations it is 1 °C/min.
- *Cooling Rate.* The result of high cooling rates for semicrystalline polymers is decreased crystallinity, or sometimes quenching to an amorphous state. Lower cooling rates will promote higher crystallinity. The cooling rate will also influence the shape of the glass transition on the subsequent reheating (e.g., enthalpy relaxation or hysteresis peaks).
- *Starting or Ending Isothermal Baselines.* These should be recorded only when and if specific heat capacity data are needed. The isothermal baselines should be stable (i.e., horizontal in appearance, indicating thermal equilibrium) and without evidence of chemical or physical processes. Erroneous heat losses lead to sloping baselines. But baseline slope can also be observed in the case of thermal degradation of the polymeric sample, secondary crystallization, and additional cure.
- *MTDSC Experiments.* The selection of experimental parameters is especially important when running MTDSC experiments; the underlying

heating rate even on the newest TA Instruments Q series DSC should never exceed 10 °C/min. When selecting the type of MTDSC experiment (heat–cool or heat only), the software is capable of selecting the amplitude and modulation period in the newer Q series DSCs, while a table summarizes the necessary values for the older 2920 DSCs. The sample mass is especially important; with samples of larger masses the modulation amplitude and frequency must be decreased. In MTDSC experiments it is critical to decrease the external thermal resistivity as much as possible. Thus, use of film samples is preferred and TA Instruments' Tzero press must be used for crimping the samples to ensure good thermal contact. When possible, standard DSC pans must be used, since these have higher contact area with the cell.

- *Running Fiber Samples.* When running fiber samples in constrained mode, the steel plate on which the fiber is wound up should be as thin as possible. With thick steel plates, excessive baseline sloping will occur.

2.14.5. Evaluation of DSC Traces

These days only computer evaluation is carried out. Thus, of course, the operator is limited to the software provided by the instrument companies. The calculation software products of the various companies have become very similar. From typical DSC runs, most often the following results are reported: the glass transition temperature (T_g), the extrapolated starting and end temperatures of the glass transition (T_1 and T_2 in Fig. 2.20), the heat capacity increase at the glass transition, the peak temperature of melting (T_{mp}), the melting point T_m, the starting temperature of crystallization (T_{c0}), the peak temperature of crystallization (T_{cp}), the heat of fusion (ΔH_f) and the heat of reaction ΔH_{rxn}. For evaluations not provided by the instrumental company, the raw data may be imported to an Excel file, and used for calculations. Here we need to emphasize again the importance of using the sigmoidal baseline for the heat of fusion determination of crystalline polymers.

2.14.6. Hardware Considerations

It is advisable to use a surge protector with your DSC. This may filter out unwanted spikes from the DSC traces. Other hardware considerations are as follows:

1. The heart of the DSC is the sample holder or the DSC cell. This should always be kept clean, and from time to time it should be cleaned and conditioned. The power compensation DSCs should be regularly "burned off"; that is, the sample holder should be opened in air (the purge gas should be shut off), and the temperature should be raised to 700–720 °C for ~10–15 min to burn off organic contaminations. The purge gas helps remove part of the contamination during the DSC measurements, but not all the contamination.

However, prior to the burn off procedure, any suspected inorganic contamination should be physically removed from the sample holder, because these may melt or react with the material of the sample holder, and this may permanently damage the cell. This procedure is also effective with heat flux DSCs, but should be used with more caution. The sample holder of the heat flux DSCs is constructed from several metals having different thermal expansivities, and excessive temperatures for prolonged periods may cause excessive stress and damage the sample holder. The manufacturers have suggestions how to clean the sample holders. TA Instruments recommends and sells special brushes, sometimes the contamination can be removed with solvents on a Q-tip. (Turn off the instrument before treating the cell with a solvent-soaked Q-tip, and make sure that the solvent is fully evaporated before turning it back on!)

2. The instrument baseline should be checked regularly. It is recommended to optimize the baseline at least for a new sample holder, before using the baseline subtraction routine. On the somewhat older types of Perkin-Elmer DSCs, the baseline may be optimized with the ΔT "Balance" and "Slope" knobs. Once the baseline is optimized for a new sample holder, it may not be necessary to do this any more; the baseline subtraction routine handles this task. It is recommended to make a baseline run for each sample run, because the baseline may shift as a result of contamination, environmental temperature variations, and so on. Conditioning ("burning off") the sample holder regularly may ensure reproducible baselines. It is suggested to use identical pan masses for the sample and reference. In heat flux DSCs, this is less of a problem; the baseline is more linear, since only one heater is employed. Thus, in heat flux DSCs a worsening baseline (flatness and slope) may indicate that the sample holder is wearing out, and replacement may be necessary.

3. The quality of the baseline is also important in isothermal measurements. In isothermal crystallization measurements the sample is melted at a temperature beyond the equilibrium melting point, held isothermally at this temperature (T_1), and then quickly cooled to the temperature of isothermal crystallization (T_c), and held there until the crystallization process is completed. A sloping isothermal baseline may indicate the absence of thermal equilibrium. This may take place, for example, if one of the cell covers is tilted, resulting in erroneous heat loss. In such a case, the baseline is tilted at both T_1 and T_c. On the other hand, if the baseline is horizontal at T_1, but slopes at T_c after the crystallization is completed, secondary crystallization may take place (see Section 2.7).

4. The instrument manufacturers suggest not to turn off the instruments regularly. However, when you turn off the instrument, and turn it back on again, make sure that it is warmed up before performing measurements. An hour waiting time is usually sufficient. Similarly, when using a mechanical cooling accessory that has been turned off, make sure that it is in operation for at least 2 h before starting an actual measurement.

5. For low-temperature (i.e., subambient) measurements, the DSC cell should be well isolated to prevent condensation. Several manufacturers sell gloveboxes for this purpose.

6. The cell should be covered to prevent any unnecessary heat losses. This is especially important in instruments in which a lid separately closes the sample and reference holders, as in DSCs of the Perkin-Elmer Corporation. It is important that the lid on the sample and reference holders reproducibly and horizontally covers the holders. Any erroneous lid position leads to unwanted baseline slope even in isothermal measurements, as mentioned above.

2.15. INSTRUMENTATION

As described in Section 2.3, there are two main types of DSC design: heat flux and power compensation. Most DSC equipment is based on the heat flux design, with each manufacturer offering a number of accessories. These accessories, in general, include controlled cooling, autosampler, pressure cell, or photocalorimeter. A current trend in manufacture is toward a modular design so that future enhancements can be made to fit one's needs and budget. In the following synopses, descriptions of various manufacturer's DSC instrumentation are given from vendor information. For a better understanding of this section, the authors first suggest reading Section 2.3. The major instrument manufacturers and their products are described below.

Instrument Specialists (ISI) offers the I Series DSC based on a heat flux design. The ISI DSC is capable of program rates of 0.1–200°C/min with a temperature range of –150–725°C. A pressure cell option is available that allows for pressures up to 1000 psi (approximately 7 MPa or 68 atm). For more information, see www.instrument-specialists.com.

Mettler Toledo (Mettler) now markets the DSC 1 unit, but at the time that this instrumental brief was written, marketed only the DSC823^e (its design is further discussed in Section 2.3, The Basics of Differential Scanning Calorimetry). For both units, Mettler now offers a two-sensor option based on a thermopile construction embedded in a chemically inert and corrosion resistant ceramic material. The sensors differ structurally in their design, although the primary difference is the number of thermocouples incorporated into each thermopile. The performance specifications are given in Table 2.8.

The Mettler DSC has a number of accessories, including a 34-position autosampler. The controlled cooling options include air cooling, a cryostat (circulating bath), a selection of refrigeration units, and liquid nitrogen cooling.

Mettler's pressure DSC is a separate module, the HP DSC827^e. It has a maximum pressure specification of 10 MPa (~100 atm) and a temperature range from ambient to 700°C. The UV unit is an add-on accessory to both

TABLE 2.8. Performance Characteristics of DSC823^e of Mettler Toledo

Performance Feature	FRS5 Sensor	HSS7 Sensor
Number of thermocouples	56	120
Temperature accuracy based on metal standards (°C)	±0.2	±0.2
Temperature precision based on metal standards (°C)	±0.02	±0.02
Signal time constant (s)	1.7	3.9
Indium peak, height to width, 1 mg at 10°C/min	17.0	6.9
TAWN resolution	0.12	0.30
TAWN sensitivity	11.9	56.5
Resolution (μW)	0.04	0.01

TABLE 2.9. Performance Characteristics of DSC 204 F1 Phoenix from Netzsch

Performance Feature	τ Sensor	μ Sensor
Range (°C)	−180 to 700	−180 to 450
Temperature accuracy (°C)	<0.1	<0.1
Temperature precision (°C)	<0.1	<0.1
Calorimetric sensitivity (μW)	<0.3	<0.02
Calorimetric precision (%)	<0.5	<0.5

types of DSC modules. For more information on both DSC units, see www.mt.com.

Netzsch offers the DSC 204 F1 Phoenix as well as a quality control model, the DSC 200 F3 Maia. The Phoenix is based on a heat flux design and has two user-interchangeable sensor options. The τ sensor consists of a pair of constantan disk sample and reference platforms surrounded by a silver plate. The μ sensor uses a doped silicon wafer and is designed for improved sensitivity. The module has a temperature range of −180°C to either 450°C or 700°C depending on the sensor. Its performance features are summarized in Table 2.9. Cooling accessories include compressed air, mechanical cooler, and liquid nitrogen. It has a 64-position autosampler and a UV attachment, as well as baseline correction software. The Maia has a temperature range of −150–600°C with an optional autosampler and various cooling options. Netzsch also offers the DTA 404 PC Ëos® high temperature DTA capable of heating to 1550°C and a pressure DSC. For more information, see www.e-thermal.com.

PerkinElmer Life and Analytical Sciences (PKI, Perkin-Elmer, Perkin-Elmer) offers two types of DSC modules. The first is constructed on a heat flux design, while the second design is based on the power compensation principle. Table 2.10 compares the characteristics of the two DSCs:

TABLE 2.10. Performance Characteristics of Perkin-Elmer DSCs

Performance Feature	Jade DSC	Diamond DSC
DSC type	Heat flux	Power compensation
Sensor construction	Chromel alloy (90% nickel/10% chromium)	Platinum
Range (°C)	−180 to 450	−170 to 730
Temperature accuracy (°C)	±0.1	±0.1
Temperature precision (°C)	±0.02	±0.01
Calorimetric accuracy (%)	±2	<±1
Calorimetric precision (%)	±0.1	<±0.1
Indium height/width (mW/°C) (1.0 mg heated 10°C/min)	Not available	17.6

1. *Jade DSC.* The Jade DSC is Perkin-Elmer's heat flux design. It has disk sensors that are made from a hardened nickel chromium alloy and sit above the sample and reference thermocouples. The block temperature is monitored by platinum resistance thermometers. A low-mass alumina-coated aluminum furnace provides heating rates up to 100°C/min. Various cooling units can be selected, which include a liquid coolant circulator, mechanical cooling, or liquid nitrogen cooling capability. The Jade DSC has a 45-position autosampler and integrated mass flow meters for precise purge flow control and gas switching capability.

2. *Diamond DSC.* The Diamond DSC is Perkin-Elmer's power compensation module. Unlike the heat flux design, this unit consists of two separate holders, each with its own heater and sensor. The holders are made of a platinum/iridium alloy with a platinum resistance thermometer at its base. The calorimeter measures the heat flow based on the power compensation principle. The holders are of very low mass, approximately 1 g each. This allows for very fast heating and cooling rates, both up to 500°C/min. Since the cooling rate is dependent on the lower temperature limit, it has a claimed performance of 400°C/min to −10°C using liquid nitrogen cooling with a helium purge. Table 2.11 lists the different cooling rate conditions for controlled cooling. A photocalorimeter accessory (DSC under UV-irradiation) and a high pressure cell kit are also available for the Diamond DSC. A 44-position autosampler can be purchased as an accessory.

The specifications for the Jade and Diamond DSCs are given in Table 2.10. For more information on both DSCs see www.perkinelmer.com.

Care should be taken with the older Perkin-Elmer units that have a front console of light indicators. The control light on the front console may not indicate that the instrument is in true control during cooling. At cooling rates exceeding 10°C/min, the program cooling rate and the sample cooling rate can

TABLE 2.11. Controlled Cooling Rates for Perkin-Elmer DSCs

Cooling Rate (°C/min)	Liquid Nitrogen Coolant with Helium Purge, Controlled Cooling to
400	−10 °C
300	−80 °C
200	−85 °C
100	−135 °C
50	−165 °C
10	−170 °C

TABLE 2.12. Performance Characteristics of EXSTAR 6000 DSC Line of Seiko Instruments

Performance Characteristic	DSC6100	DSC6200	DSC6300
Type	High-sensitivity	Standard	High-temperature
Temperature range (°C)	−150 to 500	−150 to 725	Ambient to 1500
Maximum heating rate (°C/min)	20	100	40

be considerably different, although the control light is often on in these experiments (Barrau and Judovits, 1999). This evaluation was done through the use of a Perkin-Elmer DSC7 module that allowed access to a sample temperature signal. Another method for evaluating the loss of controlled cooling is to monitor for sudden nonlinearities in the heat flow response. Listings of temperatures where the loss of controlled cooling can occur are given in the Section 2.15 for the current module, the Diamond DSC.

Seiko Instruments offers the EXSTAR 6000 DSC line. The line consists of a high-sensitivity DSC, a standard DSC, and a high-temperature DSC. Accessories include an autosampler, liquid nitrogen cooling, and an UV photocalorimetric attachment. The modules available are listed in Table 2.12. For more information, see www.sii.co.jp or the North American distributor *RT Instruments* at www.rtinstruments.com. *RT Instruments* also sells previously owned, refurbished thermal analysis systems.

Setaram offers four different lines of DSC instrumentation based on a heat flux design using either a heat leak disk or a Calvet detector. The first DSC line, the DSC131, uses a heat leak or what Setaram terms "a plate" construction. This unit has an operating range of −170–700 °C. The unit can be operated with 30- or 100-µL crucibles.

The *Sensys* DSC line is based on the Calvet detector construction. This construction is a three-dimensional array of thermocouples that surround tubular sample and reference holders. The array consists of 10 rings, each containing 12 thermocouples in series. The holders are long tubes that bisect a surrounding furnace. The unit can be operated in either a horizontal or a vertical mode. In the horizontal mode, the sample is inserted into the center of the tube where the sensors are located or, for the vertical mode, suspended. The horizontal position is the most common mode and allows the use of a 48-position autosampler. Loading a crucible at one end, centering it, and then ejecting at the other end does this. Crystallization kinetics can be investigated in this mode by preheating outside the sample tube followed by a quick insertion. The vertical mode is recommended for gas adsorption studies (not a polymer application). Its operational range is –120–830 °C with heating rates of ≤30 °C/min. The maximum sample volume is 250 µL with controlled pressure possible of ≤400 bar (=40 MPa, close to 400 atm). Calibration of the Calvet sensor is performed using the Joule effect, which is an electrically introduced known amount of heat (and can be used to calibrate all Setaram Calvet sensors). Setaram also has a high-temperature Calvet DSC, denoted MHTC (part of the "96-line" thermal analyzers). The MHTC DSC has a temperature range of ambient to 1600 °C and crucibles that range up to 450 µL. The last line of DSC instrumentation offered by Setaram is for biological applications. These units include the Micro DSC III and the Micro DSC VII. Both have a Calvet detector with Peltier heating and cooling for precise control (use of thermocouples to heat or cool instead of as a sensor). Both modules are limited in temperature range and heating rate but are sufficient for most biological applications. The Micro DSC III has a temperature range of –10–120 °C with scanning rates of 0.001–2 °C/min. The temperature range for the Micro DSC VII is –45–120 °C with rates of 0.001–1.2 °C/min. Performance specifications for the Setaram lines are given in Table 2.13. For more information see www.setaram.com.

Shimadzu offers the DSC60 and is based on a heat flux design. It has a temperature range of –150–600 °C. A liquid nitrogen bath enables low temperature measurements. For more information, see www.ssi.shimadzu.com and search on "dsc."

TA Instruments (TAI) makes their current line of DSC's under the Q series designation. The Q series consists of three different series designations, the Q20, Q200, and Q2000. The simplest design with the lowest cost is the Q20. The Q20 family is available in three models: the Q20, with no autosampler; the AQ20, with an autosampler; and the Q20P, which has the normal cell design

TABLE 2.13. Performance Characteristics of Setaram DSCs

Performance Feature	DSC131	MHTC[a]	Sensys	Micro DSC III	Micro DSC VII
Temperature range (°C)	–170 to 700	≤1600	–120 to 830	–20 to 120	–45 to 120
Cell volume (µL)	30, 100	≤450	250 maximum	850	850
Calorimetric resolution (µW)	0.4	5	0.4	0.03	0.04

[a]DSC version (because Setaram also has a high-temperature DTA).

TABLE 2.14. Performance Characteristics of Q Series DSCs of TA Instruments

Performance Feature	Q2000	Q200	Q20	AQ20	Q20P[a]
Temperature range, °C	Ambient to 725 °C	Ambient to 725 °C	Ambient to 725 °C	Ambient to 725 °C	Ambient to 725 °C
With cooling accessories (°C)	−180 to 725	−180 to 725	−180 to 725	−180 to 725	−130 to 725
Temperature Accuracy (°C)	±0.1	±0.1	±0.1	±0.1	±0.1
Temperature precision (°C)	±0.01	±0.05	±0.05	±0.05	±0.05
Calorimetric reproducibility (indium metal) (%)	±0.05	±0.1	±1	±1	±1
Calorimetric precision (indium metal) (%)	±0.05	±0.1	±0.1	±0.1	±0.1
Baseline curvature (−50 to 300 °C) (W)	10	10	<150	<150	NA
Baseline reproducibility (μW)	±10	±10	±40	±40	NA
Sensitivity (μW)	0.2	0.2	1.0	1.0	1.0
Indium height/width (mW/°C) (1.0 mg heated 10 °C/min)	60	30	8.0	8.0	NA

[a]NA = data not available.

replaced with a dedicated pressure cell. Characteristics of these DSCs are listed in Table 2.14. TAI positons the Q20 family as entry-level and employ the Tzero cell design (described in Section 2.3) as a standard component. Accordingly, the operational software for the Q20 family does not include the cell asymmetry correction, the correction for pan thermal resistance, or the capability to perform modulated temperature experiments. The pressure cell has a temperature range of −130 to 725 °C (under ambient pressure its lowest achievable temperature range is −180 °C) and its maximum pressure specification is 7 MPa (≈1000 psi or 69 atm). The pressure cell can also be used as an accessory for the more advanced Q2000 module.

Whereas the Q20 consists of a set platform, the Q200 and Q2000 are flexible platforms that can be configured by the user. The Q200 is an intermediate DSC, which can be expanded with various options, while the Q2000 is TAI's top-of-the-line module. The Q2000 is sold with most features included except the differential photocalorimeter (DPC), the pressure cell, and the cooling accessories. Although the Q200 and Q2000 offer expandable platforms, differences exist between the two modules, primarily in the technology package and

TABLE 2.15. Controlled Cooling Rates for Q Series DSCs of TA Instruments

Controlled Rate (°C/min)	RCS40 from 400°C to	RCS90 from 550°C to	LNCS from 550°C to
100	—	300°C	200°C
50	175°C	120°C	0°C
20	40°C	−20°C	−100°C
10	0°C	−50°C	−150°C
5	−15°C	−75°C	−165°C
2	−40°C	−90°C	−180°C

in some of the hardware features. The Q200 includes the $T4$ baseline correction but does not correct for the pan thermal resistance pan mass differences. For the Q200, the 50-position autosampler and the modulated temperature capability are optional, but the pressure cell is not available. For the Q2000, the autosampler and the modulated temperature capability are included as well the $T4P$ correction for the cell asymmetry and pan thermal resistivity pass mass differences. A differential photocalorimeter is an available option for both platforms.

The Q series includes a number of enhanced designed elements. The autolid assembly has an improved thermal isolation of the cell. Use of internal digital mass flow controllers ensure the accurate metering of the purge gas. To enhance cooling performance, the furnace is associated with a cooling assembly that consists of symmetrically arranged, thermally conductive rods. These separate the heat sink from the furnace and sample chamber, and provide for a uniform transfer of heat to and from the cell.

Three automatic low-temperature options are available: RCS40, RCS90, and the LNCS. The RCS40 and RCS90 are different mechanical cooling options, while the LNCS is a liquid-nitrogen-based cooling system. Table 2.15 summarizes the controlled cooling rates that can be obtained for each cooling unit. For more information, which includes a comprehensive product brochure, see www.tainstruments.com.

Other suppliers of DSC systems are *Bähr Thermoanalyse* at www.baehr-themo.de., *Linseis* at www.linseis.net., *Rigaku* at www.rigaku.com, and *Theta Industries* at www.theta-us.com.

2.15.1. DSC Resolution and Precision

Most manufacturers provide some performance values for their instruments. Two sample-based tests that are now used are (1) the TAWN protocol and (2) the ratio of the height to the width at half-height for a metal standard. The Dutch Society for Thermal Analysis (TAWN) has suggested the use of the transitions of 4, 4′-azoxyanisole to evaluate both resolution and sensitivity of a DSC (van Ekeren et al. 1997). The resolution is evaluated by the ability to

DIFFERENTIAL SCANNING CALORIMETRY (DSC)

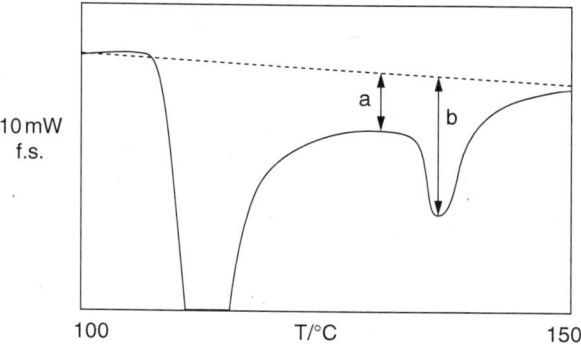

Figure 2.125. This figure shows determination of the TAWN temperature resolution. The resolution is calculated as the ratio of heights a/b. The abbreviation f.s. is used to denote full scale (van Ekeren et al. 1997; reprinted with permission from Akadémiai Kiadó, Hungary).

separate two closely spaced transitions in temperature, while the sensitivity is evaluated by how well a DSC can detect transitions in the heat flow. The crystal→nematic transition at approximately 117 °C and the nematic→isotropic transition at about 134 °C are used. The TAWN resolution is the calculation of the ratio of the minimum height between the two transitions to the baseline divided by the height of the nematic→isotropic transition main melting transition at a moderate heating rate (20 °C/min is recommended). This is shown in Fig. 2.125, where the TAWN resolution is the quotient of "height a" divided by "height b." The smaller the TAWN resolution value, the better the temperature resolution.

The TAWN sensitivity is calculated from the ratio of the nematic→isotropic transition peak height over the baseline noise. The sensitivity is evaluated at a slow rate, with 0.1 °C/min recommended. This is demonstrated in Fig. 2.126, where the TAWN sensitivity is calculated by dividing "height c" by the height of the "noise d." The larger this value, the better the TAWN sensitivity.

Another method used to evaluate instrument performance is the height:width ratio for a high-purity metal standard (usually indium). This method incorporates both elements of *sensitivity* (by the height measurement) and *resolution* (by the width measurement). Aubuchon et al. (2006) suggest a protocol.

2.15.2. Microbalances

It is very important to accurately weigh the samples for DSC measurements. To achieve an accuracy of better than 1% in heat capacity measurements, for instance, the sample must be weighed to an accuracy of at least 0.1%. For these purposes the use of electronic microbalances is recommended. The major

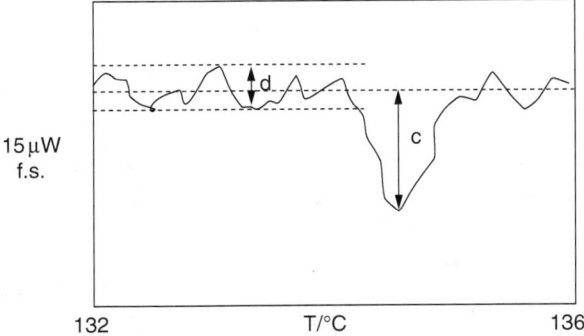

Figure 2.126. This figure shows determination of the TAWN heat flow sensitivity. The sensitivity is calculated as the ratio of heights c/d. The abbreviation f.s. is used to denote full scale (van Ekeren et al. 1997; reprinted with permission from Akadémiai Kiadó, Hungary).

manufacturers of these microbalances are Cahn Instruments (www.cahn.com), Mettler Toledo (www.us.mt.com), Perkin-Elmer (www.perkinelmer.com), and Sartorius (www.sartorius.com). The use of a microbalance with at least a 1 μg sensitivity is recommended; but a balance with 0.1 μg sensitivity is preferred.

APPENDIX

Standards for Differential Scanning Calorimetry issued under the jurisdiction of ASTM International Committee E37 on Thermal Measurements are listed below for reference. For more information, see website www.astm.org.

- E473, *Standard Terminology Relating to Thermal Analysis and Rheology*
- E698, *Standard Test Method for Arrhenius Kinetic Constants for Thermally Unstable Materials*
- E1142, *Standard Terminology Relating to Thermophysical Properties*
- E1952, *Standard Test Method for Thermal Conductivity and Thermal Diffusivity by Modulated Temperature Differential Scanning Calorimeter*
- E2069, *Standard Test Method for Temperature Calibration on Cooling for Differential Scanning Calorimeters*

ABBREVIATIONS

Symbols

A	area; preexponential factor, preexponential constant
α	crystallinity; conversion in thermoset cure
α_{gel}	α at gel point

$d\alpha/dt$	rate of conversion
β_T	isothermal compressibility
C_1	thermal lag constant
C_p	heat capacity at constant pressure
ΔC_p	heat capacity jump at glass transition
$C_{p,\text{rev}}$	reversing heat capacity (also expressed as $\text{rev}C_p$)
C_v	heat capacity at constant volume
dH/dt	heat flow
E	activation energy
F	Helmholtz free energy; fraction of material melted to temperature T_s
Δg_f	specific free-enthalpy change on melting
$_{\text{gel}}T_g$	T_g for thermoset at the gel point with $\alpha = \alpha_{\text{gel}}$
G	Gibbs free energy (free enthalpy)
γ	coefficient of volumetric thermal expansion
ΔH_f	heat of fusion
ΔH_f°	equilibrium heat of fusion
h	amplitude of the DSC heat flow signal
H	enthalpy
ΔH_{res}	residual heat of reaction
ΔH_{rxn}	heat of reaction
ΔH_t	heat generated up to time t
ΔH_{ult}	ultimate heat of reaction $\leq \Delta H_{\text{rxn}}$
k_1	catalyzed rate constant
k_2	autocatalyzed rate constant
k_a	overall rate constant
k_T	Arrhenius rate constant for chemical reaction
k_d	diffusion rate constant
K	rate constant of crystallization
K_{DSC}	instrument constant
l	lamellar thickness
λ	thermal conductivity
m	mass; reaction order
M_n	number average molecular mass
M_w	mass average molecular mass
n	order of reaction; Avrami exponent
p	pressure
q	heating rate
Q	heat
$\dot{Q}$	heat flow rate
R	thermal resistance; universal gas constant
$\text{rev}C_p$	reversing heat capacity
ρ	density
S	entropy
σ	surface free energy of lamellae

t	time
t_{gel}	time to gelation, gel point
t_{vit}	time to vitrification
t_1	time to reach constant conversion at T_1
t_2	time to reach constant conversion at T_2
T	temperature
T_1	an operatio mode of the Q100, Q200, Q1000 and Q2000 DSCs of TA instruments
T_1	specific temperature (extrapolated onset of the glass transition)
T_1	a running mode in the Q100, Q200, Q1000 and Q2000 DSCs of TA Instruments
T_2	specific temperature (extrapolated glass transition end temperature)
T_2	specific temperature (extrapolated end temperature of the glass transition)
$T4$	a running mode in the Q100, Q200, Q1000 and Q2000 DSCs of TA Instruments
$T4P$	a running mode in the Q1000 and Q2000 DSCs of TA Instruments
T_{ann}	annealing temperature
T_{av}	average temperature
T_b	beginning temperature of glass transition
T_{bl}	block temperature
T_c	crystallization temperature
T_{c0}	starting temperature of crystallization
T_{cp}	peak temperature of crystallization
T_{cure}	cure temperature
T_e	end temperature of glass transition
T_g	glass transition temperature
T_{g0}	T_g for uncured thermoset with $\alpha = 0$
$T_{g\infty}$	T_g for fully cured thermoset with $\alpha = 1$
T_i	temperature associated with constant conversion
T_{ind}	indicated temperature (temperature displayed by DSC)
T_m	melting point
T_m°	equilibrium melting point
T_r	reference temperature
T_{real}	real temperature
T_s	sample temperature
T_{sur}	temperature of the heat sink (surroundings)
ΔT	temperature difference between sample and reference sensors; correction factor in temperature calibration
ΔT_{sbl}	temperature difference between sample and block
τ	time constant of calorimeters; relaxation time
U	internal energy
V	volume

X	amplifier gain signal
x	alternate symbol for conversion
x_{imp}	mole fraction of impurity
χ_1	polymer–solvent interaction parameter
W	work
W_{av}	power from average amplifier
W_d	power from differential amplifier

Acronyms

ASA	an acrylate-styrene-acrylonitrile terpolymer
ASTM	Aerican Society for Testing and Materials
CR	cooling rate
DDS	diaminodiphenylsulfone
DEA	dielectric analysis
DF	diffusion factor
DGEBA	diglycidyl ether of bisphenol A
DMA	dynamic mechanical analysis
DPC	differential photocalorimetry
DSC	differential scanning calorimetry
HR	heating rate
IS	impact strength
MDA	methylenedianiline
MDSC®	modulated DSC
MF	mobility factor (in thermoset cure)
MTDSC	modulated temperature DSC
NHF	nonreversing heat flow
ODSC	oscillating DSC
PAN	polyacrylonitrile
PBT	poly(butylene terephthalate)
PC	bisphenol A polycarbonate
PE	polyethylene
PEEK	poly(ether–ether ketone)
PEN	poly(ethylene-2,6-naphthalene dicarboxylate)
PEO	poly(ethylene oxide)
PES	poly(ether sulfone)
PET	poly(ethylene terephthalate)
PLLA	poly(L-lactide)
PMMA	poly(methyl methacrylate)
POM	polyoxymethylene
PP	polypropylene
PPS	poly(phenylene sulfide)
PS	polystyrene
PTFE	polytetrafluoroethylene
PVA	poly(vinyl alcohol)
PVAc	poly(vinyl acetate)
PVC	poly(vinyl chloride); pigment volume concentration

PVF	poly(vinyl fluoride)
PVDF	poly(vinylidene fluoride)
RHF	reversing heat flow
RIPS	reaction-induced phase separation
RT-IR	real-time infrared spectroscopy
SSDSC	step scan DSC
TGA	thermogravimetric analysis
TGMDA	tetraglycidylmethylene dianaline
THF	total heat flow
TOPEM®	Mettler Toledo's modulated temperature DSC version
UHR	underlying heating rate
VARTM	vacuum-assisted resin transfer molding

REFERENCES

Acitelli, M. A., Prime, R. B., and Sacher, E. (1971), *Polymer* **12**, 333–343.

Adamovsky, S. and Schick, C. (2004), *Thermochim. Acta* **415**(1–2), 1.

Adamovsky, S., Minakov, A. A., and Schick, C. (2003), *Thermochim. Acta* **403**(1), 55.

Arakawa, T. and Wunderlich, B. (1965), *in Macromolecular Chemistry*, Prague (J. Polym. Sci. C **16**), Wichterle, O. and Sedlacek, B., eds, Interscience, New York, 1967, p. 653.

Arnauts, J., De Cooman, R., Vandeweerdt, P., Koningsveld, R., and Berghmans, H. (1994), *Termochim. Acta* **238**, 1.

Aubuchon, S. R., Chin, P., and Hassel, R. L. (1997), *Proc. 25th North American Thermal Analysis Soc. Conf.*, p. 78.

Aubuchon, S. R., Caufield, P. A., and Blaine, R. L. (2006), *Proc. 34th North American Thermal Analysis Soc. Conf.*

Avrami, M. (1939), *J. Chem. Phys.* **7**, 1103.

Avrami, M. (1940), *J. Chem. Phys.* **8**, 212.

Avrami, M. (1941), *J. Chem. Phys.* **9**, 177.

Bair, H. E., Huseby, T. W., and Salovey R. (1968), in *Analytical Calorimetry*, Porter, R. S. and Johnson, J. F., eds., Plenum Press, New York, p. 31.

Bair, H. E., Boyle, D. J., and Kelleher, P. G. (1980), *Polym. Eng. Sci.* **20**, 995.

Bair, H. E., Shepherd, L., and Boyle, D. J. (1981), in *Thermal Analysis in Polymer Characterization*, Turi, E. A., eds., Heyden, Philadephia, p. 114.

Bair, H. E. and Warren, P. C. (1981), *J. Macromol. Sci.* **B20**(3), 381.

Bair, H. E. and Blyler, L. L. (1985), *Proc. 14th North American Thermal Analysis Soc. Conf.*, pp. 392–398.

Bair, H. E., Blyer, L. L., and Simoff, D. A. (1993), *Polym. Mater. Sci. Eng.* **68**, 268.

Bair, H. E. (1994), in *Assignment of the Glass Transition* Seyler, R. J., ed., STP1249, ASTM, Philidelphia, pp. 50–74.

Bair, H. E., Hale, A., and Popielarski, S. R. (2003), US Patent 6,586,258.

Bair, H. E. (1997), in *Thermal Characterization of Polymeric Materials*, Vol. **2**, Turi, E. A., ed., Academic Press, San Diego, pp. 2263–2420.

Bair, H. E. (2007), *Proc. 35th North American Thermal Analysis Doc. Conf.*, p. 9.

Bares, V. and Wunderlich, B. (1973), *J. Polym. Sci., Polym. Phys. Ed.* **11**, 861–873.
Barrau, S. and Judovits, L. (1999), *Proc. 27th North American Thermal Analysis Soc. Conf.*, p. 284.
Barton, J. M. (1974a), *J. Macromol. Sci., Chem.* **A8**, 53–64.
Barton, J. M. (1974b), in: *Polymer Characterization by Thermal Methods of Analysis*, Chiu, J., ed., Marce Dekker, New York, pp. 25–32.
Barton, J. M. (1983), *Thermochim. Acta* **71**, 337–344.
Barton, J. M. (1985), *Adv. Polym. Sci.* **72**, 111–154.
Bates, F. S. and Fredrickson, G. H. (1990), *Annu. Rev. Phys. Chem.* **41**, 525.
Beatty, C. L. and Karasz, F. E. (1979), *J. Macromol. Sci., Rev. Macromol. Chem.* **C17**, 37.
Bilyeu, B. and Brostow, W. (2002), *Polym. Compos.* **23**, 1111.
Blaine, R. L. and Marcus, S. M. (1998), *J. Therm. Anal.* **54**, 467.
Blaine, R. L. (not dated), Du Pont Company, Instrument Products, Scientific and Process Div., Wilmington, DE 19898: THE CASE FOR A GENERIC DEFINITION OF DIFFERENTIAL SCANNING CALORIMETRY.
Boller, A., Jin, Y., and Wunderlich, B. (1994), *J. Therm. Anal.* **42**, 307.
Boller, A., Schick, C., and Wunderlich, B. (1995), *Thermochim. Acta* **266**, 97.
Booth, A. and Hay, J. N. (1972), *Br. Polym. J.* **4**, 9.
Breitigam, W. V., Bauer, R. S., and May, C. A. (1993), *Polymer* **34**, 767.
Brennan, W. P., DiVito, M. P., Fyans, R. L., and Gray, A. P. (1984), "An overview of the calorimetric purity measurements," in *Purity Determinations by Thermal Methods*, Blaine, R. L. and Schoff, C. K., eds., ASTM STP 838, pp. 5–15.
Buchler, F. U. and Seferis, J. C. (2000), *Thermochim. Acta* **348**, 161.
Cassel, B., Scotto, P., and Sichina, B. (1999), *Proc. 27th North American Thermal Analysis Soc. Conf.*, p. 33.
Cassettari, M., Salvetti, G., Tombari, E., Veronesi, S., and Johari, G. P. (1993), *J. Polym. Sci. Pt. B* **31**, 199.
Chen, W., Moon, I.-K., and Wunderlich, B. (2000), *Polymer* **41**, 4119.
Chern, C. S. and Poehlein, G. W. (1987), *Polym. Eng. Sci.* **27**, 788.
Chuah, H. H., and Porter, R. S. (1986), *Polymer* **27**, 1022.
Clausius, R. (1865), *Annal. Phys.* **125**, 353.
Clough, S. B. (1970a), *J. Macromol. Sci., Phys.* B**4**(1), 199.
Clough, S. B. (1970b), *Polym. Lett.* **8**, 519.
Clough, S. B. (1971), *J. Appl. Polym. Sci.* **15**, 2141.
Collins, G. L. and Menczel, J. D. (1992), *Polym. Eng. Sci.* **32**(17), 1270.
Couchman, P. R. and Karasz, F. E. (1978), *Macromolecules* **11**, 117.
Couchman, P. R. (1980), *Macromolecules* **13**, 1272.
Crivello, J. V. (1975), US Patent 4,058,401.
Crivello, J. V. and Lam, J. H. (1977), *Macromolecules* **10**, 1307.
Crivello, J. V., and Lee, J. L. (1990), *ACS Symp. Series* **417**, 398–411.
Crivello, J. V. and Liu, S. J. (2000), *J. Polym. Sci., Pt A: Polym. Chem.* **38**, 389–401.
Danley, R. L., *Thermochimi. Acta* (2003), **395**, 201.
Danley, R. L., *Thermochimi. Acta* (2004), **409**, 111.

Davidson, T. and Wunderlich, B. (1969), *J. Polym. Sci. A2* **7**, 2043.

Davis, G. J. and Porter, R. S. (1969), *J. Therm. Anal.* **1**, 449.

Decker, C. (1992), in *Radiation Curing, Science and Technology*, Pappas, S. P. ed., Plenum, New York, pp. 135–179.

Decker, C. and Moussa, K. (1988), *Makromol. Chem.* **189**, 2381.

Deng, Y. and Martin, G. C. (1994), *Macromolecules* **27**, 5141–5446.

De Groot, S. R. and Mazur, P. (1962), *Non-Equilibrium Thermodynamics*, North Holland, Amsterdam.

DiBenedetto, A. T. (1987), *J. Polym. Sci., Pt. B: Polym. Phys.* **25**, 1949.

Doyle, C. D. (1961), *Anal. Chem.* **33**, 77–79.

Dreezen, G., Groeninckx, G., Swier, S., and Van Mele, B. (2001), *Polymer* **42**, 1449.

Ebert, M., Garbella, R. W., and Wendorff, J. H. (1986), *Makromol. Chem., Rapid Commun.* **7**, 65.

Ehrenfest, P. (1933), *Proc. Kon. Nederl. Akad. Wetensch* **36**, 153.

Fischer, E., Kunze, W., and Stapp, B. (1988a), *Kunstoffe* **9**, 870–873.

Fischer, E., Kunze, W., and Stapp, B. (1988b), *Perkin-Elmer Anal. Tech. Report* 60E.

Flory, P. J. (1951), *Trans. Faraday Soc.* **55**, 848.

Flory, P. J. (1953), *Principles of Polymer Chemistry*, Cornell Univ. Press, Ithaca, NY.

Flynn, J. H. (1978a), in *Thermal Methods in Polymer Analysis*, Shalaby, S. W., eds., Franklin Inst. Press, Philadelphia, pp. 163–186.

Flynn, J. H. (1978b), in Jellinek, H. H. G., ed., *Aspects of Degradation and Stabilization of Polymers*, Elsevier, Amsterdam, pp. 573–603.

Flynn, J. H. and Wall, L. A. (1966a), *J. Polym. Sci., Pt. B* **4**, 323.

Flynn, J. H. and Wall, L. A. (1966b), *J. Res. Natl. Bur. Stand., Sect. A* **70**, 487.

Folkes, M. J., and Hope, P. S., eds. (1993), *Polymer Blends and Alloys*, Blackie, Glasgow.

Fox, F. G. (1956), *J. Appl. Bull. Am. Phys. Soc.* **1**, 123.

Freeberg, F. E. and Alleman, T. C. (1966), *Anal. Chem.* **38**, 1806–1807.

Fuller, L. C. and Judovits, L. (2002), in *Materials Characterization by Dynamic. and Modulated Thermal Analytical Techniques*, Riga, A. T. and Judovits, L., eds., STP1402, 89.

Gaur, U. and Wunderlich, B. (1980), *Macromolecules* **13**, 445.

Gaur, U. and Wunderlich, B. (1981), *J. Phys. Chem. Ref. Data* **10**, 1051.

Gaur, U., Lau, S. F., and Wunderlich, B. (1983), *J. Phys. Chem. Ref. Data* **12**, 65.

Geil, P. H., Anderson, F. R., Wunderlich, B., and Arakawa, T. (1964), *J. Polym. Sci. A* **2**, 3707.

Gill, P. S., Sauerbrunn, S. R., and Reading, M. (1993), *J. Therm. Anal.* **40**, 931.

Gillham, J. K. (1986), *Polym. Eng. Sci.* **26**, 1429–1433.

Gillham J. K. and Enns J. B. (1994), *Trends Polym. Sci.* **2**, 406.

Gordon, M. and Taylor, J. S. (1952), *J. Appl. Chem.* **2**, 493.

Gotro, J. and Prime, R. B. (2004), "Thermosets," in *Encyclopedia of Polymer Science & Technology*, 3rd ed. J. Kroschwitz, ed., Wiley, Hoboken, NJ.

Gray, A. P. (1966a), *Thermal Analysis Newsletter 5*, The Perkin-Elmer Corp., Norwalk, CT.

Gray, A. P. (1966b), *Thermal Analysis Newsletter 6*, The Perkin-Elmer Corp., Norwalk, CT.

Gray, A. P., and Fyans, R. L. (1973), *Thermal Analysis Applications Study 10*, The Perkin-Elmer Corp., Norwalk, CT.

Gunaraine, L. M. W. K. and Shanks, R. A. (2005a), *Thermochim. Acta* **430**, 183.

Gunaraine, L. M. W. K. and Shanks, R. A. (2005b), *Eur. Polym. J.* **41**, 2980.

Hale, A. (1992), in *Handbook of Thermal Analysis and Calorimetry*, Vol. **3**, Cheng, S. Z. D., ed., Elsevier, Amsterdam, pp. 295–354.

Hale, A. and Bair, H. E. (1997), "Fibers polymer blends and block copolymers," in *Thermal Charactaerization of Polymeric Materials*, 2nd ed., Turi, E., Academic Press, San Diego, pp. 745–886.

Hale, A., Macosko, C. W. and Bair, H. E. (1989), *J. Polym. Sci.* **38**, 1253.

Hale, A., Macosko, C. W., and Bair, H. E. (1991), *Macromolecules* **24**, 2610.

Harget, P. (1987), Private communication to Menczel, J.

Heijboer, J. (1968), *J. Polym. Sci., Pt C* **16**, 3755.

Hellmuth, E. and Wunderlich, B. (1965), *J. Appl. Phys.* **36**, 3039.

Hemminger, W. and Höhne, G. (1984), *Calorimetry Fundmentals and Practice*, Verlag Chemie.

Hildebrand, J. H. (1915), *J. Am. Chem. Soc.* **37**, 970.

Hildebrand, J. H. (1918), *J. Am. Chem. Soc.* **40**, 45.

Hillmyer, M. A., Lipic, P. M., Hajduk, D. A., Almdal, K., and Bates, F. S. (1997), *J. Am. Chem. Soc.* **119** (11), 2749.

Hodgkin, J. H., Simon, G. P., and Varley, R. J. (1998), *Polym. Adv. Technol.* **9**(1), 3.

Hoffman, J. D. and Weeks, J. J. (1962), *J. Res. Natl. Bur. Stand.* **66**A, 13.

Hourston, D. J., Song, M., Hammiche, A., Pollock, H. M., and Reading, M. (1997), *Polymer*, **38**, 1.

Hourston, D. and Song, M. (2005), *Ann. Proc. 33rd North American Thermal Analysis Soc. Conf.*

Iguchi, M. (1976), *Makromol. Chem.* **177**, 549.

Ikeda, M. (1967), *Kobunshi Kagaku* **24**, 378.

Illers, K. H. (1970), *Angew. Makromol. Chem.* **12**, 89.

Illers, K. H. (1977), *Polymer* **18**, 551.

Illers, K. H. (1980), *Colloid Polym. Sci.* **258**(2), 117.

Ishikiriyama, K. and Wunderlich, B. (1997), *J. Therm. Anal.* **50**, 337.

Jaffe, M. and Wunderlich, B. (1967), *Kolloid Z.-Z. Polym.* **216–217**, 203.

Jaffe, M. and Jones, S. (1985), in *High Technology Fibers*, Lewin, M. and Preston, J., eds. Marce Dekker, New York, Vol. III, Part A, p. 349.

Jaffe, M., Menczel, J. D., and Bessey, W. E.(1997), "Fibers," in *Thermal Charactaerization of Polymeric Materials*, 2nd ed., Turi, E., ed. Academic Press, San Diego, pp. 1767–1954.

Jin, Y., Boller, A., and Wunderlich, B. (1993), *Proc. 22nd North American Thermal Analysis Soc. Conf.*, p. 59.

Judovits, L. (1997), *NATAS Notes* **29**(1), 33.

Judovits L. and Crabb, C. C. (2005), *Ann. Proc. North American Thermal Analysis Soc.*

Judovits, L., Menczel, J. D., and Leray, A.-G. (1998), *J. Therm. Anal. Calorim.* **54**, 605.

Kehl, T. and van der Plaats, G. (1991), US Patent 5,033,866, Mettler Toledo.

Kenny, J. M. and Trivisano, A. (1991), *Polym. Eng. Sci.* **31**, 1426–1433.

Kenny, J. M., Trivisano, A., Frigione, M. E., and Nicolais, L. (1992), *Thermochim. Acta* **199**, 213–227.

Kissinger, H. E. (1957), *Anal. Chem.* **29**, 1702.

Kloosterboer, J. G., van de Hei, G. M. M., Gossink, R. G., and Dortant, G. C. M. (1984), *Polym. Commun.* **25**, 322.

Kubota, Y., Fukao, K., and Saruyama, Y. (2005), *Thermochim. Acta* **431**, 149.

Kurdikar, D. L. and Peppas, N. A. (1994), *Polymer* **35**, 1004.

Kurnakov, N. S. (1904), *Z. Anorg. Chem.* **42**, 184.

Kwei, T. K. (1984), *J. Polym. Sci., Polym. Lett. Ed.* **22**, 307.

Lacey, A. A., Price, D. M., and Reading, M., (2006), in *Modulated-Temperature Differential Scanning Calorimetry Theoretical and Practical Applications in Polymer Characterisation*, Hot Topics in Thermal Analysis and Calorimetry Series, Reading, M. and Hourston, D. J., eds. Springer, Vol. **6**.

Lacey, A. A., Nikolopoulos, C., and Reading, M. (1997), *J. Therm. Anal.* **50**(1–2), 279.

Lau, S.-F., Suzuki, H., and Wunderlich, B. (1984a), *J. Polym. Sci., Polym. Phys. Ed.* **22**, 379.

Lau, S.-F., Wesson, J. P., and Wunderlich, B. (1984b), *Macromolecules* **17**(5), 1102.

Lauritzen, J. I.Jr, . and Hoffman, J. D. (1960), *J. Res. Natl. Bur. Stand.* **64**A, 73.

Le Chatelier, H. (1887), *Z. Phys. Chem.* **1**, 396.

Levin, R., Hale, A., Harris, A. L., Levinos, N. J., and Schilling, F. C. (1995), *Polym. Mat. Sci. Eng.* **72**, 524.

Levy, P. F., Nieuweboer, G., and Semanski, L. C. (1970), *Thermochim. Acta* **1**, 429.

Lide, D. R. (1998–1999), *CRC Handbook of Chemistry and Physics*, 79th ed., CRC Press, Boca Raton, FL.

Lipic, P. M., Bates, F. S., and Hillmyer, M. A. (1998), *J. Am. Chem. Soc.* **120** (35), 8963.

Lopes, C. M. A. and Felisberti, M. I. (2004), *Polym. Test.* **23**, 637.

Lu, X. and Weiss, R. A. (1992), *Macromolecules* **25**, 3242.

Lu, X. and Weiss, R. A. (1993), *Macromolecules* **26**, 248.

MacKnight, W. J. and Karasz, F. E. (1989), in *Comprehensive Polymer Science*, Aggarwal, S. L. ed., Vol. **7**. p. 111.

Mandelkern, L. (1964), *Crystallization of Polymers*, McGraw-Hill.

Marcus, S. M. and Blaine, R. L. (1994), *Thermochim. Acta* **243**, 231.

Matsuoka, S., Quan, X., Bair, H. E., and Doyle, D. J. (1989), *Macromolecules* **22**, 4093.

May, C. A., ed. (1983), *Chemorheology of Thermosetting Polymers*, ACS Symp. Series **227**, American Chemical Society.

Menczel, J. and Varga, J. (1983), *J. Therm. Anal.* **28**, 161.

Menczel, J. and Wunderlich, B. (1986), *Polym. Prepr., J. Am. Chem. Soc., Div. Polym. Chem.* 255.

Menczel, J., and Wunderlich, B. (1980), *J. Polym. Sci., Polym. Phys. Ed.* **18**, 1433.

Menczel, J. D. and Wunderlich, B. (1981), *J. Polym. Sci., Polym. Lett. Ed.* **19**, 261.

Menczel, J. D. and Leslie, T. M. (1990), *Thermochim. Acta* **166**, 309 (1990).

Menczel, J. D. and Leslie, T. M. (1993), *J. Therm. Anal.* **40**, 957 (1993).

Menczel, J. D. (1994), *Thermal Analysis News, Worldwide Perkin-Elmer Thermal Analysis Customer Newsletter* (Sept.), pp. 6–7.

Menczel, J. D., Jaffe, M., Saw, C. K., and Bruno, T. P. (1996), *J. Therm. Anal.* **46**, 753.

Menczel, J. D. (1997), *J. Therm. Anal.* **49**, 193.

Menczel, J. D., Collins, G. L., and Saw, S. K. (1997a), *J. Therm. Anal.* **49**, 201.

Menczel, J. D., Jaffe, M., and Bessey, W. E. (1997b), "Films," in *Thermal Characterization of Polymeric Materials*, Turi, E. A., eds., Academic Press, San Diego, pp. 1955–2089.

Menczel, J. D. (1999), *J. Therm. Anal. Calorim.* **58**, 517.

Menczel, J. D. and Jaffe, M. (2006, 2007), *Proc. 34th North American Thermal Analysis Soc. Conf.; J. Therm. Anal. Calorim.* **89**(2), 357.

Menczel, J. D. to be published.

Meng, Y. and Simon, S. L. (2005), *Thermochim. Acta* **437**, 179.

Mettler Toledo (1991), US Patent 5,033,866.

Mettler Toledo (2004), Stare System Manual, Chapter 8.

Mettler Toledo (2005), DSC823^e module brochure.

Mijovic, J., Shen, M. Z., Sy, J. W., and Mondragon, I. (2000), *Macromolecules* **33**(14), 5235.

Minakov, A., Mordvintsev, D., and Schick, C. (2004), *Proc. 32nd North American Thermal Analysis Soc. Conf.*, p. 23.

Miyagi, A. and Wunderlich, B. (1972), *J. Polym. Sci., Polym. Phys. Ed.* **10**, 1401.

Moore, W. J., *Physical Chemistry*. 4th ed., Prentice-Hall, Englewood Cliffs, NJ, 1972, p. vi (Preface).

Murthy, N. S., Correale, S. T., and Kavesh, S. (1990), *Polym. Commun.* **31**, 50.

Nakagawa, K. and Ishida, Y. (1973a), *Kolloid Z.-Z. Polym.* **251**, 103.

Nakagawa, K. and Ishida, Y. (1973b), *J. Polym. Sci.* **11**, 2153.

Neag, C. M. and Prime, R. B. (1991), *J. Coat. Technol.* **63**(797), 37.

Neff, B. L. and Barsotti, R. J. (1999), *Proc. 27th North American Thermal Analysis Soc. Conf.*, p. 646.

Nielson, L. E. (1969), *J. Macromol. Sci.* C**3**, 69.

Noel, D., Hechler, J. J., Cole, K. C., Chouliotsis, A., and Overbury, K. C. (1988), *Thermochim. Acta* **125**, 191.

Noshay, A. and McGrath, J. E. (1977), *Block Copolymers: Overview and Critical Survey*, Academic Press, New York.

Olsson, R. T., Bair, H. E., Kuck, V., and Hale, A. (2002), *ACS Polym. Preprints* **42**(2), 797.

Olsson, R. T., Bair, H. E., Kuck, V., and Hale, A. (2003), in *Photoinitiated Polymerization*, Belfield, K. D. and Crivello, J. V. eds., ACS Symp. Series **847**, pp. 317–328.

Olsson, R. T., Bair, H. E., Kuck, V. and Hale, A. (2004), *J. Therm. Anal. Cal.* **76**, 367.

O'Neill, M. J. (1964), *Anal. Chem.* **36**, 1238.
Onsager, L. (1931a), *Phys. Rev.* **37**, 405.
Onsager, L. (1931b), *Phys. Rev.* **38**, 2265.
O'Reilly, J. M., Karasz, F. E., and Bair, H. E. (1963), *J. Polym. Sci, Pt. C* **6**, 109.
Ozawa, T. (1965), *Bull. Chem. Soc. Jpn.* **38**, 1881–1886.
Ozawa, T., J. (1970), *Therm. Anal.* **2**, 301–324.
Ozawa, T. (1971), *Polymer* **12**, 150.
Pan, R., Varma-Noir, M., and Wunderlich, B. (1989), *J. Therm. Anal.* **30**, 229.
Papageorgiou, G. Z., Achilias, D. S., Karayannidis, G. P., Bikiaris, D. N., Roupakias, C., and Litsardakis, G. (2006), *Eur. Polym. J.* **42**, 434.
Papir, Y. S., Kapur, S., Rogers, C. E., and Baer, E. (1972), *J. Polym. Sci., Pt. A2* **10**, 1305.
Pascault, J. P. and Williams, R. J. J. (1990), *J. Polym. Sci., Pt. B: Polym. Phys.* **28**, 85–95.
Pascault, J. P., Sautereau, H., Verdu, J., and Williams, R. J. J. (2002), *Thermosetting Polymers*, Marcel Dekker, New York.
Paul, D. R. and Barlow, J. W. (1982), *MMI Press Symp. Ser.* **2**, 1.
Penn, L. and Larsen, F. (1979), *J. Appl. Polym. Sci.* **23**, 59.
Penning, A. J. (1967), *Proc. Int. Crystal Growth Conf., Boston, 1966*, p. 389.
Perkin-Elmer (1970), *Thermal Analysis Newsletter* 9.
Perkin-Elmer (1976), *Instruction Manual of the DSC-2*.
Pielichowski, K., Flejtuch, K., and Pielichowski, J. (2004), *Polymer*, **45**, 1235.
Pijpers, T. F. J., Mathot, V. B. F., Goderis, B., Scherrenberg, R. L., and van Der Vegte, E. W. (2002), *Macromolecules* **35**, 3601.
Prigogine, I. (1945), *Acad. Roy. Belg. Bull. Cl. Sc.* **31**, 600.
Prigogine, I. (1954), *Proc. 3rd Symp. Temperature*, Washington, DC.
Prigogine, I. (1967), *Introduction to Thermodynamics of Irreversible Processes*, Wiley, New York.
Prigogine, I. and Mayer, G. (1955), *Acad. Roy. Belg. Bull. Cl. Sc.* **41**, 22.
Prime, R. B. and Wunderlich, B. (1969), *J. Polym. Sci., Polym. Phys. Ed.* **7**, 2061.
Prime, R. B., Wunderlich, B., and Melillo, L. (1969), *J. Polym. Sci., Polym. Phys. Ed.* **7**, 2091.
Prime, R. B. (1981), "Thermosets," in *Thermal Characterization of Polymeric Materials*, Turi, E. A., eds., Academic Press, New York, pp. 435–569.
Prime, R. B. (1997), "Thermosets," in *Thermal Characterization of Polymeric Materials*, 2nd ed., Turi, E. A. ed., Academic Press, San Diego, pp. 1379–1766.
Prime, R. B., Michalski, C., and Neag, C. M. (2005), *Thermochim. Acta* **429**, 213.
Pyda, M. and Wunderlich, B. (2000), *J. Polym. Sci., Pt. B: Polym. Phys.* **38**(4), 622.
Rabinowitch, E. (1937), *Trans. Faraday Soc.* **33**, 1225–1233.
Reading, M. (1993), *Trends Polym. Sci.* **1**, 248.
Reading, M., Elliott, D., and Hill, V. (1992), *Proc. 21st Nortth American Thermal Analysis Soc. Conf.*, p. 145.
Reading, M., Hahn, B., and Crowe, B. (1993a), US Patent 5,224,775.
Reading, M., Elliot, D., and Hill, V. L. (1993b), *J. Therm. Anal.* **40**, 949.

Reading, M., Luget, A., and Wilson, R. (1994), *Thermochim. Acta* **238**, 295.

Reading, M., Price, D. M., and Orliac, H. (2001), "Measurement of crystallinity in polymer using modulated temperature differential scanning calorimetry," in *Material Characterization by Dynamic and Modulated Thermal Analysis Techniques, American Society for Testing and Materials*, Riga, A. T. and Judovits, L. H., eds., ASTM, STP 1402.

Reading, M. and Hourston, D. J. (2006), *Modulated-Temperature Differential Scanning Calorimetry*, Springer, Dordrecht.

Riccardi, C. C. and Williams, R. J. J. (1986), *J. Appl. Polym. Sci.* **32**, 3445.

Richardson, M. J. (1993a) "The glass transition region," in *Calorimetry and Thermal Analysis of Polymers*, Mathot, V. B. F., ed., Hanser, Munich, pp. 169–187.

Richardson, M. J. (1993b), in *Calorimetry and Thermal Analysis of Polymers*, Mathot, V. B. F., ed., Hanser, New York, pp. 189–206.

Riesen, R. (1998), *Proc. 26th North American Thermal Analysis Soc. Conf.*, p. 235.

Riga, A. and Sisk, B. (1999), *Proc. 27th North American Thermal Analysis Soc.* p. 9.

Roberts-Austen, W. C. (1899), *Metallographist* **2**, 186.

Sacher, E. (1974), *J. Macromol. Sci., Phys.* **B9**(1), 163.

Sakai, Y. and Miyasaka, K. (1988), *Polymer* **29**, 1608.

Sakai, Y. and Miyasaka, K. (1990), *Polymer* **31**, 51.

Sakai, Y., Umetsu, K., and Miyasaka, K. (1993), *Polymer* **34**, 3362.

Samuels, R. J. (1975a), *J. Polym. Sci., Polym. Phys. Ed.* **13**, 1417.

Samuels, R. J. (1975b), *Appl. Polym. Symp.* **27**, 205.

Sanchez, M. S., Ribelles, J. L. G., Sanchez, F. H., and Mano, J. F. (2005), *Thermochim. Acta* **430**, 201.

Sasaki, T., Yamauchi, N., Irie, S., and Sakurai, K. (2005), *J. Polym. Sci., Pt. B: Polym. Phys.* **43**, 115.

Sauerbrunn, S. (2006), private communication.

Saw, C. K., Menczel, J., Choe, E. W., and Hughes, O. R. (1997), *SPE ANTEC '97*, April 27–May 2, 1997, Toronto, Vol. **II**, *Materials*, p. 916.

Sawatari, C. and Matsuo, M. (1985), *J. Appl. Polym. Sci.* **263**, 783.

Sbirrazzuoli, N. and Vyazovkin, S. (2002), *Thermochim. Acta* **388**, 289.

Schawe, J. E. K. (1995), *Thermochim. Acta* **260**, 1.

Schawe, J. E. K. (2002), *Thermochim. Acta* **388**, 299–312.

Schawe, J. E. K. and Hutter, T. (2005), *Proc. 33rd North American Thermal Analysis Soc. Conf.*

Scherrenberg, R., Mathot, V., and Steeman, P. (1998), *J. Therm. Anal.* **54**, 477.

Schubnell, M., Heitz, Ch., Hutter, Th., and Schawe, J. E. K. (2005), *Proc. 33rd North American Thermal Analysis Soc. Conf.*

Seferis, J. C., Salin, I. M., Gill, P. S., and Reading, M. (1992), *Proc. Acad. Greece* **67**, 311.

Shen, S. and Torkelson, J. M. (1992), *Macromolecules* **25**, 721.

Sichina, W. J. (2000), PETech-22, *Perkin-Elmer Technical Letters*, http://www.thermal-instruments.co.uk/Applications/PETech-22.PDF.

Siegmann, A. and Harget, P. J. (1973), *Bull. Am. Phys. Soc. II* **18**, 346.

Siemens, R. (2001), *Proc. 29th North American Thermal Analysis Soc. Conf.*, p. 246.
Simon, S. L. and Gillham, J. K. (1993), *J. Appl. Polym. Sci.* **47**, 461–485.
Simon, S. L. and McKenna, G. B. (2000), *Thermochim. Acta* **348**, 77.
Simon, S. L. (2001), *Thermochim. Acta* **374**, 55.
Sircar, A. K. (1997), "Elastomers," in *Thermal Characterization of Polymeric Materials*, 2nd ed., Turi, E. A., ed., Academic Press, San Diego, pp. 1379–1766.
Slough, G. (2006), TA Instruments, private communication.
Smook, J. and Pennings, J. (1984), *Colloid Polym. Sci.* **262**, 712.
Song, M., Hammiche, A., Pollock, H. M., Hourston, D. J., and Reading, M. (1995), *Polymer* **36**, 3315.
Song, M., Hourston, D. J., Reading, M., Pollock, H. M., and Hammiche, A. (1999), *J. Therm. Anal. Calorim.* **56**, 991.
Sourour, S. and Kamal, M. (1976), *Thermochim. Acta* **14**, 41.
Starkweather, H. W. Jr., Zoller, P., Jones, G. A., and Vega, A. J. (1982), *J. Polym. Sci., Polym. Phys. Ed.* **20**, 751.
Struik, L. C. E. (1978), *Physical Aging in Amorphous Polymers and Other Materials*, Elsevier, New York.
Struik, L. C. E. (1990), *Internal Stresses, Dimensional Instabilities and Molecular Orientations in Plastics*, Wiley, New York.
Stutz, H., Mertes, J., and Neubecker, K. (1993), *J. Polym. Sci., Part A: Polym. Chem.* **31**, 1879–1886.
Swarin, S. J. and Wims, A. M. (1976), *Anal. Calorim.* **4**, 155–171.
Swier, S., Pieters, R., and Van Mele, B. (2002), *Polymer* **43**, 3611.
Swier, S. and Van Mele, B. (2003a), *Polymer* **44**, 2689.
Swier, S. and Van Mele B. (2003b), *J. Polym. Sci., Pt. B* **41**, 594.
Swier, S., Van Assche, G., and Van Mele, B. (2004), *J. Appl. Polym. Sci.*, **91**, 2814.
TA Instruments (1993), DSC2920, *Differential Scanning Calorimeter Operator's Manual*.
TA Instruments (n.d.), *Modulated DSC™(Compendium Basic Theory & Experimental Considerations*.
TA Instruments (1994), TAI Modulated DSC Workshop.
TA Instruments (1997), Modulated DSC Training Workshop, McLean, VA, Sept. 11, 1997.
TA Instruments (1998), *DSC 2920CE Operator's Manual*, Appendix C.
TA Instruments (2005), *Modulated Differential Scanning Calorimetry® (MDSC®)*, TA Instruments Customer Training, June 2005.
TA Instruments (2007), *announcement at the 35th North American Thermal Analysis Soc. Conf., East Lansing, MI*, Aug. 26–28, by Terry Kelly, CEO of TA Instruments.
Tant, M. R. and Wilkes, G. L. (1981), *Polym. Eng. Sci.* **21**, 874–895.
Thomas, L. C. (2006), *Modulated DSC Technology*, TA Instruments (publisher).
Thomas, L. C., Tzero™ MDSC® Benefits and Signal Calculations, private communication.
Toda, A., Oda, T., Hikosaka, M., and Saruyama, Y. (1997a), *Thermochim. Acta* **293**, 47.

Toda, A., Oda, T., Hikosaka, M., and Saruyama, Y. (1997b), *Polymer* **38**, 2849.
Toda, A., T. Oda, T., Hikosaka, M., and Saruyama, Y. (1997c), *Polymer* **38**, 231.
Toda, A., Tomita, C., Oda, T., and Hikosaka, M. (1997d), *Proc. 25th North American Thermal Analysis Soc. Conf.*, p. 307.
Todoki, M. and Kawaguchi, T. (1977a), *J. Polym. Sci., Polym. Phys. Ed.* **15**, 1067.
Todoki, M. and Kawaguchi, T. (1977b), *J. Polym. Sci., Polym. Phys. Ed.* **15**, 1507.
Trouton, F. (1883), *Nature* **27**, 292.
US Patent 5,033,866, assignee Mettler Toledo AG, Greifensee, Switzerland, July 23, 1991, inventors. Thomas Kehl and Gosse van der Plasts.
Van Assche, G., Van Hemelrijk, A., Rahier, H., and Van Mele, B. (1995), *Thermochim. Acta* **268**, 121–142.
Van Assche, G., Van Hemelrijck, A., Rahier, H., and Van Mele, B. (1996), *Thermochim. Acta* **286**, 209.
Van Assche G., Van Hemelrijck, A., Rahier, H., and Van Mele, B. (1997), *Thermochim. Acta* **304/305**, 317.
van Ekeren, P. J., Holt, C. M., and Witteveen, A. J. (1997), *J. Therm. Anal.* **49**, 1105.
van Herwaarden, A. W. (2005), *Thermochim. Acta* **432**(2), 192.
van Krevelen, D. W. (2003), *Properties of Polymers*, 3rd ed., Elsevier, Amsterdam, pp. 149–151.
Van Mele, B., Van Assche, G., Van Hemelrijck, A., and Rahier, H. (1995), *Thermochim. Acta* **268**, 121–142.
Van Mele, B., Rahier, H., Van Assche, G., and Swier, S. (2004), in *The Characterization of Polymers Using Advanced Calorimetric Methods*, Reading, M. and Hourston, D. J., eds., Kluwer Academic, UK.
Van Mele, B., Rahier, H., Van Assche, G., and Swier, S. (2006), in *Modulated-Temperature Differential Scanning Calorimetry Theoretical and Practical Applications in Polymer Characterisation*, Hot Topics in Thermal Analysis and Calorimetry Series, Reading, M. and Hourston, D. J., eds., Springer, Vol. **6**.
Varga, J., Menczel, J., and Solti, A. (1979), *J. Thermal Anal.* **17**, 333.
Varga, J., Menczel, J., and Solti, A. (1981), *J. Thermal Anal.* **20**, 23.
Venditti, R. A. and Gillham, J. K. (1997), *J. Appl. Polym. Sci.* **64**, 3.
Vyazovkin, S. (1997), *J. Comput. Chem.* **18**, 393.
Vyazovkin, S. (2001), *J. Comput. Chem.* **22**, 178.
Vyazovkin, S. and Sbirrazzuoli, N. (1999), *Macromol. Chem. Phys.* **200**, 2294.
Weese, R. K. (2005), *Thermochim. Acta* **429**, 119.
Widmann, G. (1975), in *Thermal Analysis*, Vol. **3**, Buzas, I., ed., Heydon, London, p. 359.
Williams, R. J. J., Rozenberg B. A., and Pascault, J. P. (1997), *Adv. Polym. Sci.* **128**, 95.
Wisanrakkit, G. and Gillham, J. K. (1990a), *J. Appl. Polym. Sci.* **41**, 2885.
Wisanrakkit, G. and Gillham, J. K. (1990b), *J. Coat. Technol.* **62**(783), 35.
Wisanrakkit, G., Gillham, J. K., and Enns, J. B. (1990), *J. Appl. Polym. Sci.* **41**, 1895.
Wisanrakkit, G. and Gillham, J. K. (1991), *J. Appl. Polym. Sci.* **42**, 2453.
Wu, Z. Q., Dann, V. L., Cheng, S. Z. D., and Wunderlich, B. (1988), *J. Therm. Anal.* **34**, 105.

Wunderlich, B. (1960), *J. Phys. Chem.* **64**, 1052.

Wunderlich, B. and Cormier, C. M. (1967), *J. Polym. Sci., Polym. Phys. Ed.* **5**, 987.

Wunderlich, B. (1976), *Macromolecular Physics, Vol. 2, Crystal Nucleation, Growth*, Academic Press, New York.

Wunderlich, B. and Czornyj, G. (1977), *Macromolecules* **10**, 906.

Wunderlich, B. (1980), *Macromolecular Physics*, Vol. 3, *Crystal Melting*, Academic Press, New York.

Wunderlich, B. (1990), *Thermal Analysis*, Academic Press, Boston.

Wunderlich, B. (1997a), "The basis of thermal analysis," in *Thermal Characterization of Polymeric Materials*, Vol. **1**, Turi, E. A., ed., Academic Press, San Diego.

Wunderlich, B. (1997b), in *Modulated DSC Practical Applications*, TA Instruments, Sept. 10, 1997.

Wunderlich, B. (2005), *Thermal Analysis of Polymeric Materials*, Springer, Berlin, pp. 90–91.

Wunderlich B. (2006), *Thermal Analysis of Polymeric Materials*, optical disk.

Wunderlich, B., Jin, Y., and Boller, A. (1994), *Thermochim. Acta* **238**, 277.

Wunderlich, B. and Okazaki, I. (1997), *J. Therm. Anal.* **49**, 57.

Wunderlich, B., Okazaki, I., Ishikiriyama, K., and Boller, A. (1998), *Thermochim. Acta* **324**, 77.

Ye, P. (2006), Private communication to L. Judovite.

Zhang, Y. and Vyazovkin, S. (2005), *Macromol. Chem. Phys.* **206**, 1084.

CHAPTER 3

THERMOGRAVIMETRIC ANALYSIS (TGA)

R. BRUCE PRIME
IBM (Retired)/Consultant, San Jose, California

HARVEY E. BAIR
Bell Laboratories (Retired)/Consultant, Newton, New Jersey

SERGEY VYAZOVKIN
University of Alabama at Birmingham, Birmington, Alabama

PATRICK K. GALLAGHER
The Ohio State University (Emeritus), Columbus, Ohio

ALAN RIGA
Cleveland State University, Cleveland, Ohio

3.1. INTRODUCTION

Thermogravimetric analysis (TGA) or thermogravimetry (TG) is a technique where the mass[1] of a polymer is measured as a function of temperature or time while the sample is subjected to a controlled temperature program in a controlled atmosphere (Earnest 1988). Temperature ranges for commercial TGAs are typically ambient to 1000 °C or more, a sufficient upper limit for polymer applications. A purge gas flowing through the balance creates an atmosphere that can be inert, such as nitrogen, argon, or helium; oxidizing, such as air or oxygen; or reducing, such as forming gas (8–10% hydrogen in nitrogen). With polymers, a reducing atmosphere is rarely needed. The moisture content of the purge gas can vary from dry to saturated.

Polymers generally exhibit mass loss, although mass gain may be observed prior to degradation at slow heating rates in an oxidizing atmosphere (see

---

[1] As discussed in Chapter 1, the property measured by TGA is now referred to as *mass* and is no longer referred to as *weight*. While we adhere to this terminology, some figures in this chapter are reproduced with their original ordinates, which may be *weight* or *weight percent*.

Thermal Analysis of Polymers: Fundamentals and Applications, Edited by Joseph D. Menczel and R. Bruce Prime
Copyright © 2009 by John Wiley & Sons, Inc.

Fig. 3.30). Mass loss may be categorized as volatile components such as absorbed moisture, residual solvents, or low-molecular-mass additives or oligomers that generally evaporate between ambient and 300 °C; reaction products, such as water and formaldehyde from the cure of phenolic and amino resins, which generally form between 100 °C and 250 °C; and generation of volatile degradation products resulting from chain scission that generally require temperatures above 200 °C but not more than 800 °C. All of these mass loss processes may be characterized by TGA to yield information such as composition, extent of cure, and thermal stability. The kinetics of these processes may also be determined to model and predict cure, thermal stability, and aging due to thermal and thermooxidative processes.

In this chapter we present background principles of thermogravimetric analysis and the various measurement modes; issues associated with and recommendations for calibration; measurement and analysis methods for characterizing polymeric materials, including recommendations for how to perform a TGA experiment; and kinetics. At the end of the chapter we present selected applications of actual industrial problems where TGA has been instrumental in their solution.

3.2. BACKGROUND PRINCIPLES AND MEASUREMENT MODES

3.2.1. The Thermobalance

The heart of the thermogravimetric analyzer is the thermobalance, which is capable of measuring the sample mass as a function of temperature and time. The relationship between the components of a thermobalance varies from one instrument to another. A schematic representation as shown in Figs. 3.1a and 3.1b indicates typical thermocouple placements relative to the sample. The three standard sample and furnace positions relative to the balance are depicted in Fig. 3.1a. Figure 3.2 shows actual examples of currently available commercial instruments.

The bottom- and side-loading thermobalances shown in Figs. 3.2a and 3.2b utilize conventional resistive heating elements, while the bottom-loading example selected in Fig. 3.2c utilizes radiant heating elements necessitating a transparent tube to constrain the atmospheric flow.

The three different configurations necessitate different sample support systems. The sample holder is rigidly attached to the end of the balance arm in the horizontal configuration, and there is no added mass due to a suspension harness. In the top-loading arrangement, the suspension arm must be rigid with a weight below to keep it vertical. This added mass reduces the total mass range available for the sample. Suspension below the balance may be less rigid and does not require the compensating weight and thus detracts less from the available range. Note that top- and bottom-loading instruments have a vertical orientation while the side-loading TGA has a horizontal orientation.

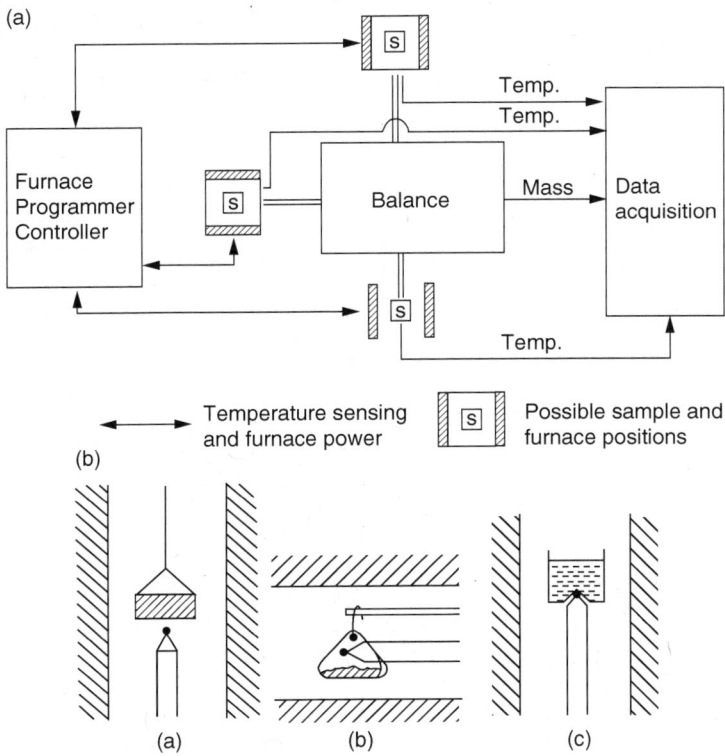

Figure 3.1. (a) General arrangement of components in a thermobalance [from Gallagher (1993) with permission of Elsevier]; (b) typical location of thermocouples: configuration (a) top-loading, configuration (b) side-loading, and configuration (c) bottom loading [from Gallagher (1997) with permission of Elsevier].

3.2.2. Factors Affecting TGA

Table 3.1 lists some of the major experimental factors influencing TGA measurements. Disturbances in the measurement of mass arise from three general sources: atmosphere effects, secondary reactions, and electrical considerations.

In thermogravimetric analysis *buoyancy* is the upward force on the sample produced by the surrounding atmosphere, which will affect the apparent mass during a TGA experiment. The buoyancy phenomenon occurs as the density of the atmosphere in the balance decreases with increasing temperature, resulting in an apparent mass gain. Consequently, it depends on the volume of the sample and its supports, and the density of the atmosphere. Both may vary with temperature and rate of temperature increase, so any buoyancy correction may change during the course of the experiment. Effects of buoyancy can be quite apparent at the start of a heating segment, and when an atmosphere is deliberately switched, such as, from nitrogen to air, owing to

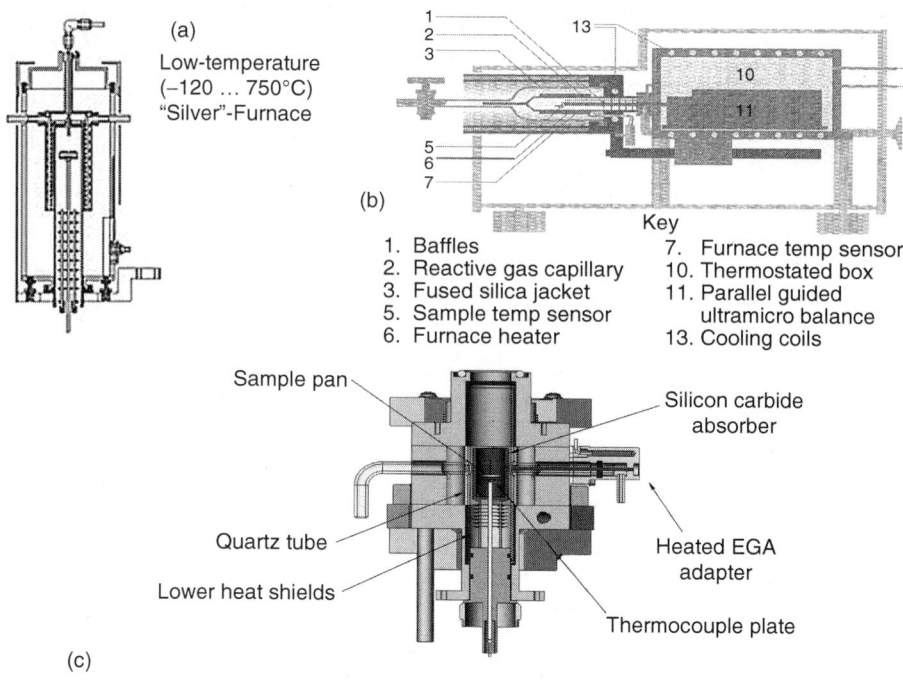

Figure 3.2. Three examples of commercial thermobalances, including typical thermocouple placement: (a) a top-loading model (courtesy of Netzsch Instruments); (b) a side loading model (courtesy of Mettler-Toledo); (c) a bottom-loading model (courtesy of TA Instruments).

TABLE 3.1. Major Factors Affecting TGA Results

Mass	Temperature
Buoyancy and thermal expansion	Heating rate
Atmospheric turbulence	Thermal conductivity
Condensation and reaction	Enthalpy of the processes
Electrostatic and magnetic forces	Sample–furnace–sensor arrangement
Electronic drift	Electronic drift

differences in gas properties such as density and flow rate. Modern thermobalances have design features that minimize or compensate for these effects. In general no corrections are required, except for the most sensitive experiments involving very small changes in mass.

One approach to compensate for buoyancy and general baseline issues encountered with very small mass loss is to perform a second experiment with a closely matching inert sample and subsequently subtracting this "blank" experiment. A series of such experiments can be stored for future use. An excellent example of this technique was the determination of the coating thickness on fused-silica fibers (Gallagher 1992). Fibers were initially heated in an oxidizing atmosphere to burn off the very thin polymer coating. They were then immediately reheated under identical conditions and the second experiment subtracted from the initial one to determine the mass loss curve associated with the very small loss in mass resulting from combustion of the thin coating.

Other minor problems introduced by the changing temperature are related to thermal expansion of the thermobalance components. Unequal expansion of the balance lever arms will distort the mass signal and the underlying baseline. The horizontal configuration shown earlier will minimize this problem because both lever arms are in the same furnace environment. Controlling the temperature of the balance chamber by thermostatting the balance compartment or, more practically, by using simple water-cooled baffles in the vertical configurations, will solve the problem. Thermal expansion of the sample support, that is, the mechanism on or from which the sample rests or is suspended, will move the relative sample position within the furnace and may result in slightly different thermal environments or worse yet, disturb the balance signal by contact of the sample with another component such as the temperature sensor or furnace wall.

All thermobalances effectively utilize baffling to reduce the thermal convection currents that would otherwise disturb the measurement of mass and heat the balance compartment. Optimization of the atmospheric flow pattern and the baffling are critical to reduction of thermal noise and drift in the measurement of the mass. Many models will impose a water-cooled plate between the balance compartment and the furnace. When long-term stability is required, it is advisable to consider actually thermostatting the balance compartment. Since heat normally rises, vertical configurations are most susceptible. The horizontal arrangement leads to less interference from the flow patterns arising from control of the atmosphere.

Clearly it is essential that the sample, any volatile species, and any reactive atmosphere not react with the sample holder or other components with which they may come in contact. This is seldom an issue with polymers; however, the interaction of phosphorous containing polymers with platinum above 900 °C is one exception. Condensation of volatile products on the suspension in cooler portions of the furnace may be avoided by using a sufficient gas flow or longer hot zone. The flow pattern of the atmosphere, the baffling, and the thermobalance configuration all affect the likelihood of such a problem.

Many of the factors discussed above pertain to considerations in choosing the technique and specific instrumentation. Others are concerned with the nature of the particular material or process under investigation (enthalpy, reversibility, potential condensation, etc.) and an awareness of the potential caveats associated therewith. Finally, there is the selection of the specific experimental conditions (sample size, temperature program, atmosphere and its flow rate, data analysis, etc.).

A routine for the calibration of mass and temperature should be established (see Section 3.3). Under typical heating and gas flow conditions, the accuracy of the mass measurement is decreased by about an order of magnitude over the static specification for the balance at room temperature. The accuracy in temperature is affected by many factors (sensor and/or mass placement, enthalpy of the process, and the many conditions that control thermal transport). Calibration and actual measurements should be conducted under as similar conditions as possible.

3.2.3. Controlled Rate TGA

Controlled rate TGA or CRTGA operates in at least three modes. The first is to establish preset rates of change in mass with time and utilize the measured dm/dt values to control the furnace temperature. Figure 3.3b shows the results

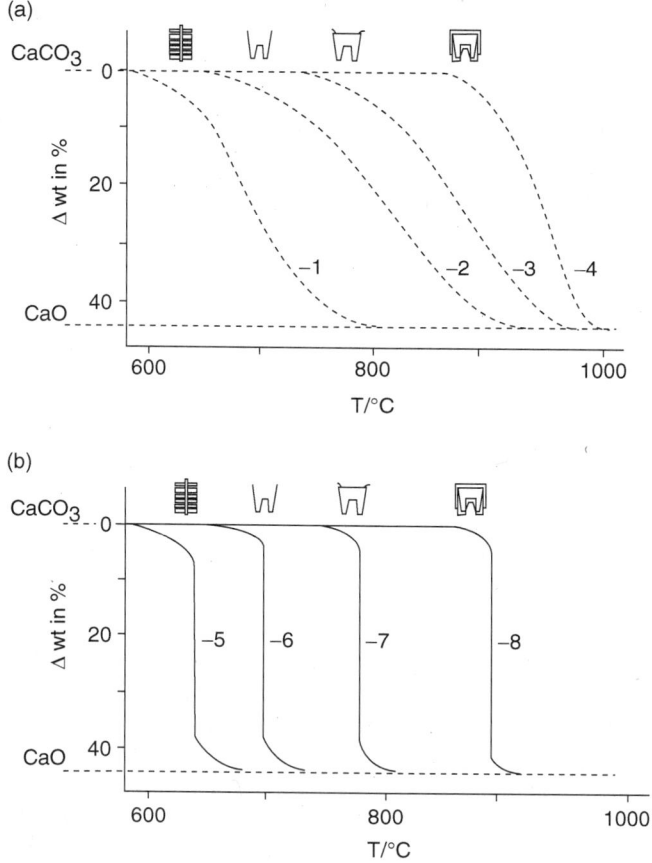

Figure 3.3. TGA curves of the thermal decomposition of $CaCO_3$ in air: (a) linear heating rates increasing numerically; (b) CRTGA. Sample configurations that control the equilibrium decomposition temperature are illustrated. The sample holders, from left to right, are multiple thin layers, open crucible, covered crucible, and pinhole labyrinth crucible—providing decreasing exposure to the flowing atmosphere in this order. [From Paulik and Paulik (1986) with permission of Elsevier.]

of this approach for the thermal decomposition of $CaCO_3$ (Paulik and Paulik 1986). Comparison with the curves in Fig. 3.3a shows how the decompositions here appear to occur at a constant temperature, the equilibrium decomposition temperature associated with the particular partial pressure of CO_2 determined by the sample holder and atmospheric flow rate. An excellent practical example of this method is the slow controlled burnout of the polymeric and organic binders, surfactants, and lubricants from a green pressed ceramic body. Ceramic powders are preformed into shapes by compression molding, casting, and other procedures. The organic compounds must be removed prior to the sintering process, or they will lead to cracks and local deformations.

A second mode of CRTGA is essentially a stepwise isothermal approach in which upper and lower limits or thresholds are set for the rate of mass change (Sichina 1993). This method is also referred to as *autostepwise*, and several instrument suppliers provide the necessary software. The sample is heated at a preselected linear rate until the upper rate of mass change is achieved. At that point the temperature is maintained constant until the lower rate of change is reached, that is, until the mass loss step is essentially completed, after which the linear heating rate is reapplied and the process continues to the next mass loss step. A typical upper limit would be 10–100 times greater than the lower limit. Figure 3.39 compares this approach vis-à-vis 10°C/min heating for an ethylene–vinyl acetate (EVA) copolymer. The stepwise isothermal approach gives the highest resolution, but it takes time to properly establish the upper and lower thresholds.

A closely related third approach is referred to as dynamic rate Hi-Res™ TGA (Gill et al. 1992). This methodology is based on adjusting the heating rate during a TGA run in order to improve separation between closely occurring mass loss events. The dynamic rate approach provides a good compromise between improved resolution and analysis time. In dynamic rate Hi-Res TGA, the heating rate of the sample is dynamically and continuously varied in response to changes in the sample's rate of mass loss. This allows very high heating rates to be used in regions where no mass changes are occurring but prevents the temperature from overshooting a thermal event. Two experimental parameters need to be selected to run a dynamic rate Hi-Res TGA experiment—the initial heating rate (usually 20–50°C/min) and the resolution index, a user-selected parameter that defines the range of dm/dt values over which the system will respond by adjusting the heating rate. Another variable (sensitivity) is available for further fine tuning if required. Figure 3.4 compares the results of this approach with traditional TGA at 5°C/min for the thermal decomposition of $CuSO_4 \cdot 5H_2O$ (Gill et al. 1992). Note the five distinct mass loss steps in the dynamic Hi-Res measurement. The steps associated with the dehydration are much more clearly distinguished in this Hi-Res mode.

Besides achieving better resolution, the overall elapsed time of the experiments may be reduced depending on the choice of parameters. The three

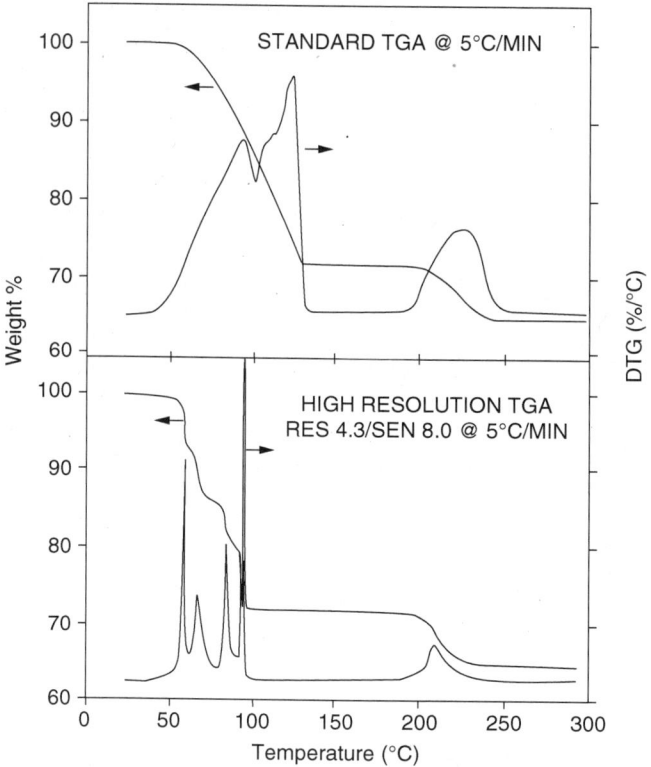

Figure 3.4. A comparison of dynamic Hi-Res and traditional TGA for the thermal decomposition of $CuSO_4 \cdot 5H_2O$ (courtesy of TA Instruments, also see Gill et al., 1992).

CRTGA methods described have been compared by Barnes et al. (1994). It is wise to look at the resulting time–temperature profile in order to achieve or confirm what has transpired. The experimenter should be aware that in rare cases the resulting CRTGA temperature program may lead to a different mechanism and products than does linear heating or isothermal experiments (Arthur and Redfern 1992).

3.2.4. Modulated TGA

Modulated TGA (MTGA™) is a methodology for obtaining continuous kinetic information for processes that produce volatile reaction products, such as cure of phenolics and decomposition of polymers (Blaine and Hahn 1998). Like modulated temperature DSC, MTGA superimposes a sinusoidal temperature modulation on the traditional underlying heating profile. MTGA may be used under quasi-isothermal conditions to characterize kinetics of a single mass loss, or it may be combined with linear heating or controlled rate TGA to analyze materials with single or multiple mass loss steps. In a single experiment it can provide results similar to the ASTM E1641 method (discussed in Section 3.5 of this chapter) that requires a minimum of three experiments.

3.2.5. Other Techniques and Measurements

Thermogravimetric analysis provides a quantitative measurement of the mass change, but does not indicate the nature of the material lost. Hence, it is often used in a combined manner with evolved gas analysis (EGA) measurements on the same sample. EGA determines the specific nature of the changes in the gas phase and may be done in a qualitative or quantitative manner. There are several modes of operation. Gas samples can be trapped and analyzed periodically either during or at the end of the experiment. More common and more useful is to monitor the atmosphere and correlate its chemical nature with the observed mass change in a near-simultaneous mode, with negligible delay between mass loss in the TGA and gas detection by EGA.

The resolution of the chemical analysis with the temperature and time of the TGA measurement will depend on a number of factors, importantly the heating rate, the required analysis or scan time of the EGA measurement, and the delay caused by the transfer time between the TGA source and the EGA detector. The latter is a function of the atmosphere's flow rate and the length of the path prior to analysis. The temperature (typically 200–230 °C, high enough to prevent condensation but not so high as to promote degradation) and nature of the transfer path can influence the results through condensation or reaction of the products.

Any technique for gas analysis can be applied to EGA. The most frequently used methods are mass spectroscopy (MS) and Fourier transform infrared spectroscopy (FTIR). Many instrument manufacturers provide the ability to interface their TGAs with MS or FTIR (see Section 3.7, on instrumentation). Temporal resolution between the TGA and the MS or FTIR detector is an important feature, for example, in distinguishing absorbed water from water as a reaction product and in assigning a decomposition product to a specific mass loss. Each method has its experimental requirements, limitations, and advantages. Mass spectroscopy is a very sensitive technique that identifies volatile species by their mass-to-charge ratio, referred to as m/z. The evolution of the sum of all m/z species can be plotted and compared with the derivative TGA plot to ensure temporal resolution between the TGA and the mass spectrometer. The evolution of a specific m/z, associated with species such as water or formaldehyde, can show the distinct evolution of these compounds. The most common ionization is by 70 eV electron impact (EI), which operates under high vacuum. Because the ion source filaments are sensitive to oxygen, TGA is normally restricted to pyrolysis in an inert atmosphere. EI tends to fragment larger molecules that are formed during thermal degradation, creating ambiguity in the precise identification of the degradation products. However, the fragmentation patterns can be an aid in identifying the class of degradation product, such as saturated hydrocarbons, polyethers, or carbonyl-containing species. Mass spectrometers that employ atmospheric pressure chemical ionization (APCI) simplify interfacing to ambient pressure TGA and allow analysis of processes that occur in air (Prime and Shushan 1989), but they are more expensive than EI mass spectrometers.

Many polymers evolve relatively simple gases on heating, such as water, ammonia, acetic acid, hydrogen chloride, and oxides of carbon, nitrogen, and sulfur. These gases are easily identified in the purge gas by FTIR, since they have characteristic absorption frequencies and band contours. More complex products are not always totally identifiable, but the observed group frequencies are often enough to define the class of product. An evolution profile for the entire gas stream can be compared with the DTG plot to ensure temporal resolution between the TGA and the FTIR. Specific gas profiles such as water, CO_2, or even the carbonyl group can also be plotted. FTIR gas analysis is an ambient pressure process that can accommodate both reactive and inert atmospheres within the TGA. An option to maintaining the FTIR gas cell at elevated temperature to prevent condensation is to purposely trap the higher-molecular-mass condensable products, which tend to foul the gas cell and complicate the IR spectra, on a membrane for subsequent analysis while allowing the gaseous products to be analyzed in real time (Khorami et al. 1986). Lewis et al. (1989) used a fiber-reinforced composite as both the sample and the membrane and were able to measure diffusion of volatile degradation products through the sample. Thus, the selection of an EGA detector is based on the specific experimental conditions and goals. See Prime (1997a) for a detailed discussion of TGA/MS and TGA/IR. Besides the common texts on thermal analysis, there are good reviews of the subject for consideration (Mullens 1998; Gallagher 1997). The pulsed-gas techniques make quantitative measurements more reliable (Gallagher 1997; Mullens 1998; Maciejewski and Baiker 2008).

Energetic information supplied through DTA or DSC also complements the information from TGA. They not only provide information on the enthalpy of the process but also detect transitions and reactions that do not result in a change in mass of the system (see Chapter 2). Simultaneous TGA/DTA or TGA/DSC has several advantages over separate experimentation. In addition to a savings in time and some costs, the primary advantages are concerned with the alleviation of potential ambiguities in the interpretation of the results that might arise from differences in samples, atmospheric conditions, temperature, and other variables. The combination of the calorimeter and quartz crystal microbalance (Smith and Shirazi 2005; Smith et al. 2004, 2006) is a more recent novel application of such measurements.

3.3. CALIBRATION AND REFERENCE MATERIALS

3.3.1. Mass

The drifts in stability of modern electronic devices over short periods of time are minimal, particularly if the environmental temperature is maintained stable. Frequent calibration procedures serve to protect against long-term changes depending on one's experience with a particular instrument. Most important in TGA is the linearity of the mass measurement, for instance, that mass losses of 5%, 50%, and 95% be measured with equal reproducibility and accuracy. At room temperature it is relatively simple to verify the mass using standard calibrated weights. Verifying accuracy under changing temperature conditions, however, is another matter. Such standards are constantly being sought and investigated. Mixtures of liquids with markedly different vapor pressures or well-characterized stable compounds, such as hydrates and carbonates that have reliable decomposition patterns, are utilized. TA Instruments developed and sells three certified reference materials with nominal mass losses of 2%, 50%, and 98% at 160°C (Blaine and Rose 2004).

Undesired electromagnetic fields can be a problem when magnetic samples are involved. These may include potentially magnetic sample holders, magnetic recording materials such as tape and disk media, and binder burnout in polycrystalline magnets. Current flow through the resistive elements of a furnace can generate substantial magnetic fields, particularly during the initial heating surge. Manufacturers are well aware of the problem and have generally wound their furnaces in a bifilar mode or arranged the elements in other ways to cancel the resulting field. Electrostatic disturbances, often on the quartz furnace tube, can cause the sample to become attracted to nearby components such as a furnace wall. This phenomenon is most prevalent during loading of the sample and closing of the furnace prior to a TGA run. Antistatic sprays and coatings are used or an ionizing element, such as a weakly radioactive polonium (Po) source, may be used to counteract this problem. In the Perkin-Elmer TGA a curtain of charged particles can be activated to neutralize static buildup on sample and furnace walls. This problem is exacerbated at low humidity, and increasing the local humidity will diminish the severity of the problem.

A general conservative guideline is that the accuracy of the mass in a TGA experiment while heating or at elevated temperature is an order of magnitude less than that at static room temperature conditions. Both precision and accuracy can be improved markedly through appropriate use of tares and careful baseline subtraction techniques.

3.3.2. Temperature

The preceding section has indicated the uncertainties in determining the change in mass. Establishing the exact temperature to associate with the mass, however, is an even more difficult task. The flow of heat to or from the heat source, sample, and temperature sensor is subject to many factors. The thermal conductivity of the appropriate pathways changes markedly with the nature and flow of the atmosphere. The thermal conductivity of the atmosphere depends primarily on its average molecular mass (see Table 3.2). The differences between air, nitrogen, and oxygen are trivial while for argon the difference is small. However, the use of helium, hydrogen, or carbon dioxide will require recalibration of temperature and heating rates because of their significantly higher thermal conductivities. In addition, the temperature dependence of this heat transfer will vary with temperature as the primary mechanism for heating changes among convection, conduction, and radiation.

In the bottom-loading configuration the temperature sensor may be in direct contact with the sample pan. (See Fig. 3.1b for illustrations of different balance configurations.) In the side- and top-loading arrangements, however, the temperature sensor is nearby but not touching in order not to disturb the mass signal. In these latter configurations the lack of direct contact makes temperature calibration even more critical. The location of the temperature sensor touching or near the sample rather than the heat source increases the lag time for temperature control. Manufacturers may employ an additional sensor near the heating elements for smoother control of the temperature program.

Because of the thermal transport conditions, the heating rate has a strong influence on the apparent temperature of events. The inability to instantly heat or cool the sample means that there is inevitably a lag between the sample's response and the temperature program. This lag obviously increases with increasing scan rate.

Two other factors give rise to the departure of the sample's actual temperature from the desired programmed temperature. One is associated with the enthalpy of the event. An endothermic reaction will consume the heat supplied rather than raise the temperature of the sample. Conversely, an exothermic reaction can lead to the temperature getting ahead of the intended temperature. If the sample combusts, the temperature can get far ahead. The

TABLE 3.2. Thermal Conductivity of Purge Gases in W/(m·K) at 20°C

Purge Gas	Thermal Conductivity
Argon	0.018
Nitrogen	0.023
Air/oxygen	0.024
Carbon dioxide	0.087
Helium	0.138
Hydrogen	0.172

Source: Young (1992).

second factor comes into play only for reversible reactions. Under such conditions the equilibrium temperature of the reaction will be influenced by the concentration of the products and is dependent on the ability of the flowing atmosphere to sweep the products away. If the partial pressure of a product is allowed to increase, then the sample temperature must increase in order to continue reacting.

Figure 3.3 clearly indicates these effects for the reversible decomposition of $CaCO_3$. The reaction is highly endothermic, so significantly greater demands are made on the rate of heat supply during the decomposition period. Because the reaction is reversible, the decomposition temperature depends on the partial pressure of CO_2 and, therefore, the ability of the flowing atmosphere to sweep the product away from the reacting interface. Faster heating rates further accentuate the lags between the temperature of the sample, sensor, and program. All the shifts observed in Fig. 3.3 reflect those effects.

Two approaches are utilized for the calibration of the temperature axis in TGA. One method employs thermomagnetometry (TM) to detect well-defined magnetic transition temperatures (Norem et al. 1970; Garn et al. 1980; Gallagher et al. 2003). A permanent magnet is used to induce a small magnetic attraction, equivalent to ~16% of the nickel sample mass shown in Fig. 3.5. The *transition temperature* is defined as the disappearance of the magnetic attraction, so the endpoint of the apparent change in mass is assigned to the known transition temperature, as illustrated in Fig. 3.5. The current set of *ICTAC Certified Curie Temperature Reference Materials* may be obtained through TA Instruments, Inc. Perkin-Elmer provides a set of Curie temperature reference materials that are not certified.

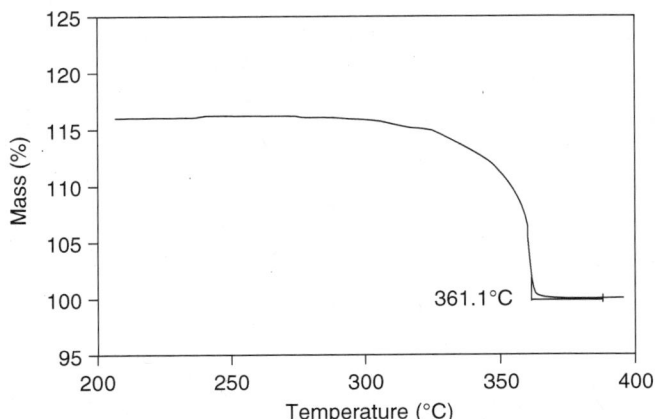

Figure 3.5. Thermomagnetometry experiment with nickel, illustrating the magnetic transition temperature as the extrapolated endpoint. Note that the magnetically induced mass represents 16% of the sample mass. Compare the measured temperature of 361.1°C with the Curie temperature of 358.2°C shown in Table 3.3. (Courtesy of TA Instruments.)

The second approach utilizes the melting points of high-purity metals as defined in the International Temperature Scale (Quinn 1990). For simultaneous TGA/DTA or TGA/DSC instruments, the melting temperature T_m is determined by the DTA or DSC component as described later and in Section 2.5. In the case of a standalone TGA it is necessary to obtain at least a momentary change in mass at the well-defined melting points. This can be achieved using the fusible-link technique advocated by McGhie (McGhie, 1983; McGhie et al. 1983). A thin link of the pure metal is placed on the sample holder near the sample position. A weight suspended from this link is designed to fall onto the pan or through an opening in the pan to create the momentary change in observed mass. Figure 3.6 illustrates possible configurations. Fusible links are not available for purchase, but indium fibers and thin films are available from Alfa Aesar, www.alfa.com. Indium films are recommended because they are less fragile than the fibers and can be cut into strips of appropriate size. Care must be taken with the atmosphere so as not to oxidize the pure metals. A very low partial pressure of oxygen is needed at more elevated temperatures in order to prevent such oxidation. Both the magnetic and fusible-link methods

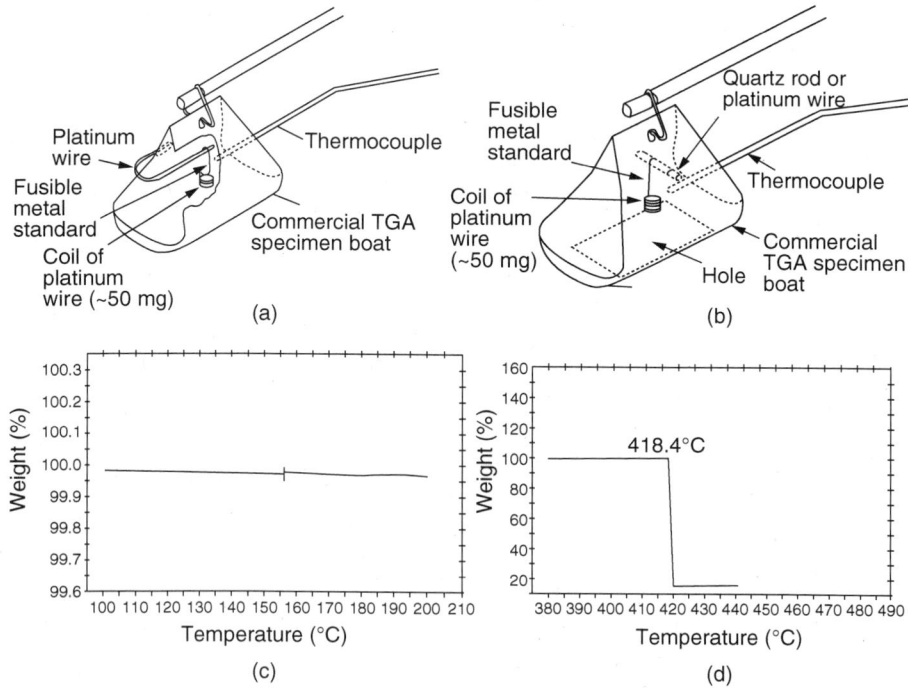

Figure 3.6. An example of the fusible-link approach to temperature calibration: (a) solid pan, (b) open pan, (c) mass–temperature plot for (a); (d) mass–temperature plot for (b). (Adapted from McGhie, 1983 with permission of the American Chemical Society and from McGhie et al., 1983 with permission of Elsevier.)

were found to work equally well in a comparison study (Gallagher and Gyorgy 1986).

Figure 3.7 shows the use of both magnetic and melting point standards in a simultaneous TGA/DTA instrument. The melting points were used initially to establish the magnetic transition temperatures for subsequent independent use to calibrate standalone TGA units. Table 3.3 lists some of the recommended materials used to calibrate the temperature axis of TGA instruments.

In summary, no general values can be listed for the precision or accuracy to be expected in a TGA experiment, even when well calibrated. As with the

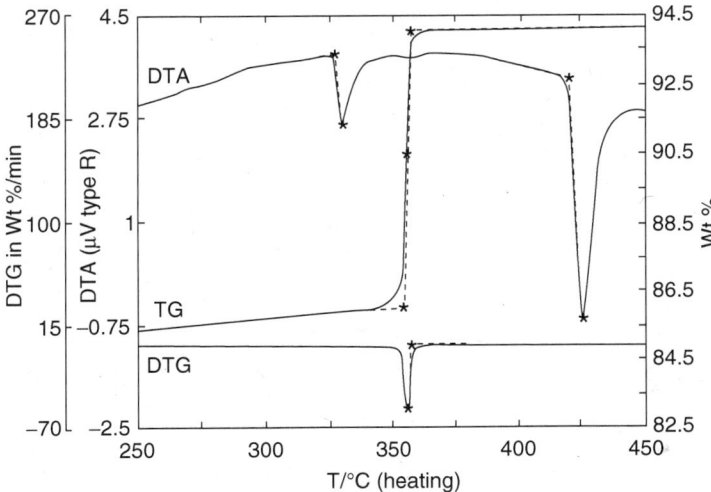

Figure 3.7. Simultaneous TGA/DTA curves show the melting of zinc and lead from the DTA signal and the magnetic transition of nickel, intermediate between the melting points, in the TGA and DTG signals; 10 °C/min heating rate in flowing nitrogen. [From Gallagher et al. (1993). Reprinted with permission of Akadémiai Kiadó, Hungary.]

TABLE 3.3. Common Materials Used for Temperature Calibration of TGA Instruments in Range 25–1000 °C

Fusible-Link Calibration		Magnetic Calibration[a]	
Material	Melting Temperature (°C)	Material	Curie Temperature (°C)
Indium	156.6	Alumel	152.6 ± 2.0
Tin	231.9	Nickel	358.2 ± 2.1
Lead	327.5	$Ni_{0.83}Co_{0.17}$	554.4 ± 4.3
Zinc	419.5	$Ni_{0.63}Co_{0.37}$	746.4 ± 3.1
Aluminum	660.3	$Ni_{0.37}Co_{0.63}$	930.8 ± 1.9
Silver	961.8	—	—

[a]ICTAC *Certified Curie Temperature Reference Materials*, available from TA Instruments.

mass calibration, a general conservative guideline is that the accuracy of the temperature in a TGA experiment while heating or at elevated temperature is an order of magnitude less than at static room temperature conditions. The precision and accuracy of both temperature and mass measurements are strongly influenced by the heating or cooling rate, thermal transport, and the chemical or physical processes occurring at that time. Naturally, sample size and shape will also have an effect. Consequently, the error limits will vary during the course of most experiments. In addition, the degree of match between the conditions used for calibration and the current experimental conditions is clearly a factor. For example, a simple isothermal experiment at 250 °C under a steady modest flowing atmosphere with no chemical reaction or physical transformation would be expected to have the limits imposed by the calibration, of the order of a degree or less or a small number of micrograms (depending on the type of balance). On the other hand, an experiment at 20 °C/min that is undergoing a reaction or transformation involving substantial enthalpy and evolving volatiles may easily have errors an order of magnitude greater. Two relevant ASTM standards are E2040, *Standard Test Method for Mass Scale Calibration of Thermogravimetric Analyzers* and E1582, *Standard Practice for Calibration of Temperature Scale for Thermogravimetry*.

3.4. MEASUREMENTS AND ANALYSES

The purpose of this section is to provide guidance to the reader in the general design and implementation of a TGA experiment as well as to provide insight and practical examples of specific types of measurement, such as thermal stability, isothermal mass loss, and material composition.

3.4.1. Designing and Performing a TGA Experiment

Thermogravimetric analysis enables one to continuously monitor the mass of a sample as a function of temperature and/or time. Unfortunately, the instrument cannot identify the volatile products that are evolving as temperature is increased unless it is coupled to another analytical tool such as a mass spectrometer (MS) or Fourier transform infrared spectrometer (FTIR), as discussed in Section 3.2.5. For standalone TGA it is important to gather as much information as possible about a sample from prior work, from the supplier, or from the open literature. In some instances, a more complete characterization of a sample may be possible using complementary experiments, including other thermal or analytical techniques.

Before performing the TGA experiment, it is important to have a clear understanding of the objectives and how best to meet those objectives. In terms of precision and accuracy, the needs of many experiments will be satisfied by a properly calibrated TGA over wide temperature and mass ranges. If the temperature of a thermal event is especially important, then concentration on an even higher degree of accuracy over a limited temperature range may be warranted. The best possible control of isothermal temperature is often necessary, but in some situations, such as quality control, precision may be more important than accuracy. On the other hand, temperature accuracy is key to obtaining good kinetic data. In most cases the accuracy of mass percent or $\Delta(\text{mass}\%)$, for example, in the measurement of material composition, is important. Often these mass measurements are made at elevated temperature where accuracy is reduced, but measuring the mass after the TGA has cooled to room temperature can improve the accuracy. We recommend that readers refer to the ASTM standard methods listed in the Appendix (at the end of this chapter) for specific measurements and where documented results are required. The precise and accurate measurement of very small mass losses (e.g., <1%) will typically benefit from baseline subtraction routines. In both isothermal and multiple heating rate kinetic studies, the accuracy of the temperature and the mass or conversion will determine the accuracy of not only the kinetic results but also the predictions made with these results. How the exothermic or endothermic nature of the reaction being investigated affects the accuracy of the kinetic data needs to be considered.

The main choices one has to decide on in performing a TGA experiment are the sample pans, the sample size, the temperature program, including possible isothermal steps; and the gas environment—either inert or oxidative. Occasionally, it may be helpful to add moisture to the TGA's gas stream to measure moisture sorption/desorption properties (see Section 3.4.4 and Fig. 3.15) or to test a polymer's susceptibility to hydrolysis. Several manufacturers sell accessories that allow the operator to control the room temperature humidity of the purge gas (see Section 3.7 on instrumentation). A fritted glass bubbler can also be employed to saturate the purge gas with water.

The pans used for analyzing polymers are usually made of platinum to withstand temperatures to 800 °C or more and often in an oxidizing atmosphere. Disposable aluminum pans can be used for measurements that leave a stubborn residue provided the test can be completed below the 660 °C melting temperature of aluminum. Examples of TGA pans that are commercially available are shown in Fig. 3.8. Ceramic pans should be used for high-temperature TGA studies with phosphorus-containing chemicals, additives, or polymers. At temperatures above 900 °C phosphorus will react with platinum resulting in significant deterioration of the platinum pans. Care must be taken in reusing platinum pans. Usually, placing the platinum pan in a torch flame and heating the pan until it is red in color (>1100 °C) will remove most organic or polymer residues. However, residues resistant to this heat treatment can prevail. At that time it is best to replace the platinum pan and recycle the platinum.

A typical, commercial TGA can sense in a reproducible fashion mass changes as small as a few tenths of a microgram. With this level of sensitivity, it is seldom necessary to analyze large samples. Obviously, smaller samples will equilibrate with the furnace temperature faster than will larger samples.

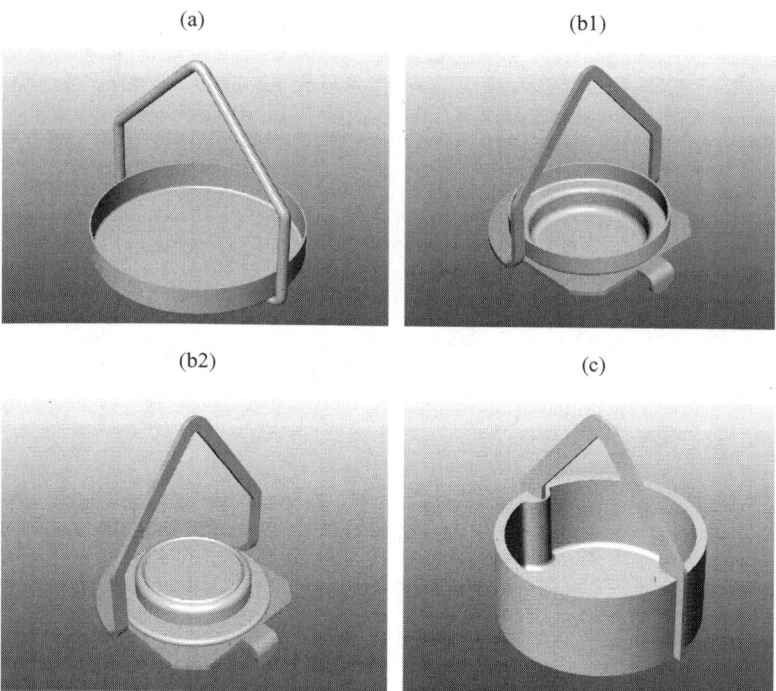

Figure 3.8. Sample pans: (a) platinum, 50 and 100 µL; (b) aluminum, 80 µL, can be open (b1), sealed (b2), or sealed and punched (not shown); (c) ceramic, 250 µL (courtesy of TA Instruments).

In addition, during pyrolysis a heavy sample (>10 mg) as compared to a relatively light sample (<5 mg) will result in more degradation products being trapped on the furnace itself and/or on the inside of the TGA's relatively cool glass walls, and may cause turbulence inside the TGA, leading to noise. The furnace and sample holders may be kept clean by periodically heating to 800 °C in air for 10–15 min. Condensed products may be removed from the furnace walls by using a modest amount of solvent on a long Q-tip or pipe cleaner. When this fails, the unit's glassware will need to be baked in air or oxygen at 600 °C or higher, and perhaps afterward one should consider hand rubbing the instrument's inner glass surfaces with an abrasive, fine-grit sandpaper in order to clean the system prior to the next run. Since most TGA practitioners resist keeping their glassware clean, we suggest that it is time for a cleanup when the sample is partially obscured by condensation products on the glass walls or static charge forces between the sample and its environment become significant. Starting when the TGA is new and following these cleaning procedures, a baseline should be run from ambient to 800 °C. A comparison of the current heating run with the new-instrument baseline, possibly using a baseline subtraction routine, will provide a measure of the cleanliness of the instrument. Pyrolysis runs with smaller samples mean that more samples can be run without the need to institute lengthy cleanup procedures. Thus, during degradation runs, preparing samples that weigh only a few milligrams can minimize many of the difficulties noted above. Of course, using TGA where samples are not being degraded and only minute mass changes occur will benefit from using larger samples. An extreme case that demonstrates the utility of large sample mass are TGA water desorption experiments from polyethylene (PE) (McCall et al. 1984). This work established the solubility and diffusion of water in hydrophobic PE. In these studies bars of PE weighing roughly 2.2 g were placed in a thermogravimetric analyzer with its internal furnace removed in order to accommodate such a large sample. All but 100 mg of the sample's mass was tared away. Under these conditions water mass losses were detected with a precision of 0.5 ppm (Bair et al. 1983).

It is always a good idea to control the size and geometry of the sample. It is typical for rates of simple evaporation to be controlled by the surface area where additional volume merely serves as a reservoir. It is also common for more complex mass loss processes to show a dependence on sample geometry, such as the area: volume ratio. Mass loss rates will depend not only on the formation of volatile products but also their diffusion to the surface where mass loss occurs. In oxidative degradation processes the diffusion of oxygen into the sample must also be considered. Examples of these effects are described in Sections 3.4.2 and 3.4.3 for the thermooxidative degradation of poly(vinyl methyl ether) (PVME) and in Section 3.4.5 and Fig. 3.19 for the cure of polymer thick films (PTF) that involves evaporation of a high-boiling solvent as well as crosslinking. For these reasons it is always good practice to investigate the effects of sample size or geometry on mass loss behavior to see how rigidly these parameters need to be controlled.

The operator may choose from a variety of purge gas options, depending on the purpose of the experiment. High-purity dry nitrogen is a good default and should be used when it is important to avoid oxidative processes. High-purity dry air may be used if the purpose of the experiment calls for it or if oxidation is not a concern. If the source of the purge gas is compressed air, it is important to use a filter to remove any pump oil. The thermal conductivity of the purge gas may be a consideration, for instance, high-thermal-conductivity gases will benefit experiments with fast heating rates or large samples. Values for most common purge gases are listed in Table 3.2. Air, nitrogen, and argon are low-conductivity gases, while helium has high thermal conductivity. Gas-switching accessories allow the purge gas to be changed during the experiment, for example, when the first part of an experiment is conducted in nitrogen and then finished in air. Prime et al. (1988a) held a magnetic coating in nitrogen at 230 °C to remove the solvents and cure the thermoset binder, and then switched to air to quantify a critical oxidative mass loss (Fig. 3.21). Carbon fiber or carbon-black content can often be measured by a similar technique (see Section 3.4.6). The flow rate of the purge gas is important, and it should be set as suggested by the manufacturer; 50 mL/min is a typical value. In the actual TGA experiments, the purge gas and the flow rate should be the same as that used for calibration.

The heating program is based on the user's information needs, discussed in some detail in the following subsections. Constant heating rates are typically in the 5–20 °C/min range. Faster rates like 20 °C/min take less time and are good for survey runs. Slower rates give better separation of overlapping thermal events. Hi-Res and autostepwise techniques, described in Section 3.2.3 and illustrated in Figs. 3.4 and 3.39, respectively, provide an attractive approach to separating distinct thermal processes. As illustrated in Fig. 3.30, fast heating rates in air can mask mass gain due to oxidation observed at slower rates. For isothermal experiments, heating rapidly (e.g., ≥100 °C/min) to the isothermal temperature is useful to minimize evaporation or reaction during heating to temperature. The upper-limit temperature in a TGA experiment depends on the objectives: low for solvent or moisture evaporation or phenolic cure (100–300 °C), intermediate for most polymer decomposition (500–600C°), and high for filler content and thermally stable polymers (650–1000 °C).

The derivative TGA, or DTG, curve is useful in many ways, for example, to distinguish overlapping mass loss events, to identify shapes and maxima of mass loss processes, and to help identify minor mass loss steps (see Section 3.6.4 and Fig. 3.40). The DTG peak maxima indicate the maximum rate of mass loss. Each peak in the DTG represents a separate event. The first derivative may indicate better than the mass loss curve itself that more than one process, chemical reaction, or physical phenomenon, is taking place. The DTG curve is a sensitive measurement and supplies the analyst with information of relative rates of volatilization and polymer decomposition. The DTG curve generally has more noise than a DSC curve does and should therefore not be used for kinetic analyses.

The following subsections present examples and additional guidelines for specific types of TGA experiments. While the primary objective is the experiment itself, an attempt has been made to use common, well-known polymers where appropriate examples were available.

3.4.2. Polymer Thermal and Oxidative Stability

Probably the most frequent and classical use of TGA involves the pyrolysis of a polymer at a constant heating rate that typically falls within 5–20 °C/min. When carried out in an inert gas atmosphere, this type of experiment is useful in ranking the relative thermal stability of a group of polymers (a technique for estimating the lifetime of a polymer at service temperatures is described in Section 3.5.4 and illustrated in Section 3.6.1). An example of this type of TGA run is shown in Fig. 3.9, where the mass loss behavior of poly(vinyl chloride) (PVC), poly(methyl methacrylate) (PMMA), low-density polyethylene (LDPE; polytetrafluroethylene (PTFE), and polyimide (PI) are plotted against temperature. These samples were run in nitrogen at a heating rate of

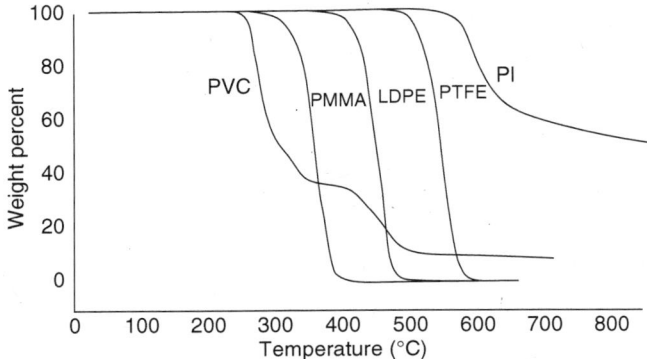

Figure 3.9. The mass loss curves for a series of polymers pyrolyzed in an inert atmosphere [from Chiu (1966) with permission by John Wiley & Sons, Inc.].

5 °C/min (Chiu 1966). Sample mass in each case was 10 mg. Clearly, PVC is the least stable, while PI is observed to be the most stable.

In general, the degradation mechanisms of polymers are free-radical processes initiated by bond dissociation at the temperature of pyrolysis (Wampler 1995). The specific pathway followed is related to the bond strengths and structure of the polymer. These mechanisms are generally grouped into three categories: random scission, unzipping to monomer, and sidegroup elimination. The pyrolysis of polyethylene is an example of random chain scission where the primary fragments produced are a series of pieces of the original molecule. Some polymers such as PMMA or polyoxymethylene (POM) simply unzip at pyrolysis temperatures, generating mostly momoner. For some polymers the bonds attaching sidegroups to the mainchain are the weakest link, in which case these groups are stripped from the chain before it is broken into smaller pieces. Examples of the sidegroups eliminated include PVC (hydrogen chloride), poly(vinyl methyl ether) or PVME (methanol), and ethylene–vinyl acetate copolymer or EVA (acetic acid).

It is well known that polymers based on linear aromatic and other cyclic structures promote char formation on degradation in nitrogen (Bair 1997). In Fig. 3.9 the polyimide retains nearly half of its original mass when heated above 800 °C, while at this high temperature the degradation of PVC yields a stable char of about 8 mass%. The remaining polymers completely decompose into volatile products above 600 °C.

Note in Fig. 3.9 that PVC's mass was reduced by 60–70% after heating to almost 400 °C. By far the major volatile products are hydrogen chloride (HCl) and benzene, which were shown by TGA/MS to evolve simultaneously between 250 °C and 400 °C in air (Khorami and Prime 1988). The mechanism in air is predominately nonoxidative for this first stage of degradation and therefore similar to the pyrolysis in nitrogen (Khorami and Prime 1988). Approximately

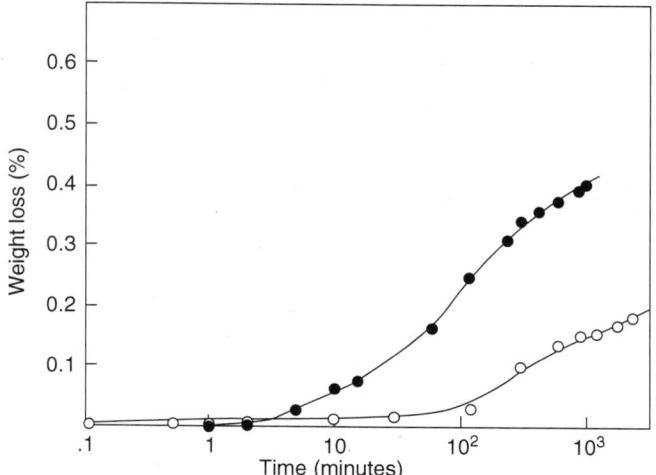

Figure 3.10. PVC mass loss at 78 °C plotted against log time: (○) 0 Mrad (i.e., no irradiation); (●) 25 Mrad dose of ionizing radiation [from Salovey and Bair (1970) with permission by John Wiley & Sons, Inc.].

95% of the theoretical HCl is liberated during decomposition, equivalent to ~55% by mass of the polymer (Woolley 1971; Chang and Salovey 1974). The amount of volatile HCl is independent of molecular mass and polymer structure, while the amount of benzene liberated is dependent on these properties and varies from about 2% to 25% of the original polymer. This type of TGA analysis plus analyses of gaseous decomposition products demonstrates that below 300 °C PVC's mass loss is due primarily to sidegroup elimination, in this case dehydrochlorination (Stromberg et al. 1959).

Under the conditions employed in running the PVC sample in Fig. 3.9, there is no sign of any mass loss until the sample is heated above 200 °C. However, this behavior is reproducible only as long as the same heating rate of 5 °C/min is employed. With slower heating rates the first sign of mass loss will shift to lower temperatures. The maximum, positive effect on sensitivity to mass loss occurs if PVC is held for long periods of time under isothermal conditions. This type of TGA experiment is demonstrated in Fig. 3.10, where a neat PVC film, 0.31 mm thick with a mass of about 7 mg, is held at 78 °C for approximately one day (Salovey and Bair 1970).

In this investigation samples were heated to their isothermal temperature of either 78 °C, 126 °C, or 155 °C at a rate of 320 °C/min in order to minimize degradation during the heating segment of the experiment. The isothermal temperatures were set according to calibrations with Curie temperature reference materials monel, alumel, and nickel (with Curie temperatures of 65 °C, 153 °C, and 358 °C, respectively) at five heating rates within 0.31–80 °C/min and extrapolated to zero heating rate. Mass losses in excess of 0.3 μg were reproducibly determined. Note that mass losses well below 1% were observed after several thousand minutes of aging at 78 °C and that mass loss is not a linear function of time. The glass transition temperature T_g for commercial PVC with no plasticizer present as measured by DSC at a heating rate of 20 °C/min is 89 °C (Bair and Warren 1981). Note that at the lowest isothermal temperature, the PVC sample lost mass while in the glass transition region, where physical aging was also occurring (see discussion of physical aging in Section 2.7).

Poly(vinyl chloride) was exposed to ionizing radiation to check whether high-energy, 1-MeV electrons would enhance degradation (Fig. 3.10). From this work it was determined that the later stages of mass loss follow third/second-order kinetics for either thermal or radiolytic decomposition (Salovey and Bair 1970). The results suggest that TGA in this isothermal, high-resolution form may be useful in testing and evaluating stabilizing additives for PVC.

Some polymers are particularly susceptible to thermooxidative degradation when exposed to oxygen. Poly(vinyl methyl ether) (PVME), used as a film-forming additive in coatings and in model blend studies with polystyrene, is such a polymer. The oxidative thermal degradation of PVME was investigated by TGA, TGA/MS, DMA, FTIR, and NMR techniques (Prime et al. 1988b). Its susceptibility to oxidative attack can be readily observed in Fig. 3.11, which compares TGA at 10 °C/min in nitrogen and air. As can be seen, pyrolysis in nitrogen occurs in a single step between 350 °C and 475 °C. In air degradation starts much earlier and is considerably more complex with three major mass loss processes. While significant mass loss begins at about 180 °C in Fig. 3.11 signs of oxidation have been observed at much lower temperatures, such as the appearance of carbonyl (C=O) functionality near 120 °C (Naito and Kwei 1979) and a small mass gain starting at 88 °C in TGA at 0.3 °C/min in air prior to the onset of mass loss at 110 °C (Prime, unpublished). Adding to the complexity of its thermooxidative degradation PVME was observed to gel (i.e., crosslink) at ~20% mass loss by isothermal DMA in air. Thus the three mass loss steps in air in Fig. 3.11 were attributed to degradation of PVME to form an oxygenated crosslinked polyene structure, degradation of this structure to form a char, and finally decomposition of the char (Prime et al. 1988b).

In Fig. 3.9 low-density polyethylene (LDPE) is shown to be a relatively stable polymer in nitrogen where significant degradation is not detected at temperatures below 500 °C. However, when LDPE is exposed to air or oxygen at elevated temperatures, its stability is dramatically reduced, as shown in the TGA experiment illustrated in Fig. 3.12. In this instance a few milligrams of polyethylene was heated at 320 °C/min to the test temperature, and the rate of evolution of low-molecular-mass volatile fragments formed by the degradation of this unprotected LDPE was measured in both nitrogen and oxygen. The degradation products are detected thermogravimetrically as a loss of mass, and the derivative of that mass loss (dm/dt) or volatilization rate is plotted against absolute temperature [see Eq. (3.7)]. Activation energies of

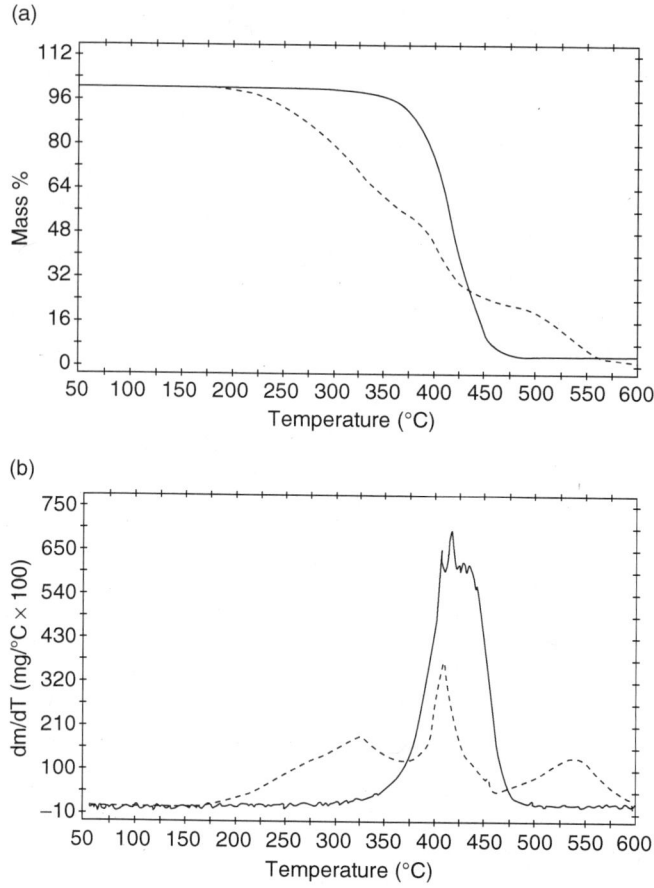

Figure 3.11. TGA of poly(vinyl methyl ether) (PVME) in nitrogen (solid line) and air (dashed line) at 10°C/min: (a) mass loss versus temperature; (b) derivative or DTG versus temperature [from Prime et al. (1988b), with permission of the Society of Plastics Engineers].

113 and 255 kJ/mol were calculated for degradation in oxygen and nitrogen, respectively (Bair 1981). At 260°C the degradation of LDPE is approximately 100,000 times more rapid in oxygen than in nitrogen. For this reason, stabilizer systems have been designed to inhibit its oxidation (Bair 1997). These results demonstrate the need to determine not only the thermal stability of a polymer but also its thermooxidative stability, particularly if it is to be used in air.

Usually the thermooxidative stability of an unknown polymer is evaluated by initially performing a TGA scan of the material in an air or oxygen environment. Comparison with TGA in nitrogen can be a good indicator of ther-

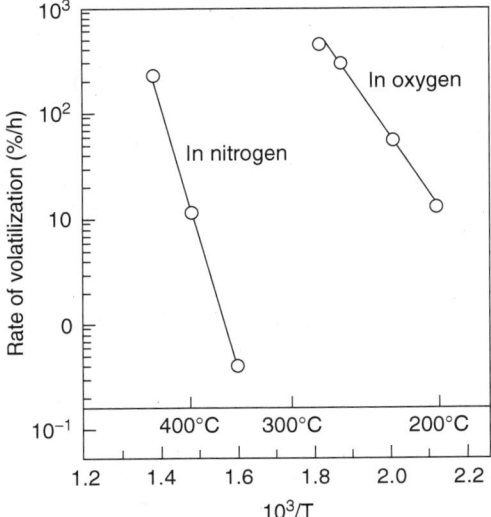

Figure 3.12. Mass loss of degradation products from LDPE as the polymer is held under isothermal conditions at elevated temperatures in nitrogen and in oxygen; there are no antioxidant additives in these LDPE samples [from Bair (1997); reproduced with permission from Elsevier Ltd.]

mooxidative stability. Note that with nominal heating rates such as 10 °C/min, mass loss processes can obscure or mask the mass gains associated with oxidation, but when slower heating rates (e.g., ≤1.5 °C/min) are selected, the mass gaining oxidation step can be readily observed (see Fig. 3.30 as an example). Note that software packages such as dynamic Hi-Res or the autostepwise technique, which are designed to slow down or temporarily stop heating to increase resolution of overlapping thermal events will not function properly if mass loss is masked by mass gain as is shown in Fig. 3.30 for the case of benzocyclobutene heated in oxygen at 10 °C/min.

Thermogravimetric analysis can also be used to determine the oxidation induction time (OIT) by heating a sample of less than 10 mg to an isothermal temperature (typically 200 °C) in oxygen and waiting until the sample begins to gain mass as oxidation starts. This induction period between the time to reach the isothermal temperature and the first sign of mass gain is the sample's OIT (Bair 1997). The OIT of a sample with an unknown level of a specific antioxidant can be determined by constructing a calibration curve from the OIT for samples with known amounts of the particular antioxidant. This TGA technique can detect antioxidants reproducibly at concentrations above 0.001 mass% and has been used in numerous ways to follow oxidation in polyolefins (Bair 1997).

3.4.3. Isothermal TGA Experiments

Isothermal experiments are often appropriate when one wants to focus on a specific mass loss process such as moisture sorption or desorption (see Section 3.4.4 and Fig. 3.15), cure where volatile reaction products such as water and/or formaldehyde are formed during the cure of a phenolic resin (Tabaddor et al. 1997), the liberation of hydrogen chloride as PVC degrades (Salovey and Bair 1970), or the kinetics of thermal and thermooxidative degradation of PVME (Prime 1988). Isothermal measurements can yield excellent kinetic data for these processes. It should be noted that unlike DSC, where fast scans (e.g., 150–500 °C/min) can yield increased sensitivity to small transitions (see Section 2.12), TGA is best suited to standing isothermally at a constant temperature and allowing the process to proceed toward completion. Heating rates at or above 100 °C/min are usually best suited for suppressing mass loss processes from occurring during the heating of a sample to temperature. Obviously, it is advantageous in working with an unknown sample to initially heat the material rapidly (e.g., at 20 °C/min) to gain an overview of its thermal behavior before setting parameters for future TGA work.

A thermoplastic/thermoset adhesive that consists of a reactive phenol–formaldehyde resin and an unreactive acrylonitrile–butadiene copolymer not only liberates heat during cure but also yields volatiles. The evaporation of these volatile cure products are endothermic and cause significant noise to develop during DSC curing scans, which competes with the exothermic reaction and makes quantitative tracking of the cure by DSC difficult if not impossible (Tabaddor et al. 2000). However, a TGA scan to 300 °C yields 10% loss of low molecular mass reaction products that result from the cure of the phenolic resin (see Section 3.6.4 and Fig. 3.40). If the 10% loss of volatiles from the dried adhesive film represents the sum of condensation products associated with 100% curing of the phenolic, the extent of reaction during a TGA experiment (α) is given by

$$\alpha = 0.1 \times \text{mass loss\%} \tag{3.1}$$

Note that the extent of reaction in percent is 100α. See also Eq. (3.3) in Section 3.5 for the general definition of conversion α in a TGA experiment.

In these isothermal TGA curing studies each adhesive sample weighed 5 mg and was stored in a dry box prior to its TGA run. The T_g of the phenolic phase for the as-received adhesive blend and later when fully cured was 75 °C and 235 °C, respectively, as determined by DSC scans at 15 °C/min. Mass loss behavior was recorded in a dry nitrogen environment at five isothermal temperatures: 100 °C, 125 °C, 150 °C, 200 °C, and 225 °C. Each sample was held at 23 °C for 5 min and then heated at an appropriate rate to arrive at the selected cure temperature in exactly 3 min in an attempt to give each sample the same amount of time to cure before the isothermal study begins. For example, to reach 125 °C in 3 min, a heating rate of 33.33 °C/min was selected. In Fig. 3.13

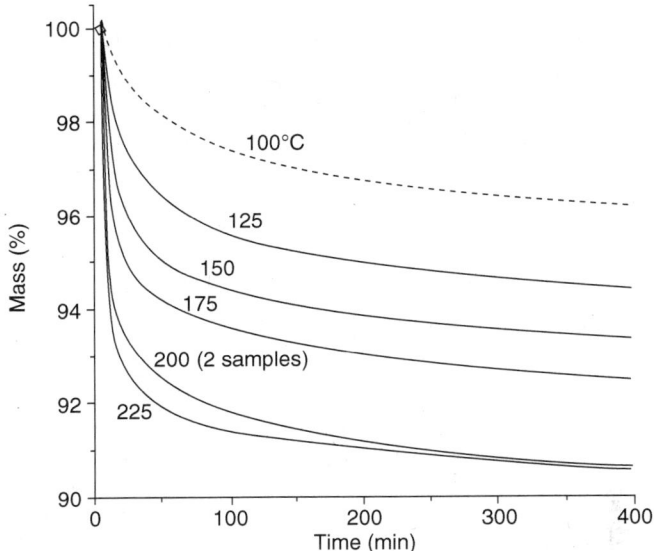

Figure 3.13. Isothermal TGA mass loss curves associated with the curing of a thermoplastic/thermoset adhesive [from Tabaddor et al., (2000) with kind permission from Springer Science and Business Media].

there is less than 0.2% mass loss (2% reaction) during the heating ramp of any isothermal run, indicating that no significant curing has occurred during the initial dynamic heating phase of these experiments. No check was made to see if sample size had any significant effect on the sample's mass loss rate.

At each curing temperature the reaction appears to follow two stages. Initially the cure proceeds rapidly when the reaction temperature is above the phenolic's glass transition temperature, but later, when T_g reaches and exceeds the reaction temperature, the process becomes diffusion-controlled and the cure rate slows noticeably (see Section 2.10 for discussion of the shift from chemical control to diffusion control on vitrification).

Using Eq. (3.1) and data from Fig. 3.13, conversion of the phenolic after 5 min at 100 °C, 125 °C, 150 °C, 200 °C, and 225 °C is estimated to be 0.05, 0.14, 0.24, 0.36, 0.55 and 0.63, equivalent to 0.5%, 1.4%, 2.4%, 3.6%, 5.5%, and 6.3% mass loss, respectively. Note that in Fig. 3.13, 5 min of isothermal cure occurs after 13 min have elapsed, which includes 5 min of conditioning at room temperature, a 3-min heating ramp, plus 5 min at the given isothermal temperature. These data support and complement the DMA investigations of the chemical conversion and T_g development in this thermoset/thermoplastic resin (Tabaddor et al. 2000).

The thermal and thermooxidative stability of poly(vinyl methyl ether) (PVME) was described in Section 3.4.2. PVME was a component in some polymer-based magnetic coatings where its phase separation and subsequent decomposition during cure played an important role in developing key properties of these coatings (Burns et al. 1989). Among the characteristics investigated were the degradation kinetics in both nitrogen and air. Flow rates of both gases were 100 mL/min. Sample size and geometry were kept constant at 2.4 ± 0.1 mg, equivalent to an area density of 9.6 ± 0.4 mg/cm^2 in pans 0.28 cm in radius, as these parameters were found to strongly influence the observed mass loss rate. This level of reproducibility was achieved with the use of a microsyringe to rigidly control the amount of sample. To ensure uniform film formation, 1.8–2.2 mg of heat-cleaned glass fabric was first preweighed into the standard sample pan. TGA experiments were run in two steps. The first step was to heat to 150 °C in nitrogen for 20 min to evaporate solvent. For isothermal experiments, the second step was a rapid heat at 100 °C/min to temperature followed by isotherms of varying duration. Multiple heating rate experiments were conducted at rates between 0.3 °C/min and 30 °C/min.

Conversion (α) was computed according to Eq. (3.3) in Section 3.5. Multiple heating rate data for the pyrolysis in nitrogen were analyzed by the Flynn–Wall method, described in Section 3.5.3, and showed that the activation energy E_α increased with conversion. However, between 40% and 65% conversion E_α was constant with a value of 231 ± 1 kJ/mol. Between 20% and 85% conversion E_α increased from 215 to 235 kJ/mol with a small positive slope. As shown in Fig. 3.14, the isothermal data between 20% and 80% conversion gave a good fit to the second-order kinetics equation [see Eq. (2.84) in Chapter 2] with an intercept of ~1 and a slope of k. An activation energy of 219 kJ/mol was obtained from an Arrhenius plot of the rate constants versus absolute

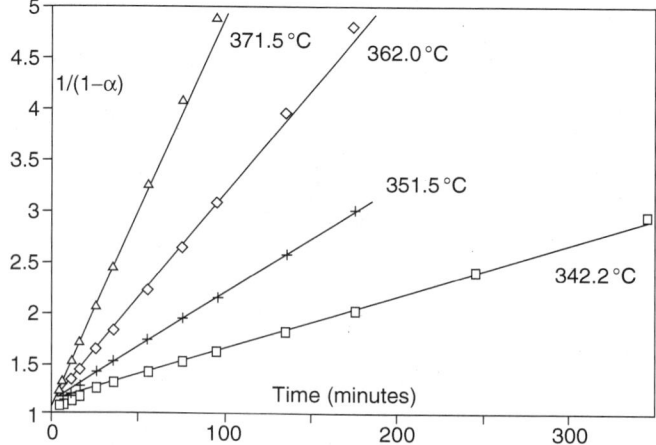

Figure 3.14. Second-order plot of isothermal mass loss of PVME in nitrogen [from Prime (1988b)].

temperature according to Eq. (3.6). Thus the multiple heating rate data showed the activation energy to be approximately constant over a wide conversion range, allowing a detailed description of the kinetics of decomposition in nitrogen from isothermal measurements.

As described earlier, the thermooxidative degradation of PVME was complex, with three major mass loss steps. Multiple heating rate experiments gave low activation energies for the first step of 54–67 kJ/mol from 10% to 30% conversion, increasing to high activation energies for the second step and reaching a maximum above 300 kJ/mol near the end of that step, and reverting back to lower activation energies for the third step. Unique to the thermooxidative degradation, slower heating rates (0.3–3 °C/min) gave lower activation energies than the faster rates (3–30 °C/min). In air isothermal decomposition is initially very rapid but then slows considerably, becoming quite sluggish beyond 50% mass loss. This behavior is attributed to crosslinking, which was observed to begin at about 20% mass loss. The large changes in activation energy with conversion precluded a detailed kinetic analysis from isothermal data.

3.4.4. Low-Molecular-Mass Components in Polymers

Many polymers contain low-mass additives that evaporate at relatively modest temperatures, and their composition in the polymer can often be measured by TGA. The most common small molecule, water, is soluble to some degree in all polymers, and its sorption/desorption properties can be studied by TGA. Typically, a hydrophilic polymer like poly(vinyl acetate) (PVAc) will absorb in excess of 5 mass% water at room temperature. Initially the water binds to groups on the polymer chain (bound water), causing the glass transition temperature of PVAc to decrease, but as more water is added to the polymer, T_g remains constant. In this latter case, the water is found clustered in small cavities created within the PVAc (Bair et al. 1981). In contrast to this behavior, water and a hydrophobic polymer such as polyethylene are essentially incompatible. However, the presence of trace amounts of impurities and groups formed by oxidation, such as carbonyl (C=O) along the polymer chain, can cause water solubility to become relatively large compared to pure PE. In this latter case water binds to the oxygen-containing groups on the polymer chain (McCall et al. 1984). The chemical nature and number of hydrophilic sites found in polyethylene controls the quantity of water and the speed at which it can move through the polyolefin. The amount of water absorbed by a silica-filled, uncured epoxy novolac at 23 °C was found to be a reversible function of humidity except for a small amount that appears to be silica-bound and can be removed only at temperatures above 100 °C (Bair 1992). These studies show that even the interaction of a simple compound like water with a polymer can be complex and require diligence on the part of the thermal analyst as well as the application of several techniques to gain a more complete understanding of its behavior in a particular polymer system.

Absorption and desorption of water in and out of a polymeric material can be readily monitored in a TGA that has a nitrogen gas stream whose humidity level is adjustable. Some instrument manufacturers sell accessories to control humidity. But the moisture level of the gas can also be controlled by the relative flow of "wet" and dry nitrogen gas lines that are mixed before they enter the thermobalance. Using this kind of experimental approach, it was found that the desorption of 0.53% by mass water from an epoxy (DGEBA) required about 600 min of storage at 23 °C in the 0% relative humidity nitrogen atmosphere of the TGA before equilibrium was reached (Fig. 3.15).

After 1250 min the relative humidity (RH) of the nitrogen gas inside the TGA at 23 °C, as detected by an RH sensor placed near the sample, is changed from 0% to 64%. Initially, the dried epoxy begins to gain mass rapidly. After about 1350 min of aging in 64% RH, the epoxy resin has gained 0.86 mass% water but still has not reached equilibrium. The presence of moisture in this DGEBA epoxy not only will lower its T_g but also, if present during cure, may cause the cured epoxy's ultimate properties to be diminished from what they would have been if curing had been carried out in a dry state (Prime 1997b). For example, if ~1.7 mass% water is present in an epoxy Novolac portion of a molding compound prior to curing, after cure T_g will be ~107 °C, 95 °C below the epoxy's ultimate T_g of 202 °C. After postbaking, this epoxy's glass transition is raised, but only to 141 °C. Obviously, in this case the water not only plasticized the resin but also altered its structure irreversibly (Bair 1992).

In addition to water, other low-mass compounds such as residual solvent, additives, or contaminants may be present in polymers, and their amounts can often be determined by TGA. In an inert atmosphere high-molecular-mass polymers typically do not lose significant mass below 300 °C unless low-molecular-mass materials are initially present in the form of additives or contaminants. However, there are exceptions, such as the degradation of PVC

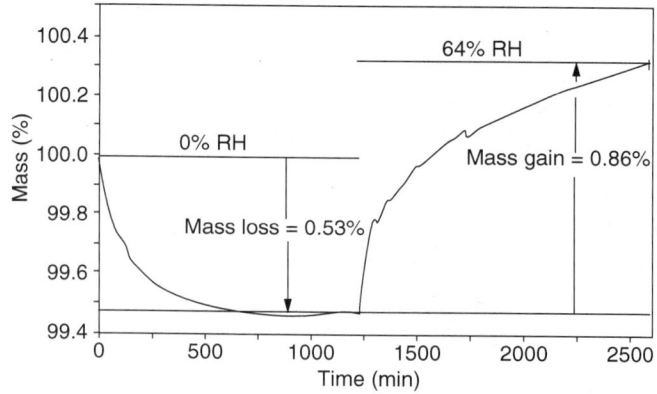

Figure 3.15. Water desorption and absorption by an epoxy in a TGA flushed with dry (0% RH) and wet (64% RH) nitrogen at 23 °C. (Bair, unpublished results, 1997).

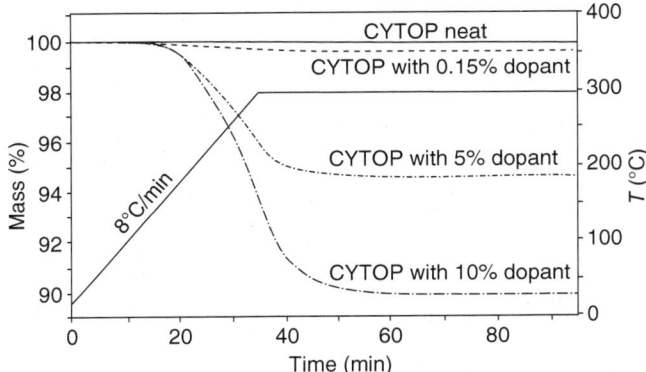

Figure 3.16. Mass loss behavior of Cytop compounds after heating to and holding at 290 °C for one hour in nitrogen (Bair, unpublished results, 1994).

discussed earlier. Hence, if a single known low-molecular-mass component is present and sublimes or vaporizes below 300 °C and the polymer host is stable, one can track the mass loss quantitatively as a function of time in a TGA. For example, in Fig. 3.16 the neat resin, Cytop®, a thermally stable fluorinated polymer, is heated in nitrogen at 8 °C/min to 290 °C and held there for one hour. Within 30 min of reaching 290 °C the compound's mass stabilizes with a loss of mass of 0.27%. However, when the supplier blended 0.15%, 5%, and 10% by mass of a low-molecular-mass dopant into the neat resin, TGA analysis yielded net mass losses (blend mass loss minus neat resin mass loss) of 0.18, 5.18, and 9.94 mass%, respectively. Obviously, this thermoanalytical routine could be used as a quality control method to check the incoming manufacturer's dopant level in their product.

The major sources of error in mass as detected by TGA are due to three effects: buoyancy, electrostatic forces, and electronic drift. The buoyancy phenomenon, discussed in Section 3.2.2, occurs as the density of the nitrogen in the balance decreases with increasing temperature, resulting in an apparent mass gain (when heating Cytop samples to 290 °C an apparent gain of ~0.05 mass% was detected). These effects are small and usually are taken into account only in the most demanding situations. With these samples temperature effects can be eliminated by simply weighing samples at room temperature both before heating to 290 °C and again afterward when the sample has cooled back to room temperature.

3.4.5. Loss of Volatile Reaction Products

Cure of materials like phenolics and polyimides produce volatile products that result in mass loss that generally occurs between 100 °C and 300 °C, and may overlap with the loss of volatiles initially present. The chemical reactions that lead to chain growth and crosslinking split out products such as water or formaldehyde. Polyimide precursors are often dissolved in a high-boiling(-point) solvent such as *n*-methyl pyrolidone (NMP), whose evaporation will contribute to the mass loss. Solvent-based systems like screen-printable polymer thick films (PTF) are dissolved in high-boiling solvents that moderate evaporation at ambient conditions and help stabilize the rheology of these systems. In these materials cure involves solvent evaporation in addition to crosslinking, and residual solvent after cure is common. The *drying* of coatings and paints is another example. Reactants or catalysts, such as in thermosetting systems, may be blocked with a thermally labile adduct. The unblocking of these components generally occurs over a specific temperature range and is accompanied by the release and evaporation of the blocking agent, often an alcohol. The loss of hydrogen chloride from poly(vinyl chloride) described earlier (see Figs. 3.9 and 3.10) can also be considered as the loss of a volatile reaction product.

Figure 3.17 shows mass loss and its derivative or DTG from uncured and partially cured phenolic bonding compound (Cassel 1976, 1980). The uncured phenolic (A curves) exhibits the apparent loss of a very small amount of absorbed water below ~125 °C (observable only in the DTG curve) and a two-step mass loss associated with cure overlapping the absorbed water loss up to 300 °C. Total mass loss on heating to 300 °C is 3%. Sample B, previously cured for 1 min at 160 °C, shows no absorbed water and predominantly the higher temperature mass loss associated with cure. By comparison with the uncured sample and by analogy with Eq. (3.1), the 1.8% mass loss suggests about 40% cure for this sample. Sample C, cured for 1 min at 180 °C, shows a total mass loss of 0.3%, which is exclusively the higher-temperature mass loss step. By a similar analogy, the degree of cure determined for this sample was ~90%.

Isothermal TGA of a phenolic-containing system, where mass loss on cure was measured directly in the TGA, was described earlier (see Fig. 3.13). With

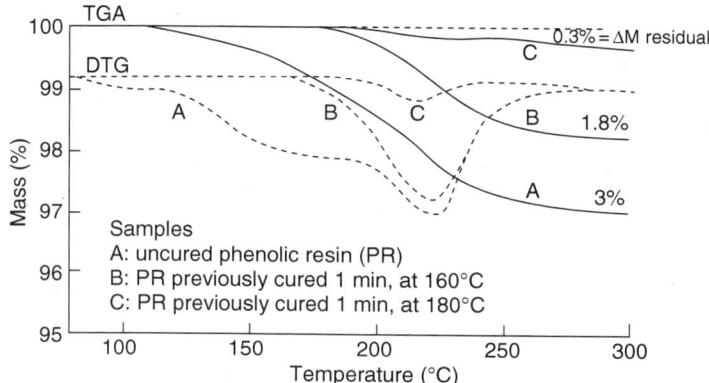

Figure 3.17. Residual mass loss and derivative TGA of uncured and partially cured phenolic bonding resin [from Cassel (1976, 1980), courtesy of Perkin-Elmer].

this technique, which is analogous to DSC isothermal method 1 described in Chapter 2 (Section 2.10.1.3), the entire cure process is monitored in the TGA. Greenberg and Kamel (1977) used similar techniques to study the cure kinetics of a poly(acrylic acid)–alumina composite. This system cured by anhydride bond formation, where each crosslink was accompanied by the elimination of a water molecule. When using this method quantitatively, it is recommended that experimenters establish that the assumed mass loss is indeed the one being measured. By a mass spectroscopic technique the authors showed that the volatile reaction products consisted of water and carbon dioxide in an approximate ratio of 12:1. Full cure was established to be equivalent to a mass loss of 6.0% for the composite system. Using isothermal techniques similar to those described in Section 2.10.1.3, the authors found the reaction to be second-order up to about 40% conversion.

Techniques analogous to DSC isothermal method 2 have also been reported, where residual cure is measured by TGA. One example is the phenolic bonding compound shown in Fig. 3.17. TGA and DMA were used to measure composition and cure of dielectric and conducting polymer thick films (Prime 1992). With polymer thick films (PTF) it is possible to screen-print conductive and resistive elements onto suitable substrates such as rigid printed circuit boards or flexible polyimide film. Applications include videogames, inexpensive calculators, and under-dashboard electronics in automobiles. To create an electrical connection, first a dielectric layer is deposited and cured, followed by a conducting layer, and finished with another dielectric layer. The PTF materials studied here were screenable polymer pastes with significant quantities of butyl carbitol, a high-boiling solvent (boiling point = 231 °C). The relative amounts of solvent, polymer matrix, and filler were measured by heating 3-mg samples of dielectric PTF and 10-mg samples of the highly filled conducting PTF in disposable aluminum pans at 5 °C/min in air to 630 °C. The silver-filled conducting PTF was found to contain 70% filler after removal of the solvent (Fig. 3.18), from which a pigment volume concentration (PVC) of 43% was estimated [see Prime et al. (1988) and Fig. 3.21 for determination of pigment volume concentration by TGA].

Results for the dielectric PTF, which contained more solvent and was therefore the more difficult to cure, were reported in this study. Residual solvent and percent coating were measured on specimens of approximately 25 and 100 µm thicknesses coated onto stainless-steel mesh of similar thicknesses. Samples were heated rapidly in the TGA to 220 °C in nitrogen and held for 60 min to evaporate all residual solvent and other volatiles, and then ramped at 10 °C/min to 630 °C to pyrolyze the polymer binder, leaving the nonconducting filler and the mesh. In this way the amount of solvent-free coating, residual solvent, and filler content could be measured.

The T_g (by DMA) and residual solvent (by TGA) were measured on different pieces of the same mesh samples, where time, temperature, and thickness were controlled. On the basis of the manufacturer's recommended cure of 20 min at 200 °C, initial samples were cured in a convection oven at 165 °C,

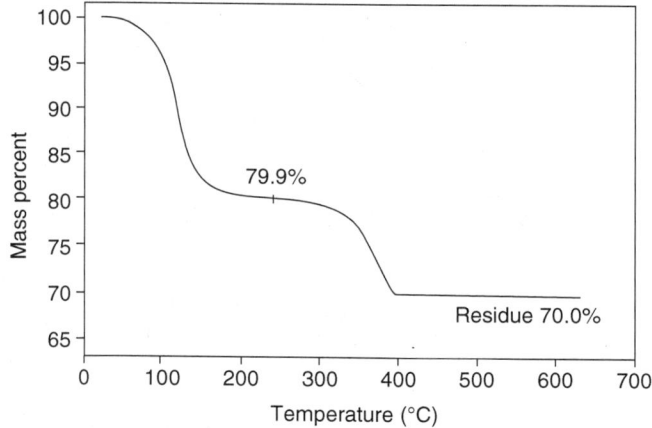

Figure 3.18. TGA in air of silver-filled conductive polymer thick film (PTF) ink; 5 °C/min in air at 100 mL/min flow rate [from Prime (1992), with permission of the Society of Plastics Engineers].

185 °C, and 200 °C for various times. Activation energy for cure was determined by the time–temperature superposition techniques described in Section 2.10.3 (Neag and Prime 1991). Additional samples were cured in an infrared (IR) conveyor oven, following the PTF manufacturer's recommendation of 3 min at 200 °C for IR cure. Note the implication that IR cure was faster or more efficient than convection-oven cure. Time–temperature profiles for IR cure were converted to equivalent times at a reference temperature of 185 °C using the same time–temperature superposition techniques. A strong correlation was found between T_g and residual solvent, demonstrating the important role of residual solvent in the cure process. As expected, due to diffusion to the surface and evaporation of the high-boiling solvent, sample thickness was also found to play a significant role, as shown in Fig. 3.19. Even highly cured specimens were found to contain more than 2% residual solvent. No difference could be discerned between convection and IR cure. Only time and temperature determined degree of cure irrespective of how specimens were heated.

3.4.6. Compositional Analyses

Thermogravimetric analysis can provide valuable information, often quantitative, on the composition of polymeric materials. If a multicomponent material contains low-molecular-mass compounds, polymeric material, and inorganic additives, the three groups can be separated by temperature. The quantitative determination of a low-mass additive in a polymeric material by isothermal TGA was described in Section 3.4.4 (see Fig. 3.16). This general principle of molecular fractionation by temperature is common to most linear polymers, where rupture of carbon–carbon bonds typically occurs at temperatures

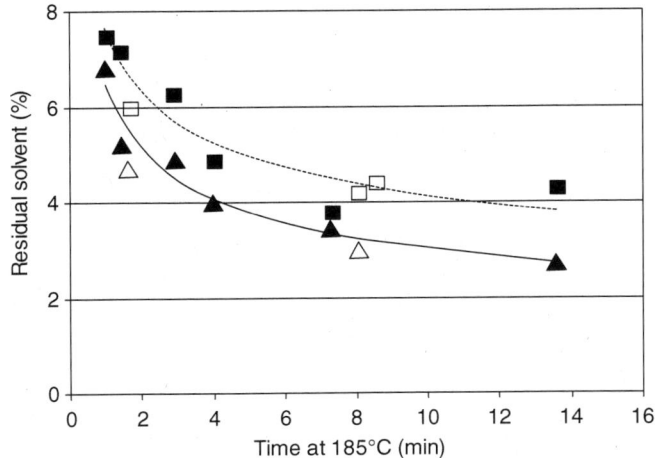

Figure 3.19. Time–temperature–superposition master curves of residual solvent by TGA versus time at a reference temperature: solid line—25-μm-thick samples, dashed line—100-μm-thick samples, open symbols—convection oven cure, filled symbols—IR cure [plotted from data in Prime (1992)].

between 500 °C and 550 °C. Inorganic additives are usually stable in an inert atmosphere to 900 °C or higher, although fillers like $CaCO_3$ will convert to CaO and CO_2 between 600 °C and 700 °C. In fact, the mass loss of CO_2 measured can help quantify the amount of calcium carbonate in the original composition. Hence, a TGA scan at a heating rate of 10 °C/min in nitrogen of an unknown polymeric material can often show mass loss below 300 °C as the loss of low-molecular-mass components such as water, additives, or possibly reaction products if present and mass loss between 300 °C and 550 °C as macromolecular degradation occurs. Residues that remain or mass loss steps above 600 °C are normally associated with inorganic compounds such as silica particles, glass fibers, or calcium carbonate. Obviously, even in an inert atmosphere some polymers degrade at temperatures below 300 °C such as PVC as shown in Fig. 3.9. If the thermal stability of a polymer is unknown, it must be determined and taken into account before attempting to carry out compositional analysis by TGA.

Often the amount of carbon black (CB) in a sample can be determined by switching the gas in the TGA from nitrogen to air or oxygen at 500 °C to 550 °C such that CB converts to CO_2. The carbon-black surface area can be approximated from the temperature of its decomposition, often taken as T15 or T20, the temperature at which 15% or 20% of combustion is complete [see Mauer (1981); Sircar (1997); and Fig. 3.20]. This CB analysis assumes all polymeric material has been volatilized prior to the introduction of oxygen. Beware of polymers that can form a char, such as those that contain aromatic structures, for they can cause erroneous CB determinations unless the pyrolysis of their

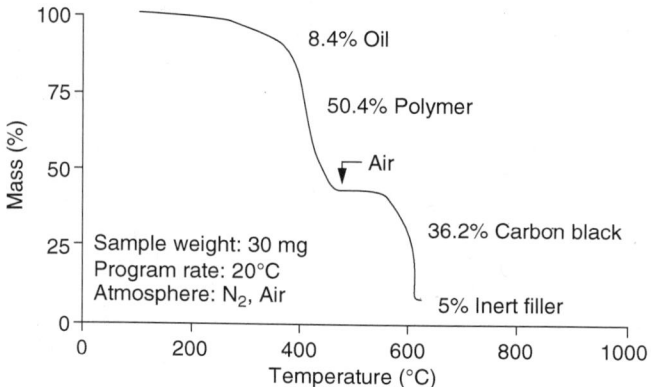

Figure 3.20. Compositional analysis of rubber products by TGA: styrene–butadiene rubber (courtesey of TA Instruments).

char is well separated from that of the carbon black. These analyses are often aided by methods where, following the initial pyrolysis in nitrogen, the TGA is cooled to ~400 °C before switching gases. This may afford a more discernible baseline prior to the onset of decomposition in air. Dynamic Hi-Res and autostepwise TGA techniques, described in Section 3.2.3, that slow or stop the rate of temperature increase in response to an accelerating rate of mass loss, allowing isothermal or near-isothermal mass loss steps, can be a valuable aid to compositional analyses.

Compositional analysis of polymeric rubber products by TGA has been used for many years to determine the quality and content of various rubber products (Kau 1988; Sircar 1997). A polystyrene butadiene rubber composition is illustrated in Fig. 3.20. The protocol for this analysis includes the following: heating rate 20 °C/min, 30 mg sample mass, and switching the purge gas from nitrogen to air at 500 °C. The results are illuminating for a 30-min determination of composition: 8.4 mass% oil, which evaporates prior to the onset of polymer decomposition just below 400 °C; 50.4 mass% polymer, whose decomposition is complete by 500 °C, at which point the purge gas is switched to air; 36.2 mass% carbon black, taken as the mass loss in air; and 5.0 mass% residue, either ash or mineral filler plus ash.

Thermogravimetric analysis can provide precise and accurate compositional analysis that can be used for quality and process control, as illustrated in the following example. Prime et al. (1988) describe a TGA measurement of percent solids and pigment volume concentration (PVC) of magnetic coating ink, consisting of solvents, magnetic iron oxide and aluminum oxide particles or "pigment," and thermosetting resins. This technique was used in the manufacturing process to control magnetic performance of the media in hard-disk drives. Figure 3.21 shows the TGA technique. Rigid control of sample size (8.00 ± 0.25 mg) and sample temperature in step 2 (±1 °C), made

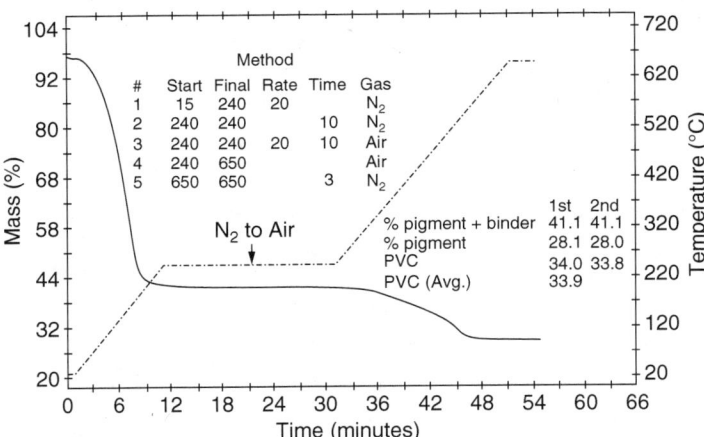

Figure 3.21. TGA method for measuring percent solids and pigment volume concentration (PVC) in a coating ink [from Prime et al. (1988), with permission of the Federation of Societies for Coatings Technology].

possible by frequent temperature calibration with a nickel standard, was required to achieve the required precision. The first step in the analysis is solvent removal and cure in nitrogen. Next, the furnace and balance are allowed to fill with air, which is necessary to completely pyrolyze the organic binder, leaving only the inorganic oxides or pigment. The TGA PVC was calculated from

$$\text{PVC} = \frac{V_{\text{pigment}}}{V_{\text{pigment}} + V_{\text{binder}}} \tag{3.2}$$

after converting mass percent to volume percent using the appropriate densities. Note the reproducibility between the two runs in the tabular display in Fig. 3.21. Disposable aluminum pans were used for these measurements.

3.5. KINETICS

3.5.1. Overview of Kinetic Principles

Kinetic information is crucial for evaluating the times and temperatures associated with the processing, service lifetimes, and storage of materials. It is also of value for understanding the mechanisms of thermal processes. In a pragmatic sense the objective of kinetics is often to provide a mathematical relationship between time, temperature, and conversion.

At a constant (e.g., ambient) pressure, the rate of many thermally activated processes can be described as a function of two variables: the temperature (T)

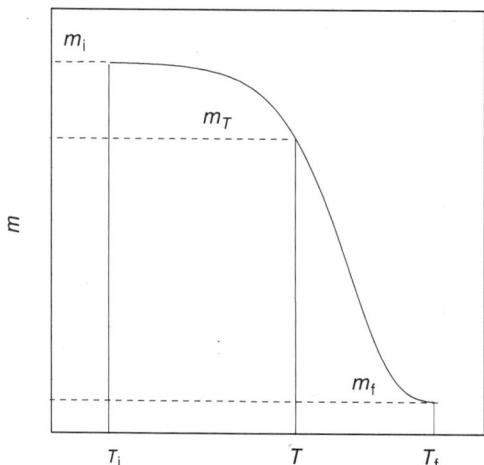

Figure 3.22. Use of TGA mass loss curve for determining the extent of conversion by Eq. (3.3).

and the extent of conversion (α). The temperature is controlled by a TGA instrument in accord with a program set by a user. The extent of conversion is conveniently determined from mass loss measurements. At a given temperature, α is defined as

$$\alpha = \frac{m_i - m_T}{m_i - m_f} \tag{3.3}$$

where m_T is the mass at temperature T and m_i and m_f are respectively the initial and final masses for a given step of mass change (Fig. 3.22). A common application of TGA is to monitor thermal degradation of polymers. In this case, the α–T curve represents the kinetics of conversion of a polymer to volatile degradation products. However, TGA can also be used to measure the kinetics of polymerization or cure if the reaction product(s) released during the reaction are volatile and there is no other contribution to the mass loss. An example of such a process is the condensation of hydroxymethyl phenols with phenol to form Novolac resins, where each bond formed is accompanied by the release of one water molecule. The cure of phenolic resin systems, such as those illustrated in Figs. 3.13, 3.17, and 3.40, that release water and/or formaldehyde is another example.

In the simplest case of the so-called single-step process, the rate equation has the following form

$$\frac{d\alpha}{dt} = k(T) f(\alpha) \tag{3.4}$$

where $k(T)$ is the rate constant and $f(\alpha)$ is the reaction model. A process can generally involve multiple steps, so its rate equation may be more complex.

For instance, the rate of a process involving two parallel reactions can be described as follows:

$$\frac{d\alpha}{dt} = k_1(T)f_1(\alpha) + k_2(T)f_2(\alpha) \tag{3.5}$$

The temperature dependence is almost universally described by the Arrhenius equation

$$k(T) = A\exp\left(\frac{-E}{RT}\right) \tag{3.6}$$

where A and E are Arrhenius parameters (the preexponential factor and the activation energy, respectively) and R is the universal gas constant. Substitution of Eq. (3.6) into Eq. (3.4) gives

$$\frac{d\alpha}{dt} = A\exp\left(\frac{-E}{RT}\right)f(\alpha) \tag{3.7}$$

The reaction rate depends on the conversion, which can be represented by a reaction model [$f(\alpha)$] which may take various mathematical forms, depending on the physical mechanism assumed in the mathematical derivations (Brown 2001). A combination of the activation energy, preexponential factor, and reaction model is sometimes called a "kinetic triplet." For a process involving more than one step, the individual steps are likely to have differing kinetic triplets. For example, a two-step kinetic equation involves two kinetic triplets: $A_1,E_1,f_1(\alpha)$ and $A_2,E_2,f_2(\alpha)$.

In most general terms, the conversion dependence of the reaction rate can be one of three types—accelerating, decelerating, and autocatalytic (i.e., passing through a maximum)—as illustrated in Fig. 3.23. Accelerating models represent processes whose rate monotonically increases with the extent of conversion. A typical example is the so-called power-law models that are described by the following general equation

$$f(\alpha) = n\alpha^{(n-1)/n} \tag{3.8}$$

where n is a constant. Accelerating kinetics are typically observed in the thermal degradation of polymers in an oxidative atmosphere such as oxygen or air. These processes commonly demonstrate a slower initial stage or an induction period associated with the formation and accumulation of peroxides. Decelerating models describe processes whose rate monotonically decreases with the extent of conversion, for example, as the reactants are depleted. Reaction order models of the general form

$$f(\alpha) = (1-\alpha)^n \tag{3.9}$$

(where n denotes reaction order) provide the most common example of decelerating processes. This type of kinetics is common for the thermal degradation of polymers in an inert atmosphere such nitrogen or argon. The second-order decomposition of poly(vinyl methyl ether) (PVME) in nitrogen, described in Section 3.4.3 and Fig. 3.14, is an example. A maximum in the rate observed between 0% and 100% conversion is characteristic of autocatalytic processes that can be described, for example, by the Avrami–Erofeev models

$$f(\alpha) = [n(1-\alpha)][-\ln(1-\alpha)]^{n-1/n} \qquad (3.10)$$

where n is a constant. The kinetic behavior of this kind may be observed when the thermal degradation of a polymer yields a highly reactive product capable of accelerating degradation as in the case of the formation of NO_2 in the degradation of nitrocellulose. Under isothermal conditions, these three types of processes can be recognized by the characteristic appearances of the respective kinetic curves illustrated in Fig. 3.23.

An objective of kinetic analysis is to determine kinetic parameters such as the activation energy, preexponential factor, and reaction model. The experimentally determined parameters are then widely used for mechanistic interpretations and kinetic predictions.

3.5.2. Kinetic Analyses Based on Isothermal Runs

Isothermal TGA data are commonly analyzed in terms of the reaction models. The analysis starts by transforming mass–time to conversion–time plots. For transformation one can use Eq. (3.3) by replacing m_T with m_t, which is the mass in a given moment of time. Figure 3.24 shows an α–t plot for the thermal

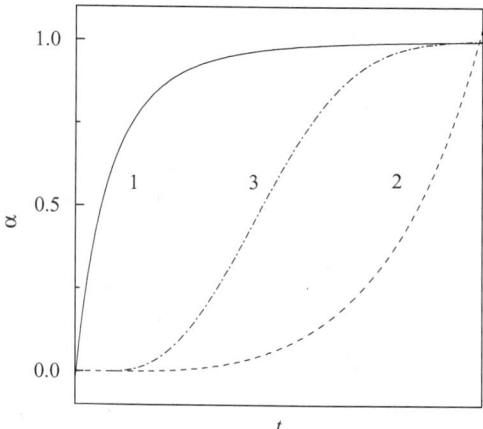

Figure 3.23. Characteristic kinetic curves (conversion vs. time) for major types of reaction models: 1—decelerating, 2—accelerating, 3—autocatalytic.

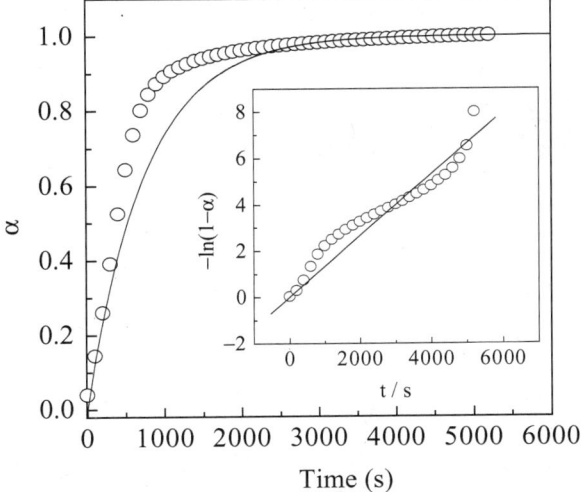

Figure 3.24. Experimental α–t data for isothermal degradation of PEN at 420 °C (circles) and a simulated α–t curve (solid line) obtained by assuming first-order kinetics [Eq. (3.14)]. Inset shows fitting α–t data to the straight line, $-\ln(1-\alpha) = k(T)t$, to determine the rate constant.

degradation of poly(ethylene naphthalate) (PEN) under nitrogen at 420 °C. It is seen that the experimental curve is of the decelerating type shown in Fig. 3.23, which means that one can attempt to describe it in terms of the reaction order model [Eq. (3.9)]. Integral data are conveniently described by the integral form of Eq. (3.4)

$$g(\alpha) \equiv \int_0^\alpha \frac{d\alpha}{f(\alpha)} = k(T)t \qquad (3.11)$$

where $g(\alpha)$ is the integral form of the reaction model. For the reaction order model [Eq. (3.9)], $g(\alpha)$ can take one of two forms:[2]

$$g(\alpha) \equiv \int_0^\alpha \frac{d\alpha}{(1-\alpha)^n} = \frac{1-(1-\alpha)^{1-n}}{1-n} \quad \text{for } n \neq 1 \qquad (3.12)$$

$$g(\alpha) \equiv \int_0^\alpha \frac{d\alpha}{1-\alpha} = -\ln(1-\alpha) \quad \text{for } n = 1 \qquad (3.13)$$

According to Eq. (3.11), substitution of experimental values of α into $g(\alpha)$ and plotting $g(\alpha)$ against the respective values of t should yield a straight line whose slope is the rate constant at a given temperature $k(T)$. However, the actual plots of $g(\alpha)$ versus t commonly show deviation from a straight line.

---

[2] Integrated forms of first-, second-, and nth-order ($n \neq 1$) rate equations are given in Eqs. (2.83)–(2.85) (in Chapter 2).

Figure 3.24 provides an example of such a plot for the thermal degradation of PEN at 420 °C. A first-order reaction model [Eq. (3.13)] has been found to fit the experimental data better than reaction order models of other n values. Although the $g(\alpha)$–t plot is characterized by a reasonable correlation coefficient ($r = 0.97$), it demonstrates obvious systematic deviations from the straight line. The reason is that actual degradation processes tend to be more complex than the simplistic models widely used for their description. The rate constant $k(T)$, obtained from the slope of the $g(\alpha)$ versus t, is $0.00133\,\text{s}^{-1}$. The value of $k(T)$ can then be used to simulate the experimental α–t data by solving Eq. (3.11) with respect to α. For a first-order reaction model [Eq. (3.13)], the solution is given by

$$\alpha = 1 - \exp[-k(T)t] \qquad (3.14)$$

Substituting the experimental value of $k(T)$ into this equation leads to the simulated α–t curve in Fig. 3.24. Note that the systematic deviations of the $g(\alpha)$ vs. t plot from the straight line give rise to systematic deviations of the simulated α–t curve from the experimental one.

The aforementioned procedure of estimating the rate constant can be applied to α–t datasets obtained at different temperatures. The resulting dependence of $k(T)$ on the temperature can then be used to determine the values of the preexponential factor and the activation energy. This is easily accomplished by substituting the respective values of $k(T)$ and T into the equation

$$\ln k(T) = \ln A - \frac{E}{RT} \qquad (3.15)$$

which is the logarithmic form of Eq. (3.6). A plot of $\ln k(T)$ versus T^{-1} is typically linear or almost linear, which permits the respective values of $\ln A$ and E to be estimated from the intercept and the slope of the line.

In order to obtain the values of $\ln A$ and E, one needs to carry out TGA runs at at least three temperatures, but the use of four or five temperatures is recommended to improve reliability. A choice of the temperatures can be challenging and is usually limited to a rather narrow interval of 20–30 °C. The use of higher temperatures is limited by the length of the sample warmup period; it may take 1–3 minutes before a sample placed in the TGA furnace reaches the preset experimental temperature. Obviously, fast heating rates are recommended. If the temperature is too high and/or the heating rate is too slow, the sample may undergo significant reaction during the ramp before isothermal conditions are reached. As a rule of a thumb, the highest temperature should be chosen so that the entire decomposition process takes about one hour. Other temperatures can then be selected by decreasing the temperature in increments of 5–10 °C. It should be kept in mind that decreasing the temperature by 10 °C will typically increase the duration of the TGA run

2–3 times depending on the value of the activation energy. Thus, if the process takes 1 h at the highest temperature, it could take close to 10 h at a temperature 20 °C below the highest temperature. The use of lower temperatures has the disadvantage of extending the length of the TGA experiment, but overnight and weekend runs can extend the practical temperature range. Also note that it is not necessary to take the reaction to completion and that valuable information can be gained from the early portion of the reaction, as suggested by the data in Fig. 3.13. Note that the temperature range is 125 °C.

The second-order decomposition of poly(vinyl methyl ether) (PVME) in nitrogen described in Section 3.4.4 is a good example of isothermal kinetic analysis. Isothermal TGA runs were conducted at four temperatures with 10 °C steps. As shown in Fig. 3.14, these data gave a good fit to second-order kinetics where $g(\alpha) = [(1/1 - \alpha) - 1]$ [cf. Eq. (3.12) with $n = 2$]. Note that it took 100 min to reach 80% conversion at the highest temperature of 372 °C and 350 min to reach 70% conversion at the lowest temperature of 342 °C.

3.5.3. The Flaw of Kinetic Analyses Based on a Single Nonisothermal Run

Many kinetic measurements are performed under nonisothermal conditions that allow for faster runs over a wider temperature range. The runs are typically carried out at a constant heating rate:

$$q = \frac{dT}{dt} \qquad (3.16)$$

For such conditions, Eq. (3.7) can be written as follows:

$$\frac{d\alpha}{dT} = \frac{A}{q} \exp\left(\frac{-E}{RT}\right) f(\alpha) \qquad (3.17)$$

It should be noted that unlike the isothermal kinetic curves in Fig. 3.23, the shape of nonisothermal curves does not provide any clear indication of the type of the reaction model. In nonisothermal runs the temperature increases, causing the reaction rate to continuously accelerate. As a result, under nonisothermal conditions all processes exhibit kinetic curves of a sigmoidal shape.

Many methods have been developed for evaluating a kinetic triplet by fitting rate equations to data obtained at a single heating rate. For example, rearranging Eq. (3.17) gives

$$\ln\left(\frac{d\alpha}{dT} \frac{q}{f(\alpha)}\right) = \ln A - \frac{E}{RT} \qquad (3.18)$$

Differential Eq. (3.18) requires the values of $d\alpha/dT$. TGA runs produce α–T curves that can be converted to $d\alpha/dT$ versus T by using numerical differentiation. However, one should be aware that this procedure typically results in reaction rate data with considerable noise, whose use may diminish the precision and accuracy of kinetic evaluations. Care must also be exercised when using instrument manufacturers' software that combines numerical differentiation with smoothing. Although the latter procedure reduces the noise in the rate data, it may also affect the shape and position of the $(d\alpha/dT)$–T peak.

Alternatively, one can use α–T data in combination with the integrated form of Eq. (3.17). The general integral form of Eq. (3.17) is as follows

$$g(\alpha) \equiv \int_0^\alpha \frac{d\alpha}{f(\alpha)} = \frac{A}{q}\int_0^{T_\alpha} \exp\left(\frac{-E}{RT}\right)dT = \frac{A}{q}I(E,T) \qquad (3.19)$$

where $g(\alpha)$ is the integral form of the reaction model. The temperature integral $I(E,T)$ in Eq. (3.19) does not have an analytical solution. It has been solved by using various approximations that resulted in a number of integral kinetic methods. Among these methods, the one by Coats and Redfern (1964) appears to be the most popular. The temperature integral approximation used by Coats and Redfern leads to the following simple equation:

$$\ln\left[\frac{g(\alpha)}{T^2}\right] = \text{const} - \frac{E}{RT} \qquad (3.20)$$

In principle, both differential Eq. (3.18) and integral Eq. (3.20) allow one to determine a kinetic triplet from a nonisothermal run performed at a single heating rate. In reality, the results of such kinetic analyses tend to be very confusing because one usually finds that the same experimental curve (i.e., α vs. T or $d\alpha/dT$ vs. T) is satisfactorily described by differing kinetic triplets. The flawed nature of kinetic methods that use a single heating rate was stressed in discussions (Maciejewski 2000; Burnham 2000) of the results of the Kinetics Project (Brown et al. 2000) of the International Confederation of Thermal Analysis and Calorimetry (ICTAC). This project led to the general recommendation that single heating rate methods should be avoided. Instead, one should employ data obtained with multiple temperature programs, for example, at multiple heating rates and/or multiple isothermal temperatures. Subject to this condition, reliable kinetic analyses can be performed by fitting appropriate kinetic models [see, e.g., Eqs. (3.4) and (3.5)] as well as in a model-free way by using isoconversional methods, described below.

3.5.4. Kinetic Analyses Based on Multiple Nonisothermal Runs

Kinetic analyses of multiple nonisothermal, also referred to as *multiple heating rate*, runs are most commonly performed by using the methods of Friedman, Ozawa, and Flynn and Wall. However, application of the Kissinger method (Kissinger 1957) is discouraged because the method yields a single value of the activation energy (or a single kinetic triplet in general) for any process, even if the kinetics is determined by more than one step [see Eq. (3.5)]. Instead, the use of the methods of Friedman, Ozawa, and Flynn and Wall are recommended. The TGA kinetics software provided by many instrument companies is based on the Flynn–Wall method as embodied in ASTM E1641 (see Appendix). These are called *isoconversional methods* because they originate from in the isoconversional principle, which states that at constant conversion, the reaction rate is a function of temperature only; thus

$$\left[\frac{d\ln(d\alpha/dt)}{dT^{-1}}\right]_\alpha = -\frac{E_\alpha}{R} \qquad (3.21)$$

where the subscript α denotes values related to a given extent of conversion. Activation energy E_α is assumed to be constant with temperature but may vary with conversion. Equation (3.21) is derived from the single-step kinetic Eq. (3.7). However, the fundamental assumption of the isoconversional methods is that a single Eq. (3.7) is applicable only to a single extent of conversion and to the temperature region (ΔT) related to this conversion (Fig. 3.25). In other words, isoconversional methods describe the kinetics of the process by using multiple single-step kinetic equations, each of which is associated with a certain extent of conversion (Fig. 3.26). Because of this feature isoconversional methods allow complex multi-step processes to be detected by a variation of E_α with α. Conversely, the absence of a variation of E_α with α is a sign of a single-step process.

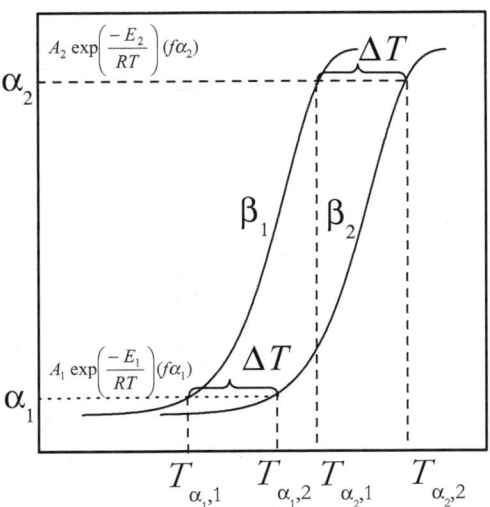

Figure 3.25. In isoconversional methods a single-step rate equation is applied to a narrow temperature region ΔT that changes with the extent of conversion [adapted from Vyazovkin and Sbirrazzuoli (2006) with permission of Wiley-VCH].

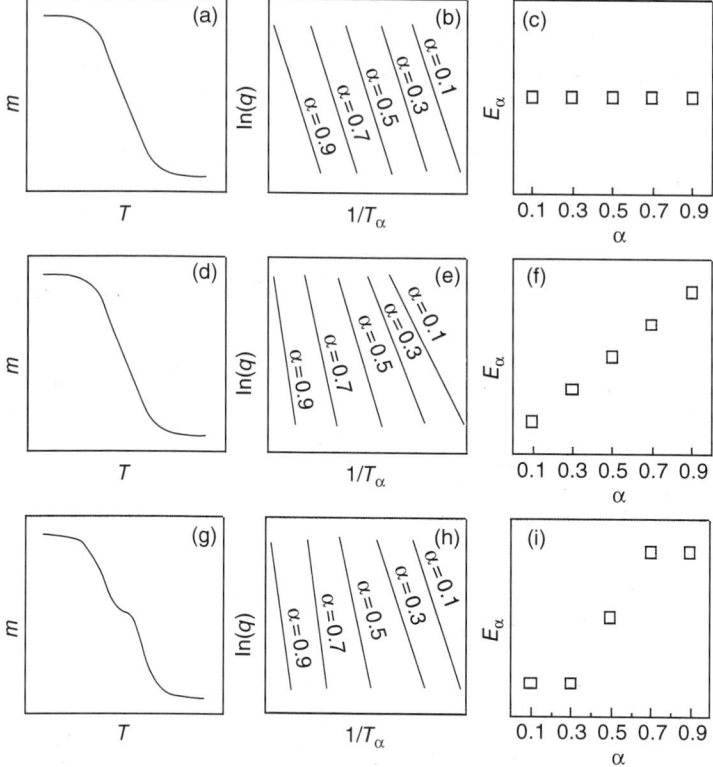

Figure 3.26. (a) single-step mass loss; (b) isoconversional plots are parallel; (c) E_α does not vary with α; (d) single-step mass loss; slope of isoconversional plots (e) and value of E_α (f) change systematically with α; (g) two-step mass loss; (h) slopes of isoconversional plots (h) and value of E_α (i) changes at the transition between the steps.

Simple rearrangement of Eq. (3.18) gives rise to the following equation

$$\ln\left[q_i\left(\frac{d\alpha}{dT}\right)_{\alpha,i}\right] = \ln[A_\alpha f(\alpha)] - \frac{E_\alpha}{RT_{\alpha,i}} \qquad (3.22)$$

which is a foundation of the differential isoconversional method of Friedman (1964). Because Eq. (3.22) is differential, its application to TGA data is associated with the aforementioned problems of numerical differentiation. These problems can be avoided by using integral isoconversional methods that make use of approximations of the temperature integral in Eq. (3.19) or solve it numerically. One of the simplest approximations (Doyle 1962) gives rise to the following equation

$$\ln(q_i) = \text{const} - \frac{E_\alpha}{RT_{\alpha,i}} \qquad (3.23)$$

which is used in the popular isoconversional methods of Flynn and Wall (1966a,b) and Ozawa (1965). Because of inaccuracies in the Doyle approximation, Eq. (3.23) tends to misestimate the values of E_α that then need to be divided by a correction factor. The value of the correction factor increases with decreasing values of E_α/RT. The use of a more accurate approximation (Coats and Redfern 1964) yields

$$\ln\left(\frac{q_i}{T_{\alpha,i}^2}\right) = \text{const} - \frac{E_\alpha}{RT_{\alpha,i}} \quad (3.24)$$

Both Eqs. (3.23) and (3.24) enable the E_α value to be calculated from the slope of a linear plot of the respective left-hand sides (LHSs) of these two equations against $T_{\alpha,i}^{-1}$.

Figure 3.25 shows how the values of T_α are determined for two different heating rates. Actual kinetic analyses require at least three heating rates, but four or five are recommended to improve reliability of the results. The values of T_α should be determined for a wide range of α (e.g., $\alpha = 0.05$–0.95) with a step of 0.05 or smaller. The use of a wide range of α values is needed to reliably establish whether E_α varies with α.

Application of the ASTM E1641, *Standard Test Method for Decomposition Kinetics by Thermogravimetry* (see Appendix), is described in the following section. This method, which makes use of the Flynn–Wall method [Eq. (3.23)], recommends evaluating E_α at the early stages of degradation for estimating lifetimes. However, the software provided by instrument suppliers can make evaluations over the entire decomposition range, which is strongly recommended. Using E_α evaluated at the early stages of degradation to predict the entire degradation may lead to significant errors in predicting the thermal degradation kinetics at larger conversions (see Section 3.5.5 and Fig. 3.29). Evaluating a dependence of E_α on α for a wide region of α can be very beneficial in obtaining helpful clues about degradation mechanisms as well as in making reliable kinetic predictions.

Several situations are typically encountered when applying an isoconversional method to TGA data on polymer degradation. For a process consisting of a single mass loss step (Figs. 3.26a and 3.26d), plotting $\ln(q)$ [see Eq. (3.23)] or $\ln(q/T_\alpha^2)$ [see Eq. (3.24)] against T_α^{-1} for different α values may result in two possible cases. The first is when the respective plots are practically parallel (Fig. 3.26b). This means that E_α does not vary with α (Fig. 3.26c); that is, the process can be described by a single-step rate equation [Eq. (3.4)] throughout the whole range of conversions and temperatures (Fig. 3.25). In this case, the whole process behaves as a single reaction step that can be represented by a single kinetic triplet, that is, a single activation energy, preexponential factor, and reaction model. This does not necessarily mean that the process includes only one single step. The process may involve multiple steps, one of which plays the rate-determining role. Often one encounters the second case when the slopes of $\ln(q)$ or $\ln(q/T_\alpha^2)$ against T_α^{-1} plots change with α (Fig. 3.26e) so that E_α varies with α (Fig. 3.26f) (Vyazovkin and Sbirrazzuoli 2006). This is an indication that the process does not obey a single-step rate equation. It means that the process involves multiple steps, at least two of which provide equally important contributions to the overall rate. A simple example of this situation would be an aforementioned process involving two parallel reactions [Eq. (3.5)], having differing kinetic triplets. Also, it is not uncommon when a degradation process demonstrates two or more consecutive mass loss steps (Fig. 3.26g). This case is illustrated in Figs. 3.38 and 3.39 for the two-step decomposition of ethylene–vinyl acetate (EVA). In this situation, one may observe that the plots of $\ln(q)$ or $\ln(q/\ln(q/T_\alpha^2))$ against T_α^{-1} are parallel to each other within the α region related to an individual mass loss step. Typically, the slope of these plots would change sharply on transition from one step to another (Fig. 3.26h). Such a change in slope is equivalent to a step change in the value of E_α (Fig. 3.26i).

Note that the simpler integral isoconversional equations [e.g., Eqs. (3.23) and (3.24)] have been derived by assuming that the value of E_α is constant in $I(E,T)$ [Eq. (3.19)] throughout the whole interval of integration, from 0 to α. This assumption introduces a systematic error in E_α if its actual value varies with α (Vyazovkin 2001). Normally, the error is not large, but in the case of strong variations can reach 20–30%. The error does not appear in the differential method of Friedman and can be easily eliminated in the advanced isoconversional methods (Vyazovkin and Dollimore 1996; Vyazovkin 1997) that use numerical integration as a part of the E_α evaluation. This is accomplished by carrying out the $I(E,T)$ integral in Eq. (3.19) in small steps $T_{\alpha-\Delta\alpha} - T_\alpha$ so that the constancy of E_α is assumed only for small intervals of conversion, $\Delta\alpha$.

More recent advances in isoconversional methods are associated with designing an integral method suitable for arbitrary temperature programs such as deviation from linearity in the heating rate that frequently occurs in samples as a result of strong self-heating or self-cooling or such as the use of consecutive isothermal and nonisothermal segments in a single run. The data obtained under such temperature programs cannot be treated by typical integral isoconversional equations such as Eqs. (3.23) and (3.24) because they have been derived assuming q to be constant (i.e., not deviating from the linear heating program) and positive (i.e., for heating only). These assumptions are not made in the differential method of Friedman that makes it applicable to the respective temperature programs. However, integral isoconversional methods can be adjusted to an arbitrary temperature program $T(t)$ by replacing integration over temperature in Eq. (3.19) with integration over time (Vyazovkin 2001, 1997):

$$g(\alpha) = \int_{t_{\alpha-\Delta\alpha}}^{t_\alpha} \exp\left[\frac{-E_\alpha}{RT(t)}\right] dt = AJ[E_\alpha, T(t_\alpha)] \quad (3.25)$$

Here, $T(t)$ represents the actual sample temperature and J represents the integral with respect to the actual sample temperature $T(t)$. The resulting advanced isoconversional method (Vyazovkin 2001) employs a set of n experiments carried out under different temperature programs, $T_i(t)$. The E_a value is determined as the value that minimizes the function

$$\Phi(E_\alpha) = \sum_{i=1}^{n} \sum_{j \neq i}^{n} \frac{J[E_\alpha, T_i(t_\alpha)]}{J[E_\alpha, T_j(t_\alpha)]} \tag{3.26}$$

When using multiple heating rate methods, one should be careful about selecting the heating rates. First, it should be kept in mind that increasing the heating rate results in shifting a TGA curve to higher temperatures. Therefore, faster heating rates may be preferred when one is interested in the kinetics of reactions that occur at higher temperatures, and vice versa. As an example, the activation energy for the thermooxidative degradation of PVME was found to be lower at heating rates within 0.3–3 °C/min than within 3–30 °C/min (Prime 1988). Modern TGA instruments provide a very wide range of heating rates. The use of slower heating rates has a disadvantage of longer experimental times. The use of fast heating rates may result in significant temperature gradients within the sample. Because polymers have a relatively low thermal conductivity, establishing a uniform temperature distribution throughout a sample may take appreciable time so that at fast heating rates the average sample temperature may lag behind that of the furnace. For this reason, it is always a good idea to secure good thermal contact between the sample and the sample holder (i.e., pan or crucible). The thermal contact is improved by maximizing the contact area between the sample and the bottom of the sample holder. This is efficiently accomplished in samples prepared as thin powders, melt or solvent cast films, pressed tablets, or similar. Using a high-thermal-conductivity purge gas like helium will be beneficial for studying thermal processes. Unfortunately there is no high-thermal-conductivity purge gas for oxidative processes (see Table 3.2). In addition, degradation of polymers is accompanied by a thermal effect that causes self-heating of the sample with exothermic processes or self-cooling with endothermic processes. The resulting deviations in the sample temperature increase significantly with increasing heating rate. Significant temperature deviations would invalidate the application of numerous kinetic methods that are based on the assumption that the sample and furnace temperatures are identical. In order to minimize such deviations, a good practice is to use small samples. Typically 3–5-mg samples are adequate for TGA runs. However, in the case of composite materials a representative sample may have to be of a larger size. It should be noted that many modern TGA instruments provide an estimate of the sample temperature that is measured throughout a run by using a thermocouple placed in close proximity to the sample. It is, therefore, wise to check that the sample temperature does not deviate from the reference (furnace) temperature by more than 1 °C. If this cannot be accomplished by decreasing the sample size and heating rate, one should consider employing kinetic methods that allow for using the actual sample temperatures, such as the method of Friedman

[Eq. (3.22)] or an advanced isoconversional method [Eq. (3.26)]. Finally, heating rates should be selected so that the respective α–T curves are well resolved, that is, that they shift to higher temperatures with increasing heating rate, but without intersecting each other. We recommend rates within 0.1–20 °C/min.

3.5.5. Applications

Typical applications of kinetic analyses are associated with gaining mechanistic insights and making kinetic predictions. The reader is also referred to the practical applications of kinetic analyses in Sections 3.4.3 and 3.6.1 in this chapter, which complement the examples presented in this subsection. Although kinetic analysis may provide only limited insights into reaction mechanisms, these are frequently the only insights available, especially in the case of complex polymeric materials. The shape of the isothermal curve may provide insight into the mechanistic model governing the process. As discussed earlier (Fig. 3.26), independence of E_α on α indicates the presence of a single rate-determining step; a step increase in E_α indicates the presence of two consecutive rate-determining steps (commonly observed by polymers such as EVA; see Figs. 3.38 and 3.39, whose degradation begins by sidegroup elimination); and variation of E_α with α suggests that the process rate is simultaneously determined by several steps. By analyzing changes in the E_α dependences that occur when modifying experimental conditions and/or reacting materials, one can obtain further mechanistic clues (Vyazovkin and Sbirrazzuoli 2006).

For instance, thermal degradation of polymers is often initiated at weak link sites inherent to the polymer chain. The process has relatively low activation energy and takes place at lower temperatures. Typical weak sites include head-to-head links in a typical head-to-tail polymer, and hydroperoxy and peroxy structures. These sites are where thermal degradation typically starts. Alternatively, initiation can occur as random scission of the polymer chain. As this process has a greater activation energy, its contribution is small at lower temperatures but becomes dominant at higher temperatures (Fig. 3.27) and/or when all weak links have given rise to initiation. This means that the two initiation processes tend to overlap so that the overall degradation rate will be determined by initiation at the weak links at lower temperatures and by random scission at higher temperatures. As a result, the experimentally derived activation energy will be increasing throughout the process of degradation. Indeed, lower values of the experimental activation energy at lower temperatures and extents of degradation are frequently found for a number of polymers (Vyazovkin and Sbirrazzuoli 2006). For example, the application of an isoconversional method to TGA data on the thermal degradation of polyethylene and polypropylene results in increasing E_α with increasing extent of degradation (Peterson et al. 2001) as shown in Fig. 3.28.

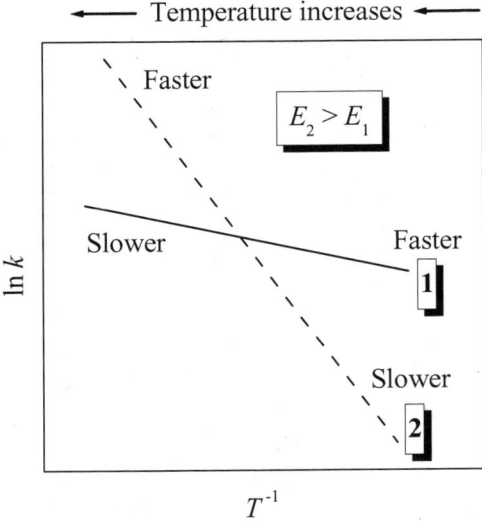

Figure 3.27. Arrhenius plots (ln k vs. T^{-1}) for two simultaneously occurring reactions (1 and 2). Reaction 2 (e.g., random scission) has a larger activation energy than reaction 1 (e.g., scission of weak links). At lower temperatures the rate constant for reaction 1 is larger than for reaction 2; that is, reaction 1 is faster. At higher temperatures the rates invert.

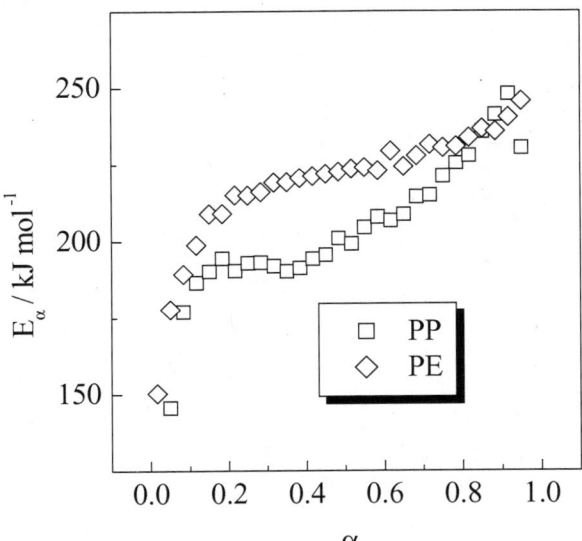

Figure 3.28. E_α dependencies derived by an isoconversional method from TGA data on thermal degradation of polypropylene (PP) and polyethylene (PE) [adapted from Peterson et al (2001) with permission of Wiley-VCH].

It should be pointed out that the experimental values of the effective activation energy may depend on the choice of the heating rates used for conducting TGA runs. As mentioned earlier, the use of faster heating rates shifts TGA curves to higher temperatures. This means that the respective average temperatures for a set of slow heating rates (e.g., all below 5 °C/min) and a set of faster heating rates (e.g., all above 10 °C/min) may differ significantly, perhaps by more than 20 °C. For this reason, the experimental activation energy derived from the slower-heating-rates will be determined primarily by the activation energies of the steps that have a faster rate at lower temperature. On the other hand, the use of faster heating rates will yield experimental activation energies that are related to the activation energies of the steps that have faster rates at higher temperatures. For example, for a thermal degradation process the use of slower heating rates would be likely to yield lower values of the activation energies associated with the initiation at the weak links, whereas the large values associated with random scission would be obtained from faster heating rates. As mentioned earlier, E_α for the thermooxidative degradation of PVME was found to be lower at heating rates within 0.3–3 °C/min than at rates within 3–30 °C/min, although no dependence of E_α on heating rate was observed for the nitrogen pyrolysis (Prime 1988). Therefore, the choice of experimental heating rates should be tailored to the practical task at hand. If the task concerns the high-temperature behavior of a polymeric material, such as in the case of exploring fire resistance, the kinetics derived from faster heating rates would be more relevant. On the other hand, the slower-heating-rate kinetics would be more representative of the thermal behavior at moderate conditions of exploitation, such as service life at elevated temperatures.

Another important practical objective of kinetic analysis is predictions. Their purpose is to evaluate the kinetic behavior of materials under temperature conditions that are different from those used in the actual experimental runs but important for practical applications. A typical example is the use of nonisothermal TGA runs for estimating thermal stability of a material at a certain temperature. Thermal stability can be evaluated as the time to reach a specific but low extent of conversion at a given temperature. Integration and rearrangement of Eq. (3.7) gives

$$t_a = \frac{g(\alpha)}{A\exp(-E/RT_0)} \quad (3.27)$$

where t_α is the time to reach the selected extent of conversion α at a given temperature (T_0), in an isothermal run. It should be emphasized that Eq. (3.27) simply tells one how long it would take to reach a certain fraction of mass loss. Care must be exercised when correlating this time with degradation of various physical properties of a polymer. As an example, degradation of mechanical properties can be detected at very early stages of thermal degradation when the polymer chains just start breaking, that is, well before forming low-molecular-mass volatile fragments that cause the mass loss. It should also be remembered that degradation of a polymer is not associated exclusively with its mass loss. For instance, thermal degradation in the presence of oxygen may result in the formation of polymer oxidation products that cause the sample mass to increase in the early stages of the process and mask initial mass loss. This is more likely to be observed at lower temperatures and slower heating rates (see Section 3.6.1 and Fig. 3.30).

As mentioned earlier, kinetic analyses based on single heating rate runs are not recommended. Problems associated with single heating rate methods can be overcome by combining Eq. (3.27) with the results of kinetic analyses of multiple heating rate runs. One example of such a solution is the ASTM E1641 *Standard Test Method for Decomposition Kinetics by Thermogravimetry* (see Appendix) that uses the following predictive equation to estimate polymer lifetimes

$$t_\alpha = \frac{-\ln(1-\alpha)}{A\exp[-(E/RT_0)]} \qquad (3.28)$$

where the value of E is determined by the Flynn–Wall method [Eq. (3.23)]. Equation (3.28) assumes that the process obeys first-order kinetics, that is, that $g(\alpha)$ in Eq. (3.27) is replaced with $-\ln(1-\alpha)$ [see Eq. (3.13)]. The same assumption is made in ASTM E1641 to determine the preexponential factor in Eq. (3.29)

$$A = -\frac{\bar{q}R}{E_r}\ln(1-\alpha)10^a \qquad (3.29)$$

where $\bar{q}$ is the mean of the experimental heating rates used to determine E by the Flynn–Wall method. The E_r value is the corrected value of the activation energy that is obtained by dividing the experimental value of E by the correction factor. Values for both the correction factor and a in Eq. (3.29) are tabulated in the ASTM method.

The ASTM E1641 method estimates E_α from Eq. (3.23) at conversions selected by the user, from which the constancy of E_α can be determined over the entire degradation process. However, for predicting polymer lifetimes the ASTM method recommends that conversions not exceed 10% (i.e., $\alpha \leq 0.1$) or, in certain cases, 20%. Calculation of E_α should be made at several decomposition levels, for instance, from 2.5% to 20% in 2.5% increments. Low E_α at very low mass loss may indicate evaporation of impurities, additives, or oligomers that immediately precede decomposition. The mass loss chosen for lifetime estimations should be in an area of consistent E_α versus α. A common value used to estimate service lifetimes is 5% conversion. The ASTM method also recommends heating rates within 1–10 °C/min. However, modeling can be done only with the first-order kinetic model and assuming constant activation energy, regardless of the actual reaction path. This can lead to erroneous predictions beyond the early stages of degradation as illustrated in Fig. 3.29. For this reason it is recommended that use of the ASTM method be limited

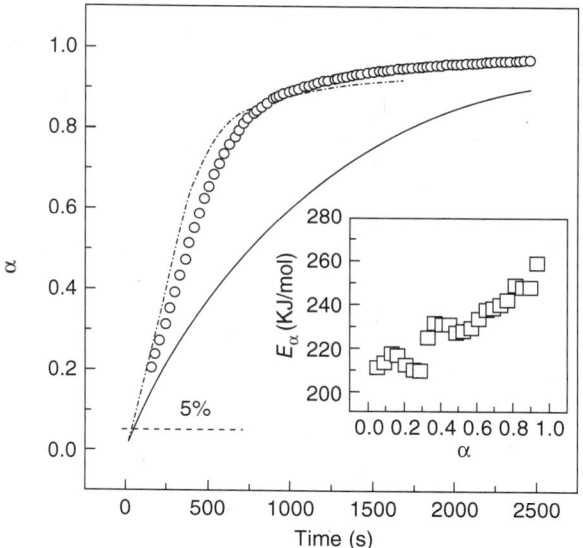

Figure 3.29. ASTM (solid line) and model-free (dashed–dotted line) predictions of the thermal degradation of PEN at 420 °C (circles; initial portion is not shown to avoid overcrowding); inset shows E_α dependence derived by an isoconversional method from TGA data (S. Vyazovkin, K. Chen, and I. Dranca, unpublished, 2006).

to the initial stages of decomposition unless it can be demonstrated that E_α is independent of α and that the degradation follows first-order kinetics.

The limitations of the ASTM method are avoided in the model-free predictions (Vyazovkin 1996), which make use of the dependence of E_α on α determined by an isoconversional method. The predictive equation was originally proposed in the following form:

$$t_\alpha = \frac{1/q \int_0^{T_\alpha} \exp(-E_\alpha/RT)dT}{\exp(-E_\alpha/RT)} \qquad (3.30)$$

This equation was later modified to employ data from arbitrary heating programs, as follows:

$$t_\alpha = \frac{J[E_\alpha, T(t_\alpha)]}{\exp(-E_\alpha/RT_0)} \qquad (3.31)$$

The respective predictions are called "model-free predictions" because they eliminate the reaction model $g(\alpha)$ in the numerator of Eq. (3.27).

Model-free predictions are typically superior to predictions based on the ASTM method. Figure 3.29 provides an example of using Eqs. (3.28) (ASTM prediction) and (3.31) (model-free prediction) for modeling the thermal degradation of PEN at $T_0 = 420\,°C$. Both predictions are based on kinetic analysis of the same nonisothermal TGA data obtained for a ~5-mg sample of PEN at five heating rates (5, 7.5, 10, 12.5, and 15 °C/min) under nitrogen. As can be seen in Fig. 3.29, both methods give similar results at 5% conversion. However, the ASTM prediction shows a systematic error that increases markedly with increasing extent of degradation. This example emphasizes the importance of accounting for reaction complexity when making kinetic predictions. In the model-free predictions [Eqs. (3.30) and (3.31)], this complexity is accounted for by using an isoconversional dependence of E_α on α determined for the degradation process (inset Fig. 3.29). In these equations each value of t_α is predicted by using the respective value of E_α that varies with α from ~210 to ~260 kJ/mol. The ASTM predictions rely on a single value of E obtained by Eq. (3.23) that treats any process as a single-step reaction that can be represented by a single value of the activation energy. For the process considered, the corrected E_r value estimated from the Flynn–Wall method is 221 kJ/mol. This value corresponds to $\alpha = 0.05$ and is used in Eq. 3.28 to predict all the t_α values for the whole range of α. The ASTM method predicts the actual decomposition data well at 5% conversion but begins to significantly underestimate the data soon afterward. The model-free prediction represents the data well over the whole degradation range.

3.6. SELECTED APPLICATIONS

In this section we present selected applications of actual industrial problems where TGA has been instrumental in their solution. The purpose is to present sufficient detail that readers may apply in adapting these techniques to their problems.

3.6.1. Thermal Stability and Lifetime Estimates

Attempting to estimate the lifetime of a polymer is often a challenging if not nearly impossible problem to deal with because of the interactions between various chemical and physical effects. Consider the statement "the effect of temperature, stress, humidity, light, contaminating molecules, and processing or design flaws, acting either alone or in combination on a polymer that is part of a product can cause the plastic to fail prematurely" (Bair 1997). Hence, it is necessary to examine a long list of potential factors that can have a deleterious effect on the performance of a given polymer, both during its processing and during its lifetime in the field. Usually it is wise to investigate each potential threat to a polymer's longevity separately and decide what inherent flaw, if any, may seriously contribute to the plastic's premature failure in the field, and then attempt to design around this flaw through the use of additives or other engineering means. In some cases, only by selecting a different polymeric material can an environmental problem be circumvented.

Numerous polymers are known to have inherent weaknesses in their basic structures that render them susceptible to thermal or thermooxidative degradation. For example, in the presence of oxygen and at relatively low temperatures polyolefins are susceptible to a random chain scission process, with outdoor weathering ABS (acrylonitrile-butadiene-styrene) terpolymers and polycarbonates are weakened when light falls on their unprotected surfaces, and poly(vinyl chloride), even in the absence of oxygen, can simply degrade from the heat generated during its processing. These many destructive influences on certain plastic families have given rise to the development of various additives that attempt to extend a polymer's lifetime by various chemical and physical means (Bair 1997). Although thermal analysis techniques can be used to help understand the complex breakdown mechanisms that may confront commercial plastics, we will limit our discussion here to the use of TGA to study the degradation of polymers due to heat alone or heat plus an oxidizing atmosphere.

With a new polymeric material, a major concern is often its thermal stability in inert and oxidizing atmospheres. TGA is well suited to enable one to gain information about the tendency of a polymer to degrade under these conditions. A few cases are shown here.

Perhaps the simplest and fastest way to investigate the stability of a polymer is to heat it in a TGA apparatus and observe at what temperature the material begins to yield volatile fragments. Unfortunately, misleading indications of a polymer's stability in the presence of oxygen can occur if an inappropriate heating rate is selected that allows the polymer's mass gaining oxidation process to be offset by the loss of volatiles. For benzocyclobutene, this behavior appears when a heating rate of 10 °C/min is chosen (Fig. 3.30, solid line),

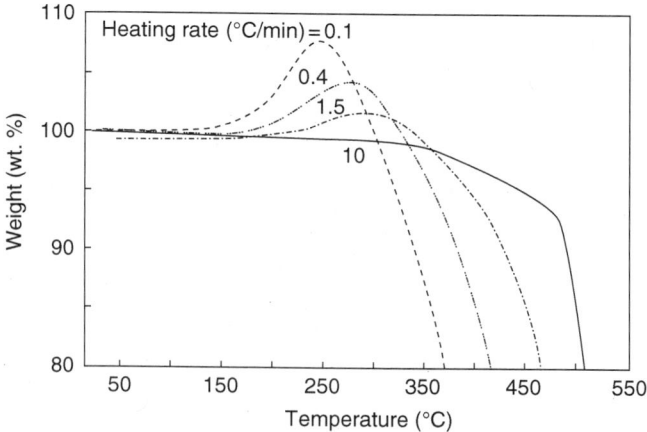

Figure 3.30. Effect of heating rate on shape of cured benzocyclobutene's TGA curves in oxygen [from Bair (1997); reproduced with permission from Elsevier].

for it appears as if the sample is stable until a temperature of about 330 °C is reached. In practice this overestimation of the stability of this material as a dielectric coating first appeared on the manufacturer's datasheet, where it stated that cured benzocyclobutene could tolerate temperatures in excess of 300 °C in oxygen (Stokich et al. 1991). However, at slower heating rates oxidation was detected at 150 °C or lower (Bair et al. 1991). Note that at a heating rate of 0.4 °C/min a mass gain for the benzocyclobutene of approximately 8% developed before the first signs of fragments evolving as a mass loss were detected. A mass gain was also observed in the degradation of poly(vinyl methyl ether) (PVME) in air at heating rates below 1 °C/min (Prime, unpublished results). A similar phenomenon occurs in isothermal polyethylene oxidation studies if the temperature is set too high; then, only mass loss events are recorded (Foster 1990). Thus, in isothermal or dynamic oxidation studies it is important to set the heating rate or isothermal temperature low enough to allow the oxidation process to be observed.

Most thermal analysis methods for studying polymeric stabilizer systems are based on the antioxidant's ability to delay the oxidation process. Usually a sample is heated to a specified temperature and the induction time, or period of time before the onset of rapid thermal oxidation, is determined [see discussion of oxidative induction time (OIT) in Section 3.4.2 of this chapter]. The end of the induction period is marked by an abrupt increase in the sample's temperature, evolved heat, or mass and can be detected by DTA, DSC or TGA, respectively (Bair 1997). The effect of antioxidant structure and its concentration on prolonging a sample's induction period can be used to determine the most effective antioxidant system for a polymer such as polyethylene. Extensive data have shown that thermal information such as this can be used successfully to estimate the lifetime of polyethylene at processing temperatures (Bair 1997).

Unfortunately, any attempt to predict the long-term stability of polyethylene based on an Arrhenius plot of high-temperature oxidative induction times measured above the melting point fails when projected to lower temperatures where polyethylene is a semicrystalline solid (Bair 1973; Chan et al. 1978). The reasons for this nonlinear behavior appear to be associated with complex chemical and physical interactions that behave differently in the solid state than in the melt.

One problem that can occur in the solid state that does not occur in the melt is that a large portion of the antioxidant that was soluble in the melt becomes insoluble in the solid state, where it blooms to the surface and is lost to its surroundings. This process is controlled by the solubility and diffusion of the antioxidant in the polymer, and it can be characterized by a TGA technique developed by Roe et al. (1974). The Roe–Bair–Gieniewski (RBG) method involves analyzing a concentration profile across a stack of polyethylene sheets through which the antioxidant has been forced to diffuse (Bair 1997). The level of antioxidant in the polyethylene sheets was determined by TGA based on an induction time calibration curve (Bair 1973). In this study the TGA was found capable of detecting quantitatively the level of phenolic antioxidants in polyethylene at concentrations as low as 0.001 mass%. From these studies it was shown that at 20 °C nearly all of a typical commercial phenolic antioxidant exudes to the surface of the polyethylene in about 2 weeks. On the polymer's surface the antioxidant can be removed by wind, rain, or sublimation processes. At 60 °C a typical antioxidant can sublime away in as little as 5 years.

Hopefully, reviewing the results of these prior studies will make the reader aware of some of the potential difficulties that can arise in trying to estimate the lifetime of a polymer from data collected in the melt when these data are extrapolated to much lower temperatures where the polymer is transformed into a solid phase. However, often the only practical way to acquire data quickly about the degradation of a polymer is to carry out tests at elevated temperatures and then attempt to apply some safeguards on how to use the data. A few examples of these kinds of thermal degradation studies are reviewed here.

The thermal stability of Cytop, a fluorinated polymer, would appear to be significantly better than most polymers simply on the basis of its fluorocarbon structure. This amorphous polymer's glass transition temperature is 108 °C by DSC when heated at 15 °C/min. However, how long it can survive in oxygen at 250 °C without degrading prior to extrusion can be predicted on an acceptable experimental timescale only if a series of dynamic runs are made by heating Cytop samples to above 500 °C as shown in Fig. 3.31. Four runs were made in oxygen on samples weighing about 8 mg each at heating rates of 0.5, 1.0, 3.0, and 6.0 (°C/min). Specimens were prepared from 125-µm-thick Cytop

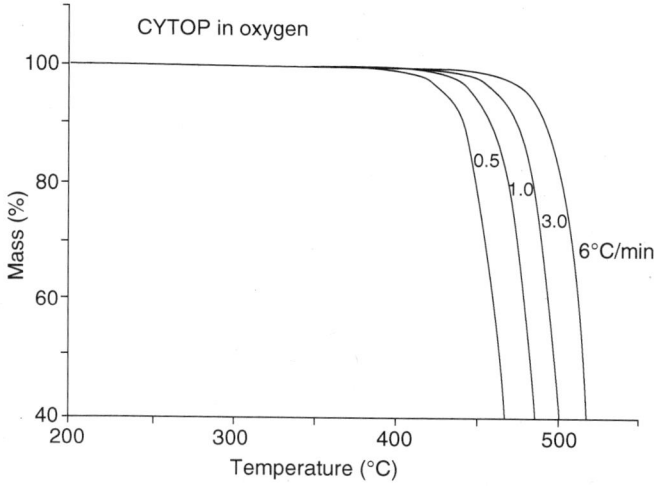

Figure 3.31. TGA scans of Cytop heated at 0.5 °C, 1.0 °C, 3.0 °C, and 6.0 °C/min in oxygen (Bair, unpublished, 1996).

films. At the slowest heating rate the first signs of degradation begin near 400 °C, and then the sample volatilizes rapidly and completely. As the heating rate increases, the shape of the degradation curve remains the same but shifts to higher temperatures.

The decomposition kinetic analysis of the mass loss curves was carried out using standard TGA kinetics software that is based on the Flynn–Wall method (Flynn and Wall 1966a,b) and described in Section 3.5.5. In this technique a plot of the logarithm of the heating rate (log q) against the reciprocal of the absolute temperature [see. Eq. (3.23)] was made where the conversion or mass loss equaled 4%, 5.5%, 8%, 12%, 18% and 26%. From the slope of this plot (Fig. 3.32) the activation energy for decomposition (E) was calculated from Eq. (3.32), which is based on Eq. (3.23):

$$E = \frac{-4.35 \, d(\log q)}{d(1/T)} \quad (3.32)$$

Note that all curves in Fig. 3.32 have the same slope, indicating that E is constant over the conversion range studied (cf. Fig. 3.26b). The activation energy for decomposition was found to be 217 ± 3 kJ/mol.

Once the kinetic analysis is completed, lifetimes may be projected assuming first-order reaction kinetics, which should apply to early mass loss behavior as discussed in Section 3.5.5. If one selects a very low mass loss such as 0.01% to represent the start of degradation, then, using the measured activation energy, a time–temperature plot can be constructed for this low onset level of decomposition. Note that while times to reach such a low level of degradation can be predicted from the kinetic analysis, they cannot actually be measured accu-

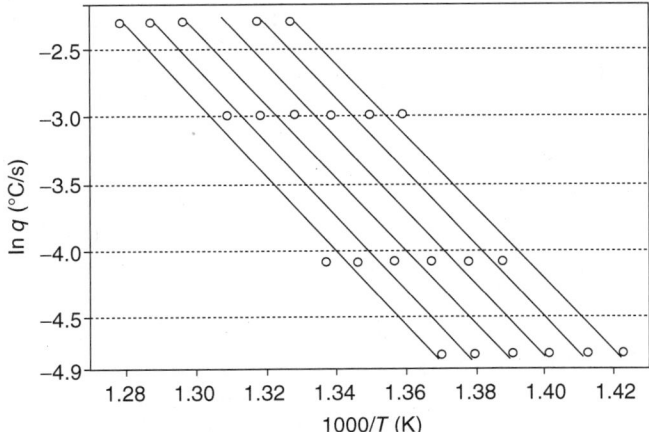

Figure 3.32. Plot of the heating rate against the reciprocal absolute temperature for Cytop in oxygen where the mass loss equals 4 (rightmost curve), 5.5, 8, 12, 18, and 26 mass% (leftmost curve) (Bair, unpublished, 1996).

rately since, among other factors, evaporation other components will compete with the loss of degradation products. The software allows one to calculate the time to reach 0.01% mass loss for a selected temperature with the previously determined value of E. The plot shown in Fig. 3.33 was prepared from data obtained in this manner using an Excel program.

From this analysis the stability of Cytop in oxygen is estimated to be about 20 days at 250 °C and 417 days at 220 °C. In addition, the thermal stability of Cytop in nitrogen was determined in a similar manner and the time to the onset of decomposition at 250 °C was shifted by more than an order of magnitude beyond the value found for degradation in oxygen. Actual extrusion trials of this material in the melt indicate these predicted lifetimes are reasonable estimates of Cytop's stability.

What happens if the estimate of time to reach a certain level of degradation causes one to question the results? The quickest way to gain confidence in the projected results of a TGA kinetic analysis is to run a check on the predicted time to obtain a certain level of degradation at a specific isothermal temperature. An example of this approach is shown for a silicone gel polymer that was analyzed by dynamic TGA scans similar to Fig. 3.31 but in nitrogen. The kinetics software predicted the silicone would lose 6% of its mass in 1388 min (23.1 h) and 417 days if aged isothermally at 260 °C and 150 °C in nitrogen, respectively. Hence, the more practical timescale for degradation at the higher isothermal temperature was selected as a check on the estimated time to 6% mass loss.

Figure 3.34 shows results for a sample of the silicone that was run in nitrogen from 23 °C to 260 °C at a rate of 15 °C/min and then held at 260 °C for 2000 min. In the initial run the sample appears to lose about 2.2 mass% of trapped low-molecular-mass components and finally begins to degrade after about 1000 min of aging. This sample was next cooled to room temperature and then reheated (run 2) and held at 260°C for an additional 1100 min. During the second run the mass loss rate stabilizes at 0.0043%/min between 500 and 1000 min, after the trapped species created during the cool down at

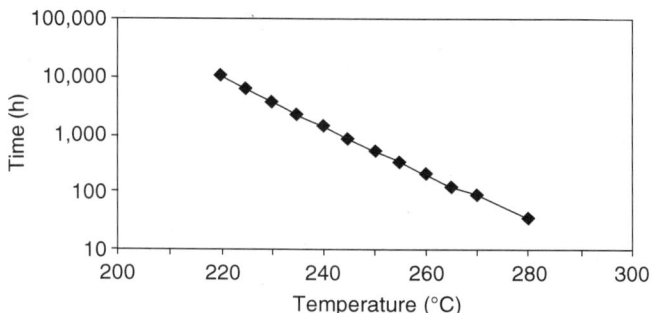

Figure 3.33. Time to onset of decomposition at 0.01% mass loss for Cytop in oxygen plotted against temperature (Bair, unpublished, 1996).

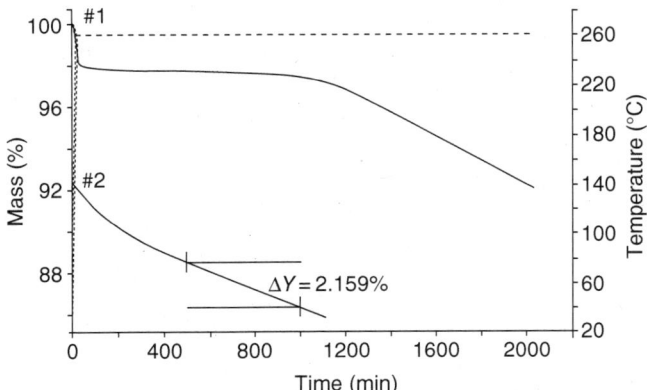

Figure 3.34. Silicone gel heated to 260 °C in nitrogen at 15 °C/min and then held there for 2000 min (#1, solid line) before cooling to room temperature and again heating and holding at 260 °C (#2, solid line); temperature is shown as dashed line. (Bair, unpublished, 1998).

the end of run 1 were lost. Using the latter rate, one calculates that 6% of the sample mass would be lost in 1388 min. The predicted loss from the dynamic data matches the actual degradative loss observed in the isothermal experiment. Thus, we conclude that the kinetic analysis is able to reasonably predict the isothermal degradation behavior of the silicone gel at elevated temperatures.

Several examples of the pros and cons of using TGA experiments to probe the thermal stability of polymeric materials have been reviewed. In particular, the difficulty of trying to apply an Arrhenius plot to data collected in the melt for antioxidant protected polyethylene in order to predict the polymer's lifetime in the solid state at service temperatures below 80 °C was shown to be impossible because of the antioxidant's solubility and diffusion behavior in the partially crystalline polymer. Another problem that can occur is if the selected heating rate is so fast that it enables the mass loss process to be masked by the oxidative, mass-gaining step. In such cases one can be fooled into thinking the tested polymer has far greater thermooxidative stability than is the actual case. Slower heating rates should eliminate the latter difficulty. Finally, the application of commercial software to kinetically analyze dynamic TGA mass loss curves to calculate activation energy values and predict lifetimes was demonstrated.

3.6.2. Permeability of Water and Water Vapor Transport

Sometimes it is impractical to measure the movement of water through a film using the standard ASTM E96/E96M method, where the minimum film size recommended is at least 3000 mm^2 or 4.65 in.^{2} (ASTM 2005). Often research-grade films of a new material may not be available in an amount large enough to meet ASTM's film size requirements. Also, the ASTM technique calls for making intermittent mass measurements by removing the sample and its container from a controlled environment and placing it on an analytical balance. This latter step not only unnecessarily lengthens each water vapor transport experiment but also causes the sample to be removed from its controlled environment during weighing.

To overcome these difficulties, a TGA-size metal capsule was developed to enable one to make rapid, in situ water vapor transport (WVT) measurements through polymeric films or membranes inside a TGA as a function of humidity and temperature (Bair et al. 2003). This capsule has an opening about 200 times smaller than that recommended by ASTM; nevertheless, it yields permeability data that are in reasonable agreement with past measurements by other techniques (Holcomb and Bair 2004, 2005).

A film of the polymer or membrane to be tested is positioned inside a circular lid and held in place via a rubber O-ring. With the film in place the lid is then placed atop the impermeable metal receptacle as shown in Fig. 3.35. The top edges of the container's walls are bent toward the center of the receptacle. Before sealing the capsule with water in it, the capsule's mass is tared in the TGA. Then, about 50mg of water or any other liquid of interest is placed inside the capsule using a syringe before the lid is sealed in place.

The capsule's lid is positioned across the top edges of the container's sidewalls, and the two parts are squeezed together. The lid compresses the O-ring onto the lower half of the capsule, sealing it against the bent top edges of the receptacle and the polymer film above it. Finally, the mass of the water is recorded in the TGA as the initial or starting mass.

"Wet" and dry nitrogen gas streams are mixed to control the humidity inside the TGA to any value between 0% and 90%. In this manner a number of WVT measurements can be carried out quickly at a variety of humidities and temperatures. Note that the O-ring will have a small but measurable leak rate, particularly at elevated temperatures. In a separate experiment the leak rate as a function of temperature can be determined by making a run with a

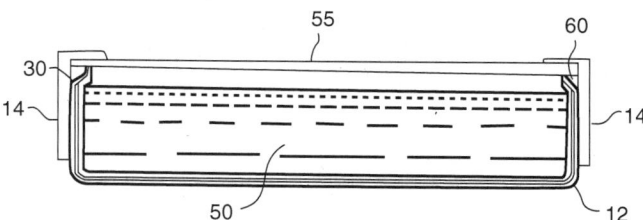

Figure 3.35. Cross-sectional view of a water vapor transport (WVT) capsule showing (12) receptacle, (14) lid, (30) top edges of side wall, (55) polymer film, (50) liquid permeant or water, and (60) O-ring [from Holcomb and Bair (2005); reprinted with permission from the American Chemical Society].

lid that has no hole in it. These O-ring leak rate values can be used as a correction term in subsequent experiments when films are in place.

Permeability (P) is calculated as

$$P = \frac{\Delta m * d}{At(p_1 - p_2)} \quad (3.33)$$

where Δm is the mass change for the sealed receptacle, d is the thickness of the polymer film, A is the area of the hole in the receptacle lid, t is time, p_1 is the water vapor pressure in the sealed capsule, and p_2 is the vapor pressure inside the TGA chamber.

An amorphous polycarbonate (bisphenol A polycarbonate, Lexan®) film with a thickness of 30.5 μm was placed inside the lid of a WVT capsule. The lid had a 4.25-mm-diameter hole. The capsule was sealed with about 53 mg of water present. The TGA was held at 23 °C during the mass loss experiment. In this case the relative humidity inside the TGA was zero (i.e., $p_2 = 0$). The vapor pressure above the water drop inside the capsule (p_1), was 21.1 mm (2.11 cm) of mercury as found in a table of water vapor pressure values (Weast 1973). In Fig. 3.36 the mass loss of water in the sealed receptacle is plotted as a function of time. From these data it was determined that water vapor moved through the amorphous polycarbonate film at a constant rate of 0.062 ± 0.001 mg/h for more than 275 min. This is equivalent to $\Delta m/At = 0.44$ mg/(cm²·h) in Eq. (3.33).

On the basis of this water vapor transport (WVT) data, the water permeability of Lexan polycarbonate at 23 °C can be calculated as ~1.8 × 10⁻⁹ (g·mm)/(cm²·s·cmHg). Converting the mass of water to its volume at standard temperature and pressure (STP) yields a value of ~2.2 × 10⁻⁶ (cm³·mm)/(cm²·s·cmHg) (STP). In 1963 Norton measured a value of 1.4 × 10⁻⁶ (cm³·mm)/

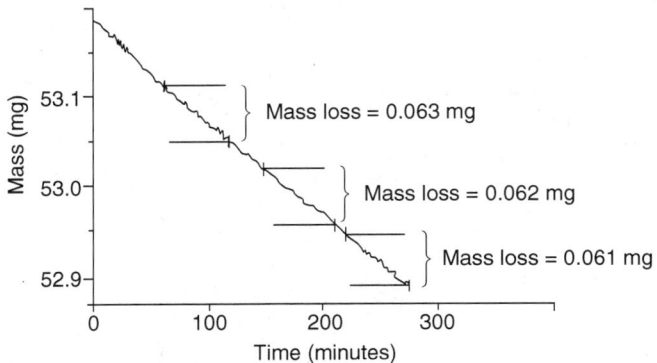

Figure 3.36. Water vapor loss through a polycarbonate film at 23 °C by TGA (mass changes indicated are in one hour intervals) [from Holcomb and Bair (2005); reprinted with permission from the American Chemical Society].

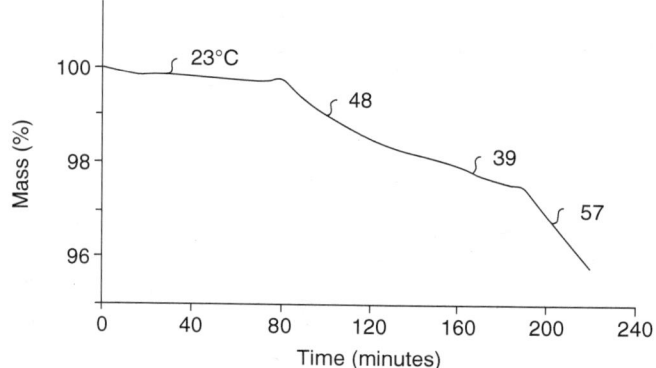

Figure 3.37. Water vapor loss through a silicone film at 23 °C, 48 °C, 39 °C, and 57 °C [from Holcomb and Bair (2005); reprinted with permission from the American Chemical Society].

(cm^2·s·cmHg) (STP) on polycarbonate at 25 °C using a mass spectrometer as a detector of the water vapor.

Using a syringe a drop of water was placed inside a stainless-steel WVT capsule. Next a 0.32-mm-thick film of cured GE silicone, RTV-615, was secured inside the capsule's lid, which had a 4.25-mm hole. The sealed capsule was placed in the center of the TGA platinum sample cup, and the TGA chamber was closed. The rate of vapor transport across the cured film was found to be 1.2, 9.3, 4.5 and 16.7 mg/(cm^2·h) at 23 °C, 48 °C, 39 °C, and 57 °C, respectively (Fig. 3.37). Observe how quickly the film adjusts to temperature changes in the TGA and steady-state water loss values are obtained. Note, in particular, how rapidly the capsule adapts to cooling from 48 °C to 39 °C.

Using this TGA/WVT capsule technique, vapor transport data have been collected not only for glassy PC and rubbery silicone films as reviewed here but also for crystalline films of poly(ethylene terephthalate) (PET) and a microporous Gortex® membrane (Holcomb and Bair 2005). With a small modification this WVT capsule can be employed to confine volatile reactive liquids such as the UV-curable materials studied in DPC work at elevated temperatures (see Section 2.11, on differential photocalorimetry).

3.6.3. Quantitative Analysis of Ethylene–Vinyl Acetate Copolymers

It is possible to use a TGA to carry out quantitative analysis on certain blends and copolymers if the different polymer segments in the blend or copolymer have significantly different thermal stabilities. Chiu (1966) showed how TGA could be used to separate quantitatively an 85/15 blend of polychlorotrifluoroethylene (PCTFE) and polytetrafluoroethylene (PTFE) by heating the blend at 5 °C/min in nitrogen from room temperature to above 600 °C.

Ethylene–vinyl acetate (EVA) copolymers are widely used as hot-melt adhesives and as insulation over heavy-duty power cables. When EVA is thermally degraded, acetic acid is generated by sidegroup elimination during the early stages of degradation in an inert atmosphere, whereas the remainder of the copolymer breaks apart at much higher temperatures. A TGA multiple heating rate methodology for multistage degradation of EVA revealed that in nitrogen the first stage of degradation had an activation energy of 175 kJ/mol, while the second stage had an activation energy of 260 kJ/mol (Nam and Seferis, 1991). Compare E_α versus conversion in Fig. 3.26i.

This decomposition behavior is shown in Fig. 3.38 (Hale and Bair 1997). The mass loss near 350 °C is associated with the evolution of acetic acid, while the remainder of the copolymer begins to degrade at about 430 °C. The amount of vinyl acetate in the copolymer can be determined from the mass loss between 300 and 400 °C, assumed to be entirely acetic acid, using the following equation (Chiu 1966):

$$\% \text{ vinyl acetate} = \% \text{ mass loss} \times 1.43 \qquad (3.34)$$

Although the evolution of acetic acid is evident in this temperature range in Fig. 3.38, the end of the process cannot be readily found since the TGA curve doesn't completely level off before the next stage of the degradation begins. One solution would be to heat at a slower rate. But the end of the initial degradation can be better defined by employing controlled rate TGA (CRTGA), described in Section 3.2.3 of this chapter. CRTGA can automatically switch the TGA to an isothermal mode when the mass loss rate exceeds a set limit (autostepwise), or it can dynamically adjust the heating rate in response to an increase or decrease in the mass loss rate (dynamic Hi-Res).

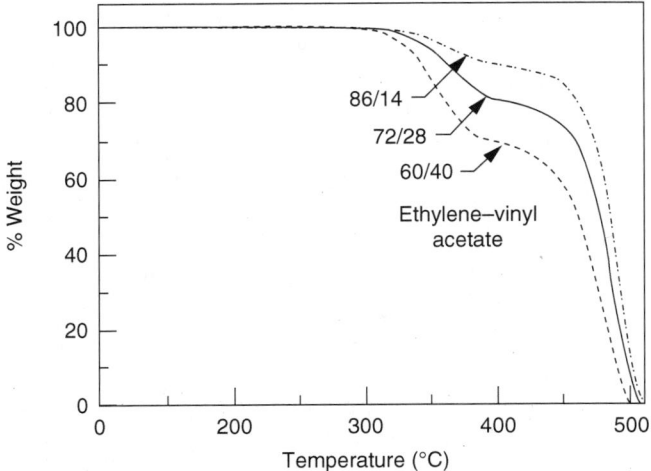

Figure 3.38. Mass loss curves of several EVA copolymers in nitogen when heated at 10 °C/min; the copolymer composition is placed next to each curve [from Hale and Bair (1997); reproduced with permission from Elsevier).]

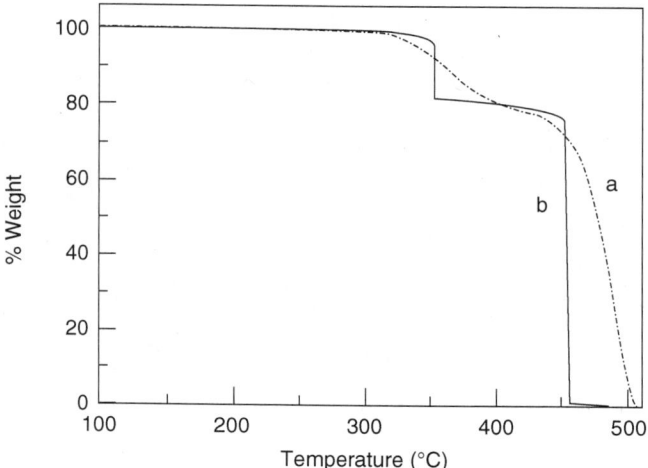

Figure 3.39. Mass loss curves of a 72–28 ethylene–vinyl acetate copolymer. Curve (a) was obtained by heating at 10 °C/min. Curve (b) was produced using autostepwise software, which automatically switched the 10 °C/min rate to an isothermal mode when the mass loss rate increased to a preselected value and returned to the initial heating rate when the mass loss rate slowed to an also preselected small value. [From Hale and Bair (1997); reproduced by permission from Elsevier).]

The autostepwise method is shown in Fig. 3.39 for the degradation of an EVA sample.

In this case the initial 10 °C/min heating rate switches automatically to an isothermal mode near 350 °C, when the first degradation step associated with acid loss exceeds a rate of 3 mass%/min, and after 5.8 min returns to the starting heating rate of 10 °C/min when the degradation rate falls below 0.20 mass%/min.

This software method results in a mass loss curve where the end of acid loss is clearly defined as shown in Fig. 3.39. In this example the acid mass loss is 18.4% and from Eq. (3.34) corresponds to a vinyl acetate content of 26.3%. The vinyl acetate level for this EVA copolymer reported by the manufacturer was 28%.

3.6.4. Analysis of Adhesive Blend

Even in cases where one cannot unambiguously identify the components of a complex blend by its individual degradation pattern, it still may be possible to use the blend's TGA degradation curve as a quality control tool. For example, a commercial blend of an acrylonitrile–butadiene (ACN/BD) copolymer and a phenolic resin yields a tough thermoplastic/thermoset adhesive when cured that can hold brake linings in place and meet other demanding industrial applications. Isothermal cure of this adhesive was described in Section 3.4.3.

DSC studies of the as-received blend show two glass transitions at 7 °C and 75 °C (Tabaddor et al. 2000). The lower T_g is associated with the rubbery ACN/BD phase, and the other glassy phase is due to the uncured phenolic resin.

When initially heated at 15 °C/min from 23 °C to 300 °C, this complex adhesive blend begins to lose low-molecular-mass reaction products such as water and formaldehyde as soon as the phenolic devitrifies above 75 °C (Fig. 3.40, solid line). At 300 °C the reaction appears to cease when about 10 mass% gaseous products have evolved. Cooling the cured material to 23 °C and rescanning the adhesive blend to 700 °C did not show any additional sign of mass loss until a temperature of ~400 °C was reached (Fig. 3.40, dashed line). Above 400 °C, degradation of the polymeric material began with the evolution of low-molecular-mass degradation products and terminated near 700 °C.

About 2 mass% of residual ash remains at 700 °C and is attributed to a zinc oxide additive referred to in the manufacturer's datasheet. Hence, in this manner one can often separate a commercial polymeric material into its small molecules, macromolecules, and inorganic parts by simply heating the material in an inert atmosphere from room temperature to above 700 °C at a nominal rate of 10–15 °C/min.

In general, if the polymeric portion of a material is relatively stable, only low-molecular-mass components will be lost below 300 °C in nitrogen, between 300 °C and 600 °C most polymers will degrade and volatilize, and finally above 600 °C only the inorganic compounds are stable enough to remain. However, some polymers based on aromatic and other cyclic structures will not completely convert to gaseous products when heated to 600 °C in nitrogen but instead will form a char residue (carbonaceous structure). By switching to oxygen at about 550 °C, one finds that the char will degrade, forming carbon-containing gases. A method for measuring CB content in a polymeric material was described in Section 3.4.6, where the sample was heated to 500–600 °C in

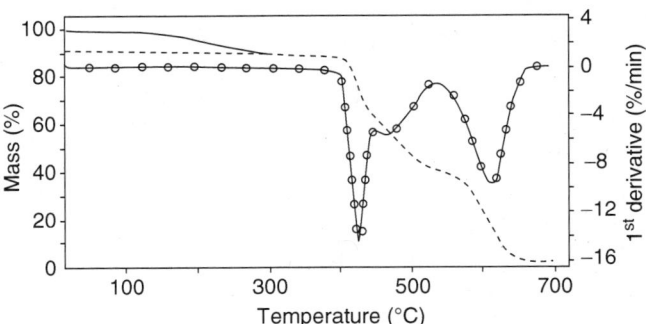

Figure 3.40. TGA mass loss and derivative DTG curves of a thermoplastic/thermoset adhesive. Sample was predried prior to the first run. Heating rate was 15 °C/min in nitrogen. [From Tabaddor et al. (2000) with kind permission from Springer Science and Business Media.]

nitrogen (i.e., until no more mass loss was observed) and then the gas was switched to air or oxygen to pyrolyze the carbon black.

The derivative mass loss curve of the rerun sample is depicted in Fig. 3.40 (line with circles). Three maxima occur in the derivative curve at 420 °C, 470 °C, and 620 °C and appear from IR analysis to be associated with degradation of the three major components of the blend: butadiene, acrylonitrile, and phenolic, respectively. Unfortunately, the IR analysis of the evolved gases during the rerun also indicates that some breakdown of the phenolic resin is occurring during the degradation of the ACN/BD copolymer in the temperature range 370–500 °C. Thus, the TGA analysis cannot provide a direct measure of the amount of each component in the blend. Nevertheless, this TGA "fingerprint" is reproducible and can be used as a quality control check for the consistency of the incoming product.

These several examples of TGA mass loss curves have shown how useful this thermal instrument can be in kinetic analysis of thermal stability and in quantifying the amount of low-molecular-mass products in a polymeric material as well as determining the level of components in a copolymer. Even in the case where degradation of a complex blend cannot be carried out quantitatively, it was demonstrated how the TGA fingerprint can be used to check the consistency of the supplier's incoming product.

3.7. INSTRUMENTATION

In this section we briefly review commercial thermogravimetric analyzers in alphabetical order of manufacturer. Please visit the respective supplier Websites for current offerings and more information. See Section 3.2.1 for descriptions of balance configurations. While most suppliers offer autostepwise software to control the furnace temperature (see Fig. 3.40), kinetics software such as the ASTM E1641 Flynn–Wall method described in Section 3.5.5, and the ability to interface to a Fourier transform infrared spectrometer (FTIR) or mass spectrometer (MS), the reader interested in these applications is advised to confirm the features of particular TGA systems.

Bähr Thermoanalyse offers a top-loading balance (TGA 502) and a side-loading balance (TGA 503) for use in various temperature ranges. Both balances have a 1-g sample capacity and can heat at rates of 0.01–100 °C/min. The TGA 502 has temperature ranges of approximately −160–600 °C and ambient to 1500 °C and higher, while the TGA 503 has ranges of −160–700 °C and ambient to 1500 °C. Temperature ranges depend on how the system is configured. The TGA 503 can cool from 1500 °C to 100 °C in 15 min. It is possible to connect an FTIR and MS to both systems, but evolved gas analysis is better suited to the small-volume and horizontal configuration of the 503. For more information, see www.baehr-thermo.de.

Cahn VersaTherm bottom-loading thermogravimetric analyzers have up to 100 g sample capacity and up to 35 mL sample volume. Built-in flow meters can perform gas switching and mixing. Nominal range is ambient to 1100 °C with vacuum capability to 5×10^{-5} torr, but high-temperature capability up to 1700 °C is available. Samples can be studied under corrosive or reactive gas environments. These TGAs have the capability to be coupled with FTIR or mass spectrometry. For further information, see www.thermo.com and click on "tga."

Instrument Specialists offers the iSeries TGA with a vertically oriented bottom-loading balance. With different configurations the temperature range is ambient to 1000 °C or 1500 °C with heating rates of ≤300 °C/min, and with maximum sample capacities of 40 or 400 mg. Evolved gas analysis capability is optional. Upgrades to existing thermal analysis systems are available. For more information, see www.instrument-specialists.com.

Linseis offers several thermogravimetric analyzers, including the L81-II economy line and the L81-I research line. Exchangeable measuring systems are available for TGA, TGA/DTA, and TGA/DSC. Horizontal as well as vertical balance configurations are available in both lines. The L81-I research series all have sample capacity up to 25 g. Temperature ranges vary from −150 °C to 500 °C and ambient to 1000 °C, and to 2400 °C in several stages. All members of this series are rated for 10^{-5} torr vacuum. Coupling to FTIR and MS instruments is available. For more information, see www.linseis.net.

Mettler Toledo offers the TGA/DSC1 with a horizontal balance and multiple furnace configurations. Three calorimetric sensors are available that allow simultaneous measurement of TGA and DTA or DSC. Three furnace types are available that can accommodate sample volumes of 100 µL (~100 mg) or 900 µL (~1 g) and with upper temperature limits of 1100 °C or 1600 °C. Heating rates are from 0.01–10 °C/min to 250 °C/min. Cooling from 1000 °C to 100 °C can be accomplished in 20–25 min. All of these properties depend on the particular furnace chosen. Optional accessories include a 34-position autosampler with sealed-pan punching capability, programmable gas switching with mass flow control, and capability to be coupled with FTIR or mass spectrometry (MS). The TGA can be converted to a TGA sorption analyzer that allows samples to be analyzed under controlled conditions of relative humidity and temperature. An advanced model-free kinetics (MFK) package is available. For more information, see www.mt.com.

Netzsch offers several thermogravimetric analyzers, including the TG 209 F3 Tarsus® and the TG 209 F1 Iris® TGAs. The F3 is a top-loading TGA with a 2 g capacity that operates between ambient and 1000 °C with heating rates of 0.001–50 °C/min. The furnace will cool from 1000 °C to 100 °C in 20–25 min. Options include a calculated DTA signal (c-DTA®) useful both for calibration and for information on endothermic and exothermic processes, and a 20-pan autosampler. The F1 operates between 10 and 1000 °C with heating rates of 0.001–100 °C/min, and has an automatic evacuation and gas-filling system. A 64-pan autosampler with sealed-pan punching capability is optional. The F1 supports the simultaneous coupling of a mass spectrometer (MS) or Fourier transform infrared (FTIR) spectrometer. Simultaneous

TGA-DTA/DSC measurements are available up to 2400 °C with a series of top-loading configurations. For more information, see www.e-thermal.com.

Perkin-Elmer offers two thermogravimetric analyzers. The Pyris 6 is a top-loading balance design with a capacity of 1.5 g. The temperature range is ambient to 1000 °C with heating rates of 0.1–100 °C/min. The furnace will cool from 1000 °C to 100 °C in less than 10 min. It comes with a 45-pan autosampler. The Pyris 1 is a bottom-loading instrument with two furnace options. The standard furnace has a range from subambient to 1000 °C, can heat at rates of 0.1–200 °C/min, and can cool from 1000 °C to 40 °C in less than 15 min. The high-temperature furnace has a range of 50–1500 °C, can heat at rates of ≤50 °C/min, and can cool from 1500 °C to 100 °C in less than 30 min. A curtain of charged particles can be activated to neutralize static buildup on sample and furnace walls. Optional accessories include a 20-position autosampler with sealed-pan punching capability, and capability to be coupled with FTIR or MS instrumentation. The top-loading STA6000 provides simultaneous TGA-DTA/DSC measurements from 15 °C to 1000 °C. A model-free kinetics (MFK) package is available. For more information, see www.perkinelmer.com.

Rigaku offers the TG/DTA 8120 system. The standard model heats from ambient to 1100 °C, and there is an IR furnace model that heats to 950 °C. An autosampler and a humidity control device are available. For more information, see www.rigaku.com.

Seiko offers the EXSTAR6000 TGA/DTA series with dual-beam, horizontally oriented balance configuration. The TG/DTA6200 has an ambient–1100 °C range, and the TG/DTA6300 has an ambient–1500 °C range. Both systems have a maximum sample capacity of 200 mg, can heat from 0.01 to 100 °C/min, and will cool from 1000 °C to 50 °C in less than 15 min. A 30-pan autosampler is optional. For more information, see www.sii.co.jp or the North American distributor *RT Instruments* at www.rtinstruments.com. *RT Instruments* also sells previously owned, refurbished thermal analysis systems.

Setaram offers thermogravimetric analyzers with ambient–1600 °C capability (Labsys TGA), ambient–2400 °C capability (SETSYS Evolution TGA), and ambient–2100 °C with large-volume capability (96-line TGA). The SETSYS Evolution TGA is a modular TGA with two weighing capacities of 35 or 100 g. Simultaneous TGA-DTA measurements up to 2400 °C and TGA-DSC up to 1600 °C can be made. Coupling to a gas analyzer is also possible. The 96-line TGA can weigh samples up to 100 g, and simultaneous TGA-DTA/DSC measurement is available. A gas humidity generator accessory is available. An advanced thermokinetics software package is available. For more information, see www.setaram.com.

Shimadzu offers the TGA-50/50H with 1 g sample capacity and the TGA-51/51H with 10 g sample capacity. All systems have bottom-loading balances. Standard systems heat from ambient to 1000 °C, while the "H" systems heat from ambient to 1500 °C. All four systems can heat from 0.002 to 100 °C/min and support vacuum to 10^{-3} torr. The TGA-50 series has smaller volume and is recommended for evolved gas analysis. For more information, see www.ssi.shimadzu.com and click on "tga."

TA Instruments offers three ranges of vertically oriented bottom-loading thermogravimetric analyzers. The Q50 and Q500 use resistance-wound heating with a range of ambient to 1000 °C, heat at rates of 0.1–100 °C/min, and will cool from 1000 °C to 50 °C in less than 12 min. Both have a mass range up to 1 gram. The Q500 contains options such as a 16-pan autosampler and Hi-Res TGA that are not available on the Q50. Four different variable heating rate algorithms are provided with Hi-Res TGA, which can be used either alone or in combination. They are dynamic rate, constant reaction rate, constant heating rate, and autostepwise. The Q5000IR uses IR heating with a range of ambient to 1200 °C, heats at rates of 0.1–500 °C/min, and will cool from 1200 °C to 35 °C in less than 10 min. Features include Hi-Res TGA, modulated TGA or MTGA™, a 25-pan autosampler with sealed-pan punching capability, and vacuum capability to 10^{-3} torr. TGA/MS is optional on all models. A Q5000SA model is available that is designed for sorption analysis under controlled temperature (5–85 °C) and humidity (0–98%) conditions. The SDT Q600, a simultaneous TGA/DSC, can heat at ≤100 °C/min to 1000 °C or 25 °C/min to 1500 °C. For more information, see www.tainstruments.com.

Theta Industries offers the Gravitronic II—1600 °C TGA, a top-loading instrument with a maximum sample capacity of 100 g. This system will support a corrosive environment and vacuum of 10^{-6} torr. Also offered is the Gravitronic IV—1100 °C DTA/TGA, a vertically oriented bottom-loading system with a maximum sample capacity of 5 g. For more information, see www.theta-us.com.

APPENDIX

Standards for thermogravimetry issued under the jurisdiction of ASTM International Committee E37 on Thermal Measurements are listed below for reference. For more information, see Website www.astm.org.

E1131, *Standard Test Method for Compositional Analysis by Thermogravimetry*

E1582, *Standard Practice for Calibration of Temperature Scale for Thermogravimetry*

E1641, *Standard Test Method for Decomposition Kinetics by Thermogravimetry*

E1868, *Standard Test Method for Loss-on-Drying by Thermogravimetry*

E1877, *Standard Practice for Calculating Thermal Endurance of Materials from Thermogravimetric Decomposition Data*

E1970, *Standard Practice for Statistical Treatment of Thermoanalytical Data*

E2008, *Standard Test Method for Volatility Rate by Thermogravimetry*

E2040, *Standard Test Method for Mass Scale Calibration of Thermogravimetric Analyzers*

E2402, *Standard Test Method for Mass Loss and Residue Measurement Validation of Thermogravimetric Analyzers*

E2403, *Standard Test Method for Sulfated Ash of Organic Materials by Thermogravimetry*

E2550, *Standard Test Method for Thermal Stability by Thermogravimetry*

E2551, *Standard Test Method for Humidity Calibration (or Conformation) of Humidity Generators for Use with Thermogravimetric Analyzers*

ABBREVIATIONS

Symbols

A	pre-exponential factor, pre-exponential constant, area
α	conversion
d	thickness
$d\alpha/dt$	rate of conversion
dm/dt	rate of mass change
Δ_α	interval of conversion
E	activation energy
E_α	activation energy for conversion α
$f(\alpha)$	reaction model
$g(\alpha)$	integral form of reaction model
I	temperature integral
J	integral with respect to the actual sample temperature
$k(T)$	rate constant at temperature T
m	mass, sample mass
m_i	initial mass
m_f	final mass
m_T	mass at temperature T
n	reaction order, constant
p	water vapor pressure
P	permeability
Φ	function
q	heating rate
t	time
t_α	time to reach conversion α
T	temperature
T_α	temperature associated with conversion α
T_0	given temperature
T_g	glass transition temperature

Acronyms

ACN/BD	acrylonitrile/butadiene
APCI	atmospheric pressure chemical ionization
ASTM	American Society for Testing and Materials
CB	carbon black
CRTGA	controlled rate TGA
DGEBA	diglycidyl ether of bisphenol-A
DMA	dynamic mechanical analysis
DSC	differential scanning calorimetry
DTA	differential thermal analysis
EVA	ethylene-vinyl acetate copolymer
FTIR	Fourier transform infrared spectroscopy
GC	gas chromatography
ICTAC	International Confederation of Thermal Analysis and Calorimetry
IR	infrared
LDPE	low density polyethylene
MFK	model-free kinetics
MS	mass spectroscopy
NO_2	nitrogen dioxide
OIT	oxidative induction time
PC	polycarbonate
PCTFE	polychlorotrifluoroethylene
PE	polyethylene
PEN	poly(ethylene naphthalate)
PET	poly(ethylene terephthalate)
PMMA	poly(methyl methacrylate)
POM	polyoxymethylene
PP	polypropylene
PS	polystyrene
PTFE	polytetrafluoroethylene
PVAc	poly(vinyl acetate)
PVC	poly(vinyl chloride)
PVME	poly(vinyl methyl ether)
PTF	polymer thick film
RTV	room temperature vulcanate
STP	standard temperature and pressure
TA	thermal analysis
TGA	thermogravimetric analysis
TGA/IR	thermogravimetric analysis coupled with FTIR
TGA/MS	thermogravimetric analysis coupled with mass spectroscopy
WVT	water vapor transport

REFERENCES

Arthur, J. K. and Redfern, J. P. (1992), *J. Therm. Anal.* **38**, 1645.

ASTM (2005), E96/E96M-05, *Standard Test Methods for Water Vapor Transmission of Materials*, West Conshohocken, PA (see also www.astm.org).

Bair, H. E. (1973), *Polym. Eng. Sci.* **13**, 435–439.

Bair, H. E. (1981), in *Thermal Characteriazation of Polymeric Materials*, Turi, E. A., ed., Academic Press, New York, pp. 845–909.

Bair, H. E. (1992), *Proc. 21st NATAS Conf. Preprints*, p. 60.

Bair, H. E. (1997), in *Thermal Characterization of Polymeric Materials*, 2nd ed., Vol. 2, Turi, E. A., ed., Academic Press, San Diego, pp. 2263–2420.

Bair, H. E., Hale, A., and Popielarski, S. (2003), US Patent 6,586,258 (July 1).

Bair, H. E. and Warren, P. C. (1981), *J. Macromol. Sci. Phys.* **B20**(3), 381–402.

Bair, H. E., Johnson, G. E., Anderson, E. W., and Matsuoka, S. (1981), *Polym. Eng. Sci.* **21**, 930–935.

Bair, H. E., Blyler, L. L., Jr., Johnson, G. E., and McCall, D. W. (1983), *Proc. 12th North American Thermal Analysis Soc. Conf. Preprints*, pp. 550–552.

Bair, H. E., Pryde, C. A., Johnson, G. E., and Han, B. J. (1991), *Proc. 20th North American Thermal Analysis Soc. Conf.*, pp. 352–358.

Barnes, P. A., Parkes, G. M. B., and Charsley, E. L. (1994), *Anal. Chem.* **66**, 2226.

Blaine, R. L. and Hahn (1998), *J. Therm. Anal.* **54**, 695–704.

Blaine, R. L. and Rose, J. E. (2004), *Proc. 33rd North American Thermal Analysis Soc. Conf.*

Brown, M. E. (2001), *Introduction to Thermal Analysis*, Kluwer, Dodrecht.

Brown, M. E., Maciejewski, M., Vyazovkin, S., Nomen, R., Sempere, J., Burnham, A., Opfermann, J., Strey, R., Anderson, H. L., Kemmler, A., Keuleers, R., Janssens, J., Desseyn, H. O., Li, C.-R., Tang, B. T., Roduit, B., Malek, J., and Mitsuhashi, T. (2000), *Thermochim. Acta.* **355**, 125.

Burnham, A. K. (2000), *Thermochim. Acta.* **355**, 165.

Burns, J. M., Prime, R. B., Barrall, E. M., and M. E. Oxsen (1989), in *Polymers in Information Storage Technology*, Mittal, K. L., ed., Plenum, New York, pp. 237–256.

Cassel, B. (1976), Perkin-Elmer Thermal Analysis Application Study (TAAS) 19.

Cassel, B. (1980), *Polym. News* **6**(3), 108–115.

Chan, M. G., Gilroy, H. M., Heyard, I. P., Johnson, L., and Martin, W. M. (1978), *Soc. Plast. Eng. Tech. Pap.* **24**, 381–383.

Chang, E. P. and Salovey, R. (1974), *J. Polym. Sci., Polym. Chem. Ed.* **12**, 2927.

Charsley, E. and Dunn, J. G. (1981), Plastics and Rubber Processing and Applications, 1, 3–33.

Charsley, E. and Warrington, S. B. (1988), "Industrial Application of Compositions Analysis by Thermogravimetery", in *Compositional Analysis by TG*, ASTM STP 997, M. Earnest ed., ASTM, West Conshohocken, PA, pp. 19–27.

Chiu, J. (1966), *Appl. Polym. Symp.* **2**, 25.

Coats, A. W. and Redfern, J. P. (1964), *Nature* **201**, 68.

Doyle, C. D. (1962), *J. Appl. Polym. Sci.* **6**, 639.

Earnest, C. M. (1988), Compositional Analysis by Thermogravimetry, ASTM STP 997. ASTM International, Philadelphia.

Flynn, H. and Wall, L. A. (1966a), *J. Res. Natl. Bur. Stand.* **70A**, 487.

Flynn, J. H. and Wall, L. A. (1966b), *Polym. Lett.* **4**, 323.

Foster, G. N. (1990), in *Oxidation Inhibition in Organic Materials*, Pospisil, J. and Klemchuk, P. P. eds., CRC Press, Boca Raton, FL, pp. 299–346, Vol. 2.

Friedman, H. (1964), *J. Polym. Sci.* **C6**, 83.

Gallagher, P. K. (1992), *J. Therm. Anal.* **38**, 17–26.

Gallagher, P. K. (1993), *Adv. Anal. Geochem.* **1**, 211–257.

Gallagher, P. K. (1997), in *Thermal Characterization of Polymeric Materials*, 2nd ed., Vol 1, Turi, E. A., ed., Academic Press, San Diego, Chapter 1.

Gallagher, P. K. and Gyorgy, E. M. (1986), *Thermochim. Acta.* **109**, 193–206.

Gallagher, P. K., Blaine, R., Charsley, E. L., Koga, N., Ozoa, R., Sato, H., Sauerbrunn, S., Schultze, D., and Yoshida, H. (2003), Magnetic Task Group, ICTAC Committee for Standardization, *J. Therm. Anal. Calorim.* **72**, 1109–1116.

Gallagher, P. K., Zhong, Z., Charsley, E. L., Mikhail, S. A., Todoki, M., Taniguahi, K., and Blaine, R. L. (1993), *J. Therm. Anal.* **40**, 1423–1430.

Garn, P. D., Menis, O., and Wiedemann, H. G. (1980), in *Thermal Analysis*, Wiedemann, H. G., ed. Birkhäeuser, Basel, Vol. 1, pp. 201–206.

Gill, P. S., Sauerbrunn, S., and Crowe, B. S. (1992), *J. Therm. Anal.* **38**, 255.

Greenberg, A. R. and Kamel, I. (1977), *J. Polym. Sci., Polym. Chem. Ed.* **15**, 2137–2149.

Hale, A. and Bair, H. E. (1997), in *Thermal Characterization of Polymeric Materials*, 2nd ed., Vol. 1, Turi, E. A., ed., Academic Press, San Diego, pp. 745–886.

Holcomb, B. D. and Bair, H. E. (2004), *Proc. 32nd North American Thermal Analysis Soc. NATAS, Conf.* p. 210.

Holcomb, B. D. and Bair, H. E. (2005), in *New Polymeric Materials*, Korugic-Kanasz, L. S., MacKnight, W. J., and Martuscelli, E., eds., ACS Symp. Series **916**, p. 350.

Kau, H.-T. (1988), in *"Compositional Analysis by Thermogrowimetry"* (C. M. Earest, ed.), ASTM STP 997. ASTM International, Philadelphia.

Khorami, J., Chauvette, G., Lemieux, A., Manard, K., and Jolicoeur, C. (1986), *Thermochim. Acta.* **103**, 221–230.

Khorami, J. and Prime, R. B. (1988), *Proc. 17th NATAS Conf.*, pp. 86–92.

Kissinger, H. E. (1957), *Anal. Chem.* **29**, 1702.

Lewis, M. L., Nam, J.-D., Seferis, J. C., and Prime, R. B. (1989), *Soc. Plast. Eng. (Proc. Annu. Tech. Conf.)* **35**, 1092–1095.

Maciejewski, M. (2000), *Thermochim. Acta.* **355**, 145.

Maciejewski, M. and Baiker, A. (2008), in *Handbook of Thermal Analysis and Calorimetry: 2: Recent Advances and Techniques*, Brown, M. E. and Gallagher, P. K., eds., Elsevier Science.

Mauer, J. J. (1981), In *"Thermal Characterization of Polymeric Materials"* (E. A. Turi, ed.), pp. 571–708. Academic Press, New York.

McCall, D. W., Douglass, D. C., Blyler, L. L., Jr., Johnson, G. E., Jelinski, L. W., and Bair, H. E. (1984), *Macromolecules.* **17**, 1644–1649.

McGhie, A. R. (1983), *Anal. Chem.* **55**, 987–991.

McGhie, A. R., Chiu, J., Fair, P. G., and Blaine, R. L. (1983), *Thermochim. Acta* **67**, 241–247.

Miller, T. M., Neenan, T. X., Zayas, R., and Bair, H. E. (1991), *Proc. 20th NATAS Conf.*, pp. 624–632.

Miller, T. M., Neenan, T. X., Zayas, R., and Bair, H. E. (1992), *J. Am. Chem. Soc.* **114**, 1018.

Mullens J. (1998), "Evolved gas analysis," In *Handbook of Thermal Analysis and Calorimetry: Principles and Practice*, Elsevier Science, pp. 509–546.

Naito, K. and Kwei, T. K. (1979), *J. Polym. Sci., Chem. Ed.* **17**, 2935–2945.

Nam, J. and Seferis, J. C. (1991), *J. Polym. Sci., Part B: Polym. Phys.* **29**, 601–608.

Neag, C. M. and Prime, R. B. (1991), *J. Coat. Technol.* **63**(797), 37.

Neenan, T. X., Miller, T. M., Kwock, E. W., and Bair, H. E. (1994), in *Advances in Dendritic Molecules*, Vol. 1, Newkome, G. R., ed., JAI Press, Greenwich, CT., pp. 105–132.

Norem, S. D., O'Neill, M. J., and Gray, A. P. (1970), *Thermochim. Acta.* **1**, 29–34.

Norton, F. J. (1963). *J. Appl. Polym. Sci.* **7**, 1649–1659.

Ozawa T. (1965), *Bull. Chem. Soc. Jpn.* **38**, 1881.

Paulik, F. and Paulik, J. (1986), *Thermochim. Acta.* **100**, 23–59.

Peterson, J. D., Vyazovkin S., and Wight, C. A. (2001), *Macromol. Chem. Phys.* **202**, 775.

Prime, R. B. (1988), *Proc. 17th NATAS Conf.*, pp. 707–713.

Prime, R. B. (1992). *Polym. Eng. Sci.* **32**, 1286–1289.

Prime, R. B. (1997a), in *Thermal Characterization of Polymeric Materials*, 2nd ed., Vol. 2, Turi, E. A., ed., Academic Press, San Diego, pp. 1537–1554.

Prime, R. B. (1997b), in *Thermal Characterization of Polymeric Materials*, 2nd ed., Vol. 2, Turi, E. A., ed., Academic Press, San Diego, pp. 1677–1679.

Prime, R. B., Burns, J. M., Karmin, M. L., Moy, C. H., and Tu, H.-B. (1988a), *J. Coat. Technol.* **60**(761), 55.

Prime, R. B., Whelihan, E. F., and Burns, J. M. (1988b), *Soc. Plast. Eng. (Proc. Ann. Technol Conf.)* **34**, 1268.

Prime, R. B. and Shushan, B. (1989), *Anal. Chem.* **61**, 1195–1201.

Quinn, T. J. (1990), *Temperature*, 2nd ed., Academic Press, San Diego.

Roe, R. J., Bair, H. E., and Gieniewski, C. (1974), *J. Appl. Polym. Sci.* **18**, 843–856.

Salovey, R. and Bair, H. E. (1970), *J. Appl. Polym. Sci.* **14**, 713–721.

Sichina, W. J. (1993), *Am. Lab.* (Fairfield, CT), **25**, 45.

Sircar, A. K. (1997), In *"Thermal Characterization of Polymeric Materials"* 2nd Ed. (E. A. Turi, ed.), pp. 887–1378. Academic Press, San Diego.

Smith, A. L. and Shirazi, H. M. (2005), *Thermochim. Acta.* **432**, 202–211.

Smith, A. L., Mulligan, Sr. R. B., and Shirazi, H. M. (2004), *J. Polym. Sci., Pt. B: Polym. Phys.* **42**, 3893–3906.

Smith, A. L., Ashcraft, J. N., and Hammond, P. T. (2006), *Thermochim. Acta.* **450**, 118–125.

Stokich, T. M., Jr., Burdeaux, D. C., Mohler, C. E., Townsend, P. H., Warmington, S. C., Tou, J. C., Han, B. J., Pryde, C. A., Bair, H. E., and Johnson, G. E. (1991), *Spring Meeting, Mater. Res. Soc., Final Program and Abstracts*, Anaheim, CA, p. 253.

Stromberg, R. R., Straus, S., and Achammer, B. G. (1959), *J. Polym. Sci.* **35**, 355.

Tabaddor, P. L., Aloisio, C. J., Bair, H. E., Plagianis, C. H., and Taylor, C. R. (1997), *J. Therm. Anal. Calorim.* **59**, 559–570.

Tabaddor, P. L., Aloisio, C. J., Bair, H. E., Plagianis, C. H., and Taylor, C. R. (2000), *J. Therm. Anal. Cal.* **59**, 559–570.

Vyazovkin, S. (1996), *Int. J. Chem. Kinet.* **28**, 95.

Vyazovkin, S. and Dollimore, D. (1996), *J. Chem. Inform. Comput. Sci.* **36**, 42.

Vyazovkin, S. and Sbirrazzuoli, N. (2006), *Macromol. Rapid. Commun.* **27**, 1515.

Vyazovkin, S. (1997), *J. Comput. Chem.* **18**, 393.

Vyazovkin, S. (2001), *J. Comput. Chem.* **22**, 178.

Vyazovkin, S. and Dollimore, D. (1996), *J. Chem. Inform. Comput. Sci.* **36**, 42.

Vyazovkin S. and Wight, C. A. (1999), *Thermochim. Acta.* **340**/341, 53.

Wampler, T. P., ed. (1995), *Applied Pyrolysis Handbook*, Marcel Dekker, New York.

Weast, R. C., ed. in chief, (1973), *Handbook of Chemistry and Physics*, 54th ed., CRC Press, Cleveland, OH, Section D, pp. 158–160.

Woolley, W. D. (1971), *Br. Polym. J.* **3**, 186.

Young, H. D. (1992), *University Physics*, 7th ed., Addison-Wesley, Reading, MA, Table 15-5.

CHAPTER 4

THERMOMECHANICAL ANALYSIS (TMA) AND THERMODILATOMETRY (TD)

HARVEY E. BAIR
Bell Laboratories (Retired)/Consultant, Newton, New Jersey

ALI E. AKINAY AND JOSEPH D. MENCZEL
Alcon Research, Ltd., Fort Worth, Texas

R. BRUCE PRIME
IBM (Retired)/Consultant, San Jose, California

MICHAEL JAFFE
New Jersey Institute of Technology, Newark

4.1. INTRODUCTION

Fabricated polymers, plastics, fibers, and films are a group of materials used in our daily lives. Applications include simple food containers, kitchenware coatings, medical devices, materials for controlled drug delivery, blood storage bags, and sutures, as well as plastic devices such as composite parts for automobiles and aerospace devices. This list can be easily expanded to areas of specific polymer applications. In all these areas, determination and understanding of process–structure–property relationships are of primary importance because of their effect on end-use performance. For instance, a multilayer food package can keep food fresh for a longer period of time, because it acts as a barrier against oxygen permeation. A multicomponent–multilayer tire can serve under extreme mechanical and temperature conditions without delaminating.

Unlike metals and ceramics, polymers are viscoelastic—that is, their properties have both viscous and elastic components, which control the mechanical response as a function of time and temperature (see Section 4.2.3 in this

Thermal Analysis of Polymers: Fundamentals and Applications, Edited by Joseph D. Menczel and R. Bruce Prime
Copyright © 2009 by John Wiley & Sons, Inc.

chapter and, for a detailed description of viscoelasticity, see Chapter 5, on dynamic mechanical analysis). Therefore, the accurate characterization of polymeric materials is extremely important for prediction of performance and service lifetime. Among the thermal characterization techniques described in this book, themomechanical analysis (TMA) has its own distinct advantages, as will be shown in this chapter.

Thermomechanical analysis measures changes in sample length or volume as a function of temperature or time under load at atmospheric pressure. The load is usually static or variable. This technique is also referred to as *thermodilatometry* (TD), if sample dimensions are measured under negligible load (Mackenzie 1979; IUPAC 1989). TMA is an especially useful thermal analysis technique, because measurements can be carried out on actual manufactured parts of a variety of shapes, provided they are small enough to fit in the TMA apparatus. TMA experiments are generally carried out under static load with a variety of probes for measurement of sample dimensions in compression, tension, or flexure. The most important TMA measurements include determination of the coefficient of linear thermal expansion (CLTE, α) and the glass transition temperature, T_g. But several other measurements can be made by applying special modes and various attachments. These include stress relaxation, creep, tensile properties of films and fibers, flexural properties, dimensional stability (reversible and irreversible dimensional change), parallel-plate rheometry, and volume dilatometry. The goal of this chapter is to explain the basis of the TMA techniques and their application to polymeric materials and shaped plastic products. A number of problem solving applications are presented in the final section of this chapter.

4.2. PRINCIPLES AND THEORY

4.2.1. Principles

Thermomechanical analysis measures the change in sample length in compression or tension under constant or variable force as a function of temperature or time. For thermal expansion measurements, the applied force must be negligible, just enough to keep the probe in contact with the sample. Some confusion exists about the terminology. *Thermodilatometry* (TD) and *thermomechanical analysis* (TMA) are essentially identical techniques. According to ASTM E473 (see Appendix), the International Union of Pure and Applied Chemistry (IUPAC, 1989) and the International Confederation for Thermal Analysis and Calorimetry (ICTAC, 1980), "thermodilatometry is a technique in which a dimension of a substance under negligible load is measured as a function of temperature while the substance is subjected to a controlled temperature program in a specified atmosphere. Linear TD and volume TD are distinguished on the basis of the dimension measured." On the other hand, TMA is "a technique in which the deformation of a substance under nonoscillatory (static) load is measured as a function of temperature or time while the substance is subjected to a controlled temperature program in a specified atmosphere." Thus, a researcher who uses an instrument that was marketed as a "thermomechanical analyzer" puts a minimum load on the probe to measure the coefficient of thermal expansion essentially carries out a dilatometry measurement. Nevertheless, the vast majority of publications talk about TMA (and not TD) even when the load on the sample is negligible. Although this is incorrect in principle, in publications the word *dilatometry* is often used only for measurements in which the volume change of the sample is measured in some liquid, as in a mercury dilatometer.

The relative change of the length in one direction (x, y, or z) is expressed as

$$\frac{\Delta L}{L_0} = \frac{L_T - L_0}{L_0} \qquad (4.1)$$

where L_T is the length at temperature T and L_0 is the initial length of the sample at the starting temperature, or zero time of the measurement in the case of isothermal experiments. In the expansion mode, one of the most common techniques is to measure the linear isobaric expansivity, also called the *coefficient of linear thermal expansion* (CLTE). CLTE, denoted by the symbol α, is the slope of $\Delta L/L_0$ with respect to temperature at constant pressure P and has the units of K^{-1} (°C^{-1} and ppm/°C are also used). If the expansion of the sample is linear in a certain temperature range of ΔT, then

$$\alpha = \frac{1}{L_0} \left(\frac{\Delta L}{\Delta T} \right)_P \qquad (4.2)$$

For isotropic materials the coefficient of linear thermal expansion is identical along all three axes, $\alpha_x = \alpha_y = \alpha_z$. From these values, the volumetric expansivity, also called the *coefficient of volumetric expansion* (γ), can be calculated as

$$\gamma = \alpha_x + \alpha_y + \alpha_z = 3\alpha \qquad (4.3)$$

A schematic curve of a TMA run is shown in Fig. 4.1. The glass transition temperature T_g, is defined as the point of intersection of the glassy and rubbery (or melt) expansivities. The change in the slope of the expansion curves below and above T_g is related to the expansion of free volume. The actual volume of the molecules (hard-core volume) has similar values below and above T_g, but the free volume increases starting at T_g.

The available free volume is one of the most important physical quantities influencing the glass transition and the thermal expansion. Changes in free volume (v_f), can be measured by determining the volumetric change of the polymer, and this can be done with TMA experiments (Menard 1996; Bird

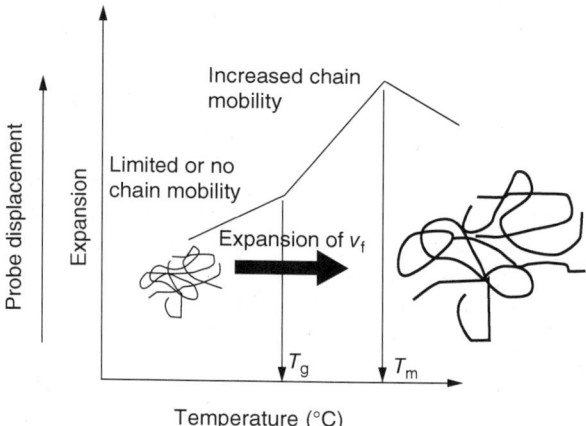

Figure 4.1. A schematic representation of a TMA curve in an expansion-mode experiment.

et al. 1987). Several polymer properties are strongly influenced by their free volume. These include viscoelastic (energy dissipative–energy accumulating) properties (Ferry 1980; Flory 1955, 1970; Aklonis and McKnight 1983; Litt 1976), physical aging (Struik 1978, 1986; Matsuoka 1992), solvent and gas diffusion through polymers (Vrentas et al. 1986; Litt 1986) and impact properties (Litt and Tobolsky 1967; Litt 1976; Brostow and Macip 1989).

In addition to thermal expansivity, Young's modulus (also referred to as *elastic modulus* or *tensile modulus*) can also be measured by TMA (within a limited range of modulus and sample dimensions). Young's modulus is defined as

$$E = \frac{\sigma}{\varepsilon} = \frac{L(\Delta f/\Delta L)_T}{A} \qquad (4.4)$$

where f is force and A is the cross-sectional area of the sample, so that $\Delta f/A$ represents the change in stress and $\Delta L/L$, the change in strain.

4.2.2. Free-Volume Theory

Free volume (v_f) plays an important role in determining the properties of polymers as mentioned in the previous section. *Free volume* can be defined as

$$v_f = v - v^* \qquad (4.5)$$

where v is the total specific volume or occupied volume (in units of cm^3/g), and v^* is the characteristic volume (also known as *hard-core* or *incompressible* *volume*) corresponding to extremely high pressure or zero thermodynamic temperature. Knowledge of free volume is important, because it is the property that eventually determines what kinds of molecular motion can exist in the material.

Some key processes that occur in a polymeric material under the influence of mechanical forces and/or temperature are as follows:

- Transmission of energy to neighboring chains depending on the presence of entanglements, crosslinks, and crystal phases
- Conformational rearrangements within the chains
- Elastic energy storage resulting from bond stretching and bond angle changes.

Sufficient free volume is essential for these processes. An increase in v_f results in an increase in the number of conformational changes in the chains enhancing the chain relaxation capability (CRC) (Brostow and Macip 1989; Doolittle 1951; Akinay et al. 2001). Brostow and Macip (1989) defined CRC as "the amount of external energy dissipated by a relaxation process in a unit of time and unit mass of the polymer." The excess energy that cannot be dissipated by relaxational processes can go into destructive processes (irreversible changes in response to an applied force and/or temperature such as bond fracture and plastic deformation). The theory of the chain relaxation capability is discussed in detail in several references (Akinay et al. 2002; Brostow and Kubat 1996; Brostow 2000; Brostow et al. 2000; Brostow et al. 1999a, b).

The Doolittle viscosity equation (Doolittle 1951) was the pioneering equation for most free-volume theories

$$\eta = A \exp \frac{Bv}{v_f} \qquad (4.6)$$

where η is the viscosity, v and v_f are the occupied and the free volumes, respectively, and A and B are constants. This equation provides the theoretical background for the Williams–Landel–Ferry (WLF) equation (Williams et al. 1955; Ferry 1980), which has the following form:

$$\log a_T = \frac{-C_1(T - T_g)}{C_2 + (T - T_g)} \qquad (4.7)$$

where a_T is the temperature shift factor (also related to the change in viscosity η or the relaxation time). These authors found that C_1 and C_2 were similar for many amorphous polymers with $C_1 = 17.4$ and $C_2 = 51.6$ in the temperature range between T_g and $T_g + 100\,°C$. This equation is referred to as the "universal" WLF equation when C_1 and C_2 take these values. For this case, the constant C_1 ($\approx 1/2.3 v_{ff}$, where v_{ff} is the fractional free volume at $T_g \approx v_f/v$), indicates that v_{ff} at $T_g \approx 0.025$; that is, the fractional free volume at the glass transition temperature is about 2.5%. The constant C_2 ($\approx v_{ff}/\gamma_G$, where γ_G is the expansion coefficient in the glassy state) suggests that the volumetric expansion coefficient of the free volume (v_f) would be about $0.042 \times 10^{-6}/°C$. However, the parameters C_1 and C_2 should not be treated as strictly universal, because polymer–polymer variations were reported, especially those with rigid backbones or those containing groups with hindered rotation in the mainchain (Ferry 1980). Therefore, the parameters C_1 and C_2 should be considered as empirically adjustable parameters, the values of which are determined by a fit to experimental values of a_T. In addition, as discussed by Ferry (1980), the WLF equation, which assumes a linear relation between the mechanical property measured and the temperature, is expected to apply in the temperature range between T_g and $T_g + 100°C$. Several authors discussed the possible expansion of this equation to a broader temperature range (Akinay et al. 2001, 2002; Brostow 1985; Brostow et al. 1991, 1994; Boiko et al. 1995; Akinay and Brostow 2001). An application of a_T is given in this chapter in section 4.6.9.

Figure 4.2a schematically shows a practical means of estimating v_f assuming that all of the volume changes come from the expansion of free volume. Figure 4.2b shows an actual TMA trace for a cured epoxy polymer (Turi et al. 1988). In Fig. 4.2a the glassy and rubbery expansivities are extrapolated to absolute zero. The extrapolated difference in volume between glassy (v_G) and rubbery (v_R) expansivities is defined as the free volume at 0K. Simha and Boyer (1962); Simha and Wilson (1973) suggested that T_g is observed when the difference ($v_G - v_R$) reaches ~$0.113 cm^3/g$.

4.2.3. Viscoelasticity

Polymers are viscoelastic materials, whose mechanical behavior exhibits characteristics of both solids and liquids. See Chapter 5, on dynamic mechanical analysis, for a detailed description of viscoelasticity. The timescale and temperature of observation are critical to the relative degree of solidlike and liquidlike behavior exhibited by viscoelastic materials. Generally, they will behave more solidlike at lower temperatures or over shorter timescales, but more liquidlike at higher temperatures or longer timescales. As will be described later in this chapter, TMA can be used to measure the viscoelastic phenomenon of creep and, in some cases, stress relaxation. But it should be pointed out that whenever a load is imposed on a polymeric material, the material will respond in a viscoelastic fashion; specifically, its penetration or elongation will exhibit time dependence. When the load is small, the temperature is low and/or the time is short, these effects (such as penetration of the probe into the polymer and creep) will be negligible. But when the load is large, the temperature is high, and/or the time is long, these effects may be significant and may in fact be necessary to observe the desired effect.

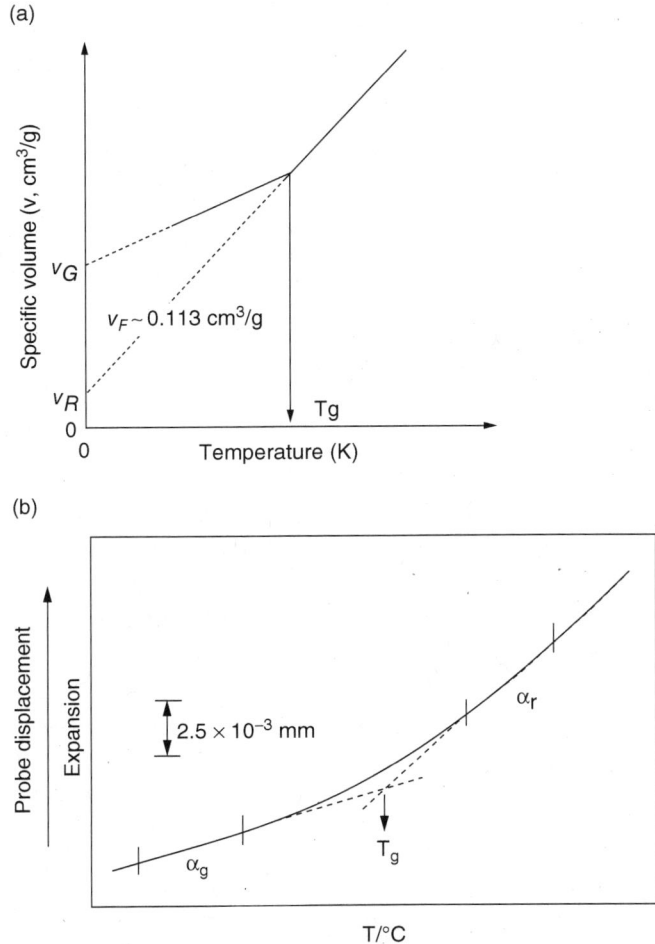

Figure 4.2. (a) The temperature dependence of free volume in an amorphous polymer; (b) a typical TMA experiment of a cured epoxy thermoset (probe displacement is proportional to expansion) for determination of T_g; symbols α_g and α_r represent glassy and rubbery CLTEs, respectively [Turi et al. (1988); reprinted with permission of VCH Publishers].

4.2.4. Physical Quantities Measured by TMA

Some important physical quantities that can be measured by TMA are listed below:

Creep—for a viscoelastic material when the stress is held constant, the strain will increase with time. This phenomenon is called *creep*. It is a relatively slow, progressive deformation of a polymer under constant stress. Creep can be characterized by creep compliance ($J(t) = \varepsilon(t)/\sigma$) (Kaye et al. 1998). See Sections 4.6.7 and 4.6.9.

Stress relaxation—when the strain is held constant, the stress will decrease with time. This phenomenon is called *stress relaxation*. Stress relaxation is characterized by the stress relaxation modulus ($E(t) = \sigma(t)/\varepsilon$). See Section 4.6.7.

Coefficient of linear thermal expansion (CLTE)—this is the slope of the relative change in the length ($\Delta L/L_0$) with respect to temperature [CLTE = α = $(1/L_0)(\Delta L/\Delta T)$ (K^{-1} or $10^{-6}/°C$ or 10^{-6} ppm/°C)]. See Eq. (4.2).

Glass transition temperature (T_g)—the point of intersection of the glassy and rubbery (or melt) expansion curves (°C; see Fig. 4.1).

Young's modulus—the slope of the stress–strain curve in the initial elastic region [$E = \sigma/\varepsilon$ (Pa)]; also referred to as the *elastic modulus* or *tensile modulus*.

Shear modulus—the ratio of shear stress to the shear strain (Pa).

Flexural stress—the ability of a material to resist deformation under load in a three-point bending test (Pa).

Softening point and heat deflection temperature—the temperature at which the probe penetrates to a certain depth.

Gel time—the time at fixed temperature for a thermosetting material to reach the gel point, where the material transforms from a viscoelastic liquid to a crosslinked rubber.

Degree of swelling—increase in thickness of the sample with absorption of a solvent as measured by the expansion probe. See Section 4.6.5 and Eq. (4.8).

Hard-core volume—also known as *incompressible volume* corresponding to extremely high pressure or zero thermodynamic temperature conditions. See Section 4.2.2 and Eq. (4.5).

4.3. INSTRUMENTAL

4.3.1. Instrumental Components

A wide variety of commercial instrumentation is available. Table 4.1 lists some of the current TMA instruments and their manufacturers, including contact information.

Figure 4.3 is a simple diagram of a typical vertical design TMA instrument. Almost all TMA instruments have similar components: a sample platform surrounded by a furnace (for controlled heating and cooling), a sample probe (one end of which touches the sample while the other end is connected to the

TABLE 4.1. Some Current TMA Instruments

Manufacturer	Model	Contact Information
Bähr-Thermoanalyse GmbH	TMA801/813	www.baehr-thermo.de
Linseis	PT10	www.linseis.net
	PT1000EM	
	PT1000	
	PT1600	
Metler Toledo	TMA/SDTA840	www.mt.com
Netzsch	TMA202	www.e-thermal.com
Perkin-Elmer	DMA 8000	www.perkinelmer.com
Rigaku	TMA8310	www.rigaku.com
Shimadzu	TMA-50/50H	www.ssi.shimadzu.com
Setaram	SETSYS TMA	www.setaram.com
	96 line TMA	
Seico/RT Instruments	TMA/SS6100	www.sii.co.jp
	TMA/SS6200	www.rtinstruments.com
	TMA/SS6300	
TA Instruments	TMAQ400	www.tainst.com

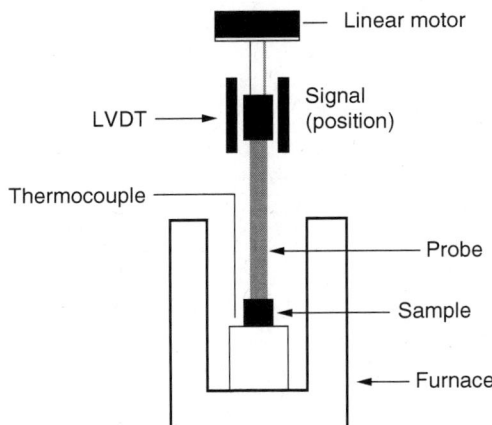

Figure 4.3. Schematic diagram of a typical vertical design TMA instrument (see www.anasys.co.uk).

position transformer), and a sensitive position transformer used to measure probe displacement [in this case a linear voltage differential transformer (LVDT)]. Optical and mechanical transducers are also used as position transformers. A schematic diagram of an LVDT is shown in Fig. 4.4. The output signal (voltage) of the LVDT is calibrated against a height displacement standard (see Section 4.4, on calibration). The direction of the motion of the

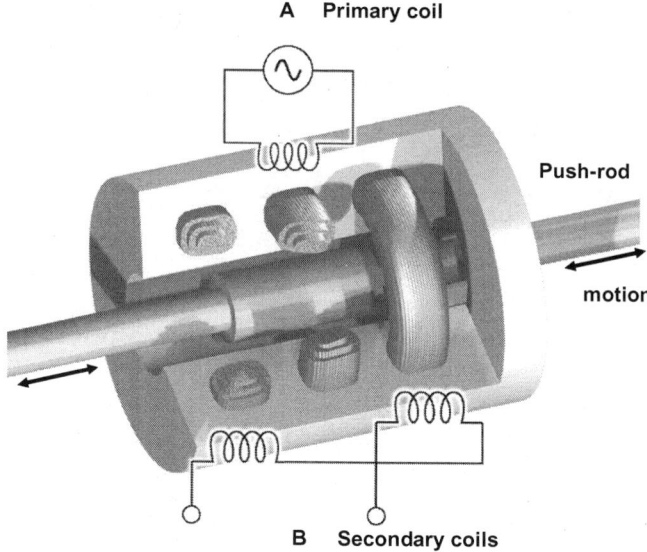

Figure 4.4. Schematic diagram of an LVDT. Current is driven through the primary coil at A, causing an induction current to be generated through the secondary coils at B (see http://en.wikipedia.org/wiki/Linear_variable_differential_transformer).

push-rod determines the sign of the output voltage, so that expansion and compression measurements are distinguishable from each other. Dimensional changes less than 0.01 μm can be detected by an LVDT. Because the output voltage depends on temperature, the LVDT is thermally isolated in the TMA instrument.

The sample temperature is measured with a thermocouple, located in close proximity to the sample. The sample platform and the probe are made from a material such as quartz, which has a low, reproducible, and accurately known coefficient of thermal expansion and also has low thermal conductivity to isolate the LVDT from the temperature changes in the furnace.

A purge gas is used during the TMA measurements. Its purpose is to ensure continuous laminar gas flow, in order to prevent formation of air turbulence as the temperature is being increased, to prevent deposition of degradation products inside the various parts of the instrument, to increase heat transfer to the sample, and to prevent oxidation in high temperature measurements. Helium is preferred for this purpose because of its high thermal conductivity. [The thermal conductivity of He and N_2 are 0.138 and 0.023 W/(m·K) at 20 °C, respectively. See Table 3.2 (in Chapter 3), which lists thermal conductivities of several common purge gases.] The load on the sample probe may be applied by static weights or more commonly by a force motor. Computerized force motors can precisely apply force to the sample in the range of 0.001–10 N (equivalent to weights of 0.1 g–1 kg). See Section 4.4 for force calibration.

Table 4.2 lists some of the current TMA instruments and their specifications such as temperature and force ranges. Common to all instrument offerings are four probe types (expansion, penetration, three-point bending, and tension) and purge systems that accommodate common gases like nitrogen, helium, and air. As noted, some TMA systems can operate in vacuum or reactive gas environments. Data were collected from current instrument catalogs.

4.3.2. Modulated Temperature TMA

In a modulated temperature TMA (MTTMA) experiment, the sample is exposed to a sinusoidal temperature modulation overlaid on a linear underlying heating, cooling, or isothermal profile, similar to modulated temperature DSC (see Section 2.13). Of course, the dependent physical property measured is length for this technique, and the temperature modulation with its resulting modulated length can be noted in Fig. 4.5.

As with MTDSC, MTTMA determines the reversing signal by the use of a Fourier transform of the amplitude of the modulated length divided by the amplitude of the modulated temperature multiplied by the calibration constant. The total length is the average of the modulated length at a particular temperature, or, if modulated at one temperature, quasiisothermally, then time, which is also generated by a Fourier transform. The nonreversing signal is calculated by subtracting the reversing signal from the total length signal. This allows for effects such as the coefficient of thermal expansion and glass transition to separate into the reversing signal while kinetic phenomena such as stress relaxation and heat shrinking would be observed in the nonreversing signal.

A modulated temperature experiment provides a valuable tool for separating the thermally reversing effect of thermal expansion from irreversible dimensional changes. An example of irreversible dimensional change is shrinkage in oriented fibers and films that lose their orientation on heating (see Fig. 4.6). Creep is another example of irreversible dimensional change. The relaxation of residual stresses induced by processing of thermoplastics and thermosets are often observed on the first heating through the glass transition, obscuring T_g and necessitating its measurement from uncomplicated thermal expansion on a second heating. In such cases MTTMA allows separation of these processing effects from the underlying thermal expansion without the necessity of a heat–cool–heat protocol that can add unwanted effects, such as additional crosslinking in a partially cured thermoset. MTTMA measurements are less sensitive to the initial load in compression, for example, where creep occurs. The results of detailed MTTMA measurements can be found in the works of Price (1998a,b,c, and 2000). In a typical modulated temperature experiment the modulation period ranges from 60 to 600 s with an amplitude of 1–5 °C and underlying heating rates of 0.2–2 °C/min. Separation of the signal into reversing and nonreversing components provides the thermal expansion coefficient and irreversible dimensional changes in a single TMA experiment.

TABLE 4.2. Some Available TMA Instruments and Their Specifications

Manufacturer and Model	Temperature Range	Maximum Force /Force Resolution (N)	Displacement Sensitivity	Maximum Sample Size (L)	Options and Attachments
Bähr: TMA801, TMA813	−160–2400 °C	2.5, 0.005	10 nm	25 mm	Controlled gas flow, vacuum
Linseis: PT10, PT1000, PT1600	−150–1600 °C	5.7, 0.001	12.5 nm	30 mm	Volumetric dilatometer, reactive gas capability
Metler-Toledo: SDTA840	RT–1100 °C	10, 0.0013	10 nm	20 mm	Controlled gas flow, reactive gas capability
Netzsch: TMA202, TMA402	−150–500 °C, RT–1000 °C	2, 0.001	10 nm	50 mm	Gas flow meter, temperature calibration kit, vacuum, reactive gas capability
Perkin-Elmer: DMA 8000	−190–600 °C	10, 0.002	10 nm	52 mm	Fluid bath (−196–150 °C), humidity generator, reactive gas capability, vertical and horizontal operation, additional probes: compression, shear, single/dual cantilever, parallel plate, powder pocket
Seico EXSTAR6000	−150–600 °C, RT–1500 °C	5.8		25 mm	Gas flow meter
Shimadzu TMA60	−150–600 °C, RT–1500 °C	5.0	±3%		
TA Instruments: TMA Q400	−150–1000 °C	2, 0.001	15 nm	26 mm	Modulated TMA (see next section), controlled gas flow, reactive gas capability, additional probes: compression, shear, parallel plate, single/dual cantilever

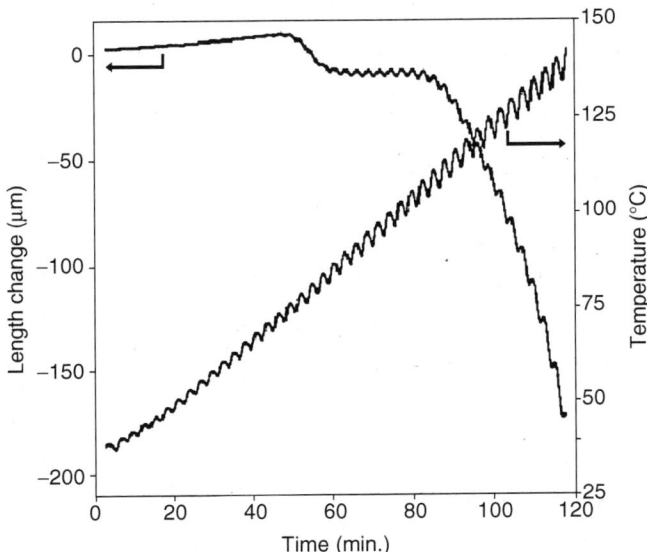

Figure 4.5. Raw MTTMA signals; both the modulated temperature and its resulting effect on length can be noted [from Price (1998c); reprinted with permission from Elsevier Ltd.].

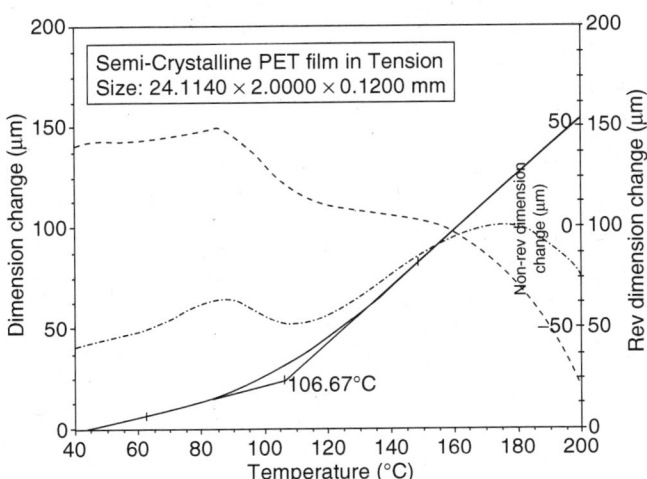

Figure 4.6. Modulated Temperature TMA run on an oriented PET film sample in the tensile mode. Total dimensional change (dot-dashed line) is equivalent to a standard TMA curve. Thermal expansion below and above T_g is observed in the reversing signal (solid line). The nonreversing signal (dashed line) indicates that shrinkage takes place in two stages, at the glass transition and at temperatures above 150 °C. Modulation conditions: ±5 °C/300 s, underlying heating rate 1 °C/min. (Courtesy of TA Instruments.)

TA Instruments provides the only commercial modulated temperature TMA (see TA Instruments *Application Notes TA311* at www.tainstruments.com).

4.3.3. Sample Probes

In general, the output of a TMA measurement is a plot of sample dimension or dimensional change versus temperature or time recorded in expansion, compression, tension, or flexure. Expansion, compression, and tension experiments are the most common measurements. Table 4.3 summarizes some key TMA applications in different measurement modes using different sample probes. Details of these measurements are discussed further in Sections 4.6 and 4.7 of this chapter. In addition to these basic types of measurement shear and torsional modulus of films, fibers, laminates, and adhesives can be measured using specially designed probes.

4.4. CALIBRATION

The TMA instrument must be calibrated in height (or position), force, eigendeformation (calibration of the probe when large forces are applied), temperature, and expansion (cell constant). As in any thermal technique, the purpose of these calibrations is to minimize the difference between the measured values of these parameters and their true values. These calibrations are instrument-specific, and they are described in detail in the respective manufactures' instruction manuals.

During the position calibration procedure, the displacement transducer used to measure position is calibrated against a height displacement standard (usually made from a hard material such as sapphire) with a precisely known height. Current TMA instruments can measure probe position changes as small as 10 nm (see discussion of LVDT in Section 4.3.1).

Before performing a force calibration, the weight of the probe itself must be tared. Then, the force motor used to apply the static force on the sample is calibrated using a weight standard (such as a 50 or 100 g gram-force calibration standard weight). Force loads as small as 1 mN (0.1 g-force) and as high as 10 N (1 kg g-force) can be applied with a resolution of 0.001 N.

Eigen-deformation calibration should be performed after height and force calibrations. Eigen-calibration is used to calibrate the very small movement of the probe when large forces are applied (e.g., forces exceeding 500 mN or 50 g-force). Following this calibration, when a large force is applied to the probe touching the empty sample platform, zero displacement will be measured by the instrument.

There are a number of approaches to temperature calibration. The accuracy of the temperature values in TMA, as in other thermal analysis techniques, is especially important. The next section describes temperature calibration of TMA instruments.

TABLE 4.3. TMA Probes and Key Applications

Mode	Probe	Applications
Compression	Expansion	Expansion probes are used primarily to measure CLTE and T_g under negligible load, 2.5–3 mm in diameter; the same probe can be rounded to form a hemispherical probe; some suppliers also provide a macroprobe with ~6 mm diameter
	Volumetric expansion	Irregular-shaped sample expansion behavior can be directly measured using a volumetric dilatometer by embedding the sample (a few grams or more) in aluminum oxide powder or placed it into a buoyancy fluid in a volumetric dilatometer made of fused quartz or sapphire; an expansion probe is used to measure the difference in the expansion of the dilatometer with and without sample as a function of temperature
	Penetration	Penetration probes apply the load to a small area to measure softening and melting points, ~1 mm in diameter (they can be applied to coatings and thin or thick films); this probe does not measure T_g, but the temperature at which the probe penetrates the sample to a certain depth under a specific load, referred to as the *softening temperature*; for amorphous polymers, softening is a consequence of the glass transition, so in this case the softening point is close to T_g; for semicrystalline polymers, the softening point is between T_g and the melting point
Tension	Tension	Tension probes measure tensile properties such as stress and strain of thin films and fibers; in addition, creep and stress relaxation measurements can be carried out in the tensile mode; CLTE and hygroscopic expansion can also be measured in this mode (Prime et al. 1974).
Flexure	Flexure	Flexural probes measure deflection properties in three-point bending, such as flexural strength and modulus of stiff materials and fiber-reinforced composites
Shear		Two samples are sandwiched between two fixed plates and both sides of the drive shaft; the shear mode is especially useful for lower-modulus materials such as elastomers; shear modulus can be measured

4.4.1. Temperature Calibration

The purpose of temperature calibration is to match the thermocouple readings to the true temperature. In a TMA instrument, a single thermocouple is used to control and measure the temperature of the sample in the furnace. The conditions at which the temperature calibration is carried out [i.e., the measurement mode (compression, tension, etc.), heating rate, flow rate of the purge gas, the purge gas itself, etc.] should be identical to those of the measurements on actual samples. Also, the geometry of the calibration standard should be as similar as possible to the geometry of the samples. This means that for optimum calibration, film standards must be used for measurements on films, pellets for pellets, and fibers for fibers (Lotti and Canevarolo 1998; Mano and Cahon 2004).

In heating experiments, the most popular temperature calibration methods are based on the melting of high-purity metal standards, such as indium (In, T_m=156.6°C), tin (Sn, T_m = 231.9°C), lead (Pb, T_m = 327.5°C), and zinc (Zn, T_m = 419.7°C) (Earnest and Seyler 1992). For fiber and film measurements, indium and other high-purity metal films and fibers are commercially available (e.g., from Alfa Aesar, http://www.alfa.com). Metal fibers tend to break easily even in the solid state, so it is advised to use thin metal film strips instead. In addition to metal standards, distilled water (T_m = 0°C) and organic substances (in liquid and powder forms) with sharp melting points (e.g., toluene, T_m = −95°C; acetic anhydride, T_m = −73°C; benzophenone, T_m = 48.1°C; phenoxybenzene, T_m = 26.9°C, and naphthalene, T_m = 80.2°C) can be used as temperature calibration standards (Lotti and Canevarolo 1998; Charsley et al. 2006). Sigma-Aldrich (http://www.sigmaaldrich.com) is a source for these materials. If organic standards are used, care should be taken to control their moisture content. The temperature scale of the instrument should be linear. To check the linearity, it is recommended to use at least three calibration standards with transition temperatures that cover the range of interest.

ASTM E1363, test method for *Temperature Calibration of Themomechanical Analyzers*, provides a standard method for calibrating TMA instruments (Seyler and Earnest 1991 and 1992). In the recommended ASTM method, three standard materials (phenoxybenzene, Pb, and In) are used.

Special calibration standards are often used for fiber measurements. In such cases hollow fibers or fibers absorbing large amounts of liquids are employed, and the calibration should be carried out in the tension mode. Nonwoven polymer matrices (fabrics) can be used for calibration in compression measurements. These matrices must be capable of absorbing large amounts of a liquid standard. Nonwoven fabrics are flat, porous sheets made of fibers or plastic films as sheet or web structures. The elements of the fabrics are bonded together by entangling fibers or perforating films mechanically, thermally, or chemically (disposable diapers, filters, and dishwashing cloths are some of the products made of nonwoven fabrics). In the actual temperature calibration, the liquid or powder standards should be absorbed into the nonwoven polymer matrix and used in the tensile or compression mode to measure the melting point of the absorbed standard. The advantage of these calibration methods is the similar geometry of the standard and the sample (Lotti and Canevarolo 1998; Mano and Cahon 2004).

4.4.2. Position Calibration

The thermal expansion value displayed by the instrument represents the total thermal expansion of the system (sample + the sample holder + the probe). However, the operator is interested in thermal expansion of the sample only, so the thermal expansion curve of the sample holder and probe should be subtracted from the sample run. In most cases, the sample holder and the probe are made of fused quartz, with CLTE = 0.6 ppm/°C. This correction should be applied to materials with CLTE below 30 ppm/°C but can be ignored for materials with higher expansivities.

The expansion response of the TMA instrument can be checked with a standard reference material of defined thermal expansion. The *Polymer Handbook* (Brandrup 1999) and the *Handbook of Chemistry and Physics* (Lide 1998) provide tables of CLTEs versus temperature for numerous materials. Prime (1997) also lists CLTE values for three calibration standards (lead, aluminum, and copper) in the temperature range from −100°C to 180°C. ASTM E831 describes the standard test method for measuring CLTE of solid materials by TMA.

4.5. HOW TO PERFORM A TMA EXPERIMENT

4.5.1. Preliminary Preparations

Before starting a TMA experiment, the instrument should be warmed up as recommended by the manufacturer. If the instrument is used frequently, it should be kept on continuously. The calibration and the performance of the instrument should be regularly checked. Standard materials such as borosilicate glass (NIST standard reference material 731L1) and copper (NIST 736L1) can be used to check the performance in expansion measurements (available from NIST, www.nist.gov). The probe weight should be tared before starting a calibration procedure or an actual measurement. Cooling of the furnace is an important feature. It can be accomplished with a mechanical cooling accessory, liquid nitrogen, or a dry ice/organic solvent cooling medium. Liquid nitrogen is capable of cooling the furnace to temperatures as low as −180°C. Furnace temperatures down to −90°C can be achieved with a mechanical cooling accessory. If a mechanical cooling accessory is used, it should be turned on at least 2 h prior to calibration or an actual measurement. The advantage of such mechanical cooling devices is their continuous operation. Cooling media have cooling capabilities similar to those of refrigerating mechanical units [liquid nitrogen/toluene (−95 °C) and dry ice/isopropanol (−78 °C)], but the cooling medium needs to be refreshed periodically.

4.5.2. Atmosphere

The TMA sample cell should be continuously purged by an inert dry gas to increase heat transfer to the sample, and to prevent oxidation (Perkin-Elmer PETech-122). Helium (vs. nitrogen or argon) is recommended for this purpose because of its considerably higher thermal conductivity (see Table 3.2 in Chapter 3, for thermal conductivities of common purge gases). However, for economical and practical reasons (the cost of nitrogen is considerably less than that of helium and may be readily available in the laboratory as a house gas), nitrogen may be the preferred purge gas. Because of the lower thermal conductivity of nitrogen compared to helium, a ramp rate reduction is recommended in TMA runs when nitrogen is used as a purge gas (see Section 4.5.4 on heating and cooling rates). The flow rate of the purge gas is usually 20–50 mL/min, and should be set as suggested by the instrument manufacturer.

4.5.3. Sample Requirements

One advantage of TMA is that measurements can be carried out on actual manufactured parts with a variety of shapes, or on small pieces chipped off of a part. When using flat samples in compression and flexure, sample thicknesses between 0.5 and 2.5 mm are recommended in order to limit temperature gradients in the sample to ~1 °C (Prime 1997). The heating rate should not exceed 5 °C/min, and for thicker samples as low as 1 °C/min is recommended. Ideally, the minimum sample thickness should be around 0.5 mm, but if a thinner sample is used in the compression mode, the empty cell baseline should be subtracted from the actual run (Sircar and Chartoff 1994).

In some cases, there may be undesired physical interactions (such as tackiness) between the sample and the sample substrate. This so-called substrate effect may adversely affect CLTE measurements, especially when soft elastic materials (such as silicone gels) are measured as demonstrated in Fig. 4.7. The tacky surface of the approximately 1-mm-thick silicone gel film adheres to the flat quartz bottom of the TMA probe and floor of the sample holder. During heating this adhesion restricts the sample's lateral expansion and causes the film's vertical increase in height to be anomalously high. Thus, a single layer of the film has an unexpectedly high CLTE value of about 520 ppm/°C. However, if silicone films are stacked together, when a height of ~6 mm (six layers) or greater is reached, a consistent CLTE value of ~330 ppm/°C is obtained. Another example of this phenomenon, the effects of adhesion on expansion of dual coated optical fibers, is reviewed in Section 4.7.7 of this chapter.

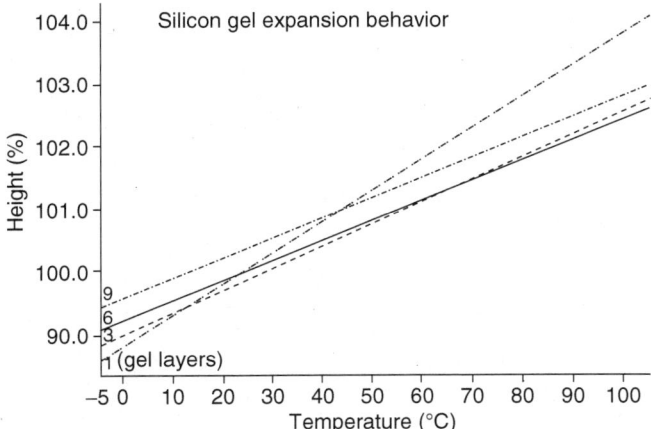

Figure 4.7. Linear expansion of one (1-mm-thick), three (3-mm-thick), six (6-mm-thick), and nine (9-mm-thick) layers of silicone gel films heated at 5 °C/min from −5 °C to 105 °C in the expansion mode under a probe load of 141 Pa with helium as a purge (Bair, unpublished).

In the tension mode, film thickness or fiber diameter from 20 to 200 μm is preferred. The sample width or diameter should be chosen consistent with probe dimensions (see Section 4.5.6 for the required load levels).

Irregularly shaped samples can also be studied after flattening them. Flattening can be accomplished by heating the sample under a high load using a compression probe. However, it is important to remember that flattening by heating affects the thermal history of the sample (see Section 4.5.5 on sample preparation and conditioning). If flattening is not possible just by heating (e.g., in crosslinked polymers), the expansion behavior of an irregularly shaped sample can be studied using a volumetric dilatometer probe. For this, usually several pellets can be embedded in aluminum oxide powder or placed into a buoyancy fluid in the volumetric dilatometer (see Table 4.3). The difference in the expansion of the dilatometer with and without the sample provides the volumetric changes of the material as a function of temperature. Irregularly shaped samples can also be studied in compression without flattening, but only a softening temperature is obtained.

4.5.4. Heating and Cooling Rates

In TMA experiments, scanning rates should be sufficiently low to prevent excessive thermal lag in the sample. Heating rates of 1–5 °C/min are recommended depending on the desired sensitivity, sample size, and purge gas. In general, when larger sample size and/or lower thermal conductivity purge gas such as nitrogen are used, a ramp rate reduction is recommended. The flow rate of the purge gas is usually 20–50 mL/min.

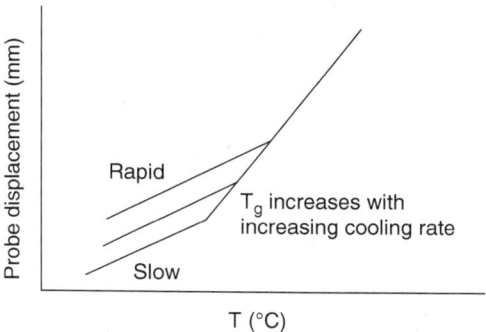

Figure 4.8. The cooling rate dependence of the volume expansion of an amorphous polymer at various cooling rates [from Gedde (2001); reprinted with permission of Cambridge University Press].

The glass transition temperature is often measured on a second run following relaxation of residual stresses (see Section 4.5.5). This requires heating the sample to T_g or somewhat higher and then cooling to a temperature well below T_g. Depending on the experimental conditions and the purpose of the experiment, the sample can be heated to the molten state before cooling to measure T_g. It should not be forgotten that relaxations, shrinkage, and additional cure can take place above T_g. Therefore, various heating and cooling rates can be selected depending on the operator's goal. Once the experimental conditions (i.e., heating up to a certain temperature above T_g, heating and cooling rates) are set, they have to be applied to all the samples. Otherwise, as displayed in Fig. 4.8, different cooling rates will lead to inconsistent T_g measurements. Figure 4.8 shows the effect of different cooling rates on the glass transition temperature: when a material is heated above its T_g and then cooled to below T_g, the glass transition temperature will shift to lower values at slower cooling rates (Ngai 2003; Gedde 2001). It should be noted that in Fig. 4.8 the subsequent heating rate should be the same as the cooling rate.

4.5.5. Sample Preparation and Conditioning

In expansion- or compression-mode TMA measurements, the sample surface should be flat to ensure good contact with the probe. If the sample is punched from a sheet, the burred edge, if it exists, must be removed. Alternatively, a probe with a diameter smaller than that of the sample may be used so that it fits within the edges. The CLTE for irregularly shaped polymers can be directly measured using a volumetric probe assembly (see Table 4.3), or these materials can be flattened using a hot or cold press, although this may introduce stresses that will need to be relieved by heating to T_g.

The measured properties in thermal analysis of polymers (e.g., T_g, T_m, crystallization temperature, CLTE) are often a function of the thermal and mechanical history, sample preparation, and storage conditions, as well as the heating or cooling rates, since polymers in the solid state are far from equilibrium. To measure inherent material properties, the thermal history of the samples needs to be erased first. This procedure is often called "conditioning" the sample. Conditioning can be done in the thermal analysis instrument by heating the sample to above its glass transition temperature, as described in Section 4.5.4. For semicrystalline polymers, heating to above the melting point is recommended for introducing reproducible thermal history. However, for some semicrystalline polymers, such as polypropylene, owing to its helical structure, it is not easy to erase the thermal history even if heated above the melting point. In general amorphous thermoplastic and thermoset polymers should be heated to or only slightly above their T_g. After conditioning, the TMA measurement should start well below T_g.

In addition to thermal conditioning, attention should also be paid to the moisture content of the sample. Varying moisture content will lead to inconsistent results, because water often plasticizes the polymers and reduces T_g. Furthermore, evaporation of water from the sample during measurement can lead to erroneous and irreproducible results. Therefore, prior to the measurement, the samples can be dried in a vacuum oven at an appropriate temperature for a sufficient period of time. They can also be dried at elevated temperature (but below T_m for crystalline polymers) at ambient pressure, with care taken to avoid thermal degradation.

4.5.6. Applying the Initial Load to the Sample

In TMA experiments, the probe in contact with the sample measures the response of the sample to the applied load (as in tension and penetration experiments) and/or temperature (as in expansion experiments). Therefore, some initial load must be applied to the sample. For expansion measurements the initial load should be large enough for the probe to contact the sample and to keep them in contact, but not so large that the material's properties are altered by the load. Figure 4.9 shows the dimensional changes (expansion) in the glass transition temperature region of an amorphous polymer (polystyrene in this example) when the initial applied load is too high. When the viscosity of the sample in the rubbery plateau state above T_g is reduced to a certain value, the probe will indent into the sample (curve A in Fig. 4.9). This undesired indentation makes the T_g measurement ambiguous. Curve B in Fig. 4.9 was recorded under a considerably smaller force, so T_g could be measured as the intersection of the glassy and rubbery curves. Therefore, it is recommended to apply a small force (3–10 mN) to ensure that the probe is in contact with the sample, but will not influence its expansion. A force of 10–30 mN (1–3 g) is recommended by ASTM E831 for coefficient of thermal expansion measurements. However, the ASTM method covers a wide variety of materials, so the initial load should be selected on the basis of properties of the specific sample.

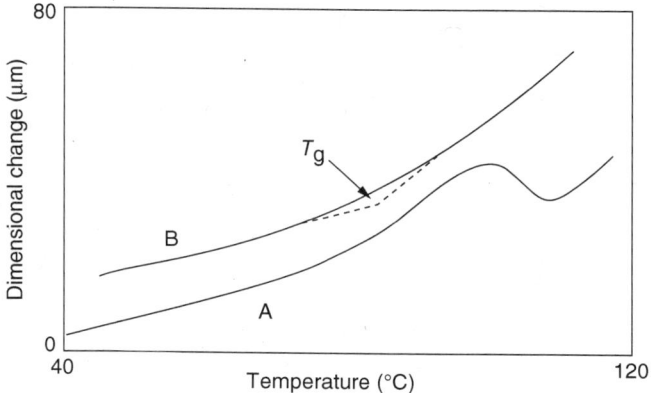

Figure 4.9. Dimensional changes due to overload (curve A) and correctly selected low load (curve B) (see www.anasys.co.uk).

It is also recommended that the initial load be applied after the polymer equilibrates at a temperature below its T_g for a certain period of time (usually 10–15 minutes). Sircar and Chartoff (1994) demonstrated poor reproducibility of coefficient of thermal expansion with a number of elastomeric samples when the load was applied at room temperature.

In tension-mode experiments, essentially zero force should be applied to the sample (such as films or fibers) to measure the tensile properties. However, a small initial force (≤ 2 mN) may be applied to the film or fiber sample to keep the sample straight. Care should be taken not to apply more than the necessary initial force, because this can cause creep and sometimes unnecessary orientation. Too high a load may also reduce the level of shrinkage during heating.

4.6. KEY APPLICATIONS

Various TMA measurements and the information that can be acquired by TMA are summarized in Table 4.4.

4.6.1. Glass Transition Temperature, Thermal Expansion, and Melting Temperature

Two major application areas of TMA in the polymer field are determination of the coefficient of thermal expansion and measurement of T_g and sometimes the melting point (T_m) of semicrystalline polymers. These quantities can be measured in a single TMA run, as shown schematically in Fig. 4.1 and for a cured epoxy resin in Fig. 4.2b (Turi et al. 1988). As mentioned earlier, the glass transition temperature is determined as the point of intersection of the

TABLE 4.4. Key Applications of TMA

Sample Type	Probe Type/Load	Measured Physical Quantity	As a Function of	Information Acquired
Flat bulk sample	Expansion probe Small load	Linear expansivity	Temperature	CLTE (α) and T_g
	Progressive load	Linear displacement	Force	Compression, elastic modulus and crosslink density
	Loading/unloading cycles	Strain/strain recovery	Time/temperature/load	Creep behavior
Cylindrical sample	Large-area PPR probe Constant load or loading/unloading cycles	Strain or strain/strain recovery	Time/temperature/load	Creep behavior, cure
Irregular-shaped bulk sample	Volumetric probe Small load	Volumetric expansion	Temperature	CTE (γ) and T_g
Irregular-shaped bulk sample	Expansion probe Sample smaller than probe, significant load	Linear displacement	Temperature	Softening temperature
Flat bulk sample	Penetration probe Significant load	Linear displacement (depth of penetration)	Temperature Time	Softening and melting temperatures; creep and cure behavior
Thin film or fiber	Film/fiber tension probe Significant load[a]	Linear displacement (uniaxial extension or shrinkage in the direction of measurement)	Temperature	T_g, T_m
	Progressive load	Strain/strain recovery	Force	Elastic (Young's) modulus, crosslink density
	Loading/unloading cycles		Time	Creep behavior
Flat bulk sample, bars	Flexural probe (three-point bend) Significant load	Linear displacement (bending)	Temperature	Softening temperature, heat deflection temperature (HDT)
			Time	Creep behavior

[a]The load, although significant, cannot exceed a certain value: for example, it should not change the original draw ratio (this is especially important for as-spun fibers and fibers with small draw ratio).

expansion curves of the glassy and rubbery (or melt) phases. During expansion measurements of amorphous and crystalline polymers that are not oriented, a sudden increase in expansion rate is observed above T_g. In the case of crystalline polymers only, a further increase in temperature may result in penetration of the probe into the polymer, even with a negligible load on the sample. This sudden decrease in the probe position (onset temperature), as schematically represented in Fig. 4.1, can in most cases be assigned to melting, and the temperature of the break in the curve represents the melting point (T_m).

The glass transition temperature (and the melting point) as measured by DSC and TMA are often different, because the volume change is not always parallel with the enthalpy change. But close correlation exists in the T_g and T_m values determined by these two techniques. Figure 4.10 shows this correlation for several engineering plastics (Riga and Collins 1991).

The melting point can also be determined in a penetration-mode measurement, which is important in the characterization of multilayer polymer films and coatings. Multilayer thin films are used primarily as packaging materials in order to control the end-use properties, such as selective atmosphere permeability or toughness. TMA in the penetration mode can characterize multilayer films in terms of layer thickness as shown in Fig. 4.11. Here, a multilayer film coated on a foil substrate was characterized in a penetration-mode experiment. Two distinct changes due to the penetration of the probe are clear at 103°C and 258°C, suggesting that two layers of film were coated onto the metal substrate. The thicknesses of the layers were measured as 93 μm (polymer 1) and 15 μm (polymer 2). These temperatures are close to the melting points

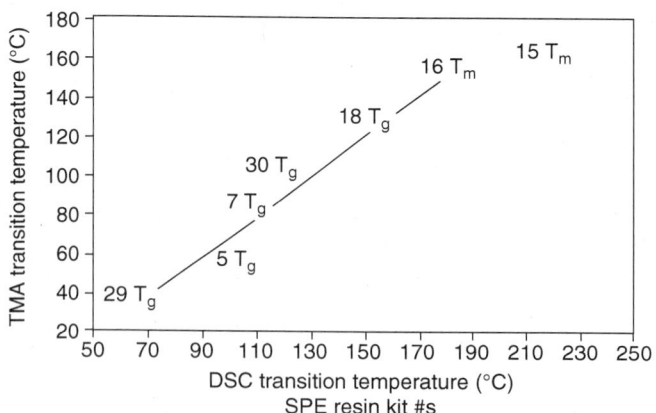

Figure 4.10. Correlation of the T_g and T_m values as measured by TMA and DSC for engineering plastics from the Society of Plastic Engineers (SPE) resin kit (#1043), including polystyrene, poly(methyl methacrylate), polycarbonate, polytetrafluoroethylene, and polypropylene [From Riga and Collins (1991); reprinted with permission of ASTM].

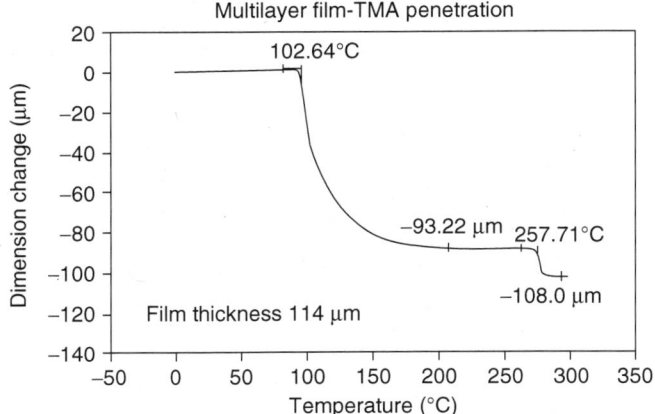

Figure 4.11. Results for a TMA penetration-mode experiment on a multilayer film. Total film thickness was 114 μm. A heating rate of 5 °C/min was used, and a penetration probe was employed under 0.1 N loading. (TA-108, TA Instruments *Thermal Analysis Application Notes*; courtesy of of TA Instruments).

of low-density polyethylene (LDPE) and poly(ethylene terephthalate) (PET), as identified by similar experiments carried out on films made of LDPE and PET (TA 108).

In practice TMA is mostly used for measurement of the glass transition temperature by measuring CLTE and on occasion volume expansion. Therefore, it is appropriate to review those chemical factors (effects of pendant groups, molecular mass, tacticity, crystallinity, crosslinking, etc.) that have the largest influence on the glass transition temperature and the expansion of a polymer.

4.6.1.1. Effect of Pendant Groups and Molecular Mass

In general, stiffer chains lead to higher T_g values. When the chain is made more rigid, the segmental length increases, restricting translational motion, and T_g increases. For example, when the mobile units in the mainchain are restricted by the presence of a rigid pendant group, T_g increases with the size and number of the pendant groups, as summarized in Table 4.5. Clearly larger and/or more rigid pendant groups hinder the motion in the backbone, resulting in higher glass transition temperatures (Eisenberg 1993; Gedde 2001).

As expected, the influence of flexible side groups is different from that of rigid substituents. The increase in length of the flexible alkyl sidechains (e.g., in acrylates or methacrylates and polystyrene) causes the glass transition temperature to decrease (Mark et al. 1993; Gedde 2001).

The glass transition temperature of commercial polymers changes little with the molecular mass beyond a certain value (polymers by definition have a molecular mass higher than 10,000 g/mol). However, there is a significant

TABLE 4.5. Effect of Rigid Pendant Groups on T_g in $(CH_2\text{-}CHX)n$ Polymers

Subsituent X	T_g (°C)
Methyl	−10
Phenyl	100
ortho-Methylphenyl	115
α-Naphthyl	135
Biphenyl	145
Methyl and phenyl on same carbon	175
α-Substituted naphthyl attached to two backbone carbons	264

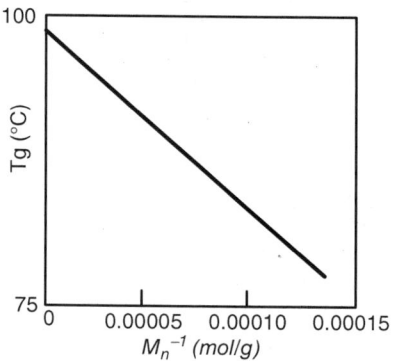

Figure 4.12. T_g versus $1/M_n$ for polystyrene [from Gedde (2001); reprinted with permission of Kluwer Academic Publishers].

increase of T_g with molecular mass for smaller molecules in the oligomeric region, as shown in Fig. 4.12 (Gedde 2001). The free-volume approach presents a simplified way to understand this. It is assumed that chain ends, at any temperature, move faster than do similar chemical moieties in the chain backbone. This is because a chain end has only one covalent bond to tie it down, while the interior segments of the chain are tied down on both ends. The concentration of chain ends increases with decreasing molecular mass, leading to an overall increase in mobility and decrease in T_g.

4.6.1.2. Crystallinity, Stereochemistry, and Crosslinking Crystallinity has an effect on the glass transition temperature of the amorphous fraction. In most cases, T_g increases with increasing crystallinity, because crystallites immobilize the amorphous chains (they behave as physical crosslinks). But there are rare cases where T_g decreases with increasing crystallinity, as observed in bisphenol A polycarbonate (Grimau et al. 1997).

The stereochemistry of a polymer may significantly influence T_g. Most polymers having only one substituent on every second carbon atom of the mainchain do not show any tacticity effects (e.g., acrylates or styrenes). However, polymers containing both a methyl group and a pendant ester on the same carbon atom may have different tacticities that effect T_g. As an example, syndiotactic poly(methyl methacrylate) has a glass transition temperature of 115 °C, whereas the isotactic polymer has a T_g of 45 °C (Ngai 2003). The asymmetric double sidegroups on alternating carbon atoms restrict the mainchain mobility.

Similarly, an increase in intermolecular interactions (ionic interactions, π bonding, H bonding) of different groups in different tacticity polymers also restricts mainchain motions, thus increasing the glass transition temperature— And, of course, T_g increases with crosslink density because of restrictions of mobility in the mainchain. The relationship between T_g and conversion in thermosets is discussed in Section 2.10 (of Chapter 2).

4.6.1.3. Thermal Expansion of Polymeric Materials

In addition to the glass transition and melting temperature measurements, the other major application of TMA is determination of the thermal expansion of polymeric materials. The coefficient of linear thermal expansion (CLTE or α) is the slope of the relative change in the length ($\Delta L/L_0$) with respect to the temperature range of interest, as defined by Eq. (4.2) (see Fig. 4.2b). CLTE measurements are important for many reasons. For example, mismatch in CLTE values may cause dimensional change or delamination in composites and adhesives due to the thermal stresses.

It should be noted that, in addition to traditional TMA, other techniques for measuring CLTE include capacitance change, laser interferometry, dielectric analysis, and elipsometry. These techniques, summarized by Menczel et al. (1997), are especially useful for very thin films, where the traditional TMA instruments reach their sensitivity limits.

4.6.2. Determination of Softening Point

The softening point or temperature is an important practical characteristic of plastic materials. Physical testing laboratories of chemical companies frequently determine the VICAT softening point or the heat deflection temperature (HDT). ASTM defines HDT as the temperature at which a sample with specified dimensions under a specified load (4.6 kg/cm^2) will exhibit a given deflection (ASTM D648). The VICAT softening temperature is the temperature at which a circular probe of 1.0-mm^2 cross section, under a load of 1000 g, penetrates 1.0 mm into a sample 12.7 mm thick (ASTM D1525). By TMA, the softening point of polymers can be measured in the compression mode using a penetration probe, the bending mode using a three-point bending fixture and the tension mode (TA Instruments TA-138). The softening points measured by these methods are known to be a function of load, depending on both the T_g and the viscosity of the material. Therefore, a clear distinction must be made between T_g and the softening point, and the experimental conditions must always be given when reporting results of such measurements (Neag 1991 and 1995). The softening temperature is often similar to the T_g measured in the expansion mode. Figure 4.13 summarizes the types of TMA information that can be obtained from free expansion, compression, and tension experiments for neoprene rubber (Sircar 1997). Figure 4.14 illustrates that T_g determined from expansion data and the softening point measured in the penetration mode correlate well for polycarbonate (Chartoff 1997). However, this is not always the case, as shown in Fig. 4.15, which compares T_g value obtained on a biaxially oriented PET film in expansion and softening point measured in penetration (Moscato and Seyler 1994): the softening point measured in penetration, and T_g determined from expansion experiments, are different. There is often a reasonable correlation between T_g and the softening point for amorphous polymers. However, for semicrystalline polymers the softening point is between T_g and T_m; the value will depend on the test conditions such as heating rate and applied force as well as structure of the sample (crystallinity, crystallite size and orientation, and probably the content of the *rigid amorphous phase* (see Section 2.7.1.1 in Chapter 2) in addition to thermal history of the sample). The presence of the crystalline phase influences the softening point, because crystallites play the role of physical crosslinks.

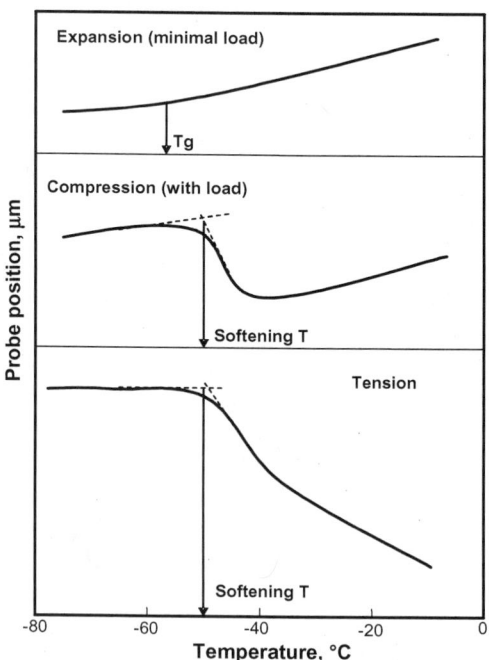

Figure 4.13. Expansion-, compression-, and tensile-mode TMA curves of a neoprene rubber (heating rate = 5 °C/min, 50 mN load in penetration and tension) [From Brazier and Nickel (1978); reprinted with permission of Elsevier, Inc.].

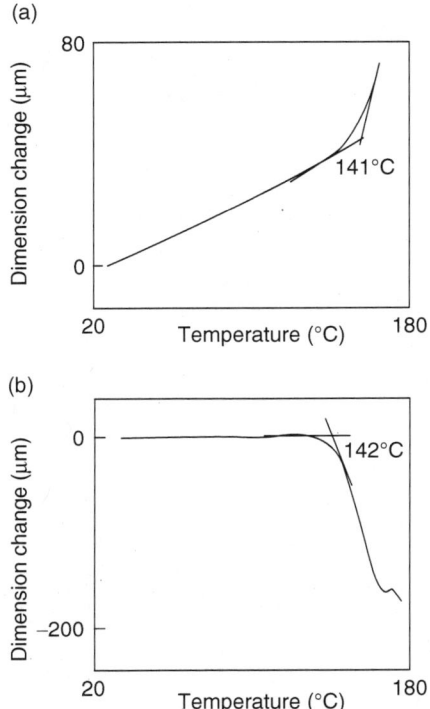

Figure 4.14. (a) TMA curves of amorphous bisphenol A polycarbonate carried out in expansion, 2°C/min, 0.5 g force, and (b) in penetration, 5°C/min, 1.0 g load [From Chartoff (1997); reprinted with permission of Elsevier, Inc.].

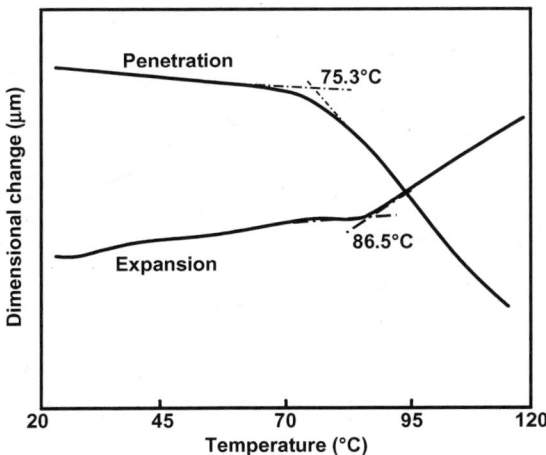

Figure 4.15. Expansion and penetration measurements on a biaxially oriented PET film. In the expansion measurement, negligible force was applied using a 3.5 mm-diameter quartz probe. In the penetration measurement, 300 mN force was applied by a 1-mm diameter probe. Reprinted from Moscato and Seyler (1994) by permission of ASTM.

4.6.3. Tensile Properties

Properties such as Young's modulus and in some cases tensile strength and strain for thin films and fibers can be measured by TMA in the tension mode. A typical stress–strain curve for a polymer is given in Fig. 4.16. Tensile strength, also called "ultimate stress," is an important property for materials that will be stretched such as films and fibers. The *tensile strength* (see Fig. 4.16) is defined as the upper limit of the tensile stress before the material ruptures (therefore, it has units similar to those for stress, i.e., force/area in pascals). *Elongation* is the tensile deformation and the elongation or strain at ultimate stress is defined as the ultimate strain.

Young's modulus is the initial slope of the stress–strain curve in Fig. 4.16, and its magnitude is a measure of the stiffness of a material. In this linear portion of the stress–strain curve, polymers behave as elastic materials and strain is assumed to be recoverable on removal of the stress. Most polymers show inelastic deformation with increasing stress, and they begin to yield (Fig. 4.16). Beyond the yield point, chains continue to deform and absorb energy.

Toughness is the energy absorbed at failure, and it corresponds to the area under the stress–strain curve (units of J/m^3). In general, rigid (glassy) plastics and highly crosslinked networks have high modulus with low elongation at break, and elastomers (such as rubbers) have low modulus with high elongation. General types of stress–strain curves are summarized in Fig. 4.17. Of course, all these stress–strain behaviors depend on the timescale of the test (how fast the test is run), the temperature at which the test is run, the presence of fillers, and even sample geometry. It should be noted that TMA in the tensile mode has some limitations for measuring tensile properties such as

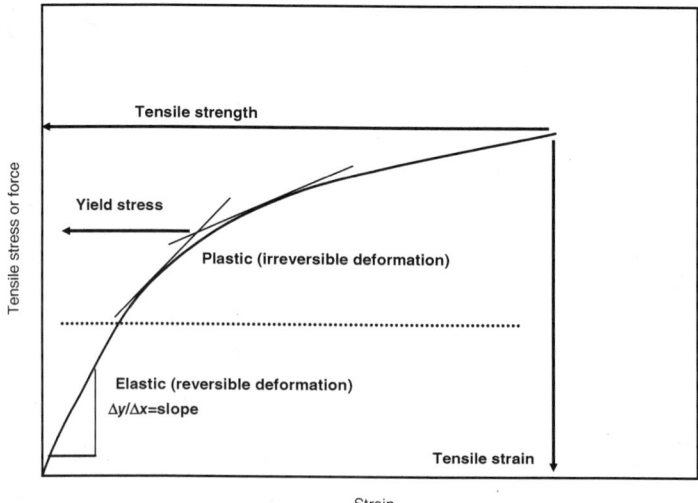

Figure 4.16. Typical stress–strain curve.

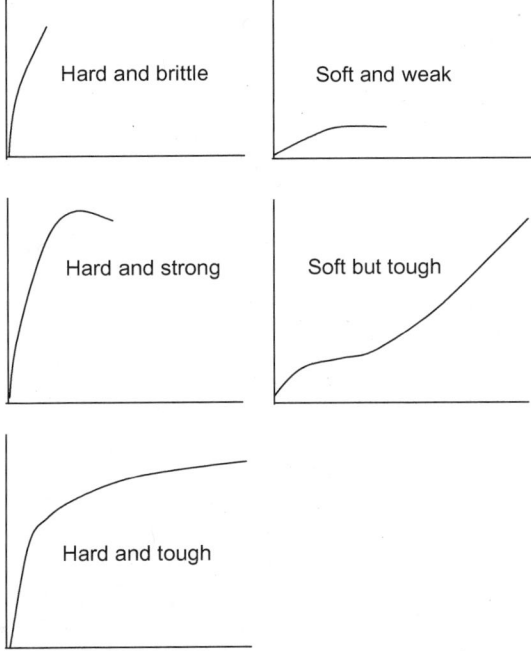

Figure 4.17. Types of stress–strain curves.

tensile strength and elongation. One limitation is the maximum force (depending on the instrument, about 10 N, equivalent to ~1 kg) that can be applied to a film or fiber sample. The other limitation is the probe travel distance, which may not allow stretching samples over 100% strain. One way to overcome these limitations is to adjust the sample dimensions.

During a stress–strain measurement, a small amount of initial stress called *pretension* (e.g., 1–3 mN) may be applied to the film or fiber sample in order to keep it straight. However, it is better to start from zero stress to prevent creep, so one can see on the curve where the slack is taken up.

Thermomechanical analysis in the tension mode can be employed in the production of films and fibers to follow the consistency of the product during manufacturing, and to characterize properties of the product, such as CLTE, thermal shrinkage and shrinkage force, and physical properties like T_g and T_m. All these properties are important, and they determine the end-use possibilities. For example, matching fiber-matrix expansion properties is necessary to avoid delamination of components, such as in tire cords. Shrinkage (i.e., irreversible contraction) is important to avoid in textiles for safe ironing, and it is also an important property in heat-shrink films (Jaffe et al. 1997).

The following paragraph summarizes a typical shrinkage phenomenon for a drawn, partially crystalline fiber. Figure 4.18 schematically shows four regions

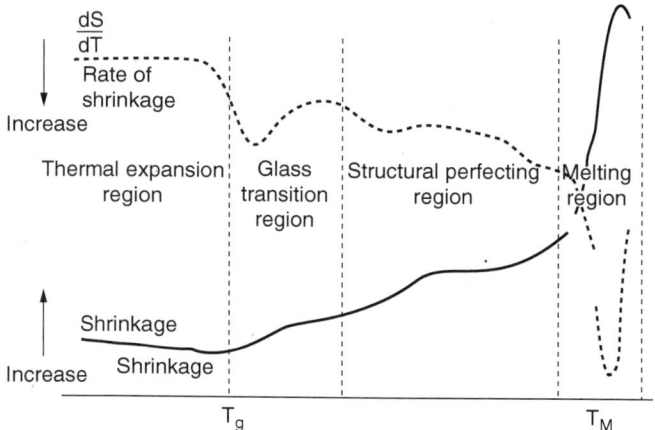

Figure 4.18. Schematic representation of a shrinkage TMA curve of a drawn semicrystalline fiber [From Jaffe et al. (1997); reprinted with permission of Elsevier, Inc.].

of fiber shrinkage. First, at temperatures below T_g, the sample exhibits reversible thermal expansion. Some irreversible shrinkage may also be possible in this region, if solvent or water evaporate from the fiber. Next is the glass transition region. At T_g the fiber shows a rapid irreversible shrinkage because of the relaxation of oriented amorphous chains. In the temperature region between T_g and T_m, reorganization and recrystallization may take place. Finally, at the temperatures just below T_m, rapid shrinkage occurs as a result of relaxation of the tie chains connecting the crystallites. A schematic representation of the molecular chains in the three higher-temperature regions is given in Fig. 4.19. Depending on the process conditions and structure, variations and overlap in the four regions described above can be expected.

Such thermal shrinkage schemes have been investigated by Keum and Song (2005), Wu et al. (1998), and Sichina (2000) for high-speed melt-spun PET fibers. Various degrees of chain orientation, crystallinity, and chain conformation can be obtained by controlling the spinning and quenching conditions [see Jaffe et al. (1997) for details of fiber production].

Thermomechanical analysis in tension is a sensitive method to measure shrinkage force, which is an important parameter for fibers because of various end-use possibilities. The requirements for textile fibers are different from those for fibers used for load-bearing applications such as belts used in machines. The textile fiber must have dimensional stability at ironing temperatures, while fibers used in load-bearing applications should have dimensional stability in the temperature range of their application. Ultimately the expansion and shrinkage properties can be tailored by adjusting processing parameters such as draw ratio, fiber-spinning speed, and cooling conditions, and characterizing the final product by appropriate techniques such as TMA. The *shrinkage force* is the level of force that a film or fiber exerts during shrinkage.

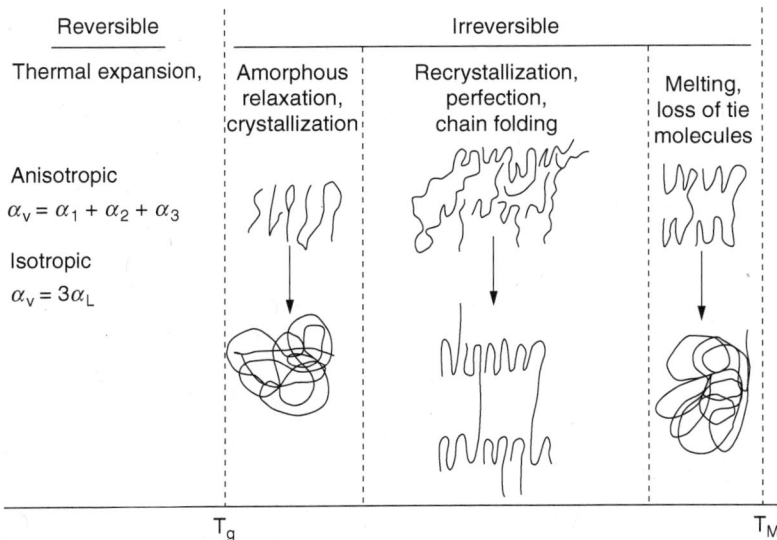

Figure 4.19. Schematic representation of molecular chains and processes taking place in a semicrystalline fiber during heating [From Jaffe et al. (1997); reprinted with permission of Elsevier, Inc.].

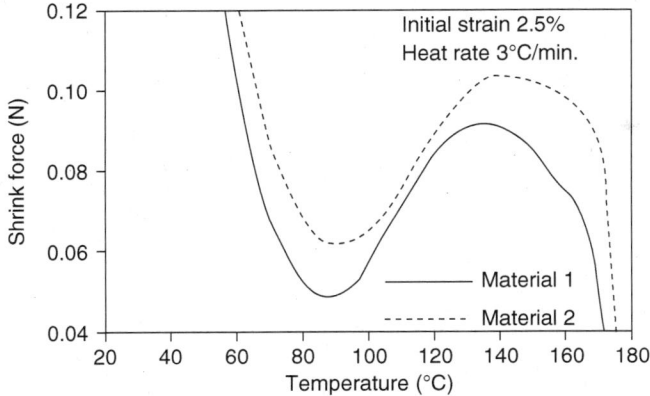

Figure 4.20. Shrinkage force curves for two polypropylene copolymer fibers. Measurements were taken at a heating rate of 3°C/min under 50 mN tensile load [from TA Instruments Thermal Analysis Application Note TA-208, courtesy of TA Instruments].

Figure 4.20 shows the force curves measured during thermal expansion and shrinkage for two different polypropylene fibers (Foreman et al., TA Instruments Application Note 208 reference). A strain is imposed on the fiber and held constant. As the fiber is heated, the force is measured. The force decreases when the fiber expands and increases when the fiber shrinks. The fibers depicted in Fig. 4.20 expand up to 85 °C. On further heating, they begin to shrink, and this process lasts up to the melting temperature at around 160 °C.

TMA data on ultradrawn HDPE fibers (Chartoff 1997; Zachariades et al. 1979; Grebowicz et al. 2001) indicate that the shrinkage of these fibers is a function of draw ratio and spinning speed as shown in Fig. 4.21. Negative CLTEs represent shrinkage. Orientation and crystallinity play an important role in the shrinkage. An increase in crystallinity will spatially fix the oriented amorphous regions through their partial immobilization, and thus will reduce shrinkage.

It should be noted that the properties of oriented films show considerable dependence on the processing conditions (e.g., biaxially oriented films are obtained by blow molding, while extrusion produces uniaxially oriented films). This is important because the mechanical properties of oriented specimens are considerably different in the flow and transverse directions of the films as shown in Fig. 4.22 (measurements were done by a traditional tensile tester). When the film samples in the form of strips are prepared for TMA measurements, the strips should always be cut from the same orientation (Menczel et al. 1997). When preparing such samples, film widths of ≥2 mm are recommended in order to minimize edge effects.

Figure 4.23 schematically shows TMA curves in the machine and transverse directions for film stretched in one dimension to produce uniaxial orientation. In the machine or stretch direction, the film gradually shrinks up to the melting

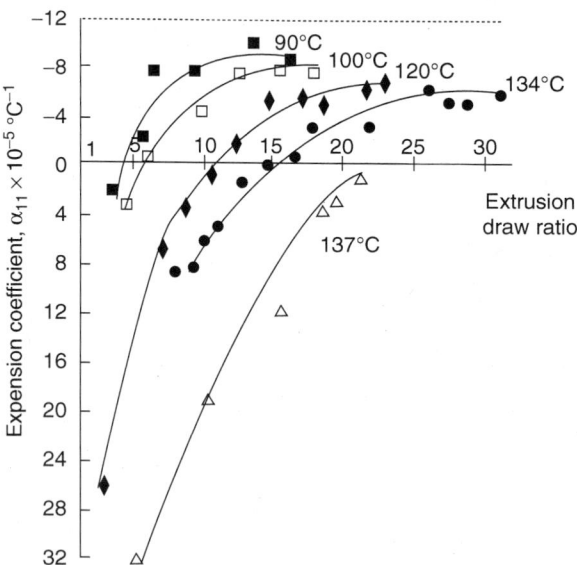

Figure 4.21. Variation of coefficient of linear thermal expansion of HDPE fibers with extrusion draw ratio; measurement made in axial direction [from Zachariades et al. (1979), reprinted with permission of the authors].

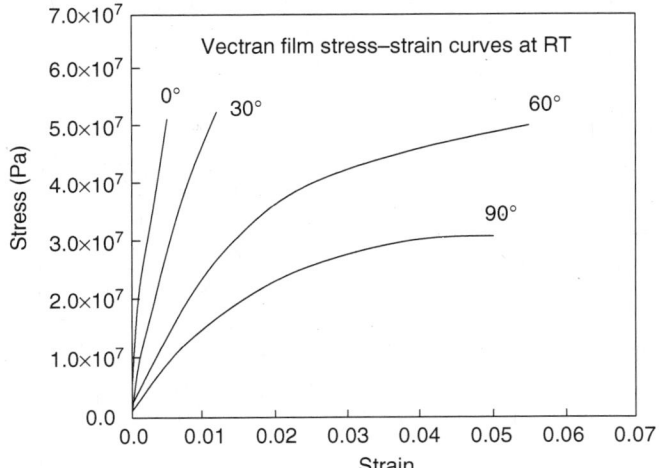

Figure 4.22. Stress–strain curves of Vectran film samples cut at four different angles relative to the direction of molecular orientation, as indicated for each curve. (Menczel, unpublished results).

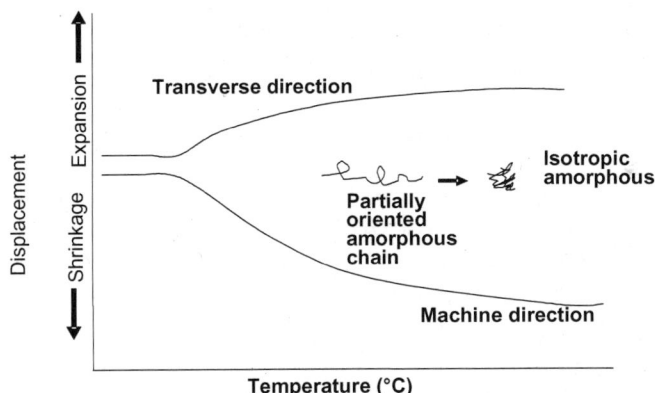

Figure 4.23. Schematic tensile-mode TMA curves of a uniaxially oriented polymer film in the machine and transverse directions (Akinay, unpublished, 2007).

temperature, because partially oriented noncrystalline chains lose their orientation. The extent of shrinkage increases with increasing noncrystalline chain orientation as a result of increased stretching (Menczel et al. 1997). In the transverse direction a gradual expansion is observed with increasing temperature.

Commercially available liquid crystal polymers (LCPs) (e.g., Vectra and Kevlar) exhibit excellent chemical resistance, low flammability, high elastic modulus, and low expansivity (depending on the mainchain orientation direction). Their classification in terms of molecular structure was summarized by Brostow (1990). The liquid crystalline behavior produces anisotropy during processing (such as injection molding or extrusion). Therefore, similar to nonliquid crystalline drawn films and fibers, LCP products will have different thermal expansion in different directions. LCPs have very small negative expansion (i.e., they contract a little) in the drawing or molding direction and low positive expansion in the transverse directions (Akinay et al. 1993; Brostow et al. 1999a,b, 2000).

In summary, anisotropic samples such as polymer fibers and films, oriented bulky polymeric objects, and products from polymer liquid crystals will have different properties, such as thermal expansion, depending on the direction of measurement. TMA, in both the tension and expansion modes, is an important tool for characterizing the effect of orientation on thermal properties of these materials (Jaffe et al. 1997).

4.6.4. Flexural Properties

When a structural polymeric material is subjected to bending, it must have sufficient flexural strength (the level of stress at break or yield in flexural deformation; see www.matweb.com) and flexural modulus (the ratio of stress to strain in flexural deformation; see Chapter 5 for a more detailed discussuon). These properties can be conveniently measured by TMA using the three-point bending probe. For practical purposes, the flexural modulus can be regarded as an estimate of the tensile modulus made from measurements taken in a flexural configuration. For isotropic materials, the flexural and tensile moduli are usually nearly identical, but there may be large differences for anisotropic materials such as composites. Flexural strength and modulus of many commercially available polymers can be found at www.matweb.com.

4.6.5. Swelling

Expansion-mode TMA can be used to measure the swelling behavior of various elastomeric polymers. Swelling is especially important for hydrogel polymers (a lightly crosslinked, highly hygroscopic network). Their applications extend from sanitary products (such as disposable diapers) to medical products (e.g., contact lenses, medical electrodes, controlled drug release matrices). End-use properties of these hydrogels depend on several factors such as the degree of swelling, morphology, chemical structure, interactions with water, and crosslink density. Swelling ratio and swelling kinetics can be measured using the expansion probe when the sample is immersed in a liquid (usually water) at a constant temperature. In an isothermal experiment, the increase in thickness of the sample with swelling is measured. The degree of swelling (S_ε) can be defined as

$$S_\varepsilon = \frac{\ell_t - \ell_0}{\ell_0} \times 100\% \tag{4.8}$$

where l_t and l_0 are the length of the sample at hydration time t and zero, respectively. One disadvantage of this measurement is that there must be a small amount of force on the material to ensure contact of the probe with the sample. Even though the force is negligible (e.g., between 3 and 5 mN), swollen polymers are soft and may be compressed, adversely affecting the swelling measurements. To avoid this complication, Nakamura et al. (2000) suggested measuring the swelling rate under various loading conditions as a function of time. Extrapolating swelling rate versus load provides the zero-load swelling value.

Mechanical measurements on hydrogels can easily be performed by TMA in compression in a liquid environment at various temperatures. In one study, the swelling behavior of hydrogels made from polysaccharide, calcium pectin, and cellulose ethers was studied by TMA (Iijima et al. 2005; Ford and Mitchell 1995). In these studies, samples with varying crosslink densities were immersed in water, and creep behavior was characterized as a function of immersion time and temperature using a compression-type probe. As expected the creep increased with increasing swelling temperature and swell ratio.

Hygroscopic expansion can be viewed as "reversible" expansion and contraction due to the sorption and desorption of water. It is much slower than thermal expansion and is best measured on thin samples in the tensile mode (Prime et al. 1974). Note that any polymeric material with some absorbed water will show simultaneous effects of thermal expansion and hygroscopic contraction when heated (Prime 1997).

As mentioned above, basic mechanical properties, such as compression modulus and tensile modulus of hydrogels, or any solvent-swollen polymer, can be studied by TMA in a liquid environment. The modulus data obtained can be used to calculate effective crosslink density, water–polymer interactions, and thermodynamic properties (enthalpy, entropy, and Gibbs free energy or free enthalpy) of solvent and polymer in a network system (Flory and Rehner 1943; Flory 1955; Rabek 1977; Gedde 2001; Sideridou et al. 2004). These are important parameters for swollen systems, because they characterize the pore size and structure as well as interactions between the liquid and the polymer. Ultimately, diffusion (e.g., ions and gases), controlled release of small-molecular-mass substances (e.g., controlled drug release), and the amount of absorbed liquid (water in hydrogels) all depend on these parameters. The effective crosslink density (v_e/V_0), which is the number of moles of elastically effective crosslinks per unit volume of polymer, can be calculated for a solvent swollen polymer using the following relationship

$$\frac{v_e}{V_0} = \frac{E}{3\phi_p RT} \tag{4.9}$$

where ϕ_p is the volume fraction of the solvent-swollen polymer, T is the absolute temperature in kelvins, and R is the universal gas constant [8.314 J/(mol·K)]. This testing requires that a flat polymer sample be swollen in a solvent at a constant temperature. The elastic modulus (E) of the swollen polymer can be measured in a compression- or extension-mode experiment while the sample is still immersed in the solvent. The amount of absorbed solvent can be measured gravimetrically, and the density of the polymer should be known in order to calculate its volume fraction. It should be noted that the "effective" crosslink density is a measure of both chemical and physical attachment of the chains; that is, it includes physical entanglements and hydrogen bonds as well as covalent crosslink bonds between the polymer chains. Of course, covalent bonds are much stronger and more permanent than entanglements and hydrogen bonds.

Knowing the crosslink density, it is possible to calculate the Flory–Huggins solvent–polymer interaction parameter χ (Flory 1955) using Eq. (4.10) as suggested by Rabek (1977).

$$\ln(1-\phi_2) + \phi_2 + \chi\phi_2^2 + \upsilon_e V_1 \frac{\phi_2^{1/3} - 2\phi_2}{f} = 0 \qquad (4.10)$$

where V_1 is the molar volume of the solvent, f is the functionality of the crosslinks, and ϕ_2 is the molar volume fraction of the polymer.

Ferrer et al. (2004), Johnson et al. (2002), and Sanchez et al. (2004) studied these properties for poly(2-hydroxyethyl acrylate) (PHEA). In these studies, the extent of water–polymer interactions was investigated as a function of temperature and water content in varying crosslink density samples. Average water–water and polymer–polymer interactions were found to be stronger than the water–polymer interactions. These authors concluded that water interacts primarily with hydroxyl groups of the polymer at even the lowest water contents. Changes in the extent of water–polymer interactions with varying chemical structure may contribute to a better understanding of the fundamental behavior and properties of these systems.

4.6.6. Curing

Thermomechanical analysis can be employed in a number of ways to evaluate cure-relevant properties such as gelation and extent of cure (e.g., by invoking the T_g-conversion relationship; see Section 2.10) through the use of expansion, compression, penetration, or tension measurements. *Gelation* is an irreversible transformation from a liquid to a crosslinked system, and the *gel point* may be estimated as the time required to reach a fixed high viscosity at a constant temperature. TMA, especially in the parallel-plate mode, is a good method for determining gelation parameters, which are critical to processing, because after the material has gelled, it is not able to flow (TA Instruments *Application Notes*, TA126). The factors that may influence the cure reaction

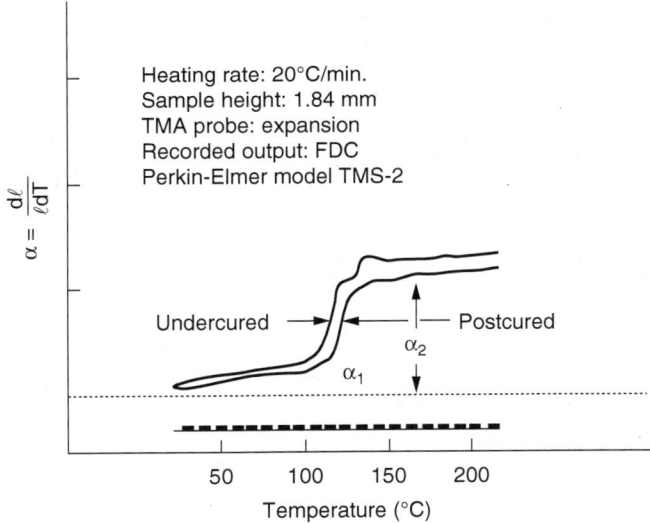

Figure 4.24. Expansion-mode TMA measurement on samples with different degrees of cure for thermosets [From Cassel (1976, 1980), courtesy of Perkin-Elmer.].

and gelation process, such as type and quantity of hardener, filler, and pigments, can be evaluated by TMA. More detailed discussion of gel time determination can be found in Cessna and Jabloner (1974) and Prime (1997).

Figure 4.24 illustrates a correlation between the expansion coefficient and degree of cure for a typical thermoset. In general, the glass transition temperature increases and CLTE decreases in both the glassy and rubbery states as the degree of cure increases (Prime 1997; Cassel 1976 and 1980).

The degree of cure can also be monitored with penetration measurements. For example, as the rubbery modulus above T_g increases with increasing cure, the temperature of the probe penetration also increases (the material softens at a higher temperature), but the probe penetrates to lesser depths as a result of increased crosslinked density as illustrated in Fig. 4.25 (Prime 1997; Sircar 1997; Olivier 2006; Cassel 1976 and 1980).

4.6.7. Viscoelasticity, Creep, and Stress–Relaxation Behavior

Polymers are viscoelastic materials that exhibit characteristics of both solids and liquids. Their mechanical response is both time and temperature dependent, with what is often described as time-dependent elastic properties. As described in Section 4.2.3, viscoelastic materials will undergo creep and stress relaxation. Viscoelasticity is discussed in more detail in Chapter 5.

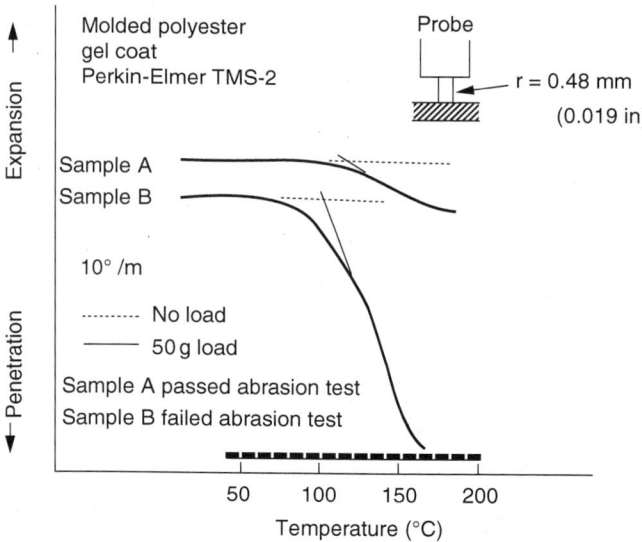

Figure 4.25. Penetration-mode TMA measurement to determine degree of cure for thermosets (as crosslinking increases, the softening temperature increases and the probe penetration decreases) [From Cassel (1976, 1980), courtesy of Perkin-Elmer.].

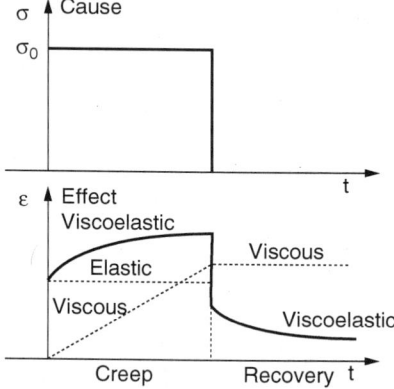

Figure 4.26. A schematic representation of creep and recovery [From Lakes (1999); see http://silver.neep.wise.edu/~lakes/].

Thermomechanical analysis instruments are ideally suited to measure creep. In these experiments the increase in strain is measured with time following the application of a constant stress to the sample, followed by the recovery of the strain when the stress is removed. Figure 4.26 shows a typical TMA creep–recovery curve. In these experiments, an instantaneous compression or tensile stress is applied to the sample, and the time-dependent strain is measured at constant temperature. During the loading cycle, the resultant creep curve shows (1) an instantaneous elastic response, which can be completely recovered on removal of stress; (2) a delayed or time-dependent elastic response; and (3) viscous flow for materials that are not crosslinked. If the load is removed, a strain recovery curve can be recorded. Both a time-dependent recovery due to the viscoelastic nature of polymers and, in some cases, non-recoverable elastic response due to viscous flow (sometimes referred to as *permanent set*) will be observed. In this manner, TMA allows the viscoelastic response of a polymer to be easily measured (Lakes 1999).

Prime (1985) used TMA in the parallel-plate compression mode to measure creep compliance of toner, largely thermoplastic polymers and carbon black, which is used in electrophotographic printers. Measurements at several temperatures allowed characterization of the toner from the glassy state through the rubbery plateau and into the viscous flow region. Using time–temperature superposition, a creep compliance master curve was generated that allowed modeling of the toner fusing process. Measurements of a narrow molecular mass polystyrene, for which accepted literature data were available, demonstrated that this TMA technique is capable of generating quantitative rheological data.

In stress relaxation measurements an instantaneous strain is imposed on the sample and maintained constant, and the time-dependent stress is measured at constant temperature. Stress relaxation can be measured by TMA, provided the instrument is capable of maintaining a constant strain and continuously monitoring the stress.

4.6.8. Plasticizer Efficiency

Plasticizers, such as low-molar-mass substances (e.g., phthalates), are often added to a brittle polymer to reduce its T_g and thus improve its flexibility. Plasticized PVC is a common example. The plasticizer effect may be treated semiquantitatively with a free-volume approach as shown by the following equation

$$v_{ff} = 0.025 + \gamma(T - T_g) \quad (4.11)$$

where v_{ff} is the fractional free volume estimated to be 0.025 (2.5%) at T_g (see Section 4.2.2) and γ is the coefficient of volumetric expansion. This equation can be further generalized to describe the fractional free volume at any temperature for a polymer–diluent system by assuming that the equation is valid for both the polymer and the diluent. The following equation shows the change of T_g as a function of plasticizer content (Kelly and Bueche 1961):

$$T_g = \frac{\gamma_{fp} V_p T_{gp} + \gamma_{fd}(1-V_d)T_{gd}}{\gamma_{fp} V_p + \gamma_{fd}(1-V_d)} \quad (4.12)$$

where p and d indicate polymer and diluent respectively, and V_p and V_d indicate the respective volume fractions. The glass transition temperature (T_{gd})

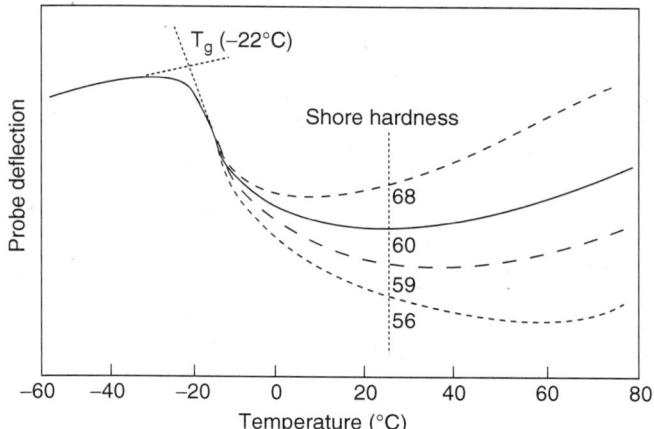

Figure 4.27. Effect of plasticizer on the indentation of crosslinked nitrile rubber: 5 °C/min, no plasticizer (medium dashed line); 10 phr coumarone indene resin (solid line); 10 phr hydrocarbon resin (long dashed line); 10 phr hydroxy resin (short dashed line) [from Brazier and Nickel (1978); reprinted with permission of Elsevier].

and the free-volume expansion of the diluent (γ_{fd}) at T_g are usually not known. However, if these are treated as adjustable parameters, the fit to Eq. (4.12) can be useful to predict T_g of polymer containing plasticizer at any plasticizer concentration.

As just mentioned, making soft PVC from rigid PVC is a well-known example of plasticization. The efficiency of the plasticizer can be easily characterized by TMA simply by measuring T_g. Adding a plasticizer, such as diethylhexyl phthalate with $T_g \sim -90\,°C$, to PVC decreases T_g from $\sim -90\,°C$ (Bair and Warren 1981) to about $-20\,°C$ at a 60% plasticizer concentration (Ceccorulli et al. 1983; Gedde 2001).

In some cases (especially in crosslinked elastomers), TMA penetration measurements can provide useful information about the effectiveness of plasticizers as demonstrated in Fig. 4.27 (Sircar 1997). The shape of the thermal curves vary with hardness resulting from the type and concentration of plasticizer. Although the softening point of the polymer does not change, the magnitude of the indentation increases with decreasing hardness of the polymer.

4.6.9. Service Performance of Polymeric Materials: Lifetime Predictions

Prediction of long-term behavior on the basis of short-term tests is one of the most important tasks in polymer characterization. Viscoelasticity provides tools for modeling the time–temperature behavior of mechanical properties of polymers and predicting their long-term service performance. Williams,

Landel and Ferry (1955) performed the pioneering work in this area, which lead to the time–temperature superposition (TTS) principle [see Eq. (4.7)]. Using TTS, mechanical measurements made at several temperatures can be used to build a master curve for a chosen reference temperature (T_{ref}) (Prime 1985; Akinay et al. 2002). Such master curves may cover time intervals of many decades that are not accessible by experiment. The shifting of individual curves to create the master curve provides experimental temperature shift factors a_T. These a_T values can then be used to create master curves at temperatures other than the reference temperature. When the temperature range of the master curve covers the use temperature of the polymer, the master curve can be used to predict service lifetimes. TTS can be applied to experimental data such as TMA creep and stress relaxation measurements made at several different temperatures. Figure 4.28 shows the creep compliance of a magnetic tape measured at various temperatures (TA Instruments TA287). An increase in creep is observed with increasing time and temperature. Figure 4.29 shows an example of a master curve constructed from the data of Fig. 4.28 for a reference temperature of 30°C. Note that the master curve covers a timespan of 0.1 to 10^{12} s (ca. 300 centuries). C_1 and the C_2 in Eq. (4.7) can be derived from the curve fitting by using the experimental a_T values obtained from superimposing the curves in Fig. 4.28. As a last resort in the absence of experimental data, the so-called universal values of C_1 and C_2 can be used (17.4 and 51.6, respectively).

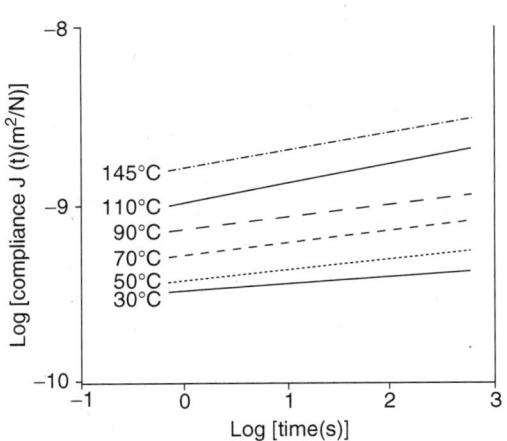

Figure 4.28. Selected creep compliance curves for a magnetic tape at a series of temperatures. The width and thickness were 8mm and 8μm, respectively. A tension film clamp was used in creep mode. In the original experiment, the temperature was stepped in 5°C increments from 30°C to 165°C. A 10MPa creep stress was applied for 10min at each temperature. (Courtesy of TA Instruments.)

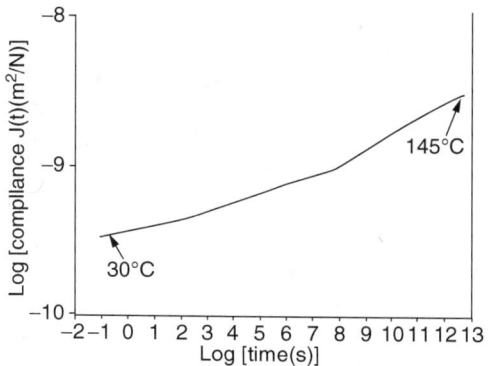

Figure 4.29. Time–temperature superposition (TTS) master curve for $T_{ref} = 30\,°C$ obtained using the data in Fig. 4.28. Individual compliance curves recorded at each temperature shifted along the log[time] axis starting from 30 °C up to 145 °C created the master curve. (courtesy of TA Instruments.)

4.6.10. Determination of Characteristic Parameters, Free Volume, and T^*

Coefficient of linear thermal expansion measurements can also be used to estimate some important characteristic parameters of polymers, such as free volume (v_f), hard-core volume (v^*), and magnitude of intermolecular interactions (T^*) in the materials (Brostow and Szymanski 1986). As temperature increases, polymers expand and their conformational entropy, a measure of disorder in polymer chains, increases. Theoretically, at a sufficiently high temperature, volume expansion allows each polymer chain to relax individually without interacting with its neighbors. This temperature is defined as T^*. In polymers, T^* is an extrapolated value above the degradation temperature, and for many commercial polymers T^* is above 500 °C (Matsuoka 1997). Variables such as v_f, v^*, and T^* can be evaluated by the application of an equation of state such as the Hartman equation shown below (where $V' = v/v^*$, $T' = T/T^*$, and $P' = P/P^*$). (Hartman and Haque 1985a, b):

$$P'V'^5 = T'^{3/2} - \ln V' \qquad (4.13)$$

It has been reported (Akinay et al. 2001, 2002) that the Hartman equation works well for both polymeric solids and melts, including amorphous and crystalline polymers and their blends. Moreover, it also works well in more complicated multiphase systems, such as liquid crystalline polymers (LCPs). Use of the Hartman equation is relatively simple, and its variables can be measured experimentally. Several other equations of state are cited in the *Polymer Handbook* (Brandrup 1999). Some of these do not permit direct measurement of parameters or contain universal constants that complicate determination of the required parameters. Another advantage of the Hartman equation is that it can be easily applied to TMA expansion data. CLTE and specific volume (v) are obtained at atmospheric pressure (about 0.1 MPa); thus $P'V'^5$ in Eq. (4.13) becomes a very small value. If $P'V'^5$ is assumed to be zero, the Hartman equation can be further simplified to the following form assuming isotropic behavior, i.e. identical expansivity in all three dimensions [for derivation, see Brostow et al. (1991)]:

$$3\ln\left(\frac{l}{l_0}\right) = \left(\frac{T}{T^*}\right)^{3/2} + 3\ln\frac{l^*}{l_0} \qquad (4.14)$$

where the ratio l/l_0 indicates normalized length. The slope of a plot of $3\ln(l/l_0)$ versus $T^{3/2}$ gives T^*, and v^* can be calculated from Eq. (4.13).

4.7. SELECTED INDUSTRIAL APPLICATIONS (WITH DETAILS OF EXPERIMENTAL CONDITIONS)

4.7.1. Background on TMA of Films, Fibers, and Composites

Heating a polymer usually causes the sample to expand, whereas cooling results in contraction. An abrupt increase in slope in a TMA expansion curve as temperature is increased is normally taken as evidence of the glass transition, particularly if the slope of the expansion curve above the transition is approximately 3 times its value below the transition for a single-component amorphous polymer. The temperature of intersection of the expansion curves of the glassy and rubbery states defines the glass transition temperature. The optimal way of determining the glass transition temperature T_g is by cooling the sample from its equilibrium state above T_g as shown in Fig. 4.37. Note that T_g experiments at faster or slower cooling rates will give higher or lower values of T_g, respectively.

Unfortunately, in addition to the glass transition, other phenomena can occur to alter this typical expansion behavior. Perhaps the simplest case occurs during heating when too large a compressive load is placed on the sample, causing it to flow or deform when the viscosity of the sample becomes low enough (McKenna and Simon, 2002. see Fig. 4.9). This experimental difficulty is not normally a problem at temperatures well below the glass transition; however, as T_g is approached or exceeded, it can become an issue. Usually commercial TMAs require a small load on the probe to ensure that contact between the sample and the probe is maintained during a run. Normally only a load of 300 Pa (2 mN on a 3-mm-diameter probe) or less is needed. On the other hand, when the sample is held in tension and the force is too great, the sample will elongate irreversibly above T_g. Hence, the tensile load must be minimized to around 300 Pa or until there is just enough force to keep the sample extended. One indication of a good TMA run on a stable sample is to cool the sample back to room temperature and see if its height or length has returned to its original value.

364 THERMOMECHANICAL ANALYSIS (TMA) AND THERMODILATOMETRY (TD)

Physical or chemical events such as stress relief, orientation recovery, gain or loss of moisture or solvents, partial crystallization, curing shrinkage, or adhesive forces that constrain the sample's movement can not only complicate the interpretation of TMA curves but also provide important clues about the sample's past history or enable processing conditions to be established. When these kinds of phenomena arise in TMA work, one is usually well advised to use complementary techniques such as DSC, TGA, DMA, or modulated TMA to gain additional insight into the thermal behavior of a specimen. Also, it is useful to cycle a sample to the highest temperature of interest several times in order to separate irreversible processes from reversible ones. A good example of this cycling technique is shown in Fig. 4.36, where a coated fiber's thickness is under a small compressive load and it must be cycled several times above and below T_g before a stable height or fiber thickness is established. Several other examples of TMA studies being used to examine the expansion and contraction behavior of film, fiber, and composite samples are presented here.

4.7.2. Measurements on Thin Bonded Films

A light load of about 0.2 g on a flat 3-mm-diameter TMA probe (2 mN, equivalent to ~300 Pa pressure) will not normally sink into a polymeric film at or for ~30 °C above T_g, but neither will it produce an unambiguously detectable break in the expansion curve at T_g if the film that is being examined is extremely thin. Hence, large loads such as 5 g (~7000 Pa on the same probe) that cause the probe to indent or penetrate the sample's bulk during heating to and beyond the glass temperature is probably the best way to locate an unknown glass transition of a sub-micrometer-thick polymer film bonded onto a substrate.

In Fig. 4.30 a sub-micrometer-thick film of a cured, unknown network-forming polymer that was adhering to an ~750-µm-thick silicon wafer was placed in the TMA's quartz enclosure. The supplier reported the film's thickness after baking at 200 °C was of the order of 400 nm. An 0.8-mm-diameter quartz rod with a flat bottom was placed on top of the film with a load of approximately 10^5 Pa. The sample chamber was purged with helium, and a heating rate of 5 °C/min was employed.

In the first run the expansion curve of the coated wafer has an unexpected negative slope (height decreasing with increasing temperature), and heating was stopped arbitrarily at 210 °C with the hope that a positive expansion would occur on the second heating scan. On rezeroing and reheating the sample, a positive expansion behavior is displayed until near 190 °C, where the probe sinks 0.45 µm into the film (note that the amount of penetration is close to the film's reported thickness). This latter temperature is assumed to be the unknown's T_g. In the third experiment, a bare silicon wafer was heated from 20 °C to 260 °C, and a linear increase in sample height was observed over the entire temperature range. Also note that between 20 °C and 180 °C the slopes

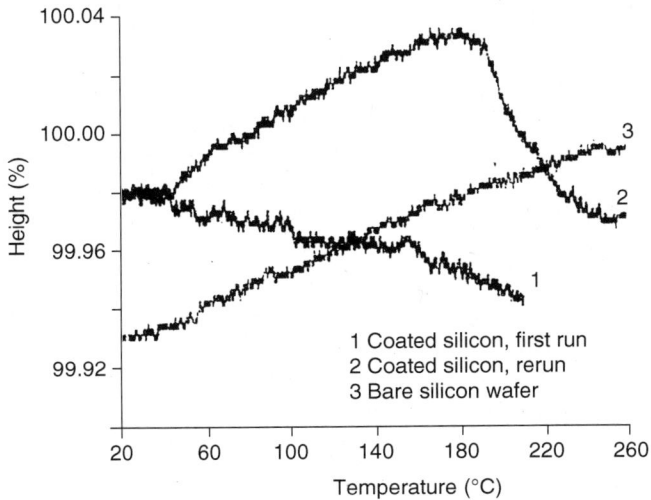

Figure 4.30. Normalized heights of a thin, flat polymer-coated and bare silicon wafer under a 0.1 MPa load at a heating rate of 5 °C/min in a Perkin-Elmer TMA-7 in a compressive mode with helium flowing at 22 mL/min (Bair, unpublished, 2002).

of the expansion for curves 2 and 3 are similar. From the slope of scan 3, a linear coefficient of thermal expansion (CLTE) of $\sim 3.4 \times 10^{-6}/°C$ was calculated. This latter value of CLTE is about 13% above the accepted value for silicon. In each run the TMA was resting on an air table, and the noise level of each curve is approximately 30 nm (Fig. 4.30).

In the first run heating the sample to above 200 °C results in a slight diminution of the sample's apparent height. Since prior to the initial scan the thin film had been baked at 200 °C, it is unlikely that any devolatilization occurred during this run. The probable cause of the apparent contraction in height during the first run is believed to be due to a tiny pocket of air that was trapped between the optically flat surfaces of the TMA's quartz floor bottom and the silica substrate sitting on it. Presumably heating to 210 °C enabled the trapped air to escape. Also, under these conditions it should be noted that the thin polymer's expansion is too low to be detected apart from the expansion of the 750-μm silicon wafer (the film's substrate).

It is always advisable to try to minimize trapped air when placing a thin flat sample in the TMA by sliding the sample back and forth across the floor of the TMA with a load of a few hundred millinewtons on the sample, and then cycle the sample to an elevated temperature above T_g before attempting to collect usable data. Finally, it is important to remember that the slope of a TMA curve with any thin coating should always be similar to the bare substrate, except for a transition region where the probe penetrates most of the sample, since the substrate represents the bulk of material even with the coating present.

4.7.3. Thermal Cure Shrinkage

Important quantitative shrinkage information can be gleaned from careful TMA measurements during the heating of network-forming polymers. In this example it is shown how the thickness of a reactive film decreases under a given load and temperature as curing proceeds and T_g advances. This type of TMA experiment is shown in Fig. 4.31, which depicts the thermomechanical effect of cycling a photoimageable polyimide (PI) and where the temperature is cycled between 0°C and 120°C, 20°C and 200°C, 20°C and 300°C, and finally between 20°C and 360°C.

The prepolyimide resin is supplied in a solvated form. Its processing normally involves spin-coating, soft-baking at ≤110°C, exposure, developing, and hard-baking above 200°C. The sample used in Fig. 4.31 was soft-baked at 80°C, and TGA/FTIR analysis indicates about 12 mass% of N-methylpyrrolidone (NMP), the solvent (boiling point 204°C), is lost. NMP will continue to outgas until the sample reaches 200°C, and then imidization will begin. Note that on heating, the softening temperature occurs when the expansion of the material ceases and the probe begins to sink into the material. In the initial cycle the softening point is detected near 75°C. On reaching 120°C, the sample is cooled back to 20°C and its height has decreased by 7%. Subsequent heating cycles to 120°C, 200°C, and 300°C reveals the softening point (which is close to T_g) at the temperature where expansion stops and penetration begins to be near 110°C, 180°C, and 295°C, respectively, as the sample loses the high-boiling solvent and imidization proceeds. The ultimate softening point for this PI is ~305°C. Note that after cooling to room temperature

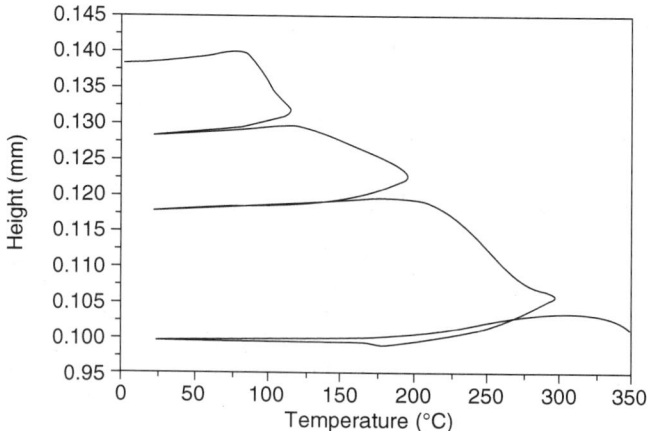

Figure 4.31. Expansion and contraction of a polyimide film plotted against temperature in the direction perpendicular to the plane of the film; the soft-baked PI sample was under a compressive load of 10,300 Pa and heated in steps from 0°C or 20°C to 120°C, 200°C, 300°C, and 360°C at 5°C/min in helium (Bair, unpublished, 2000).

from 300°C the sample's height or thickness has decreased by 27%. In this manner a process can be developed to create an appropriate sample thickness as a function of pressure, conversion, and temperature.

4.7.4. Photoinitiated Polymerization Shrinkage

Photopolymers are valuable materials for a number of optical technologies because of their low cost, high sensitivity, and ease of preparation. One potential limitation to their use in holographic data storage applications is the significant shrinkage that often accompanies their cure. Progress toward understanding this shrinkage behavior is limited by the lack of tools to evaluate dimensional changes during irradiation. It has been shown that photopolymerization-induced shrinkage of acrylate polymers could be determined in a DMA modified for optical access (Bair et al. 1998).

The DMA can be operated in a dynamic or static mode. In the dynamic mode, a probe with an oscillating frequency enables the entire photoinduced polymerization to be followed with real-time monitoring of shrinkage, sample viscosity, and modulus. Alternatively, if it is operated in a static mode, the DMA essentially becomes a TMA. The dynamic-mode DMA measurements are possible but difficult to carry out since the probe must essentially "float" on the initially uncured liquid and hence, is not discussed here. Instead, TMA-like measurements in a static mode are described, as illustrated in Fig. 4.32. Bair et al. (1998, 2000). This static-mode technique can be applied to samples with a wide range of shrinkages, from <0.1% to >10%. The detection limit for sample shrinkage in 500-μm-thick specimens using this static DMA method is

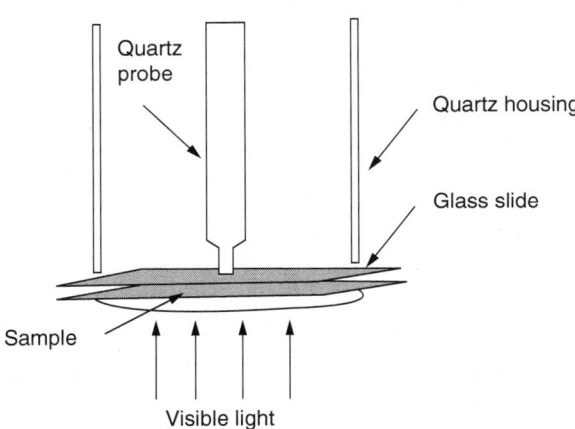

Figure 4.32. Schematic of a DMA quartz sample chamber used for shrinkage measurements at 23°C. The DMA is operated in a TMA-like static mode. Typically, polymerizable samples were 0.5 mm thick and the 3.5-mm-diameter probe carried a load of 5 mN (equivalent to a pressure of 520 Pa). [From Bair et al. (1998); reprinted with permission from the American Chemical Society.]

quite low in practice, where dimensional changes on the order of 50 nm were readily observable. An added value of this approach is that the samples can be stored between quartz plates and analyzed in an either unreacted or partially cured state in the same way as they are used in storage media applications.

All measurements were made in a Perkin-Elmer DMA7e, whose sample chamber was replaced with a large quartz housing designed to allow samples held between glass slides as wide as 4.5 cm to be placed on a flat quartz floor. In the static mode, uncured liquid or partially reacted photosensitive materials are placed between quartz plates and the probe rests with a load of 5 mN on the top plate during the photoreaction and provides an accurate measure of the sample thickness perpendicular to the plane of the glass plates or in the transverse direction.

The photopolymer materials shown in Fig. 4.33 contain acrylate and vinyl monomers blended with an oligomer capped with urethane acrylate endgroups ($M_n \sim 2000$) and 1% by mass photoinitiator for light-induced polymerization. Formulation ACR contained 40 mass% acrylate monomer and 60 mass% oligomer. In the second formulation, VACR (vinyl monomer-containing polymer), 10% of the acrylate monomer was replaced with a vinyl monomer while the amount of oligomer remained unchanged. Prior DSC studies of ACR and VACR show that the components are miscible and the cured T_g values are 16 °C and 26 °C, respectively (Bair et al. 1997).

The viscous liquid samples were layered between two quartz plates (4.5 × 7.5 cm) with double-sided sticky tape attached around the periphery of the plates. The tape does not interfere with the sample's shrinkage since the reactive material adheres strongly to the glass plates and causes the tape to

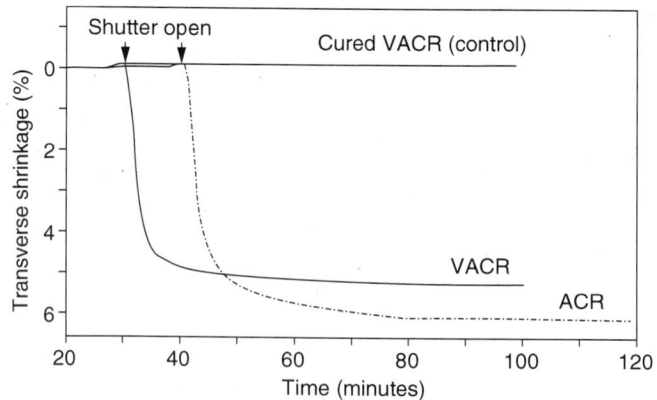

Figure 4.33. Sample thickness versus irradiation time for uncured, liquid samples of ACR (broken line, shutter opened at 40 min) and VACR (solid line, shutter opened at 30 min) as compared to a fully cured VACR control sample. Nitrogen purge. [From Bair et al. (1998); reprinted with permission from the American Chemical Society.].

flex as curing and shrinkage proceeds. Prior to placing the sample between the plates, the thickness of the two glass plates was measured and added to the TMA zero position. These unreacted liquids were found to slowly equilibrate and sometimes required about 10h to reach a stable height. Hence, it is convenient to leave this type of sample in the TMA overnight and monitor its height while waiting for it to stabilize before initiating irradiation. In contrast to this behavior, for a partially reacted sample, the probe stabilized in 15 min or less. Samples were exposed to a $10 \pm 0.5\,\text{mW/cm}^2$ halogen lamp filtered to provide light in the visible part of the spectrum between 540 and 850 nm, where the photoinitiator responds. Figure 4.33 shows the TMA scans for the ACR and VACR materials from the uncured liquid state to the fully cured solid state. For ACR after 40 min of equilibration, a total shrinkage of 6.2% was measured during 50 min of irradiation. Similarly, for VACR, a total shrinkage of 5.3% was determined. An important issue in this measurement is that the probe detects changes only in the polymer dimension that is perpendicular to the glass plates.

These TMA shrinkage measurements were carried out without the instrument furnace in place, that is, at ambient temperature. During a typical 60-min run an estimate of the ambient thermal fluctuations of the surroundings was made and found to be of the order of $\pm 0.2\,°C$. For a 0.6-mm-thick photopolymer, these temperature changes would result in only a 0.004% change in sample height and would be a negligible source of error in the experiments.

The actual amount of volumetric shrinkage expected for these two samples is not known. However, a model exists that can be used to predict the shrinkages (Pezron and Magny 1996). With this approach, calculated shrinkages in volume for ACR and VACR are similar and approximate 6%. Another acrylate formulation was prepared with a calculated shrinkage of 14%, which may be compared with a value of 12.4% measured by TMA. The agreement between the calculated total shrinkage for each sample with the shrinkage determined in the static mode suggests that the polymer films were not contracting to any significant degree in the lateral plane. If shrinkage occurred in all directions, the transverse shrinkage in thickness would account for only about one-third of the total shrinkage. Thus, in this sample configuration the transverse shrinkage is a measure of the volumetric shrinkage. In support of this contention, the VACR's CLTE when it was constrained between glass slides was 455 ppm/°C, or nearly 3 times the CLTE of a free film of VACR or 154 ppm/°C.

4.7.5. Compositional Heterogeneity of Polymers: UV-Stabilized Polycarbonate

Polycarbonate (PC) is susceptible to photooxidation when aged outdoors. In order to reduce the effects of weathering, UV stabilizers such as benzotriazoles are blended into PC. Then, UV stabilizer-enriched sheets of PC are coextruded with unprotected polycarbonate. TGA vaporization studies at 180°C of a commercial coextruded, 3.4-mm-thick PC sheet show that 5–7%

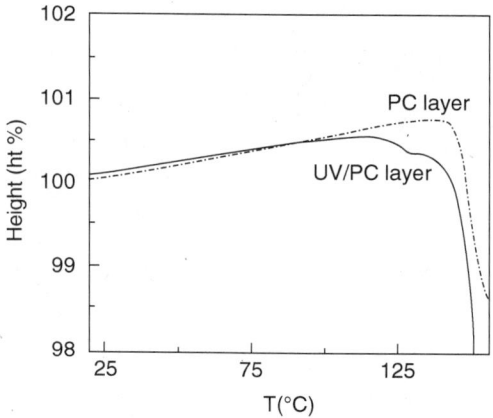

Figure 4.34. Penetration by a rounded-nose TMA probe versus temperature (compressive load 15 g on a maximum 3-mm-diameter probe, helium purge) into a UV-enriched PC layer compared to the neat polycarbonate layer [from Bair (1997); reproduced with permission from Elsevier].

by mass of a benzotriazole-type UV absorber is present in the top layer of the product (the neat PC is thermally stable under these conditions). By DSC, it is found that the UV absorber is soluble in the PC and lowers T_g by 14 °C below the 147 °C glass temperature of neat PC (Bair 1997).

This difference in T_g values can be detected mechanically as shown in Fig. 4.34. Above 100 °C the probe begins to sink into the plasticized layer (solid line, Fig. 4.34). Between 120 °C and 130 °C the quartz probe continues to sink into the UV-protected layer. Above 135 °C the penetration rate accelerates again. On reversing the sample sheet (broken line, Fig. 4.34), only a single, smooth penetration step is noted above 140 °C. Hence, the two-step penetration found on the first side corresponds to the UV-rich PC layer followed by the neat polycarbonate layer.

4.7.6. Orientation Recovery in a Cable Jacket

Suddenly and for unknown manufacturing reasons, many vinyl outer jackets in the assembly of an optical undersea cable system ruptured at their point of entry into an optoelectronic device during a solder reflow operation at elevated temperature. As part of an investigation into the cause of failure, several thermal analysis experiments, including DSC, DMA, and TMA, were conducted. A tensile-mode TMA study is shown in Fig. 4.35. The cable's outer jacket (OJ) was removed. This tubular structure's outer diameter measured about 1.6 mm with a wall thickness of about 0.20 mm. OJ samples were cut to a length of about 23 mm and clamped between the TMA's steel clamps. A slight positive tension was maintained as the tubular sample was heated at

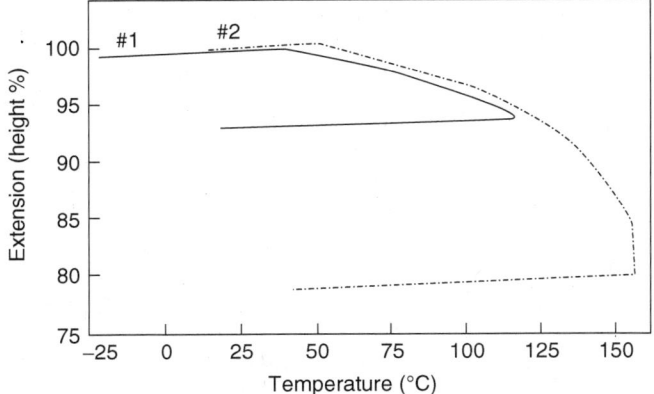

Figure 4.35. The length of an optical cable's vinyl outer jacket plotted against temperature for two tubular PVC samples; tensile mode, with a heating rate of 6 °C/min and a helium purge (Bair, unpublished, 1995).

6 °C/min in helium from −20 °C to 120 °C [Fig. 4.35, line 1, (solid line)]. Between −20 °C and 35 °C the vinyl jacket expands at a rate of 104 ppm/°C, which equals its CLTE. Near 40 °C the OJ begins to shrink with increasing temperature until it is about 7% shorter at 120 °C than its initial length. On cooling to 20 °C and reheating to 85 °C, it contracts and expands without any sign of additional shrinkage. However, if a second sample is heated to higher temperatures, it recovers more of the orientation induced during manufacture as shown by the broken line (Fig. 4.35, line 2). Note that above 100 °C the sample's contraction rate accelerates. When the OJ sample reaches 158 °C, it is 17% shorter than its starting length but the vinyl jacket compensates by thickening in its radial direction. The elevated temperature recovery of these "frozen in" strains produces sufficient energy to cause the outer jacket to fail. These latter data were determined by a DMA measurement of the force that was required to keep the OJ at a fixed length as the temperature was raised. Finally, steps were taken to minimize and control orientation and the potential for shrinkage during the manufacture of the vinyl outer jacket used in this commercial optical cable system.

4.7.7. Adhesion in Dual-Coated Optical Fibers

Thermomechanical analysis measurements have proved to be useful tools in understanding the behavior of polymeric coatings used on silica optical fibers. In a typical dual-coated fiber the inner primary coating has a T_g below room temperature and acts to cushion the fiber against microbending losses, while the secondary or outer coating has a glass transition temperature above room temperature and provides a tough protective cladding for the fiber (Blyler et al. 1987). It is instructive to compare the expansion behavior of the two coupled UV coatings in situ on the fiber with the two freestanding films. Note that the fiber sample consists of four layers, two above the inner silica fiber and two below it. The goal of this work is to gain insight into the effect of adhesion at the coating–silica interface on the dual coating's expansion behavior.

A single, 8 mm length of a dual-coated optical fiber is placed on the flat quartz floor of the TMA sample chamber. Then a flat, circular 3-mm-diameter quartz probe is lowered gently onto the fiber with a load of 1 mN or 0.1 g. Note that the probe's exact contact area atop the fiber is unknown. The overall thickness of the fiber is about 0.233 mm (Fig. 4.36). The sample chamber should be purged with helium to minimize temperature gradients. As noted earlier thermal conductivity of helium at 20 °C is about 6 times greater than that of nitrogen (see Table 3.2 in Chapter 3). Heating and cooling rates of 2–4 °C/min are recommended. Also, the sample should be heated initially to above T_g to allow the probe to flatten out any high spots on the fiber surface. When the dual-coated fiber is run in the expansion mode, the preliminary heating step to 135 °C is critical because the probe will flatten the 233-µm-thick fiber (inner and outer polymer coatings are ~30 µm thick) by about 5 µm during the seating process.

An example of this "seating" process is shown in Fig. 4.36, where a single strand of optical fiber is cycled several times between −50 °C and 135 °C at 3 °C/min until the curves nearly collapse into a single line. Note that on the initial heating from 20 °C to 135 °C the probe starts to flatten the fiber near 50 °C and its initial height of 0.233 mm is diminished by about 2 µm but above 80 °C expansion resumes. However, some additional flattening occurs as the fiber is cycled from 135 °C to −50 °C and back to 135 °C. Subsequent tempera-

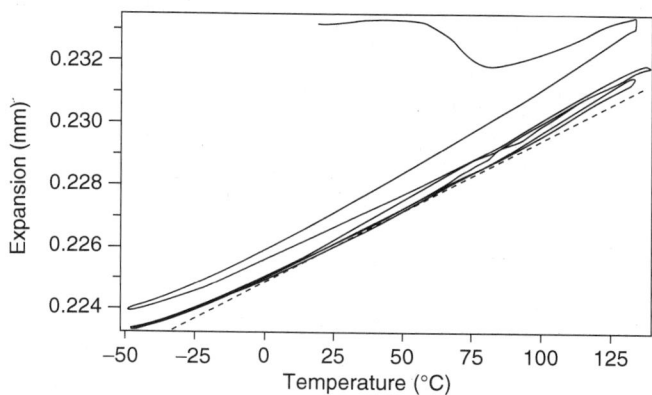

Figure 4.36. Expansion and contraction of a dual-coated optical fiber plotted against temperature at a rate of 3 °C/min; initially heated from 20 °C to 135 °C and then cycled three times between −50 °C and 135 °C with a probe load of 1 mN (0.1 g) in an expansion mode with a helium purge (Bair, unpublished, 1999).

ture cycles reveal that the flattening process stopped when the optical fiber's height reached 0.226 mm at 25 °C. Additional temperature cycling does not change the sample's thickness at 25 °C, or, in other words, the sample is now "seated".

Once this condition is reached, the primary coating's glassy expansion curve from −50 °C to −30 °C and the secondary coating's rubbery expansion curve from 125 °C to 90 °C can be extrapolated upward and downward, respectively, until they meet the linear line extending from the expansion curve between 0 °C and 75 °C (dashed line, Fig. 4.36). This process yields glass transition temperatures near −20 °C and 87 °C. DSC measurements at 15 °C/min on this same optical fiber sample indicated that glass transition temperatures occur at −31 °C and 78 °C. In normal practice with the probe "seated," the final stable heating run would be selected and isolated in a separate plot for determination of T_g and CLTE values.

All CLTE measurements were corrected for the presence of the silica fiber and the baseline behavior of the TMA sample chamber and probe. To achieve this, a fiber was heated to 750 °C in oxygen to remove the double coating. Measurements were then taken of the bare fiber when it was heated over the same temperature range as that used with the coated fibers. This stored instrumental baseline was subtracted from all subsequent tests on coated fibers.

Two 0.4-mm-thick free films from the primary and secondary coating materials were run individually in the expansion mode in the TMA (Fig. 4.37). The in situ T_g of both coatings was taken from the intersection of the linear extrapolations of the expansion behavior below and above each glass transition. Note the dashed lines in Fig. 4.37, which help to delineate the beginning

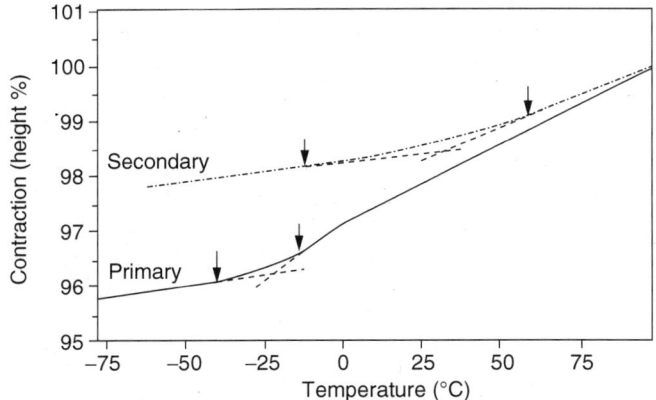

Figure 4.37. Contraction of 0.4-mm-thick primary and secondary coatings in film form during cooling from 100 °C at 4 °C/min in helium (compression mode, 0.1 g on a 3-mm-diameter probe (141 Pa pressure) in helium; arrows depict the onset and end of each glass transition (Bair, unpublished, 1998).

and end of each glass transition. Separate contraction curves for a primary film and a secondary film are depicted in Fig. 4.37, which shows results for samples cooled from 100 °C to −75 °C at 4 °C/min. Note that the primary coating used in the film is identical to that used in the optical fiber shown in Fig. 4.36 but the secondary coating on the fiber is different, and this results in a lower T_g but a CLTE similar to that of the secondary coating used on the optical fiber. The primary coating begins to vitrify as denoted by its reduced rate of contraction near −10 °C, whereas the secondary coating begins to vitrify near 59 °C. As shown by the arrows in Fig. 4.37, the breadth of each transition is clearly marked and shows the widths of the glass transitions to be about 28 °C and 70 °C for the primary and secondary coatings, respectively. The glass transition temperature was placed at −23 °C for the primary coating and 33 °C for the secondary coating, respectively. Above T_g, the coefficient of linear thermal expansion for these primary and secondary photocured materials is 284 and 220 ppm/°C, respectively. In the glassy state CLTE values for the primary and secondary coatings are 86 and 67 ppm/°C, respectively.

The adhesion between the primary coating and the silica fiber is known to decrease as the temperature is increased. Unexpectedly, evidence for this effect can be seen by comparing CLTE measurements on a dual-coated fiber to the freestanding films of the coatings. The CLTE values measured in situ on an optical fiber containing an adhesion promoter were 214, 428, and 506 ppm/°C at three 10 °C intervals centered at −40 °C, 25 °C, and 115 °C, respectively. In contrast, the corresponding values for the freestanding films without an adhesion promoter and at the same temperature intervals were 77, 175, and 252 ppm/°C. Note that the latter values are simply the arithmetic averages of the CLTEs of the primary and secondary free films. The ratios of the CLTE values from the fiber divided by the free-film numbers are 2.8, 2.4, and 2.0 °C at −40 °C, 25 °C, and 115 °C, respectively.

The results show that the coating is constrained in its longitudinal and radial directions by adhesion of the primary coating to the silica glass and cohesion of the primary and secondary coatings. Last, and perhaps most important, as the temperature is increased, the CLTE ratio decreases, indicating a reduction in adhesive constraints.

Not only does a properly engineered optical fiber need good adhesion at service temperatures but also, if the fiber must be stripped, the ability to reduce adhesion by heating the fiber is desirable. In this manner CLTE measurements can play an important role in engineering an optimum type and level of adhesion promoter for an optical fiber system.

4.7.8. Bowing in Integrated Circuit Packages

Because of the thermal contraction mismatch of a silicon chip, metal leadframe and silica-filled epoxy molding compound integrated circuit (IC) packages bow or warp when cooled to room temperature after manufacture. The magnitude of the bow in an IC package can be determined quantitatively by

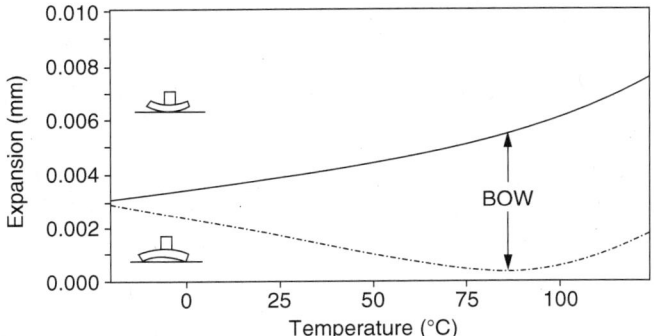

Figure 4.38. Thin integrated circuit package cooled from 120 °C to –20 °C at 3 °C/min with its bow up (solid line) and down (dashed line) and the probe on the package center (0.2 g ≈ 2 mN on a 3-mm-diameter probe in helium (Bair, unpublished, 1988).

TMA measurements using a comparative technique. First, the sample is scanned with its bow up (Fig. 4.38, solid line) and next with its bow down (Fig. 4.38, dashed line). In the lower curve the negative slope ceases when the bow has flattened out near 90 °C, and then expansion occurs at a rate similar to the upper curve as the IC package is heated to 120 °C. Hence, when the package is cooled from 90 °C to –20 °C, a 5 µm bow is formed.

Comparison of the predicted bowing for an IC package with the experimentally determined bow can be used to evaluate adhesion between components in a package. When the predicted bow (Suhir 1993) matches the actual experimental value, good adhesion is assumed. A bow smaller than predicted indicates that poor adhesion is present in the package. Good adhesion is highly desirable in a microelectronic package in order to distribute stresses evenly and minimize the possible buildup of stress at the corner of a chip that can lead to delamination or cracks inside the package.

4.7.9. Thermoplastic Composite

DICLAD® 880 is a copper-clad composite of high-performance fibers with a thermoplastic matrix that can be used for printed wiring board (PWB) applications. The product is sold in a sheet that is about 0.60 mm thick while its outer copper skin is about 0.02 mm thick. By DSC analysis, the matrix was found to be partially crystalline with a melting temperature of 327 °C, and near room temperature the polymer undergoes two first-order phase transitions at 23 °C and 33 °C. This thermal fingerprint identifies the polymeric matrix as polytetrafluoroethylene (PTFE). Heating the product in nitrogen minus the Cu cladding in a TGA to 900 °C removes the Teflon and leaves bundles of fibers that were arranged in flat beltlike bundles that measured 0.50 mm across and 0.02 mm thick. Individual fibers were about 5 µm in diameter. Thus, an individual fiber belt is composed of hundreds of long fibers woven into an array somewhat like the webbed backing found in an outdoor chair. These fiber mats are layered in the product's x–y plane with a spacing of approximately 0.15 mm between fiber planes. Hence, four of these webbed layers are evenly spaced through the thickness (z direction) of DICLAD 880.

A TMA study was carried out in the expansion mode in the x, y, and z directions and with the outer copper layers removed. As seen in Fig. 4.39, between 10 °C and 40 °C, the DSC reveals two solid–solid phase transitions associated with the onset of internal motion in the polymer's crystal phase (Bunn and Howells 1954; Kimmig et al. 1994). This structural change manifests itself as an abrupt thickening (0.64%) in the PTFE/fiber composite, but with the restraints imposed by the fibers, no anomaly is seen in the expansion of the material in the x direction (y-direction expansion is not shown but was found to be the same as in the x direction). Below the phase transitions CLTE values of 24, 25, and 163 ppm/°C were found for the x, y, and z directions, respectively; and above the transitions, CLTEs of 20, 20, and 186 ppm/°C were determined for the material. Thus, the fibers restrict expansion on heating in the x and y directions of the material but not in the z direction. Teflon itself has isotropic CLTE values around 114 ppm/°C at both 0 °C and 45 °C, while in this temperature range copper's CLTE is 17 ppm/°C. Thus, there is a reasonable match with temperature changes between the expansion and contraction of the Cu cladding and the Teflon below it. However, movement induced by temperature cycling near room temperature in the Cu/Teflon/fiber composite in the z direction could possibly cause delaminating forces to develop near holes in a printed wiring board.

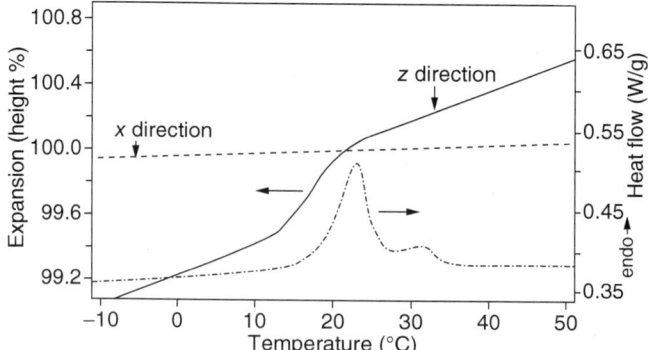

Figure 4.39. Expansion of a composite of poly(tetrafluoroethylene) and fibers in two directions (x and z, dashed and solid lines, respectively) plotted against temperature at a heating rate of 4 °C/min in the expansion mode in helium (0.1 g on a 3-mm-diameter probe, equivalent to 141 Pa pressure on the sample); DSC scan at 15 °C/min in nitrogen of the same fluoropolymer as run in the TMA (Bair, unpublished, 1995).

4.7.10. Glass-Reinforced Thermoplastic

Glass-filled (GF) polyphenylene sulfide (PPS) is a thermoplastic with many desirable properties such as inherent flame resistance, low creep, excellent resistance to chemicals and stress cracking, and low water absorption. In applications where dimensional stability is needed, it is useful not only to select a resin with a high glass filler loading (>40% by mass) but also to carry out injection-molding processes in a manner that will cause the composite's crystallinity to exceed 50%. At this level of crystallinity and higher, the molecular motions in the glassy regions of GFPPS are restricted (T_g increases) and the material's resistance to softening and warping are improved. By measuring the amount of heat that is required to melt PPS crystals in a DSC, the level of crystallinity in a molded PPS part can be determined (see Section 2.7). In this manner one can check the influence of crystallinity on the mechanical deformation of PPS.

An example of how the TMA can be utilized to find the heat distortion temperature (HDT) of partially crystalline GFPPS (Fortron® 4184, Hoechst) with 49 mass% glass filler is shown in Fig. 4.40. In determining the DSC crystallinity for this resin, a heat of fusion of 38.3 J/g was used. In a TMA three-point bend test, a bar of rectangular cross section (1 mm thick × 3 mm wide) was placed on two parallel knife supports that were 10 mm apart, and then a load was applied via a knife-edged, quartz probe at the center of the bar to yield a maximum fiber stress of 1820 kPa (264 psi). These conditions of stress are analogous to those used in the ASTM D648 heat distortion test, except their recommend sample size is much larger. Samples were heated from 25°C to 284°C and the deflection under load was monitored. Typically, in the ASTM test the heat deflection temperature is taken as the temperature where deflection reaches 0.90 mm. In Fig. 4.40 the GFPPS sample (as-molded, 53%

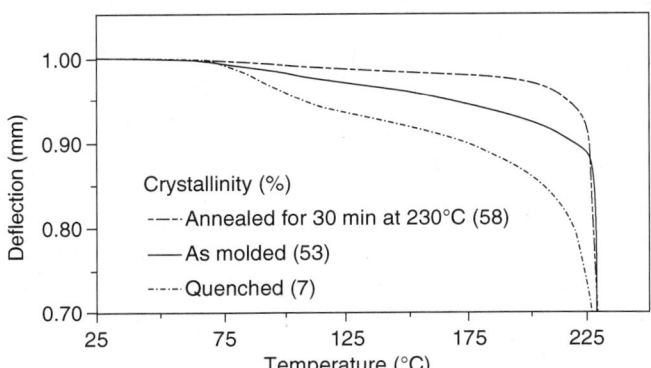

Figure 4.40. Effect of crystallinity on the deflection of glass-filled poly(phenylene sulfide) under an 1820 kPa load with increasing temperature; TMA three-point bending mode at 3°C/min in helium (Bair, unpublished, 1999).

crystalline) begins to bow slightly above 70 °C (T_g by DSC is at 87 °C) until near 230 °C, when deflection occurs at a rapid rate. The as-molded sample was annealed for 30 min at 230 °C, which increased its crystallinity to 58%. This modest increase in crystallinity resulted in a significant improvement in its deflection under load at elevated temperatures (broken line with dots, Fig. 4.40) as compared to the as-molded material. For example, at 200 °C the deflections of the as-molded and annealed bars were 0.071 and 0.026 mm, respectively. Finally, the GFPPS was melted and quenched. This process produced a sample with a large crystallinity gradient whose outer surfaces were amorphous. The quenched bar shows significant deformation under load at temperatures above T_g (Fig. 4.40, broken line with dashes). At 200 °C the quenched PPS sample deflected 0.135 mm below its initial position at room temperature or 5 times more than the annealed GFPPS.

These and other complementary thermomechanical tests have been carried out on injection-molded GFPPS components in order to set processing conditions and, thereby, maximize the plastic's mechanical stability. These GFPPS components are part of a telecommunication system's optoelectronic package (Bair and Osenbach 2004).

In summary, the examples presented above show how thermomechanical analysis can be used to establish processing conditions or troubleshoot problems as needed in various industrial products. In particular, TMA techniques were shown to be useful in detecting the softening temperature of an amorphous, sub-micrometer thick, unknown film, in determining imidization-induced linear shrinkage of a polyimide film under the effect of pressure and temperature, in carrying out volumetric shrinkage measurements induced by photopolymerization, in detecting film layers via penetration tests, and in analyzing orientation recovery as a function of temperature in a vinyl tubular jacket. In addition, TMA comparisons of CLTE data on free films of UV-cured coatings versus the same materials on silica fiber showed the effect of adhesive constraints on the expansion behavior of polymeric claddings in an optical fiber cable. Additionally, a TMA method was described for measuring the amount of bowing in a microelectronic package. Finally, heat distortion measurements were used to set crystallinity conditions to enhance mechanical stability in a glass-filled poly(phenylene sulfide) optoelectronic device.

APPENDIX

Standards for thermomechanical analysis issued under the jurisdiction of ASTM International Committee E37 on Thermal Measurements are listed below for reference (for more information, see Website www.astm.org).

E228, *Test Method for Linear Thermal Expansion of Solid Materials with a Vitreous Silica Dilatometer*

E473, *Definition of Terms Relating to Thermal Analysis*

D696, *Coefficient of Linear Thermal Expansion of Plastics*

E831, *Test Method for Linear Thermal Expansion of Solid Materials by Thermomechanical Analysis*

E1363, *Test Method for Temperature Calibration of Thermomechanical Analyzers*

E2092, *Standard Test Method for Distortion Temperature in Three-Point Bending by Thermomechanical Analysis*

D3418, *Transition Temperature of Polymers by Thermal Analysis*

STP23594S, *The Use of Thermomechanical Analysis as a Viable Alternative for the Determination of the Tensile Heat Distortion Temperature of Polymer Films*

D1525, *Standards to Determine Vicat Softening Point*

D648, *Deflection Temperature of Plastics under Flexural Load*

TABLE 4A.1. CLTE and T_g of Commercially Available Plastic Materials as Measured by TMA

Polymer	CLTE $\times 10^6$/°C Test Method ASTM D696	DSC T_g °C Test Method ASTM D3418
Rigid PVC	9–18	81
Flexible PVC[a]	13–45	Varies
PTFE	99	−100
HDPE	100–200	N/A
LDPE[b]	250	−35
LLDPE[c]	250	−35
ULDPE[c]	250	−38
PP	90	−20
HIPS[d]	90	93–105
SAN[e]	67	115
ABS[f]	67	110
PC[d]	68	230
Nylon 6[e]	81	220
Nylon 66[d]	81	255
PMMA[e]	50–90	105
Silicone[g]	10–19	—

[a]Range includes plasticizers and fillers.
[b]General-purpose for film and molding.
[c]For blown film.
[d]General-purpose grade.
[e]Injection-molding grade.
[f]30% glass-filled.
[g]Flexible casting resin.
Source: Brandrup et al. (1999).

ABBREVIATIONS

Symbols

$\varnothing$	excess free volume per chain end
α	coefficient of linear thermal expansion (CLTE)
γ	coefficient of volumetric thermal expansion
Δ	difference
χ	Flory–Huggins polymer–solvent interaction parameter
ε	strain
σ	stress
η	viscosity
A	cross-sectional area
a	thickness
a_T	temperature shift factor
d	density
E	Young's modulus
f	force
J	creep compliance, Joules
L	length
M_n	number average molecular mass
P	pressure
P^*	characteristic pressure
r	diameter
S	shrinkage
S_ε	degree of swelling
T	temperature
T^*	characteristic temperature
T_g	glass transition temperature
T_m	melting temperature
v	specific volume
v^*	hard-core volume
v_f	free volume
v_{ff}	fractional free volume

Acronyms

ABS	acrylonitrile–butadiene–styrene copolymer
ASTM	ASTM International
CLTE	coefficient of linear thermal expansion
CRC	chain relaxation capability
CTE	coefficient of thermal expansion (usually volume)
DEA	dielectric analysis
DMA	dynamic mechanical analysis
DSC	differential scanning calorimetry
FDC	first derivative computer

GF	glass-filled
GFPPS	glass-filled polyphenyl sulfide
HDPE	high-density polyethylene
HDT	heat distortion (or deflection) temperature
HIPS	high-impact polystyrene
IC	integrated circuit
ICTAC	International Confederation for Thermal Analysis and Calorimetry
IUPAC	International Union of Pure and Applied Chemistry
LDPE	low-density polyethylene
LVDT	linear voltage differential transformer
NIST	National Institute of Standards and Technology
NMP	N-methyl pyrrolidone
OJ	outer jacket
PC	polycarbonate
PET	poly(ethylene terephthalate)
PHEA	poly(2-hydroxyethyl acrylate)
PI	polyimide
PMMA	poly(methyl methacrylate)
PP	polypropylene
PPS	poly(phenylene sulfide)
PTFE	polytetrafluoroethylene
PVC	poly(vinyl chloride)
PWB	printed writing board
RT	room temperature
SAN	styrene–acrylonitrile–styrene
TD	thermodilatometry
TGA	thermogravimetric analysis
TMA	thermomechanical analysis
TTS	time–temperature superposition
ULDPE	ultra-low-density polyethylene
WLF	Williams–Landel–Ferry

REFERENCES

Akinay, A. E. and Brostow, W. (2001), *Polymer* **42**, 4527.

Akinay, A. E., Brostow, W., and Maksimov, R. M. (2001), *Polym. Eng. Sci.* **41**, 977.

Akinay, A. E., Brostow, W., Castano, V. M., Maksimov, R., and Olszynski, P. (2002), *Polymer* **43**, 3593.

Akinay, A. E., Brostow, W., Ertepinar, H., and Lopez, B. (1993), *POLYCHAR-3 Conf. Proc.*

Aklonis, J. J. and McKnight, W. J. (1983), *Introduction to Polymer Viscoelasticity*, 2nd ed., Wiley, New York.

Bair, H. E. and Warren, P. C. (1981), *J. Macromol. Sci.*, Part B, **20**, pp. 381–402.

Bair, H. E. (1997), "Thermal analysis of additives in polymers," in *Thermal Characterization of Polymeric Materials*, 2nd ed., Turi, E. A., ed., Academic Press, New York, 2263–2420.

Bair, H. E., Colvin, V. L., and Schilling, M. L. (1997), *Proc. 25th NATAS Conf.* pp. 374–381.

Bair, H. E., Hale, A., and Schilling, M. L. (2000), US Patent 6,016,696.

Bair, H. E., Schilling, M. L., Colvin, V. L., Hale, A., and Levinos, N. J. (1998), *Polym. Mater. Sci. Engi* **78**, 230.

Bair, H. E. and Osenbach, J. W. (2004), US Patent 6,770,224 B2.

Bird, R., Curtis, C., Armstrong R., and Hassenger, O. (1987), *Dynamics of Polymer Fluids*, 2nd ed., Wiley, New York.

Blyler, L. L., Jr., DiMarcello, E. V., Hart, A. C., and Huff, R. G. (1987), Series 346, *Polymers for High Technology*, Bowden, M. J. and Turner, S. R., eds., ACS Sym. Series 346, pp. 410–416.

Boiko, T. M., Brostow, W., Goldman, A. Y., and Ramamurthy, A. C. (1995), *Polymer* **36**, 1383.

Brandrup, J., Immergut, E. E., and Grulke, E. A. eds. (1999), *Polymer Handbook*, 4th ed., Wiley, New York, pp. 159–169.

Brostow, W. (1985), *Mater. Chem. Phys.* **13**, 47.

Brostow, W. (1990), *Polymer* **31**, 979–995.

Brostow, W. (2000), *Chain Relaxation Capability in Performance of Plastics*, Hanser/Gardner, Munich/Cincinnati, Chapter 5.

Brostow, W. and Macip, M. A. (1989), *Macromolecules* **22**, 2761.

Brostow, W. and Szymanski, W. J. (1986), *Rheology* **30**, 877.

Brostow, W. D'Souza, N. A., and Gopalanarayanan, B. (2000), *Polym. Eng. Sci.* **40**, 490–498.

Brostow, W., Faitelson, E. A., Kamensky, M. G., Karkhov, V. P., and Rodin, Y. P. (1999a), *Polymer* **40**, 1441–1449.

Brostow, W., D'Souza, N. A., Kubat, J., and Maksimov, R. D. (1999b), *Intl. J. Polym. Mater.* **43**, 233–248.

Brostow, W., Duffy, J. V., Lee G. F., and Madejczyk, K. (1991), *Macromolecules* **24**, 479.

Brostow, W., Hess, M., and Lopez, B. E. (1994), *Macromolecules* **27**, 2262.

Brostow, W. and Kubat, M. J. (1996), *Mechanical Properties in Physical Properties of Polymers Handbook*, Mark, J. E. ed., American Institute of Physics, Woodbury, MN, Chapter 23.

Bunn, C. W. and Howells, E. R. (1954), *J. Appl. Phys.* **174**, 549.

Cassel, B. (1976) *Perkin-Elmer Thermal Analysis Application Study (TAAS)* 19.

Cassel, B. (1980) *Polym. News* **6**(3), pp. 108–115.

Cecccorulli, G., Pizzoli, M., Scandola, M., and Pezzin, G. (1983), *Polym. Commun.* **24**, 107.

Cessna, L. C., Jr. and Jabloner, H. (1974), *J. Elastom. Plast.* **6**, 103–113.

Charsley, E. L., Laye, P. G., Palakollu, V., Rooney, J. J. and Joseph, B. (2006), "DSC studies on organic melting point temperature standards, Thermochim. *Acta* **446**, pp. 29–34.

Chartoff, R. P. (1997), *Thermoplastic Polymers in Thermal Characterization of Polymeric Materials*, 2nd ed., Turi, E. A., ed., Academic Press, New York, Chapter 3.

Doolittle, A. K. (1951), *J. Appl. Phys.* **22**, 1741.

Earnest, C. M. and Seyler, R. J. (1992), "Temperature calibration of thermomechanical analyzers: Part I—The development of a standard method," *J. Test. Eval.* **20**, 430.

Eisenberg, A. (1993), *The Glassy State and the Glass Transition in Physical Properties of Polymers*, 2nd ed., Mark, J. E., Eisenberg, A., Graessley, W. W., Mandelkern, L., Samulski, E. T., Koening, J. L., and Wignall, G. D. American Chemical Society (ACS), Washington, DC, Chapter 2.

Ferrer, G. G., Pradas, M. M., and Sanchez, M. S. (2004), *Polymer* **45**, 6207–6217.

Ferry, J. D. (1980), *Viscoelastic Properties of Polymers*, 3rd ed., Wiley, New York.

Flory, P. J. (1955), *Trans. Faraday. Soc.* **51**, 848.

Flory, P. J. (1970), *Faraday Soc. Discuss.* **49**, 7.

Flory, P. J. and Rehner, J. (1943), *J. Chem. Phys.* **11**, 512–526.

Ford, J. L. and Mitchell, K. (1995), *Thermochim. Acta* **248**, 329–345.

Gedde, U. W. (2001), *The Glassy Amorphous State in Polymer Physics*, 4th ed., Kluwer Academic, Amsterdam, Chapter 5.

Grimau, M., Laredo, E., Bello, A., and Müller, A. (1997), *Am. Phys. Soc., Annual March Meeting*, March 17–21, Abstract J32.42.

Grebowicz, J. S., Brown, B., Chuah, H., Olvera, J. M., Wasiak, A., Sajkiewicz, P., and Ziabicki, A. (2001), *Polymer*, **42**(16), pp. 7153–7160.

Hartman, B. and Haque, M. A. (1985a), *J. Appl. Phys.* **58**, 2831.

Hartman, B. and Haque, M. A. (1985b), *J. Appl. Polym. Sci.* **30**, 1553.

Iijima, M., Hataketama, T., and Hatakeyama, H. (2005), Swelling behavior of calcium pectin hydrogels by thermomechanical analysis in water, *Thermochim. Acta* **431**, 68–72.

IUPAC (1989), *Pure Appl. Chem* **61**, 769 (IUPAC "macromolecular nomenclature").

Jaffe, M., Menczel, J. D., and Bessey, W. E. (1997), *Thermal Characterization of Polymeric Materials*, 2nd ed., Turi, E. A., ed., Academic Press, New York, Chapter 7.

Johnson, B., Niedermaier, D. J., Crone, W. C., Moorthy, J., and Beebe, D. J. (2002), *Society for Experimental Mechanics, 2002 SEM Annual Conf. Proc.*

Kaye, A., Stepto, R. F. T., Work, W. J., Aleman, J. V., and Malkin, A. Ya. (1998), *Pure Appl. Chem.* **70**(3) 701–754.

Kelly, F. N. and Bueche, F. J. (1961), *J. Polym. Sci.* **50**, 49.

Keum, J. K. and Song, H. H. (2005), *Polymer* **46**, 939.

Kimmig, M., Strobl, G., and Stuhn, B. (1994). *Macromolecules* **27**, 2481.

Lakes, R. S. (1999). *Viscoelastic Solids*, CRC Press, Boca Raton, FL.

Lide, R. D. (1998), *Handbook of Chemistry and Physics*, 79th ed., CRC Press, New York, pp. 6–175.

Litt, M. H. (1976), "Free volume and its relationship to the temperature effect on zero shear melt viscosity: A new correlation," *J. Rheol.* **20**, 47–64.

Litt, M. H. (1986), "Free volume concepts connecting PVT behavior and gaseous diffusion through polymers," *J. Rheol.*, **30**, 853–868.

Litt, M. H. and Tobolsky, A. V. (1967), "Cold flow of glassy polymers. II: Ductility, impact reistance, and unoccupied volume," *J. Macromol. Sci. Phys.* **B1**(3), 433–443.

Lotti, C. and Canevarolo, V. (1998), "Temperature calibration of a DMA thermal analyser," *Polym. Tes.* **17**, 523–530.

Mackenzie, R. C. (1979), *Thermochim. Acta* **18**, 421.

Mano, J. F. and Cahon, J. P. (2004), "A simple method for calibrating the temperature in dynamic mechanical analysers and thermal mechanical analysers," *Polym. Test.* **23**, 423–430.

Mark, J. E., Eisenberg, A., Graessley, W. W., Mandelkern, L., Samulski, E. T., Koening, J. L., and Wignall, G. D. (1993), *Physical Properties of Polymers*, 2nd ed., ACS, Washington, DC.

Matsuoka, S. (1992), *Relaxation Phenomena in Polymers*, Hanser, New York.

Matsuoka, S. (1997), "Entropy, free volume, and cooperative relaxation," *J. Res. Natl. Bur. Stand. Technol.* **102**, 213–228.

McKenna, G. B. and Simon, S. L. (2002), in *Handbook of Thermal Analysis and Calorimetry*, Vol. 3, *Applications to Polymers and Plastics*, Cheng, S. D., ed., Elsevier Science B. V., New York, pp. 49–109.

Menard, K. P. (1996), *Perkin-Elmer Thermal Analysis Application Note: Thermomechanical Analysis Basics*, Part I.

Menczel, J. D., Jaffe, M., and Bessey, W. E. (1997), *Thermal Characterization of Polymeric Materials*, 2nd ed., Turi, E. A., ed., Academic Press, New York, Chapter 8.

Moscato, M. J. and Seyler, R. J. (1994), *Assignment of the Glass Transition*, Seyler, R. J., ed., ASTM STP 1249, Baltimore, p. 239.

Nakamura, K., Kinoshita, E., Hatakeyama, T., and Hatakeyama, H. (2000), *Thermochim Acta* **352–353**, 171–176.

Neag, M. (1991), "Thermomechanical analysis in materials science," in *Materials Characterization by TMA*, Neag, R, ed., ASTM STP1136, Philadelphia.

Neag, M. (1995), "Coating characterization by thermal analysis," *ASTM Manual* **17**, p. 841.

Ngai, K. L. (2003), "The glass transition and the glassy state," in *Physical Properties of Polymers*, 3rd ed., Mark, J., Ngai, K. L., Graessley, W., Mandelkern, L., Samulski, E., Koening, J. and Wignall, G. eds., Cambridge Univ. Press, Chapter 2.

Olivier, P. A. (2006), *Composites, Pt. A* **37**, 602–616.

Perkin-Elmer Thermal Analysis Application Application Note: Effect of Temperature Calibration and Purge Gas on TMA CTE Analysis, PETech-122.

Peȝron, E. and Magny, B. (1996), *UV/EB Conf. Proc. RadTech*, pp. 99–106.

Price, D. M. (1998a), "Modulated temperature thermomechanical analysis," *Proc. 26th NATAS Conf.*

Price, D. M. (1998b), *J. Therm. Anal.* **51**, 231–236.

Price, D. M. (1998c), *Thermochim. Acta* **315**, 11–18.

Price, D. M. (2000), "Modulated temperature thermomechanical analysis," *Thermochim. Acta* **357–358**, 23–29.

Prime, R. B. (1985), "Creep compliance measurements by TMA-parallel plate rheometry," *J. Therm. Anal.* **30**, 1001–1011.

Prime, R. B. (1997), *Thermosets in Thermal Characterization of Polymeric Materials*, 2nd ed., Turi, E. A., ed., Academic Press, New York, Chapter 6.

Prime R. B., Barrall, E. M. II, Logan, J. A., and Duke, P. J. (1974), *Proc. 17th AIP Conf.*, pp. 72–83.

Rabek, J. F. (1977), Basics of Physical Chemistry of Polymers. Publishing of Polytechnic Institute of Wroclaw, Wroclaw, Poland. *Podstawy Fizykochamii Polimerow*, Wyndawnictwo Politecnici Wroclawskiej, Wroclaw.

Riga, A. T. and Collins, E. (1991), *Material Characterization by Thermomechanical Analysis: Industrial Applications*, ASTM STP1136, pp. 71–83.

Sanchez, S. M., Pradas, M. M., and Ribelles, J. G. (2004), *Eur. Polym. J.* **40**, 329–334.

Seyler R. J. and Earnest, C. M. (1992), "Temperature calibration of thermomechanical analyzers: Part II—An interlaboratory test of the calibration procedure," *J. Test. Eval.* **20**, 434.

Seyler, R. J. and Earnest, C. M. (1991), "Temperature calibration of thermomechanical analyzers: Part II—An interlaboratory test of the calibration procedure," in *Materials Characterization by Thermomechanical Analysis*, Riga, A. T. and Neag, M., eds., ASTM Publications, STP1136, Philadelphia.

Sichina, W. J. (2000), *Perkin-Elmer Thermal Application Notes* 28.

Sideridou, I., Achilias, D. S., and Kyrikou, E. (2004), *Biomaterials* **25**, 3087–3097.

Simha, R. and Boyer, R. F. (1962), *J. Chem. Phys.* **37**, 1003.

Simha, R. and Wilson, P. (1973), *Macromolecules* **6**, 908.

Sircar, A. K. (1997), in *Elastomers in Thermal Characterization of Polymeric Materials*, 2nd ed., Turi, E. A., ed., Academic Press, New York, Chapter 5.

Sircar, A. K. and Chartoff, R. P. (1994), in *Assignment of the Glass Transition*, Seyler, R. J., ed., ASTM STP1249, pp. 226–238.

Struik, L. C. E. (1978), *Physical Aging in Amorphous Polymers and Other Materials*, Elsevier, New York.

Struik, L. C. E. (1986), *Failure of Plastics*, Brostow, W. and Corneliussen, R. D., eds., Hanser, New York.

Suhir, E. (1993), *J. Reinforced Plast. Compos.* **12**, 953.

TA Instruments, *Thermal Analysis & Rheology Application Note* TA-138.

TA Instruments, *Thermal Analysis Aplication Note*, TA108.

TA Instruments, *Thermal Analysis Application Note*, TA208.

TA Instruments, *Thermal Analysis Aplication Note*, TA287.

TA Instruments, *Thermal Analysis Aplication Note*, TA311.

Turi, E. A., Khanna, Y. P., and Taylor, T. J. (1988), in *A Guide to Materials Characterization and Chemical Analysis*, Sibilia, J. B., ed., VCH, New York, Chapter 9.

Vrentas, J. D., Duda, J. L., and Huang, J. W. (1986), *Macromolecules* **19**, 1718.

Williams, M. J., Landel, R. F., and Ferry, J. D. (1955), *J. Am. Chem. Soc.* **77**, 3701.

Wu, G., Yoshida, T., and Cuculo, J. A. (1998), *Polymer* **39**, 6473–6482.

Zachariades, A. E., Mead W. T., and Porter, R. S. (1979), in *Ultrahigh Modulus Polymers*, Ciferi, A. and Ward, J. M., eds., Applied Science, London.

CHAPTER 5

DYNAMIC MECHANICAL ANALYSIS (DMA)

RICHARD P. CHARTOFF
The University of Dayton, Dayton, Ohio

JOSEPH D. MENCZEL
Alcon Laboratories, Fort Worth, Texas

STEVEN H. DILLMAN
Western Washington University, Bellingham, Washington

5.1. INTRODUCTION

Polymers are viscoelastic materials, whose mechanical behavior exhibits characteristics of both solids and liquids. Thermal analysts are frequently called on to measure the mechanical properties of polymers for a number of purposes. Of the different methods for viscoelastic property characterization, dynamic mechanical techniques are the most popular, since they are readily adapted for studies of both polymeric solids and liquids. They are often referred to collectively as *dynamic mechanical analysis* (DMA). Thermal analysts often refer to the DMA measurements on liquids as *rheology* measurements.

Dynamic mechanical analysis involves imposing a small cyclic strain on a sample and measuring the resulting stress response, or equivalently, imposing a cyclic stress on a sample and measuring the resultant strain response. In most commercial DMA instruments strain is the controlled input, while the resulting stress is measured. This is shown in Fig. 5.1. DMA is used both to study molecular relaxation processes in polymers and to determine inherent mechanical or flow properties as a function of time and temperature. Applications for which DMA is used are listed in Table 5.1. The primary use of these techniques for the study of polymeric solids and liquids is well documented. Excellent general discussions covering the subject are provided in several well-known references [e.g., McCrum et al. (1967); Murayama (1978); Ferry (1980); Nielsen and Landel (1994)].

Thermal Analysis of Polymers: Fundamentals and Applications, Edited by Joseph D. Menczel and R. Bruce Prime
Copyright © 2009 by John Wiley & Sons, Inc.

- An oscillatory (sinusoidal) input (stress or strain) is applied to a sample
- The material response (strain or stress) is measured
- The phase angle δ, or phase shift between the input and response, is measured

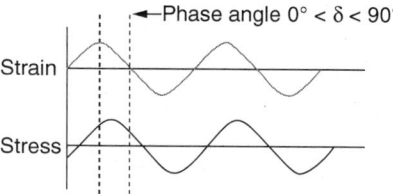

Figure 5.1. Basic principle of DMA operation; a sinusoidal strain is applied to a sample, and the resulting sinusoidal stress is measured. The strain and stress are out of phase (courtesy of TA Instruments).

TABLE 5.1. Applications of DMA and Structure-Property Characterization

Dynamic Mechanical Analysis

1. Detect transitions arising from molecular motions or relaxations
2. Determine mechanical properties, i.e., modulus and damping of viscoelastic materials over spectrum of time (frequency) and temperature
3. Develop structure–property or morphology relationships

Polymer Structure–Property Characterization

1. Glass transition
2. Secondary transitions
3. Crystallinity
4. Molecular mass/crosslinking
5. Phase separation (polymer blends, copolymers, polymer alloys)
6. Composites
7. Aging (physical and chemical)
8. Curing of networks
9. Orientation
10. Effect of additives (plasticizers, moisture)

Linear amorphous polymers exist in a number of characteristic physical states depending on the timescale of the measurement and temperature. These are illustrated in Fig. 5.2 in terms of an arbitrary modulus function and are classified as glassy, leathery, rubbery, rubbery flow, and viscous (Tobolsky 1960; Collins et al. 1973). All linear amorphous polymers exhibit these five physical states when they are observed over a wide range of time or temperature. Materials of this type are typical of amorphous thermoplastics, such as polystyrene (PS), poly(methyl methacrylate) (PMMA), or polycarbonate (PC) polymers. Polymers that are either crosslinked or crystalline do not exhibit the rubbery flow and viscous liquid responses as illustrated. Crystalline polymers, however, will exhibit a viscous response at temperatures above the melting transition.

INTRODUCTION

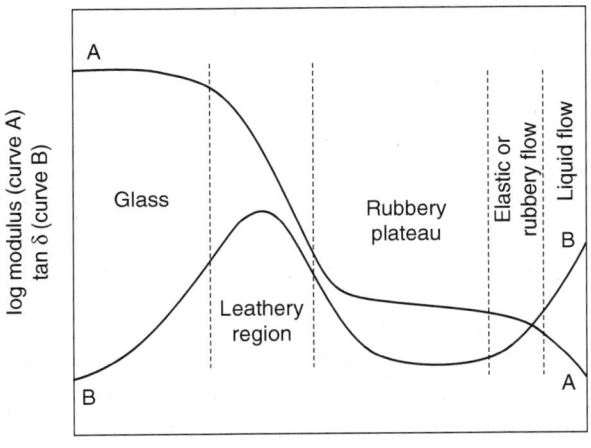

Figure 5.2. Illustration of loss tangent (damping) and storage modulus characteristics for a linear amorphous polymer illustrating various distinct physical states that the polymer assumes with varying temperature or frequency (Collins et al. 1973). Reprinted with permission of John Wiley and Sons, Inc.

The left region in Fig. 5.2 corresponds to the glassy state. The elastic modulus value (curve A, indicating stiffness) in this region is fairly constant at ~10^9 Pa. This is followed by the glass transition, where the polymer changes from a glassy to rubbery consistency. The glass transition serves as an important benchmark or corresponding state for viscoelastic response. In the transition region the modulus decreases by fully three orders of magnitude to ~10^6 Pa. Starting out as a rigid glass, the polymer then softens, varying from a stiff leathery consistency to a soft, pliable consistency. Above the glass transition in the rubbery plateau region, the polymer has the truly elastic characteristics that we associate with rubbers. Modulus values for this region are ~10^6 Pa. At higher temperatures (or lower frequencies) the polymer exhibits elastic behavior but a noticeable tendency to flow, termed "rubbery flow." In this region modulus values are ~10^5 Pa. To the right of the diagram the material assumes the characteristics of a liquid with modulus values falling below 10^5 Pa. Because the glass transition corresponds to the physical change from glassy to rubbery behavior, it is frequently referred to in the literature as the "glass-to-rubber transition."

The particular physical state that the polymer exhibits during a DMA test depends on the ratio of the timescale of the measurement (in this case the inverse of frequency) to the speed of molecular motion. In general, the modulus decreases either with a decreasing frequency or an increasing temperature. There is a strong relationship between time and temperature, as will be discussed later in this chapter. The modulus values cited above for each of the five regions are fully characteristic of each physical state (Tobolsky 1960;

Collins et al. 1973; Ferry 1980; Nielsen and Landel 1994). They serve as a quantitative measure of the physical state in which the polymer exists.

Since DMA is a technique for viscoelastic characterization, understanding the basics of DMA is facilitated by first considering the behavior of ideal elastic solids and ideal viscous liquids, which are briefly described below, followed by discussion of the characterization of viscoelastic behavior and the role of DMA in relation to other test methods. Following this, a major portion of this chapter is concerned with the interpretation of DMA data, citing numerous examples of what DMA data look like and how the data can be used effectively to solve important problems in polymer applications. Finally, some of the key factors that influence viscoelastic properties are described, along with other important issues in conducting DMA measurements and determining accurate data.

5.1.1. Ideal Elastic Solids

If a solid of constant cross section A_0 and an initial length l_0 is subjected to a tensile force F, it will deform in the direction of the force, as shown in Fig. 5.3. The stress σ, having units of force per area, is defined as

$$\sigma = \frac{F}{A_0} \tag{5.1}$$

and engineering strain ε, having units of length per length, is defined as

$$\varepsilon = \frac{l - l_0}{l_0} \tag{5.2}$$

where l is the length of the sample at any moment.

From a mechanical perspective an *ideal solid* may be defined as a material in which strain is proportional to stress in conformance to Hooke's law

$$\sigma = E \cdot \varepsilon \tag{5.3}$$

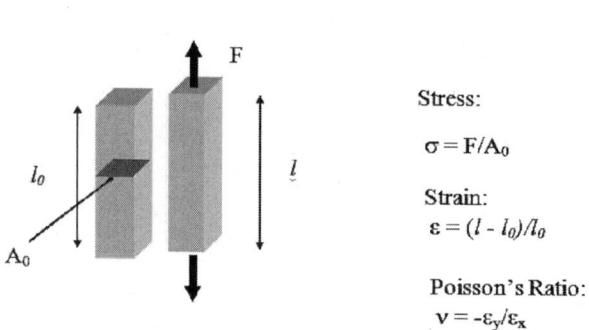

Figure 5.3. Sample under tensile load showing the deformation in the direction of applied force and the corresponding definitions of stress and strain.

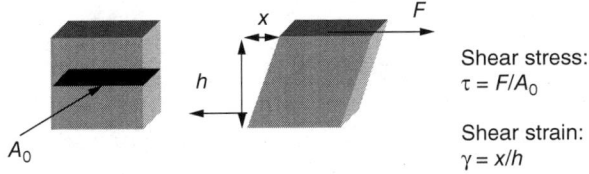

Figure 5.4. Sample under shear load showing the resulting deformation in the direction of applied force and the corresponding definitions of stress and strain.

where the constant of proportionality E, having units of force per area, is known as the *modulus of elasticity*, *Young's modulus*, the *elastic modulus*, the *tensile modulus*, or simply the *modulus*.

Similarly, for a material subjected to a shear force as shown in Fig. 5.4, a shear stress τ and shear strain γ may be defined as follows:

$$\tau = \frac{F}{A_0} \tag{5.4}$$

$$\gamma = \frac{x}{h} \tag{5.5}$$

In a direct analogy to the tensile case, a Hooke's law of shear may be used to define the elastic shear modulus:

$$\tau = G \cdot \gamma \tag{5.6}$$

For isotropic materials the shear modulus and tensile modulus are related by the expression

$$E = 2G(1+\nu) \tag{5.7}$$

where ν is the Poisson ratio of the material, the ratio of lateral contraction to longitudinal expansion due to the tensile force. For soft elastic solids $\nu = 0.5$; for glassy polymers, ν is in the range 0.3–0.4—often about 0.33.

In addition to tension and shear, experiments may be done in a flexural mode of deformation, as shown in Fig. 5.5. This is sometimes referred to as *three-point bending* by thermal analysts. For a sample of length L, width w, and thickness h, loaded as shown in the figure with a load P, the flexural modulus (E_f), is defined by

$$E_f = \frac{L^3 P}{4wh^3 \delta} \tag{5.8}$$

where δ is the linear deflection at the point of loading. This corresponds to the deflection that would be observed in an ideal isotropic material with tensile modulus E_f under identical loading conditions. Thus, for practical purposes the

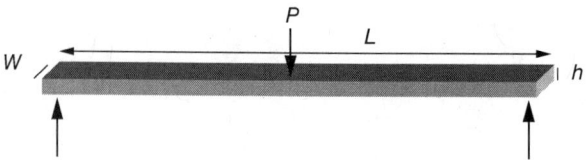

Figure 5.5. Sample under flexural load; force is applied at a central point and the sample is supported at the ends on pivot points.

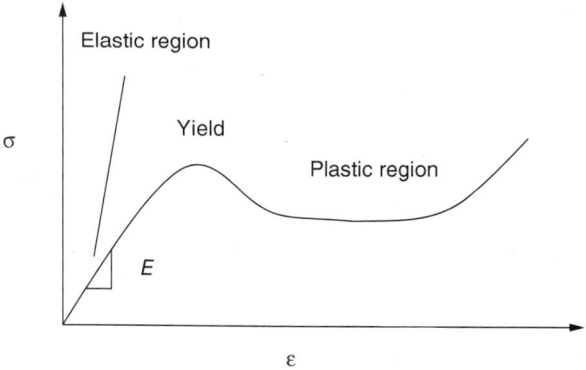

Figure 5.6. Typical stress–strain behavior for a ductile material showing initial elastic behavior (linear) followed by plastic deformation beyond the yield point and fracture of the specimen at the ultimate strain limit.

flexural modulus can be regarded as an estimate of the tensile modulus made from measurements taken in a flexural configuration. For isotropic materials the flexural and tensile moduli are usually nearly identical, but there may be large differences for anisotropic materials such as composites.

For an ideal elastic solid the modulus is a constant, independent of strain, strain rate, time, and temperature. Thus, a tensile stress–strain plot is a straight line passing through the origin with slope E as shown by the initial linear portion of the curve in Fig. 5.6. As the figure illustrates, real solids show ideal behavior over only limited ranges of strain, strain rate, time, and temperature. While extremely brittle materials (such as thermosets) may exhibit linear behavior up to the point of fracture, most materials that are normally regarded as elastic solids will exhibit linear stress–strain behavior only for small deformations of up to a few percent. At larger deformations they generally deviate from linearity. Other materials, such as elastomers, have only very small linear regions. In cases such as these the modulus is considered to be constant only at very small strains, above which it becomes a function of the strain magnitude. In addition, real solids often exhibit time-dependent behavior, with the observed modulus increasing with increasing rate of strain. Finally, the modulus of a real solid is temperature-dependent. For most materials the modulus decreases with increasing temperature, but the modulus of elastomers often increases as temperature increases in accord with the theory of rubber elasticity (Tobolsky 1960; Sperling 2006).

5.1.2. Ideal Liquid Behavior

If a fluid is placed between two large flat plates, one of which is moving at a velocity v_x relative to the other (as illustrated in Fig. 5.7), a force will be required to keep the two plates in relative motion. The shear stress can be defined exactly as in Eq. (5.4). The strain changes with time because the plates are moving relative to one another, but as long as the velocity remains constant the rate of change of strain $\dot{\gamma}$ is constant:

$$\dot{\gamma} = \frac{d\gamma}{dt} = \frac{dv_x}{dy} \tag{5.9}$$

The strain rate is commonly referred to as the *shear rate*.

An ideal fluid (referred to as a *Newtonian fluid*) is defined as one that conforms to *Newton's law of viscosity*, in which the viscosity is independent of the strain rate. Newton's law states that the shear stress and strain rate are proportional, with the constant of proportionality, called the *viscosity*, μ:

$$\tau = \mu\dot{\gamma} \tag{5.10}$$

Many real fluids, including most gases and many liquids, are Newtonian over normal ranges of strain rate and temperature. However, polymers are generally non-Newtonian, except over narrowly defined ranges of strain rate. The viscosity of a polymer is usually given the symbol η in the viscosity equation to distinguish it from μ, since it is a function of strain rate:

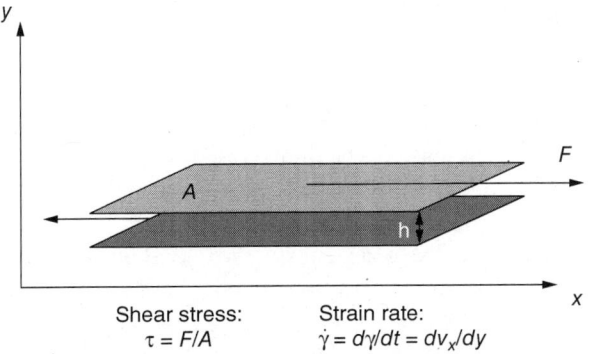

Shear stress: $\tau = F/A$
Strain rate: $\dot{\gamma} = d\gamma/dt = dv_x/dy$

Figure 5.7. Illustration of two parallel plates separated by a fluid under shear, indicating the corresponding definitions of shear stress and strain rate.

$$\tau = \varphi(\dot{\gamma})\dot{\gamma} \tag{5.11}$$

where $\varphi(\dot{\gamma}) = \eta$ is the strain rate dependent viscosity.

Numerous mathematical models exist that describe the strain rate dependence of viscosity (Carreau et al. 1997), but they are not important to the understanding of DMA.

5.1.3. Viscoelastic Materials

Viscoelastic materials, such as polymers, exhibit behavior that is intermediate between that of an ideal solid and that of an ideal liquid, showing characteristics of both. A classic example is "silly putty," which can be rolled into a ball that will bounce elastically like a solid, but if placed on a table will slowly flow over a period of several hours into a puddle like a liquid. When it is stretched slowly, it elongates continuously, but when stretched quickly, it fractures like a brittle solid. The timescale and temperature of observation are critical to the relative degree of solid- and liquidlike behavior exhibited by viscoelastic materials. Generally, they will behave more solidlike at lower temperatures or over shorter timescales, but more liquidlike at higher temperatures or longer timescales.

5.2. CHARACTERIZATION OF VISCOELASTIC BEHAVIOR

There are three fundamental test methods for characterization of the viscoelastic behavior of polymers: creep, stress relaxation, and dynamic mechanical analysis. Although the primary focus for this chapter is DMA, it is useful first to discuss the fundamentals of creep and stress relaxation, not only because they are conceptually simpler but because most DMA instruments also are capable of operating in either a creep or stress relaxation mode. All three of the methods are related, and numerical techniques are available for calculating creep and stress relaxation data from dynamic mechanical data (Ferry 1980).

5.2.1. Creep Testing

In a creep test a sample is placed under a constant stress, and strain is recorded as a function of time. If an ideal elastic solid is subjected to a creep test, it will exhibit an immediate elastic strain in accordance with Hooke's law, but the strain will remain constant thereafter until the stress is removed, when the sample will return elastically to zero strain. An ideal liquid responds to a creep test quite differently. There is no initial elastic response, and there will be a continuously increasing strain with a slope inversely proportional to the viscosity; the strain rate will remain constant. When the stress is removed, there is no elastic recovery—the liquid simply stops flowing; in other words, the strain rate returns to zero.

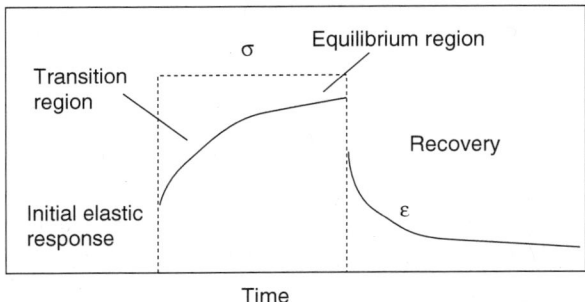

Figure 5.8. Typical stress (----) and strain (—) curves as a function of time during a creep experiment.

We noted that a viscoelastic material exhibits behavior intermediate between those of an ideal solid and an ideal liquid. Figure 5.8 shows an idealized creep curve for a viscoelastic material. As indicated in the figure, there are four distinct regions on the curve: the initial *elastic response region*, in which the strain is given by Hooke's law using the zero-time modulus; a *transition region*, where a nonlinear increase in strain occurs; an *equilibrium region*, in which strain increases linearly with time, characteristic of viscous flow; and, on removal of the stress, a *recovery region* that is characterized by an initial elastic retraction followed by a strain decay. Note that the strain never returns to zero, but asymptotically approaches a nonzero value. This is known as *permanent set*, and is another example of liquidlike behavior. The amount of permanent set increases with the time over which the stress is applied.

Because the strain in a creep test is time-dependent (while the stress is constant), creep data are commonly reported as a time-dependent creep compliance:

$$J(t) = \frac{\varepsilon(t)}{\sigma} \tag{5.12}$$

Note that (to a first approximation) the compliance is simply the inverse of the modulus.

5.2.2. Stress Relaxation

In a stress relaxation test a sample is quickly placed under a strain that is then held constant, and the resulting stress is recorded as a function of time. The response of an ideal elastic solid in stress relaxation is a stress that remains constant with time, while an ideal liquid responds with an immediate return to zero stress as soon as the test strain is imposed. Viscoelastic materials respond with a stress that decays with time. Stress relaxation data are commonly reported as a time-dependent stress relaxation modulus:

$$E(t) = \frac{\sigma(t)}{\varepsilon} \qquad (5.13)$$

5.2.3. Dynamic Mechanical Analysis

In contrast to creep or stress relaxation, DMA involves applying a sinusoidal strain (or stress) to a sample and measuring the corresponding stress (or strain) developed. Both approaches are theoretically identical. Consider what happens if a sinusoidal strain is applied to an ideal elastic solid:

$$\varepsilon(t) = \varepsilon_0 \cdot \sin(\omega t) \qquad (5.14)$$

At any point in time the stress will be proportional to the strain in accordance with Hooke's law:

$$\sigma(t) = E \cdot \varepsilon(t) = E \cdot \varepsilon_0 \cdot \sin(\omega t) = \sigma_0 \cdot \sin(\omega t) \qquad (5.15)$$

Thus, for an ideal solid the stress will be a sinusoidal function in phase with the strain, as shown in Fig. 5.9, and the ratio of amplitudes of stress and strain will be the modulus of the material:

$$E = \frac{\sigma_0}{\varepsilon_0} \qquad (5.16)$$

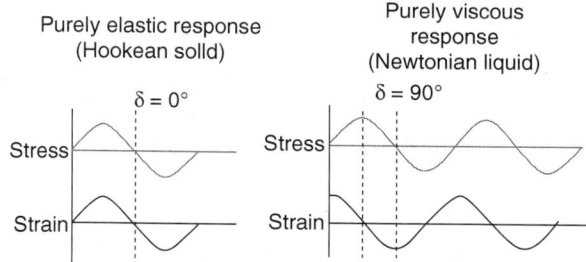
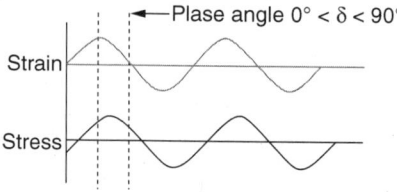

Figure 5.9. DMA sinusoidal stress–strain response curves for ideal elastic, ideal viscous, and viscoelastic materials illustrating that the phase shift between stress and strain for a viscoelastic material lies between those for the ideal materials (courtesy of TA Instruments).

CHARACTERIZATION OF VISCOELASTIC BEHAVIOR

Now consider what happens if a sinusoidal shear strain is applied to an ideal liquid:

$$\gamma = \gamma_0 \cdot \sin(\omega t) \tag{5.17}$$

At any point in time the stress will be proportional to the *strain rate* in accordance with Newton's law of viscosity:

$$\tau(t) = \eta \cdot \dot{\gamma}(t) = \eta \cdot d\gamma(t)/dt = \eta \cdot \gamma_0 \cdot \omega \cdot \cos(\omega t) = \eta \cdot \gamma_0 \cdot \omega \cdot \sin\left(\omega t + \frac{\pi}{2}\right) \tag{5.18}$$

Thus, for an ideal liquid the stress will be a sinusoidal function 90° out of phase with the strain as shown in Fig. 5.9.

This 90° phase difference between sinusoidal stress and strain in liquids is the key to the use of DMA as a tool for the characterization of viscoelastic materials. Since a viscoelastic material has properties intermediate between those of an ideal solid and an ideal liquid, it exhibits a phase lag somewhere between 0° (ideal solid) and 90° (ideal liquid), also shown in Fig. 5.9. Thus, DMA applies a given strain and measures the resulting stress as well as the relative amplitudes of stress and strain (the modulus) and the phase lag, which is a measure of the relative degree of viscous character to elastic character.

When a sinusoidal strain is initially applied to the sample, there is a delay of at least one cycle before the stress attains the desired steady-state response. Data taken before steady state is achieved are not reliable. However, unless the frequency of the measurement is extremely low, the time required to attain steady state under a given set of conditions is much shorter than the time required to perform a creep or stress relaxation test under a single set of conditions. This short measurement time is one of the most important advantages of DMA over creep and stress relaxation.

5.2.3.1. The Complex Modulus
Dynamic mechanical analysis data are most commonly reported using a quantity known as the *complex modulus*.

The complex modulus evolves from a complex variables treatment of the sinusoidal deformation, which is well documented in various references on DMA [e.g., McCrum et al. (1967); Murayama (1978); Ferry (1980); Nielsen and Landel (1994)].

The complex modulus may be defined as the ratio of the sinusoidal stress to strain:

$$E^* = \frac{\sigma(t)}{\varepsilon(t)} = \frac{\sigma_0 \cdot e^{i(\omega t + \delta)}}{\varepsilon_0 e^{i\omega t}} = \left(\frac{\sigma_0}{\varepsilon_0}\right) e^{i\delta} = \frac{\sigma_0}{\varepsilon_0}(\cos(\delta) + i\sin(\delta)) \tag{5.19}$$

The complex modulus is a characteristic property of the material that changes only when the material changes. It is a function of time only, since DMA experiments are performed under conditions of very small strain. Under these conditions the material response is in the *linear viscoelastic* range. This means that the magnitudes of the stress and strain are linearly related and the deformation behavior is completely specified by the complex modulus function. The theory is developed assuming deformation under isothermal conditions and temperature does not appear (nor is implicit) as a variable. There is no well-developed theoretical treatment for the temperature dependence of viscoelasticity, but, as will be shown later in this chapter, there is a well-established empirical relationship relating the time and temperature response of viscoelasticity known as the *time–temperature superposition principle*.

The complex modulus may be divided into real and imaginary components:

$$E^* = E' + iE'' \quad (5.20)$$

where
$$E' = \frac{\sigma_0}{\varepsilon_0}\cos(\delta)$$

$$E'' = \frac{\sigma_0}{\varepsilon_0}\sin(\delta)$$

E' is known as the storage modulus and is a measure of the elastic character or solidlike nature of the material; E'' is known as the *loss modulus* and is a measure of the viscous character or liquidlike nature of the material. The larger E' is relative to E'', the more of the energy required to deform the sample is elastically recoverable. The larger E'' is relative to E', the more of the deformation energy is viscously dissipated as heat. For an ideal elastic solid, the phase lag is zero. Thus, E' is simply Young's modulus of the material and E'' is zero. For an ideal viscous liquid the phase lag is 90° or $\pi/2$ radians. Thus, the storage modulus is zero and the loss modulus is related to the viscosity of the material, as will be described later in the chapter. In a physical sense, the storage modulus is related to the stiffness of the material and the loss modulus is reflected in the damping capacity of the material.

In addition, a third quantity may be defined by taking the ratio of the loss modulus to the storage modulus:

$$\frac{E''}{E'} = \frac{\sin(\delta)}{\cos(\delta)} = \tan\delta \quad (5.21)$$

This quantity is known as the *material loss factor* or *loss tangent*, or, more commonly, "tan δ" (pronounced "tan delta"). Tan δ ranges from zero for an ideal elastic solid to infinity for an ideal liquid. It represents the ratio of energy dissipated to energy stored per cycle of deformation. When the mode of deformation is shear, the shear storage modulus, (G'), shear loss modulus (G''), and tan $\delta = G''/G'$ are used. Their definitions are directly analogous to those above, using shear stress and shear strain as in the equations cited above. This is discussed in the following section on "rheology" (Section 5.2.3.2).

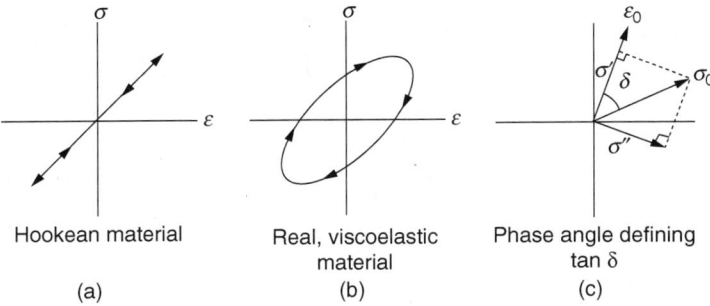

Figure 5.10. DMA sinusoidal stress–strain response curves for a typical viscoelastic material illustrating the phase shift between strain and stress (Sperling 2006): (a) vectorial representation; (b) hysteresis effect for a viscoelastic material—area within the ellipse represents viscous heat loss; (c) an ideal elastic material shows no heat loss. (Reproduced with permission of John Wiley & Sons.)

Data are sometimes reported simply as the magnitude of the complex modulus, defined as

$$|E^*| = |E'^2 + E''^2|^{1/2} = \frac{\sigma_0}{\varepsilon_0} \quad (5.22)$$

This offers the advantage of being a single real quantity describing the stiffness of the material, but information about the relative viscous and elastic character of the material is lost. Use of the complex modulus magnitude $|G^*|$ is particularly prevalent in data characterizing measurements on fluids in shear.

Based on the complex variables treatment of the dynamic mechanical modulus, the relationships between stress and strain can be considered on a complex plane in terms of vectors (Sperling 2006) as shown in Fig. 5.10a. The applied strain of magnitude ε_0 is out of phase with the resulting stress of magnitude σ_0 by an amount represented by angle δ. The stress components of σ^* are resolved as σ', in phase, and σ'', 90° out of phase with $\varepsilon^* = \varepsilon_0$.

Further, Fig. 5.10b illustrates that the phase shift between the cyclic stress and strain for a viscoelastic material results in a hysteresis loop, the size of which reflects the rate of viscous dissipation or heat generation during the cyclic deformation. For an ideal Hookean material, there is a zero-area loop or line (Fig. 5.10c). But as tan δ increases, so does the size of the hysteresis loop. This notion is particularly important in applications such as the damping of vibration and sound.

5.2.3.2. Dynamic Mechanical Analysis and "Rheology"

The principles of DMA are equally applicable to liquids. In fact, dynamic mechanical methods are an important way to determine (to a first approximation) flow viscosity data for thermoplastics in the melt state, in lieu of using more expensive and more time-consuming rheological methods such as capillary rheometry. This is discussed further in the following paragraphs. In DMA an ideal liquid will simply respond with a phase lag of 90° between the stress and strain, and a viscoelastic liquid will have a phase lag less than 90°. Rheological measurements are nearly always done in a shear mode of deformation, so assuming the application of a sinusoidal shear strain (γ) and the measurement of a shear stress (τ), we can derive a complex shear modulus (G^*), using the same procedure used to define the complex tensile modulus:

$$G^* = G' + iG'' \tag{5.23}$$

where
$$G' = \frac{\tau_0}{\gamma_0} \cdot \cos(\delta)$$

$$G'' = \frac{\tau_0}{\gamma_0} \cdot \sin(\delta)$$

Since viscosity is the quantity of interest in rheology, it is a simple matter to define a complex viscosity (η^*) in a dynamic rheological test based on a complex form of Newton's law of viscosity

$$\tau(t) = \eta^* \cdot \dot{\gamma}(t) \tag{5.24}$$

$$\eta^* = \frac{\tau(t)}{\dot{\gamma}(t)} \tag{5.25}$$

where η^* is a complex quantity

$$\eta^* = \eta' - i\eta'' \tag{5.26}$$

with
$$\eta' = \frac{\tau_0}{\omega \gamma_0} \cdot \sin(\delta)$$

$$\eta'' = \frac{\tau_0}{\omega \gamma_0} \cdot \cos(\delta)$$

Examination of the real and imaginary components of the complex viscosity reveals a simple relationship between the complex viscosity and the complex shear modulus:

$$\eta' = \frac{G''}{\omega} \tag{5.27}$$

$$\eta'' = \frac{G'}{\omega} \tag{5.28}$$

$$\eta^* = \frac{G^*}{i\omega} \tag{5.29}$$

$$|\eta^*| = |\eta'^2 + \eta''^2|^{1/2} = \frac{|G^*|}{\omega} \tag{5.30}$$

The loss modulus, representing the viscous nature of a material, is related to the real component of the complex viscosity, and the storage modulus, representing the elastic nature of the material, is related to the imaginary component. For an ideal liquid η' will simply be the viscosity of the liquid and η'' will be zero. For an ideal solid η' will be zero and η'' will be the ratio of the shear modulus to frequency. Thus, in DMA there is no real distinction between viscosity and modulus, and knowledge of either quantity along with the measurement frequency can be used to calculate the other.

Dynamic viscosity data can be used to approximate the steady shear viscosity by taking advantage of an empirical relationship known as the *Cox–Merz rule* (Cox and Merz 1958), which relates the magnitude of the complex viscosity at frequency ω to the steady shear viscosity at a shear rate $\dot\gamma$ equal to ω:

$$|\eta^*(\omega)| = \eta(\dot\gamma)|_{\dot\gamma=\omega} \tag{5.31}$$

Dynamic viscosity measurements generally are easier to perform over a wide range of frequencies than are steady shear measurements over a wide range of shear rates, while steady shear viscosity is of more value in the analysis of polymer processing. Thus, the Cox–Merz rule is frequently assumed and η^* data are used in place of steady shear viscosity data.

5.3. THE RELATIONSHIP BETWEEN TIME, TEMPERATURE, AND FREQUENCY

If an amorphous polymer is subjected to a stress relaxation experiment over a very long period of time, the modulus–log time curve will be similar to that shown in Fig. 5.11. There is a striking similarity in shape between the curves

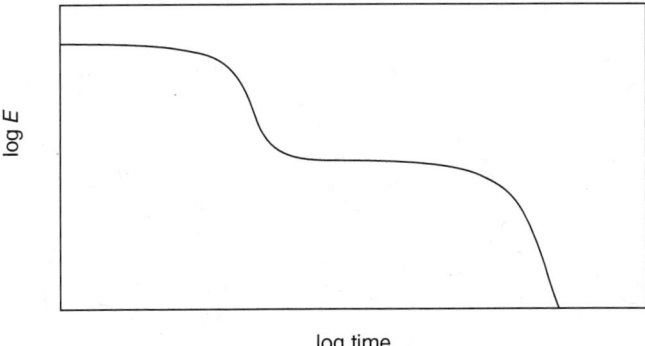

Figure 5.11. Illustration of modulus versus time for a creep or stress relaxation experiment on a typical amorphous polymer (T = constant). Note the similarity to the plot of DMA storage modulus versus temperature. The reciprocal of creep compliance, $1/J(t)$, would be used to generate the modulus curve derived from the creep response.

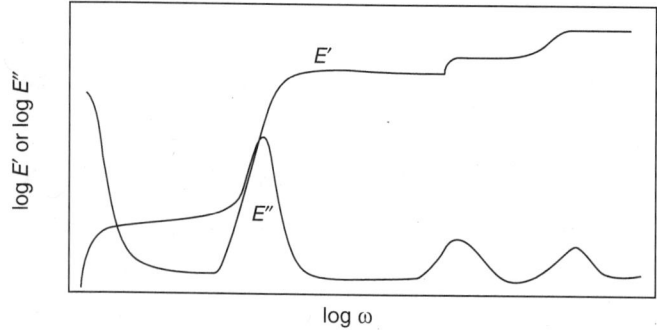

Figure 5.12. Schematic of the storage and loss moduli from a DMA frequency scan for an amorphous polymer showing multiple viscoelastic transitions (T = constant).

in Figs. 5.11 and 5.2. The same five regions of viscoelasticity displayed in Fig. 5.2 are found in Fig. 5.11, except in time rather than in temperature. This leads to a concept known as the principle of time–temperature equivalence; that is, mechanical response at low temperature is equivalent to response at short times and response at high temperature is equivalent to response at long times. There is a fundamental basis for this concept, which is not covered in detail here [the reader is referred to Ferry (1980)].

As the frequency is increased, the time allowed for molecular motion in a given cycle decreases. Thus, frequency may be regarded as inverse time, consistent with the units of frequency. Rather than performing a DMA scan in temperature at constant frequency, a DMA scan can be performed in frequency at constant temperature. Such a scan is shown schematically in Fig. 5.12. If a sufficiently broad range of frequencies is scanned, all the transitions observed in a temperature scan are found, but in reverse order (i.e., the curve is a mirror image). However, the range of frequencies that it is practical to measure is limited. Very low frequencies are difficult to measure because of the time required to complete a period, and very high frequencies are difficult to measure owing to instrumental limitations. Thus, it is generally possible to observe a wider range of viscoelastic behavior with a temperature scan than with a frequency scan.

Thermal analysts more commonly perform temperature scans covering a modest range of frequencies. Figure 5.13 shows the effect of changes in frequency on an idealized dynamic mechanical temperature scan. Transitions shift to higher temperatures at higher frequencies. The shift to higher temperatures is a direct consequence of time–temperature equivalence. As the frequency is increased, only shorter timescale motions are possible; thus the polymer responds more as if it were at a lower temperature than a sample run at a lower frequency but the same temperature. Consequently, higher temperatures are required for a sample to achieve an equivalent mechanical state

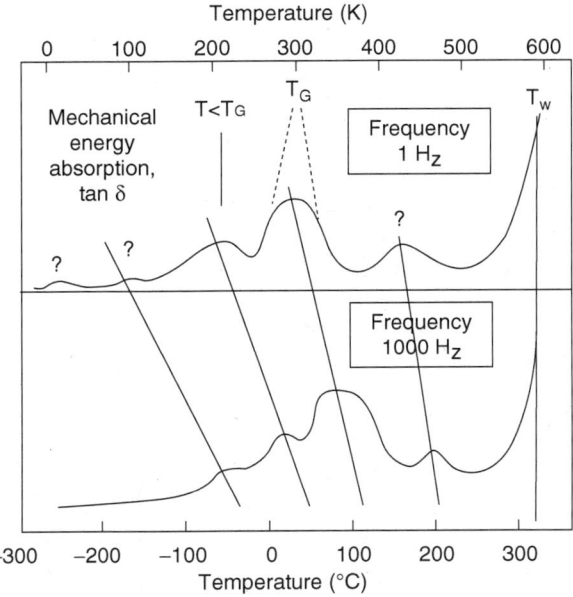

Figure 5.13. Illustration of the effect of changes in frequency on an idealized dynamic mechanical temperature scan. The loss peaks for various transitions shift to higher temperatures with increasing frequency. Better resolution of the transitions is favored at lower frequencies. Note that each transition shifts differently, since the activation energies differ (from Boyer, 1968, with permission of the Society of Plastics Engineers).

at higher frequencies, and the transitions shift to higher temperatures. It should be noted that different transitions in the same polymer will be shifted by differing amounts. This is discussed in more detail in a later section. The magnitude of the shift is a function of the activation energy of the transition in question. Thus, glass transitions, which have relatively high activation energies, shift less than do lower-temperature transitions, which have lower activation energies (McCrum et al. 1967; Nielsen and Landel 1994). In general, the activation energies for the lower-temperature transitions decrease with temperature in descending order. In addition to the temperature shift, the magnitudes of the peaks in loss modulus and tan δ usually decrease as frequency increases, and the peaks broaden.

5.3.1. Time–Temperature Superposition

While there is a relationship between time and temperature, the theories of viscoelasticity (Ferry 1980) do not deal with the temperature dependence. However, there is an empirical relationship referred to as the *time–temperature superposition* (TTS) principle, which provides a useful, practical application of the principle of time–temperature equivalence. The rationale behind time-temperature superposition is as follows. If the complete modulus–temperature curve and the modulus–log time curve have the same shape, then data taken over a narrow range of time at any temperature will duplicate a portion of this complete curve. If data are taken at different temperatures over the same timespan, they will duplicate a different portion of the curve. Thus, we can construct the curve over any broad range of time by taking data at several temperatures over a more limited range of time (or frequency) and shifting the curves horizontally on the log time axis until they fit together and exactly align in a single continuous curve. TTS, then, simply is based on the observation that the curves representing the viscoelastic properties of a single material, determined at several different temperatures, are similar in shape when plotted against log t or log ω.

This principle applies to all of the viscoelastic functions defined previously for stress relaxation, creep, and dynamic mechanical experiments. A most comprehensive discussion of TTS is provided in the text by Ferry (1980). An example of TTS is shown in Fig. 5.14 for the modulus data of Fig. 5.15, which were determined at various temperatures (Mercier et al. 1965) by both stress relaxation and creep measurements. The creep compliance data were converted to modulus by the relationship $E(t) = 1/J(t)$. The single curve formed by superposition of the various curves shifted on the time axis to the given reference temperature (T_R) is referred to as a "master curve." In this case

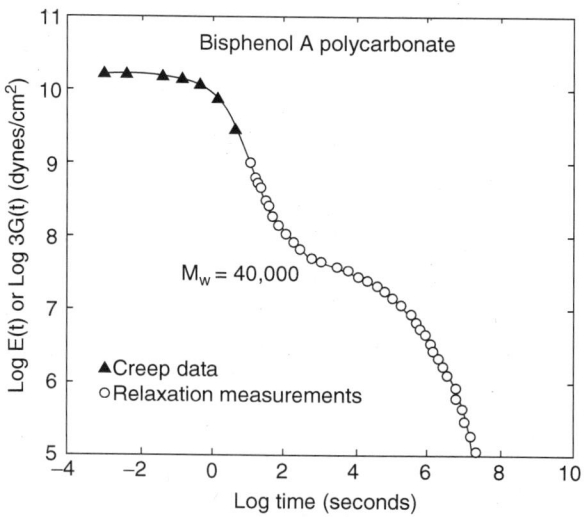

Figure 5.14. Time–temperature superposition master curve for modulus versus time formed from the polycarbonate stress relaxation and creep data shown in Fig. 5.15; the master curve covers 10 decades on the log timescale. Reference temperature is 150 °C. [From Mercier et al. (1965); reprinted with permission of John Wiley and Sons, Inc.]

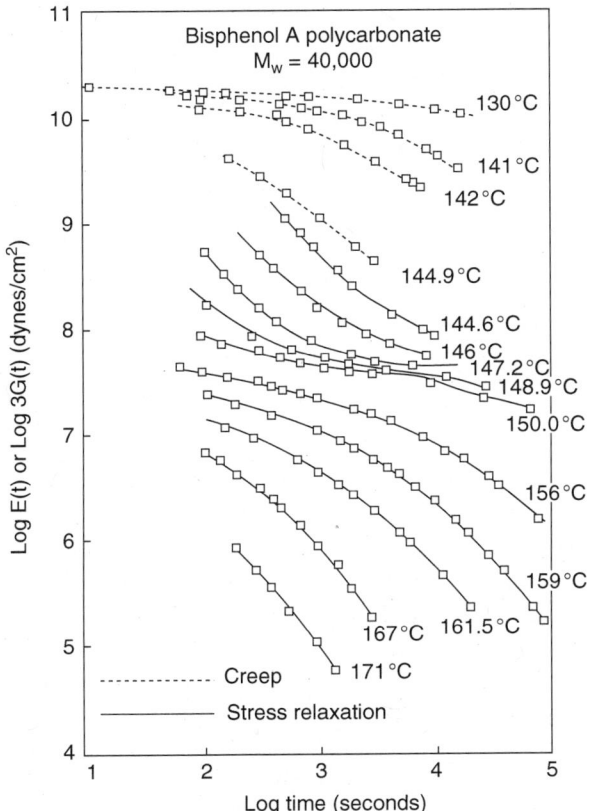

Figure 5.15. Set of individual stress relaxation and creep curves taken at various temperatures over a narrow time range; these form the basis for the master curve of Fig. 5.14. [From Mercier et al. (1965); reprinted with permission of John Wiley and Sons, Inc.]

$T_R = 150\,°C$. The master curve clearly displays the five different characteristic viscoelastic regions or physical states noted in Fig. 5.2, with respect to the time or frequency axis. The superposition process underscores the fact that the viscoelastic properties of polymers depend more strongly on temperature than time, with changes in modulus and other properties over several orders of magnitude occurring over a modest temperature range (in this case ~40 °C).

Figure 5.16 shows in more detail how the shifting procedure actually is accomplished. In the simulated example depicted there the reference temperature is chosen as T_1. The data taken at T_1 are not shifted on the time axis. The other curves are shifted to the right as shown along the time axis until they fit together, superimposing to give a single continuous curve. The power of this technique comes from its ability to predict behavior over periods of time that are impractical to measure. For example, using the data in Fig. 5.15

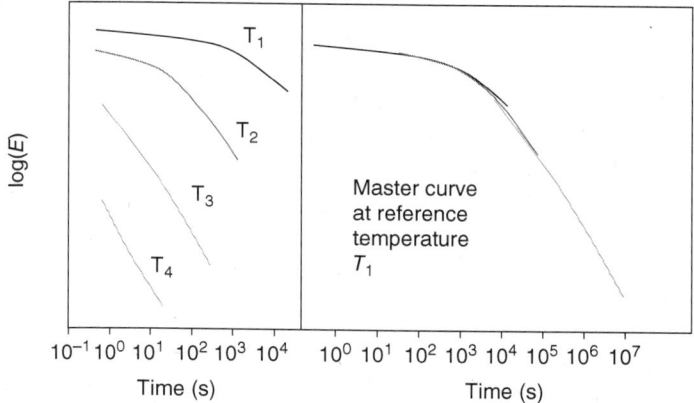

Figure 5.16. Illustration of formation of a master curve by shifting several sets of simulated modulus data taken at different temperatures to superpose; reference temperature is T_1. In shifting, the T_1 curve is fixed and the others are shifted to the right.

taken over periods of time of 10^4 s (2.8 h), we are able to predict behavior to 10^7 s, or over 115 days. In practice, by choosing even a wider range of temperatures over which to collect data, behavior may be predicted for longer periods of time.

However, predictions for very long time periods should be approached with a degree of caution because some of the factors that render TTS invalid as described below may come into play. These include anything that changes the chemical and/or thermodynamic state of the material, such as oxidation, cross-linking, crystallization, or physical aging, all of which may be influenced by the stress field imparted to the material. [*Physical aging* is a spontaneous change in the thermodynamic state or properties of a polymer that results in changes in its physical and mechanical properties (Matsuoka 1992; Chartoff 1997). While the rate of physical aging depends on the temperature and the amount of stress applied, the driving force for physical aging is the difference in chemical potential that exists between the current thermodynamic state of the material (which is nonequilibrium) and the equilibrium state. Please refer to Chapter 2 in this book on DSC and the references cited above for additional information on the topic.]

When TTS was originally implemented, the distance that a curve must be shifted in order to transpose it from the temperature of observation T to the reference temperature T_R was determined by the simple empirical curve-fitting procedure illustrated above. Pioneering research on this aspect of TTS was carried out by Leaderman (1941, 1943) and Tobolsky (1960), who also modified the procedure to account for the proportionality of modulus to absolute temperature. This has the effect of creating a slight vertical shift in the data. Williams et al. (1955) further modified time–temperature superposition to account for changes in density at different temperatures, which has the effect of creating a small additional vertical shift factor. The effect of the density/temperature ratio on modulus is frequently ignored, however, since the ratio generally is close to unity.

While time–temperature superposition may be performed by actually shifting the curves until the best alignment is found, it is useful to quantify the shift by defining a quantity known as the *temperature shift factor*, a_T (a function of temperature)

$$a_T = \frac{t_T}{t_{T_R}}. \tag{5.32}$$

where t_T is the time required to reach a given value of the stress relaxation modulus or creep compliance at temperature T and t_{T_R} is the time required to reach the same value of the modulus or compliance at the reference temperature T_R. (see also Section 2.10).

Determination of the shift factor by aligning curves taken at multiple temperatures provides shift factor values only for temperatures at which data are taken. Often it is useful to be able to shift the master curve to any arbitrary temperature, which requires knowledge of the shift factor as a continuous function of temperature. This is generally accomplished by fitting the empirically obtained shift factors to an appropriate function. The most common function used for this purpose is the Williams–Landel–Ferry (WLF) equation (Williams et al. 1955):

$$\log a_T = \frac{-C_1(T - T_g)}{C_2 + (T - T_g)} \tag{5.33}$$

Williams, Landel, and Ferry found that C_1 and C_2 were similar for many amorphous polymers with $C_1 = 17.44$ and $C_2 = 51.6$ in the temperature range between T_g and $T_g + 100\,°C$. The equation is referred to as the "universal" WLF equation when C_1 and C_2 assume these values. While this equation is not truly universal, it was developed from a large database for various polymers. When the equation is written in this form, it is clear that T_g serves as a corresponding state for viscoelastic behavior. A plot of $\log a_T$ versus $(T - T_i)$ for the data of Fig. 5.15 is shown in Fig. 5.17; here T_i is the inflection point in the modulus curve at T_g. Each DMA vendor has software available that automates the TTS operations of curve shifting, determining values of a_T, and fitting the data to the WLF equation.

Ferry (1980) later acknowledged that "It is evident that the actual variation from one polymer to another is too great to permit use of the 'universal' values except as a last resort in the absence of other specific data." Thus, usually C_1 and C_2 are treated as empirically adjustable parameters, the values of which are determined by a fit to experimental values of a_T. Furthermore, the equation is sometimes applied over temperatures outside the recommended temperature range, including polymers below T_g. As with any empirical data fit,

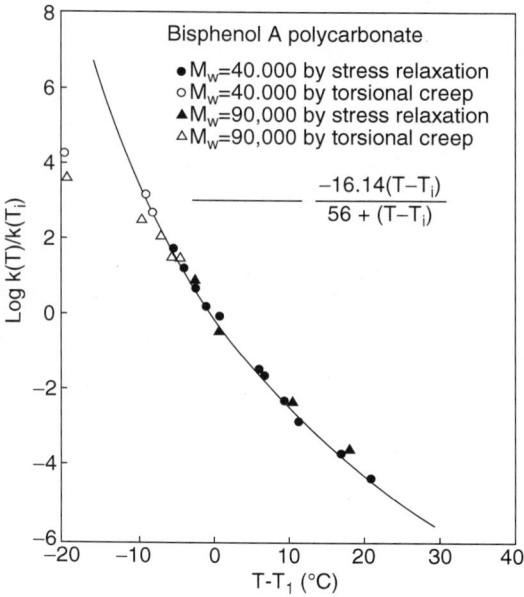

Figure 5.17. WLF shift factor curve for the data of Fig. 5.15; reference temperature T_i (the inflection temperature) is equivalent to T_g. Note that C_1 and C_2 here are not the "universal" values but are specific for the polycarbonate polymer. [From Mercier et al. (1965); reprinted with permission of John Wiley and Sons, Inc.]

the WLF equation may be used with greater confidence for interpolation of shift factors between experimentally determined values. Thus caution should be used in extrapolating with the equation.

Another popular equation for fitting shift factors is based on the Arrhenius equation

$$\ln a_T = \ln\left(\frac{t}{t_g}\right) = \left(\frac{E(T_g - T)}{RT_g T}\right) \quad (5.34)$$

where E is a relaxation activation energy, R is the universal gas constant, and T and T_g are absolute temperatures. With the Arrhenius approach there is only one adjustable parameter rather than two. Vendor software for TTS also includes curve fitting to the Arrhenius equation. Generally the WLF equation will give a better fit over a wide range of temperatures because of the two adjustable parameters. For this reason it is recommended that the WLF equation be used.

We noted previously that most DMA instruments can measure in a stress relaxation or creep mode. The discussion above relates directly to this. When applied to DMA data, time–temperature superposition relates temperature and frequency rather than temperature and time. Data are taken over a range of frequencies at multiple temperatures. The data may be shifted onto a single master curve by treating frequency as inverse time ($f = 1/t$) and determining shift factors as described above. The master curve represents the complex modulus as a function of frequency at the reference temperature. Modulus versus frequency at another temperature may then be determined by using the shift factor to shift the master curve to the temperature of interest. In addition, the master curve may be used to predict behavior at frequencies outside the practical limitations of the test equipment.

While time–temperature superposition is very useful, it will not work in all cases. Predictions based on TTS conform well to the observed behavior of many polymers, but others exhibit behavior inconsistent with TTS. A number of assumptions inherent in the principle of time–temperature equivalence (Ferry 1980) are incorrect for many polymers. For example, implicit in TTS is the assumption that the effect of temperature on the relaxation time spectrum, is consistent for the entire spectrum, but this is frequently in error (Dealy and Wissbrun 1990).

First, we have noted that relaxations in different temperature regions have different activation energies, so they will not superpose by a single set of a_T values. Also, the polymer must be homogeneous. An example in which this clearly does not hold is the case of multiphase blends or block copolymers. These materials will have multiple relaxations that are unrelated, and therefore may be expected to have differing shift factors for any given reference temperature, even when each phase in isolation conforms to the WLF equation. In addition, TTS assumes that the chemical structure and morphology of the polymer is constant. Any polymer that undergoes a chemical change with time, such as oxidation, UV degradation, additional polymerization, or crosslinking, will be expected to undergo a change in properties in a manner inconsistent with TTS. Similarly, if a polymer undergoes a change in morphology as a result of crystallization, partial melting, or crystal refinement during annealing, its relaxation spectrum will be altered and it will not follow TTS. Thus, caution is warranted when using TTS to predict properties for times and temperatures outside the range over which data were taken and for systems that are nonhomogeneous. However, as a general rule, good alignment (i.e., a smooth fit of the curves), when shifting the data taken at different temperatures, is a good indicator that the assumptions inherent in TTS are valid for the system of interest.

In summary, the TTS principle has great practical significance, since it allows data taken at various temperatures over a conveniently narrow time or frequency range to be used for estimating viscoelastic properties over an extended time range. When the appropriate shift factors are known it also allows the estimation of properties at various temperatures from the master curve at a single reference temperature.

5.4. APPLICATIONS OF DYNAMIC MECHANICAL ANALYSIS

5.4.1. Viscoelastic Transitions or Relaxations

Viscoelastic transitions or relaxations in amorphous polymers include T_g and several secondary transitions resulting from relaxation processes that are observed at temperatures below T_g. The secondary transitions are not observed in DSC experiments because the associated C_p changes are too small, and there are no heat associated with these transitions. In semicrystalline polymers there often is an additional secondary relaxation between T_m and T_g. These relaxations are of continuing scientific interest and engineering importance. They are intimately associated with the stiffness of engineering thermoplastics, impact toughness, diffusion rates of low-molecular-weight vapors and liquids, and the effectiveness of plasticization, among others. The three points of greatest interest in relation to characterizing relaxation processes are (1) determining in what phase they originate, (2) describing the molecular processes underlying them, and (3) relating relaxation behavior to engineering properties and practical applications. DMA has proved most useful in helping to address these issues as noted in the review by Chartoff (1997).

For constant-frequency (sometimes referred to as *isochronal*) experiments on amorphous polymers, the highest temperature relaxation is the glass transition, also referred to as α, while secondary β and γ relaxations are commonly observed. There also may be a δ process. In this case all the relaxations are associated with the amorphous phase. The γ and δ relaxations (and in certain cases also the β relaxation) are below −100 °C. These relaxations would be missed if the only cooling device available were a mechanical chiller that would cool to only −100 °C. Figure 5.18 illustrates the various loss peaks associated with the transitions in amorphous and crystalline polymers.

For semicrystalline polymers in the temperature range between T_m and liquid nitrogen temperature [−196 °C (77 K)], at least three relaxation processes are often found. The high-temperature $α_c$ process, when present, is a secondary relaxation related to the crystalline fraction. The β process in semicrystalline polymers that have an $α_c$ process represents the glass transition, which is related to the amorphous phase. This is different from the case of amorphous polymers, where the α relaxation is the glass transition. The low temperature γ process is generally considered to originate in the amorphous phase but may also have an important component associated with the crystalline phase. There also may be a δ process or even in certain instances lower-temperature ε and ζ relaxations as for Nylon 6 (Papir et al. 1972), which are associated with the amorphous phase.

As we will illustrate later in the chapter, the various relaxations are influenced by polymer structure, morphology, and environmental factors. For example, the intensity, temperature position, and temperature range of the glass transition in semicrystalline polymers depends on the degree of crystallinity and morphology of the polymer. These will vary depending on the pro-

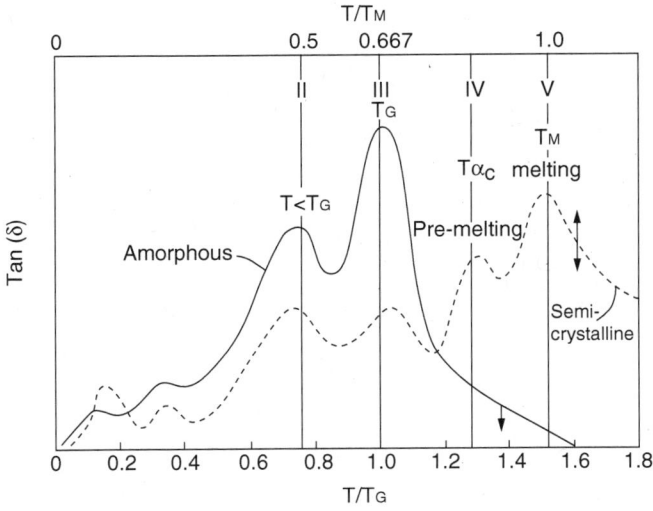

Figure 5.18. This diagram illustrates the loss peaks associated with the various transitions found in semicrystalline and amorphous polymers. [From Chartoff (1997); reprinted with permission of Elsevier.]

cessing history of the material. Secondary transitions are influenced by diluents, moisture, fillers, and pigments, among other factors. In crosslinked polymers the degree of crosslinking and the nature of the crosslink network are important influences. Changes in these will systematically alter the transitions. Thus DMA can be used as an important analytical tool to study such property variations.

5.4.2. Determining T_g by DMA

Dynamic mechanical analysis methods are most popular among thermal analysts for measuring mechanical properties at various temperatures. Identifying the glass transition and how various system modifications affect T_g is a major application for DMA. The glass transition generally is easily identified from dynamic mechanical data because of the sharp decrease in storage modulus E' (or shear storage modulus G'), and the corresponding loss dispersion in E'' (shear loss G'') or tan δ that occur at T_g as shown in Fig. 5.19. These data are for a typical amorphous polymer (Nielsen 1962). It is evident that there is latitude in how the value of T_g is chosen from a set of DMA data, and this often leads to confusion in the literature. Most often the criterion for selection of T_g from DMA data is either the peak in loss modulus E'' or peak in tan δ. Either one is valid. However, the values are different, as we explain in the following paragraph, and arguments are forwarded favoring the peak loss modulus.

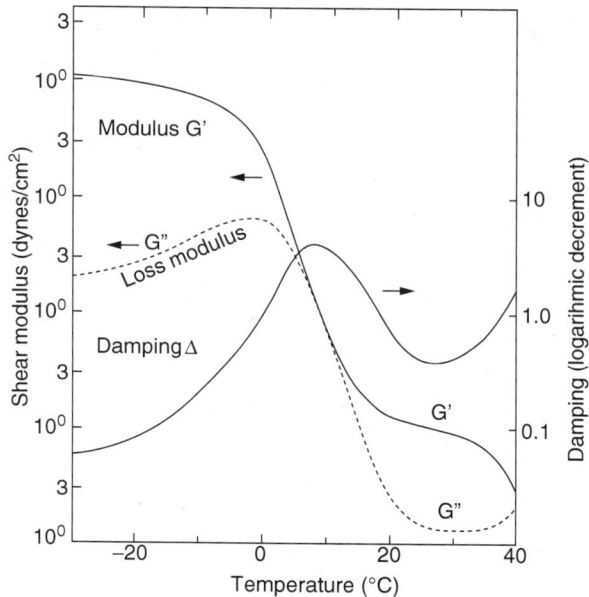

Figure 5.19. DMA E′, E″, and tan δ (△) curves for a butadiene–styrene copolymer; the data illustrate that E″ and tan δ have different maxima (Nielsen 1962).

The T_g value obtained from the peak tan δ is several degrees higher than that from the peak E''. The tan δ peak corresponds more closely to the transition midpoint or inflection point of the decreasing log E' curve, while the loss modulus peak more closely denotes the initial drop of E' from the glassy state into the transition. In this respect the E'' peak T_g value is generally close to the intersection of the two tangents to the log storage modulus curve originating from both the glassy region and the transition region, the so-called onset temperature. For polymers that have unusually broad glass transitions (e.g., crosslinked, semicrystalline, and heterophase systems), peak tan δ will occur at a much greater ΔT. Furthermore, as noted earlier, T_g is a kinetic quantity, so it changes with frequency.

While the tan δ peak is often cited as T_g in the literature, in general the maximum loss modulus is the more appropriate value. This is the method for specifying T_g prescribed by ASTM E1640-07 (see Appendix). It is a reasonable criterion from molecular and practical or engineering perspectives because the upper use temperature of many amorphous polymers is the "softening" point. The onset point is related to the initiation of segmental molecular motions and loosening of the polymer structure associated with T_g. It is clear that at the transition midpoint or inflection point (peak tan δ), the softening point already has been exceeded. In this region substantial molecular relaxation involving increased cooperative segmental motions of the polymer chains occurs. Thus the temperature for tan δ maximum is more sensitive to parameters such as molecular weight, crosslink density, filler content, and morphology. Related to this is the observation that for many polymers the tan δ damping peak maximum increases by about 7°C for each decade increase in frequency while the loss modulus maximum rises only about 3°C (Nielsen and Landel 1994).

For most linear amorphous polymers the transition region is rather narrow, covering around 15–20°C. In these cases the distinction between peak E'' and tan δ temperatures is not substantial. The tan δ peak may occur 10–15°C higher in temperature than the E'' peak. There are cases, however (e.g., crystalline polymers, crosslinked thermosets, or heterophase polymers) where the transition region is very broad and neither the peak E'' nor tan δ may be entirely suitable for specifying T_g. This is illustrated in Fig. 5.20 (Chartoff et al. 1994) for a crosslinked acrylate photopolymer. The polymer is a heterogeneous mixture of phases with overlapping glass transitions. The difficulty in specifying T_g from these data is discussed further in a subsequent section.

In selecting a criterion to be used for specifying T_g, the experimenter may take into account the major application for use of the T_g data; for example, whether it is to be used as (1) a material property to measure material consistency; (2) to evaluate the effects of processing, as in the curing of thermosets where T_g–conversion relationships are important; or (3) as an engineering property where the T_g value has significance as a structural property. If mea-

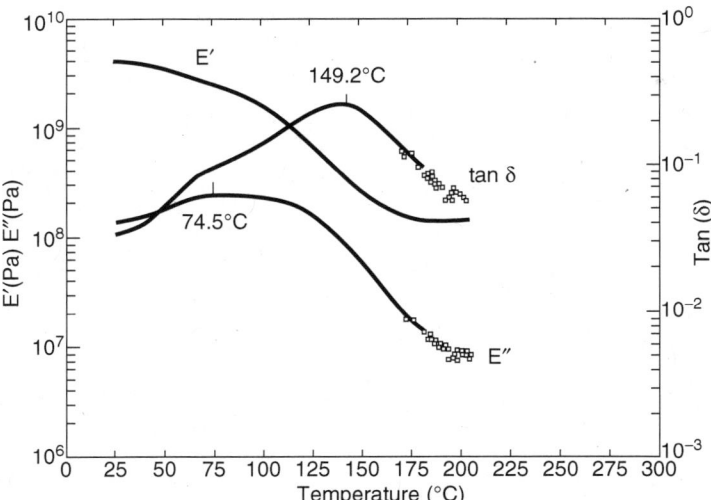

Figure 5.20. DMA data at 1 Hz for a crosslinked photopolymer formed from a mixture of several acrylate monomers; the T_g region is very broad, making it difficult to pick a specific T_g value. There are at least two distinct transitions that are apparent in tan δ but are not resolved in E'' (from Chartoff, et al., 1994, reprinted, with permission, from *STP 1249-Assignment of the Glass Transition*, copyright ASTM International, 100 Barr Harbor Drive, West Conshohocken, PA 19428).

surement precision is foremost, the choice of either E'' or tan δ might be taken depending on which appears sharper. It also should be noted that one can sharpen the data by plotting modulus or tan δ on a linear rather than log scale.

For both highly crystalline polymers and highly crosslinked polymers the glass transition is a less prominent event than for linear amorphous polymers and may be difficult to observe by DSC because the ΔC_p change will be small and occur over a broad temperature range. However, generally the glass transition in these polymers may be observed more clearly with DMA than by DSC because the baseline shift ($\Delta E'$) at T_g is much larger than the ΔC_p measured in DSC. The change in modulus at T_g is often of the order of 10–10^3 Pa, while the ΔC_p change will be of the order of 10–30% of the baseline value.

5.4.3. Factors Influencing the Measurement of T_g by DMA

Many factors influence the measurement of T_g by DMA methods. These can be grouped as: (1) instrumental factors, (2) test frequency, and (3) material characteristics. The various material characteristics that influence T_g are discussed in detail later in the chapter. Here, our discussion provides a summary of these effects. The reader is also referred to a more comprehensive review on the subject of determining T_g by DMA methods (Chartoff et al. 1994).

5.4.3.1. Instrumental Factors
Several instrumental factors affect determination of the glass transition from DMA data, such as error in the thermocouple or thermocouple placement, temperature gradients within the sample chamber (or oven) or within the sample proper, and exceeding the compliance limitations of the machine. These factors are specific to each individual instrument. Some instruments allow the user to input software corrections to correct errors in thermocouples but fail to allow their proper electronic calibration by the user. Each instrument has a different size and shape oven, employs different heating and cooling mechanisms, and has a variety of gas flow rates. Finally, the instruments are constructed with various rigidities and offer a variety of sample clamping configurations. The sample's modulus, the sample's dimensions, and the fixture all combine to determine whether the compliance limits of the machine have been exceeded. All of these factors are highly specific to a given instrument and should be noted by the analyst when interpreting DMA data. These aspects are discussed in more detail in Section 5.7.

In addition, changing the heating rate also will affect the position of T_g on the temperature scale. If a ramp heatup is used, an increase of heating rate generally will cause an elevation of T_g. When the heating rate exceeds $2\,°C/min$ (for most samples), the sample temperature may lag the DMA oven temperature so that a time lag in the apparent T_g will be observed and the transition will appear to occur at a higher temperature. This is also discussed more fully in a later section of this chapter.

5.4.3.2. Test Frequency Since the glass transition is inherently kinetic, it is strongly influenced by the rate or frequency of mechanical energy input. We noted that substantial molecular relaxation involving cooperative segmental motions of the polymer chains occurs in the region of T_g. The rate of this segmental motion depends on temperature, so that if the test frequency is increased, the relaxations associated with the glass transition have difficulty in keeping up with the mechanical strain input, and the polymer appears to be more rigid. The segmental motions associated with T_g then can occur only at a higher temperature. Thus T_g increases with frequency as illustrated in Figs. 5.21 (Becker 1955) and 5.22 (Chartoff 1997).

Characteristic features of the transition with increasing frequency, as illustrated by these data, are a general decrease in the intensity of tan δ or E'', a broadening in the peak, and a decrease in the slope of the storage modulus curve in the transition region. This is related to a broadening of the relaxation spectrum in the T_g or α relaxation vicinity for PVC, coupled in many cases with the merging of relaxations associated with the β relaxation, which is below T_g in temperature. Cases where the α and β relaxations are well separated include high-T_g aromatic chain polymers such as polycarbonate and polysulfone. The temperature position of the E'' peak at various frequencies for the α and β transitions in poly(methyl acrylate) is shown in Fig. 5.23 (McCrum

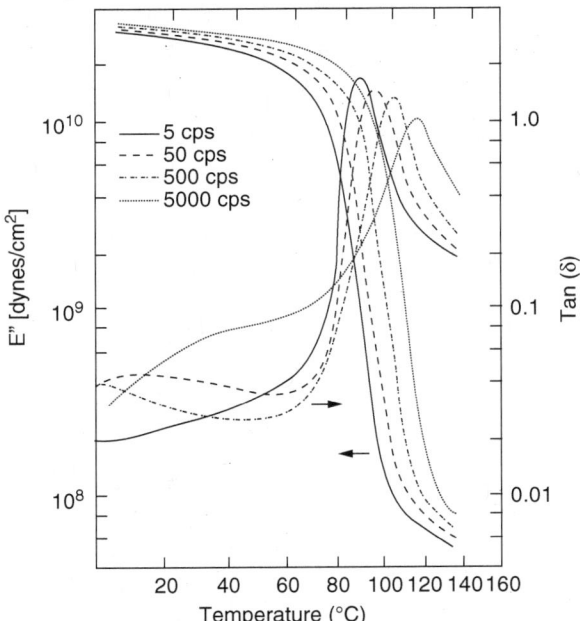

Figure 5.21. Experimental DMA data for PVC at various frequencies; a general decrease in the intensity of tan δ or E'', a broadening of the loss peak, and a slight decrease of the slope of the storage modulus curve in the transition region occur with increasing frequency (from Becker, 1955, with permission of Springer Science and Business Media).

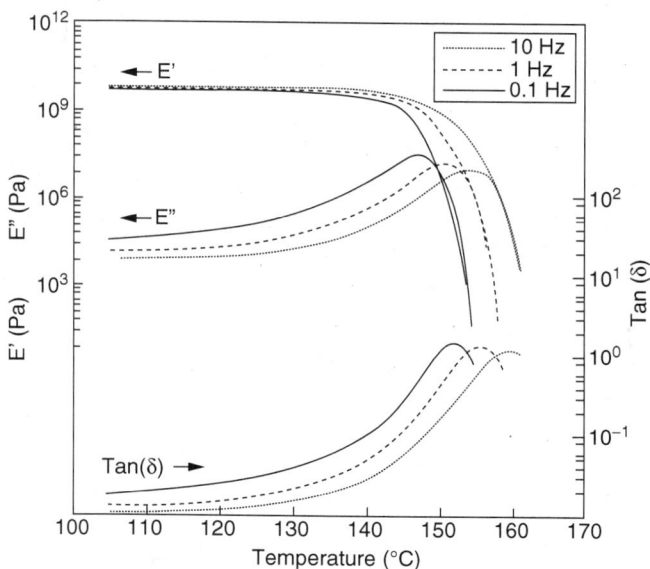

Figure 5.22. DMA curves for polycarbonate at several frequencies. [From Chartoff (1997); reprinted with permission of Elsevier.]

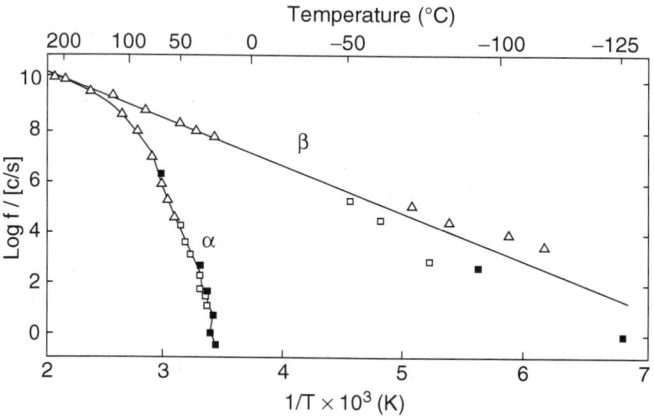

Figure 5.23. Arrhenius plots for the α and β transitions in poly(methyl acrylate) indicating the different temperature dependences for the two transitions. [From McCrum et al. (1967); reprinted with permission of John Wiley and Sons, Inc.]

et al. 1967). This is a composite diagram that includes both DMA and dielectric data. Note that the α transition curve is nonlinear and does not follow the Arrhenius equation. Thus the variation in T_g increases with frequency as shown. Analysis indicates that the variation actually follows the WLF equation. Only over limited frequency ranges (one to two decades) will the frequency variation of T_g with $1/T$ approximately follow an Arrhenius dependence.

In general, however, linear (Arrhenius) relationships over a wide frequency range are found for secondary relaxations, including those associated with the crystalline phase. Only glass transitions exhibit nonlinear (WLF) relationships in these plots. It is also noted that the glass transition is less sensitive to frequency than is the case for secondary relaxations. Thus the glass transition has a higher apparent activation energy. The subject of sub-T_g relaxations is covered in more detail below in Section 5.4.4 on secondary viscoelastic relaxation processes.

Because of the frequency dependence of T_g, the convention adopted for assignment of the glass transition temperature is an important consideration. Traditionally, a frequency of 1 Hz has been used as a standard value. This is based on the historic precedence, since the torsion pendulum was the most widely used DMA technique in the early days of viscoelastic property measurements. The torsion pendulum is a free vibration technique with a natural frequency of approximately 1 Hz. The 1 Hz value also is reasonably close (within 10 °C) to the T_g values determined by other widely used methods such as DSC, dilatometry, and TMA. The relation between DMA and DSC T_g values is considered further at the end of this chapter (Sircar and Drake 1990). Because of the ambiguity inherent in the kinetic nature of T_g, it is most important that the test frequency be reported along with any T_g value determined by a DMA technique.

5.4.3.3. Material Characteristics

5.4.3.3.1. Crystallinity. In semicrystalline polymers the fact that the amorphous and crystalline phases are intimately interconnected has a significant effect on relaxation at the glass transition, resulting in a broadening of the transition region. In polymers that develop a high degree of crystallinity, the glass transition is suppressed so that the glass transition loss dispersion appears as a relatively minor event and in some cases may be difficult to detect at all. The subject of relaxation processes in crystalline polymers is treated in an excellent review paper by Boyd (1985).

An example of the broadening of the glass transition in a moderately crystalline polymer (PET) is illustrated by the data shown in Fig. 5.24 (Takayanagi 1965). The low-crystallinity sample exhibits the glass transition relaxation at 80 °C, while the higher crystallinity samples show the α relaxation around 105 °C. As the crystallinity increases, the relaxation broadens and the loss peak shifts to higher temperatures, and assignment of the T_g value becomes more difficult.

Figures 5.25 (Sha et al. 1990) and 5.26 (McCrum 1959) illustrate the suppression of relaxation at T_g in highly crystalline polymers. Although identification of the value of T_g in polyethylene is still a controversial subject, substantive arguments point to the β (−30 °C) relaxation as the glass transition (Sha et al. 1990; Davis and Eby 1973; Gaur and Wunderlich 1980; Sauer et al. 1995). There is an α_c relaxation in polyethylene below T_m. In this case the α_c relaxation is

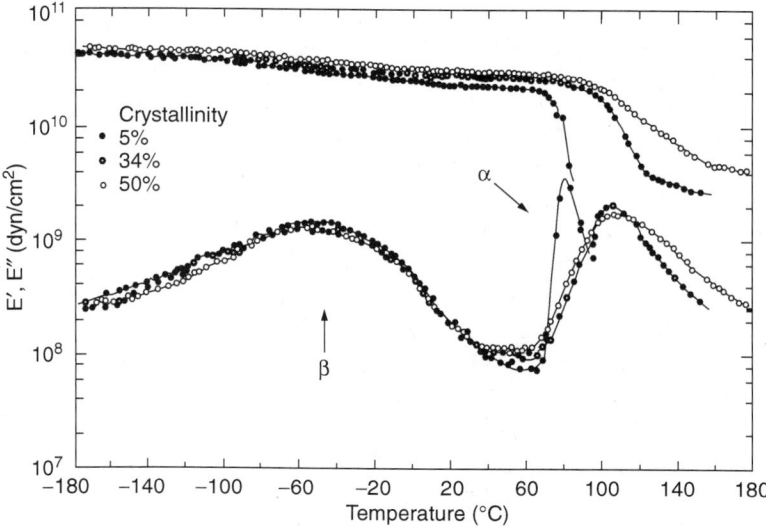

Figure 5.24. An example of the broadening of the glass transition in semicrystalline PET; the low-crystallinity sample exhibits glass transition relaxation at 75 °C, while the higher-crystallinity samples show the relaxation around 105 °C. As crystallinity increases, the relaxation broadens and the loss peak shifts to higher temperatures (from Takayanagi, 1965).

not considered a crystalline relaxation itself, but it does require the presence of the crystalline phase (Boyd 1984).

Regarding the effect of crystallinity on the intensity of the glass transition, consider Fig. 5.25. It is well known that the crystallinity of branched (low-density) polyethylene is much less than that of linear (high-density) polyethylene. This is due to the disrupting effect of the branching on the crystallites. The data shown in Fig. 5.25 (Sha et al. 1990) indicate that with decreasing the amount of branching, that is, increasing crystallinity the intensity of the β relaxation diminishes until it is hardly detected. Similarly, in PTFE (Fig. 5.26) (McCrum 1959), the intensity of the glass transition, noted as α at around 135 °C (408 K), decreases as the degree of crystallinity increases. The α relaxation in PTFE is the commonly accepted glass transition for PTFE. The reader also should be advised that there is reference in the literature to the possibility that the γ relaxation at −73 °C (200 K) is the glass transition of PTFE (Lau et al. 1984a,b).

The difficulty in assigning T_g for highly crystalline polymers is that the glass transition relaxation in these materials is a relatively minor process and there is often more than one relaxation below T_m that may be influenced by crystallinity. Thus the assignment of the particular type of molecular motion associated with the relaxation and relating it solely to the amorphous fraction is not a straightforward matter.

From a practical view, the physical and mechanical properties of highly crystalline polymers are dominated by the presence of crystallinity, and their

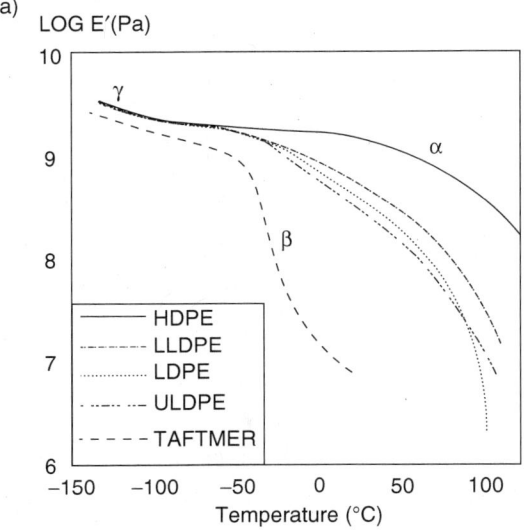

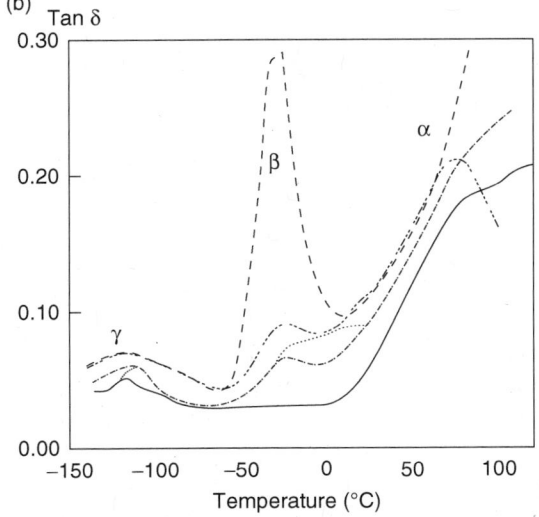

Figure 5.25. Suppression of relaxation at T_g in a highly crystalline polymer for a series of polyethylenes with different amounts of short-chain branching; as crystallinity increases, the intensity of the β-loss peak diminishes. HDPE = high-density polyetnylene (PE), LLDPE is linear low-density PE, LDPE is low-density PE, ULDPE is ultra low-density PE, and TAFTMER is a specially synthesized low-low density PE. [From Sha et al. (1990); reprinted with permission of the North American Thermal Analysis Society.]

upper use temperature limit is T_m, not T_g. Thus, the glass transition has less significance in this class of materials than it does in amorphous polymers, where T_g generally represents the maximum use temperature. It should be noted that below T_g crystalline polymers tend to be more brittle than they are

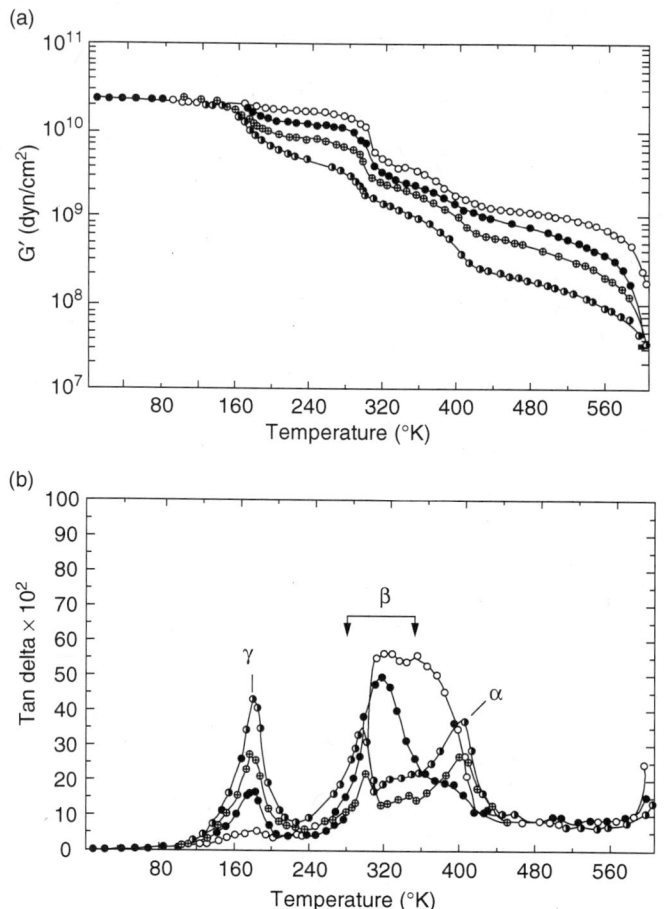

Figure 5.26. In PTFE the intensity of the glass transition [noted as α at around 135 °C (408 K)] decreases as the degree of crystallinity increases. Crystallinity: 92% (○), 76% (●), 64% (⊕), 48% (◐). (From McCrum, 1959, with permission of John Wiley & Sons.)

above T_g. This can lead to problems due to cracking under stress during low-temperature applications.

A complication that arises in measuring the glass transitions in crystallizable polymers involves possible crystallization during the measurement at temperatures below T_m. Crystallizable polymers that are quickly cooled from the molten state often have poorly developed or no crystallinity. When such materials are heated above their T_g, crystallization occurs spontaneously at temperatures between T_g and T_m. This process is referred to as *cold crystallization*. Thus attempts to study the α_c relaxation will be affected by cold crystallization. This is the case when amorphous or slightly crystalline modifications of crystallizable polymers are being considered. Data obtained for such samples must be interpreted with caution. Time–temperature superposition (TTS) is certainly not valid for such materials, since the material is changing during the test, which is contrary to the rules for applying TTS. The application of TTS to semicrystalline polymers is discussed in detail by Van Krevelen (1990).

5.4.3.3.2. Polymers Containing Particulate Fillers. Particulate fillers can cause noteworthy changes in the linear viscoelastic properties of amorphous polymers in the vicinity of T_g. It is well established that rigid fillers increase the storage modulus E' and shift the T_g to a higher temperature; usually the increase in E' is greatest above T_g in the transition and plateau regions of the viscoelastic spectrum (Lee and Nielsen 1977; Chartoff 1986). All of these features are illustrated in Fig. 5.27 (Lee and Nielsen 1977). At the same time, the glass transition damping peak (tan δ) broadens and the peak position shifts to higher temperatures (not shown). The amount of broadening of the transition increases as the volume fraction of filler increases or as the particle size of the filler decreases for a given volume fraction, when the particles are spherical. Depending on the type of filler and the interfacial bonding between the filler and the matrix, however, other effects have been noted (Chartoff 1986, 1988).

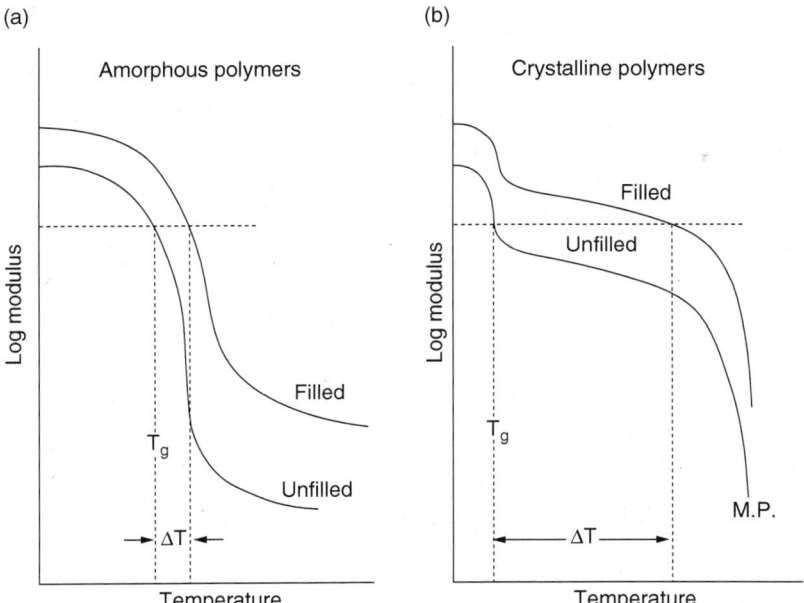

Figure 5.27. Schematic illustration of the effect of rigid particulate fillers on the storage modulus of amorphous and crystalline polymers in the vicinity of T_g; fillers increase the storage modulus E' and shift the T_g to a higher temperature; the glass transition tan δ peak (not shown) broadens and the peak position shifts to a higher temperature. [From Lee and Nielsen (1977); reprinted with permission of John Wiley and Sons, Inc.]

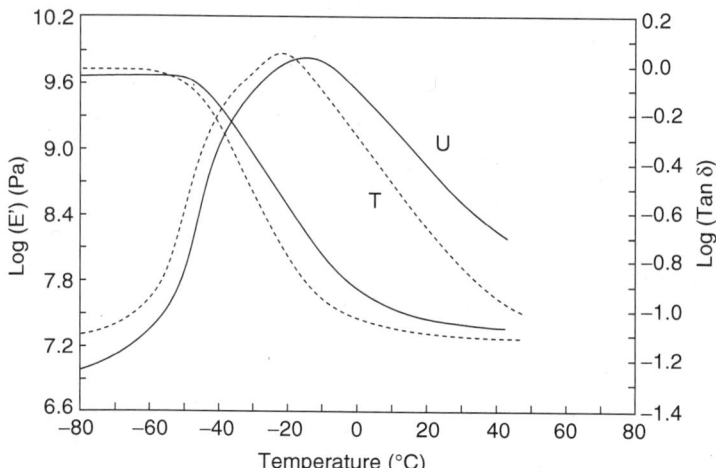

Figure 5.28. Composites with platelet-shaped graphite filler particles that are poorly bonded to the polymer matrix have broad T_g regions (U); when the filler is treated with a coupling agent to create stronger adhesion, the glass transition damping peak narrows and the loss peak temperature decreases (T) (from Chartoff, 1986, with permission of de Gruyter).

Platelet fillers such as mica and graphite are more efficient than spherical particles in broadening the damping peak with no reduction in damping peak intensity at filler volume fractions as low as 5–10%. First, platelet-shaped particles have a higher specific surface area. In addition, these particles have effects on viscoelastic properties that are more pronounced as the filler particle size increases. Chartoff (1986, 1988) cites evidence that two additional damping mechanisms that affect the relaxation spectrum are introduced in the case of platelet particles: (1) particle–matrix slippage and (2) polymer shear between filler particles that is operative when the platelets are oriented and parallel to each other. Chartoff notes that natural graphite has a slippery surface, so that normally there is poor adhesion between the platelet graphite particles and the matrix polymer. When the graphite is treated with a silane coupling agent to create stronger adhesion, the glass transition damping peak narrows and the peak temperature decreases as shown in Fig. 5.28 (Chartoff 1986, 1988). For a more detailed review of the effects of particulate fillers on the glass transition, the reader is referred to the reviews presented by Lee and Nielsen (1977), Chartoff (1986), and Nielsen and Landell (1994).

5.4.3.3.3. Plasticizers and Moisture. The T_g values of polymers are seriously affected by the presence of diluents. Liquid organic plasticizers may be added specifically to lower T_g. The most important example among commercial polymers is poly(vinyl chloride), which can vary from very rigid to soft and flexible products, depending on the plasticizer content. In the case of polar polymers (particularly nylons and epoxy resins) moisture acts as a plasticizer and significantly lowers the glass transition. More specific effects of plasticizers are considered in greater detail later in this chapter.

5.4.4. Secondary Transitions in Amorphous and Crystalline Polymers

The nomenclature for various transitions or relaxations in polymers was considered in a previous section. In isochronal experiments on amorphous polymers, for the temperature interval between the melt and liquid nitrogen temperature, each relaxation process commonly has distinct characteristics related to the specific motions associated with it. The highest-temperature or α process in amorphous polymers is the glass transition, involving the cooperative motion of several molecular segments or subunits. The glassy state β and γ relaxations take place in polymers either with or without flexible sidegroups and are believed to be due to local rotational motions of the mainchain and/or motions of pendant groups with differing degrees of cooperativity. The sub-T_g relaxations are characterized by an Arrhenius-type temperature dependence. The more complex are the cooperative motions associated with a relaxation process, the greater are the activation energies. Thus each relaxation process has its own activation energy. This is discussed further below.

In semicrystalline polymers there are at least two and often three relaxation processes observed below the melting point. In polymers that show three processes, the high-temperature $α_c$ process commonly is associated with the crystalline fraction in the material. However, while the mechanical $α_c$ relaxation requires the presence of the crystalline phase and is influenced by lamellar thickness, it is an amorphous phase process and involves deformation and cooperative motion of the latter. The nature of the $α_c$ molecular process was considered by Boyd (1984, 1985). The β process in these polymers originates in the amorphous phase and is the glass–rubber transition, short "the glass transition". In polymers where there is no secondary $α_c$ process above T_g, the highest temperature transition is the glass transition and is referred to as α. The low-temperature process, γ (or β when α is the glass transition), is generally thought to originate in the amorphous phase but may also be affected by the presence of the crystalline phase.

Some generalizations can be made about these types of relaxation behavior (Boyd 1984, 1985). Inherently low crystallinity polymers show no crystalline high-temperature relaxation process (but do possess a well-developed amorphous fraction glass transition, in this case denoted α). Inherently easily crystallizable, high-crystallinity polymers show both $α_c$ and β relaxations, where β is the glass transition. However, in these materials the β process does not tend to be very prominent because the amorphous phase is the minor phase. All crystalline polymers show the low-temperature γ process (referred to as β when α is the glass transition).

The study of relaxation processes in semicrystalline polymers is a subject of technological interest because of its practical importance. This is based on the observation that the modulus or stiffness of typical semicrystalline engineering thermoplastics at room temperature may be only 20–34% that of the same polymer at a low temperature. The decrease in modulus takes place in regions of temperature associated with a relaxation process.

Similarly, in amorphous polymers such as polycarbonate, the drop in modulus accompanying sub-T_g relaxations can have an important effect on engineering properties, particularly brittleness and impact resistance. This is discussed in Section 5.4.2.

5.5. EXAMPLES OF DMA CHARACTERIZATION FOR THERMOPLASTICS

5.5.1. Effects of Plasticizers and Moisture

In amorphous polymers the glass transition and the relaxation behavior associated with it are very sensitive to the addition of small amounts of diluents. As the diluent is added, the relaxation is shifted to a lower temperature at constant frequency or higher frequency at constant temperature. The usual explanation is that since the diluent molecules are small and mobile, they act to effectively increase the available free volume for segmental motion and hence speed it up. Similar plasticizing effects on the glass–rubber transition in semicrystalline polymers are observed. One such example is the case of poly(vinyl alcohol), which is water-soluble (Takayanagi 1965). Other examples of semicrystalline polymers where the effects of moisture are observed are aliphatic polyamides such as Nylon 6–6 (Starkweather 1980), Nylon 6–10 (Boyd 1959; Woodward et al. 1960), and Nylon 12 (Varlet et al. 1990).

There are three major relaxations in Nylon 6–6 (McCrum et al. 1967; Starkweather 1980; Schmieder and Wolf 1953; Starkweather 1973a). The α relaxation is the glass transition reflecting motion in fairly long-chain segments in amorphous regions. The β relaxation has been attributed to motion of labile amide groups that hydrogen-bond with water and may be absent in dry, annealed specimens. The γ relaxation has been assigned to motion of short polymethylene segments with some involvement of adjacent amide groups. The addition of small amounts of water or alcohols (which readily dissolve in the amorphous fraction by virtue of hydrogen bonding) moves the α loss peak to lower temperatures (or higher frequencies at constant temperature) and narrows it. Moisture also reduces the temperatures of both the β and γ relaxations but to a lesser degree than the α relaxation.

The sorption of water by nylon has a major effect on the mechanical properties of engineering importance. For example, in molded Nylon 6–6 at room temperature, as the water content is increased from dryness to saturation, the modulus decreases by a factor of ~5, the yield stress decreases by more than half, and there are major increases in the tensile elongation and energy to break (Starkweather 1973a,b). These changes are associated with the glass transition shifting from above room temperature to below it. The reported properties of nylons are frequently those of samples containing some amount of absorbed water. Thus, it is important to specify the water content or the relative humidity with which the polymer is in equilibrium.

The dependence of the relaxation temperatures on the level of absorbed water in Nylon 6-6 is known from dynamic mechanical studies (Starkweather 1980, 1973a; Prevorsek et al. 1971) as well as dielectric studies (Starkweather and Barkley 1981). The temperature variations with sorbed moisture of the loss modulus peaks for the three relaxations are shown in Fig. 5.29 (Starkweather 1980). The test frequency for the three relaxations varies slightly but is around 1 Hz. The data indicate that the temperature of the α relaxation at a given frequency decreases by about 100 °C (to below ambient) between dryness and saturation. The β relaxation is also shifted to lower temperatures and higher frequencies by absorbed water, while the temperature of the γ relaxation is only slightly affected, shifting to somewhat lower temperatures and higher frequencies.

In addition, it was found that moisture increases the intensity of the β relaxation and reduces its activation energy while decreasing the peak height of the γ relaxation with little change in activation energy. The decrease in γ intensity has been referred to as an "antiplasticization" effect, since it is associated with an increase in modulus and decrease in impact strength. This is

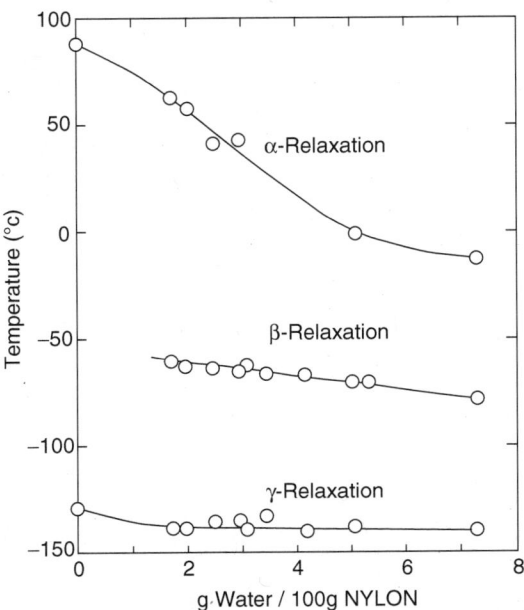

Figure 5.29. Temperature variations with the level of absorbed moisture for the loss modulus peaks of the three relaxations in Nylon 6-6 (from Starkweather, 1980, reprinted with permission from ACS Symposium Series 127, 433. Copyright 1980 by the American Chemical Society).

discussed further below (Section 5.5.2) in reference to impact behavior. The increase in β intensity has been associated with the motion of water molecules hydrogen-bonded with carbonyl and amide groups about the hydrogen bond axis (Woodward et al. 1960; Starkweather and Barkley 1981).

Moisture is an effective plasticizer for highly polar, phenylene polymers with α relaxations in excess of the 100 °C boiling point of water. This is particularly noted in polyamides and polyimides, where strong hydrogen bonding prevents the loss of moisture even at high temperatures up to T_g. An example is the polyamide derived from hexamethylenediamine and a 70/30 mixture of isophthalic and terephthalic acids (Nylon 6–I/T; 70/30), which exhibits an α peak temperature decreasing from 154 °C at 0% RH to 60 °C at 100% RH as a linear function of relative humidity (Avakian et al. 1991). Discussions on the absorption of moisture by polyimides are provided by Lim et al. (1973) as well as Chartoff and Chiu (1980). It is known that in order to effectively dry polyimides, the application of heat and vacuum at elevated temperatures for an extended time period is required. For a polymer that contains both amide and imide groups, this effect would be even more pronounced.

Dynamic mechanical analysis data for samples of an amorphous aromatic polyamid–imide polymer containing various amounts of moisture are shown in Fig. 5.30 (Hiltz and Keough 1991). These data indicate a decrease in T_g of >100 °C at 4.25% moisture sorption. The dry T_g value is around 285 °C. Each α loss dispersion has the same shape, indicating that moisture was not lost during the measurement. (If moisture were lost, the loss dispersion would broaden as the result of a gradual shift in the relaxation to higher temperatures during the measurement.) Since the α relaxation in this polymer occurs at

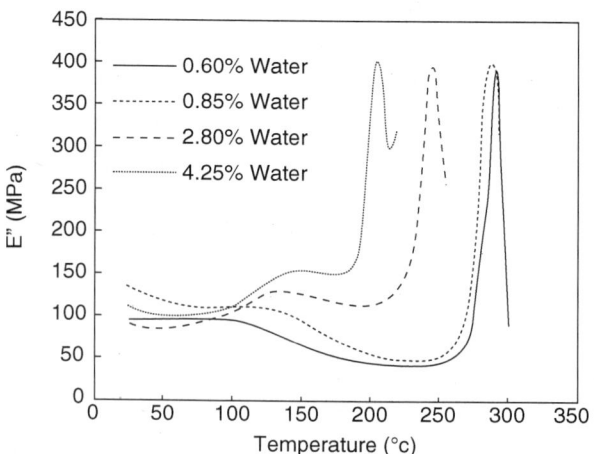

Figure 5.30. DMA data for samples of an amorphous aromatic polyamid–imide polymer containing various amounts of moisture; the data indicate a decrease in T_g of over 100 °C with 4.25% moisture sorption. [From Hiltz and Keough (1991); reprinted with permission of the North American Thermal Analysis Society.]

temperatures well in excess of 100 °C, the data show that water is indeed tightly bound in such systems. A broad β relaxation at around 100 °C is seen to sharpen, increase in intensity, and shift upward to around 140 °C in the wet specimens.

The effect of moisture on the β relaxations in amorphous phenylene polymers is well documented (Allen et al. 1971; Lim et al. 1973; Chung and Sauer 1971). For example, Allen et al. (1971) determined that both the mechanical and dielectric β relaxations in polysulfone, polycarbonate, poly(phenylene oxide), and poly(ether sulfone) were dependent on the water content of the samples as illustrated for polysulfone in Fig. 5.31. In addition, the amount of water absorbed depended on the polarity of the molecule. The results indicate that the absorbed water hydrogen bonds to polar groups along the polymer chain and, as such, takes part in the molecular processes that give rise to the β relaxation. This is consistent with the other examples cited previously.

The polymers studied by Allen et al. (1971) are modestly polar with maximum moisture uptakes of less than 1%. In more polar phenylene polymers such as polyimides (Lim et al. 1973) and polyamid–imides (Hiltz and Keough 1991) the mechanical β relaxation shows a more significant effect because of the larger amounts (ranging from 1.5 to >4%) of moisture absorbed.

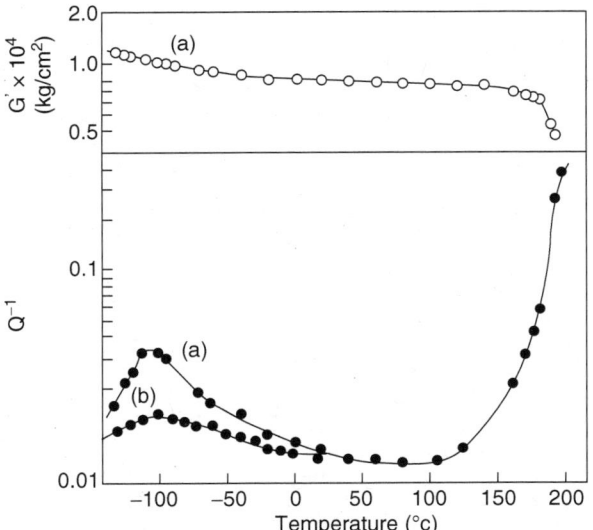

Figure 5.31. Effect of absorbed H_2O on the storage modulus and mechanical damping behavior (Q^{-1} or tan δ) of poly(ether sulfone) at 1 Hz; curves (a) 0.31% H_2O, curve (b) 0.07% H_2O; the β relaxation in this polysulfone is dependent on the water content of the samples as illustrated; the relaxation intensity increases with the amount of moisture present. [From Allen et al. (1971); reprinted with permission of Elsevier.]

As in other crystalline polymers, the glass–rubber (β) transition in linear or high-density polyethylene (HDPE) is sensitive to plasticizing diluents. CCl$_4$ is effective in this regard (Illers 1972; Arai and Kuriyama 1976; Sha et al. 1990). Loss data for HDPE containing absorbed CCl$_4$ such as those of Fig. 5.32 (Arai and Kuriyama 1976), show a prominent E'' β peak in the range from −60 °C to −80 °C compared to a very weak (or nonexistent) relaxation peak in the range from −30 °C to −50 °C for the unplasticized material. It is evident that the diluent is effective in removing constraints on the amorphous segments, thus increasing their mobility (causing a reduction in $T_β$) and in making more conformations available to reorientation (resulting in an increase in β intensity).

5.5.2. Impact Behavior

The fracture behavior of polymers under large-scale deformations is of considerable interest for many thermal analysts who are interested in specifying polymers for use in critical load-bearing applications. It is important in material selection to have some practical criteria for deciding what material to use. Such criteria are discussed by Menges and Boden (1986), who indicate how linear viscoelastic DMA data can be of benefit as an aid in decisionmaking.

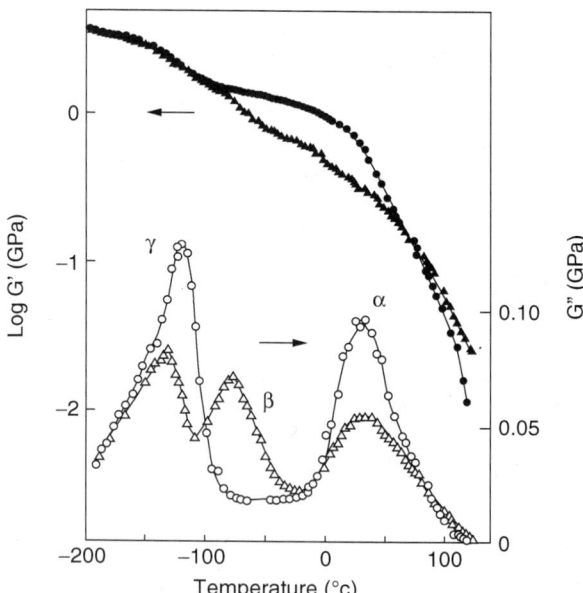

Figure 5.32. DMA shear storage and loss moduli for linear polyethylene (LPE) unswollen (○,●) compared to LPE swollen with CCl$_4$ (11 wt %) (△,▲); the loss modulus data clearly show the effect of the absorbed CCl$_4$ plasticizer on the glass transition relaxation (β relaxation) (from Arai and Kuriyama, 1976, with permission of Springer Science and Business Media).

The cause of brittle fracture in polymers is crack formation due to the inability of the material to quickly dissipate by molecular relaxation and related processes the internal stresses generated as a result of the imposed deformation. Brittle fracture occurs when the time to failure is of the same order of magnitude (or faster) than the speed of the relaxation process that dominates the mechanical behavior in the temperature range of interest. The relevant relaxation processes are the first secondary relaxation below T_g (β or γ). A qualitative criterion for determining whether the relaxation will significantly influence the impact properties is if the isochronal storage modulus (on a log modulus–T plot) exhibits a distinct step at the damping maximum. The size of the step is related to the relaxation intensity. It has been known for some time that polymers with relatively intense low-temperature relaxations may be expected to be "tough" at room temperature (Boyer 1968; Retting 1969; Hartmann and Lee 1979).

Thus the occurrence of a secondary relaxation loss peak by itself does not ensure improved impact strength (Menges and Boden 1986; Boyer 1968). Additional information such as actual impact test data at temperatures above and below the relaxation, should be used to verify whether a given relaxation process has a significant influence on the impact behavior.

Evidence of the importance of secondary (sub-T_g) relaxations in determining the impact strength of polymers is provided by numerous examples of changes in these relaxations by the addition of diluents that act as antiplasticizers. The term *antiplasticizer* relates to the fact that, while these diluents lower T_g, they also cause increases in storage (and tensile) modulus and tensile strength but lower impact strength and elongation at break. Figures 5.33a and 5.33b (Robeson and Faucher 1969) show storage modulus and tan δ curves for antiplasticized polysulfone and polycarbonate. As noted previously, both polymers in the pure state have well-defined β relaxations at about $-100\,°C$ (173 K). It is readily apparent that at high diluent levels these relaxations are greatly reduced and virtually eliminated.

The presence of an antiplasticizer in these systems and the resulting near elimination of the β relaxation leads to a higher modulus value at temperatures above the relaxation range up through room temperature. This also results in higher tensile strength, lower elongation at break, and reduced impact strength as noted in Table 5.2. Specific volume data suggest a qualitative explanation of antiplasticization in these systems, where the antiplasticizer molecules fill polymer free volume and restrict the local-mode molecular motions associated with the β relaxation.

Similar observations are found in the case of PVC (Bohn 1963; Pezzin et al. 1967), where the addition of normal plasticizers actually results in "antiplasticization" until the level of 15–20% plasticizer is reached. At this point modulus and tensile strength begin to decrease because the T_g has been reduced to below room temperature. Pezzin et al. (1967) show that this antiplasticization effect is closely related to a reduction in the intensity of the β relaxation centered around $-40\,°C$. The ability of polymers to be antiplasti-

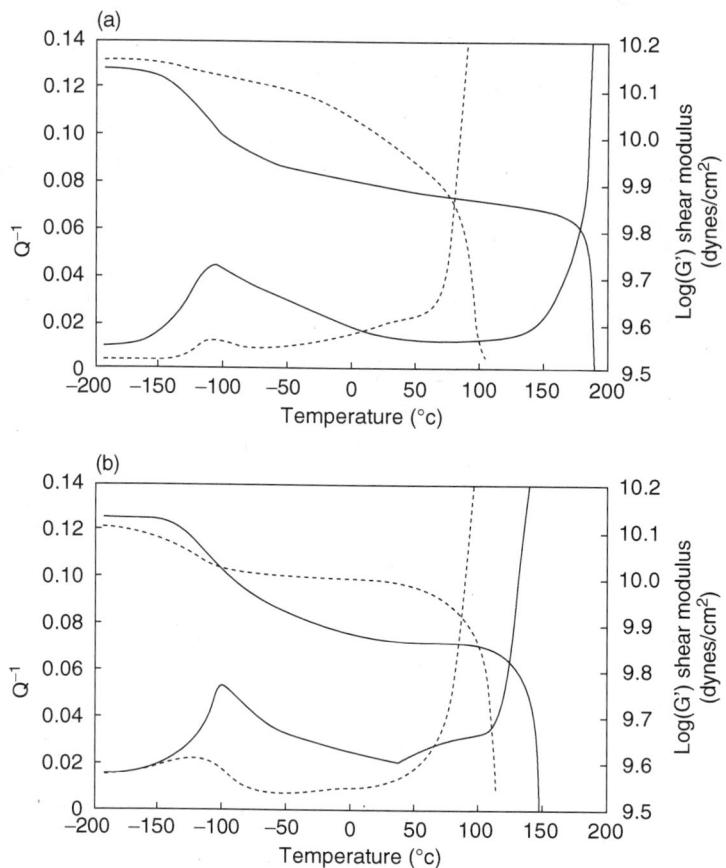

Figure 5.33. Storage modulus and tan δ curves for antiplasticized polysulfone (a) and polycarbonate (b); both polymers in the pure state have well-defined β relaxations at about −100 °C (173 K); at high diluent levels these relaxations are greatly reduced and virtually eliminated. [From Robeson and Faucher (1969); reprinted with permission of John Wiley and Sons, Inc.]

cized will be related to the strength of their secondary relaxations. Thus polymers with small secondary relaxations should be unaffected by antiplasticizers (Bohn 1963).

We noted previously that various workers (Starkweather 1980; Chung and Sauer 1971; Varlet et al. 1990; Prevorsek et al. 1971) have reported similar effects for nylons. At temperatures below the α relaxation, water is an antiplasticizer for nylon, increasing the modulus and tensile strength and reducing the intensity of the γ relaxation. Absorbed water reduces the activation entropy for the γ relaxation from a substantial positive value in the dry state to zero at saturation (Avakian et al. 1991; Starkweather 1988). This could be explained by water-forming bridges between amide groups and the speculation that

TABLE 5.2. Mechanical Property Data for Various Antiplasticized Aromatic Polymer Systems[a]

Polymer/Antiplasticizer	Antiplasticizer (wt%)	Tensile Modulus (psi)	Tensile Strength (psi)	Elongation at Break (%)	Izod Impact (ft·lb/in. of Notch)	T_g (°C)
Polysulfone/N-phenyl-2-naphthylamine	0	370,000	10,900	50	1.5	189
($T_g \cong -16°C$)[b]	10	450,000	12,850	7.8	0.5	137
	20	461,000	12,350	7.5	0.4	110
Polysulfone/4,4'-dichlorodiphenyl sulfone	10	452,000	12,380	9.2	0.9	148
($T_g \cong 0°C$)[c]	20	469,000	11,700	6.7	0.5	125
	30	479,000	11,240	6.0	0.3	102
Bisphenol A polycarbonate/Arochlor 5460	0	330,000	9,080	50–100	12–16	147
($T_g \cong 55°C$)	10	418,000	11,400	30.0	1.30	131
	20	444,000	12,400	6.2	1.00	120
	30	480,000	12,200	9.0	0.50	112

[a]Data from Robeson and Faucher (1969).
[b]Determined by measurement on torsion pendulum.
[c]Determined by extrapolation. Data from Robeson and Faucher (1969).

these bridges reduce the mobility of the aliphatic segments adjacent to the amide groups. It is known that local-mode relaxations of the aliphatic groups in cooperation with the amide groups give rise to the γ relaxation in nylon.

5.6. CHARACTERISTICS OF FIBERS AND THIN FILMS

Fibers and thin films generally are characterized by DMA in the tensile mode using fixtures specially designed for gripping them. Further, they usually consist of thermoplastics that have a high degree of orientation, which strongly affects their mechanical properties. Before presenting a discussion of DMA measurements of fibers and films, we briefly describe some basic aspects of their preparation. In general, fiber-forming polymers are semicrystalline. Fibers from synthetic polymers are made by spinning, which means that the molten (or dissolved) polymer is continuously extruded or pushed through narrow holes. The filaments formed in this way are cooled rapidly, and this process leads to the formation of so-called as-spun fibers. For polymers that can be easily quenched (such as PET), the as-spun fibers have low crystallinity. Since these fibers usually are unoriented (with little or no crystallinity), they are then stretched significantly at some optimum temperature to develop the necessary orientation, so that their modulus and a tenacity increase (*tenacity* is a fiber industry term that refers to the specific stress at break; it is a measure of the fiber's resistance to static forces). These fibers are then called *drawn fibers*, and are characterized by the *draw ratio*, which is a measure of the degree of stretching during the orientation of a fiber, expressed as the ratio of the cross-sectional area of the undrawn material to that of the drawn material.

A special method for fiber preparation is gel spinning and gel drawing (Pennings 1967). When the polymer chains are disentangled from each other in a dilute solution, this dilute solution can be spun into a coagulation bath, and in this way a solid gel is formed. Then, these gel-spun fibers can be drawn to very high draw ratios (since the individual polymer chains are disentangled from each other) to impart extremely high orientation. By this method, ultra-high-molecular-weight polyethylene can be drawn up to a draw ratio of 300×, and polypropylene to 57×. Extremely high tenacities can be achieved with gel-spun and gel-drawn fibers.

Single filaments (known as *monofilaments*) cannot be used directly in fabric products. For fabrics a certain number of monofilaments is first put together, and then twisted in order to increase the friction between the individual monofilaments. These interlocked fibers are called *yarns*. It is important to always report whether DMA measurements were made on single fibers or yarns; the mechanical properties of a yarn are not simply the sum of the mechanical properties of single fibers.

Films are usually prepared from a polymer melt or solution by slit-die extrusion followed by one- or two-dimensional stretching; thus a film may possess uni- or biaxial orientation. A special case of film extrusion is film blowing, which produces biaxially oriented films in one step, but the level of orientation is low. Similar to the case of gel spinning of fibers, precursor films of low chain entanglement can be produced, which, after drawing and annealing, provide films of very high modulus and tenacity. Films also may be produced by casting from polymer solutions.

A special method, called *zone drawing–zone annealing*, was developed by Kunugi et al. (1986, 1988, 1991) to prepare high-modulus and high-tenacity fibers and films from various polymers [polyethylene, polypropylene, PET, poly(ether ether ketone) (PEEK), nylons, poly(vinyl alcohol)]. In this method the fiber is drawn in a small heated zone, which is created by a spot-heating device. More recently, a CO_2 laser has been used as the heating device. This provides a uniform heat zone that produces fibers with high draw ratios and, therefore, high tenacity (Okamura et al. 2004; Uddin et al. 2004).

Measurements on fibers and films using DMA yield information about the various transitions using the temperature dependence of the storage and loss moduli. In order of decreasing temperatures, these transitions are

- *Cold crystallization* (discussed previously in this chapter and Chapter 2), where crystallization takes place during heating to above T_g of a polymer that is amorphous for kinetic reasons (i.e., a polymer that had been quenched) but is able to crystallize. PET and PEN [poly(ethylene-2,6-naphthalene dicarboxylate)] are examples of polymers that exhibit this phenomenon. In such cases, when the samples are heated, the modulus undergoes a sudden drop of two or three orders of magnitude, due to the glass transition followed by a sudden modulus increase in the cold crystallization temperature range above T_g. Remember that the modulus of a semicrystalline polymer is always higher than the modulus of a similar amorphous polymer above the glass transition temperature.
- *The α_c transition*, a relaxation process that takes place in some cases in the amorphous phase but is influenced by the crystallites or in other cases it occurs within the crystallites; this relaxation takes place at temperatures below the melting point, but higher than the glass transition temperature.
- *The amorphous α relaxation* (or β-relaxation when an α_c relaxation is present), which is the glass transition.
- *Various second-order transitions* taking place in either the amorphous or crystalline phase (or both of them) below T_g.

DMA measurements can provide information about the orientation of the fiber or film through the magnitude of the storage modulus or by the glass transition temperature.

An important topic is the effects of crystallinity and orientation of the fiber on the glass transition temperature. Several research groups have noted that

T_g of fibers at first increases as crystallinity increases, then, with further increase in crystallinity, T_g begins to decrease (Dumbleton and Murayama 1967; Dumbleton et al. 1968; Illers and Breuer 1963; Thompson and Woods 1956). This phenomenon was explained by the change in the number and size of the crystallites in the fiber. The effect of orientation is more straightforward; in most cases, T_g of a fiber increases with increasing orientation. The reasons for this increase are a decrease of the free volume with increasing orientation and the fact that amorphous phase molecular segments are constrained in an elongated condition. However, the dependence of properties on orientation can be complex, if the crystallinity of the fibers changes simultaneously with orientation. Figure 5.34a shows how the storage modulus of PET yarn changes as the draw ratios of the fibers constructing the yarn increase. It is obvious that the storage modulus increases with increasing draw ratio (and therefore orientation) of these fibers. The orientation and the crystallinity also have an effect on the glass transition, as can be seen on the tan δ–temperature curves in Fig. 5.34b: the yarn from the as-spun fibers has a sharp glass transition, because it is unoriented and nearly amorphous.

Figure 5.35 shows the dynamic mechanical curves for as-spun and drawn poly(ethylene-2,6-naphthalene dicarboxylate) (PEN) monofilaments. The as-spun fiber, which is unoriented and amorphous, exhibits a slow modulus decrease with increasing temperature, then beyond 100 °C the modulus suddenly drops by two or three orders of magnitude because of the glass transition of the polymer. At higher temperatures, the modulus increases again as a result of cold crystallization. The modulus increase above the glass transition is approximately one order of magnitude, which is typical for semicrystalline polymers that are quenched to an amorphous state. Finally, with a further temperature increase, the storage modulus decreases again, this time because of melting of the PEN crystallites. The drawn fibers do not exhibit such a modulus increase, because they are already crystalline, so no cold crystallization can take place.

Garrett and Grubb (1988) described an excellent example of how a crystalline relaxation can affect the drawing conditions of a fiber. Working with highly drawn gel-spun poly(vinyl alcohol) (PVA) fibers, they observed an intense crystalline $α_c$ relaxation. The temperature of this relaxation (which was determined from the peak temperature of the tan δ–T curves) depended on the draw ratio of the fiber. At a draw ratio of 1 (i.e., for undrawn fiber), $T_{α_c} = 160°C$, and it moves up to 220 °C at a draw ratio of 38× (see Fig. 5.36). It was suggested that in this polymer the hydrogen bonding becomes weaker beyond the $α_c$ relaxation and this weakening is responsible for making drawing possible. But at very high draw ratios, the temperature of the $α_c$ relaxation increases and moves toward the melting point of PVA, thus rendering gel drawing impossible.

Crystal → crystal transitions sometimes can appear on the DMA curves. Figure 5.37 shows the storage modulus–temperature data for an as-spun monofilament and a drawn monofilament prepared from poly(2-methylpenta-

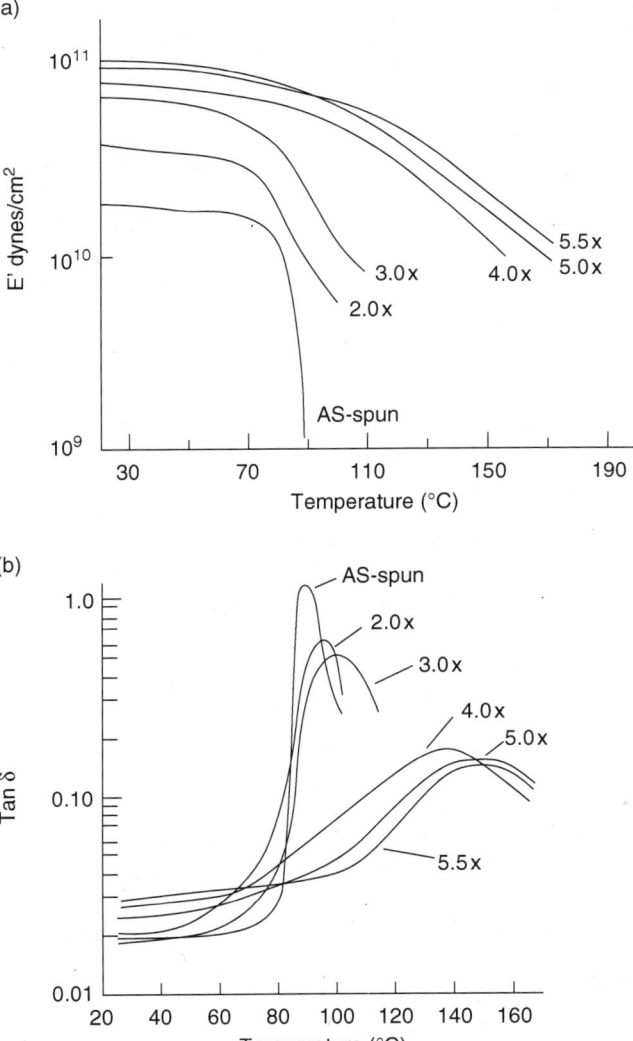

Figure 5.34. (a) Storage modulus of PET yarn from fibers drawn to various draw ratios. The storage modulus increases with increasing draw ratio (note that 1 dyn/cm^2 = 0.1 Pa); the draw ratio is indicated at each curve; (b) Loss tangents for PET yarns from fibers drawn to various draw ratios. As the orientation (i.e., draw ratio) and crystallinity increase, the glass transition broadens and becomes more shallow; the draw ratio is indicated at each curve. [Plots (a) and (b) both from Miller and Murayama (1984); reprinted with permission of John Wiley and Sons, Inc.]

methylene terephthalamide) [Nylon M5T] [compare with the DSC curves in Chapter 2 (Menczel et al. 1996)]. The storage modulus of the drawn fiber exhibits a sudden drop of about an order of magnitude due to the glass transition at ~160 °C, then a further slow modulus decrease as T increases owing to

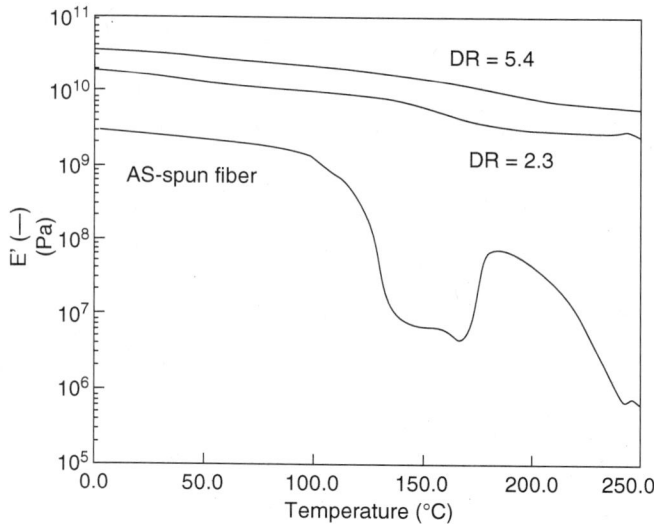

Figure 5.35. DMA storage modulus curves for as-spun and drawn poly(ethylene-2,6-naphthalene dicarboxylate) (PEN) monofilaments; the as-spun fiber, which is unoriented and amorphous, exhibits a modulus decrease due to the glass transition of the polymer; then, the modulus increases again due to cold crystallization. [From Saw et al. (1997); reprinted with permission of Society of Plastics Engineers.]

a gradual melting process. The drawn Nylon M5T fibers do not exhibit a modulus increase, due to cold crystallization. The curve for the as-spun fiber, which is unoriented and amorphous, is quite different. At the glass transition, which now occurs at ~145 °C, the storage modulus drops by approximately three orders of magnitude, then it suddenly increases because of cold crystallization. This is followed by a decrease in modulus due to melting and an increase due to recrystallization. As temperature is increased, two more recrystallization–melting events are observed. Compare with the description of thermal behavior of Nylon M5T shown in Chapter 2.

Dynamic mechanical analysis data for thin films are similar to, but somewhat more complex than those for fibers; there is an additional variable in these measurements—the angle between the molecular chain orientation of the film and the measurement strain direction. Changing this angle can drastically change the results. Figure 5.38 (Menczel et al. 1997a) shows storage modulus data for a series of Vectran liquid crystalline polymer film samples. It is clear that the modulus is greatest when the strain direction is parallel to the chain orientation in the film. This situation is analogous to the case of unidirectional fiber-reinforced composites discussed in a section 5.8.5.2.

It is expected that in general, the tensile storage modulus of more conventional polymeric films also should be greater in the draw direction than in the direction perpendicular to the draw (i.e., $E_{\parallel} > E_{\perp}$). When these moduli are

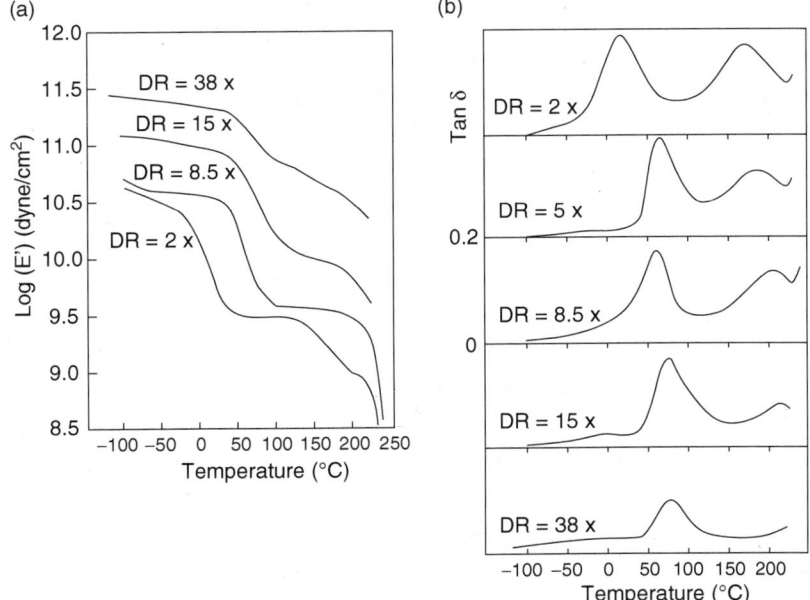

Figure 5.36. (a) Storage modulus of PVA fibers gel drawn to various draw ratios. The draw ratio is indicated at each curve; (b) loss tangent of PVA fibers gel drawn to various draw ratios; the temperature of the α_c relaxation (the higher-temperature relaxation on each curve) shifts to higher temperatures with increasing draw ratio moving toward the melting point of PVA so that it disappears at DR > 38. [From Garrett and Grubb (1988); reprinted with permission of John Wiley and Sons, Inc.]

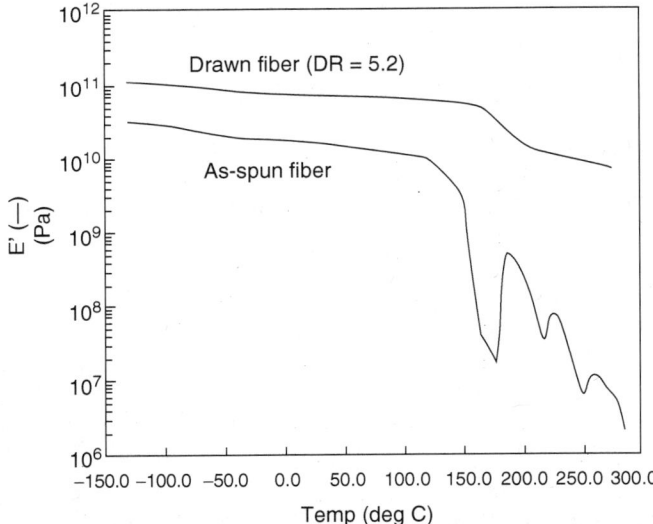

Figure 5.37. Tensile storage modulus curves of as-spun and drawn Nylon M5T fibers as a function of temperature (from Menczel et al., 1996, with permission of Springer).

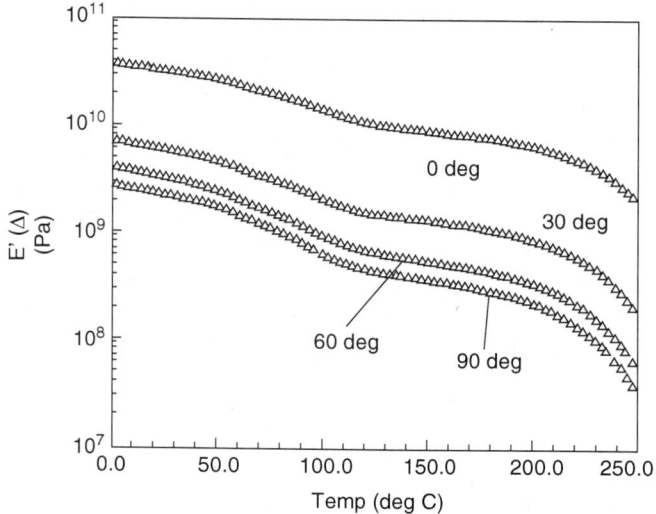

Figure 5.38. Tensile storage modulus of Vectran liquid crystalline film samples as a function of temperature; the number at each curve indicates the angle between the measurement strain direction and the direction of polymer chain orientation (from Menczel et al., 1997a, with permission of Akadémiai Kiadó, Hungary).

compared for cold-drawn polyethylene, polypropylene, polyoxymethylene, poly(ethylene oxide), and Nylon 66 films, $E_{\parallel}$ is indeed greater than $E_{\perp}$ (Dumbleton and Murayama 1970).

5.7. DMA CHARACTERIZATION OF CROSSLINKED POLYMERS

5.7.1. The Glass Transition

Highly crosslinked polymers are an important class of materials used as adhesives, coatings, and matrices for composites, to name just a few applications. Many of this class are referred to as *thermosets*, and almost all are completely amorphous (an exception is the case of liquid crystal thermosets). The material presented here is complemented by the presentation on thermosets in an earlier chapter (see Section 2.10). The glass transition temperature is a most significant parameter for thermosets, since it is a material property related to the degree of crosslinking and is an important parameter for engineering applications. In thermosets a difficulty arises in specifying T_g because of the effect of crosslinking on relaxation in the transition region (Chartoff et al. 1994). The general effect on the glass transition from increasing crosslink density is shown in Fig. 5.39. The transition loss dispersion decreases in intensity, broadens, and shifts to higher temperatures. Also, the transition slope of

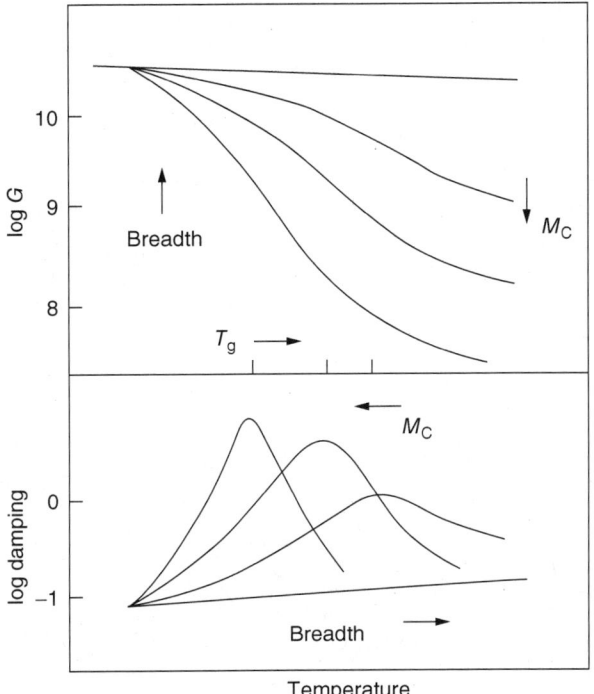

Figure 5.39. Diagram showing the general effect on glass transition by increasing crosslink density in a typical thermoset; the transition loss dispersion decreases in intensity, broadens, and shifts to higher temperatures; the drop in modulus decreases in magnitude and occurs over a broader temperature range. At a very high crosslink density, the loss peak may be entirely suppressed so that a glass transition is not apparent. Mc is the molecular weight between crosslinks, which is inversely proportional to crosslink density. All arrows are in the direction of increasing values.

the storage modulus decreases. The broadening of the relaxation spectrum is further enhanced by network heterogeneity and microphase separation introduced into systems formed from mixtures of reactive monomers that have different functionalities. The data of Fig. 5.20 for a UV cured acrylate polymer are representative of such a system.

The highly crosslinked acrylic represented in Fig. 5.20 has a very broad (possibly two-step) transition region that clearly illustrates the problem involved in assigning T_g in such cases. The issue to consider is what criterion should be used to specify T_g? The loss dispersion in Fig. 5.20 covers a span of approximately 100 °C and is very flat, with no well-defined maximum value. The E'' maximum value chosen (computer-selected, by numerically taking the largest E'' value) is 70 °C lower than the peak tan δ. Furthermore, the E'' peak appears to occur in the transition region above the temperature where E' begins to decrease gradually from the glassy state value. Our recommendation in such cases, where the T_g range is unusually broad, is based on practical considerations. Since network polymers are frequently used in structural applications as adhesives and composite matrix materials, the point where E' begins to drop from its glassy value is the most appropriate T_g assignment, based primarily on the criterion of retention of structural integrity. This is most easily specified by the onset point, defined by intersection of the tangents to the storage modulus glassy state curve and transition region curve.

5.7.2. The Extent of Thermoset Cure as Characterized by DMA

From the information on crosslinking and T_g noted above, DMA can be used for the cure characterization of thermosets. There are two aspects of this:

1. As the chemical conversion (extent of cure) and the degree of crosslinking advance, T_g increases. Thus the T_g measured by DMA correlates with the cure state of the material. In cases where T_g is not overly broad and a well-defined maximum in the loss modulus is evident, the T_g value obtained is related to crosslink density. Similarly, peak tan δ values may be used to indicate T_g as long as one is consistent. In this case reproducibility is the main criterion and either the peak in E'' or tan δ may be taken as appropriate measures of T_g.
2. The magnitude of the plateau in storage modulus above T_g increases with crosslink density. This can be used to estimate an average crosslink density (or molecular weight between crosslinks) for the network. We present a brief review of these issues in the following paragraphs.

It has been shown that in many cases for thermoset systems there is a quantitative relationship between the chemical conversion of the thermoset and its T_g value, independent of the time–temperature cure history (Hale et al. 1991; Venditti and Gillham 1997). This can be useful from an applications standpoint because T_g is often a simpler and more sensitive measure than conversion itself. This implies that either the molecular structure at a given degree of conversion is the same, regardless of the reaction path, or that differences in the structures produced for different reaction paths do not affect T_g.

In dealing with thermoset cure, T_g–conversion data should be generated as part of any cure characterization study. The easiest way to develop a T_g–conversion relationship is via DSC. Where degree of conversion cannot be measured directly, the T_g serves as a good means for evaluating cure advancement. This will be useful even if there is no direct proportionality between T_g and conversion.

It should be noted that, while a T_g–degree of cure relationship has been found valid for many epoxy–amine and other thermosets, this correlation has not been observed in a limited number of cases, such as some epoxy–DICY (dicyandiamide), cyanate–ester, and phenolic systems, or for certain resin systems cured by two different methods such as thermal and microwave (Wei et al. 1995) or thermal and electron beam cures (Klosterman et al. 2003).

It is likely that cured molecular networks that have a significant degree of heterogeneity due to composition or molecular weight variations in the initial monomer, or from phase separation, or even the partial formation of a liquid crystal phase during polymerization will not have a well-defined T_g–degree of cure relationship. In the case of the epoxy–DICY and cyanate–ester systems, composition heterogeneity may arise because the crosslinking agents are added initially as crystalline powders. In such cases there is always some degree of incomplete mixing of these components so that the crosslinked network is heterogeneous, with multiple phases having different crosslink densities occurring on a microscale (Bobalek et al. 1964; Manson et al. 1977). In the phenolic systems the starting material is generally a mixture of oligomers, with a range of chemical functionalities. This also leads to a heterogeneous crosslinked phase distribution.

5.7.3. The Time–Temperature Transformation Cure Diagram

5.7.3.1. Characteristics of the Curing Process The curing reaction process is a most important factor that affects a thermosetting polymer's mechanical performance. Consequently, it is necessary to understand the cure process and its kinetics in order to design the proper cure schedule for obtaining the optimum network structure and performance. Excellent reviews on this topic are available in the literature (Ellis 1993; Pascault et al. 2001). Both rheology-mode and solid-mode DMA geometries develop data that are useful for cure characterization and optimization. These can be used to effectively develop cure temperature–cure time, gelation, and vitrification relationships as well as cure temperature–cure time viscosity relationships. These quantities also relate to conversion via T_g.

Isothermal time–temperature transformation (TTT) cure diagrams, as illustrated in Fig. 5.40, are a useful tool for illustrating the physical and chemical changes that take place during cure, including gelation, vitrification, complete cure, and degradation (Enns and Gillham 1983; Seferis and Nicolais 1983; Gillham 1986a,b). The TTT diagram is a generalized representation of the various events that occur during cure, where the time to gelation and vitrification are plotted as a function of isothermal cure temperature.

The S-shaped vitrification curve and the gelation curve divide the TTT plot into four distinct states related to the thermosetting cure process: liquid, gelled rubber, ungelled glass, and gelled glass. Three important temperatures are marked on the temperature axis of the TTT cure diagram: T_{g0} is the glass transition temperature of the unreacted resin mixture, $T_{g\infty}$ the glass transition temperature of the fully cured resin, and $_{gel}T_g$ is the temperature where gelation and vitrification occur simultaneously as well as the point where the vitrification and gelation curves intersect. At temperatures below T_{g0}, reaction takes place in the glassy state and is therefore slow to occur. To minimize

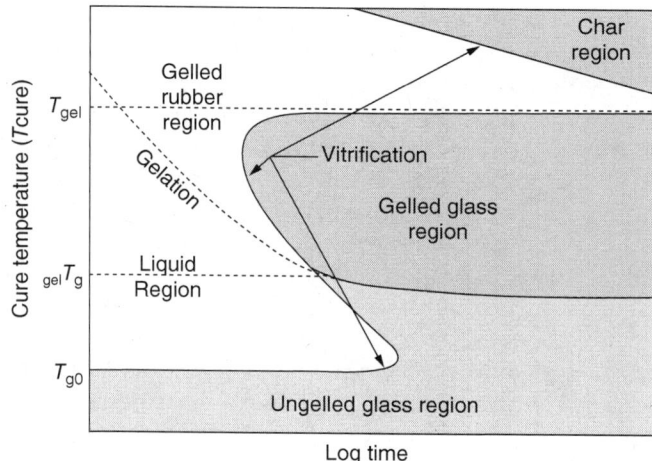

Figure 5.40. Generalized time–temperature transformation (TTT) cure diagram; a plot of the times to gelation and vitrification during isothermal cure versus temperature, delineates the regions of four distinct states of matter: liquid, gelled rubber, gelled glass, and ungelled glass (from Enns and Gillham, 1983b with permission of John Wiley and Sons).

reaction during storage, it is recommended that unreacted systems be stored deep within the glassy state at least 20–30 °C below T_{g0}. In the early stages of cure prior to gelation or vitrification, the reactions are kinetically (chemically) controlled. In the region between gelation and vitrification (rubbery region), the reaction is still chemically controlled but may change from chemical to diffusion control. When vitrification occurs, the reaction becomes diffusion controlled, and the reaction rate decreases by two or three orders of magnitude below that in the liquid or rubbery regions, so that for practical purposes the reaction is essentially quenched soon after vitrification.

The *gel point* (or *gelation*) is defined as the onset of the formation of insoluble, crosslinked polymer (gel fraction) in the reaction mixture. However, a portion of the sample is still soluble (termed the *sol fraction*). Vitrification is glass formation caused by T_g increasing from below T_{cure} to above T_{cure} as a result of the crosslinking reaction, and is defined as the point where $T_g = T_{cure}$ (Gillham 1986b). Thus the onset of vitrification occurs when the glass transition temperature of the curing sample becomes equal to the cure temperature T_{cure}. Ideally, the most useful structural thermoset would react until all starting monomers are incorporated into the crosslinked network, resulting in no soluble fraction. This occurs at 100% conversion of monomer and the T_g for this condition is $T_{g\infty}$. It is important to note that vitrification is a phenomenon completely distinct from gelation; it may or may not occur during cure depending on the cure temperature relative to the $T_{g\infty}$. Vitrification can occur anywhere during the reaction to form either an ungelled glass or a gelled glass. It can be avoided by curing at or above $T_{g\infty}$.

The gel point conversion may be calculated if the chemistry is known (Flory 1941, 1953). Molecular gelation may be detected as the onset of insolubility of the reacting resin (Flory 1941, 1953; Enns and Gillham 1983). Gelation in condensation or step-growth systems typically occurs between 50 and 80% conversion (degree of cure $\alpha = 0.5$–0.8). For free-radical initiated, chain-growth systems or high functionality step-growth systems, gelation may occur at much lower conversions (e.g., $\alpha = 0.2$–0.5).

From a practical view we are most often interested in finding the time required to reach gelation. This can be done by various DMA or rheology techniques. The literature indicates that the gel point is a frequency independent event (Winter 1987; Winter et al. 1997) as considered in the discussion below pertaining to Fig. 5.45. Further, based on multi-frequency isothermal rheology data, it has been proposed that there should be a unique point on a plot of the tan δ vs. time curves for the different frequencies where all the tan δ values are identical, and this point will give an exact measure of the gel point. While the Winter method should apply to DMA measurements in the rheology mode and possibly to some DMA modes with supported samples, the authors are not aware of such DMA measurements having been reported. The main advantage of this method would be an exact measure of the gel point, but the approximation methods discussed below are simpler to implement and will in most cases give a reasonable estimate of the gel point. The Winter method will not apply to all thermosetting materials, particularly where the time to gel is fast relative to the time to complete a frequency sweep.

The fact that the major change in rheological properties at gelation is a precipitous increase in viscosity provides a basis for the more advantageous approximation methods that are commonly used for identifying the gel point. They include 1) the time required to reach a specific viscosity value; 2) the $G' = G''$ crossover in a dynamic rheology measurement (Tung and Dynes 1982); 3) the intersection of tangents to the dynamic viscosity curve taken from the minimum and ascending portions of the cure profile (Fig. 5.42); and 4) the time associated with the damping peak that accompanies the rise in viscosity in a dynamic mechanical measurement for a supported sample. This can be either tan δ or loss modulus; but note that while either the tan δ or loss modulus peak can be used, they will each give different results (see Fig. 5.41). The damping peak is accompanied by a noteworthy increase in storage modulus, but considerably smaller than would be observed upon vitrification.

During gelation the storage modulus rises to a value typical of a rubbery material, e.g., 1 to 100 MPa. If this is followed by vitrification, G' (or E') continues to increase to above 1 GPa (Enns and Gillham 1983). But these modulus values may not be observed using a supported sample in the tensile or dual cantilever modes where the support dominates the mechanical response of the composite (e.g., Lee and Goldfarb 1981a,b). It should be pointed out that with the wire mesh method (Prime 1992, 1997; Neag and Prime 1991) the mesh does not dominate the mechanical response when the sample is run in the tensile mode with the mesh oriented at a 45° bias to the mesh fibers. The actual modulus measured for the supported sample also will depend on the resin content of the sample. The $G' = G''$ crossover (a value of tan $\delta = 1.0$) does not always give an accurate measure of the gel point (Winter 1987). As discussed in the following section, the crossover should not be used to determine the gel point for filled composite systems, particularly those on a glass or metal fiber support. Often there will not be a crossover in these cases (see Fig. 5.41). Of the methods cited above the one most frequently used to determine gel times in the composites industry is method 3), taking the intersection of tangents to the dynamic viscosity curve. This method also will work for supported samples, although the dynamic viscosity values in this case will be only relative values.

The degree of conversion at the gel point (α_{gel}) is constant for a given thermoset, independent of cure temperature; that is, gelation is isoconversional. For this reason the time to gel–reciprocal absolute temperature plot can be used to measure the activation energy for cure. Gelation seldom inhibits cure (e.g., the reaction rate remains unchanged on passing through gelation), and gelation cannot be detected by DSC, because there is no detectable change in reaction rate or heat capacity at the gel point. DSC is sensitive only to the heat effects occurring during a chemical reaction. Beyond the gel point, the reaction proceeds toward the formation of one infinite network with substantial increase in crosslink density, T_g, and ultimate physical properties.

As curing proceeds, the increases in molecular mass and crosslinking are accompanied by an increase in the viscosity of the system, which can be measured by steady shear rheometry or DMA in the rheology mode. Eventually the viscosity becomes extremely high as the material approaches the gel point. At the gel point the material first exhibits an equilibrium or time-independent modulus but additional crosslinking is required to complete the network formation. This is often accomplished as a "postcure" at a fixed temperature to enable the thermoset system to achieve greater crosslinking and better performance. Postcure is most effective at temperatures near to or greater than $T_{g\infty}$ so that maximum crosslinking can be achieved. However, it also should be noted that at temperatures sufficiently above $T_{g\infty}$, the onset of network degradation also can occur if sufficient time is involved. Thus one must be careful about potential "overcuring."

Isothermal TTT diagrams are relevant to the selection and management of cure temperatures (T_{cure}) and heating schedules in industrial practice. Often thermoset cures are carried out in industry in stages by heating at a series of isothermal holds. This is most effective for controlling the internal heat buildup due to the reaction exotherm. Since polymers have relatively poor thermal conductivity, the polymer can overheat and degrade if the reaction is carried out too quickly, resulting in charring and blistering. Also, if T_{cure} is too low, vitrification may occur before gelation and further reactions may not be completed, resulting in an incomplete network structure. This is of particular relevance in ambient cures and some radiation cures (Glauser et al. 1999).

Further, attention must be paid to the relationship between mixing of reactants and gelation, especially for fast-reacting systems. In the case of epoxy resins, for example, curing agents must be thoroughly mixed with the resin prior to gelation, since the rapid viscosity buildup as the gel point is approached inhibits homogeneous mixing of reactants, resulting in potential network and morphological inhomogeneities and defects (Dusek 1986).

Since the cure reaction is effectively quenched by vitrification, curing isothermally for times longer than the vitrification time is counterproductive. Further cure advancement is most effectively achieved by advancing the cure temperature, either by staging isothermally or through a postcure.

5.7.3.2. DMA Techniques for Developing Cure Parameters

5.7.3.2.1. Isothermal Cures. As noted, the TTT diagram may be constructed by using DMA to determine gelation and vitrification times. Several techniques already described have been developed using DMA instruments in both solids and rheology modes to determine both gel and vitrification times (Tung and Dynes 1982; Enns and Gillham 1983; Smith and Ishida 1999). As noted previously gel times are often estimated from isothermal measurements in the rheology mode by the intersection of the storage and loss moduli ($G' = G''$) (Tung and Dynes 1982). But in some cases the crossover can be in error.

An alternative (and preferred) method is the intersection of tangents to the low temperature and high temperature portions of the dynamic viscosity profile (examples of the dynamic viscosity profiles are shown in Fig. 5.42). Gel times also can be estimated for thermosetting systems on a glass or metal fiber support by a peak in the loss modulus or $\tan \delta$ accompanied by an increase in storage modulus. As pointed out previously, the increase should be to a typical rubbery value of about 1–100 MPa but will be influenced by the support. The gel point is a frequency independent event (Winter 1987; Winter et al. 1997) as noted in the discussion below pertaining to Figure 5.45. Thus the point where $\tan \delta$ appears to be independent of frequency in a multi-frequency measurement in principle should give an exact measure of the gel point by DMA. Vitrification times can be determined from the maximum in the loss modulus or $\tan \delta$ of an isothermal dynamic mechanical spectrum or alternatively from the intersection of tangents to the storage modulus curve as it approaches a glassy state modulus above ~1 GPa. The change in mechanical properties during the progress of an isothermal cure of an epoxy resin on a glass fiber support is illustrated in Fig. 5.41 (Lee and Goldfarb 1981a,b) and is discussed in the following paragraphs. It is important to note that for supported specimens the rubbery and glassy modulus values cited above are not applicable. This is also illustrated in Fig. 5.41. These modulus values usually will be higher, sometimes considerably higher than for unsupported resins.

Dynamic mechanical analysis in the rheology mode (oscillatory rheometry in cone and plate or parallel-plate geometries; see Fig. 5.50) can be used to

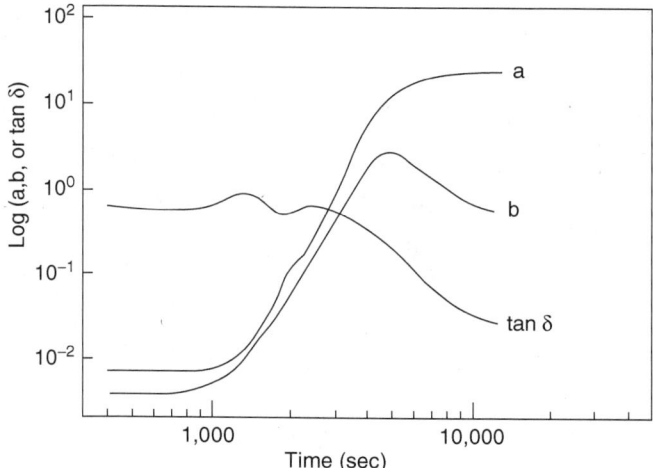

Figure 5.41. Increase in storage modulus (which parallels the rise in viscosity) from the onset of reaction for a supported sample (resin coated onto an inert substrate) of an epoxy resin. Curve (a) is the isothermal storage modulus–cure time plot; curve (b) shows loss modulus. The storage modulus rises in two steps, the first due to gelation and the second to vitrification, and then levels off at a high value characteristic of the glassy state. The loss modulus peak indicates the time to vitrify. High performance epoxy, cure at 155 °C (from Lee and Goldfarb, 1981a, b with permission of the Society of Plastics Engineers).

determine thermoset gel times by running isothermal or temperature programmed experiments that simulate actual processing conditions. An added value of the rheology mode is that actual values of the dynamic viscosity change during cure can be measured as illustrated in Fig. 5.42. The methodology for this is discussed in detail by Prime (1997). In general, when thermoset reactions proceed at a given temperature, the uncured or B-staged (partially reacted but not gelled) thermoset is initially liquid, or converted to a liquid phase if solid, before reaction begins. Once the sample is in the liquid phase, reaction occurs (and T_g increases) until the viscosity of the system advances rapidly (Fig. 5.42) and the system gels forming a rubbery solid. Reaction continues in the rubbery state until T_g increases to the cure temperature where the sample passes into the glassy state and vitrifies. At this point the reaction is inhibited due to restrictions on diffusion of reactive groups. In most instances the reaction rate decreases drastically so that the reaction appears to stop even though the chemical reaction is not complete.

For unsupported thermosets, a sample must be rigid enough to be clamped. Thus DMA in the solids mode only can be used to measure the cure process from beyond gelation through vitrification if $T_{cure} < T_{g\infty}$, or through the completion of cure if $T_{cure} > T_{g\infty}$, as well as to measure the properties of cured thermosets. For supported samples (resin coated onto an inert substrate), the

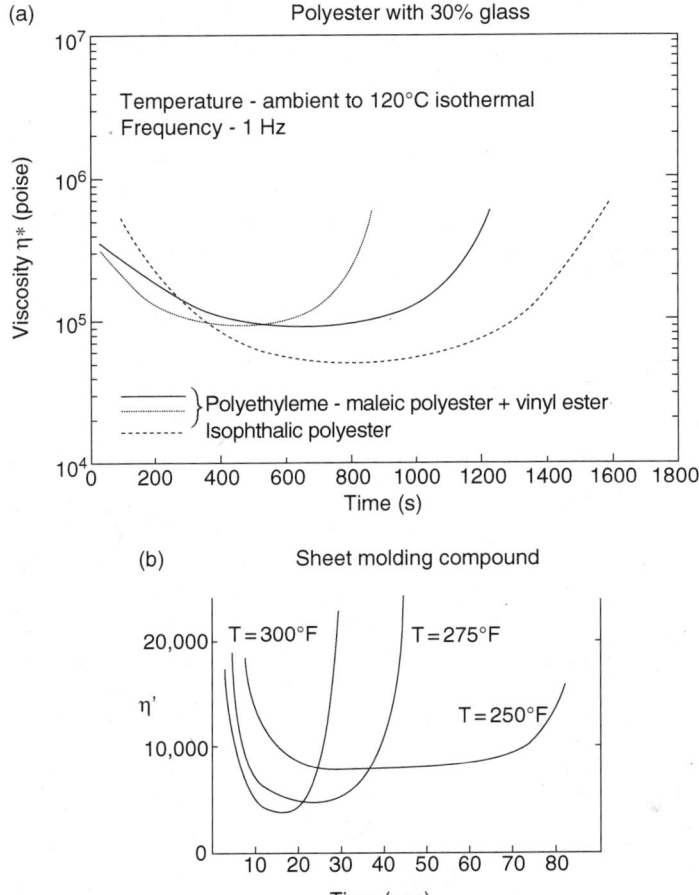

Figure 5.42. Dynamic viscosity–cure time data for two different polyester thermoset systems taken at various temperatures in cone and plate rheology mode (courtesy of TA Instruments).

increase in storage modulus (which reflects the rise in viscosity of the system) from the onset of reaction can be mapped by DMA. But the modulus values measured in this case are those of the composite. In this mode gel times cannot be determined by the intersection of the storage and loss moduli ($E' = E''$) because the actual intersection depends on the resin content of the composite. In many cases (e.g., as in Fig. 5.41), there is no intersection. Data for the isothermal cure of a supported epoxy resin sample are shown in Fig. 5.41 (Lee and Goldfarb 1981a,b). Here the storage modulus–cure time plot (curve a) indicates that the storage modulus rises in two stages, accompanied by two tan δ peaks; these events have been interpreted as gelation followed by vitrification. The storage modulus then levels off at a high value above 1 GPa, which is characteristic of the glassy state. This high modulus indicates that the sample has vitrified and that the cure temperature is below $T_{g\infty}$. If the cure temperature were above $T_{g\infty}$, the storage modulus would plateau at a lower value more typical of a rubbery material. The time to gel may also be taken from the first step change in modulus and the time to vitrify from the second. Alternatively, the vitrification time can be taken from the peak in loss modulus. In this example the loss modulus curve does not show a gelation peak, but this is not always the case and both gelation and vitrification times can often be taken from these curves. Note that when $T_{cure} > {}_{gel}T_g$, gelation will occur and later be followed by vitrification, as shown in Fig. 5.41. But when $T_{cure} < {}_{gel}T_g$, the sample will vitrify and the reaction will be effectively quenched, preventing gelation. The latter is an example of B-staging.

As noted above, DMA solid-mode experiments can be performed using liquid samples on a support, where uncured thermosets in liquid or paste form are applied to a substrate such as a metal shim (Prime 1997; Wetton 1986), wire mesh (Prime 1992, 1997; Neag and Prime 1991), or glass fiber or filter media (Prime 1997; Enns and Gillham 1983; Wisanrakkit and Gillham 1991; Prime 1986; Lee and Goldfarb 1981a,b). Using this technique, cure can be monitored from the uncured liquid state to the completion of the reaction. Isothermal gelation and vitrification times can be measured directly at several temperatures to generate TTT diagrams such as Fig. 5.40 (Enns and Gillham 1983b). Cure also can be followed by measuring T_g versus time at several temperatures such as in Fig. 5.43 (Wisanrakkit and Gillham 1991). Figure 5.44

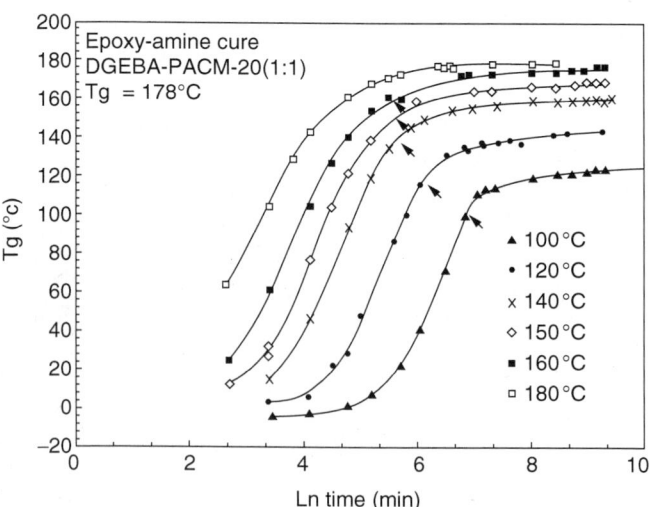

Figure 5.43. Glass transition temperature T_g-cure time curve for an epoxy system illustrating the use of correlating degree of cure with T_g. [From Wisanrakkit and Gillham (1991); reprinted with permission of John Wiley and Sons, Inc.]

illustrates the wire-mesh technique, which has been used in the cure characterization of magnetic disk coatings (Burns et al. 1989), powder coatings (Neag and Prime 1991), and dielectric and conducting polymer thick films (Prime 1992). Figure 5.45 shows the multifrequency DMA data during heating for a B-stage epoxy from below T_g through the completion of cure (Sichina and Matsumori 1991). The frequency-dependent transition starting near room

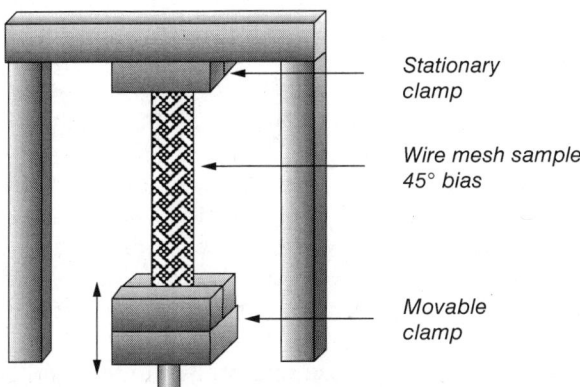

Figure 5.44. Illustration of wire mesh technique for DMA of liquid or paste materials; tensile mode is illustrated with wire mesh cut at 45° bias to minimize shear resistance of the mesh (Prime, unpublished).

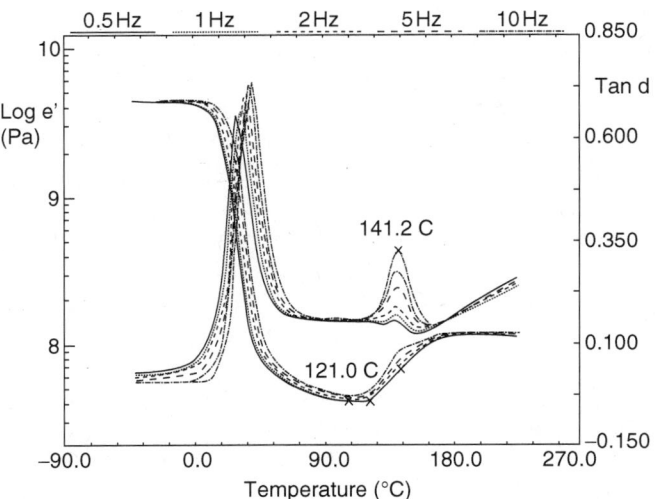

Figure 5.45. Multifrequency DMA data taken during heating at 2°C/min for a B-staged epoxy resin from below T_g through the completion of cure. [From Sichina and Matsumori (1991); reprinted with permission of the North American Thermal Analysis Society.]

temperature is the glass transition of the B-stage epoxy, which gels near 140 °C during heating at 2 °C/min, as evidenced by the small increase in storage modulus and frequency-independent tan δ peaks.

It was mentioned previously that the effective T_g of a partially reacted, vitrified polymer will be equal to or slightly greater than the cure temperature (T_{cure}), because the reaction rate slows down on vitrification. If the temperature is then raised above T_g, the curing reaction will continue, and T_g will again increase to the new reaction temperature or to $T_{g\infty}$ if T_{cure} is greater than $T_{g\infty}$. If an isochronal DMA scan is run for a partially cured sample, the storage modulus will decrease with temperature at T_g in the usual manner until the reaction starts again. Then the modulus will be observed to increase with temperature if the reaction rate is fast relative to the heating rate. This is discussed in the next section.

An alternative to DMA rheology-mode experiments is the "shear sandwich" experiment carried out with a solids mode DMA instrument. This is done with a parallel-plate fixture subjecting the sample to simple shear as illustrated in Fig. 5.4. Shear-sandwich-mode DMA data are shown in Fig. 5.46 for the cure of an epoxy resin, at frequencies from 0.1 to 100 Hz (Wetton et al. 1987).

The frequency-dependent increase in storage modulus to ~1 GPa and the frequency-dependent loss peaks signify vitrification during cure. Vitrification, as expected, is frequency-dependent since it is the cure-induced glass transition. Gelation, however, is an event that is independent of frequency. It is important to note that the shear sandwich experiment is useful primarily for B-staged resins whose viscosity is sufficient so that the resin does not flow out of the gap between the parallel-plate shear sandwich fixtures.

Dynamic mechanical analysis is also used for studying transitions in cured thermosets. The analysis of crosslinked epoxy resins typically shows, in order of decreasing temperature, an α transition corresponding to the glass transition, a β transition associated with relaxation of the glyceryl groups, and a γ transition due to methylene group motion (Charlesworth 1988). Both the β and the γ transitions, which are typically observed at −30 °C to −70 °C, and at about −140 °C, respectively, are attributed to crankshaft motions of the polymer chain segments (Boyer 1968). The appearance of other transitions between the α and β transitions is highly variable and has been attributed to segmental motions due to particular curing agents (Urbaczweski-Espuche et al. 1991).

5.7.3.2.2. Nonisothermal Cure. The concepts of gelation and vitrification as well as T_{g0} and $T_{g\infty}$ are useful for gaining an understanding of nonisothermal cure and the potential benefits of cure at slow heating rates. They are also useful for developing a qualitative interpretation of isochronal DMA (constant frequency) scans of partially cured thermosets that can be complex. Both the storage modulus and the loss modulus curves play an important role in developing this understanding. Dillman (1988) and Dillman and Seferis (1989)

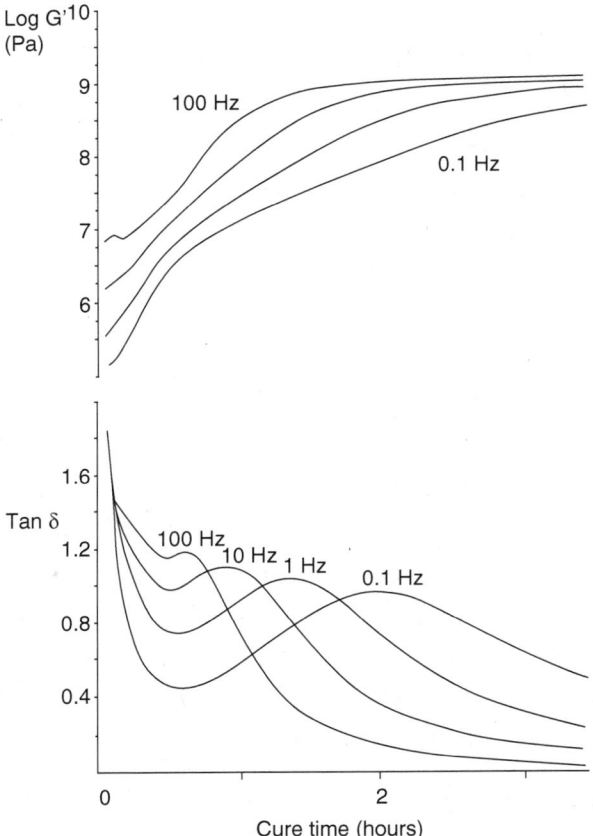

Figure 5.46. Illustration of 40 °C isothermal cure of an epoxy resin in the shear sandwich mode, which is analogous to the parallel-plate rheology mode, at frequencies from 0.1 to 100 Hz. [From Wetton et al. (1987); reprinted with permission of the North American Thermal Analysis Society.]

considered the effects of reaction on dynamic mechanical behavior during heating at various constant rates. The nonisothermal behavior is more complex than in the isothermal case because of the interplay between reaction kinetics and viscoelasticity, both of which have different temperature dependences. They termed this interplay *kinetic viscoelasticity*. It was found that the heating rate relative to the reaction rate has a significant effect on what is observed in a dynamic DMA (constant-heating-rate) scan for a curing thermoset system, even one involving a single reaction mechanism. On increasing the heating rate from low to high relative to the reaction rate, the number of loss modulus peaks observed decreases progressively from three to one. This is illustrated in Fig. 5.47 for the reaction of a partially cured TGDDM—35 phr DDS epoxy at heating rates of 2 °C, 5 °C, and 10 °C/min.

Conceptually, if we consider network formation as indicated by the change in T_g from T_{g0} to $T_{g\infty}$ during the course of reaction (dashed curve) relative to the temperature ramp rate (solid line), the events that occur can be depicted

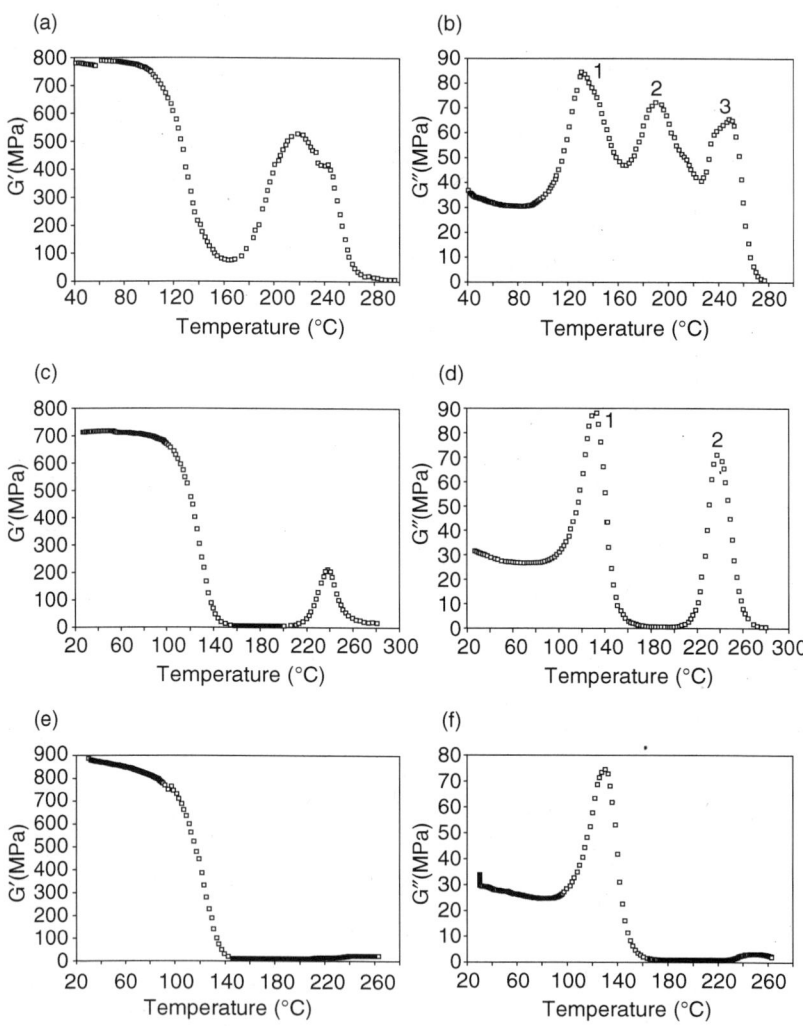

Figure 5.47. A series of dynamic DMA scans for the reaction of an epoxy resin at heating rates of 2 °C/min [top: (a), (b)], 5 °C/min [middle: (c), (d)], and 10 °C/min [bottom: (e), (f)]. On increasing the heating rate from slow to fast the number of loss modulus peaks decreases progressively from three to one (from Dillman, 1988, with permission of S. H. Dillman).

as in Fig. 5.48. Figure 5.48a illustrates a slow heating rate relative to the reaction rate, where three intersections of T_g with the experimental temperature can be observed. In order, the events represented by these intersections are devitrification at T_{g0}, or at the initial T_g, if the thermoset is partially cured; vitrification at a higher T_g (a value dependent on the heating rate); and devitrification of the fully cured thermoset at $T_{g\infty}$. This can be explained in terms of interplay between reaction rate or cure kinetic effects and viscoelastic effects. Because the polymer begins in a vitrified state, reaction rates initially

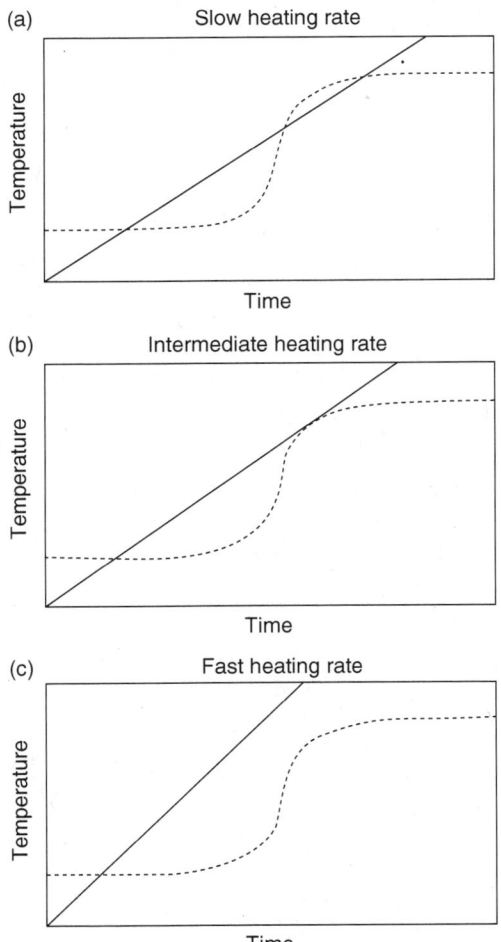

Figure 5.48. Illustration of the change in T_g from T_{g0} to $T_{g\infty}$ during the course of reaction (dashed curve) relative to the temperature ramp rate (solid line); the events that occur each time the curves cross represent a loss peak due to either devitrification, vitrification, or both (J. C. Seferis, private communication).

are close to zero. As the sample is heated, it first undergoes devitrification (peak 1 in Fig. 5.47), which allows for an increase in reaction rate, shown in Fig. 5.47 as resulting in an increase in storage modulus (due to crosslinking) and in Fig. 5.48 as an increase in the slope of the T_g–time curve. At this point the reaction rate is faster than the heating rate, allowing T_g to catch up with and slightly exceed the experimental temperature and vitrification to occur (peak 2 in Fig. 5.47). Subsequently, the reaction rate is moderated by a combination of two factors—primarily because of vitrification, but also because of the reaction nearing completion. Reaction continues in the glassy state until cure is complete. A further increase in temperature leads to a final devitrification (peak 3 in Fig. 5.47). Breitigam et al. (1993) employed a technique related to this in the manufacture of epoxy matrix composite parts. They showed that, following initial cure in an autoclave, removing the parts and heating them unrestrained (or freestanding) in an oven at rates of 0.8–1.7 °C/min ensured that further reaction occurred in the vitrified state where parts maintained a glassy modulus and retained dimensional stability. These studies, supported by dynamic mechanical measurements, showed that the time required for expensive autoclave curing could be reduced by 75% or more by completing the cure outside the autoclave.

As the heating rate is increased, vitrification peak 2 shifts to a higher temperature until it merges with devitrification peak 3. This upward shift with increased heating rate is the result of the interaction of reaction rate, reaction time, and the degree of conversion at vitrification.

The heating rate at which peaks 2 and 3 merge is the rate at which vitrification is immediately followed by devitrification at the completion of the reaction (see Fig. 5.48b). The G' curve for the intermediate heating rate in Fig. 5.47 shows a small peak that is accompanied by loss peak 2. Unlike the case where distinct vitrification or devitrification events occur, in which the loss modulus peak corresponds only to a step change in the storage modulus, this loss modulus peak corresponds to a peak in storage modulus that results from vitrification followed immediately by devitrification.

All heating rates greater than this are fast enough so that the experimental temperature is always greater than the sample T_g, thus preventing vitrification. This is true for any further increases in the heating rate. Thus, only the initial devitrification peak 1 is observed at the fast heating rate. The effect of heating rate on vitrification and cure in the glassy state, starting with uncured thermosets, has been characterized by modulated temperature DSC; this is discussed in Section 2.17 of this book. In general, heating rates below ~2 °C/min are required to observe vitrification in the nonisothermal mode.

An example relating some of the features of nonisothermal cure discussed here is provided by Patel et al. (1989). These authors describe the dynamic mechanical behavior at constant heating rate of an epoxidized Novolac–anhydride system both uncured and at various stages of cure. The different degrees of cure were imparted by curing isothermally at the temperatures indicated on each curve. Samples coated on a wire-mesh support were characterized in a TA Instruments 983 DMA, which measures resonant frequency (roughly proportional to the square root of the storage modulus) and damping where the maximum in the damping curve is equivalent to the corresponding maximum in the loss tangent curve. In Fig. 5.49 the uncured resin system can be seen to gel near 140 °C as evidenced by the small step increase in storage modulus and small damping peak. The absence of vitrification indicates that the heating rate is fast enough to keep ahead of the increasing T_g. From the preceding discussion we can better interpret the behavior of a thermoset cured to apparent completion at a series of isothermal temperatures. Note that $_{gel}T_g$ for this system was found to be 30 °C and $T_{g\infty}$ is 193 °C, from which it can be concluded that all partially cured samples had gelled and vitrified. All samples show initial devitrification that increases with cure temperature. The sample cured at 40 °C shows complete initial devitrification, while the 60 °C

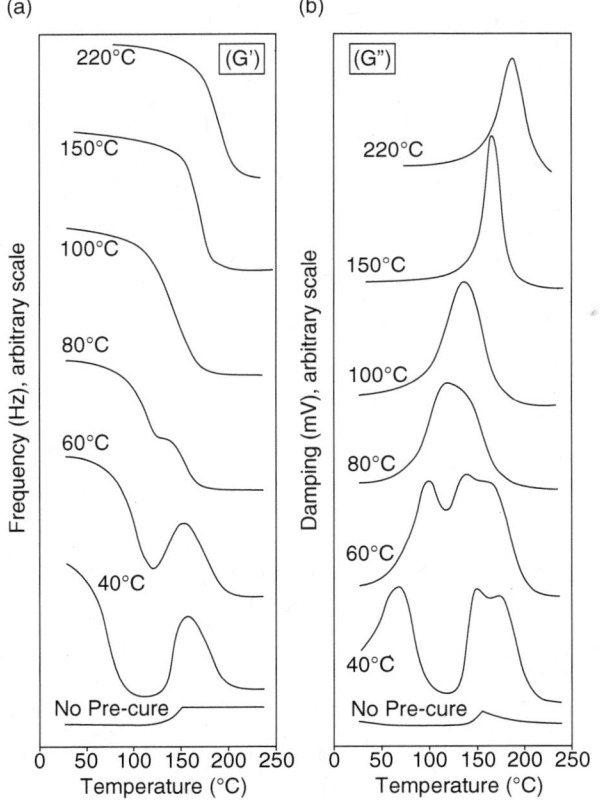

Figure 5.49. DMA data for an epoxidized novolac–anhydride system on wire mesh at 5 °C/min heating rate. Curves in (a) are resonant frequency or relative storage modulus. Curves in (b) are damping or relative loss tangent. [From Patel et al. (1989); reprinted with permission of the North American Thermal Analysis Society.]

sample shows incomplete initial devitrification. In both cases devitrification is followed by a closely spaced revitrification/devitrification, as evidenced by the peak in storage modulus and overlapping loss peaks. The 80 °C sample shows a two-step drop in modulus and broad asymmetric damping peak, suggesting a pause as increasing reaction and increasing temperature compete prior to completion of the reaction. The shallow slope of the storage modulus curve and the broad damping peak suggest that a similar competition is occurring in the 100 °C sample. The steep storage modulus slope and sharp damping peak of the 150 °C sample suggest that the material is close to complete cure. For the sample cured at 220 °C, well beyond $T_{g\infty}$, the broader damping peak not only gives an estimate of $T_{g\infty}$ but also may indicate some effects of postcure.

5.8. PRACTICAL ASPECTS OF CONDUCTING DMA EXPERIMENTS

5.8.1. Typical Modes of Operation

Table 5.3 lists the principal experimental methods used in dynamic mechanical testing. Of the experiments considered below, the thermal scan mode (method 1) is the technique most commonly used by thermal analysts. Here typical applications in quality control or processing look for differences in material batches, thermal history, different grades, reactivity, and other characteristics. The stepped isotherm (or step isothermal) experiment (method 2) is used mainly in studies involving detailed mechanical property determination for structural analysis, vibration damping applications, and for determining time–temperature superposition master curves. Method 3 (fast scan or single isotherm) is application specific.

The thermal scan method (method 1) usually is carried out at a constant frequency of 1 Hz. All relevant dynamic mechanical data are obtained from a single thermal scan completed within a few hours. The frequency is held

TABLE 5.3. Basic DMA Operation Modes

Method	Temperature Mode	Frequency Application	Typical Use and Comments
1	Thermal scan	Single or multiple	Polymer transition fingerprint, T_g
2	Stepped isotherm (or step isothermal)	Sweep of multiple frequencies (full range)	Accurate viscoelastic property dataset, better T accuracy, time–temperature–superposition
3	Rapid ramp followed by isotherm	Single or sweep	Quality control, cure studies, crystallization kinetics

constant (or possibly covers a few discrete settings) and the temperature is scanned at a constant rate, usually from low to high. This technique is suited to small specimens. The accuracy will be better for smaller, thinner samples that achieve thermal equilibrium quickly and generally for slower heating rates (typically 2 °C/min or less) for the same reason. The instrument manufacturer will recommend sample sizes for particular specimen geometries and moduli. This is discussed further in the next section. In comparative screening studies, faster rates (and the possible temperature errors that result) may be tolerated, since any thermal lag will be the same for similar samples.

In the stepped isotherm method (method 2) a number of frequencies may be scanned at a constant temperature. The temperature is then stepped to a new value and the scan repeated. This is carried out until the measurements have bracketed the temperature and material property range of interest for the sample under test. This method generates significantly more data than any other type of experiment. It can be regarded as the classical experiment for viscoelastic property determination. It is the method required for obtaining data necessary for applying time–temperature superposition. The method is more accurate, but tests require a longer time period for completion, especially when a wide temperature range is involved. For presentation purposes the data may be plotted either as multiple frequency scans on a temperature axis or multiple temperature scans on a frequency axis. TTS, however, can be performed only on the multiple temperature scan data plotted on a frequency axis.

In order to shorten the time for covering the complete temperature range of interest, the temperature step increments may be adjusted so that they are larger for regions where no transition is expected: for example, steps of 5 °C in the glassy state and steps of 2 °C in the glass transition region. The time to reach temperature equilibration will vary depending on the thickness of the sample. For thin films and fibers, 2 min should be adequate. For thicker samples, 4–5 min may be desirable. One can test the equilibration time required for thick samples by simulating an actual test run using a "dummy" sample with a thermocouple imbedded in it. It is important to note that this mode is not appropriate if the sample is reactive, will change morphology with time and temperature, or undergoes stress relaxation during the test. For DMA experiments covering an extended temperature range and multiple frequencies, automated programming of the instrument allows for overnight and weekend runs while the instrument is unattended. If subambient temperatures are involved, such measurements can be facilitated by employing a mechanical cooling device.

The viscoelastic analysis for DMA requires that the sample be in the linear viscoelastic range. In practice, this means that the strain/stress behavior is independent of the strain/stress level. Unmodified polymers, such as PMMA and PC, which are amorphous, are not likely to exhibit strain-dependent behavior as long as the strain amplitude is kept below about 0.3%. However, certain filled materials, especially carbon black or silica-filled rubbers, may demonstrate significant strain dependence in mechanical properties even at lower amplitudes down to 0.1%. Care must be observed in dealing with such samples. As a general rule, an unknown filled material should be evaluated over a range of increasing strain amplitudes at temperatures both above and below the glass transition. If the modulus values determined do not change with amplitude, then the measurement is in the linear viscoelastic range. Normally at low amplitudes it is unlikely that any strain dependence will be found, but this should be confirmed experimentally before final test conditions are decided.

5.8.2. Sample Geometries

Several sample test geometries are available with most DMA instruments. These include the following (see illustrations in Fig. 5.50):

DMA solid sample geometries

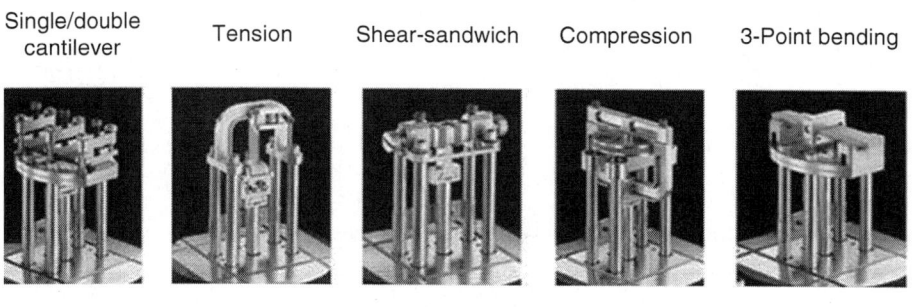

DMA rheology geometries

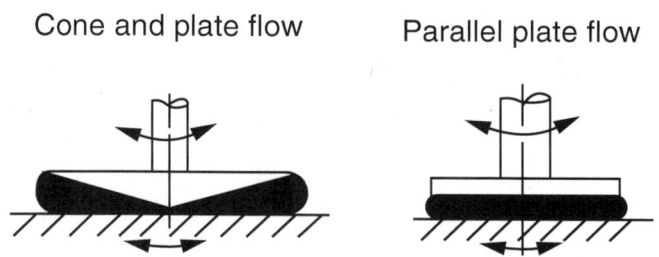

Figure 5.50. DMA sample geometries for solids and liquids: for solids—dual cantilever, tension, single cantilever, three-point bending or flexure, compression; for liquids—cone–plate flow, torsional flow; the rheometers that measure liquid-state properties also measure the properties of solid samples in torsional shear (courtesy of TA Instruments).

Tension. In the tension fixture a sample is strained in tension between a fixed clamp and a moving clamp. In tensile oscillation a static load (pretension) must be applied to prevent buckling and creep. This is discussed in Section 5.8.5.1.

Dual/Single Cantilever. In this geometry a sample is clamped at both ends and either flexed in the middle (dual cantilever) or at one end (single cantilever). Cantilever bending is a general-purpose mode for evaluating relatively stiff thermoplastics and thermosets. The dual-cantilever mode is also suitable for studying the cure of supported elastomers. Clamp corrections must be applied. These generally will be included in the software supplied with the instrument.

Three-Point Bending. Here the sample is supported at both ends but not clamped, so clamping effects are eliminated. The strain is applied at the middle by a blunt oscillating probe.

Shear Sandwich. In this geometry a rectangular sample is sheared between fixed and movable plates in a horizontal (or vertical) position. There are usually two sample sections with the movable plate situated between them. This mode of deformation is used only for very soft samples. The shear sandwich fixture is not shown in Fig. 5.49.

Compression. The sample in this geometry is sandwiched between horizontal parallel plates, and the upper plate is oscillated so as to apply a small-amplitude strain. This mode of deformation is also useful with soft samples.

Torsional Shear. Here a rectangular specimen is secured between a fixed clamp and a rotating clamp. The rotating clamp is oscillated by a small displacement.

Oscillatory Shear for Fluids. A viscoelastic melt or fluid is confined between a fixed disk and a movable disk. The movable disk is oscillated. The movable disk is flat or more commonly a narrow angle cone. The latter is referred to as a "cone and plate" geometry.

Thermal analysts sometimes refer to the oscillatory fluid experiment as a "rheology" or DMA rheology measurement. It may be possible to perform this type of measurement in the shear sandwich geometry if the fluid has a high enough viscosity that it will not leak out from between the shear sandwich fixture surfaces during the measurement. The method may be applied to thermosets using disposable tooling.

5.8.3. Frequency Effects

As noted previously, the most popular operating frequency is 1.0 Hz. This is also cited as an ASTM standard frequency for determining T_g (ASTM E1640-07). A majority of temperature sweep data are taken at this frequency. T_g values obtained from 1-Hz DMA data for E'' are often within a few degrees of DSC T_g values obtained at 10–20 °C/min heating rates (the DMA values are always higher). A method for relating DMA and DSC T_g values taken at various DMA frequencies and DSC heating rates is considered at the end of this chapter. Running at lower frequencies will shift T_g values to lower temperatures. But it also will aid in separating close transitions, thus improving resolution. The tradeoff is that lower frequencies require longer times to run, since ensuring steady-state operation of the DMA instrument requires the completion of at least one sine wave of strain. Higher frequencies always shift the transitions to higher temperatures, move different transitions closer together (thus reducing resolution, see Fig. 5.14), and reduce the time required to reach steady-state oscillation. Measurements taken at higher frequencies (up to several thousand hertz) may be used to provide modulus and loss tangent data that can be applied in designing vibration-damping devices.

5.8.4. Instrument Calibration

5.8.4.1. Modulus Accuracy A key property that a dynamic mechanical analyzer measures is stiffness or storage modulus. For a typical instrument, with uniform, accurately shaped samples that are within the instrument's measurement range, accuracy of better than ~5% should be attained. This is best tested using a three-point bending geometry and steel bar samples. The test is referred to as a *compliance calibration*, since it is used to measure the flexibility of the clamping system and calibrates the instrument to that flexibility. Typically, a set of steel bars having different thicknesses (i.e., varying stiffness) might be supplied by the instrument manufacturer. For example, this might include steel bars 60 mm long × 13 mm wide with thicknesses ranging from 0.13 to 3 mm. The compliance calibration is performed when installing a single- or dual-cantilever clamp, three-point bending clamp, or film tension clamp.

The first test should use a rigid steel sample (with known dimensions) whose stiffness falls at or slightly above the instrument's lower stiffness limit. After calibrating the instrument, a value of 200 GPa (within 5%) should be obtained for mild steel. It is better to use ordinary steel, as opposed to stainless steel, since the modulus of the latter can change markedly with varying alloying additions—ordinary steel is more consistent. In the second test, a thicker steel bar should be used, at or slightly in excess of the instrument's upper stiffness limit. If the same value is obtained all is well. If a lower value is obtained, the difference is probably explained by the compliance of the instrument itself during the measurement, which accounts for some of the measured displacement, yielding a lower apparent modulus value for the sample. This is discussed in more detail in the following paragraphs.

5.8.4.2. Errors Due to Machine Compliance Issues As noted above, DMA instruments have a measurable stiffness associated with them. When a strain is imposed on the sample, both the sample and the instrument (including clamps) are deformed. The actual sample strain is typically determined by calibrating the machine's compliance (a constant value in m/kg) with a very stiff material such as steel. The deformation of the machine is then subtracted from the total deformation, yielding the deformation of the sample. However, if the stiffness of the sample and the test geometry combine to allow deformations in the sample that are close to or less than that of the machine, significant errors can result. A sample too stiff will result in suppression of E' and E'' because of compliance limits.

Error due to compliance limits gives artificially low storage and loss moduli. The storage modulus, for example, can be suppressed to one-half or more of its actual magnitude, but the data will appear quite uniform and consistent. This can have a dramatic effect on the position and shape of the loss modulus and loss tangent peaks and as a result will change the apparent transition temperature values measured. Figures 5.51a and 5.51b are DMA data for a crosslinked acrylate polymer taken with tensile and three-point bending fixtures, respectively. For the tensile sample of Fig. 5.51a the cross section of the sample was considerably larger than the machine compliance limits allowed. The sample of Fig. 5.51b tested in the three-point bending configuration was within the machine's compliance limits. For the tensile sample, the loss modulus and loss tangent peaks are relatively sharp and rounded compared to the broad transition peaks for the sample tested in bending. Moreover, for the tensile sample the E'' peak maximum is 30 °C higher while the loss tangent maximum is 20 °C lower than the data taken with three-point bending fixture.

The operating limits are based on the stiffness or storage modulus of the sample and the sample geometry. Typically the constraints for a particular instrument would define a region on a modulus geometry factor chart as shown in Fig. 5.52. The limits on such a diagram are different for each geometry. The diagram can help to determine the proper sample shape for a given material, or to determine whether a sample of a particular size can be measured as is. However, determining the proper sample size for a specific material requires some knowledge of the approximate behavior (modulus) of the material to be tested. This is based on the operator's estimate of the physical state of the material as discussed earlier in this chapter.

5.8.4.3. Accurate Temperature Measurement—Effect of Heating Rate

Typical modern DMA instruments use two thermocouples in the oven compartment. One of these is used in a feedback loop to the computer and controls the oven temperature. The second thermocouple is positioned by the user very near the sample and recorded as the sample's temperature. The positioning should be done so that the thermocouple is as close as possible to the sample but does not touch it during the sample vibration. This will depend on the deformation mode and vibration amplitude. In general, the thermocouple should be placed near the center of the sample, and the spacing can be controlled with a sheet of paper or a metal shim. It is important that this thermo-

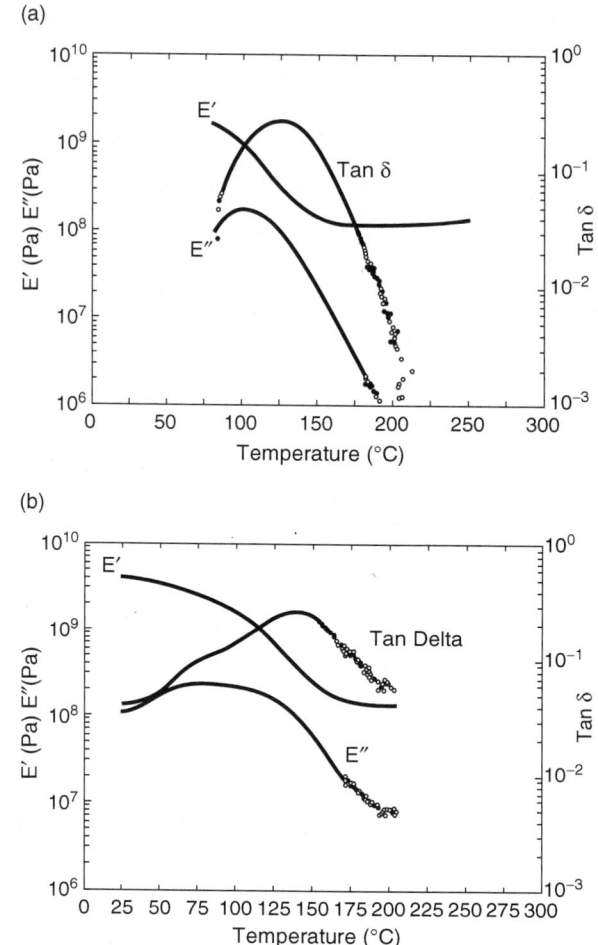

Figure 5.51. DMA data for a crosslinked acrylate polymer taken with tensile (a) and three-point bending (b) fixtures, respectively, illustrating the effect of machine compliance errors. (Chartoff et al., 1994, reprinted with permission, from *STP 1249- Assignment of the Glass Transition*, copyright ASTM International, 100 Barr Harbor Drive, West Conshohocken, PA 19428.)

couple accurately represent the sample temperature, but it should be noted that there are several issues that combine to create an overall error:

Errors in the Electronic Thermocouple Calibration. The thermocouple interfaces with the computer via an analog-to-digital conversion board. Since the thermocouple's output is in millivolts, the interface board will linearize and set the limits of the signal. The instrument usually also will

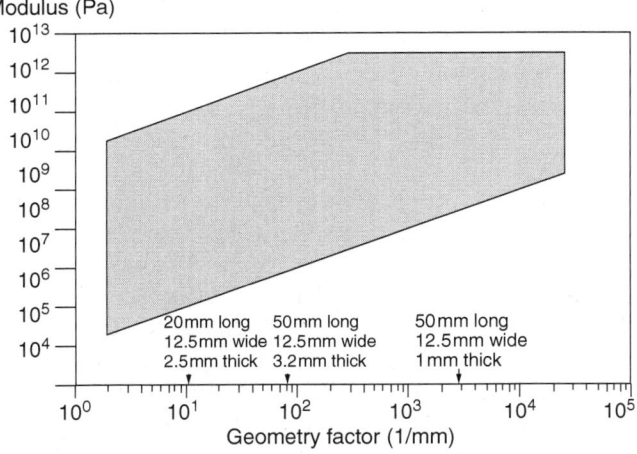

Figure 5.52. DMA machine operating limits typically define a region on a modulus–geometry factor chart as shown here (courtesy of TA Instruments).

allow the operator to use a standard thermocouple calibrator to ensure that the sample thermocouple is within specifications.

Oven Temperature Variations. This is highly specific to the oven and heating mechanism used. Variations of up to several degrees can exist, which can cause temperature gradients within the sample and can also cause the sample temperature reading to vary significantly depending on the thermocouple placement within the oven. The user is advised to check for significant temperature variations in the oven at various operating temperatures and heating rates. These variations may be reduced by changing gas purge flow rates, convection fan speeds, reducing heating rates, or a combination of these.

Actual Sample Temperature. Any polymer sample will take a finite time to reach thermal equilibrium, and this is frequently commensurate with the timescale of the experiment. Most polymers have poor thermal conductivities. Thus, the most accurate measurements are made isothermally in the step mode, since a soak period can be arranged so that the sample reaches thermal equilibrium. Poorly designed ovens can exacerbate the problem. The guideline of 2 °C/min as the maximum rate for temperature scanning in the ramp heating mode is a useful one for a DMA instrument. This is the rate specified in the various ASTM standards (e.g., ASTM E1640-07, ASTM E1867-06; see Appendix). Samples such as thermoset adhesives coated onto wire mesh, which are thin and have relatively high heat transfer rates, may be run at higher heating rates up to 5°C/min. There may be situations where very high heating rates may be desired for the purposes of shortening running time for repetitive samples. This could involve cases where it is not important to know the exact sample temperature and samples of the same geometry and similar composition are being run (e.g., for quality control comparison purposes).

Since the sample thermocouple is not in direct contact with the sample, it measures the oven's atmospheric temperature near the sample. The actual sample temperature is a function of its thermal conductivity, mass shape and dimensions, the mass and composition of the fixture in contact with the sample, the uniformity of the oven thermal environment, and the heating rate. As noted above, the result is that a sample can have a thermal gradient within it. Even if the oven temperature were completely uniform, the sample temperature could lag behind the oven temperature because of heat transfer limitations.

Thermal gradients within the sample are greatest for thick samples (with lower surface area/volume ratio), large high-mass fixtures (such as parallel-plate or cone–plate fixtures), and fast heating rates. For large, thick samples these thermal gradients can have a significant effect on the measured T_g (Mora and Macosko 1991).

It is also important to note that samples undergoing a first-order phase transition (melting, cold crystallization, or crystal–crystal transition) during a ramp heating will not be at the same temperature as the oven, which continues to increase in temperature throughout the transition, while the sample remains isothermal until the transition is completed. Another case where the sample and oven temperatures may vary is for thermosets undergoing cure. Because the cure reaction is exothermic, the sample temperature may exceed the oven temperature.

5.8.4.4. Methods of Sample Thermocouple Calibration: Above-Ambient, Subambient

Sample thermocouple calibrations can be made both above ambient and subambient by several techniques for various types of samples (this includes samples that require dynamic heating, e.g., those that contain volatile components or consist of polymers that react chemically during the test): (1) testing polymer samples of thermal conductivity and dimensions similar to those of the samples of interest and that have well-established T_g values, (2) using fiberglass or metal cloth specimens impregnated with a suitable calibration material, and (3) measuring a bar or film of a pure metal that melts sharply. In general it is desirable to establish more than one calibration point over the temperature range of interest. This is certainly true for precise quantitative work. However, for routine work over a restricted temperature range (~100°C), a single calibration point should be sufficient. As a rule, calibration on a weekly basis is a good idea. A calibration sample should be run as a normal DMA specimen with the sample thermocouple(s) placed as close as possible to the sample surface (0.1–0.2 mm) in the same place as for a normal DMA run. One calibration point that may be used is ambient temperature with the oven open so that the sample thermocouple is accurately reading the local temperature. For accurate multiple-point temperature calibration, high-purity metal films can be purchased or fabricated (In, Sn, Zn; e.g., from Alfa Aesar, as noted in Chapter 4).

A simple test when accurate temperatures are required for measured transitions is to make measurements on the same samples at different thermal scanning rates, (e.g., 1, 2, 3 °C/min). If all transition temperatures in heating experiments are coincident, then the sample is in thermal equilibrium in all tests. If higher transition temperatures are measured for the faster rates, however, then the sample is lagging behind the measured temperature. For cooling experiments lower values will be seen at the faster cooling rates if the sample temperature is lagging behind the oven temperature.

Techniques for subambient sample calibrations such as for elastomers are more difficult. Here it is most desirable to calibrate in the same temperature range as the T_g of the elastomers, which might range from slightly below room temperature to about −120 °C for silicone rubber. Calibrants that have sharp melting points may be used, such as hexane (−95 °C), cyclohexane (−83 °C), octane (−56.8 °C), decane (−29.7 °C), dodecane (−9.6 °C), water (0 °C), and tetradecane (5.9 °C). The calibrants should be of high purity. A calibrant is formed into a DMA sample by impregnating it into an inert substrate such as a rectangular strip of fiberglass cloth (or metal mesh). The fiberglass/calibrant composite is then run as a normal DMA specimen, with the melting event providing a discontinuity in the storage modulus signal at the transition.

A similar procedure uses a small hollow tube of polymer filled with a frozen calibrant. The tube material is chosen so that it is "inert" in the transition region of the calibrant and it is impervious to the calibrant. One of the authors (Chartoff) has used a PEEK tube 6 cm in length, 4.5 mm in diameter, and 0.2 mm wall thickness. The calibrant is pipetted into the tube, the ends are capped with an inert sealant such as a rapidly curing epoxy, and the calibrant is frozen. Several temperature calibration points are taken with various calibrants.

5.8.5. DMA Test Samples and Testing Mode

Test samples include various geometries, including films, fibers, sheets, or plaques and in the case of thermosets, liquid resins, B-staged resins, and prepregs, as well as pastes. Sample preparation methods for thermoplastics include pressed films, molded rectangular strips, monomer or solvent cast films, cast plaques, and fibers. For thermosets, cast or molded films or strips and prepregs are most suitable. Casting may be done in flexible silicone rubber molds, which allow for easy removal of the sample after solvent evaporation, monomer polymerization, or curing. Teflon molds also may be used, but sample removal in this case is more difficult because the mold is rigid. Samples should be uniformly shaped with smooth edges and surfaces. Any irregularities in dimensions such as nonparallel surfaces or variations in thickness, will lead to inaccuracies in the data obtained, since the calculations for moduli are based on samples with regular geometries. Cuts or cracks in a sample may lead to fracture during a measurement, thus negating an entire experiment. Liquid thermoset samples, such as those coated on wire mesh or fiberglass, and prepreg samples should be secured by wrapping their ends in aluminum foil to prevent them from bonding to the clamps.

Concerning testing modes, some materials are more easily tested than others. Those that represent the greatest challenge are samples whose moduli change the most at the glass transition temperature. Amorphous thermoplastics such as PMMA [poly(methyl methacrylate)] or PC (polycarbonate) are good reference test samples (see, e.g., Fig. 5.22). Both of these polymers show a sharp glass transition with well-known literature T_g values (PMMA, $T_g = 103\,°C$; PC, $T_g = 147\,°C$ at ~1 Hz) and a rubbery plateau modulus of approximately 1 MPa. Such samples afford the opportunity to test both the temperature accuracy and the instrument's lower stiffness limit. In the glassy region most solids test modes will work well. Glassy moduli results from tension and bending experiments should be comparable.

In the transition and rubbery regions thermoplastic samples are best run in a cantilever mode (single or dual). The three-point bending mode cannot be used for glass transition measurements of amorphous thermoplastics because the sample will be extremely deformed at the glass transition. Also, tension may not work well because large strains are required to maintain a fixed pretensioning force. The autotension feature of the instrument (see the following paragraph) will not be able to follow these deformation values, and also, the shaft of the instrument may reach its final position. However, for crosslinked (thermoset) and crystalline samples that remain stiff through the transition, three-point bending may be preferred when measuring the glass transition and into the plateau region. The torsional (shear) mode may work well for thermoplastics in all viscoelastic regions but not for brittle thermosets. Cured thermosets in general are brittle and should be measured at low strain amplitudes so as to avoid sample cracking. The shear modulus G' will be approximately one-third of the E' value obtained.

As noted above, in the tensile mode problems can arise with amorphous thermoplastics in the glass transition region when they are subjected to autotensioning. Autotensioning is applied by the instrument to maintain a selected positive force level (or pretensioning) required to eliminate sample buckling [see discussion of fibers and thin films below (Section 5.8.5.1) for more detail]. The static force (preload force) is applied to the sample prior to initiating the experiment. Typical static (preload) forces for DMA tension experiments are 0.01–0.05 N. Static (preload) force is rarely used for nontensioning clamps. It is required with tensioning clamps in order to sustain oscillation. However, when the sample softens above T_g, autotensioning may cause the sample to elongate excessively (or creep) and result in molecular orientation that affects the T_g itself. When running in the transition region the pretension level should be reduced as the sample softens to avoid this (the pretension does need to be sufficient to eliminate sample buckling). The DMA instrument can be programmed to accommodate for this. Additional comments on pretensioning are presented in the following section. Another class of materials that may cause problems in tension is semicrystalline thermoplastics, such as polyolefins. They also have a propensity to creep excessively, especially near the melting point, T_m. In order to avoid excessive elongation (or creep) during the course of a DMA experiment, the clamped single-cantilever or double-cantilever bending geometries may be a better choice.

Elastomers present a challenge for any DMA instrument. They are one of the most difficult types of samples to run in a tensile experiment because of their subambient T_g values. A recommended procedure is to load a tension sample by fastening one end at room temperature and then cool to −100 °C. At this point the sample normally will be glassy (the exceptions are some silicone elastomers), and the other end of the sample should be clamped and the appropriate pretension should be applied. This will entail opening the sample chamber and quickly completing the clamping operation. In doing so, one must take care to minimize frost buildup on the sample. Frost formation can be mitigated by contriving a way to circulate dry nitrogen or air around the sample as it is loaded for testing. The DMA measurements then should be made as fast as possible, but the heating rate should not exceed 2 °C/min and possibly should be slower if more than one frequency is being measured. A smooth curve should be obtained for tan δ through the glass transition, and the pretension should be reduced rapidly to avoid any overextension of the sample once it becomes rubbery. This can be programmed via the strain control option available with many DMA instruments. Opening the oven to clamp the specimen at subambient temperatures may be avoided if a three-point bending fixture is used. In this configuration the sample freely contracts as the temperature is lowered and strain is not applied to the sample until the experiment starts.

An option that should be considered for covering extended temperature ranges involving subambient and above-ambient temperatures is to use different test geometries to cover different temperature ranges. For example, below ambient a three-point bending fixture might be used and above ambient a tension fixture might be used with appropriate sample sizes.

5.8.5.1. Fibers and Thin Films In terms of sample preparation, it is not easy to perform DSC measurements on fibers and thin films. In DSC, experience is difficult to recognize the glass transition of highly drawn fibers, since the heat capacity increase of drawn fibers owing to the glass transition is broad and shallow. Similarly, TMA, although very useful (especially in the case of highly drawn fibers), seldom provides an easily noticeable break in the $V = f(T)$ or $l = f(T)$ curve to clearly characterize the glass transition (where V is the specific volume and l is the sample length). Therefore, DMA is the most frequently used technique in the fiber/film technology field. A single measurement will provide the temperature dependence of the modulus and the glass transition temperature and will characterize possible second-order transitions of the material. It is important to remember, however, that the glass transition temperature measured by DSC always will be different from that determined by relaxation methods, such as DMA or DEA (due to the frequency excitation effect). Details of DMA measurements of fibers and films can be found in several references [e.g., Murayama (1978), Jaffe et al. (1997); Menczel et al. (1997b)]. Here we briefly summarize the most important aspects of this type area of DMA measurement.

Dynamic mechanical analysis measurements of fibers and thin films require special attention because of the unusual mechanical properties of fibers and films originating from their unique geometry. Their special geometry allows only DMA measurements in tension; because of the small diameter of polymeric fibers and flexibility of polymeric thin films (and also often elastomeric thick films), these types of samples are unable to support necessary levels of compressive stress. Therefore, some positive static stress has to be applied to the fiber or the film sample that is greater than the peak level of the stress attained in the dynamic oscillation. If a static stress is absent during the measurements, sample buckling will occur, as shown in Fig. 5.53, and in such a case the lower part of the stress signal will be truncated (Grehlinger and Kraft 1988a,b). This additional static force is called *pretension*. As can be seen in

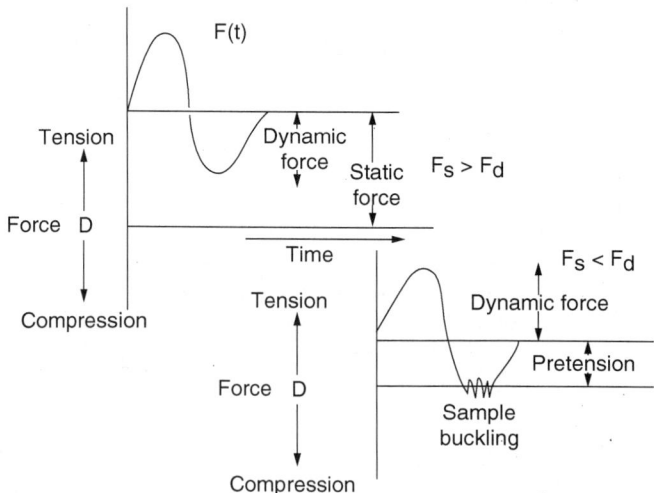

Figure 5.53. Illustration demonstrating the necessity of using a pretension in DMA tensile measurements of fibers and thin films; sample buckling will take place below a certain pretension level; F_s is the static pretension force, and F_d is the variable dynamic oscillating force imposed during the test (from Grehlinger and Kraft, 1988b, with permission of the Society of Plastics Engineers).

Fig. 5.53, the pretension raises the level of the stress sine wave, so that the originally zero level of the midpoint of the sine wave will have a positive value. The lowest level of the pretension must be higher than the half-height of the peak-to-peak value of the dynamic force to avoid sample buckling. However, it is not always easy to determine the highest possible level of the pretension, because too high a pretension can lead to permanent changes in the sample structure; for example, it can increase the orientation of the fiber. Grehlinger and Kraft (1988a,b) described the effect of pretension on the storage modulus of fibers.

Another problem with highly oriented (i.e., drawn) fibers or films is shrinkage as the temperature is being raised during the DMA measurement. Creep may also take place, and these permanent deformations in addition to the modulus changing with temperature (especially in the glass transition region), require monitoring and continuous adjustment of the pretension level. Therefore, most commercial instruments do have an automated mechanism, termed *autotension*, to deal with this problem. Usually, before the measurement, the static pretension force can be selected on the basis of one of the following:

1. It can remain constant during the entire run.
2. For every point of the measurement it can be predicted from the change of the dynamic force with temperature using the points measured at lower temperatures.
3. The static force can be predetermined by using the temperature dependence of the sample storage modulus.

Static stress–strain measurements are usually needed for estimating the appropriate pretension. In these measurements, one deforms the fiber or the film in the tension mode at a constant deformation rate, and then plots the resulting stress as a function of strain. The pretension needs to be selected such that the total stress during the subsequent DMA measurements falls onto the initial linear portion of the stress–strain curve (refer to Fig. 5.7 of this chapter—the initial portion of the curve below the yield point). It should be noted that the information presented here on pretensioning applies to tension measurements on any type of sample.

Obviously, the autotension mode will influence the DMA curve, since it may lead to different levels of shrinkage, creep, and change in orientation. Almost all presently available commercial DMA instruments have special fiber and film fixtures; sometimes the same fixture can be used for both types of measurements.

5.8.5.2. Composites: Sample end Corrections, Composite Specimen Geometry Effects

Sample geometry modes with clamped samples (shear and three-point bending are not clamped) will exhibit errors due to clamping of the sample. Generally speaking, tension suffers the least error, compression the worst, with clamped bending somewhere in between. The error will depend on sample stiffness and any correction must take this into account *during* a test, which may involve large modulus changes of up to 10^3. The three-point bending mode also requires end corrections. Correction factors for various geometries are incorporated into the analysis software supplied by the instrument vendors.

Bending modes are particularly popular for characterizing continuous-fiber composites, which represent a special case for analysis. Composite samples normally will be tested in a bending mode (three-point or cantilever) because they are too stiff for tension. However, the sample end correction here becomes a substantial contributor. A simplified analysis of this (Mullin and Knoell 1970) indicates that the total deformation in beam-type specimens tested in flexure (or bending) is mixed-mode, consisting of both a shear component and a bending component. The shear occurs at the sample ends. For unidirectional composites (the most common type of composite test specimen), the G/E ratio is the parameter of primary interest in composite beam deflection, where G is the shear modulus and E is the flexural modulus. The ratio indicates how important shear deformation is relative to flexure. G/E is relatively large for homogeneous materials ($G/E = 0.36$). But for a composite, $G/E = 0.01$–0.03. The net effect is that in calculating the relationship between the length (or span) to thickness ratio of a sample and the relative percent of shear deformation, we find the results depicted in Fig. 5.54. For homogeneous samples (curve A), shear deformation is relatively unimportant unless the sample L/t ratio is

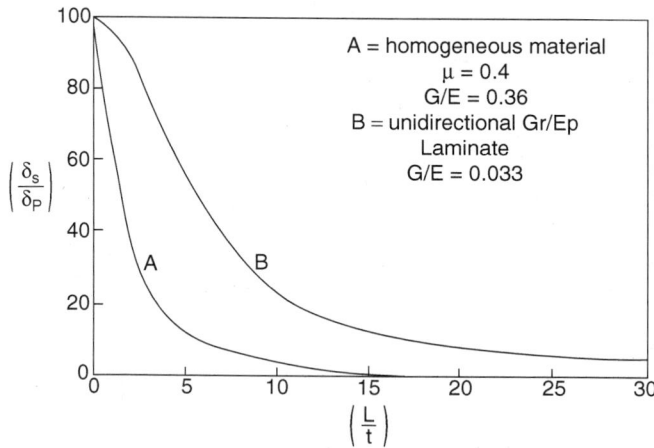

Figure 5.54. Results showing the relative percent of shear deformation due to end effects in the three-point bending experiment in relation to sample length to thickness ratio; a homogeneous sample (curve A) has a low end effect, while a unidirectional fiber composite (curve B) has a much more serious end effect. δs/δp is the relative percent of shear deformation compored to flexural deformation. (Mullin and Knoell, 1970, reprinted with permission, copyright ASTM International, 100 Barr Harbor Drive, West Conshohocken, PA 19428.)

small ($L/t < 5$). But for composites (curve B), unless L/t is greater than 20, the percent shear deformation is large. Thus, in the case of composites the normal bending analysis formulas used for calculating the modulus values will be in error except for long specimens.

Another issue with continuous fiber unidirectional composites is the test geometry relative to the fiber orientation. The storage modulus and tan δ data for an epoxy–carbon fiber composite in both three-point bending and torsional modes (Gerrard et al. 1990) are shown in Fig. 5.55. Both sets of data are correct but quite different. The torsional E' values are an order of magnitude lower than the flexure data. In torsion it is the matrix epoxy resin that primarily contributes to the measured modulus. In flexure the fibers, which are much stiffer, dominate the response. Thomason (1990) has shown that for a series of samples measured at different angles (0–90°) relative to the fiber axis (Fig. 5.56), the storage modulus decreases continuously. This is similar to the change in modulus with orientation for the Vectran liquid crystal polymer illustrated in Fig. 5.38.

An additional comment on composites is that samples must be representative of the material. This can be a challenge with composites because of the

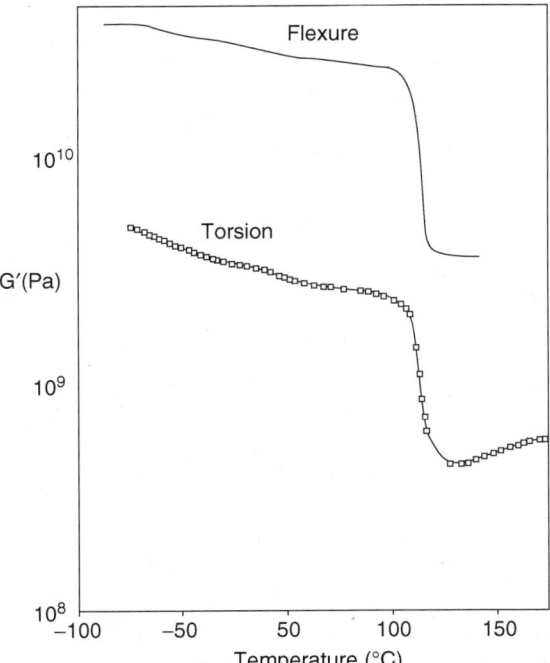

Figure 5.55. Effect on modulus measured in DMA due to fiber orientation of a unidirectional fiber composite relative to sample clamping in three-point bending and torsion (from Gerrard, et al., 1990, reprinted with permisssion of the Society of Plastics Engineers).

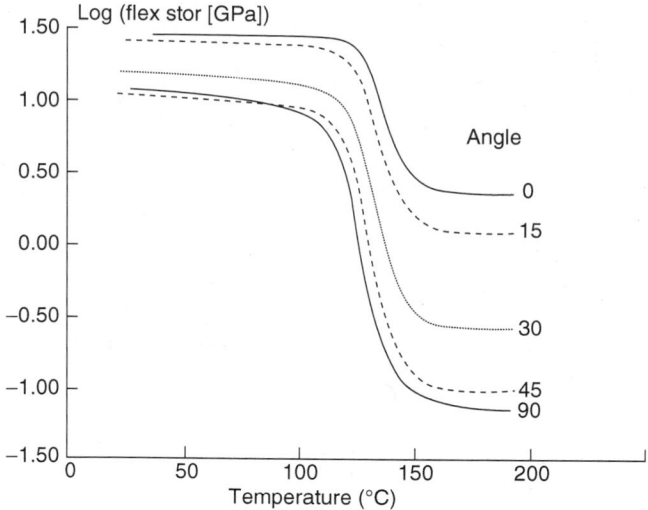

Figure 5.56. Change in storage modulus for a series of unidirectional composite samples measured at different angles (0–90°) relative to the fiber axis (from Thomason, 1990, with permission of the Society of Plastics Engineers).

scale of inhomogeneities and the small sample sizes. Care should be taken to avoid composition gradients. Thinner samples reduce thermal gradients. Residual stresses in the samples will influence modulus and damping data as will physical aging.

5.8.5.3. Thermosets Supported samples coated onto glass fiber cloth or metal fabric mesh may be used for thermoset cure studies or evaluating the properties or drying of paints and coatings. Composite prepregs are ideally suited for this. Several factors should be kept in mind. The modulus values obtained will not be quantitative for the polymer. Reasonable modulus values for the polymer can be derived by a separate calculation subtracting the contribution of the fibers, following a method discussed by Prime (1997). In order for this to work, however, care must be taken to prepare samples with regular dimensions—smooth parallel surfaces and smooth parallel edges.

The presence of the support material may change the cure process if the fibers interact with the polymer. Glass fibers should not be surface-treated with a sizing; some metals may catalyze (enhance or inhibit) the chemistry of the cure reaction (stainless steel has proved to be a good choice for a support material, since it does not readily interact). Also, it is important to prevent bonding between the sample and DMA sample clamps by covering the ends of the sample with aluminum foil so that the cured polymer will not adhere to the tooling. In the event that does occur, then the sample can be removed by heating in air to pyrolyze the polymer.

5.8.6. Temperature–Time/Frequency Analysis

5.8.6.1. Time–Temperature Superposition Software Various vendors of DMA equipment provide software that allows the user to perform time–temperature superposition (TTS) of the viscoelastic data. As noted previously, TTS treatment of the data will increase the range for predicting viscoelastic property changes versus frequency (or time) by several decades over the range covered just by the raw experimental data. TTS is an empirical curve-fitting procedure, which is best visualized by actually using it. Multiple frequency measurements must be taken at a series of discrete temperatures in order to use TTS to correlate DMA data. The TTS software allows the operator to take the E', E'', J', J'', or tan δ curves plotted from multiple oscillation experiments and shift those curves to superimpose on a single curve for a selected reference temperature (i.e., form a master curve). The shifting can be performed automatically by the software or manually by the operator. The amount of shifting required to superimpose the data onto the reference curve is subsequently fitted to different equations such as WLF or Arrhenius, and the constants in the equations are displayed. This provides the determination of which equation best describes the data and the associated shift factors. Once the master curve is obtained, it can be stored as a "data file" for subsequent recall and overlay with the master curves from similar materials (to compare differences in behavior) or other master curves for the same material in order to compare properties at different reference temperatures.

Time–temperature superposition software generally has many user-friendly features such as toolbars, icons, and point-and-click mouse interactions. In addition, the automatic shifting capability enables even an inexperienced operator to rapidly generate master curves and evaluate the alternative equations for correlating the data. Plots of the shift factors calculated for the respective equations can be compared to the actual shift factors obtained from both the automatic or the manual curve-fitting routine in the software. Also, TTS can be coupled with the instrument control software to allow completely unattended experimental evaluation and master curve generation once a sample is loaded.

In using the software, the recommended procedure is to first carry out the automatic shifting procedure. If the agreement between the computed shift factors and those obtained from automatic shifting is not satisfactory, the operator has the option to manually adjust the shift factors to agree completely with those calculated from an equation and generate a new master curve. The same shift factors should apply to each of the viscoelastic functions. This "consistency" criterion can be used to judge whether the shift factors determined need further adjustment.

5.8.6.2. The Time–Temperature Superposition Nomograph It was noted previously that, as in most methods of handling data, TTS yields more reliably accurate results when used to interpolate between two temperatures for which experimental data are available than when used to make extrapolations. However, it has been noted by Chartoff and Graham (1982) that data in the transition region obtained from low-frequency dynamic mechanical tests can be extrapolated to higher frequencies with reasonable assurance of its accuracy. This extrapolation is facilitated by use of the reduced frequency nomograph discussed below. The most important practical feature of TTS is that it permits us to condense our full knowledge of the viscoelastic properties of a given polymer over a wide temperature/frequency range into two curves—a master curve corresponding to reference temperature T_R and the shift factor–temperature curve.

When considering the coordinates of the master curve it is important to note that the appropriate timescale is a reduced time t/a_T or reduced frequency ωa_T rather than simply time or frequency. Thus t or ω can be decoupled to obtain more meaningful time/frequency or temperature data only if either values of a_T versus T or the WLF equation are known. In order to circumvent this problem and display master curve data in a more convenient format, Jones (1978) first suggested representing the dynamic mechanical master curves for E' and E'' (or tan δ) in a novel graphing format known as the *reduced-frequency nomograph*. This nomograph not only displays modulus, E' and E'' (or loss tangent), as a function of reduced frequency but also includes axes enabling the operator to read frequency and temperature data simultaneously. Figure 5.57 presents simulated data as a typical master curve displayed in the nomograph format. Note that the frequency is denoted here in hertz or 1/s.

The nomograph is created by plotting the viscoelastic function versus reduced frequency. Reduced frequency is defined as $f_j a_{T_i}$, where f_j is the

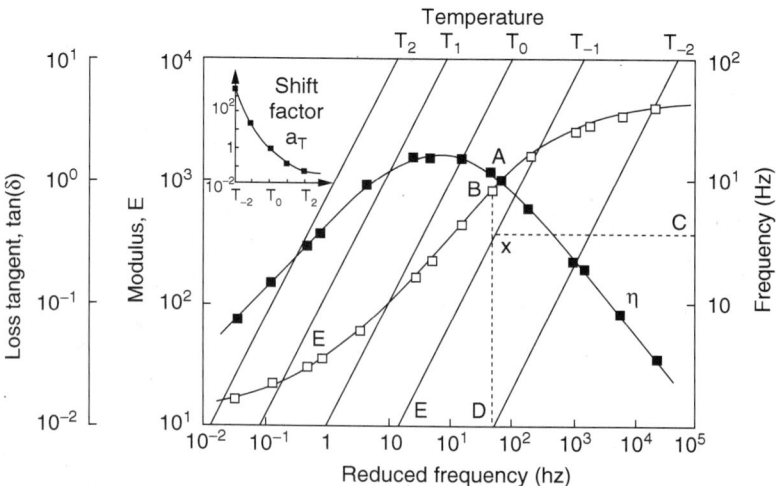

Figure 5.57. Simulated DMA data for a typical master curve displayed in the temperature–frequency nomograph format (from Jones, 1978).

frequency (in Hz) and a_{T_i} is the value of a_T versus temperature T_i. (Note that in this discussion we are using the symbol f for frequency in Hz or 1/s in order to remain consistent with the original literature. The same symbol is used for force elsewhere in this book.) An auxiliary frequency scale is then constructed as the ordinate on the right side of the graph. The values of $f_j a_{T_i}$ and f_j form a set of corresponding oblique lines representing temperature.

To illustrate the use of the nomograph, assume that we wish to find the value of E' and $\tan \delta$ at T_{-1} and some frequency f (point C in Fig. 5.57). The intersection of the horizontal line ω = constant, line CX, with $f_j a_{T_i}$ (point X) defines a value of $f a_T$ at point D, of about $4 \times 10^2 \, \text{s}^{-1}$. From this value of $f a_T$ it follows from the plot of E' and $\tan \delta$ that $E' = 10^3 \, \text{N/m}^2$, point B, and $\tan \delta = 1.2$, point A.

Construction of the nomograph can be computerized so that plots can be generated directly from dynamic mechanical data files (Weissman and Chartoff 1990). As an example of the use of the nomograph, master curves of E' and E'' for a poly(methyl methacrylate) sample are shown in Fig. 5.58. The data were reduced using the "universal" WLF constants. From the master curve one can also generate isochronal (fixed frequency) data such as that shown for a frequency of 0.1 Hz in Fig. 5.59.

Note that the isochrone generated from the master curve will represent the specific properties of the polymer used to generate the nomograph. In Fig. 5.59 the isochrone extends down to 90 °C, well below T_g. This is a nonequilib-

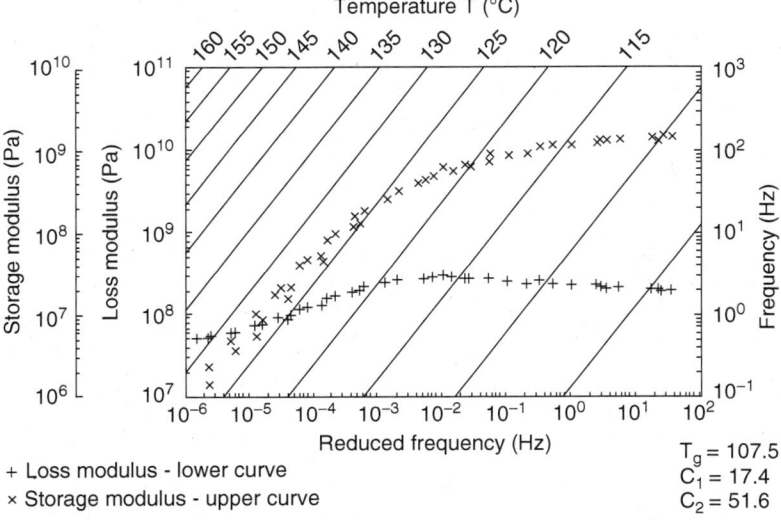

Figure 5.58. An example of the use of the temperature–frequency nomograph; master curves of E' and E'' for a poly(methyl methacrylate) (PMMA) sample (from Weissman and Chartoff, 1990, reprinted with permission from ACS Symposium Series 424, 111. Copyright 1990 by the American Chemical Society).

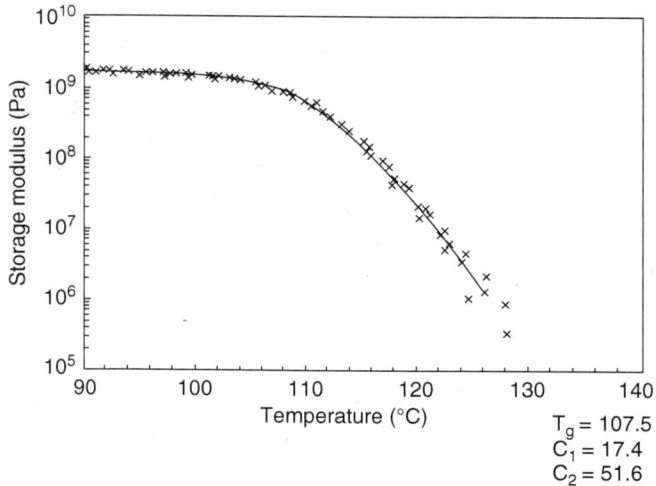

Figure 5.59. Isochronal (fixed frequency) data for a frequency of 0.1 Hz generated from the master curve for PMMA of Fig. 5.58; the solid line is from experimental data, and the crosses (x) are calculated from the master curve (from Weissman and Chartoff, 1990, reprinted with permission from ACS Symposium Series 424, 111. Copyright 1990 by the American Chemical Society).

rium state, so that the actual experimental data (solid line) will not be unique. The actual data may vary for different samples depending on their processing history.

5.8.6.3. Relating DMA and DSC T_g Values

An empirical equation based on the Arrhenius equation provides a method for estimating peak E'' or peak tan δ temperatures at various DMA frequencies from DSC T_g data. This may be useful for applications where knowledge of the loss peak temperature at various frequencies is important, such as vibration damping and noise control. Viscoelastic damping devices function optimally when tan δ is maximum. The procedure for the calculation is described by Sircar and Drake (1990). It involves using DSC T_g values either measured or obtained from the literature. The apparent activation energy, E_a, for the glass transition is assumed, taken from the literature [e.g., McCrum et al. 1967; Ferry 1980), or determined by using the Arrhenius activation energy equation for the dependence of T_g on frequency in DMA measurements.

$$f = A \exp \frac{-E_a}{RT} \quad (5.35)$$

where f = frequency, A = constant, E_a = activation energy, R = gas constant, and T = absolute temperature. Then the T_g for a given DMA frequency is calculated from

$$\ln \frac{f_{\text{DMA}}}{f_{\text{DSC}}} = \frac{E_a}{R} \left(\frac{1}{T_{g,\text{DSC}}} - \frac{1}{T_{g,\text{DMA}}} \right) \quad (5.36)$$

The reference frequency for DSC was assigned the very low value of 0.0001 Hz, which was found to yield good agreement with experimental data for the shift in peak E'' or peak tan δ temperatures at various DMA frequencies. The accuracy of the calculation varies according to the nature of the T_g data available, the value of the activation energy, and the approximation inherent in the Arrhenius equation. Nevertheless, the method provides a quick and practical method for screening candidate polymers for applications requiring a specified T_g range, such as vibration damping.

5.9. COMMERCIAL DMA INSTRUMENTATION

Several types of DMA and rheology instruments are currently available from commercial vendors. In this section we present an overview of some representative examples and their specifications. The information reported here is based on that directly provided by the vendors and is presented here with a limited amount of editing. Thus, from the information provided, it is not easy to compare every feature of the different instruments, since different vendors emphasize different features. The reader is advised that this is a dynamic field and that the instruments and their features and availability are subject to change. For more detailed information on a given instrument, available software, and the manufacturer's technical support and product line, we suggest that you visit the vendor Website and/or contact them directly. Several of the vendors mentioned here may not be familiar names, since they are not traditional thermal analysis system suppliers. All of the instruments are worthy of consideration. They are diverse, covering a wide range of different force, frequency, and temperature ranges as well as various useful test geometries and, of course, they vary in price. The available DMA systems may be classified in three categories:

1. Low-force tabletop systems with maximum force capability up to 18 N
2. Higher force tabletop or floor models; these most frequently are load frame instruments with capacities up to 450 N (or possibly greater)
3. Rotational rheometers that perform both rheology and DMA solids measurements in the torsional shear mode

Low-force tabletop instruments are offered by the traditional thermal analysis vendors Mettler Toledo, Netzsch, Perkin-Elmer, and TA Instruments as well as Seiko Instruments. Higher-force tabletop and floor model instruments are offered by Bose and 01dbMetrovib. Rotational rheometers that also perform torsional shear measurements on solid samples are marketed by Reologica

Instruments AB, TA Instruments, and ThermoFisher Scientific (Haake Rheometers).

The drive systems used for the various instruments represent approaches to obtaining precise sample positioning and minimizing system losses due to friction. This results in maximizing accuracy and precision in measuring dynamic mechanical and rheological properties over a wide range of frequencies, temperatures, and modulus values. The different drive modes represented are those involving magnetic coupling (or magnetic actuators), air bearings, and servomechanical actuators or combinations of these. All of the systems boast of high accuracy in E', $\tan \delta$ (DMA), or force and torque (rheology) of a maximum of ±1–2%. A summary of available instrumentation is presented for each manufacturer.

5.9.1. DMA Instruments from Thermal Analysis Vendors

5.9.1.1. Mettler Toledo, Inc.
1900 Polaris Parkway
Columbus, OH 43240 (USA)
Tel: 800-638-8537
http://us.mt.com

Mettler Toledo DMA SDTA 861 This is a tabletop instrument that is available in different versions with different force ranges of 12, 18 or 40 N, two different maximum frequency ranges 200 and 1000 Hz, and stiffness ranges of either four or six decades. The lower frequency limit is 0.001 Hz. It has a modular design in that the basic instrument can be upgraded in any of these categories after purchase. The instrument also has test fixtures that can be loaded externally and then inserted into the instrument. The instrument has a temperature sensor mounted very close to the sample. Mettler asserts that it provides very precise sample temperature measurements and that simultaneous sample DTA data can be obtained with the sensor system. Optional immersion, humidity control, and gas-switching systems are also available. The instrument does not perform stress relaxation or creep measurements. Because this is a dynamic field we may have omitted from this discussion some DMA instruments that are available commercially. The reader should be aware that if a specific instrument is not described here, it still should be deemed worthy of consideration.

Specifications
Temperature: –150 °C to 500 °C
Measurement modes: three-point bending, dual/single-cantilever, tension, shear, compression
Force: from 0.001 to 12, 18, or 40 N; sensitivity 1 mN
Displacement: ±1.6 mm; sensitivity 5 nm

Stiffness range: 10–10^8 N/m (depends on measurement mode); precision 0.2%

tan δ: range 0.001–100; sensitivity 0.0001

Frequency: 0.001–200 Hz or 0.001–1000 Hz

5.9.1.2. Netzsch Instruments, Inc

37 North Avenue

Burlington, MA 01803 (USA)

Tel: 781 272 5353, Ext. 112

Fax: 781 272 5225

www.nib.netzsch.us

Netzsch DMA 242 C The DMA 242 C can determine viscoelastic properties over a wide modulus range with the common types of deformation geometries: three-point bending, single- and dual-cantilever bending, compression/penetration, shear, and tension. Netzsch also offers a series of special sample holders, such as those designed, for example, for extremely stiff composite materials and metals. A stainless-steel container is available to facilitate measurements with the sample in various media (the immersion test—option).

The DMA 242 C operates in the temperature range of –170–600 °C. The low-temperature range is achieved with a low-consumption liquid nitrogen cooling system. There is minimal temperature gradient over a sample length of up to 60 mm in the bending mode. A purge gas system provides for a controlled sample atmosphere and also protects the measurement electronics from any gases evolving from the sample. Frequencies of 0.01–100 Hz can be selected and combined with defined force of ≤16 N and deformation amplitudes of 0.1–240 μm. The instrument measures viscoelastic properties covering a modulus range from 10^{-3} to 10^6 MPa. The deformation amplitude (approximately 0.1–240 μm) and the sample position are controlled independently of one another. Thus, constant contact between the push-rod and the sample is guaranteed, even when the material softens greatly. This is useful for the study of polymers where the storage modulus changes by several orders of magnitude during a measurement.

Dynamic, isothermal, and stepwise temperature control makes it possible to design flexible temperature programs. Digital filtering via Fourier analysis enhances the signal-to-noise ratio, which helps to resolve small tan δ values. Comprehensive, multidimensional calibration of the DMA system leads to reproducible test results for the modulus and damping (viscoelastic) behavior of a sample.

Specifications

Deformation modes: three-point bending; single/dual-cantilever bending; shear compression/penetration; tension

Measurement modes: oscillatory; TMA-mode creep/relaxation (optional), stress/strain sweep (optional)
Temperature range: −170° to 600 °C
Heating and cooling rates: 0.01–20 K/min
Cooling time: 10 min (from 20 °C to −150 °C)
Frequency range: 0.01–100 Hz
Controlled force range: maximum 8 N static and maximum ±8 N dynamic
Controlled strain amplitude ranges: maximum ±240 µm
Modulus range (E'): 10^{-3}–10^{6} MPa
Damping range (tan δ): 0.00006–10
Atmospheres: inert, oxidizing, static, dynamic
Immersion tests (optional)

5.9.1.3. Perkin-Elmer Life and Analytical Sciences
710 Bridgeport Avenue
Shelton, CT 06484-4794 (USA)
Tel: 800-762-4000 or 203-925-4602
www.perkinelmer.com

Perkin-Elmer DMA 8000 This is a low-cost, compact dynamic mechanical analyzer originally designed by Triton Technologies. It is recommended for both research and routine quality testing for various types of samples such as polymers, composites, pharmaceuticals, and foods.

A feature of the instrument is a rotating analysis head, which can be oriented through 180° to adjust the analysis configuration for different test types and sample geometries. In addition to operation in the dynamic mechanical mode, the DMA 8000 operates in a constant-force (TMA) mode versus time or temperature. Applications such as thermal expansion coefficient, softening and penetration, or extension or contraction in the tension geometry provide data equivalent to those obtained by many commercial standalone TMA instruments.

The cooling system operates to −190 °C in 15 min using less than one liter of liquid nitrogen. An environmental fluid bath optional accessory allows for immersion studies on a sample while measuring the dynamic mechanical properties. Another option is a humidity generator, which delivers the capability to apply and accurately control relative humidity in the sample environment. The standard furnace of the DMA 8000 is configured with a quartz window, which allows visual inspection of the sample and clamping system throughout the experiment without interrupting the temperature profile or other experimental conditions. The drive system incorporated in the DMA 8000 has low compliance with no stabilizing springs or air bearings and no shaft support system.

Specifications

Geometry: typical orientation

three-point bending: vertically up; cantilever—horizontal; tension/compression—vertically up/down; shear—any orientation

Rotating analysis head: vertically up, vertically down, horizontal (forward) and 45° in between these positions

Temperature range: standard furnace −190–400 °C

High-temperature furnace −190–600 °C

Immersion bath −196–150 °C

Scanning rates: heating rate 0–20 °C/min (standard furnace); cooling rate 0–40 °C/min (standard furnace) at midrange (100 °C), may not be achieved at elevated temperatures

Liquid nitrogen (N2) coolant consumption: −100 °C, 5 min, 0.3 liters N2; −150 °C 10 min, <1 liter N2; −190 °C, 15 min, <1 liter N2

Frequency: range 0–300 Hz (depends on sample); maximum ≤100 per experiment

Resolution: 0.001 Hz

Dynamic displacement: from 0 to ±1000 μm

Stiffness range: from 2×10^2 to 1×10^8 N/m; resolution 2 N/m

Modulus: range ~10^3–10^{16} Pa; resolution 0.0001 Pa

tan δ: resolution 0.00001

Force: Range ±10 N minimum, resolution 0.002 N

Displacement/strain: resolution 1 N·m, range ±1000 μm

Sample size: maximum 52.5 × 12.8 × 8.0 mm

Geometry options: single-cantilever bending 1.0–17.5 mm; dual-cantilever bending 2.0–35.0 mm; three-point bending 20.0–45.0 mm

Tension <10 mm; compression unlimited, <10 mm

Shear: 10-mm-diameter plate

TMA-mode measurement range ±1000 μm

5.9.1.4. TA Instruments Inc.

109 Lukens Drive
New Castle, DE 19720 (USA)
Tel: 302-427-4000
www.tainstruments.com

TA Instruments Q800 DMA This instrument incorporates unique technology to provide the ultimate in performance, versatility, and ease of use. A noncontact, linear drive motor provides precise stress control. Ultrasensitive optical encoder technology is used to measure strain, and air-bearing technology ensures virtually friction-free movement. The Q800 dynamic mechanical analysis instrument operates over a temperature range from −150 °C to 600 °C and provides multiple modes of deformation, including dual/single-cantilever and three-point bending, tension, compression, and shear. The instrument also operates in the stress relaxation and creep modes. The clamps are individually calibrated for data accuracy, and the design facilitates sample mounting. This instrument is capable of performing DMA, stress relaxation and creep measurements on samples immersed in a liquid.

Specifications

Atmospheres: static; controlled flow with air or inert gas; fluid (immersion); humidity

Dimensions: depth 17 in. (43.2 cm); width 26 in. (66 cm); height (furnace open) 28 in. (71 cm), (furnace closed) 22 in. (56 cm)

Weight: ~85 lb (~38.6 kg)

Sample size: length 2 in. (50 mm) maximum; width 0.6 in. (15 mm) maximum; thickness 5–10 mm

Displacement range: 1.0 in. (25 mm)

Force: 0.001–18 N

Temperature range: −150–600 °C

Programmed heating rate: 0.1–20 °C/min

Cooling rate: 0.1–10 °C/min

Temperature precision: ±2 °C

Isothermal stability: ±0.1 °C above 50 °C; ±1.0 °C below 50 °C

Modulus range: from 10^3 Pa to 3×10^{12} Pa

Modulus precision: ±1 %

Frequency range: 0.01–200 Hz

tan δ range: 0.0001–10; tan δ resolution 0.00001

Dynamic deformation: ±0.5 to 10,000 μm

Strain resolution: 1 nm

5.9.1.5. Seiko Instruments (US distributor)
RT Instruments, Inc.
10 N. East Street, Suite 106,
Woodland, CA 95776 (USA)
Tel: 530-666-6700
Fax: 530-662-2875
www.rtinstruments.com

EXTAR 6000 Dynamic Mechanical Spectrometer This instrument applies various deformations, such as bending, tension, compression, and shear, to a solid sample and operates in the oscillatory mode as well as the static mode for stress relaxation and creep. For dynamic measurements, a new "synthetic" oscillation mode has been added to the existing high-precision sine wave oscillation mode. The synthetic oscillation mode can measure multiple frequencies at an extremely fast rate, which allows the instrument to measure samples with extremely rapid elastic modulus transformations. Measurements from $-150\,°C$ are fully automatic using the automatic gas cooling unit.

Specifications

Deformation modes: bending, tension, shear, film shear, compression, three-point bending

Deformation mode moduli limits: 10^5–10^{12} Pa (bending), 10^5–10^{12} Pa (tension), 10^3–10^9 Pa (shear), 10^7–10^{11} Pa (film shear), 10^5–10^9 Pa (compression), 10^5–10^{12} Pa (three-point bending)

Temperature range: -150–$600\,°C$; program rate 0.01–$20\,°C/min$

5.9.2. DMA Instruments from Other Vendors

5.9.2.1. Bose Corporation

ElectroForce Systems Group
10250 Valley View Road, Suite 113
Eden Prairie, M-N 55344 (USA)
Tel: 866-837-8464
952-278-3070
www.bose-electroforce.com

Bose ElectroForce (ELF) DMA Instruments The group of tabletop test systems marketed by Bose, designated ElectroForce 3100, are designed for researchers and product developers for testing various materials and full-size components. They incorporate a magnetic linear actuator that generates a linear motion with precise load and frequency characteristics. The ELF features a patented electromagnetic linear motor design. The design utilizes a fixed winding and moving magnet assembly coupled with a flexure support guidance system that eliminates friction. The magnetic frictionless bearing eliminates special needs for maintenance. This system performs a variety of material tests as well as DMA, which include the following:

- Tension/compression
- Fatigue
- Creep/stress relaxation

- Viscoelastic property measurement (DMA)
- Shear sandwich

Specifications

Actuator: load capacities—peak/maximum sine amplitude ±22 N (±5 lb)

Stroke: 5 mm (0.2 in.)

Frequency range: 100 Hz maximum, 0.00001 Hz minimum

Transducer load cell—fatigue-rated: capacity ±22 N (±5 lb); percent error <0.5% of full-scale range

Load frame: height 500 mm (19.4 in.), width 300 mm (11.5 in.), Depth 175 mm (7 in.); weight 18 kg (40 lb)

Test space: adjustable—178 mm (7.00 in.); vertical—0–200 mm (0–8.00 in.)

Temperature range: from −150 °C to 315 °C

Bose also markets a higher-capacity unit, the 3200 series, which has a much higher force range. The 3200 series has the following available: peak/maximum sine = ±225 N (50 lb) and ±450 N (100 lb); static or root mean square (RMS) (continuous) = ±160 N (35 lb) and ±320 N (70 lb).

5.9.2.2. 01dB-Metravib Instruments

Siège Social

200, Chemin des Ormeaux

F-69578 Limonest Cedex, France

Tel: +33 (0)4 72 52 48 00

Fax: +33 (0)4 72 52 47 47

USA:
  Tel: 248 592 2990
  Fax: 248 592 2991

www.01db-metravib.com

01dB-Metravib manufacturers a family of dynamic mechanical analyzers (DMAs) for measuring viscoelastic properties, and more generally, the mechanical properties of materials and their dependence on temperature. The various Metravib instruments are designated DMA-XX, where XX refers to the maximum force range of the instrument. The 01dB-Metravib DMAs cover a wide range of applications, including

- Dynamic mechanical analysis (DMA)
- Thermomechanical analysis (TMA)
- Simultaneous DMA/TMA analysis
- Creep and stress relaxation tests

- Dedicated mechanical testing
- Robotic DMA and mechanical testing

The 01db-Metravib DMAs meet multiple requirements in two major areas:

- *Basic research*—to understand the relationships between the material's molecular structure and its mechanical properties
- *Industrial research*—to handle the key performance characteristics of manufactured products in their conditions of use and ensure their quality and lifetime

The DMA50 is a desktop DMA offering a high force range and outstanding flexibility from glass transition determination to immersed tests, which makes it a general thermomechanical testing platform.

Main assets:

- High force: 50 N (peak)
- Broad frequency range: from 10^{-5} Hz to 200 Hz
- Temperature range: from −150 °C to 600 °C
- Analysis of specimens with sizes representative of the material
- High performance of autotension mode (coupled static/dynamic control) for film and three-point bending modes
- Multipurpose thermal chamber (Heating/cooling, aging)
- Possible immersion tests for all stress modes at no extra cost

Main applications:

- DMA, TMA, and simultaneous DMA/TMA tests
- Determination of glass and subglass transitions
- Testing polymers and composites
- Analysis of films and specimens with low stiffness
- Tests on materials immersed in a liquid
- R&D/quality control

In addition, Metravib offers the DMA + 100, DMA + 150, and DMA + 450, which are larger floor-model instruments that have very rigid frames and exhibit extended ranges of measurable stiffness and frequency.

5.9.3. Rotational Rheometers

5.9.3.1. TA Instruments Inc.
109 Lukens Drive
New Castle, DE 19720 (USA)
Tel: 302-427-4000
www.tainstruments.com

AR Series Rheometers These are a group of rheometer systems that measure viscoelastic data for both fluids and solids. The AR series is a sequel to the former rheometrics rheometers with updated performance and efficiency. The instruments feature sensitive low-torque performance, accurate stress, direct strain control, and intuitive software. There are three AR series models; in addition, TA offers the ARES-G2 rheometer, which is a research instrument. Each instrument has different features and torque ranges. The newest of the group, the ARG-2 instrument, has a magnetic drive that provides enhanced low-torque precision and accuracy.

The AR2000ex rheometer has a drive system with a low-inertia drag cup motor and porous carbon air bearings for accurate operation in controlled stress, direct strain, and controlled rate modes. The instrument uses an optical encoder for measuring angular displacement, and has high stiffness and low inertia, enabling it to measure a wide range of fluid and solid viscoelastic properties. Solid properties are measured in torsional shear. Disposable tooling is available for experiments with curing thermosets.

Specifications (AR2000ex)

Minimum torque: oscillation, constant stress/constant rate $0.003\,\mu N \cdot m$

Minimum torque: steady shear, constant stress/constant rate $0.01\,\mu N \cdot m$

Torque range: $0.01\,\mu N \cdot m$ to $200\,mN \cdot m$

Frequency range: 7.5×10^{-7} to $628\,rad/s$

Displacement resolution: 25 nrad

Thrust bearing: magnetic

Normal force range: $0.005\,N$ to $50\,N$

Temperature range: $-150\,°C$ to $600\,°C$

5.9.3.2. REOLOGICA Instruments Inc. US
ATS Rheosystems
52 Georgetown Road
Bordentown, NJ 08505 (USA)
Tel: 609 758 1750
www.reologicainstruments.com

Dynalyser This is a modular research-level rheometer system designed for diverse applications. The instrument's design derives from input and recommendations from rheometer users. It is used for testing diverse types of materials, including thermoplastics, thermosets, elastomers, semisolids, and fluids. It performs dynamic oscillatory measurements on solids (in rectangular torsion) and liquids as well as transient stress relaxation and creep on solids and fluids. It also measures viscosities in steady shear flow using parallel-plate, cone–plate, and Couette flow geometries, as well as normal stresses in the flow configurations. The instrument has an indirect heated and cooled oven; the oven has a heating element with a convection airflow. It covers a temperature range from $-180\,°C$ to $550\,°C$. The oven is designed for testing polymer melts, solids, and fluid materials.

The drive system has a frictionless asynchronous AC motor and inductive position sensor. The ranges are listed, accuracy is better than $\pm 10\%$ at $0.00002\,mN \cdot m$.

Specifications

Torque range: $0.00002–100\,mN \cdot m$ ($200\,mN \cdot m$ optional)

Normal force range: $0.002–50\,N$

Angular displacement: $0.00000003–6.28\,rad$ (strain is geometry-dependent)

Angular velocity: $0–320\,rad/s$ (shear rate is geometry-dependent)

Frequency range: $0.0000001–100\,Hz$ (nine decades)

Temperature range: $-180\,°C$ to $550\,°C$

Rectangular sample maximum size (solids DMA mode): length 66 mm; width 12 mm; thickness 3 mm

Cylindrical sample maximum size (rheology mode): length 66 mm; standard diameters of 0.5, 1, 1.5, 2 mm, other sizes available

5.9.3.3. ThermoFisher Scientific Materials Characterization

25 Nimble Hill Road

Newington, NH 03801 (USA)

Tel: 800-258-0830

www.thermo.com

Haake MARS Rheometer This is a modular instrument designed to allow all application assemblies, including the measuring head and electronics systems, to be interchanged. The main control and power electronics are separated from the instrument itself, thereby preventing thermal and mechanical interference from heat sources, fans, and other equipment. A variety of temperature control units are available to control temperatures from $-150\,°C$ up to $600\,°C$.

The rheometer has an optical angle encoder that has a resolution of 12 nrad, which is important when measuring delicate samples within the linear viscoelastic regime at very low amplitudes in the oscillation mode as well as the determination of the zero-shear viscosity at very low shear rates ($10^{-6}\,s^{-1}$).

The drive system is powered by a drag cup motor that applies a steady rotation or oscillatory torque to the sample. The main advantages of a drag cup motor over other motor types are its extremely smooth operation and its very low moment of inertia. The drive system also incorporates three individual air bearings: an axial air bearing that supports the motor shaft in the vertical (axial) direction and two widely spaced radial air bearings that support the motor shaft in the radial direction and prevent the shaft from tilting.

Disposable coaxial cylinders, parallel–plate and cone–plate measuring geometries, custom-tailored dimensions, and a variety of different tooling materials are available. For measurements at higher temperatures, geometries with a ceramic shaft are available.

Specifications

Minimum torque in various modes: $0.05\,\mu N \cdot m$
Maximum torque: $200\,mN \cdot m$
Torque resolution: $0.5\,nN \cdot m$
Angular resolution: 12 nrad
Oscillation frequency: minimum $10^{-5}\,Hz$, maximum 100 Hz
Normal force: minimum 0.01 N, maximum 50 N
Normal force resolution: 0.001 N
Temperature range: $-150\,°C$ to $600\,°C$
Dimensions ($W \times D \times H$) $600 \times 600 \times 870\,mm$
Weight: 55 kg

APPENDIX

ASTM Standards Relevant to Dynamic Mechanical Analysis

Standards issued under the jurisdiction of ASTM International Committee E37 on thermal analysis that relate to DMA are listed below for reference. These provide guidance to procedures for specification of the glass transition and several types of DMA instrument calibrations. We highly recommend reviewing these, particularly those who are new to DMA methods. For additional information, go to www.astm.org.

E473, *Standard Terminology Relating to Thermal Analysis and Rheology*

E1640, *Standard Test Method for Assignment of the Glass Transition Temperature by Dynamic Mechanical Analysis*

E1867, *Standard Test Method for Temperature Calibration of Dynamic Mechanical Analyzers*

E2254, *Standard Test Method for Storage Modulus Calibration of Dynamic Mechanical Analyzers*

E2425, *Standard Test Method for Loss Modulus Conformance of Dynamic Mechanical Analyzers*

E2510, *Standard Test Method for Torque Calibration or Conformance of Rheometers*

ABBREVIATIONS

Symbols

A	cross-sectional area
A_0	original cross-sectional area
a_T	WLF time–temperature superposition shift factor
C_1	constant in the WLF equation
C_2	constant in the WLF equation
C_p	heat capacity
E	elastic modulus
E^*	complex storage modulus
E'	tensile or flexural storage modulus
E''	tensile or flexural loss modulus
E_f	flexural modulus
$E_\perp$	modulus 90° from strain axis
$E_\parallel$	modulus parallel to strain axis
F	force
f	frequency (Hz)
G	shear modulus
G^*	complex dynamic shear modulus
G'	shear storage modulus
G''	shear loss modulus
$_{gel}T_g$	temperature where gelation and vitrification occur simultaneously
h	height
J	creep compliance
J^*	complex compliance
J'	storage compliance
J''	loss compliance
K	equivalent to a_T
L	length
l_0	initial length
m	mass, mass fraction
M	molecular mass
M_c	molecular weight between crosslinks
M_n	number average molecular mass
M_w	weight average molecular mass
P	pressure

Q^{-1}	tan δ
R	universal gas constant
T	temperature
ΔT	temperature difference
T_c	crystallization temperature
T_{cure}	cure temperature
T_g	glass transition temperature
T_{g0}	glass transition temperature of unreacted resin mixture
$T_{g\infty}$	ultimate glass transition for a fully crosslinked thermoset
T_i	arbitrary temperature T_1, T_2, T_3, etc.; inflection point in modulus curve at T_g
T_m	melting point
T_R	reference temperature
T_s	sample temperature
T_{α_c}	the α_c transition temperature
t	time
t_T	time at temperature T (in time–temperature superposition)
t_{T_R}	time at temperature T_R (in time–temperature superposition)
V	volume
W	width
x	degree of crystallinity
σ	tensile stress
σ_0	initial tensile stress
σ'	in-phase component of dynamic stress
σ''	out-of-phase component of dynamic stress
σ^*	complex stress
ε^*	complex strain
α	linear coefficient of thermal expansion, degree of conversion, polymer transition
α_c	transition in a crystalline polymer above T_g
α_{gel}	reaction conversion at the gel point
γ	shear strain
$\dot{\gamma}$	shear rate
γ_0	initial shear strain
δ	phase shift between strain and stress in a DMA experiment, a subambient transition
δ_s	shear deformation
δ_p	flexural deformation
ΔH_f	heat of fusion
ε	tensile strain, flexural strain
ε_0	initial tensile (flexural) strain
η	viscosity
η^*	complex viscosity
η'	in-phase component of dynamic viscosity; dynamic viscosity
η''	out-of-phase component of dynamic viscosity

λ	thermal conductivity
μ	Newtonian viscosity
ν	Poisson's ratio
ρ	density
τ	shear stress, relaxation time
ω	radian frequency
$\Delta E'$	storage modulus difference

Acronyms

DDS	diaminodiphenylsulfone
HDPE	high-density polyethylene
LLDPE	linear low-density polyethylene
LDPE	low-density polyethylene
PC	polycarbonate
PE	polyethylene
PET	poly(ethylene terephthalate)
PMMA	poly(methyl methacrylate)
PP	polypropylene
PS	polystyrene
PVA	poly(vinyl alcohol)
TTS	time–temperature superposition
TTT	time–temperature transformation
ULDPE	ultra low-density polyethylene
UV	ultraviolet radiation
WLF	Williams–Landel–Ferry

REFERENCES

Allen, G., McAinsh, J., and Jeffs, G. M. (1971), *Polymer* **12**, 85.

Arai, H. and Kuriyama, I. (1976), *Coll. Polymer Sci.* **254**, 967.

Avakian, P., Matheson, R. R., and Starkweather, H. W. (1991), *Macromolecules* **24**, 4698.

Becker, G. W. (1955), *Kolloid Z.* **140**, 1.

Bobalek, E. G., Moore, E. R., Levey, S. S., and Lee, C. C. (1964), *J. Appl. Polymer Sci.* **8**, 635–657.

Bohn, L. (1963), *Kunstoffe* **53**, 826.

Boyd, R. H. (1984), *Macromolecules* **17**(4), 903.

Boyd, R. H. (1985), *Polymer* **26**, 323.

Boyd, R. H. (1959), *J. Chem. Phys.* **30**, 1276.

Boyer, R. F. (1968), *Polymer Eng. Sci.* **8**, 161.

Breitigam, W. V., Bauer, R. S., and May C. (1993), *Polymer* **34**, 767–771.

Burns, J. M., Prime, R. B., Barrall, E. M., and Oxsen, M. E. (1989), in *Polymers in Information Storage Technology*, Plenum, Mittal, K. L., ed., New York, pp. 237–256.

Carreau, P. J., De Kee, D. C. R., and Chhabra, R. P. (1997), *Rheology of Polymeric Systems*, Hanser, New York.

Charlesworth, J. M. (1988), *Polym. Eng. Sci.* **28**, 221.

Chartoff, R. P. and Chiu, T. W. (1980), *Polym. Eng. Sci.* **20**, 245–251.

Chartoff, R. P. and Graham, J. L. (1982), in *Computer Applications in Polymer Science*, Provder, T., eds., ACS Symp. Series **197**, 367 (Chapter 8).

Chartoff, R. P. (1986), in *Polymer Composites*, Sedlácek, B., ed., de Gruyter, Berlin, pp. 89–104.

Chartoff, R. P. (1988), *Soc. Plast. Eng. ANTEC Proc.* **34**, 1143.

Chartoff, R. P. (1997), "Thermoplastics," in *Thermal Characterization of Polymeric Materials*, 2nd ed., Turi, E. A., ed., Academic Press, New York, pp. 483–743.

Chartoff, R. P., Weissman, P. T., and Sircar, A. K. (1994), "T_g determination by dynamic mechanical methods," in *Proc. ASTM Symp. Assignment of the Glass Transition*, ASTM STP1249, Seyler, R. J., ed., ASTM, Philadelphia, pp. 88–107.

Chung, C. I. and Sauer, J. A. (1971), *J. Polym. Sci., Pt. A* **9**, 1097.

Collins, E. A., Bares, J., and Billmeyer, F. W., Jr. (1973), *Experiments in Polymer Science*, Wiley-Interscience, New York.

Cox, W. and Merz, E. (1958), *J. Polym. Sci.* **28**, 619.

Davis, G. T. and Eby, R. K. (1973), *J. Appl. Phys.* **44**, 4274.

Dealy, J. M. and Wissbrun, K. F. (1990), *Melt Rheology and Its Role in Plastics Processing*, Van Nostrand-Reinhold, New York.

Dillman, S. H. and Seferis, J. C. (1989), *J. Macromol. Sci. Chem.* **A26**(1), 227–247.

Dillman, S. H. (1988), *Kinetic Viscoelasticity of Reacting Polymer Systems*, doctoral dissertation, Univ. Washington, Seattle.

Dumbleton, J. H. and Murayama, T. (1967), *Kolloid Z. Z. Polym.* **220**, 41.

Dumbleton, J. H. and Murayama, T. (1970), *Kolloid Z. Z. Polym.* **238**(1–2), 410.

Dumbleton, J. H., Murayama, T., and Bell, J. P. (1968), *Kolloid Z. Z. Polym.* **228**, 54.

Dusek, K. (1986), in *Epoxy Resins and Composites III, Advances in Polymer Science*, vol. 78, Dusek, K., ed. Springer-Verlag, Berlin, pp. 1, 59.

Ellis, B., ed. (1993), *Chemistry and Technology of Epoxy Resins*, Blackie Academic & Professional, Glasgow, UK, pp. 72, 116.

Enns, J. B. and Gillham, J. K. (1983a), *Adv. Chem. Ser.* **203**, 27–63.

Enns, J. B. and Gillham, J. K. (1983b), *J. Appl. Polym. Sci.* **28**, 2567–2591.

Ferry, J. D. (1980), *Viscoelastic Properties of Polymers*, 3rd ed., Wiley, New York.

Feve, M. (1989), *Makromol. Chem., Macromol. Symp.* **30**, 95.

Flory, P. J. (1941), *J. Am. Chem. Soc.* **63**, 3083.

Flory, P. J. (1953), *Principles of Polymer Chemistry*, Cornell Univ. Press, Ithaca, NY.

Garrett, P. D. and Grubb, D. T. (1988), *J. Polym. Sci., Pt. B: Polym. Phys.* **26**, 2509.

Gaur, U. and Wunderlich, B. (1980), *Macromolecules* **13**, 445.

Gerrard, J. F., Andrews, S. J., and Macosko, C. W. (1990), *Polym. Compos.* **11**(2), 91.

Gillham, J. K. (1986a), *Encyclopedia of Polymer Science and Technology*, 2nd ed., Wiley, New York, pp. 519, 524.

Gillham, J. K. (1986b), *Polym. Eng. Sci.* **26**, 1429.

Glauser, T., Johansson, M., and Hult, A. (1999), *Polymer* **40**, 5297, 5302.

Grehlinger, M. and Kraft, M. (1988a), *J. Plast. Film Sheeting* **4**, 3.

Grehlinger, M. and Kraft, M. (1988b), *Proc. Soc. Plastics Engineering ANTEC Conf.*, Atlanta, 1988, vol. 88, p. 1185.

Hale, A., Macosko, C. W., and Bair, H. E. (1991), *Macromolecules* **24**, 2610–2621.

Hartmann, B. and Lee, G. F. (1979), *J. Appl. Polym. Sci.* **23**, 3639.

Hiltz, J. and Keough, I. A. (1991), *Proc. 20th NATAS Conf.*, pp. 573–578.

Illers, K. H. (1972), *Kolloid Z. Z. Polym.* **250**, 426.

Illers, K. H. and Breuer, H. (1963), *J. Colloid Sci.* **18**, 1 (this source provides additional results and information on PET).

Jaffe, M., Menczel, J. D., and Bessey, W. E. (1997), "Fibers," in *Thermal Characterization of Polymeric Materials*, 2nd ed., Academic Press, 1997, pp. 1767–1954.

Jones, D. I. G. (1978), *Shock Vibration Bull.* **48**(2), 3.

Klosterman, D., Chartoff, R., Tong, T., and Galaska, M. (2003), *Thermochim. Acta* **396**, 199–210.

Kunugi, T., Hayakawa, T., and Mizushima, A. (1991), *Polymer* **32**, 808.

Kunugi, T., Ichenose, T., and Suzuki, A. (1986), *J. Appl. Polym. Sci.* **31**, 429.

Kunugi, T., Oomori, S., and Mikami, S. (1988), *Polymer* **29**, 814.

Lau, S. F., Suzuki, H., and Wunderlich, B. (1984a), *J. Polym. Sci., Polym. Phys. Ed.* **22**, 379.

Lau, S. F., Wesson, J. P., and Wunderlich, B. (1984b) *Macromolecules*, **17**(5), 1102.

Leaderman, H. (1941), *Textile Res. J.* **11**, 171.

Leaderman, H. (1943), *Elasticity and Creep of Filamentous Materials*, The Textile Foundation, Washington, DC.

Lee, B.-L. and Nielsen, L. E. (1977), *J. Polym. Sci., Polym. Phys. Ed.* **15**, 683.

Lee, C. Y.-C. and Goldfarb, I. J. (1981a), *Polym. Eng. Sci.* **21**, 390.

Lee, C. Y.-C. and Goldfarb, I. J. (1981b), *Polym. Eng. Sci.* **21**, 787.

Lim, T., Frosini, V., Zaleckas, V., Morrow, D., and Sauer, J. A. (1973), *Polym. Eng. Sci.* **13**, 51.

Manson, J. A., Sperling, L. H., and Kim, S. L. (1977), *Influence of Crosslinking on the Mechanical Properties of High T_g Polymers*, AFML-TR-77-109 (US Air Force Technical Report, Wright Patterson Air Force Base, July 1977).

Matsuoka, S. (1992), *Relaxation Phenomena in Polymers*, Hanser/Oxford, New York.

McCrum, N. G. (1959), *J. Polym. Sci.* **34**, 355.

McCrum, N. G., Read, B. E., and Wiliams, G. (1967), *Anelastic and Dielectric Effects in Polymeric Solids*, Wiley, New York (reprinted by Dover Publications, New York, 1991; contains a detailed overview of DMA).

Menczel, J. D., Collins, G. L., and Saw, S. K. (1997a), *J. Therm. Anal.* **49**, 201.

Menczel, J. D., Jaffe, M., and Bessey, W. E. (1997b), "Films," in *Thermal Characterization of Polymeric Materials*, 2nd ed., Academic Press, pp. 1955–2089.

Menczel, J. D., Jaffe, M., Saw, C. K., and Bruno, T. P. (1996), *J. Therm. Anal.* **46**, 753.

Menges, G. and Boden, H.-E. (1986), in *Failure of Plastics*, Brostow, W. and Corneliussen, R. D., eds., Hanser, Munich, pp. 169–195.

Mercier, J. P., Aklonis, J. J., Litt, M., and Tobolsky, A. V. (1965), *J. Appl. Polym. Sci.* **9**, 147.
Miller, D. R. and Macosko, C. W. (1976), *Macromolecules* **9**, 206.
Miller, R. W. and Murayama, T. (1984), *J. Appl. Polym. Sci.* **29**, 933.
Mora, E. and Macosko, C. W. (1991), *Proc. 20th NATAS Conf.* p. 506.
Mullin, J. V. and Knoell, A. C. (1970), *Mater. Res. Stand.* (MTRSA) **10**(12), 16.
Murayama, T. (1978), *Dynamic Mechanical Analysis of Polymeric Materials*, Elsevier, Amsterdam.
Neag, C. M. and Prime, R. B. (1991), *J. Coat. Technol.* **63**(797), 37.
Nielsen, L. E. (1962), *Mechanical Properties of Polymers*, Van Nostrand-Reinhold, Princeton, NJ, p. 23.
Nielsen, L. E. and Landel, R. F. (1994), *Mechanical Properties of Polymers and Composites*, 2nd ed., Marcel Dekker, New York.
Okamura, W., Ohkoshi, Y., Gotoh, Y., Nagura, M., Urakawa, H., and Kajiwara, K. (2004), *J. Polym Sci: Pt. B* **42**, 79–90.
Papir, Y. S., Kapur, S., Rogers, C. E., and Baer, E. (1972), *J. Polym. Sci., Pt. A2* **10**, 1305.
Patel, N. M., Moy, C. H., McGrath, J. E., and Prime, R. B. (1989), *Proc. 18th NATAS Conf.*, pp. 232–239.
Pascault, J. P., Sautereau, H., Verdu, J., and Williams, R. J. J. (2001), *Thermosetting Polymers*, Marcel Dekker, New York.
Pennings, A. J. (1967), *Proc. Intl. Crystal Growth Conf.* Boston, 1966, p. 389.
Pezzin, G., Ajroldi, G., and Garbuglio, C. (1967), *J. Appl. Polym. Sci.* **11**, 2553.
Prevorsek, D. C., Butler, R. M., and Reimschussel, H. K. (1971), *J. Polym. Sci.* **9**, 867.
Prime, R. B. (1986), *J. Therm. Anal.* **31**, 1091.
Prime, R. B. (1992), *Polym. Eng. Sci.* **32**, 1286.
Prime, R. B. (1997), in *Thermal Characterization of Polymeric Materials*, 2nd ed., Turi, E. A., ed., Academic Press, San Diego, pp. 1379–1766.
Retting, W. (1969), *Rheol. Acta* **8**, 259.
Robeson, L. M. and Faucher, J. A. (1969), *J. Polym. Sci., Polymer Lett. Ed.* **7**, 35.
Saw, C. K., Menczel, J., Choe, E. W., and Hughes, *Proc SEP ANTEC '97*, p. 1610.
Sauer, B. B., Avakian, P., and Starkweather, H. W. (1995), unpublished data, private communication.
Schmieder, K. and Wolf, K. (1953), *Kolloid Z.* **134**, 148.
Seferis, J. C. and Nicolais, L., eds. (1983), *The Role of Polymer Matrix in Processing and Structural Properties of Composites*, Plenum, New York, pp. 127–145.
Sha, H., Harrison, I. R., and Zhang, X. (1990), *Proc. 19th NATAS Conf.*, p. 179.
Sichina, W. J. and Matsumori, B. (1991), *Proc. 20th NATAS Conf.*, p. 41.
Sircar, A. K. and Drake, M. L. (1990), in *Sound and Vibration Damping with Polymers*, ACS Symp. Series **424**, Sperling, L. and Corsaro, R., eds., p. 132.
Smith, M. E. and Ishida, H. (1999), *J. Appl. Polym. Sci.* **73**, 593, 600.
Sperling, L. H. (2006), *Introduction to Physical Polymer Science*, 4th ed., Wiley-Interscience, New York, p. 355.

Starkweather, H. W. (1973a), *J. Polym. Sci., Polym. Phys. Ed.* **11**, 587.
Starkweather, H. W. (1973b), in *Nylon Plastics*, Kohan, M. I., ed., Wiley, New York.
Starkweather, H. W. (1980), in *Water in Polymers*, Rowland, S. P., ed., ACS Symp. Series **127**, 433.
Starkweather, H. W. (1988), *Macromolecules* **21**, 1798.
Starkweather, H. W. and Barkley, J. R. (1981), *J. Polym. Sci., Polym. Phys. Ed.* **19**, 1211.
Takayanagi, M. (1965), *Proc. 4th Intl. Congress on Rheology*, Part 1, Interscience, New York, 161.
Thompson, A. B. and Woods, D. W. (1956), *Trans. Faraday Soc.* **52**, 1383.
Thomason, J. L. (1990), *Polym. Compos.* **11**(2), 103.
Tobolsky, A. V. (1960), *Properties and Structure of Polymers*, Wiley, New York.
Tung, C. M. and Dynes, J. P. (1982), *J. Appl. Polym. Sci.* **27**, 569, 574.
Uddin, A. J., Ohkoshi, Y., Gotoh, Y., Nagura, M., Endo, R., and Hara, T. (2004), *J. Polym. Sci., Pt. B* **42**, 433–444.
Urbaczweski-Espuche, E., Galy, J., Ferard, J., Pascault, J., and Sautereau, H. (1991), *Polym. Eng. Sci.* **31**, 1572.
Van Krevelen, D. W. (1990), *Properties of Polymers: Their Correlation with Chemical Structure*, 3rd ed., Elsevier Science, Amsterdam.
Varlet, J., Cavaillé, J. Y., Perez, J., and Joharie, G. P. (1990), *J. Polym. Sci., Pt. B: Polym Phys.* **28**, 2691.
Venditti, R. A. and Gillham, J. K. (1997), *J. Appl. Polym. Sci.* **64**, 3.
Wei, J., Hawley, M. C., and DeMeuse, M. T. (1995), *Polym. Eng. Sci.* **35**, 461–470.
Weissman, P. T. and Chartoff, R. P. (1990), in *Sound and Vibration Damping with Polymers*, Sperling, L. and Corsaro, R., eds., ACS Symp. Series **424**, p. 111.
Wetton, R. E. (1986), in *Developments in Polymer Characterization*, Dawkins, J. V., eds., Elsevier, London, pp. 179–221.
Wetton, R. E., Ruff, P. W., Richmond, J. C., and Neill, J. T. (1987), *Proc. 16th NATAS Conf.*, p. 64.
Williams, M. L., Landel, R. F., and Ferry, J. D. (1955), *J. Am. Chem. Soc.* **77**, 3701.
Winter, H. H. (1987), *Polym. Eng. Sci.* **27**, 1698.
Winter, H. H. et al. (1997), in *Techniques in Rheological Measurement*, Collyer, A. A., eds., Chapman & Hall, London.
Wisanrakkit, G. and Gillham, J. K. (1991), *J. Appl. Polym. Sci.* **42**, 2453.
Woodward, A. E., Crissman, J. M., and Sauer, J. A. (1960), *J. Polym. Sci.* **44**, 23.

CHAPTER 6

DIELECTRIC ANALYSIS (DEA)

AGLAIA VASSILIKOU-DOVA and IOANNIS M. KALOGERAS
Department of Solid State Physics, Faculty of Physics, University of Athens, Panepistimiopolis, Greece

6.1. INTRODUCTION

The term *dielectric analysis* (DEA) refers to a group of techniques that measure changes in different physical properties of a polar material, such as polarization, permittivity, and conductivity, with temperature or frequency. The reorientation of dipoles and the translational diffusion of charged particles in an oscillating electric field provide the basis of the analysis based on alternating-current (AC) dielectric methods, which principally involve measurements of the complex permittivity (ε^*) in the frequency or time domain and at constant or varying temperature. The corresponding changes in the dielectric constant and polarizability of a polymer are quite large and easily detected during phase transitions (e.g., the glass transition, melting, or crystallization) and secondary transitions (localized relaxation mechanisms), as compared to the small changes in enthalpy, volume, or heat capacity measured by other more common thermal analytical techniques. Dielectric monitoring of cooperative (long-range) processes provides a link between molecular spectroscopy, which monitors the properties of individual constituents, and techniques characterizing the bulk properties of the material under investigation, especially the viscosity and the rheological behavior. In particular, dielectric analysis provides the most sensitive method to probe local motions along the chains of polar polymers, since polar bonds (such as $>C=O$, $\equiv C-OH$, and $>N-H$) are directly affected by the electric stimulus. Even though dielectric techniques are deemed inapplicable in the case of apolar materials (e.g., polyethylene), it is possible to study the relaxation dynamics even in nonpolar polymers by appropriate chemical modifications of their structures. These include partial oxidation and labeling (e.g., by chlorination or attachment of polar pendant groups), as

Thermal Analysis of Polymers: Fundamentals and Applications, Edited by Joseph D. Menczel and R. Bruce Prime
Copyright © 2009 by John Wiley & Sons, Inc.

well as the dissolution of suitable polar probe molecules (van den Berg et al. 2004).

Association of relaxation signals with movements of certain molecular/ionic species is usually a result of sufficient knowledge about the composition, the microstructure and morphology of the sample, and information from parallel nondielectric investigations (e.g., NMR and IR). A major advantage of dielectric analysis over other common thermal techniques is its applicability in a very broad frequency range (10 µHz–100 GHz), which usually requires a series of instruments for complete coverage. Nevertheless, for everyday purposes, it is generally sufficient to concentrate in a smaller frequency range (of five to eight decades in frequency) adapted to the material properties. The decomposition of the dielectric spectrum into its individual relaxation processes provides information on the relative "amplitudes" (relaxation strength) and characteristic "times" (relaxation times) of the individual motions. Of the techniques described in this book, dynamic mechanical analysis (DMA) is the most similar to isochronal (at fixed frequency) dielectric analysis. DMA measures mechanical stiffness (modulus) and energy absorption by subjecting a specimen to oscillating mechanical stress or strain within the linear viscoelastic region. In this manner, a variety of molecular motions can be traced. Direct comparisons between DMA and DEA results, obtained in the same frequency range, are usually feasible only in the case of polymers showing moderate to strong polarizability. However, considering the narrow width of the frequency window available for dynamic mechanical studies (broadest range 1 mHz–1 kHz), dielectric analysis of polar systems can add information on certain physical properties at much higher frequencies (easily extending up to the GHz region).

Since the mid-1970s, DEA has received wide recognition as a powerful thermal analysis approach for investigating condensed- and soft-matter dynamics. Intense scientific activity has been undertaken in the dielectric characterization of ion-conducting solids, biological systems, emulsions, colloids, mesophases (liquid crystals, etc.) and particularly polymers and related composites (Kremer and Schönhals 2002), in an attempt to fabricate materials for advanced engineering applications. Besides the linear chain polymers, dielectric analysis has been applied to numerous macromolecular systems with complex molecular architectures, namely, comblike and branched structures, stars, cycles, copolymers (statistical, di- or multiblock copolymers), physically or chemically bonded polymer networks, hyperbranched polymers, and dendrimers. The dynamics in such systems involves a series of relaxation processes that goes from very local motions (generally due to branching chains or side-chains) to segmental mobility exhibiting cooperativity (α relaxation or glass transition), or even large-scale relaxation processes (Rouse dynamics or reptation). In the simpler case, dielectric methods record signals ascribed to the rotational mobility of small polar molecular units (β, γ, … relaxations or secondary transitions). The thermally or electrically activated (re)orientation of chain segments, incorporating a few dozens of carbon atoms, is a much slower process that progresses cooperatively and is responsible for the signals associated with the glass → liquid or glass → rubber transition. At this temperature region the increased mobility of the polymer matrix allows for an increased mobility of intrinsic and extrinsic electric charges, which produce dielectric signals ascribed to ionic conductivity and more complex charge-trapping/liberation processes. The analysis of the high-temperature interfacial polarization modes, originating from the trapping of ions at interfaces within polymer composites, offers a sensitive tool for an indirect study of microstructural properties. Among the contributions of the application of DEA to materials science, one should make special reference to the areas described below.

6.1.1. Establishment of Relationships between a Polymer's Configuration, Dynamic Properties, and Applications

The dynamics and energetics of a polymer can be explored by monitoring structure- or composition-dependent changes in parameters (shape, position, strength, etc.) characteristic of the glass transition and various secondary signals. At the same time, the time, temperature, or pressure dependences of properties related to translational motions of charges (DC conductivity, ion mobility, trap energy depths, etc.) are sensitive to the diversity of the environment provided by the polymer chains and the bonding state of the pertinent ions. Several of these parameters can be used to appraise perturbations in the strength of inter- and intramolecular interactions of a dipolar or ionic nature. This information is fundamental in many areas of materials science, including pharmaceutical science [e.g., for controlled drug delivery (Craig 1995)] and specialized applications in biochemistry and biophysics (body fluid hydrometallurgy, protein stability, hydrophobic hydration, etc.). Interesting examples of dielectric studies on liquids, liquid crystals, biopolymers, blends, copolymers, and composites, can be found in excellent reference books [e.g., Runt (1997); Kremer and Schönhals (2002)].

6.1.2. Information on Surface and Bulk Properties in Confining Environments

Dielectric techniques have proved particularly effective in the study of polymer dynamics at interfaces and in confining geometries [polymers in ultrathin films or within nanometric pores of oxide glasses, polymer in the galleries of clays, etc. (Kalogeras 2008)] because of their enhanced sensitivity with decreasing size scale and the broad dynamic range, with an unparalleled range of frequency and temperature (Kremer et al. 2003; Hartmann et al. 2002). Probing nanoconfinement-induced effects in the segmental relaxation dynamics of the polymer, such as by monitoring shifts of transition temperatures, changes in thermodynamic parameters, and the cooperativity of the relaxation mechanism, dielectric analysis revealed the influence of entropic (e.g., chain-end segregation, density anomalies, or disentanglement) and enthalpic (hydrogen-bonding interfacial interactions) effects on chain mobility (Schönhals et al. 2004; Roth and Dutcher 2005; Alcoutlabi and McKenna 2005; Kalogeras 2005; Priestley et al. 2007). Application of dielectric methods to ultrathin films provides valuable information, since the sensitivity of the method increases with decreasing thickness. Dielectric spectroscopy has been used extensively as a probe for characterizing the phase behavior and dynamics of various polymer composites [e.g., copolymers, polymer blends, and networks (Hedvig 1977; Kalogeras et al. 2007a,b)], with special focus on information about the phase heterogeneity of a polymer composite (Runt 1997; Kalogeras et al. 2006b).

6.1.3. Online Monitoring of Chemical Reactions (Polymerization, Curing), Physical Processes (e.g., Aging), and Structure Formation

Shifts of characteristic transition temperatures and changes of the relaxation parameters (activation energy) or ion mobilities can be used to probe polymer molecular environment and changes in molecular configurations, and follow real-time reactions such as curing, aging, crystallization, phase separation, and polymerization processes (Kranbuehl 1997; Williams et al. 1999). Among other applications, the dipole reorientation and charge migration has been used to monitor in situ the development of crystallization (Fitz and Andjelić 2003) and morphology (Poncet et al. 1999) in polymers. Besides monitoring, dielectric analysis is a practical tool for the characterization and design of such processes.

6.1.4. Evaluation of Charge Transport and Ion Mobility in Semiconductors, Organic Crystals, Polymer Composites and Polymer Electrolytes (Nad et al. 2000; Karlinsey et al. 2004; Baskaran et al. 2004)

Several parameters obtained from dielectric experiments serve for probing the segmental flexibility, which is expected to be rate-determining for the mobility of the charge carriers (Jayathilaka et al. 2003). By recording the temperature dependence of the ionic conductivity in a polymeric material, one may determine the presence or absence of some degree of cooperativity between the motion of the polymer chains and that of the conducting ions. This phenomenon is particularly important in the effort to manufacture polymer conductors for specialized industrial applications, by means of chemical or physical modification of known polymers and the precise control of the conduction mechanism.

6.1.5. Analysis of Nonlinear Electrical and Optical Effects

Important technological applications are critically dependent on the influence of light on the structure and dynamics of a polymer. For example, liquid crystalline sidechain polymers are used for reversible optical data storage and dielectric analysis has been successfully applied for investigating the mechanisms involved in the storage procedure (Stracke et al. 2000).

6.1.6. Dielectric Characterization of Polymers for Engineering Applications

At an industrial level, dielectric techniques have been implemented, among others, in quality control (reliability and stability of the structure and performance, or for studying mechanical and thermal treatment effects), in the characterization of semiconductors, cable-insulating materials (polyethylene, polypropylene, polytetrafluoroethylene, etc.), and passivation layers in microchips. Dielectric heating of materials and remote sensing is another area of interest. The latter application makes use of the change in the dielectric response of certain materials when exposed to gases or liquids (Langlet et al. 2004). Dielectric results may also be used to solve questions arising from lack of compositional information, such as uncertainties pertaining to the nature, concentration, or spatial distribution of absorbed phases, dopants, or contaminants. The piezoelectric, pyroelectric, and ferroelectric (Buixaderas et al. 2004) properties of polymeric materials (e.g., polytetrafluoroethylene, polyvinylidinefluoride, and several of their copolymers) have also attracted considerable attention.

6.1.7. Prelude to the Chapter

In the following sections, an introduction to dielectric theory and various polarization/relaxation mechanisms observed in amorphous and semicrystalline polymers (Section 6.2) is presented. This will equip undergraduate and postgraduate students, as well as technicians, professionals, and academics, with the theoretical background necessary for understanding the origin of measured signals at the molecular level and for the correct interpretation of dielectric data. Section 6.3 acquaints the reader with common isothermal and nonisothermal techniques and related instrumentation, for dielectric measurements in the frequency or time domain. The discussion focuses on two groups of analytical techniques that make up the DEA family, involving dielectric measurements performed using alternating [dielectric relaxation spectroscopy (DRS)] or constant [thermally stimulated current technique (TSC)] electric fields. A detailed description of experimental aspects (e.g., recommendations for sample preparation and accurate measurements, errors), which are essential in performing good measurements and extracting meaningful results for the most common applications, is presented in Section 6.4. In Section 6.5 the reader is guided in performing routine data analysis, using thermoplastic poly(methyl methacrylate) as a reference material that is readily available. The knowledge acquired in these sections is sufficient to navigate through the topics discussed in the following sections, dealing with the application of dielectric analysis in representative thermoplastics (Section 6.6) and thermosets (Section 6.7). The examples discussed therein are illustrative of the interrelation between thermophysical properties of the testing materials, dielectric results, and information collected from other thermal methods. Section 6.8 briefly presents several up-to-date commercially available systems for dielectric measurements in the frequency, temperature, or time domain. Last, but not least, the References section, although far from being complete, provides sources detailed enough so as to guide newcomers in the field of dielectric-related thermal analysis to a deeper level of knowledge.

Since physicists and collaborating materials research laboratories publish most of the dielectric reports on polymer dielectrics, much of the experimental information presented in this chapter is plotted isothermally (with temperature as a *parameter*) as a function of frequency (the *variable*), in the form of dielectric loss $\varepsilon''(f)$ or permittivity $\varepsilon'(f)$ graphs. This presentation may seem odd for a thermal analyst with a background in chemistry, who is more used to temperature scans (*isochronal* experiments). In dynamic mechanical analysis, for example, this preference is a result of the frequency limitations and of the stronger dependence of the viscoelastic properties of polymers on temperature, than on time. So, for both reasons it is possible to observe a wider range of viscoelastic behavior with a temperature scan than with a frequency scan. However, the use of isothermal scans is essential for extracting the most detailed information from dielectric data. For example, in an isothermal dielectric experiment the measurements are always done in thermal equilibrium, in contrast to dielectric analysis based on temperature scans, including thermally stimulated current analysis. Isochronal data in the temperature domain are difficult to analyze quantitatively, because the relaxation strength and the distribution of relaxation times are all generally temperature-dependent. Therefore, the readers may need to adjust their way of thinking when reading this dielectric chapter of a thermal analysis book. Rather than being a problematic situation, this shift will most probably be highly educational for chemists, since in practice the principal visually perceptible differentiation between the two presentations is the inversion in the position of the signals. In the isochronal DEA runs, local relaxations, cooperative transitions, interfacial polarizations, conductivity, and electrode polarization signals will appear in this order as the temperature increases, and in reverse order in an isothermal measurement. Varying the temperature of the run will merely shift different mechanisms within the studied frequency window.

6.2. THEORY AND BACKGROUND OF DIELECTRIC ANALYSIS

6.2.1. Frequency-Dependent Response of Dielectrics: Basic Aspects

Careful measurement and analysis of dielectric data necessitates a well-defined environment and configuration of the material under study, along with the use of the appropriate mathematical formulation for effective data analysis. The electrical properties of a dielectric, placed in a parallel-plate capacitor—the most common sample configuration—under an alternating voltage of angular frequency ω, can be described by measuring its capacitance (C) and conductance (G), which are defined by the following equations

$$C(\omega) = \varepsilon_0 \varepsilon'(\omega) \frac{S}{d} \quad (6.1)$$

$$G(\omega) = \varepsilon_0 \omega \varepsilon''(\omega) \frac{S}{d} \quad (6.2)$$

where S and d are the sample's cross-sectional area and thickness, respectively, and ε_0 is the permittivity of vacuum. Such an experiment makes available the frequency dependence of the parameters ε' and ε'', which are, respectively, the real and imaginary parts of the (relative) complex dielectric permittivity function

$$\varepsilon^*(\omega) = \varepsilon'(\omega) - i\varepsilon''(\omega) \quad (6.3)$$

where $i^2 = 1$. The real part of the permittivity expresses the ability of the dielectric medium to store energy, and consists of the contributions of vacuum—which is necessarily real, since there can be no loss associated with vacuum—and of the real part of the susceptibility of the material medium itself. On the other hand, the imaginary component describes the energy losses due entirely to the material medium (Jonscher 1983; Macdonald 1987). The variation of these parameters with frequency—known as *dispersion*—is an essential property of all dielectric materials. The low value of the "unrelaxed" permittivity (ε_u or ε_∞; i.e., ε' measured at $f \to \infty$) is the baseline of the dielectric permittivity without the contributions of dipole orientation and free charge migration. For example, the atomic and electronic polarization are reflected in the value of the unrelaxed dielectric constant. The usually much higher value recorded for the "relaxed" permittivity (ε_r or ε_0; i.e., ε' measured at $f \to 0$) corresponds to the higher degree of dipole orientation attainable in the sample under given conditions. The difference $\Delta\varepsilon = \varepsilon_r - \varepsilon_u$ (i.e., the permittivity drop in the limits of a particular loss peak) is referred to as the *oscillator* or *dielectric strength*, or as the *dielectric increment*, and practically describes the temperature-dependent contribution of the mechanism in the overall dielectric polarizability of the material. Normally, a high population of a very mobile group of dipoles accounts for dielectric signals of high strength.

Besides frequency, time is another critical parameter for the description of dielectric phenomena in polymers. The mathematical analysis of the time-dependent response is based largely on the (macroscopic) relaxation function $\Phi(t)$, which describes the change of the system after the removal of an applied stimulus (in the present case, the electric field, in the case of DMA, the stress). Dipole orientation, which follows the application (at time $t = 0$) of a static electric field E_s to a polymer, is not instantaneous. For example, in the case of a *linear* dielectric, the dielectric induction (D) at time t can be written as

$$D(t) = \varepsilon_u \varepsilon_0 E_s + \varepsilon_0 (\varepsilon_r - \varepsilon_u) \Phi(t) E_s \qquad (6.4)$$

where the first term on the RHS of the equation expresses the instantaneous response of the system to the electric field and the second term reflects the time-dependent contribution of polarization (Fig. 6.1). Studies of the time adjustment of functions, such as the dielectric induction and dielectric polarization (P_d), after the application of the electric field, are frequently used to appraise aspects (cooperativity, interactions, distribution parameters, etc.) related to the complexity of the molecular or electric charge motions involved in the polarization phenomenon.

Dielectric experiments that involve studies of the relaxation function $\Phi(t)$ are denoted as time-domain experiments, while those related to the complex permittivity function $\varepsilon^*(\omega)$ are considered dynamic experiments. The latter have the advantage of introducing an experimental timescale ($\sim 1/\omega$), which, when compared to the different intrinsic timescales of the system (the relaxation time τ), provides useful information on the molecular level. In terms of the single-relaxation-time model of Debye (1921, 1929), the complex permittivity for a dipolar mechanism can be written as follows:

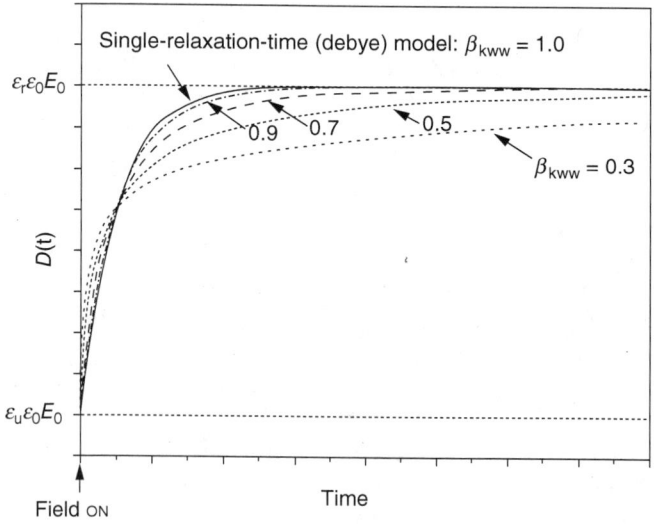

Figure 6.1. Time-dependent adjustment of the dielectric induction D, following application (at $t = 0$) of an electric field E_s to the sample, in the case of a *Debye* process, where $\Phi(t) = \exp(-t/\tau)$, or processes described by the *Köhlrausch–Williams–Watts* (KWW) stretched exponential function, where $\Phi(t) = \exp[-(t/\tau)^{\beta_{KWW}}]$ with $0 < \beta_{KWW} \leq 1$.

$$\varepsilon^*(\omega) = \varepsilon_u + \frac{\varepsilon_r - \varepsilon_u}{1 + i\omega\tau} = \varepsilon_u + \frac{\varepsilon_r - \varepsilon_u}{1 + \omega^2\tau^2} - i\frac{(\varepsilon_r - \varepsilon_u)\omega\tau}{1 + \omega^2\tau^2} \quad (6.5)$$

Direct-current (DC) conductivity contributes only to the imaginary part of permittivity, by a term $\sigma_{DC}/\varepsilon_0\omega$, where σ_{DC} is the specific DC conductivity of the material. The difference in the frequency dependence of the dipolar and the DC conductivity terms allows their experimental separation. The analysis of simple permittivity plots provides estimates for the dielectric strength ($\Delta\varepsilon = \varepsilon_r - \varepsilon_u$) and the relaxation time of each relaxation process, along with the value for σ_{DC} at the temperature of the experiment.

Equation (6.5) is sufficient to analyze relaxation processes in gases and some liquids but fails to account for the majority of experimental results in complex systems (macromolecular substances, inorganic–organic composites, etc.), characterized by broader and usually asymmetric dielectric loss peaks. The origin of the departures from the ideal single-relaxation-time response remains undetermined, despite the variety of models that have been proposed (Jonscher 1983). It is usually considered that there exists a distribution of relaxation times [$g(\log\tau)$], generally with the assumption of a distribution in the activation energies. Different forms for $g(\log\tau)$ may be used to fit experimental data, or $g(\log\tau)$ may be determined numerically from $\varepsilon'(\omega)$ or $\varepsilon''(\omega)$ data (Schäfer et al. 1996). Routinely, the dielectric response is described quite well by fitting permittivity plots by a number of empirical expressions in the frequency or time domain. In the analysis of polymers, the Havriliak and Negami (1967) (HN) expression finds the widest application:

$$\varepsilon^*(\omega) = \varepsilon_u + \frac{\Delta\varepsilon}{\left\{1 + (i\omega\tau_{HN})^{1-\alpha_{HN}}\right\}^{\beta_{HN}}} \quad (6.6)$$

This function is assumed to represent the superposition of many Debye functions [Eq. (6.5)] with various relaxation times (Bottcher and Bordewijk 1978). In terms of the Havriliak–Negami model, a complete description of a real (non-Debye) relaxation process in a polymer requires calculation of four parameters: the dielectric strength ($\Delta\varepsilon$), a parameter related to the relaxation time of the process at the temperature of the scan (τ_{HN}), and two shape parameters ($0 \leq \alpha_{HN} < 1$ and $0 < \beta_{HN} \leq 1$). The latter describe the width and the asymmetry of the loss peak, respectively, but lack a physical meaning.

6.2.2. Common Representations of Dielectric Data

The most common and meaningful representation of the dielectric results in polymers is in the form of the complex intensive parameters: permittivity $\varepsilon^*(\omega)$, electric modulus $M^*(\omega)$, conductivity $\sigma^*(\omega)$, and resistivity $\rho^*(\omega) = 1/\sigma^*(\omega)$. All these functions are important, especially because of their different dependence on and weighting with frequency. Permittivity plots are the customary representation of dielectric data. The real part of the dielectric permittivity [$\varepsilon'(\omega)$], shows a dispersion in the frequency domain, falling from ε_r to ε_u with increasing frequency (Fig. 6.2a). The experimentally determined values of ε_r, ε_u, and the relaxation time $\tau(T)$ are characteristic of each polarization mechanism. Dielectric losses [$\varepsilon''(\omega)$], in the idealized case described by Eq. (6.5)], exhibit a bell-shaped curve with a full width at half-height of 1.14 decades (in a logarithmic scale) and a maximum occurring at $\log(\omega_{max}\tau) = 0$ (Fig. 6.2a). The loss peaks for most polymeric materials are broad and distorted (asymmetric).

The dissipation factor or loss tangent {tan [$\delta(\omega)$] = $\varepsilon''(\omega)/\varepsilon'(\omega)$}, is also frequently used to characterize polymers for many electrical and engineering applications (Fig. 6.2b). δ is the phase lag between the alternating electric field $E(t) = E_0 e^{i\omega t}$ and the dielectric induction $D(t) = D_0 e^{i(\omega t - \delta)}$, and is a measure of system's polarization inertia with respect to the electric stimulus. One practical advantage of using tan δ to describe a dielectric material is its lack of dependence on sample geometry. This can be of great value in cases where the geometry of the sample is not known (e.g., semiconductor p–n junctions or electrolytic capacitors). Its values range from less than 10^{-4} for low-loss polymers to approximately 1 for high-loss materials. The tan δ function is commonly used as a measure of the sensitivity of dielectric equipment. Spectrometers with tan δ values around 10^{-3} are considered sufficient for the acquisition of data for most polymers, although instruments with tan δ as low as 3×10^{-5} and resolution of 10^{-5} are now available (see Section 6.8).

Another important representation of DEA data is the electric modulus, which has been particularly popularized by McCrum et al. (1967) in their classic book on relaxation in polymeric materials. These authors defined the electric modulus as the reciprocal of the complex relative permittivity

$$M^*(\omega) = \frac{1}{\varepsilon^*(\omega)} = M'(\omega) + iM''(\omega) = M_\infty + \frac{M_r - M_u}{1 + i\omega\tau_M} \tag{6.7}$$

in analogy with the mechanical shear and tensile moduli being the complex reciprocals of the shear and tensile compliances. For a Debye-type relaxation, the electrical loss modulus M'' gives a bell-shaped curve that peaks at $\omega_{max}\tau_M = 1$, with $\tau_M = (\varepsilon_u/\varepsilon_r)\tau$ (Fig. 6.2b). Macedo et al. (1972), and more recently Hodge et al. (2005), stressed the effectiveness of the electric modulus formalism when dealing with conducting materials, for the reason that it emphasizes the bulk properties at the expense of electrode polarization. $M''(\omega)$ plots effectively suppress large contributions of nonlocal relaxations at low frequencies. Very important is the suppression of the strong DC conductivity signal, appearing in the ε'' spectrum (Fig. 6.3a), when the electric modulus formulation is used (Fig. 6.3b). For large values of $\Delta\varepsilon$ the M'' spectrum appears at a much higher frequency than the tan δ spectrum (Fig. 6.2b), which in turn appears at an elevated frequency compared to the ε'' spectrum (Fig. 6.2a). This behavior (shift) is reversed in isochronal presentations of the dielectric results. It is useful to

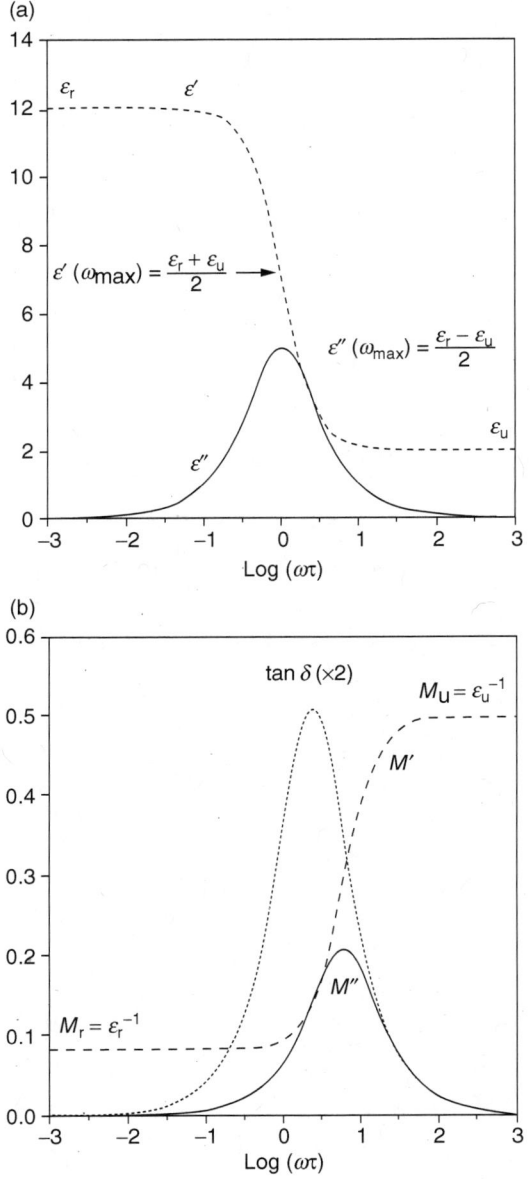

Figure 6.2. Schematic representation of the signal ascribed to a single-relaxation-time (Debye-type) mechanism ($\varepsilon_u = 2$, $\varepsilon_r = 12$, and $\tau = 1$ s): (a) ε' and ε'' versus $\log(\omega\tau)$; (b) M', M'' and $\tan\delta$ versus $\log(\omega\tau)$.

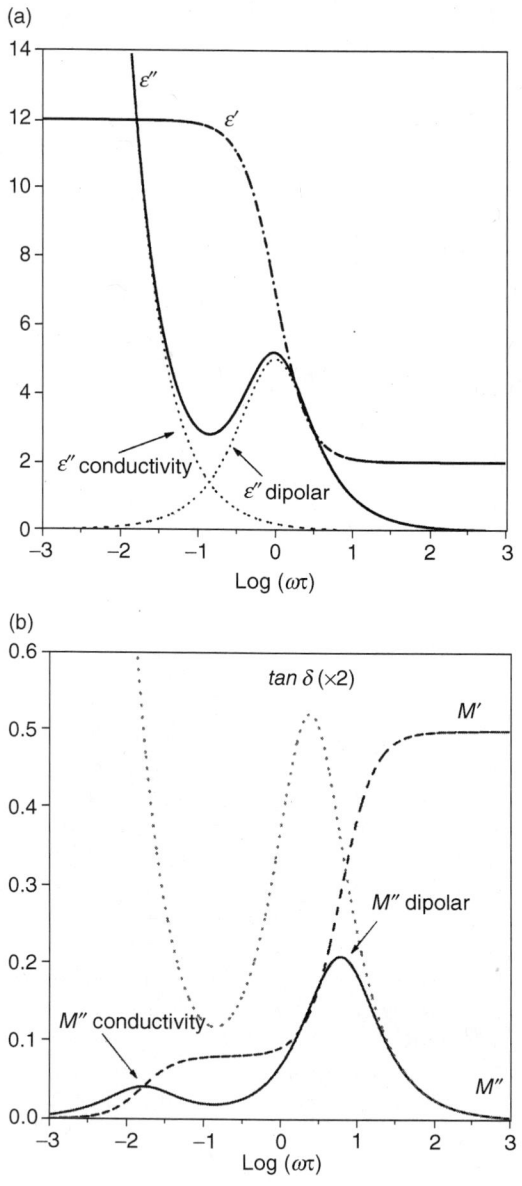

Figure 6.3. Schematic representation of (a) ε' and ε'' and (b) M', M'', and tan δ versus log($\omega\tau$) in the case of a superposition of a Debye-type mechanism ($\varepsilon_u = 2$, $\varepsilon_r = 12$, $\tau = 1\,\text{s}$) and a DC conductivity signal in its low-frequency side ($\sigma_{DC}/\varepsilon_0 = 0.2$).

note correlations between the structures of dielectric and dynamic mechanical relaxation spectra; the ε'' spectrum bears analogies with the spectrum of the imaginary part of the complex compliance J^* $[J^*(\omega) = J'(\omega) + iJ''(\omega)]$, M'' with the imaginary part of the complex modulus $E^* = 1/J^*$ $[E^*(\omega) = E'(\omega) + iE''(\omega)]$, and $\tan\delta = \omega''/\omega'$ with the mechanical loss tangent $\tan\delta = E''/E'$.

The dielectric data are usually presented either in *isothermal plots*, that is, as a function of frequency (the *variable*) at a constant temperature (the *parameter*), or as a function of temperature at a constant frequency (*isochronal plots*). It is also possible to perform step isothermal experiments for multifrequency DEA and DMA measurements on nonreactive polymer systems; in that way, the same data can be plotted as multifrequency curves versus temperature as well as multitemperature curves versus frequency. Customarily, most chemists perform temperature scans at a single frequency, or at several frequencies in step isothermal experiments, to study phase transitions in a polymer as this type of measurement provides direct information pertaining to relaxation strengths and transition temperatures (Fig. 6.4) and permits direct comparison with mechanical relaxation spectra (as well as with dilatometric and differential scanning calorimetric curves), which are essentially recorded as a function of the temperature. However, there appears to be a strong preference, among physicists and chemical engineers, in scanning frequency in the isothermal mode. Such spectra can be easily analyzed to determine relaxation characteristics and electrical properties of the polymer at the temperature of the scan (see Section 6.5).

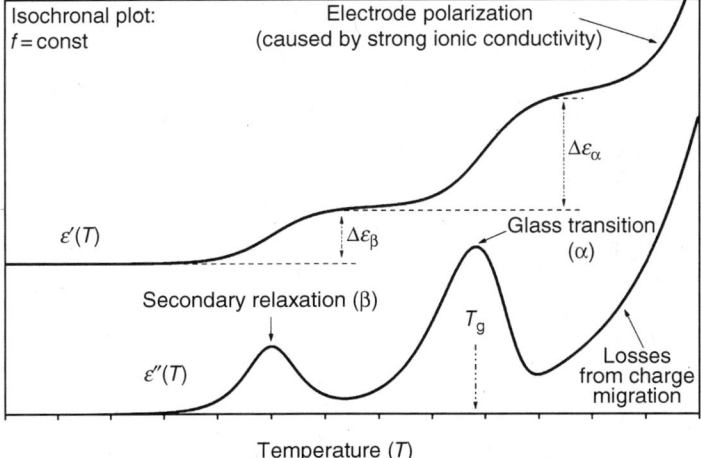

Figure 6.4. Generic behavior of temperature dependence of permittivity components (ε', ε'') recorded for an amorphous polymer with considerable ionic conductivity. The higher the relaxation time of the mechanism, the higher the temperature range at which the corresponding signal appears in this isochronal recording. The signals present in this spectrum will shift to lower temperatures by decreasing the frequency of the alternating electric field.

6.2.3. The Dielectric Relaxation Time $\tau(T)$

The relaxation time τ, probably the most important parameter in DEA studies, is connected with the transition probabilities between consecutive minimum-energy configurations of the dipoles, namely, between minima of the potential valleys (Hedvig 1977). The rotational motion of each dipole presents a characteristic relaxation time; those exhibiting rotations more effectively hindered by their environment will show higher τ, and therefore their motion will be activated and recorded at higher temperatures (Fig. 6.4). Typically, the relaxation time will be temperature-dependent, and the precise functional form of this dependence is critical information provided by DEA. Considering the case of dipoles floating in a viscous fluid [Debye model (Debye 1921)], the relaxation time at a particular temperature will be the same for each unit and strongly related to the viscosity of the environment. For relaxations described by the ideal rotational friction model of Debye (1921) or the potential barrier model of Fröhlich (1958), an Arrhenius-like equation represents the $\tau(T)$ dependence

$$\tau(T) = \tau_0 \exp\left(\frac{E}{kT}\right) \tag{6.8}$$

where τ_0 is the preexponential factor and k is Boltzmann's constant. In the case of polymers, this is true when uncorrelated rotations of sidegroups or local motions of short segments are involved (i.e., for the β- and γ-relaxation mechanisms in amorphous polymers). The apparent activation energy of the relaxation process E, related to the slope of the $\log \tau$- or $\log(f_{max})$-$1/T$ plot (Arrhenius diagram, Fig. 6.5), depends on both the internal rotation barriers and the environment of the rotating unit. For truly local processes due to isolated polar units, τ_0 should be on the order of 1 ns. Values lower than 10^{-16} s imply an activation entropy [$\Delta S^* = k \ln(kT/h) - k \ln\tau_{(0)}$, where h is Planck's constant] greater than zero, which in turn indicates more complex molecular motions (e.g., cooperativity). In the latter case, E is found in the range 100–200 kJ/mol, as frequently observed for the highly restricted conformational mainchain motions.

Relaxation mechanisms of dipoles located in dissimilar environments, or originating from complex forms of molecular or ionic motion, usually exhibit curved Arrhenius diagrams. This curvature is usually interpreted in terms of the semiempirical Williams–Landel–Ferry (WLF) equation (Williams et al. 1955)

$$\tau(T) = \tau(T_r) \exp\left(\frac{-C_1(T - T_r)}{C_2 + (T - T_r)}\right) \tag{6.9}$$

where C_1 and C_2 are the so-called WLF parameters. T_r is a reference temperature, often assigned the value of T_g. The WLF equation is a consequence of a

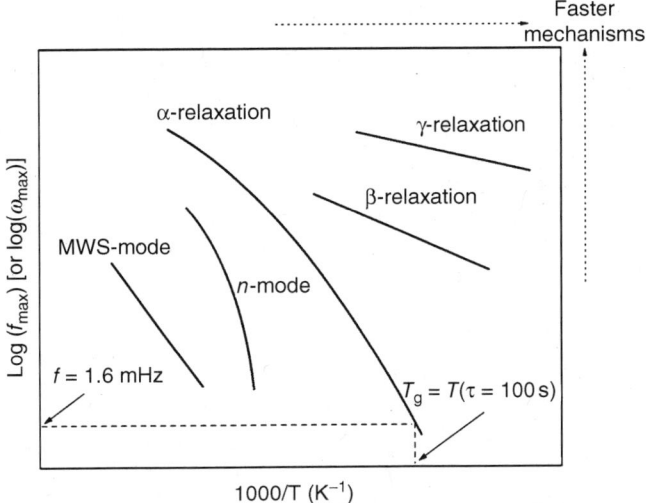

Figure 6.5. Schematic of an Arrhenius plot for mechanisms commonly observed in polymers. The lines correspond to Arrhenius [Eq. (6.8); for γ, β, and Maxwell–Wagner–Sillars (MWS) relaxations] and Vogel–Tammann–Fulcher–Hesse [VTFH; Eq. (6.10)], for α and normal-mode (n-) relaxation] temperature dependences for the relaxation time $\tau(T)$. Relaxations ascribed to small, highly mobile, dipolar units appear in the upper right side of the plot, while those originating from bulky dipolar segments, slowly moving ions, and MWS mechanisms are located in the lower-left part of the plot.

general principle, known as the *time–temperature superposition principle*, and describes, for a wide range of materials, the effect of the free volume on the relaxation behavior of polymers in the temperature range between the glass transition temperature T_g and $T_g + 100°$. For segmental (α) relaxations and similarly complex relaxation signals, the Vogel–Tammann–Fulcher–Hesse (VTFH) equation is widely used in dielectric literature (Vogel 1921; Tammann and Hesse 1926; Fulcher 1925):

$$\tau(T) = A \exp\left(\frac{B}{T - T_V}\right) \quad (6.10)$$

Here, A and B are constants and T_V is the so-called Vogel temperature, which is usually located 30–70 degrees below T_g. Despite the fact that the physical meaning of T_V has not been clearly defined [e.g., Cohen and Turnbull (1959); Adams and Gibbs (1965); Donth (1992)], the universality of the VTFH equation near T_g makes clear that T_V is a significant parameter for the dynamics of the glass transition. When dealing with the relaxational motions of long-chain segments (α relaxations) in amorphous polymers, the slope of the Arrhenius plot and the corresponding activation energy attains a maximum near T_g (van Turnhout and Wübbenhorst 2002).

6.2.4. The Origin of Dielectric Response of Polymeric Materials

The polarization of a dielectric submitted to an external electric field may occur by a number of different mechanisms (Figs 6.4 and 6.6), involving either microscopic or macroscopic charge displacement. In the case of polymer dielectrics, the net dipole moment per unit volume (i.e., the polarization) corresponds to the vector summation over all molecular dipole types in the repeating unit, the polymer chain, and over all chains in the system. As expected, the electric stimulus deforms the electronic shell with respect to the atomic nucleus (*electronic polarization*) and creates an induced dipole moment in atoms (*atomic polarization*). However, both processes have very high resonant frequencies that fall within the field of vibrational spectroscopy and are considered instantaneous in dielectric studies. Therefore, the main contribution to polarization phenomena observed in polymers arises from the rotational mobility of permanent group dipole moments (μ) (dipole moments of chemical groups in the chains that have stable configurations during thermal motion) that depend both on temperature and the stereochemical structure of the polymer. Changes in the localized charge density result from structural transformations of the chain, such as isomeric transitions, rotation of sidegroups, and segmental motions.

For long-chain molecules there are different geometric possibilities for the orientation of molecular dipole vectors with respect to the backbone. Following the notation of Stockmayer (1967), polymers are classified as type A (with dipoles fixed parallel to the mainchain, e.g., cis-1,4-polyisoprene and polyethers), type B [with dipole moments rigidly attached perpendicular to the mainchain; e.g., poly(vinyl acetate) and most synthetic polymers], or type C [with a more-or-less flexible polar sidechain; e.g., poly(*n*-alkyl methacrylate)s]. However, a polymer possessing only one type of dipole moment is an exceptional case. The timescale (speed) of each polarization (and subsequent relaxation) process will determine whether this process will be monitored by a particular dielectric technique. Characteristics and fundamental peculiarities of relaxations generally found in polymers are discussed hereafter. Note that cases where finite polarization is present even in the absence of an external field (e.g., the permanent polarization in ferroelectrics) are not considered.

6.2.4.1. Secondary Relaxation Processes in Amorphous Polymers

The relaxations representing secondary transitions in amorphous polymers β, γ, and δ, in order of decreasing temperature in isochronal plots) are made up of uncorrelated motions of molecular units involving a limited number of carbon atoms. As such, their relaxation frequencies (at a given temperature) are high enough to produce absorption bands above the glass transition signal (the α relaxation; Fig. 6.6). The main secondary transition (β relaxation) originates from localized fluctuations of parts of the mainchain, conformational changes in cyclic sidegroups, or hindered rotations of sidegroups or parts of them (Hedvig 1977; Heijboer 1978). This type of local dynamics remains active even

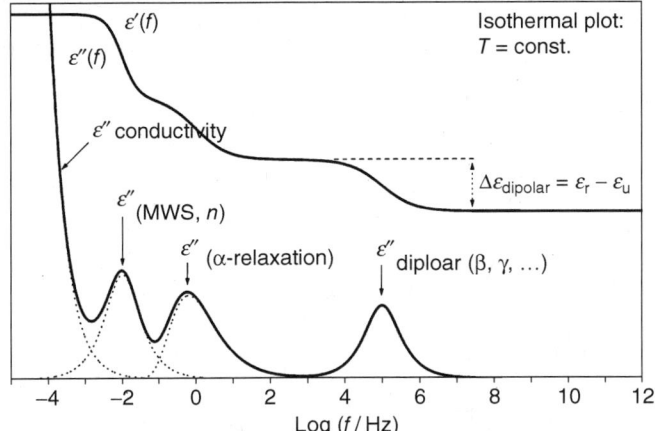

Figure 6.6. Schematic presentation of the frequency dependences of ε' and ε'' for typical dielectric relaxation modes in polymers. The traces of relaxation mechanisms with higher molecular mobility (shorter relaxation time) are recorded at higher frequencies. Compare with the isochronal dielectric spectrum of Fig. 6.4. Considering the temperature dependence of the dielectric strength ($\Delta\varepsilon$), it becomes clear that the $\Delta\varepsilon$ values measured from isothermal or isochronal isothermal spectra alone are not necessarily equal.

when the polymer is in the glassy state, that is, when the large length-scale backbone motions are frozen. Johari and Goldstein (Johari 1973; Johari and Goldstein 1970) regard the stronger secondary transition (also called the *Johari–Goldstein mode*) as a general feature of the amorphous state. There are also theoretical indications that a universal slow β relaxation may exist in all amorphous materials, as the precursor of the cooperative many-molecule dynamics of the α relaxation (Ngai 2005; Ngai and Paluck 2004). The β-relaxation mechanism is a thermally activated process characterized by an Arrhenius temperature dependence of the relaxation times [τ_β; Eq. (6.8)]. The barrier heights E_β represent the potential barrier between two possible equilibrium states (e.g., two different orientations of the polar group relative to the mainchain). These usually vary in the range 20–60 kJ/mol and are determined by the chemical composition and the stereochemical configuration of the chains. In most cases, the dielectric strength of the β relaxation ($\Delta\varepsilon_\beta$) increases with temperature. On the frequency sweep curve, in a logarithmic scale, the $\varepsilon''(\omega)$ band is broad and in most cases symmetrically shaped, with a half-width of three to six decades that generally narrows with increasing temperature. The γ relaxation, of even smaller molecular parts, can often be found at higher frequencies (or temperatures below the β relaxation in isochronal scans). A β relaxation due to isolated molecules or impurities may also be present, showing comparably higher relaxation rates.

Aspects of the analysis of the single ($\alpha\beta$) process, observed in the merging region of the α and β relaxations in several polymers [e.g., in poly(n-alkyl methacrylate)s], are discussed by Garwe et al. (1996) and Williams and Watts (1971). For example, impedance measurements in the frequency range 0.1 mHz–1 GHz were used by Garwe et al. (1996) to monitor the change (by decreasing the temperature) of the α relaxation into the local β relaxation in poly(n-butyl methacrylate) and poly(ethyl methacrylate).

6.2.4.2. The Segmental α-Relaxation Process in Amorphous Polymers

The onset of the micro-Brownian motion of chain segments gives rise to the dynamic glass transition (α) process in amorphous polymers.[1] This involves transformation of the "rigid" supramolecular glass structure to a different structure in the viscoelastic fluid state. The corresponding motional processes proceed in a cooperative manner, which means that a specific segment moves together with its environment. Therefore, the α relaxation involves both intramolecular (connectivity with the mainchain) and intermolecular (coordinated, i.e., cooperative motion with the environment) interactions. The dielectric T_g, customarily defined by convention as $\tau_\alpha(T_{g,diel}) = 100\,s$, corresponds to the temperature at which the loss factor band peaks at $f_{max} = 1.6\,10^{-3}\,Hz$ (see Fig. 6.5). As with the glass transition temperature data determined by other thermal techniques, both the dielectric values ($T_{g,diel}$) and the relaxation time at T_g [$\tau_\alpha(T_g)$] show similar dependences on the average length of the polymer chain and its architecture. For example, for linear chain and star polymers, $T_{g,diel}$ increases with weight average molecular weight (M_w) below the critical value ($\approx 10{,}000\,g/mol$), above which it is nearly constant. For polymer networks, the glass transition temperature increases with increasing crosslink density, referred to as the T_g-conversion relationship (see Sections 2.10 and 6.7), while for ring polymers the dependence is rather complex.

In most cases, the relaxation strength of the dielectric α relaxation ($\Delta\varepsilon_\alpha$) increases with temperature, but the increase is much stronger than theoretical predictions (Díaz-Calleja and Riande 1997), probably as a result of the decrease in cooperativity with increasing temperature. The $\varepsilon''(\omega)$ curve of the α-relaxation displays a relatively broad but asymmetric shape, with half-width of two to five decades (in a logarithmic scale). In isochronal recordings of the $\varepsilon''(T)$ function, the α-relaxation curve is similarly broadened. The Havriliak–Negami function provides adequate fits to the loss peak in the frequency domain. The adjusting parameters α_{HN} and β_{HN} [with $0 < (1 - \alpha_{HN})\beta_{HN} \leq 0.5$ for most polymers compared to $1 - \alpha_{HN} = \beta_{HN} = 1$ for Debye relaxations (Schönhals 1991)] show different dependences on temperature, crystallinity, and crosslink density. A stretched exponential function $\{\Phi(t) = \exp[-(t/\tau)^{\beta_{kww}}]\}$ seems to be the natural way to analyze time-domain results in all amorphous polymers and most glass-forming liquids. The strong departure of the α-relaxation mecha-

---

[1] In semicrystalline polymers the α relaxation is not always the glass transition of the amorphous phase; often it is a crystalline relaxation (also denoted as the α_c relaxation). In such cases, the glass transition is usually denoted as the β or α_a relaxation.

nism from the single-relaxation-time model of Debye ($\beta_{KWW} = 1$; Fig. 6.1) is manifested by the observation that for most polymers the β_{KWW} parameter varies in the range 0.35–0.7. The agreement between frequency-, temperature-, and time-domain results on the nonexponentiallity of the α-relaxation response is noteworthy, and provides direct experimental evidence of the intriguing complexity of the transition mechanism.

6.2.4.3. The Normal-Mode (n) Relaxation Process

The term *normal-mode relaxation* refers to the long-range motions of the end-to-end dipole moment vector along a polymer chain, and thus corresponds to the comparably slow motion of a whole chain (Adachi 1997). This relaxation mode is characteristic of polymers with dipoles fixed parallel to the mainchain (type A polymers). A representative class of such polymers are the polyethers $(-CH_2-CHR-O-)_n$, with R ≠ H [e.g., poly(propylene glycol) (Hayakawa and Adachi 2001); poly(butylene oxide) (Casalini and Roland 2005)], for which, with the exception of a few members [e.g., poly(styrene oxide) (Hirose and Adachi 2005)], a strong normal-mode relaxation signal can be resolved dielectrically. The mean relaxation time of the process, namely, $\langle \tau_n \rangle = \left(\dfrac{1}{2\pi f_{max(n)}} \right)$ [where $f_{max(n)}$ is the peak frequency of the normal-mode (n-relaxation) signal detected in isothermal scans] depends on the molecular weight of the polymer and typically shows a temperature dependence described by the VTFH equation [Eq. (6.10)].

6.2.4.4. Mechanisms Related to Long-Range Charge Migration

The migration of extrinsic (e.g., ionic impurities) and intrinsic (e.g., proton transfer along hydrogen bonds) charges lead to DC conductivity (Varotsos 2005). The relative contribution of these charges on the magnitude of the observed conductivity is determined by, for example, the chemical composition of the system, preparation conditions, sample treatment (e.g., electronic bombardment, corona discharges, and neutron or electron irradiation), ongoing chemical reactions, and physical processes (e.g., aging) in the material. Extrinsic conductivity is sufficiently suppressed in dielectric measurements of thoroughly purified materials and carefully prepared electrical contacts. Such treatments are necessary in order to reveal and study intrinsic relaxation phenomena in the material, which otherwise would be masked by the strong conductivity effects. This is particularly significant in studies of the glass transition (dielectric α-relaxation signal). The macroscopic motion of space charges is also directed to the electrodes, which may act as a total or partial barrier (blocking or non-Ohmic contacts), creating a large electric dipole (electrode polarization).

Interfacial polarization, sometimes referred to as *Maxwell–Wagner–Sillars* (MWS) polarization, is a characteristic bulk phenomenon in polymer systems with a heterogeneous structure. This kind of polarization is due to the buildup of charged layers at the interface, resulting from unequal conduction currents within the constituent phases (i.e., a *conductive* and an *insulating* phase). The general requirement for recording this effect is that the conducting phase be isolated from a direct DC path between the electrodes, that is, that the conducting phase be noncontinuously dispersed in the nonconducting matrix. Such polarizations exhibit relaxation behavior that is usually barely distinguishable from dipolar relaxations. Moreover, the position, strength, and shape of a MWS signal depend on the volume fraction, geometry (e.g., lamellar, spherical, cylindrical, or ellipsoidal shapes for the conducting inclusions) and conductivity of the dispersed phase as well as on the permittivity of the phases. In isochronal dielectric measurements [e.g., $\varepsilon''(T)$ plots at a fixed low frequency], the MWS signal is expected above the glass transition temperature. In isothermal dielectric frequency scans (Fig. 6.6), the MWS loss band appears in the very low-frequency region, usually submerged in the conductivity losses, and thus only an increase in both the low-frequency dielectric constant and loss is observed. Considering that the characteristics of the MWS relaxation are highly dependent on the morphology of the phase-separated components, the analysis of this dielectric signal can be combined with optical microscopy and related techniques for a more in-depth analysis of the microstructure. It may be considered an advantage of dielectric analysis to record this signal, as this relaxation passes unnoticed in other common thermal techniques (such as DMA and DSC).

6.2.4.5. Relaxations in Semicrystalline Polymers

If in addition to the amorphous phase, a crystalline phase is present in the polymer, the dielectric and electrical responses are dramatically influenced. In the presence of noticeable ion conductivity within the amorphous phase, the two-phase nature of the system leads to typical Maxwell–Wagner–Sillars polarization effects. The dielectric strength of the segmental relaxation, and to a lesser extent the intensity of secondary transitions, weakens with increased crystallinity, due to the dilution effect of the dielectrically weakly active or totally inactive crystals. The glass transition temperature often increases compared to the single amorphous-phase material, as a result of steric hindrances on the chains' motion at the boundaries of the phases (see examples for thermoplastic polymers; in Section 6.6). From a methodological perspective, there is a fraction of noncrystalline material—the so-called rigid amorphous phase—distinguished from crystals and from the mobile (undisturbed) amorphous material, that is not able to exhibit the relaxation characteristics of the liquidlike amorphous regions and thus does not contribute, for instance, to changes in the dielectric permittivity that occur in the glass transition (see Section 2.7 in Chapter 2). Note that in the case of the semicrystalline polymers the segmental relaxation mode is denoted as the β or α_a transition. Broadening of the α_a-relaxation signal in the frequency or time domain is a typical experimental observation, contrary to the subglass processes that show relative insensitivity to the presence of the crystal phase (Boyd and Liu 1997). The overall effect of the cooperative segmental motions that generate the glass transition phenomenon on the physical properties of semicrystalline polymers decreases with increasing crystallinity.

Crystal phase relaxations are seldom observed dielectrically because of selection rules. However, there are some examples of well-studied and well-interpreted relaxations, such as the α_c-relaxation processes in some polymers with a flexible mainchain at temperatures between T_g and the melting point; examples are found in polyethylene, polypropylene, polytetrafluoroethylene, poly(ethylene terephthalate) (Pratt and Smith 2001), and poly(vinylidene fluoride) (Neagu et al. 2002). The study of this general process is important as it is related to relevant material's properties, such as creep, crystallization, extrudability, and drawability (McGrum and Morris 1964). Furthermore, at the crystalline melting point (T_m) an order–disorder transition takes place involving a large enthalpy and entropy change due to a destruction of long-range order. The corresponding dielectric signal can be studied in parallel with its mechanical and calorimetric counterparts, in order to acquire a detailed insight into the morphology and microstructure of the material. At lower temperatures, several motional processes may also be recorded dielectrically (Hedvig 1977). These include relaxations related to the mobility of groups at the interfaces or at lattice defects, transitions of one crystal form into another, as well as relaxations due to local motions (vibration or rotation) of groups of the mainchain arranged in the crystal lattice. Among other applications, one should make reference to the application of DEA for the evaluation of isothermal crystallization on the basis of the relative decrease of the segmental-mode dielectric loss maximum (Sawada and Ishida 1975).

6.2.4.6. Effects of Moisture Absorption on Relaxation Activity of Polymers For many practical purposes (e.g., electrical insulation and outdoor applications) it is important to examine the effects of water sorption on the electrical, mechanical, and thermal behavior of a polymer. The highly polar water molecules confined in a matrix are usually either caged in structural defects or bound by hydrogen bonds. Depending on the state of bonding, water molecules are categorized as clustered (free), loosely bound, or tightly bound. The dielectric activity of water extends over wide spectral ranges, as a rule overlapping with the relaxation response of the host polymer. Free water has a relaxation time of the order of 10^{-10}–10^{-11} s, between 0 °C and 100 °C, while it increases to 10^{-2}–10^{-8} s for absorbed or bound water phases [e.g., see data for PET by Lightfoot and Xu (1993) and for Nylon 6 by Frank et al. (1996) and Laredo and Hernández (1997)]. Dielectric analysis and especially thermally stimulated current (TSC; see Section 6.3) has provided analytical insight on the state of bonding and the environment of water and its bonding states (Frank et al. 1996). Water sorbed up to a certain critical level is considered to be firmly attached (i.e., dielectrically inactive) to the macromolecule polar bonding sites, while for higher hydration levels the increase in the freedom of motion of the bound water molecules has been verified by observations of sudden discontinuities in TSC plots (Anagnostopoulou-Konsta et al. 1991).

In general, water increases the dielectric constant and the dissipation factor of a polymer, limiting in that way its potential applicability as a low-dielectric component (Shieh et al. 2004). A deterioration of several thermomechanical properties is also common; the most typical example is a decrease of the glass transition temperature (plasticization). Hydration may either facilitate or hinder (antiplasticize) local motions, depending on the position of the water molecule relative to the chain segment involved and the nature or strength of their interaction. For example, both the relaxation strength and rate of a noncooperative process are expected to increase following swelling of a polymer. Dispersion of water within a polymer causes ionization of ionic impurities (e.g., of Cl⁻), which promotes corrosion of integrated circuits incorporating polymeric components. The concomitant enhancement of conductivity, through the development of a strong protonic conduction mechanism and intensification of parallel ion conduction processes, deteriorates the electric insulation properties. Therefore, in the interpretation of dielectric results related to relaxation processes and phenomena in biomacromolecules and other biological substances, great care should be taken to separate the evolution of intrinsic properties/phenomena (crosslinking, denaturation, aging processes, etc.) from moisture transport effects (Lefebvre et al. 2006).

6.2.5. Attempting Peak Assignments in Dielectric Spectrum of a Polymer

What appears to be extremely challenging, especially for a thermal analyst new in the field of polymer dielectrics, is the assignment of the peaks in an isochronal or isothermal dielectric plot to particular types of molecular motion and, if possible, to specific relaxing polar entities. For this purpose it is worth drawing the researcher's attention to a few criteria, rules, or procedures considered necessary for categorizing the contents of a dielectric spectrum. The reader should bear in mind that there might still be a few cases (e.g., dielectric studies of very complex organic materials and multiphase inorganic–organic composites) where some of the suggested actions may perhaps fall short in providing conclusive results. Without a doubt, the first action should be the *inspection of the general structure of the recorded dielectric plot*, since the location, intensity, and number of the relaxation signals usually follow well-described trends. For instance, in an amorphous polymer the relaxations customarily appear in the following order of increasing transition temperature (or decreasing frequency): (1) several δ, γ, and/or β relaxations, depending on the number of various dipole groups (with usually symmetric $\varepsilon''(T)$ or $\varepsilon''(f)$ bands); (2) an α relaxation (usually an asymmetric band); and (3) conductivity signals (see Figs. 6.4 and 6.6). The crystalline phase within a semicrystalline polymer as a rule adds to the spectrum (4) an interfacial polarization (MWS) signal, at a temperature above the α_a transition (the amorphous-phase glass transition signal), and (5) a crystalline phase transition (α_c transition) at even higher temperatures. Most of these signals exhibit distinctive changes with variations in crystallinity. For example, the α_a-relaxation signal shifts to higher temperatures (lower frequencies) and weakens in strength with increasing crystallinity, and similar response also characterizes dielectric signals related to macroscopic charge transport (conductivity). Structural transitions are particularly sensitive to thermal pretreatment, in contrast to relaxations involving local motions of polar groups or segments that appear relatively insensitive to such treatments.

Mathematical analysis of the experimental data, such as, through the construction of Arrhenius plots (Fig. 6.5), adds valuable qualitative and quantitative information. For instance, analysis of the temperature dependence of the relaxation times of a relaxation mechanism will reveal its cooperative character [VTFH relationship; Eq. (6.10)] if the signal is related to a segmental relaxation mode (glass transition) or other structural transitions (i.e., crystalline melting). On the contrary, most relaxations in the glassy state of a polymer (at $T < T_g$), namely, relaxations involving local molecular motions, will exhibit an Arrhenius-type temperature dependence of their dielectric relaxation times [Eq. (6.8)] and a temperature-independent activation energy. In addition, while the dielectric strength of local relaxations (e.g., $\Delta\varepsilon_\beta$) generally increases monotonically with temperature (i.e., no maximum is observed at the transition temperature, T_β), the dielectric strength estimates for segmental relaxations ($\Delta\varepsilon_\alpha$) reach a maximum above the glass transition temperature (T_g or equivalently T_α).

In general, *comparisons of the dielectric spectra with published literature on related materials* are used to resolve important questions. Particularly valuable for assigning measured signals to rotational or crankshaft motions of specific structural groups (β, γ, or δ relaxations) are comparisons between dielectric spectra of polymers with analogous chemical structure or composition. This is possible since the rotational mobility of polar groups ($-OR, -COR, -OCOR, -OCH_2OH, -COOH, -CN$, etc), commonly included in the structure of different polymers, generates dielectric signals at the same or slightly different transition temperatures. Furthermore, DSC or DMA may also be used complementary to DEA to detect the glass transition temperature and thus distinguish the dielectric α-relaxation band from other signals present in a complex spectrum. No significant changes in specific volume or enthalpy are observed for local (secondary) relaxations, contrary to the behavior recorded for structural transitions. Comparisons between isochronal dielectric and dynamic mechanical spectra obtained for the material under study will be particularly useful, especially if one considers that the MWS polarization and other conductivity-related effects are absent in the DMA spectrum. Attributions of the isolated relaxation signals to specific motion at the submolecular level are also supported by IR and NMR data on the vibrational and relaxational activity of chain segments and sidegroups [e.g., by comparing spin–lattice relaxation data, analyzed through the Vogel–Tammann–Fulcher–Hesse $\tau(T)$ function, with the dielectric relaxation times]. Undoubtedly, knowledge of the chemical composition and some morphological characteristics of the material, in addition to its thermal history, play a considerable role in successfully completing this task.

6.3. DIELECTRIC TECHNIQUES

More recent developments in electronics allow the possibility of measuring complex dielectric parameters in a broad frequency range (~10 µHz–30 GHz) with a single DEA instrument. For example, spectrometers from Novocontrol Instruments are available in the ranges 1 µHz–10 MHz and 1 MHz–1 GHz and higher frequencies in the microwave region. The different measuring techniques and sample configurations (solid samples, liquids, powder or granular specimens, ultrathin films, etc.) usually require different types of dielectric cells or fixtures. With the availability of various techniques and the analogous high-precision dielectric systems, it has become possible to study both the very fast molecular motions in organic liquids, soft reacting/polymerizing systems, aqueous solutions, ferroelectrics and liquid crystals, and the very slow motions in supercooled liquids and glassy polymers (Kremer and Schönhals 2002). Nonetheless, it is important to remember that DEA is limited to materials with a reasonable concentration of dipole groups or trace ions, while the low-frequency window of the technique cannot be used to study electrically conductive materials (e.g., carbon/graphite composites). Dielectric analysis can be applied to materials with specific conductivity from 10^7 to 10^{-16} S/m.

Aspects of common DEA techniques are presented in the following paragraphs, with emphasis on the instrumentation that is likely to be used in a typical thermal analysis laboratory; examples are Novocontrol Spectrometers (BDS), TA Instruments (DEA 2970, although the production of this instrument has been discontinued), and Polymer Laboratories (DETA, this instrument is not available any more either). "Dielectrometers" (Micromet Instruments, now owned by Netzsch), used for in-process cure monitoring of thermosets, operate on similar principles. Features of the sensitive thermally stimulated current technique, and its comparison to DEA, are also discussed.

6.3.1. *Slow* Time-Domain Dielectric Spectroscopy: Isothermal Charging/Discharging Currents

The lowest frequency range (~10^{-5}–10 Hz) can be covered by DC transient methods. In the isothermal charging/discharging currents method, a step voltage (V_0) is applied to the sample and the (decreasing) charging current is recorded as a function of time, until a stationary level corresponding to the ohmic conductivity is reached. When the voltage is switched off, a similar current response of opposite sign (discharging current, I_D) is recorded. The sample's complex permittivity can be then calculated by means of a discrete Fourier-type transformation:

$$\varepsilon^*(\omega) = \frac{1}{C_0 V_0} \int_0^\infty I_D(t) e^{-i\omega t} dt \qquad (6.11a)$$

Among the alternative simplified procedures [e.g., see examples discussed by Neagu and Neagu (2000)], the one most frequently used for passing from the time-domain experimental data into a frequency-domain representation of the dielectric losses (ε'') is that originally proposed by Hamon (1952). This can be expressed as

$$\varepsilon''(\omega) \cong \frac{I_D(t)t}{0.63 C_0 V_0} \qquad (6.11b)$$

where C_0 is the geometric capacitance of the capacitor and $\omega = 0.63/t$ (Hedvig 1977). Since the reversible anomalous isothermal charging or discharging currents are most easily measured for different polymers for large times ($t > 0.1\,\text{s}$), this method is best adapted to the evaluation of the permittivity components ($\varepsilon', \varepsilon''$) at frequencies below 10 mHz, where typical AC dielectric techniques (e.g., using frequency response analysis) would be sometimes tedious or unfeasible. A disadvantage of this technique is its high sensitivity to electrical disturbance, such as noise and leakage currents. Despite the moderate accuracy of the method, it still finds wide application, due mostly to the simplicity and low cost of the apparatus. The equipment is similar to those used for the thermally stimulated current technique (see Sections 6.4.3 and 6.8.2).

6.3.2. Dielectric Analysis Based on Lumped Circuit Methods: Frequency Response and Impedance Analysis

Contrary to the previously described time-domain measurement, where a short voltage pulse excites the sample, frequency-domain dielectric analysis uses a harmonic signal for sample excitation. Frequency response analysis (FRA) (100 µHz–10 MHz) and impedance analysis (100 Hz–10 MHz) constitute the most important experimental techniques for characterizing the dynamics of molecular mobility and ion transport in polymers. Commercial analyzers used in such experiments measure the complex impedance $Z^*(\omega) = Z'(\omega) + iZ''(\omega)$ between the electrical ports of a system under test, at a given temperature, dependent on frequency $\omega/2\pi$. *Impedance* is defined as the total resistance of the system to the flow of an alternating current at a given frequency, and given its value several other important functions can be obtained [e.g., the complex permittivity, $\varepsilon^*(\omega) = -i / \omega Z^*(\omega) C_0$ (Macdonald 1987)].

The principle of the FRA measurement is illustrated in Fig. 6.7a. The AC voltage $V_1(\omega)$ is applied to the sample by the generator, and the resistor R converts the sample current $I_S(\omega)$ into voltage $V_2(\omega)$. The amplitudes and the phases of voltages $V_1(\omega)$ and $V_2(\omega)$ are measured by two phase-sensitive

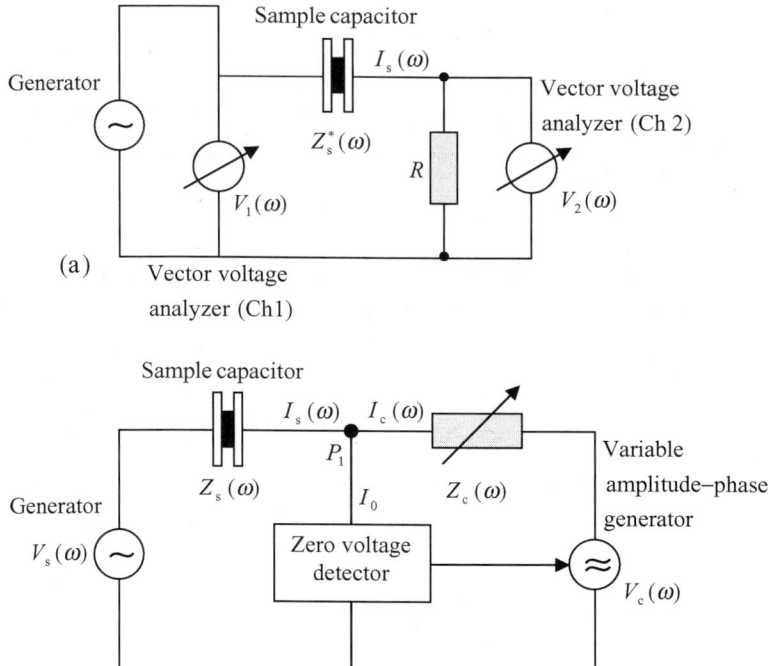

Figure 6.7. Principle of a dielectric measurement using (a) frequency response analysis or (b) AC impedance bridges.

voltage meters (vector voltmeters), and the complex impedance of the sample is calculated using the following relationship:

$$Z_s^*(\omega) = \frac{V_s(\omega)}{I_s(\omega)} = R\left(\frac{V_1(\omega)}{V_2(\omega)} - 1\right) \tag{6.12}$$

Commercial frequency response analyzers are based on either digital lock-in amplifiers (e.g., Stanford Research SR 850, frequency range 10^{-3}–10^5 Hz) or gain-phase analyzers (e.g., Solartron SI 1260A, frequency range 10^{-5}–10^7 Hz; TA Instruments DEA 2970, frequency range 3×10^{-3}–10^5 Hz). For high-accuracy measurements an active interface can also be used [e.g., the Chelsea dielectric interface (CDI) or the broadband dielectric converter (BDC) by Novocontrol]. Frequency response analysis is probably the most powerful and widespread technique, as it is insensitive to noise or nonlinear distortions. Currently available spectrometers allow comparison of each sample impedance point with a built-in variable reference capacity, which improves the accuracy in phase resolution. Errors resulting from the measuring system (e.g., accuracy of the signal analyzer and influence of the cables and sample cell) strongly depend on the frequency of the measurement and the actual sample capacity. Such limitations are usually adequately described in instruction manuals supplied by the manufacturers.

Alternating-current impedance bridges (operational range from 10 Hz to ~1 MHz and resolution $\tan\delta \approx 10^{-3}$) can be seen as an economical alternative to frequency response analysis, for applications in which low-frequency measurements are not important. General impedance analyzers (see Section 6.8.1), based on autobalancing bridges or current/voltage (I/V) methods, are convenient for polymers of sufficient conductivity. In the schematic of an AC impedance bridge, shown in Fig. 6.7b, we find the sample with impedance $Z_s(\omega)$ and the adjustable compensation impedance $Z_c(\omega)$. The generator drives the sample with the fixed and known AC voltage $V_s(\omega)$, which causes the current $I_s(\omega)$ to flow toward the right-hand side of the bridge. The variable-amplitude phase generator feeds the current $I_c(\omega)$ through the compensation impedance Z_c into P_1. The bridge will be balanced if $I_s = -I_c$ (i.e., $I_0 = 0$). Deviations are detected by the zero-voltage detector, which changes the amplitude and phase of the variable-amplitude phase generator until $I_0 = 0$. In the balanced state, the sample impedance is thus calculated by

$$Z_s(\omega) = \frac{V_s(\omega)}{I_s(\omega)} = -\frac{V_s}{V_c} Z_c(\omega) \tag{6.13}$$

The balancing operation is performed automatically over the full frequency range. The dielectric measurement is performed with a standard sample cell, connected by coaxial cables to the bridge, with temperatures controlled by dedicated systems. In the frequency range covered by impedance bridges, a disklike capacitor arrangement is a convenient sample geometry, although more specialized geometries can also be used.

As the measurement time is less that that with FRA, AC impedance bridges are particularly suitable for measurements on materials with time-dependent dielectric properties (e.g., for monitoring chemical reactions, curing and aging processes). Also, applications exhibiting considerable temperature dependence, such as characterization of phase compositions, phase transitions, and crystallization processes, can be successfully analyzed by the AC bridge methods. Of the commercially available measuring systems with AC bridges, a special reference should be made to the ARES and RSA-III rheometers offered by TA Instruments (www.tains.com). Although this and an analogous system offered by Perkin-Elmer are no longer commercially available, brief reference to their characteristics here is appropriate as these systems are still used by several thermal analysts for simultaneous DEA/DMA measurements. For example, the ARES and the RSA-III rheometers are equipped with the DETA accessory for combined computer controlled measurements of dielectric properties while the sample is at rest or undergoing deformation. The dielectric analysis option (DETA) provides simultaneous dielectric (permittivity, loss factor, $\tan\delta$) and rheological data (G', G'', $\tan\delta$) during time and temperature testing. This option includes the Agilent HP (Hewlett-Packard) 4284A bridge, with frequency range of 20 Hz–1 MHz and voltage range of 0.01–10 V. Up to 10 frequencies can be entered for a test. The electrodes are 25–40-mm-diameter stainless-steel plates. The DETA option utilizes an air or inert gas convection oven (with dual-element heaters and counterrotating airflow for optimum temperature stability, used primarily for polymer melts and solids) operating in the temperature range from −150 °C up to 350 °C, and heating rates up to 60 °C/min. Independent of this system, a remote dielectric single surface sensor that can be connected to the DEA 2970 system, has been offered by TA Instruments, allowing both DMA and DEA results to be obtained simultaneously on the same sample (although, as mentioned above, the DEA 2970 is not commercially available any more). For example, a simultaneous DMA/DEA sample configuration consists of a prepreg sample within the DMA clamps with the dielectric sensor attached on a free surface of the sample. As a result, in a single experiment complete information about the curing process can be obtained, including the glass transition temperature (to be more precise, the α-relaxation temperature) of the uncured material, the onset of cure, the gelation point, the time of minimum viscosity before cure, and the completion of cure.

6.3.3. Dielectric Analysis Based on Distributed-Circuit Methods: Coaxial-Line Reflectometry and Network Analysis

Above 10 MHz the measurement cables always contribute to the sample impedance, while above ~30 MHz standing waves arise at the line and a direct measurement of the sample impedance completely fails. At such high frequencies, where electrical wave effects become dominant, microwave techniques have to be applied. These techniques incorporate the measurement line impedance as a main part of the measured impedance. For that reason, precision lines with well-defined propagation constants are used. The network or radiofrequency (RF) impedance analyzers, with an operational frequency range from 1 MHz to ~10 GHz and $\tan\delta > 3 \times 10^{-3}$, use the transmission/reflection technique and are convenient for dielectric materials with sufficient relaxation strength, as well as relatively conductive samples (e.g., semiconducting polymers and polymer electrolytes).

Coaxial line reflectometry has to be employed at frequencies from 1 MHz to ~1 GHz. In contrast to the low-frequency techniques already described, here the sample capacitor is used as the termination of a low-loss precision coaxial line. The complex reflection factor is measured with a microwave reflectometer at the analyzer end of the line, depending on the sample impedance. For this purpose, the incoming and reflected waves are separated with two directional couplers and their amplitudes are measured. The BDS concept 70 is a commercially available system (Novocontrol Instruments) that uses the coaxial line reflectometry method. This system covers the frequency range between 1 MHz and 3 GHz with loss factor $\tan\delta$ resolution of less than 3×10^{-3}. The sample impedance determines the reflection factor of the sample capacitor, which is measured by the analyzer (e.g., an Agilent HP E4991 RF impedance analyzer; see Section 6.8.1) using the amplitudes of the incoming and reflected waves, line parameters, and sophisticated calibration procedures. To keep the waveguide as short as possible, a cryostat is directly mounted with the sample (i.e., an RF sample cell containing a parallel-plate sample capacitor) at the front end of the RF impedance analyzer. Dedicated cells, precision extension lines, and special cryostats for temperature control are supplied by Novocontrol.

Network analyzers (e.g., the Rohde & Schwarz ZVA series; see Section 6.8.1) allow impedance measurements at radio- and microwave frequencies (>1 GHz) using the transmission/reflection technique. While the RF impedance analyzers measure only the wave reflected from a sample that terminates a microwave line, in network analysis not only the reflected wave but also the wave transmitted through a sample is analyzed in terms of phase and amplitude. Although such devices are specified to work at ≤300 GHz or even higher, the calibration procedures are extremely cumbersome above 10 GHz, which seems to be the practical limit for dielectric measurements with temperature control. In the frequency range between 10 and 1000 GHz—a frequency window rarely inspected in typical thermal analysis laboratories—oversized cavity resonators or quasi-optical spectrometers are used (Kremer and Schönhals 2002).

6.3.4. *Fast* Time-Domain Dielectric Spectroscopy: Time-Domain Reflectometry

Time-domain reflectometry (TDR) dielectric spectroscopy (100 kHz–20 GHz), also referred to as *fast* time-domain spectroscopy (TDS) (Berberian and King 2002), is another well-established method for measuring the high-frequency permittivity and conductivity of materials in a liquid or a granular state. The material to be tested is inserted into a coaxial line that is itself terminated by either a short circuit, open circuit, or matched section; usually the sample is mounted in the capacitive sample cell that terminates the line. The activity of the sensor electrodes is measured not with a continuous frequency wave, but with a rapid-voltage pulse at the input to the sample cell containing a broad range of frequencies at the same time. The train of fast-rising voltage pulses, from a step generator (tunnel diode), is applied to the low-loss precision coaxial transmission line, and the waveform in the line is observed with a suitable sampling system connected to a sampling oscilloscope. The reflected signal is captured after an appropriate propagation delay, separating sensor response from connecting-line artifacts on the basis of delay time. The Fourier transforms of the incident and the reflected pulse waveforms are used to determine the relative complex permittivity of the sample. The limiting high and low frequencies are determined by the rise time and duration of the step pulse; by choice of suitable time windows and sample cells, it is possible to cover the frequency range 10^5–10^{10} Hz. Differences in the type of measuring cells and their locations along the coaxial line lead to different relationships between the values registered during the measurement and the dielectric characteristics of the material under study. Perhaps the most sensitive of the TDR techniques is the so-called precision difference method (Nakamura et al. 1982; Bone 1988), which uses a reference dielectric whose characteristics over the frequency range of interest are known. In studies of polymer solutions, in particular, the solvent of the polymeric material usually serves also as reference material. Cole et al. (1980) report successful application of this method in several dilute solutions of poly(methyl methacrylate), poly(vinyl acetate), and poly(methyl acrylate) in toluene and benzene. Mashimo (1997) has reviewed the application of TDR in biopolymers. Detailed information on cell design, procedures for accurate numerical Fourier–Laplace transforms of the waveforms, and the limitations in the implementation of the method are given by Cole and coworkers (Cole 1975, Cole et al. 1980, 1989) and by Berberian and King (2002). The main advantages of using TDR methods with respect to most frequency-domain techniques consist in the low cost, simple device handling and short measurement time.

6.3.5. Dielectric Depolarization Spectroscopy: Thermally Stimulated Current Technique

Thermally stimulated current (TSC) is a dielectric-related technique based on thermal release of stored energy (Bucci et al. 1966; Laverge and Lacabanne 1993; Gun'ko et al. 2007). The method uses molecular and ionic mobility as a probe of structural information that cannot be obtained, at least within a single temperature scan, by any other thermal technique. Properties measured by TSC include polymorphism, degree of crystallinity, purity, glass transition, relaxation phenomena, and conductivity. In a typical global TSC experiment (Fig. 6.8) a bilaterally metallized material is polarized at a sufficiently high temperature T_p (above the relaxation phenomena of interest) by applying an electric field E_p (polarizing field) for time t_p (isothermal polarization time). When the temperature of the sample is lowered to T_0 ($<<T_p$) with the electric field applied and then short-circuited at this temperature, the nonequilibrium state of the dipole system is "frozen in." During the successive heating, at a constant heating rate q, the recovery of the system is monitored by measuring the depolarization current density J_D as a function of temperature. For a single-relaxation-time mechanism the first-order kinetic function that describes the $J_D(T)$ curve has the form

$$J_D(T) = A\exp\left(-\frac{E}{kT}\right)\exp\left[-\frac{B}{q}\int_{T_0}^{T}\exp\left(-\frac{E}{kT'}\right)dT'\right] \quad (6.14)$$

The expressions for the adjustable parameters A and B depend on the nature of the process involved in the release of the current. The analysis of a dipolar

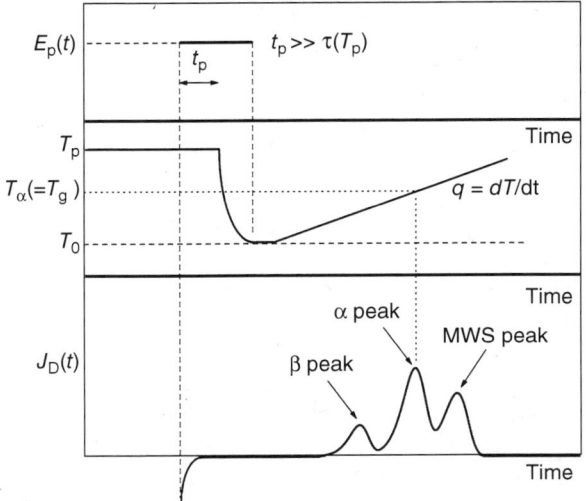

Figure 6.8. Schematic representation of the global TSC method. The location of current signals, characteristic of the relaxation activity in polymeric materials, is indicated.

mechanism ($A = P_0/\tau_0$, $B = \tau_0^{-1}$, where P_0 is the equilibrium polarization) leads to the determination of transition temperatures, relaxation parameters (activation energy E; preexponential factor τ_0) and the relaxation strength $\Delta\varepsilon$ (Garlick and Gibson 1948; Bucci et al. 1966; Laj and Berge 1966; Neagu et al. 2001). When the current is due to self-motion of space charges, analysis of the signal provides parameters such as carrier or defect mobilities, activation energies or trap depths, concentration and capture cross section of traps or "attempt to escape" frequencies, spatial distribution (when combined with surface charge measurements), and penetration depth (van Turnhout 1980).

In conventional (isochronal) dielectric or mechanical relaxation spectroscopy the timescale, which characterizes the relaxation processes at a given temperature, is well defined by the experimental frequency of the stimulus (periodic electric field or mechanical stress). In contrast, in TSC experiments the corresponding timescale (equivalent frequency, $f_{eq(TSC)}$) is intimately related to the heating rate, which is the relevant experimental variable in this case (Kalogeras 2008). The depolarization current density plot is equivalent to an isochronal $\varepsilon''(T)$ plot recorded at a low frequency ($\sim 10^{-2}$–10^{-4} Hz), but the height (i.e., the relaxation strength) and position of equivalent peaks in $\varepsilon''(T)$ and $J_D(T)$ plots may be different. Moreover, since TSC works under nonisothermal conditions, the relaxation function [$\Phi(t)$] is not directly accessible. Despite these shortcomings, there are several advantages of TSC over its isothermal or isochronal AC counterparts, described below (Sections 6.3.5.1–6.3.5.3).

6.3.5.1. Sensitive Recording of Well-Resolved Thermal Transitions

Glass transitions and other phase/structural transitions (e.g., melting) in polar polymers are easily detected with a single scan, even in cases where these transitions are not easily visible by other thermal techniques, such as the glass transition of semicrystalline samples by DSC [e.g., see Kalogeras et al. (2006a)]. TSC provides glass transition temperature estimates in excellent agreement with the results of calorimetric, dynamic mechanical, and AC dielectric studies. In TSC, T_g is selected as the temperature of the maximum current for the α-relaxation peak (T_α, Fig. 6.8) or as the temperature at which the activation energy of segmental motions attains its highest value (see Section 6.6). The α peak is selected from the available signals on the basis of several procedures (see Section 6.2.5) and the use of various experimental procedures.

6.3.5.2. Availability of Various Experimental Means for Analyzing Complex Signals

The use of curve-fitting techniques for the analysis of complex dielectric spectra is time-consuming, as it requires a choice of a specific distribution function (van Turnhout 1980), and also the use of at least two adjustable parameters. Thermally stimulated current offers a number of refined experimental procedures (thermal peak cleaning techniques) that can be applied for separating overlapping signals and identifying their origin (Vanderschueren and Gasiot 1979). Complex spectra are resolved into elementary relaxations using fractional polarization [also known as *thermal sampling* (TS), *relaxation map analysis* (RMA), or *windowing polarization* (WP)] techniques (Hino et al. 1973; Zielinski and Kryszewski 1977), or by means of computer simulations (Laredo et al. 2001) leading to the determination of relaxation features and related thermodynamic parameters. Representative TSC peak cleaning and fractional polarization techniques are described in Sections 6.5.1.2 and 6.5.1.3 (PMMA), respectively. Additional experimental processes for differentiating the various processes involve variations in the polarizing time and temperature, studies of the dependence of the peaks on field strength and polarity, use of different electrode–sample interfaces, adaptation of the resolution by testing various heating rates, controlled doping and aging processes, and variation of sample thickness.

6.3.5.3. Simplicity of Instrumentation

The low cost of the apparatus and the simplicity of the experimental procedure make TSC an efficient and time-saving tool for the characterization of novel polymer compositions and related composite materials.

6.4. PERFORMING DIELECTRIC EXPERIMENTS

6.4.1. Measuring Cells and Sample Preparation

The type of the electrode system is determined primarily by the nature of the testing material and the property or the phenomenon of interest. Several types of dielectric cells suitable for polymer samples of various forms, such as polymer films (ranging from a few nanometers to several millimeters), powders, and liquids, are now commercially available. In parallel, the availability of sensors with precise temperature control has extended the performance of dielectric spectroscopy. Many key aspects of material properties such as molecular relaxations, conductivity, phase separation, phase transitions, activation energy, glass transition temperature, rate of blending, purity, aging, and curing, can be determined when the temperature is controlled. In view of that, almost all commercial systems are equipped with suitable microprocessor-controlled temperature systems.

The parallel-plate electrode system (Fig. 6.9a) offers an easy interpretation of the measured signals in terms of the bulk dielectric properties of the material. A typical problem in this configuration is the inappropriate control of sample dimensions, namely, the sample area and thickness. The uncertainty in sample thickness increases if pressure is applied to a solid sample accompanied by strong temperature variation during the experimental run. For example, the thermal expansion or contraction taking place during heating or cooling as well as during a chemical reaction (e.g., polymerization or crosslinking) causes irregular variations in sample dimensions. In order to reduce the impact of such problems, as well as avoid piezoelectric effects, usually very little pressure is exerted on the sample when it is mounted between the electrodes. Spacers may be used to adjust the electrode spacing (Fig. 6.9b). The spacer material should have a low, frequency-independent dielectric constant and low losses (e.g., thin SiO_2 needles or precision glass spheres, or small strips of Teflon or polyethylene). The effect of the spacer on the measurement can be compensated by appropriate settings of the software used to control the measurement.

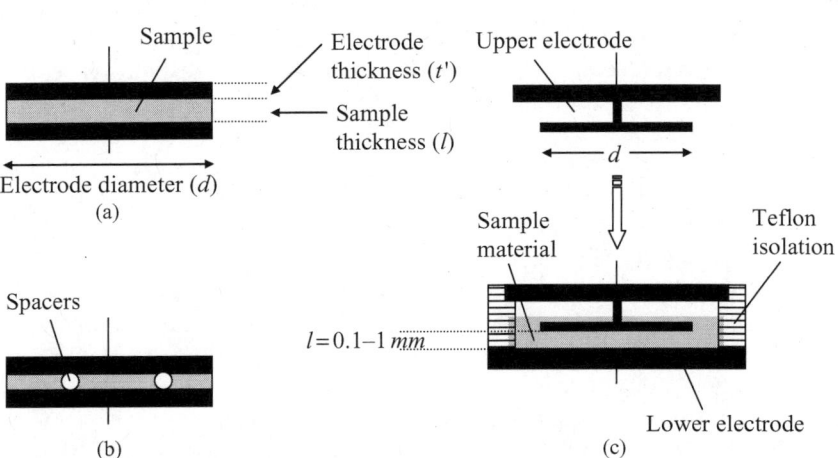

Figure 6.9. Typical parallel-plate capacitor systems for low-frequency ($f < 10\,MHz$) dielectric measurements of (a,b) solids and (c) powders, liquids, and reacting systems.

The BDS 1200 connection head (Novocontrol) is a typical sample cell with a parallel-plate electrode system and shielding unit, recommended for measurements in the low-frequency range from DC to 10 MHz and compatible with commercial temperature control systems. Several other manufacturers supply appropriate parallel-plate dielectric test fixtures for their dielectric systems [e.g., Agilent 16451B (for solids), Agilent 16452A (for liquids)]. Novocontrol also provides a high-pressure dielectric analyzer that allows scientists to study, simultaneously, the effect of temperature and pressure on a material's dielectric activity. Pressure tunability creates a whole new research landscape in polymer science and opens a new pathway to nanostructured materials. Furthermore, pressure can be used as a unique way to control crystallization processes such as nanocrystal growth on an industrial level. For such experiments, the solid sample is placed between two parallel electrodes (with diameter $d = 30$ mm), which are connected by flexible leads to high-pressure feedthrough connectors and from there by standard BNC cables to the analyzer. The electrodes are inserted into a high-pressure test cell that can be pressurized by an inert fluid such as silicon oil, and the maximum pressure that can be created by the integrated hydraulic pump is 3000 bar (i.e., 0–300 MPa).

For dielectric measurements of organic liquids (or powders), the utilization of parallel-plate capacitors is problematic. For example, mounting the sample and setting the electrodes without air bubbles is important, but not easily avoided with the classic parallel-plate configuration. The presence of bubbles can reduce the measured capacitance values and therefore be a source of error in the amplitude of the complex permittivity. Additionally, low-viscosity liquids with high vapor pressure (examples include some components of thermoset systems and antioxidants in thermoplastics) tend to evaporate from the standard capacitor, while liquid leakage may result from thermal expansion at high temperatures. Both problems can be avoided using sealed liquid or powder sample cells and sample holders with guarded electrodes. The parallel-plate shield liquid sample cell BDS 1308 (Fig. 6.9c) is a representative example (Novocontrol). As an alternative, a cylindrical liquid sample cell with guard ring (BDS 1307) is also available.

Commercially available interdigitated or comb electrodes are an alternative geometry with good reproducibility of experimental data. Such electrodes can be directly connected to the measuring device (e.g., an impedance analyzer). In a typical geometry (Fig. 6.10a), metal electrodes are fabricated on an insulating substrate (e.g., a porous polymer film, a ceramic, or a silicon integrated sensor) using photopatterning. A small quantity of sample may be placed over the electrode sensor, or the sensor may be embedded in the testing material. In these surface or fringe measurement sensors, the electrode linewidth and the spacing between the electrodes determine the approximate penetration of the fringe field into the material, and thus, the depth of the measurement. Comblike electrodes are used for online dielectric measurements in reacting systems (polymerization processes, crystallization and melting of semicrystal-

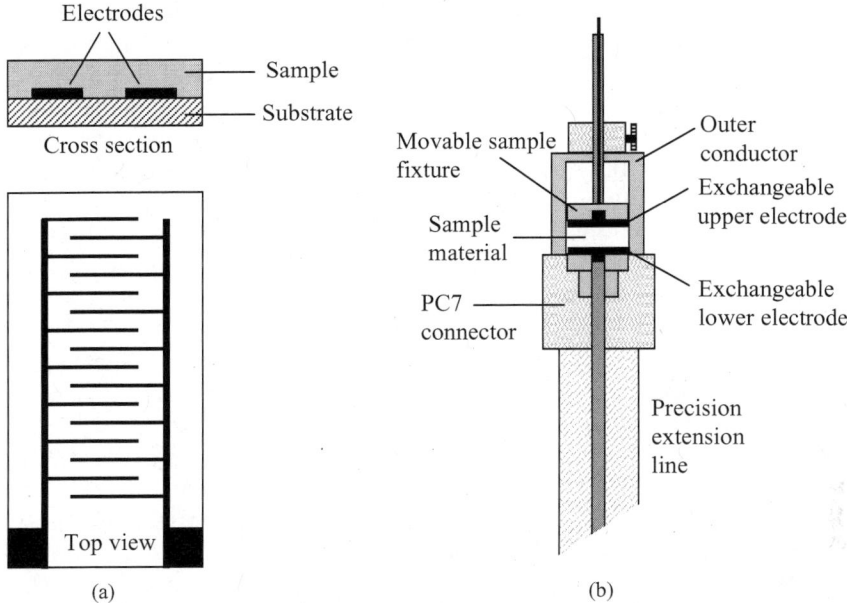

Figure 6.10. (a) The comb electrode geometry; (b) RF sample cell suitable for dielectric coaxial line reflectometry measurements in the frequency range from 1 MHz to 10 GHz.

line polymers, or curing of thermosets). The microdielectrometer sensor is a comb electrode fitted with a pair of field-effect transistors (FETs) that combines the best features of parallel-plate and comb electrodes. When this sensor is placed in the electric field, it measures the so-called complex transfer function from which dielectric permittivity is calculated. Since these electrodes are rigid and are manufactured with microelectronic precision, the calibration of the sensor is stable with respect to temperature and pressure variations. Micromet Instruments (Netzsch) offers a range of implantable sensors, such as the IDEX series sensors (e.g., the 036S, consisting of electrodes on a thin flexible polyimide substrate), for monitoring the cure of most thermosetting resins and composites. Because of their thin, flexible, and reliable nature, these sensors are suitable for use in high-pressure laminates or curved parts, as well as during manufacturing processes. Normal sensor environments are in presses, autoclaves, ovens, hotplates, and under ambient conditions.

For RF dielectric measurements, special sample cells are also commercially available (Fig. 6.10b). The BDS 2200 or 2100 RF sample cells (Novocontrol), with different maximum inner diameters (14 or 3 mm, respectively), can be used for dielectric and impedance measurements with temperature control in the range from 1 MHz to 3 or 8 GHz. Some recommendations for experimentalists engaged in the dielectric analysis and characterization of polymers and related materials are briefly given in the following paragraphs.

6.4.1.1. Control of the Sample Environment
High-vacuum conditions or constant inert gas flow are necessary for studying intrinsic relaxation phenomena in hygroscopic materials. A chemically inert atmosphere is also desirable in studies of polymer solutions, biomacromolecules, and other biological substances, where precise control of the hydration levels is vital. In isothermal scans fluctuations of the sample temperature should be as small as possible (below ~0.1 K), such fluctuations are usually controlled by most commercial temperature stabilization/regulation systems (see Section 6.4.1.2). The same recommendation applies for the heating rates typically used in nonisothermal dielectric techniques.

6.4.1.2. Selection of the Electrode Material
High-quality electrical contacts between sample and capacitor plates (i.e., perfect contact of each metal electrode with the sample surface) will reduce polarization effects at the interfacial region, which otherwise will introduce a significant error in the measured value of capacitance and permittivity. Electrode polarization, one of the most undesired effects in AC studies, renders the proper analysis of high-DC-conductivity materials almost impossible at low frequencies ($f < {\sim}1$ Hz) and high temperatures ($T > T_g$; Fig. 6.4). The electrode polarization effect is usually particularly strong in melts of even highly insulating polymers, since the loss of rigidity of the polymer network facilitates the translational motion of charged impurities. A large increase in both the real and imaginary parts of the dielectric function is a typical electrode polarization effect. The molecular origin of electrode polarization is the (partial) blocking of charges at the sample–electrode interface, which results in the formation of electrical double layers. These double layers give rise to a large capacitance in series to the conducting (bulk) sample material, which manifests itself in high apparent dielectric constants, typically in the range of 10^2–10^6. The interference of the electrode polarization signal with relaxation peaks existing in the low-frequency loss spectrum can be reduced using thicker samples (shifting in that way the signal to much lower frequencies) or using derivative presentations of the experimental data [e.g., the derivative technique proposed by Wübbenhorst and van Turnhout (2002)]. Contact problems can be solved in most cases if a highly conductive metal film (Au, Ag, or Al) is evaporated or sputtered on both sides of the sample surface. At a minimum, the solid sample should be covered with a layer of silver paint (solution of colloidal silver) or carbon black. In such cases, however, the highest temperature attained during an experiment should be below the polymer's melting point ($T_g \leq T < T_m$) in order to prevent metal diffusion in the bulk. Given that even 5–10 degrees below the melting point, 70–80% of the crystalline regions may have melted, the upper temperature limit should be well below this temperature (e.g., at a temperature where not more than ~20% of the crystals are melted). Careful selection and cleaning of the electrode material undoubtedly reduces the noise level, while permitting higher reproducibility of the measurements. Such prob-

lems may arise in experiments with metal electrodes susceptible to oxidation (e.g., silver or copper), but not in the case of most commercial instruments, which use high-purity gold-plated electrodes.

6.4.1.3. Quality of the Testing Material The quality and accuracy of the results depend considerably on the sample preparation. Uncertainties in sample geometry and poor knowledge of the chemical composition preclude acquisition of quantitative results (e.g., relaxation strengths) and their connection to specific types of molecular motion at the microscopic level. Furthermore, unknown impurities or chemicals that contaminate the sample or even influence the chemical structure of the material (residual solvents acting as plasticizers or antiplasticizers; charged impurities; environmental chemical agents causing degradation, such as hydrolysis and oxidative degradation; etc.) interact with the relaxing groups and modify intrinsic relaxation characteristics. The glass transition temperature and the melting point of the material under study change in accordance with the type and strength of the interactions, leaving an erroneous impression on the potential technological applications of the polymer. For comparison purposes between different solid polymer samples, heat treatment around the glass transition temperature, in vacuum or inert gas ovens, is usually required to diminish spurious effects related to the thermal, stress, and electrical history. In such cases, the researcher should make a judicious choice of the exact annealing temperature, depending on the property or phenomenon that is of interest. In several other cases, however, the goal of the experiment is fingerprinting (i.e., determination of thermal or mechanical history), and thus annealing prior to the dielectric experiment is useless, or even detrimental, because it changes the α-relaxation temperature. Also, in such tests, isothermal measurements are likely to be preferred over their isochronal analogs in order to keep the material in an isostructural state.

6.4.2. Measuring Conditions in AC Dielectric Experiments (DRS)

The primary purpose of everyday routine AC dielectric experiments (see Fig. 6.11 for a typical DRS experimental setup) is to measure the glass transition temperature, record the signatures of local molecular mobility mechanisms, evaluate the strength of conductivity, and characterize the crystallization of a polymer or the cure of a thermoset. In several cases, the glass transition temperature is determined in isothermal experiments using the Arrhenius plots, log τ versus $1/T$, and analyzing the data with the Vogel–Tammann–Fulcher–Hesse relaxation time function (see Section 6.5), although it would seem more logical to determine this parameter from the maximum of the corresponding band in $\varepsilon''(T)$ or tan δ–T plots at a constant frequency (*isochronal plots*). The preference for the two approaches mostly involves the depth of information that the experimentalist seeks to extract from the analysis. Bear in mind that

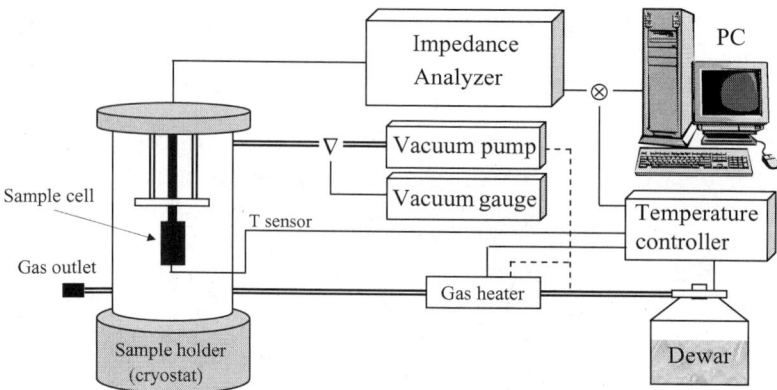

Figure 6.11. Block diagram of the components of the instrumentation, and their interconnection, typically used for dielectric relaxation spectroscopy (DRS) studies.

the kinetic character of the glass transition phenomenon is reflected in the frequency-dependent T_g, which in turn complicates comparison of the glass transition temperatures obtained from different techniques.

Modern dielectric spectrometers provide computer-controlled automated acquisition of data. Their accuracy depends on careful control of the experimental conditions, particularly if one is interested in precise values of the dielectric permittivity function. Critical steps in the preparation of the capacitor filled with the sample (in the liquid or solid form) have already been described (Section 6.4.1). Furthermore, prior to the actual measurement, the researcher should consult the instruction manual of the system operating software and adjust a number of settings. The temperature program, which usually involves a linear temperature rise, and the range of temperatures to be scanned are critical parameters for isochronal scans. The temperature control system sets the limits of the accessible temperature window. Furthermore, limitations in the selection of the upper (end) temperature, imposed by the nature of the material, have already been discussed. Again, the heating rate should be low enough to minimize thermal lag. In a step isothermal experiment, a very common method for acquiring multifrequency DEA as well as DMA data, the temperature of the frequency scan changes at a preselected mode, with complete scans performed usually every 5 or 10 degrees. In such experiments, the "waiting time" at each new temperature (i.e., the time period before the beginning of the frequency sweep) should be chosen sufficiently high—usually between 5 and 20 min, depending on the magnitude of the temperature step—to ensure true isothermal measurements. Besides temperature, the principal consideration is the range of frequencies that the frequency generator will scan in an isothermal experiment. Its limits should be determined depending on the type of the instrument and the method of measurement, as well as the conductivity of the sample and the phenomenon of interest. For example, low-frequency dielectric data are not easy to collect or interpret in practice in highly conductive polymers or at the early stages of polymerization or curing processes. On the other hand, monitoring the curing process of a thermosetting polymer requires application of a wide frequency range, with the high-frequency data of particular interest at the early stages of the reaction (see Section 6.7). For that reason, a system combining frequency response analysis or impedance analysis (Section 6.3.2) with coaxial line reflectometry (Section 6.3.3) would be desirable (although expensive; e.g., the BDS concept 80 system of Novocontrol). In performing dielectric/conductivity measurements in reacting systems via isothermal frequency sweeps, it is important to remember that, in much the same way as in DMA measurements, the time required to complete the sweep depends on the lowest frequency employed during the run. Therefore, the lowest frequency should be fast enough so as to keep the timescale of the experiment much shorter than the timescale of the changes in the polymeric network (i.e., each measurement should correspond to an *isostructural* state).

Accurate measurements of electrode diameters (d) and sample thickness (l) are essential for conversions between the different data representations (impedance, permittivity, conductivity, conductance, capacitance, etc.). Furthermore, the amplitude of the sinusoidal electric stimulus should be low enough (typically ~1 V) to ensure the linearity of the observed response. Homogeneity of the applied field $\bar{E}_0$, within common parallel-plate capacitor systems, requires electrodes with surface area $S \gg l^2$, so that fringing effects can be neglected. The influence of edge capacities, arising from the inhomogeneity of the field at the capacitor edges and the presence of a stray field outside the capacitor, can be reduced if a sample cell with a third guard electrode is used. The guard ring in the plane of the working electrode, separated from it by as narrow a gap as possible and held at the same potential as the working electrode, will increase the homogeneity of the electric field between the working electrode and the counterelectrode, transferring the edge effect to the perimeter of the guarding electrode. In cases where this configuration is not available and thick polymer samples are to be studied, it is best to estimate the ideal sample capacitance (C) by subtracting from the measured capacitance (C_{meas}) the edge capacitance (C_{edge}), using approximate formulas (Landau et al. 1984). For example, for a typical system with two round capacitor plates completely filled with the dielectric material, C_{edge} can be calculated using the relation

$$C_{edge} = \varepsilon_0 \frac{d}{2} \left[\ln \frac{8\pi d}{l} - 3 + z(x) \right] \quad (6.15)$$

where $\varepsilon_0 = 8.85 \times 10^{-12}$ (A·s)/(V·m) (the permitivity of free space), t' is the electrode thickness, and $z(x) = (1 + x)\ln(1 + x) - x \ln(x)$ with $x = t'/d$.

6.4.3. Typical Measuring Conditions In *Global* TSC Experiments

The block diagram of the apparatus for TSC and *slow* time-domain measurements is shown in Fig. 6.12. Typically, the sample is sandwiched between metal (usually gold-coated brass or chromium-coated copper) disks to which electrical contacts are made to allow polarization and current measurements. The assembly is placed in a temperature-controlled chamber, usually filled with dry N_2 or He, or simply under high-vacuum conditions. In that way, the noise of the collected current signal as well as the background current is drastically reduced. A DC high-voltage supplier is used to polarize the testing material at a carefully selected polarization temperature ($T_g \leq T_p < T_m$ or lower, depending on the information being sought). Typical voltage suppliers allow selection of voltages in a wide range with small steps. This feature allows for a precise control of the polarizing field for samples with broadly different thicknesses (typically between 10 μm and 2 mm), keeping in mind during selection of the experimental value, that the polarizing field must be kept well below the electrical breakdown of the polymer. The temperature of the sample is adjusted with special cooling/heating and temperature stabilization systems. Silicon diodes or typical copper–constantan thermocouples, placed close to the material (e.g., pinned in the grounded electrode) are used for temperature measurements (typical error below ±0.5 °C). The time-dependent discharge current of a short-circuited sample kept at T_0 levels off in a few minutes, so the warmup program can be started shortly afterward to record the short-circuit current as a function of the temperature at constant heating rate. During this final depolarization step, the current is recorded with a sensitive electrometer with a precision on the order of 10^{-16} A and internal resistance much smaller than that of the sample. Software for a personal computer allows one to control

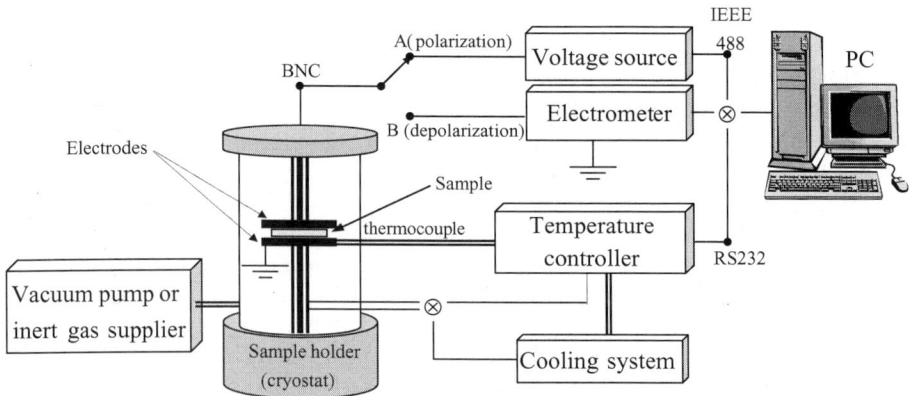

Figure 6.12. Schematic of the apparatus used for thermally stimulated current (TSC) studies. The basic concept applies also for *slow* time-domain spectroscopy measurements.

automatically the thermal cycle, to polarize the sample, and to connect it to the electrometer by activating–inactivating rotary switches, and to record the current as a function of the temperature and/or time.

Selection of the heating rate during the TSC experiment is of high importance for the acquisition of reliable results. Low-to-moderate heating rates ($0.5 \leq q \leq 6\,°C/min$) are usually necessary to avoid spurious effects that may arise from the temperature gradients that normally develop in thick samples ($l > \sim 1\,mm$). Effects arising from the nonuniformity of the temperature within the sample include erroneous estimates of transition temperatures and artificial broadening of the peaks, which results in flawed estimates of the distribution in the relaxation times. Peak intensities are reproducible only if the temperature program is rigorously fixed. A peculiarity of the TSC method, namely, the very strong dependence of the peak intensity on the rate of cooling from the polarization temperature, has been reported even from the very beginning of relevant studies of polymers. For example, Hedvig (1977) reports that by increasing the cooling rate from 2 to $13\,°C/min$, for a PMMA sample polarized at $T_p = 150\,°C$, the intensity of the α-relaxation peak (glass transition signal) decreases. Therefore, it appears that for comparison purposes one should use identical cooling rates. Furthermore, in order to calculate the dielectric strength ($\Delta\varepsilon$) of a relaxation mechanism from the area of the corresponding TSC peak, one must consider the temperature dependence of the mechanism's contribution to the static dielectric permittivity of the material. TSC-based estimates of $\Delta\varepsilon$ may be associated with an effective temperature of polarization (T_{eff}), calculated from the relation (Christodoulides et al., 1991)

$$\frac{1}{T_{eff}} = \frac{1}{T_{max}}\left[1 + \frac{1}{2\varepsilon_m} + \frac{\ln(q/q')}{\varepsilon_m + 2}\right] \qquad (6.16)$$

where q' is the sample's cooling rate and $\varepsilon_m = E/kT_{max}$ (where E is the apparent activation energy of the specific mechanism).

Most of the issues discussed in Section 6.4.1 in relation to the sample preparation apply also in this type of experiment. Moreover, TSC measurements of metallized polymers are frequently subject to a serious error, namely, the generation of a strong parasitic current at temperatures well above the glass transition temperature. This signal shows a temperature dependence that runs parallel to the true ohmic conduction current curve [e.g., the signal found in the TSC spectrum of *uncharged* PET above $120\,°C$, as reported by van Turnhout (1975)]. The signal has been attributed to a weak electrochemical potential, which arises even in the case where identical electrodes are used. The onset of this parasitic current is probably accompanied by chemical changes such as oxidation–reduction reactions, disproportionation, monomer conversion, or pyrolysis. A blank TSC (heating of an uncharged sample) will reveal the presence or absence of such effects as well as the temperature range of their appearance. In that way, the experimentalist will be able to determine the upper temperature limit of the scan, or make appropriate corrections to the data. Although such effects are seldom considered in everyday dielectric studies of most polymers and related composites, they may be determining factors for the validity of the results. For example, in the temperature range near and above the glass transition temperature of the polymeric component, parasitic currents may significantly hinder the assessment of characteristics of the conductivity mechanism. The information related to the glass transition temperature (i.e., the location of the α signal, T_α) might be severely blurred if both parasitic and true conduction currents mask the segmental relaxation. Subtraction of the parasitic current contribution from the measured signal will reveal the exact position of the intrinsic relaxation mechanisms and their dielectric strengths.

6.5. TYPICAL MEASUREMENTS ON POLY(METHYL METHACRYLATE) (PMMA)

In this section, basic procedures for analyzing dielectric spectra of polymers are described. The model sample is free-radical polymerized PMMA that has characteristics similar to those of commercial atactic materials. The focus is centered on the analysis of thermally stimulated current (TSC) spectra and the frequency dependence of complex permittivity at various temperatures. Amorphous PMMA is a material that has a number of relaxations (Hedvig 1977) corresponding to the onset of various types of short- and long-range molecular motion (Gourari et al. 1985); the δ relaxation is attributed to the rotation of the ester methyl group (β-CH_3), while the γ relaxation is assigned to the (re)orientation of low-polarity methyl groups of the mainchain (α-CH_3). The β relaxation corresponds to a 180°±20° flip of the –$COOCH_3$ sidegroup supplemented by motion of the backbone around the local-chain axis. Finally, the α relaxation appears as a result of complex forms of molecular motion near the glass transition temperature (Doulout et al. 2000). Appreciable signals related to the translational mobility of ions are also recorded at high temperatures or low frequencies.

6.5.1. Thermally Stimulated Current Analysis of Thick PMMA Films

6.5.1.1. Features of the Global TSC Spectrum A single recording of the thermally stimulated current spectrum of a thick ($8 \times 8 \times 1$ mm) PMMA film, in the broad temperature range from −263 to +160°C (Fig. 6.13), permits a qualitative appraisal of the relaxation activity in this polymer. Prior to this measurement, several procedures were followed. First, a sufficiently low-intensity polarizing electric field (1 MV/m), well below the dielectric breakdown strength of PMMA ($\approx$14 MV/m), was used so as to minimize the effect of electrode polarization and charge injection. In addition, the effects related to thermal stress and electrical history had to be removed, and for that purpose

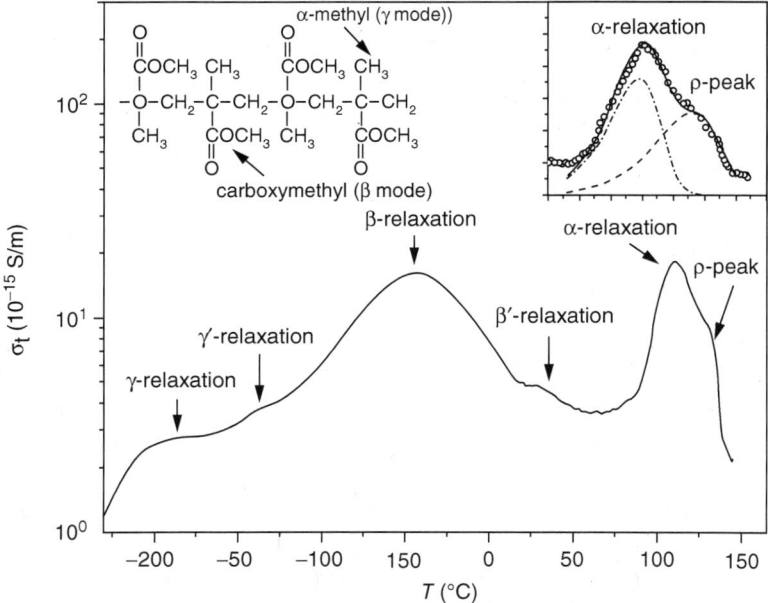

Figure 6.13. Application of the TSC technique in PMMA. Sample thickness $l = 1$ mm, $E_p = 1$ MV/m, $T_p = 100\,°C$, $t_p = 1$ h, $q = 5\,°C$/min, vacuum cell, silver electrodes provided by vacuum evaporation (Kalogeras and Vassilikou-Dova, unpublished results). Dielectric evidence for the δ relaxation is limited even in cases where the temperature window is extended below $-225\,°C$. Nonetheless, this mechanism can be resolved using other techniques [e.g., fluorescence emission; see de Deus et al. (2004)].

the sample was first annealed in a vacuum oven for 1 h at $120\,°C$ (~$5\,°C$ above T_g), and then slowly cooled to T_p. The isothermal polarization temperature (T_p) of $100\,°C$ was selected below the DSC T_g (~$115\,°C$) in order to decrease the contribution of space charge polarization (ρ peak).

In addition to the well-established molecular relaxations, the TSC spectrum shows a weak γ relaxation, ascribed to rotation of trace amounts of absorbed water, and a weak β′ relaxation, a dielectric signal that may correspond to translational mobility of charges, hindered sidechain motions, or even structural relaxation effects [see, e.g., Muzeau et al. (1995); Kalogeras (2004).[2] The α relaxation has a peak at $T_\alpha = 110\,°C$, near the DSC T_g (Kalogeras and Neagu 2004). The "intrachain" effect of the stiffness of individual chain segments is—at least in the case of PMMA and some other poly(n-alkyl metharylate)s—more important than the "interchain" effect of the cohesive (attractive) forces

---

[2] In accordance with the coarsening concept (Takahara et al. 1999), physical aging and the related structural relaxation phenomenon often observed in amorphous polymers can be treated as a result of fragmentation and aggregation of cooperatively rearranging regions, and a redistribution of the temporary high- and low-density microdomains.

between different chains in determining the values of T_g (or T_α) and to a lower extent of T_β. This effect is in part responsible for the dependence of the transition temperatures on the tacticity of the chain [$T_{g,\text{syndiotactic}} > T_{g,\text{atactic}} > T_{g,\text{isotactic}}$ (Gourari et al. 1985)]. In the inset of Fig. 6.13, the result of the curve-fitting decomposition of the high-temperature signal to the sum of a dipolar (α) and a space charge (ρ) signal is shown. Each component has been described by the approximate equation

$$\sigma_t(T) \approx A \exp\left\{-\frac{E}{kT} - \frac{T^2}{T_{\max}^2} \exp\left[\frac{E}{k}\left(\frac{1}{T_{\max}} - \frac{1}{T}\right)\right]\right\} \qquad (6.17)$$

with peak temperature of $T_\alpha \equiv T_{\max,\alpha} = 110 \pm 2\,°C$ and an activation energy of $E_\alpha = 139\,\text{kJ/mol}$ for the α relaxation, but $T_\rho \equiv T_{\max,\rho} = 132 \pm 1\,°C$ and $E_\rho = 120\,\text{kJ/mol}$ for the ρ peak.

6.5.1.2. Application of Peak Cleaning Techniques

Thermally stimulated current is a versatile technique with variants that allow experimental deconvolution of complex spectra. Isolation of selected signals is possible by applying different peak cleaning techniques, some of which are supported by the software provided with the commercial instruments. One of the most common peak cleaning techniques (Creswell and Perlman 1970) consists of the following steps (Fig. 6.14a). The material is polarized by applying an electric field at a temperature above the peak maximum of the relaxation under study (at $T_p > T_{\beta'}$) for sufficient time, and quenching at a much lower temperature. The subsequent heating cycle is interrupted at a temperature between the apparent peak maxima (at T_A: $T_\beta < T_A < T_{\beta'}$) in order to depolarize the low-temperature component only. Cooling the sample and reheating allows recording of the (nearly) pure high-temperature signal. In a variant of this technique (Bucci et al. 1966), the specimen is polarized at a temperature T_p ($T_\beta \leq T_p < T_{\beta'}$) for a short time $t_p \approx \tau_\beta(T_p) \ll \tau_{\beta'}(T_p)$, so that dipoles contributing to the low-temperature signal are polarized at close to saturation, while dipoles of the higher-temperature relaxation remain randomly oriented. Experiments performed with different T_p values (Fig. 6.14b) provide information related to the character of the signals (e.g., the presence of single or distributed relaxations; see Table 6.1).

The dielectric strength of a particular relaxation mechanism can be determined by the depolarization charge Q_{el}, obtained from the area under the related TSC peak, by

$$\Delta\varepsilon = \frac{Q_{el}}{\varepsilon_0 S E_p} = \frac{1}{h}\frac{\int_{T_0}^{T_\infty} I_D(T)\,dT}{\varepsilon_0 A E_p} \qquad (6.18)$$

The area under an isolated peak is obtained by simple graphical integration of the curve. For partially overlapping bands, it is possible to obtain rough

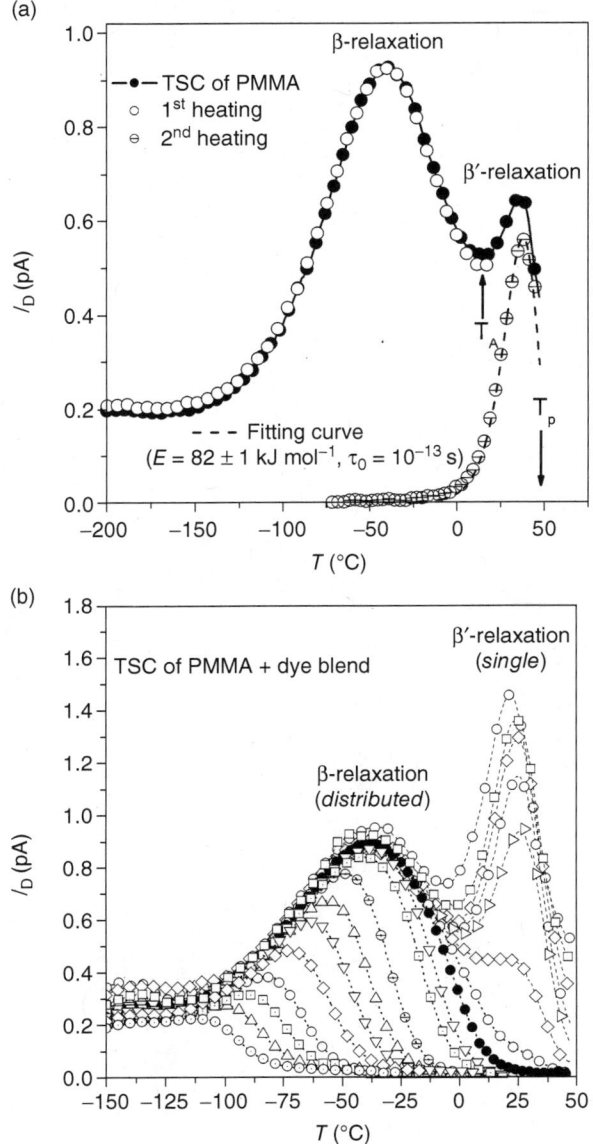

Figure 6.14. Application of peak cleaning techniques for the separation of overlapping relaxations: (a) technique introduced by Creswell and Perlman (1970), with $T_A = 10\,°C$ and $T_p = 47\,°C$; (b) variation of TSC spectrum with changes in T_p (between $-123\,°C$ and $+47\,°C$ in $10°$ increments). Filled circles correspond to the spectrum recorded with the optimum polarization temperature ($T_p = -13\,°C$) for β relaxation. [From Kalogeras (2004); reprinted with permission of John Wiley and Sons, Ltd.]

TABLE 6.1. General Trends on Effects of Experimental Parameters on Dipolar and Space-Charge-Related Thermally Stimulated Current Peaks of Polymers

Parameter	Effect on Dipolar Peaks	Effect on Space Charge (ρ) Peaks
Polarization temperature (T_p)	In relaxations with distributed parameters, T_{max} shifts if $T_p < T_{max}$ and is constant if $T_p \geq T_{max}$; single-relaxation mechanisms demonstrate peaks with position nearly independent of T_p	For polymers with intrinsic charge carriers, peak position is nearly independent of T_p, when charge carriers are not trapped at different energy levels, but its height passes through a maximum; for polymers with extrinsic charge carriers, T_p mainly influences the penetration depth of the injected carriers; penetration will increase with temperature
Polarization time (t_p)	No effect for $t_p \gg \tau(T_p)$	The effects are similar to those for T_p (i.e., the peak passes through a maximum) but t_p must change logarithmically to obtain changes of the same magnitude
Polarizing field (E_p)	Linear dependence of $\Delta\varepsilon$ on E_p (until electric saturation is reached); T_{max} is usually independent of E_p	Nonlinear dependence
Electrode material	No effect	Degree of blocking depends on electrode material, which affects peak's position and strength
Sample thickness	No effect	Weaker peak and shifted to higher temperatures for thicker polymer samples
Sample preparation	No effect (unless plasticization or antiplasticization phenomena arises in presence of residual solvents, monomers, or other chemicals)	Highly influenced by H_2O absorption, swelling agents, doping, impurities, etc.
Radiation processes	No effect (unless chain scission occurs, which significantly affects segmental relaxation modes)	Irregular dependence (usually lowering of the signal due to neutralization of carriers from radiation-induced charges)
Reproducibility	Usually good	Moderate (mechanism efficiency varies between 0% and 100%)

estimates by fitting the spectrum to a superposition of Gaussian functions (for symmetric curves; Fig. 6.15a) or approximate current functions [e.g., Eq. (6.17)]. When comparing the thermally stimulated current estimate of $\Delta\varepsilon$ with that based on Havriliak–Negami function simulations of loss peaks, the difference is usually less than $\pm 10\%$ (Fig. 6.15b). However, larger deviations may appear as a result of factors such as erroneous subtractions of the background signal, contributions from satellite peaks, incorrect estimate of sample dimensions, and differences in the equivalent temperatures of the estimates.

6.5.1.3. Experimental Analysis of Complex TSC Peaks The step or partial heating technique (Creswell and Perlman 1970) is a valuable procedure for analyzing the energy spectrum of complex or overlapping relaxations (Fig. 6.16a). In a typical run, the sample is polarized, by following an ordinary TSC protocol, and then is partially depolarized when heated up from T_0 to a temperature T_{cut} (referred to as the *cutoff temperature*). At this temperature the heating process is interrupted and the sample is cooled down to $T_L \approx T_{cut} - 50\,°C$. By repeating the heating–cooling process with a gradual increase of T_{cut} (in steps of 5–$10\,°C$), the complex signal is subjected to a moderate transformation. Analysis of the current components by means of the initial rise method provides estimates of the apparent activation energy E [obtained from the slope E/k of the $\ln I_D$–$1/T$ plots; see van Turnhout (1980)], which can then be plotted against T_{cut}. This method is the only one that can be applied for the analysis of space-charge–related signals, irrespective of the nature and characteristics of the mechanism (e.g., first- or higher-order kinetics of the process, single or distributed relaxation times, spatially uniform or nonuniform distribution of excess charges). The typical shape of the E–T_{cut} dependence for single (β') and distributed (β) relaxation time mechanisms is shown in Fig. 6.16b.

Another powerful tool for analyzing experimentally complex signals, with wide application in polymeric materials, is thermal sampling (TS). This technique consists of "sampling" the relaxation process within a narrow temperature range by polarizing at a temperature T_p and depolarizing at T_d, a few degrees lower than T_p, and by spanning the entire temperature range of the global peak with a series of TS scans (Fig. 6.17a). This thermal cycling, or equivalent protocols, is usually a batch mode in the software running most commercial TSC instruments. This technique is important because it allows experimental resolution of the complex signal and the corresponding τ distribution, without resorting to a direct analysis that would require a hypothesis on the form of the distribution function. Each quasi-non-distributed current "sample" (*i* process) is described by a set of parameters (E_i, $\tau_{0,i}$). In some cases the extrapolated plot of $\ln \tau$ versus $1/T$ of a specific relaxation mechanism shows that several Arrhenius lines converge into a single point, which is called a "compensation point" (T_{comp}, τ_{comp}) (Fig. 6.17b). At the compensation temperature (T_{comp}), all relaxation times would take the same value τ_{comp}, obeying a compensation law relationship:

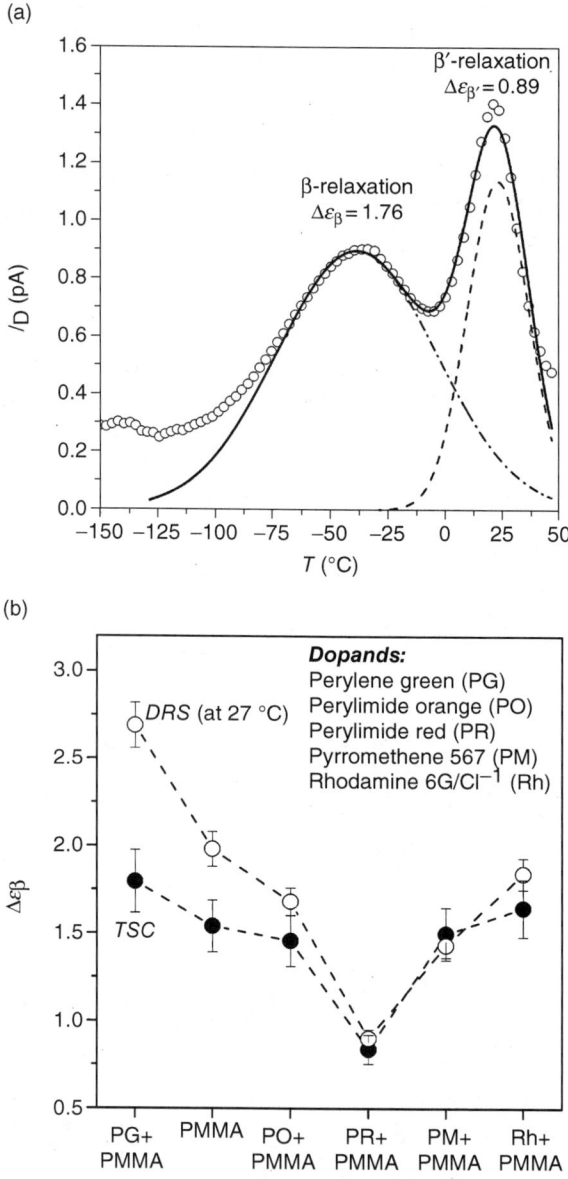

Figure 6.15. (a) Estimates of the dielectric strength of the β and β' relaxations from the area of the corresponding TSC bands of PMMA with the use of Eq. (6.17) (Kalogeras, unpublished results); (b) comparison between thermally stimulated current (TSC) and dielectric relaxation spectroscopy (DRS) estimates of the dielectric strength of the β relaxation in undoped PMMA and its blends with fluorescent dyes. [Graph (b) is reprinted with permission from Kalogeras et al. (2004). Copyright 2004, American Chemical Society.]

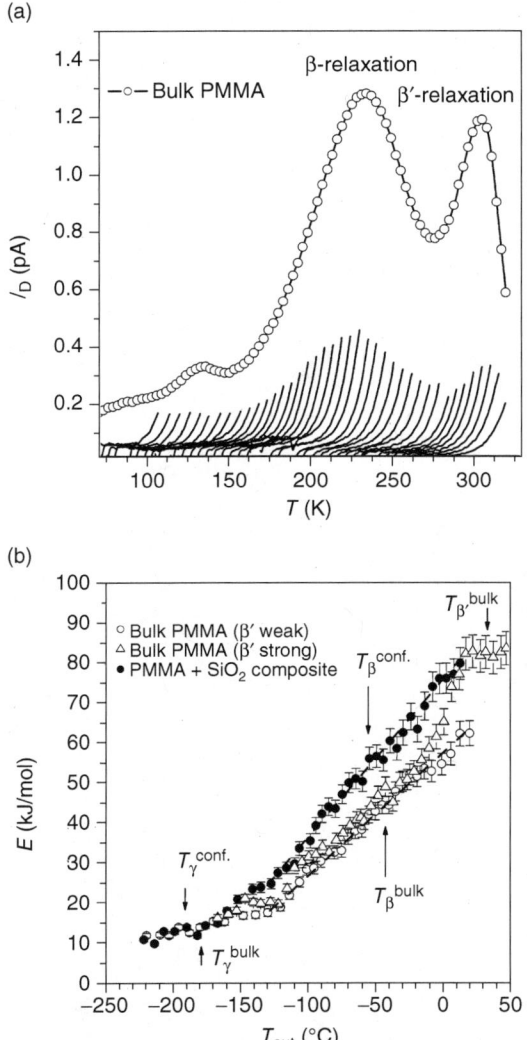

Figure 6.16. (a) Application of the partial heating technique in PMMA; (b) plot of apparent activation energies as a function of cutoff temperature recorded for PMMA samples and PMMA confined in 5 nm average pore diameter SiO_2 monoliths formed from a sol-gel process. Arrows indicate the position of the corresponding TSC peaks. [Graph 16(b) is compiled from data appearing in two publications (Kalogeras (2004, 2005).)]

DIELECTRIC ANALYSIS (DEA)

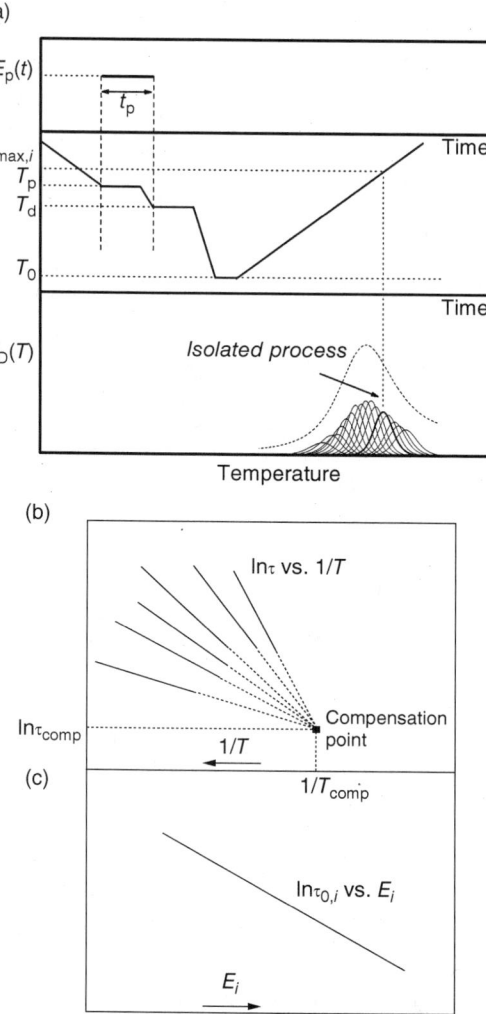

Figure 6.17. (a) Illustration of principle of the thermal sampling procedure. By varying the value of T_p by a constant step along the axis (i.e., by shifting the polarization window $T_p - T_d$ from 2° to 5°), peaks with a narrow τ distribution are recorded. (b) Schematic Arrhenius diagram of isolated processes that exhibit a compensation law behavior. (c) Schematic compensation diagram.

$$\tau_i(T) = \tau_{\text{comp}} \exp\left[\frac{E_i}{k}\left(\frac{1}{T} - \frac{1}{T_{\text{comp}}}\right)\right] \quad (6.19)$$

The variation of $\ln \tau_{0,i}$ versus E_i constitutes an alternative representation of the compensation rule (Fig. 6.17c). The intercept and slope deduced from a linear regression of $\ln \tau_{0,i}$ versus E_i allow one to obtain the compensation parameters. Observation of compensation phenomena in polymers is often considered an indication of cooperative movements (Lacabanne et al. 1994) and has been attributed to the relaxation of dipolar segments with variable lengths. In this context, the activation energy distribution reveals dipoles that are hierarchically correlated with a wide spectrum of segmental motion (Correia and Moura Ramos 2001). For segmental relaxations the lag $T_{comp} - T_g$ has been related to kinetic effects and is considered to depend on the stiffness of the polymeric chain or to be related to the width of the glass transition (Moura Ramos et al. 1997).

6.5.2. DEA Runs with Thick PMMA Films

6.5.2.1. Features of Isothermal and Isochronal DEA Spectra

The isothermal plots of the frequency dependence of permittivity (ε') and dielectric loss (ε'') of a thick PMMA sample ($10 \times 10 \times 1$ mm) are shown in Fig. 6.18. The sample was polished to obtain smooth parallel surfaces and was placed directly into the parallel-plate capacitor. Isothermal scans were performed in a nitrogen atmosphere in the frequency range of 10^{-1}–10^6 Hz and selected temperatures between $-70\,°C$ and $+190\,°C$. At temperatures below T_g ($115\,°C$ by DSC), the β relaxation dominates the response, while the α relaxation and strong conductivity losses enter the frequency window at higher temperatures. The lack of a prominent α peak in the dielectric (DRS and TSC) spectra of PMMA results primarily from the occurrence of the electric dipoles in the lateral groups (i.e., $\Delta\varepsilon_\alpha < \Delta\varepsilon_\beta$). The presentation of the DRS data will be more familiar to most thermal analysts in the form of isochronal plots, with temperature as a variable (Fig. 6.19). These can be constructed even from isothermal recordings, provided that the dielectric experiments were performed at several, closely spaced, temperatures (e.g., one scan every 5 or 10 degrees). Such plots allow direct comparisons between dielectric loss (DRS), thermally stimulated current (TSC), and dynamic mechanical analysis (DMA) spectra. The reader will notice in Fig. 6.19 the shift of the relaxation signals to lower temperatures with the gradual decrease of frequency. Because of the inherently very low value of its equivalent frequency ($\sim 10^{-3}$ Hz), the TSC recording provides the best resolution between the α and β relaxations.

The dielectric (ε'' and M'') spectra and the relaxation frequencies of the β relaxation are similar to the mechanical spectra (J'' and E'', respectively) and the corresponding relaxation rate over a wide temperatures range (Muzeau et al. 1991; Perez et al. 1999). These observations suggest that the underlying mechanisms for the local electrical and mechanical relaxation processes in PMMA are similar. Clearly, this is not always the case for polymers, since all modes of motion of a polymer chain are not dielectrically active. When rotational diffusion occurs about a variety of different axes among which only a few reorient a dipole, the shape of the relaxation and the average rates of relaxation in a dielectric measurement may and will differ from those in a mechanical test. Dielectric, dynamic mechanical, and DSC glass transition

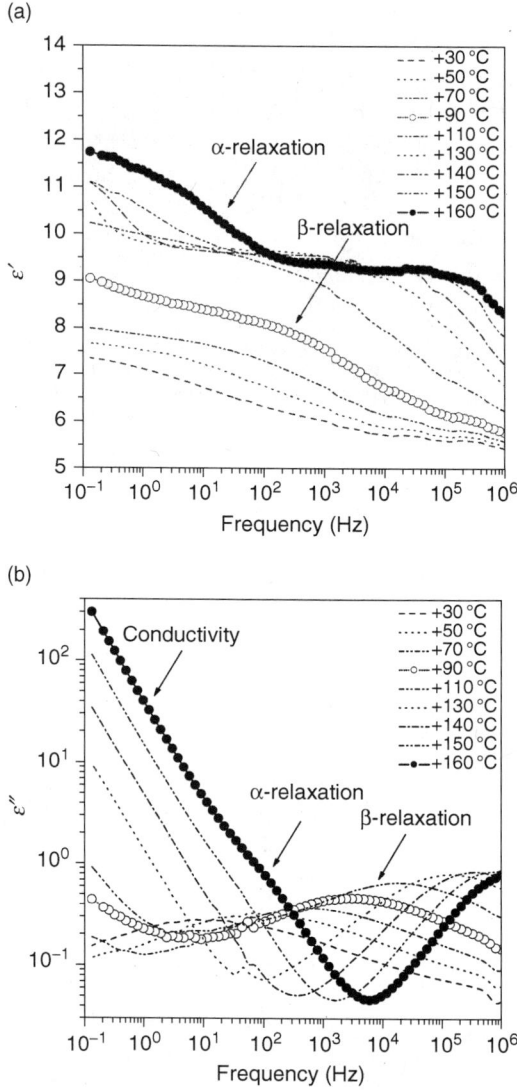

Figure 6.18. Representative permittivity (a) and dielectric loss (b) spectra of atactic PMMA recorded in the frequency range from 10^{-1} to 10^6 Hz at selected temperatures (Kalogeras and Vassilikou-Dova, unpublished results).

temperatures are seldom identical, not merely because of the equivalent frequency shifts but also because of the different nature and time response of the experimental probes (motions of side- or mainchain polar groups, or of small or large parts of the chain; electrical, mechanical, or thermal stimulation of the motion, etc). Separation of the dielectric α relaxation from the strong conduc-

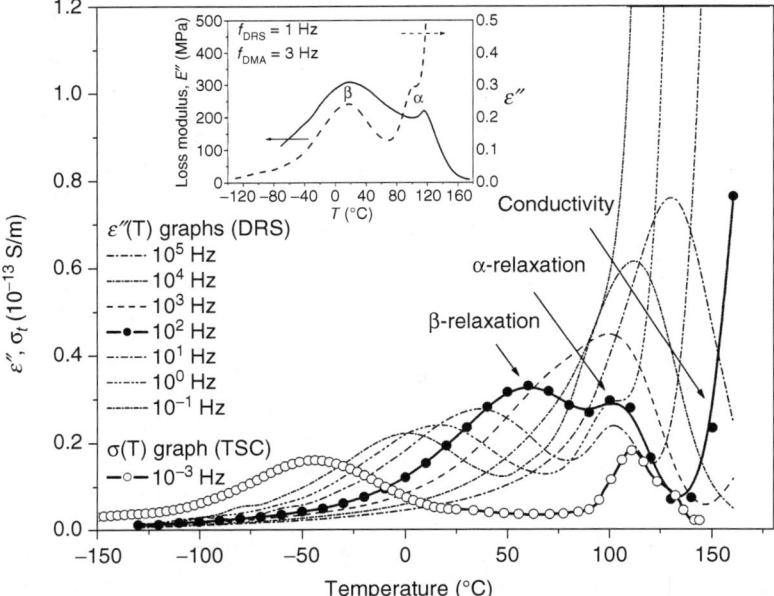

Figure 6.19. Comparison of DRS and TSC data of PMMA: selected isochronal dielectric loss spectra and the typical (global) TSC spectrum (Kalogeras and Vassilikou-Dova, unpublished results). Arrows indicate the location of the main signals in the case of the isochronal plot at 10^2 Hz. The inset shows a comparative plot of the temperature dependence of the dielectric loss (ε'') and the mechanical loss modulus (E'') for PMMA [DMA data taken from de Deus et al. (2004)].

tivity signal (e.g., in Figs 6.13, 6.18b, and 6.19) is not always feasible, contrary to the DMA technique in which the segmental relaxation is always well resolved [e.g., see Higgenbotham-Bertolucci et al. (2001), de Deus et al. (2004), also inset in Fig. 6.19].

6.5.2.2. Construction and Analysis of Arrhenius Plots

Complex permittivity data, like those shown in Fig. 6.18, are customarily analyzed by a superposition of several Havriliak–Negami functions and a conductivity term, that is

$$\varepsilon^*_{\text{exp}} = \varepsilon_\infty + \sum_k \frac{\Delta\varepsilon_k}{\left\{1+[i/(f/f_{0,k})]^{1-\alpha_{\text{HN},k}}\right\}^{\beta_{\text{HN},k}}} - i\frac{\sigma_{\text{DC}}}{\varepsilon_0(2\pi f)^s} \qquad (6.20)$$

where $k = 1, 2, \ldots$, denotes the number of transitions. For example, in the case of the frequency scan performed at 150 °C in Fig. 6.18b, the dielectric loss $\varepsilon''(f)$ spectrum can be deconvoluted into three signals. These are plotted in Fig. 6.20 using functional forms based on the Havriliak–Negami parameter estimates

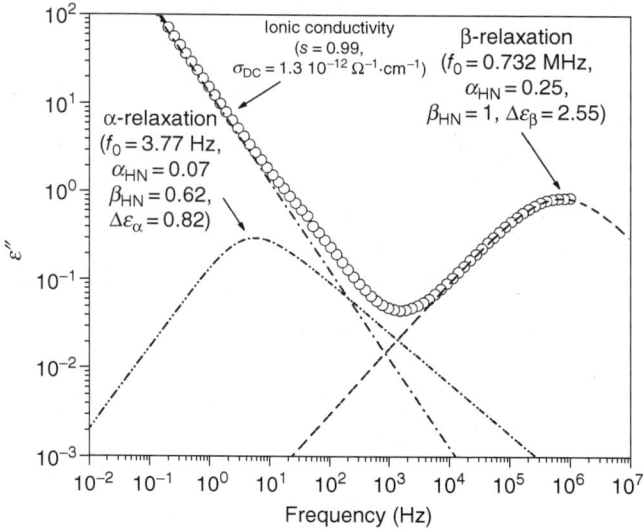

Figure 6.20. Components of the $\varepsilon''(f)$ spectrum of PMMA recorded at 150°C and the values of the fitting parameters obtained for each signal.

(Kalogeras 2008). In order to construct the Arrhenius plots from permittivity data, it is necessary to estimate the exact position of f_{max} on the frequency scale in the dielectric loss plot for each relaxation mode. This can be accomplished using appropriate relationships between f_{max} and the fitting parameters (f_0, α_{HN}, β_{HN}) [e.g., see Díaz-Calleja (2000)]. The low-temperature data shown in the Arrhenius plot for the β relaxation (Fig. 6.21) can be fitted to a straight line [Eq. (6.8), where $E = 85$ kJ/mol and $\log(\tau_0/s) = -16.5$]. It is worth mentioning that the apparent activation energies (E), as derived from the dielectric and dynamic mechanical analyses of relaxations in polymers, are rarely identical. For example, Kalogeras et al. (2001) compiled literature values for the apparent energy barrier of the α- and β-relaxation processes in PMMA phases from various thermal techniques (DRS, DMA, TSC), which suggest the following trend: $E_{TSC} < E_{DMA} < E_{DRS}$. This variation may be connected with the different modes of molecular diffusion that are coupled by the mechanical or electrical perturbation stresses. Another point of interest is the enormous scattering of the energy barriers: 50 kJ/mol $< E_\beta < 126$ kJ/mol and 37 kJ/mol $< E_\alpha < 1020$ kJ/mol. This observation emphasizes the dependence of the experimentally determined energy barriers from the tacticity of the polymer, aspects of the sample preparation (e.g., initiator type and content, dopants, contaminants, polymerization temperature, and time), and the sensitivity of each technique to specific motions. Although a direct link between such estimates may have little meaning, it may be of interest to monitor and compare the variation between materials with compositional or structural differences, as both methods disclose subtle

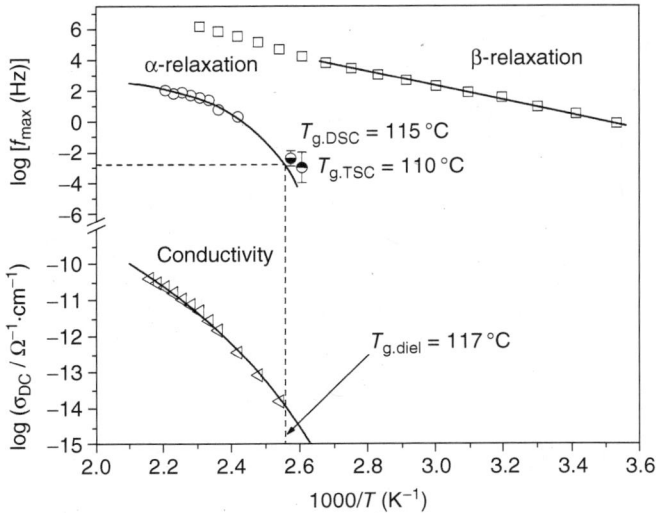

Figure 6.21. Arrhenius plot for the main relaxation mechanisms (α and β relaxations) and DC conductivity of a PMMA sample (Kalogeras, unpublished results).

structure- and composition-dependent perturbations of their relaxation dynamics.

With respect to the high-temperature signals, analysis of the Arrhenius diagram provides a wealth of information. Typically, using the Vogel–Tammann–Fulcher–Hesse (VTFH) equation for the relaxation frequency (f_{max}), one obtains several parameters describing the α-relaxation mechanism (in the present case: $\log(1/2\pi A) = 4.0 \pm 0.3$, $B = 369 \pm 31$ K, and $T_V = 366 \pm 6$ K). Apparently, the accuracy of their estimates is controlled by the number of points that are available to describe the curvature of the plot. The dielectric glass transition temperature ($T_{g,diel}$), defined by the convention $\tau(T_{g,diel}) = 100$ s, corresponds to the temperature at which the loss factor band peaks at $f_{max} = 1.6 \cdot 10^{-3}$ Hz. In the framework described above, the VTFH equation supplies a dielectric T_g estimate of 117 °C, which is obtained using the following relationship:

$$T_{g,diel} = T_V - \frac{B}{\ln(2\pi A f_{max})} \quad (6.21)$$

This value should be compared with $T_{g,TSC} \equiv T_{max,\alpha} = 110$ °C (at the equivalent frequency of ~10^{-3} Hz) and the DSC glass transition temperature of $T_{g,DSC} = 115$ °C (from second heating scans at 10 °C/min after cooling at 20 °C/min). A rough estimate of the equivalent frequency of DSC, namely, $f_{eq,DSC} = 2.6 \times 10^{-2}$ Hz, can be obtained within Donth's fluctuation model (Donth 1992) of glass transition from the relationship

$$f_{eq,DSC} = \frac{q'}{2\pi\alpha\delta T} \qquad (6.22)$$

where q' is the cooling rate, α is a constant on the order of 1, and δT is the mean temperature fluctuation (on the order of 2 °C). The reader is referred to the work of Hensel et al. (1996) for a discussion of the rules transforming cooling rates from DSC into frequencies, and to Section 2.13 (in Chapter 2), which allows measurement of the glass transition temperature in the mHz range. Note that the rate-dependent difference between the T_g values obtained from cooling and heating scans is less than 3 °C at 20 °C/min. Therefore, T_g values from slow heating scans may also be used to compose Fig. 6.21. Once the VTFH parameters are known, several other material properties and parameters can be estimated. These include the fragility index (Böhmer et al. 1993)

$$m = \left.\frac{d\log(\tau)}{d(T_g/T)}\right|_{T=T_g} = \frac{1}{\ln 10} \frac{B/T_g}{(1-T_V/T_g)^2} \qquad (6.23a)$$

which is indicative of the degree of inter- and intramolecular coupling between chains, the free-volume dilatation coefficient

$$\alpha_f = \frac{1}{B} \qquad (6.23b)$$

and the kinetic free volume fraction at T_g:

$$f_g = \frac{T_g - T_V}{B} \qquad (6.23c)$$

Finally, an informative representation of the DRS dielectric analysis data is through the real part of the AC conductivity (Fig. 6.22). At sufficiently low frequencies and high temperatures, conductivity becomes frequency-independent and the plateau value can be used to estimate the temperature variation of DC conductivity (σ_{dc}). Because of the coupling of segmental and space charge motions, the Arrhenius plot of the DC conductivity (log σ_{DC} vs. 1000/T; Fig. 6.21) of PMMA and several other polymers does not follow the classical Arrhenius-type dependence (linear), but shows a curvature accounted for by the relationship

$$\sigma_{DC}(T) = A\exp\left(\frac{B}{T-T_0}\right) \qquad (6.24)$$

with log $A = -5.2 \pm 0.9$, $B = -901 \pm 251$ K, and $T_0 = 287 \pm 18$ K. The physical meaning of parameter T_0 is ambiguous, although some evidence—not

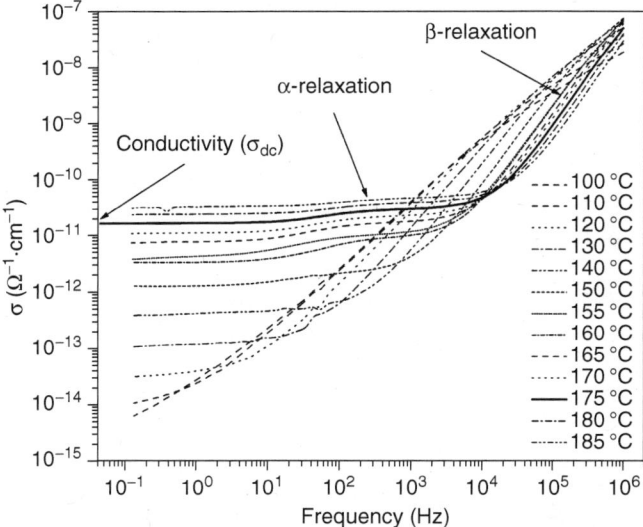

Figure 6.22. Frequency dependence of the real part of AC conductivity (σ) of PMMA recorded in the frequency range 10^{-1}–10^6 Hz at selected temperatures (Kalogeras and Vassilikou-Dova, unpublished results).

supported by the present analysis of PMMA—has been provided relating T_0 with the glass transition temperature [e.g., the data for α-PVDF; see Tuncer et al. (2005)].

6.6. DIELECTRIC ANALYSIS OF THERMOPLASTICS

Dielectric analysis of thermoplastics and related composites is a particularly active area of polymer characterization (McGrum et al. 1967; Hedvig 1977; van Turnhout 1980; Runt 1997). Parameters reflecting dielectric, electrical, or thermal characteristics (e.g., molecular interactions, conductivity mechanisms, glass transition temperatures, melting points), interesting kinetic problems (e.g., physical aging, crystallization) as well as key morphological aspects of thermoplastic systems (e.g., crystallinity, phase separation, miscibility) have been studied extensively. Furthermore, dielectric results were successfully related to results of other spectroscopic or scattering techniques, providing information with respect to biotechnological and engineering applications. This section provides an overview of dielectric information in relation to the nature, behavior, and analysis of the relaxation mechanisms and macroscopic charge transport phenomena in selected thermoplastic systems.

6.6.1. Simple Nonpolar Polymers

Dielectric analysis of nonpolar thermoplastics was made feasible by introducing a number of polar groups in the mainchain, such as carbonyls or hydroperoxy groups that are the products of oxidation. An alternative is provided by blending the polymer with small organic molecules, having a high dipole moment, which are used as "dielectric probes" or "dielectric labels." The probes sense the microviscocity in their close vicinity by exhibiting large angular fluctuations strongly coupled to the segmental (α) relaxation of the nonpolar chain. On the other hand, their relaxation behavior was reported to be almost unaffected by the local (β) relaxation dynamics or dynamics related to the crystalline phase (Kessairi et al. 2007; van den Berg et al. 2004).

The most representative example of the dielectric relaxation activity observed in this group is polyethylene (PE), which is a semicrystalline polymer. Depending on its synthesis, low-density (LDPE), high-density (HDPE), and linear low-density polyethylene (LLDPE) can be distinguished, and several types of HDPE exist. Mainly the branching is different in these polyethylenes. LDPE and LLDPE are essentially copolymers of ethylene with longer-chain olefins differing in the branch content and regularity of branching. Morphology, crystal structure, molecular motion, and dielectric properties of polyethylene have been studied extensively, in part as a result of its widespread engineering application in wire covering and high-voltage insulation. The dielectric spectra of semicrystalline PE samples typically show three relaxations, which derive from extraneous carbonyl groups and/or impurity dipoles (Das-Gupta 1994), with relaxation signal intensities dependent on the type of the polymer. In order of decreasing temperature, these are the complex α_c- (i.e., the α-crystalline, $T_{\alpha,c} \approx +100\,°C$ at $f = 10\,kHz$) relaxation signal, involving several dielectrically active processes together, the β relaxation [which is sometimes designated as α_a-(or α-amorphous) relaxation, at $T_\beta \approx 0–50\,°C$ at $f = 10\,kHz$] that is usually related to the glass transition, and the strong γ relaxation that corresponds to a crankshaft-type local motion in amorphous chains ($T_\gamma \approx -90\,°C$ at $f = 10\,kHz$) (Graff and Boyd 1994; Suljovrujic 2002).[3] Dielectric and mechanical measurements performed on different types of polyethylene demonstrate appreciable scattering in the values for the transition temperatures and the activation energy parameters [see, e.g., Suljovrujic (2002)]. This is in part a result of variations in the crystalline content and the degree of branching, orientation processes that provoke selected variations in free volume, as well as radiation processes that induce, among other effects, oxidative degradation. It is worth mentioning a review by Boyer (1973), which discusses the presence of a double glass transition [$T_g(U)$ and $T_g(L)$, upper and lower T_g, respectively] in partially crystalline linear polyethylene. Considering that the dielectric glass transition temperature corresponds to the temperature

---

[3] A crankshaft motion has been proposed in the solid state for molecular fragments consisting of three or more rotors linked by single bonds, whereby the two terminal rotors are static and the internal rotors experience circular motion (Schatzki 1962).

of the dielectric loss band at a scanning frequency of $\sim10^{-2}$ Hz, it appears that DEA gives in the case of semicrystalline polyethylene T_g estimates between $-50\,°C$ and $-20\,°C$, which lay close to $T_g(U)$, and also supports Wunderlich, who reported $-36\,°C$ as T_g of polyethylene from heat capacity measurements (Gaur and Wunderlich 1980). Of the available literature, one should also make special reference to the studies of Das-Gupta and Scarpa (1996) on the effects of aging on the polarization and dielectric relaxation behavior of electrically aged LDPE (i.e., study of polyethylene samples placed within alternating electric fields for several hours), and of Suljovrujic et al. (2001) on the effects of accelerated aging on the dielectric behavior of drawn (oriented) and γ-irradiated LDPE. In the latter report, the changes in the intensity and position of the γ- and β-relaxation signals were found to be strongly dependent on the changes in the microstructure of the amorphous phase induced by crosslinking and oxidation.

Detailed dielectric information can be found in the literature for several other nonpolar thermoplastics; examples include isotactic (semicrystalline) or atactic (mostly amorphous) polypropylene, polyisobutylene, and the highly crystalline fluorocarbon polymer poly(tetrafluoroethylene) (PTFE) (Kessairi et al. 2007; Krygier et al. 2005). In the case of PTFE, the intense dielectric activity is considered an effect of oxidation, which creates carbonyl groups and/or peroxy radicals (–COO·) in the mainchain (Hedvig 1977; Enns and Simha 1977). However, great difficulty exists in attributing the signals to specific motions in either amorphous or crystalline regions, owing their complex dependence on changes in crystallinity. Part of this arises from the impure or "labeled"—also referred to as "decorated"—PTFE samples studied by DEA methods in the past. More importantly, confusion arises because of the complexity of the relaxation dynamics, including the detection of a cooperative or high-activation-energy, low-temperature transition (i.e., the γ relaxation), as well as a more typical but difficult-to-detect high activation energy transition (the α relaxation), which has some similarities to a glass transition (McCrum et al. 1967). Starkweather and coworkers performed a series of DEA (Starkweather et al. 1992, 1994) and DMA (Starkweather 1984) studies of the plasticizing effects of several chemicals (chloroform, carbon tetrachloride, fluorocarbon-113, etc.) on the relaxation activity in PTFE. These studies revealed that the α transition [$T_\alpha \approx +110\,°C$ at $f = 1\,kHz$ (Sauer et al. 1996)] shifts to lower temperatures with volume fraction of the absorbed chemical compounds, whereas the γ transition [$(T_\gamma \approx -90\,°C$ at $f = 1\,kHz$ (Sauer et al. 1996)] does not. Besides that, the α transition in 95% crystalline solution-grown single crystals of PTFE disappears, whereas the γ transition remains, presumably because short-range local motions are still possible in the crystalline fraction. The transition temperature of the γ relaxation also remains essentially constant with crystallinity, as would be expected for a transition attributed to local motions. In contrast to the dielectric studies of "labeled" PTFEs, the AC dielectric data for undecorated poly(tetrafluoroethylene) samples show no evidence of the α_c crystal transitions (at 19 °C and 30 °C, as detected by DSC and dilatometry) since the dipolar changes associated with these crystal–crystal transitions are extremely weak. Note that these crystal transitions are clearly recorded in DMA studies, which additionally reveal the insensitiveness of this relaxation to any chemical compound absorbed in the amorphous fraction of the polymer (Starkweather 1984).

6.6.2. Polymers with Flexible Polar Sidegroups Attached to an Apolar Mainchain

Thermoplastics with apolar mainchains and polar sidegroups, such as the acrylate and the methacrylate polymers (Scheme 6.1), have been widely studied by dielectric techniques. In such polymers, and in contrast to the response observed in the dynamic mechanical spectra, the strongest dielectric signal corresponds to the subglassy β relaxation that originates from the hindered rotation of the polar sidegroups, followed by the weaker segmental (α) mode at much higher temperatures (see Fig. 6.19 for PMMA). In general, the relaxation characteristics of these molecular motions are critically determined by the size (volume and length) and the polarity of the R_1 and R_2 groups. Several trends have been reported and confirmed for the change of the dielectric β- and α-relaxation modes with structural and compositional adaptations of the thermoplastic polymer, especially for the case of the poly(n-alkyl methacrylates). For example, by increasing the size of the ester sidegroup R_2 (from methyl, to ethyl, n-propyl, n-butyl, etc.), the strength of the dielectric α relaxation increases and that of the β relaxation decreases. At the same time, the α transitions are shifted to lower temperatures ("internal plasticization" effect) and merge with the secondary signals (Hedvig 1977; Dudognon et al. 2004). Such an increase of the chain mobility (reduction of T_g) with the addition of flexible R_2 alkyl groups is similar to that induced by the addition of low-molecular-weight plasticizers (toluene, tricresyl phosphate, etc.). Furthermore, the stereochemical structure of the polymer controls the dielectric activity of the material. This becomes clear if one considers the dependence of the effective dipole moment on the tacticity (isotactic, atactic, or syndiotactic), and the differences in the physical structure, packing, and molecular mobilities due to the dissimilar stereochemistry.

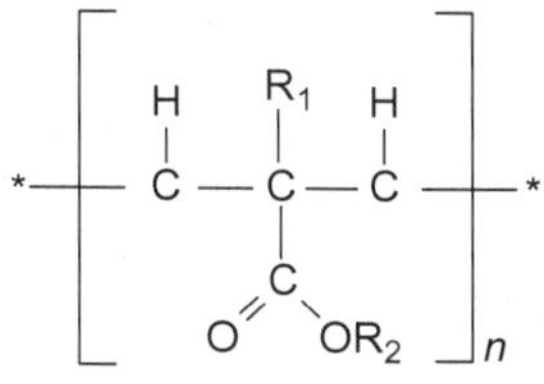

Scheme 6.1. Chemical structure of acrylate (R_1 = H) and methacrylate (R_1 = CH_3) polymers, both possessing a nonpolar mainchain and a polar sidegroup.

In addition to the analysis of the typical molecular relaxation mechanisms, more recent dielectric studies also addressed the behavior of a controversial transition, the so-called liquid–liquid transition in poly(n-alkyl methacrylates) (Dudognon et al. 2003; 2004). Early experimental evidence for the occurrence of a liquid–liquid transition in several polymers, such as polystyrene, PMMA, and polybutadiene, has been reviewed by Boyer (1979b), who suggested that this transition exhibits both kinetic (relaxational) and thermodynamic aspects (Boyer 1979a,b). Murthy (1993) indicates that this phenomenon is due to a smooth changeover of the liquid from one dynamic regime to the other and hence is not due to any real phase transition. The liquid–liquid transition typically appears at an absolute temperature (T_{LL}) between $1.1T_g$ and $1.2T_g$ [T_g (K), (Dudognon et al. 2002a,b)] and is usually considered to involve further loosening of the cohesive liquid state that persists above T_g. As the temperature increases, the chains become less entangled and begin to unfold, and their movement involves longer chain segments, producing this dielectric mode. The signal shows a narrow relaxation time distribution and is dependent on the chain structure [e.g., by showing internal plasticization with increasing alkyl group length (Dudognon et al. 2003, 2004)] or dopants [e.g., by being antiplasticized—i.e., appearing at higher temperatures—in blends with dye molecules (Kalogeras et al. 2006b, 2007a)]. Interest in this process and the underlying chain motions remains strong since a number of processes show significant characteristics at this temperature range; for example, melt processing, such as extrusion, injection molding, and compression molding, requires temperatures significantly above T_g, and the optimum melt processing temperature is usually at least $1.2T_g$.

6.6.3. Polymers with Polar Sidegroups, Rigidly Attached to an Apolar Mainchain

Representative examples of this group of thermoplastics are several halogenated polymers[4] where chlorine, bromine, or fluorine is linked to the carbon–carbon chain and the C–Cl, C–Br, and C–F groups cannot move independently of the chain. In such polymers, besides the main relaxation signal related to the segmental motions, several other relaxations may appear. The latter involve dielectrically active local motions, such as isomerization of the mainchain, but not relaxation of the rigid polar sidegroups. The corresponding dielectric signals show great sensitivity to intermolecular and intramolecular interactions between the sidegroups (Hedvig 1977). A comparison of relaxation data collected for poly(vinyl chloride) (PVC), based on dynamic mechanical measurements and dielectric relaxation spectroscopy, bring to light interesting observations with respect to the relaxational activity in this group

---

[4] Not all halogenated polymers are polar, nor can they be included in this subgroup of thermoplastic polymers (e.g., see the discussion of PTFE in Section 6.6.1). Interesting results on the dielectric properties of fluoropolymers have been reported by Avakian et al. (1997).

of polymers. Based on DMA data, conventional PVC (crystallinity < 5%, when the polymerization temperature is above room temperature) exhibits six short-scale relaxations, which are related to the rotational mobility of the CH_2 groups [$\beta(CH_2)$ and $\beta'(CH_2)$ relaxations] and polar CHCl groups [$\beta(Cl)$ and $\beta'(Cl)$ relaxations] and to the dissociation of dipole–dipole CCl $\cdots$ CCl bonds between chains, which play the role of weak physical crosslinks [$\pi(Cl)$ and $\pi'(Cl)$ relaxations]. The splitting of the transitions into two close components results from the presence of two types of microstructure in the PVC chains, isotactic and syndiotactic fragments, which affect both the dipole moments of CHCl groups and intermolecular interactions. In contrast to DMA, using DEA, one observes strong α-($T_\alpha \sim 70$–$90\,°C$) and λ ($T_\lambda \sim 140$–$160\,°C$) relaxations and only a single but broad subglassy $\beta(Cl)$ relaxation ($T_\beta \approx -50\,°C$) (Hedvig 1977). The DMA and DEA data fit a common frequency–temperature curve for the $\beta(Cl)$ relaxation (Bartenev et al. 2000). The other transitions pass unnoticed dielectrically, due to either the absence of electric dipole moments in the CH_2 groups or a low concentration of the corresponding polar (CHCl) groups or to the typically low dielectric activity of the groups participating in the physical crosslinks. For the α relaxation of relatively "free" segments (i.e., noninteracting) and the high-temperature large-scale λ relaxation (from "bound" segments; i.e., with weak C—H $\cdots$ Cl—C linkages), both DMA and DEA give comparable results. These are also well described by similar frequency–temperature dependences (Bartenev et al. 2000). The aggregate structure of PVC in conjunction with the particularly effective conductivity mechanism also produces a significant interfacial (MWS) polarization mode. Of the dielectric publications, special reference should be made to the study of internal plasticization phenomena in PVC (Elicegui et al. 1998) and the attempt to establish correlations between the polymer's stereochemical microstructure and molecular dynamic behaviors (Guarrotxena et al. 2004).

Another important member of this group is poly(vinylidene fluoride) (PVDF), of which different crystal structures (crystal forms α and β) have been studied extensively. The dielectric (DRS and TSC) spectra of the nonferroelectric, nonpiezoelectric, α-form PVDF show an intense low-temperature β-relaxation signal with several components ($T_\beta \approx -100$ to $-80\,°C$) and a high-temperature signal, the α_a relaxation [at $T_{\alpha,a} \approx -30\,°C$, $f = 1\,kHz$ (Grimau et al. 1997)], which is the dielectric manifestation of the glass transition. At much higher temperatures, the α_c process, associated with the crystalline phase, is present [at $T_{\alpha,c} \approx +100\,°C$, $f = 1\,kHz$ (Sy and Mijovic 2000)]. With removal of a fluorine atom [i.e., passing from poly(vinylidene fluoride) to poly(vinyl fluoride) (PVF)], both the temperature and the dielectric strength of the α_a-relaxation signal increase. For example, $T_{\alpha,a}$ shifts to approximately $+40\,°C$ in the case of PVF. The change in the dielectric strength illustrates the partial cancellation of the dipole moments in PVDF. Of the available results for PVDF containing the β-crystal form, it is worth referencing the dielectric information acquired for the effects of poling on the ferroelectric and piezoelectric properties of the polymer (Sencadas et al. 2005) and for the effects of the crystalline phase and uniaxial drawing on its glass transition (Gregorio and Ueno 1999).

6.4.4. Polymers with Polar Mainchain

In this class of thermoplastic polymers the polar group is directly incorporated in the main chain. Aspects of the dielectric activity of selected thermoplastics without formation of hydrogen bonds between chains (e.g., oxide polymers, polyesters, liquid crystalline polymers), are discussed in the following sections.

6.6.4.1. Polyoxides Polyoxides, such as polyoxymethylene (POM), poly(ethylene oxide) (PEO), and poly(propylene oxide) (PPO), are semicrystalline thermoplastic polymers that are distinguished by the presence of polar C—O—C bonds in their mainchains. Several relaxations have been isolated and comprehensively studied by dielectric as well as dynamic mechanical spectroscopy, and subsequently attributed (not always in a straightforward manner) to the amorphous or crystalline microphases in the material [e.g., see Sauer et al. (1997) for information on the relaxation activity of POM]. The high-temperature α_c relaxation at 130–150°C is frequently associated with chain motions at the surface of the crystallites. At somewhat lower temperatures, and only in selected polymers, the spectra show a normal-mode (n) relaxation (see Section 6.2.4.3). Typically, the isochronal dielectric spectra of oxide polymers demonstrate the α_a relaxation, also often referred to as the β relaxation, which is the dielectric manifestation of the glass transition. Its relaxation strength ($\Delta\varepsilon$) and position in an isochronal dielectric spectrum ($\approx T_g$) is highly dependent on the mass fraction of the amorphous phase, the degree of packing—usually different from polymer to polymer—and the nature and physical and chemical characteristics of the other component (when used in polymer blends, networks or composites). Crystal-induced constraints on the cooperative motions of segments located in amorphous–crystalline interfaces cause broadened glass transition signals and a considerable increase of T_g over the value in the purely amorphous material; this phenomenon is rather typical irrespective of the thermoplastic polymer considered [e.g., POM, PEO, PE, PTFE, and related linear flexible semicrystalline polymers (Sauer et al. 1997)]. Broad low-temperature γ relaxations in oxide polymers are typically related to the motion of the hydroxyl endgroups and/or to local twisting motion of the mainchain. The strength and the relaxation rate of such processes are relatively insensitive to changes in the molecular environment, although intense plasticization effects have been reported with moisture absorption (Wasylyshyn 2005; Kyritsis and Pissis 1997). Changes in the local molecular environment are possible by a number of means, including the use of high- or low-molecular-weight diluents, crystallization (McGrum et al. 1967), or interpenetrating network formation (Kalogeras et al. 2006b, 2007a). The complexity of the γ-relaxation signals in polyoxides is related to the number of participating motions in different environments that include local relaxation motions in the noncrystalline phase (main contribution) and relaxations due to dislocations or defects in the crystallites (low-temperature feature of the γ signal). From the preceding discussion, the reader will witness the close relationships between most of the relaxation signals present in polyoxides and in polyethylene (Section 6.6.1).

6.6.4.2. Polyesters and Related Thermoplastic Polymer Structures In polyesters and polycarbonates, the presence of the benzene ring makes the chain much stiffer, and thus the dielectric relaxation modes appear at temperatures much higher than those observed in polyoxides. Dielectric information for poly(ethylene terephthalate) (PET; Scheme 6.2a) and its naphthoic terephthalate analogue, poly(ethylene-2,6-naphthalate) (PEN; Scheme 6.2b), is discussed in the following paragraphs as these polymers offer characteristic examples of the relaxation activity in polyesters and its dependence on crystallinity and processing (e.g., orientation, by uniaxial or biaxial drawing performed above the glass transition temperature).

Of the various dielectric and dynamic mechanical signals observed in the two polyesters and the polycarbonate mentioned above, the best explained is the β-relaxation mode, which is assigned to local fluctuations of carbonyl groups alone, or motions of ester groups associated with flips of the aromatic phenyl ring, in PET or polycarbonate. The β relaxation was also ascribed to the onset of the naphthalene ring rotation in PEN (Wübbenhorst et al. 2001; Hardy et al. 2002). The dielectric β relaxations [in the range from −120 to −50 °C, at ~1 Hz, with $T_{\beta(PEN)} < T_{\beta(PET)}$] demonstrate an Arrhenius-type temperature dependence of the relaxation times, and occasionally the corresponding signals can be decomposed into a pair of highly overlapping relaxations. The low-temperature component (β_1) observed in isochronal dielectric plots provides the main contribution in dried samples, with an amplitude effectively influenced by changes in the draw ratio, while the high-temperature compo-

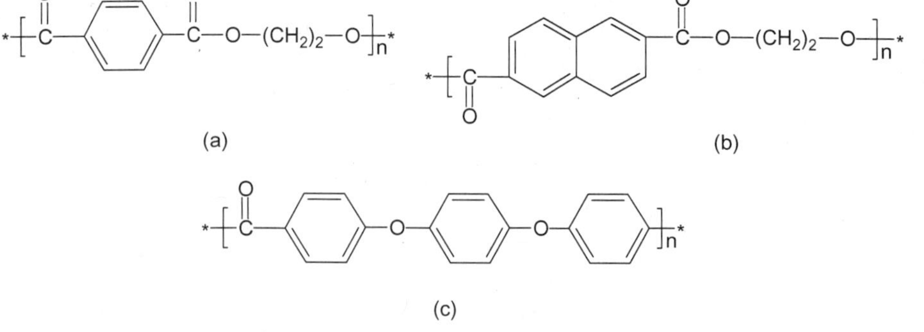

Scheme 6.2. Chemical structures of: (a) poly(ethylene terephthalate) (PET); (b) poly(ethylene-2,6-naphthalate) (PEN); and (c) poly(ether ether ketone) (PEEK).

nent (β_2) is frequently related to the presence of water, which is strong in as-received samples while it disappears almost completely on dry samples (Hakme et al. 2005). Crystallinity decreases the strength of the β-relaxation signal, with no significant change in its transition temperature. The latter shifts to lower temperatures with increasing length of the aliphatic part of the chain [e.g., by ~20 °C in going from the two methylene groups in PET to four methylene groups in poly(butylene terephthalate)], since this change enhances the mobility and the local motions along a few consecutive atoms. This behavior is consistent with the result obtained by Boyer (1977), who had observed that lengthening of the alkene unit shifts the β transition down to the value characteristic of the polyethylene secondary transition. Another signal, the β* relaxation (in the range 0–90 °C, at ~1 Hz), emerges close to the segmental relaxation (α) signal. This is the least understood relaxation process in PEN, attributed to out-of-plane rotational motions of the naphthalene group, and does not exist in the other polymers considered here. Its high activation energy implies a cooperative molecular mechanism that has prompted researchers to consider also an attribution of the β* process to motions of naphthalene aggregates and of the polar ester groups located close to them (Hakme et al. 2005) or even to local orientation order of PEN that was suggested to be considered as a mainchain liquid crystalline structure (Wübbenhorst et al. 2001).

For *completely amorphous* PET and PEN materials, the spectral position of the dielectric (DEA and TSC) signals related to the glass transition is close to the DSC T_g. For example, the TSC α peak [at $T_\alpha = 66\,°C$ at $f \sim 1\,mHz$ (Saiter et al. 2003)] and the dielectric loss α peak (ε'' peak at ~90 °C, at $f = 1\,kHz$ (Psarras et al. 2006)] in PET, are close to the midpoint of the heat capacity step; the deviation in part reflects the dissimilarity between the equivalent frequencies of the probe methods. The α relaxation is more susceptible, compared to the local β relaxation, to the influence of free-volume fraction, crystallinity and other macrofactors; an increase of crystallinity in PET broadens and diminishes the intensity of the dielectric α peak, moves the peak to higher temperatures, affects the distribution of relaxation times, and substantially reduces the apparent activation energy barrier (Pratt and Smith 2001). Some of these features are also influenced by the application of pressure, stretching, and orientation (Dargent et al. 2005). The naphthalene group in the repeating unit of PEN provides higher stiffness to the mainchain compared with PET, which has a *para*-substituted phenyl group in the backbone. This raises the DSC glass transition temperature of *amorphous* PEN to 125 °C [ε'' peak near 135 °C, at $f = 1\,kHz$ (Psarras et al. 2006)]. In spite of this shift, the overall behavior of the dielectric segmental mode in PEN, which changes with orientation and/or crystallinity, is similar to that observed in PET (Hardy et al. 2002; Hakme et al. 2005). The flexibility of the mainchains of polyesters is increased when the chain length between the rings is increased [from ethylene in poly(ethylene terephthalate), to propylene in poly(propylene terephthalate) (PPT), and butylene in poly(butylene terephthalate) (PBT), etc. (Hedvig 1977;

Pratt and Smith 2001)]. This substitution affects the packing of the chains and reduces the temperatures both of the α transition (T_g) and the secondary relaxations ($T_β$). In contrast to the segmental relaxation signal, the β relaxation, which involves localized motions, is only weakly dependent on changes in crystallinity and applied pressure (i.e., variations of free volume), although as was shown by Hakme et al. (2005), deformation increases the strength of its low-temperature ($β_1$) component.

The interesting molecular dynamics of poly(ether ether ketone) (PEEK; Scheme 6.2c) have also been extensively studied by DEA, over broad frequency and temperature ranges, allowing characterization of polymer dynamics in the glassy, semicrystalline, and molten states (Nogales et al. 1999). Attention has focused on the relaxation at the interface between amorphous and semicrystalline PEEK, identifying two amorphous phases ["mobile" or traditional and rigid amorphous phase (Huo and Cebe 1992)]. The subglassy relaxation behavior involves highly localized (noncooperative) wagging of the polar bridges (γ relaxation), which is insensitive to aging history and details of the semicrystalline morphology, and cooperative motions of polar bridges and flips of phenyl rings (β relaxation) (Goodwin and Hay 1998). The energy barriers controlling the onset of β motions are intermolecular in nature. Thus, perturbations of the environment of the relaxing moiety (water content, aging history, details of the semicrystalline morphology) drastically affect this mechanism (Kalika and Krishnaswamy 1993). The glass transition signal (α relaxation) appears in a substantially higher temperature range, compared to PET and PEN, and is related to the highly cooperative motions of segments incorporating several phenylene rings. The strength of the dielectric α-relaxation signals indicates the presence of a rigid amorphous fraction in semicrystalline samples, in which the amorphous segments, mainly at the crystal–amorphous interface are immobilized by the presence of crystallites. This rigid amorphous phase relaxes at temperatures above T_g (Menczel and Wunderlich 1981, 1986). In accordance with this scenario are reports of bimodal α and β relaxations for semicrystalline PEEK (Krishnaswamy and Kalika 1994; Huo and Cebe 1992). Following the conventional behavior of "nanoconfined" thermoplastic polymers, the glass transition temperature of completely amorphous PEEK samples ($T_g = 148\,°C$) increase with an increase in the crystalline phase content. For example, $T_α$ increases to ≈155 °C for 15% and to 170 °C for 37% crystallinity [(ε″ peak at 10 Hz (Korbakov et al. 2002)]. DEA and TSC also provide ample experimental evidences of the increase in the DEA glass transition temperature with thermal treatment or electron beam irradiation. In the latter case, the increase is a result of crosslinking of chains via radiation-induced free radicals (Shinyama and Fujita 2001; Li et al. 1999).

6.6.4.3. Liquid Crystalline polymers
Liquid crystalline polymers (LCPs) demonstrate an anisotropic liquid state with an orientational order intermediate between that of three-dimensional lattices and isotropic liquids. These polymers exhibit electro- and magnetooptical properties similar to those

observed for the low-molecular-weight liquid crystals, while also having the good film-forming qualities that are valuable attributes of the high-molar-mass polymers. Thermal-analysis-related spectroscopic studies of the molecular motions have been carried out on several LCPs, including the important groups of thermotropic sidechain and mainchain LCPs, and a number of relaxation signals have been reported and are briefly discussed in the following paragraphs. In thermotropic LCPs one or more liquid crystalline phase(s) may be detected along the temperature axis, where polar mesogenic groups, with a highly anisotropic geometry and rigid structure, show some orientational ordering. For further information on thermotropic LPCs, the reader is referred to the excellent review of Simon (1997), while the dielectric relaxation behavior of lyotropic LCPs (i.e., rigid molecular materials that show liquid-crystal properties if dissolved in a nonmesogenic solvent at certain concentrations) has been discussed in detail by Moscicki (1992).

In the sidechain LCPs, mesogen moieties are covalently linked to the polymer backbone laterally, by means of a flexible spacer. The function of the spacer is to decouple and enhance, to some extent, the motion of the mesogenic sidechain from the—usually acrylate of siloxane—mainchain, and at the same time it allows the formation of different (ordered) liquid crystalline mesophases [nematic, smectic, cholosteric, etc.; see Simon, (1997)], which are located between the glass transition temperature and the isotropization or clearing temperature (T_{cl}). The reorientational dynamics in these polymers have attracted particular interest since the response time of the structure to an external electric field is a key determinant for their application in the display industry. In the sidechain LCPs one may find, in order of increasing transition temperature, isolated (noncooperative) relaxations of short-chain units (β, γ, ...) with properties similar to those found in conventional amorphous thermoplastic polymers, the α relaxation, a new relaxation labeled δ, and at much higher temperatures (or at lower frequencies in isothermal scans) strong MWS polarization effects due to the locally inhomogeneous environment in the material. The relaxation activity in the glassy state (at $T < T_g$) involves local motions of sufficiently long spacer groups, rotational motions of terminal substituents in the mesogens, as well as internal motions within the same units (e.g., motions of the central ester moiety between the aromatic rings in the mesogen core). The molecular nature of the relaxation activity in the sidechain LCPs remains controversial, especially in the case of the α relaxation, which appears at a transition temperature well below that of the first-order nematic $\rightarrow$ smectic and the smectic $\rightarrow$ isotropic transitions (Floudas et al. 2003), but close to the glass transition temperature determined by DSC. This dielectric relaxation is usually discussed either in terms of the cooperative motions associated with the glass transition or as the manifestation of special motions, involving mainly the transverse component of the mesogenic dipole moment (Mano and Gómez Ribelles 2003). The dielectric δ-relaxation signal is situated in the LCP phase and is associated mainly with the mobility modes of the longitudinal component of the dipole moment of the mesogens, characterized by a very narrow distribution of relaxation times. This relaxation is also mechanically active, but not detected by DSC, likely because of instrument sensitivity issues. Moreover, the δ relaxation exhibits a certain degree of cooperativity (i.e., evident from correlations between motions of the relaxing units), despite involving only fluctuations on the dipolar moment of the mesogenic groups (Mano 2003). Interesting dielectric studies have been devoted to the influence of the polymer architecture (molecular weight, polymer backbone, spacer length, nature of bridging group, terminal substituent, etc.) on the relaxation dynamics of sidechain LCPs [see Simon (1997) for a review]. Furthermore, significant DEA results have been presented for the temperature and pressure dependences of the relaxation dynamics—and in particular the behavior of the glass → nematic, nematic → smectic, and smectic → isotropic phase transition temperatures—in sidechain LCPs with a poly(methyl acrylate) backbone and a (p-alcoxyphenyl)-benzoate mesogen group (Floudas et al. 2003).

The mainchain LCPs, which usually exhibit a nematic phase, find significant engineering applications in electrical components, in fibers, as reinforcing materials, and so on. A series of thermotropic mainchain liquid crystalline polyesters have been extensively studied by dielectric and other thermal techniques to clarify the influence of variations in the type of the chain segments, as well as of crystallinity and orientation, on dielectric and thermal characteristics of the material (Avakian et al. 1996; Boersma et al. 1998b). One of the best-known examples of rigid aromatic mainchain liquid crystalline copolyesters is the Vectra group. Of the commercially available members, the most widely studied are Vectra A950 [p-hydroxybenzoic acid (HBA)–6-hydroxy-2-naphthoic acid (HNA) copolyester] and Vectra B950 [HNA–terephthalic acid (TA)–aminophenol (AP) copolyesteramide]. For Vectra A950, three relaxations have been observed (Sauer et al. 1993; Collins and Long 1994; Avakian et al. 1996) and attributed, from high to low temperature, to the glass transition (α relaxation; at 100 °C at 10 Hz), rotation of the hydroxynaphthoate units (β relaxation; at ~35 °C at 10 Hz), and rotation of the hydroxybenzoate units (γ relaxation; at ~ −50 °C at 10 Hz) (Simon 1997). In the case of Vectra B950, in which there is greater hydrogen bonding between chains, analogous signals have been reported (Boersma et al. 1998a), corresponding to the glass transition (α), HNA moieties (β), TA moieties (γ), and the combined rotation of TA and AP moieties (δ relaxation). For the latter material, the combination of TSC and DMA analysis was used to distinguish between dipole and charge relaxations. TSC and DRS results were found to compare well for the low-temperature dipole relaxations, but some inconsistency appears in the case of the α relaxation, due to the influence of space charges. For random copolyester mainchain LCPs with widely varying crystallinities, including highly amorphous samples, very broad glass transition regions were observed and attributed to structural heterogeneity of the chains [e.g., see the thermal sampling method and DSC data of Sauer et al. (1993) for a moderately crystalline Vectra A950 polymer]. On the contrary, structurally regular mainchain LCPs exhibit relatively sharp glass transitions more comparable to those of ordinary isotropic amorphous or semicrystalline polymers.

Note that several LCPs exhibit ferroelectric properties in chiral tilted smectic phases (e.g., S_c^* phase). Broadband dielectric spectroscopy studies on such polymers have reported two collective relaxation modes at frequencies below 1 MHz: the Goldstone mode and the "soft" mode. The Goldstone mode is not a conventional relaxation process. It is assigned to thermal fluctuations or to modulations induced by the field, respectively, of the phase angle of the helical superstructure, which is connected to the different polarization vector of the smectic layers. The "soft" mode is attributed to amplitude fluctuations of the ferroelectric helix. In the microwave range (1 MHz–10 GHz) the β relaxation is detected, and is assigned to fluctuations in the mesogenic groups along their long molecular axis [for further information, see Kremer (1997)].

6.6.5. Polymers with Linear Hydrogen-Bonded Mainchains

Materials such as the linear or low branched polyurethanes and the polyamides fall into the category of polar mainchain polymers, but their relaxational response is somewhat distinctive, principally because of the occurrence of hydrogen bonds between their chains.

6.6.5.1. Polyurethanes Segmented polyurethanes are typical linear block copolymers of the type $(A–B)_n$, where block A consists of a relatively long and flexible soft segment (mainly ether, ester, or diene groups) and block B is a highly polar hard segment (mostly aromatic diisocyanates). Since the mid-1990s, in particular, the molecular dynamics and phase morphology of thermoplastic linear/low-branched polyurethanes have been studied in detail by TSC and DEA, for their utilization in specialized technological and biomedical applications (Frübing et al. 2002; Charnetskaya et al. 2003; Okrasa et al. 2005; Kalogeras et al. 2005b). A representative example of a TSC spectrum with the typical relaxations found in polyurethanes is shown in Fig. 6.23a. The γ (at $T\gamma \sim -165\,°C$) and β (at $T_\beta \sim -120\,°C$) relaxations (Pissis et al. 1996) arise from the crankshaft motions of $(—CH_2—)_n$ sequences (located between the hydrogen bonds) with attached dipolar groups and the rotational mobility of carbonyls, respectively. These mechanisms exhibit a spectrum of activation energies and relaxation strength strongly dependent on the packing density and the interchain interactions (Georgoussis et al. 2000b). The β relaxation shows particularly strong plasticization effects in hydrated polyurethanes but can also be effectively hindered when the carbonyls involved in this rotational relaxation strongly interact with neighboring segments. The α relaxation signal (with a peak temperature of $T_\alpha \sim -60\,°C$, near the DSC T_g) typically arises from reorientation of polar soft segments in the amorphous regions. Thermal sampling analysis of α relaxations in polyurethanes (Hsu et al. 1999; Roussos et al. 2004) reveals a typical compensation law behavior (see Section 6.5). Modification in the packing density and the morphology of the polyurethane

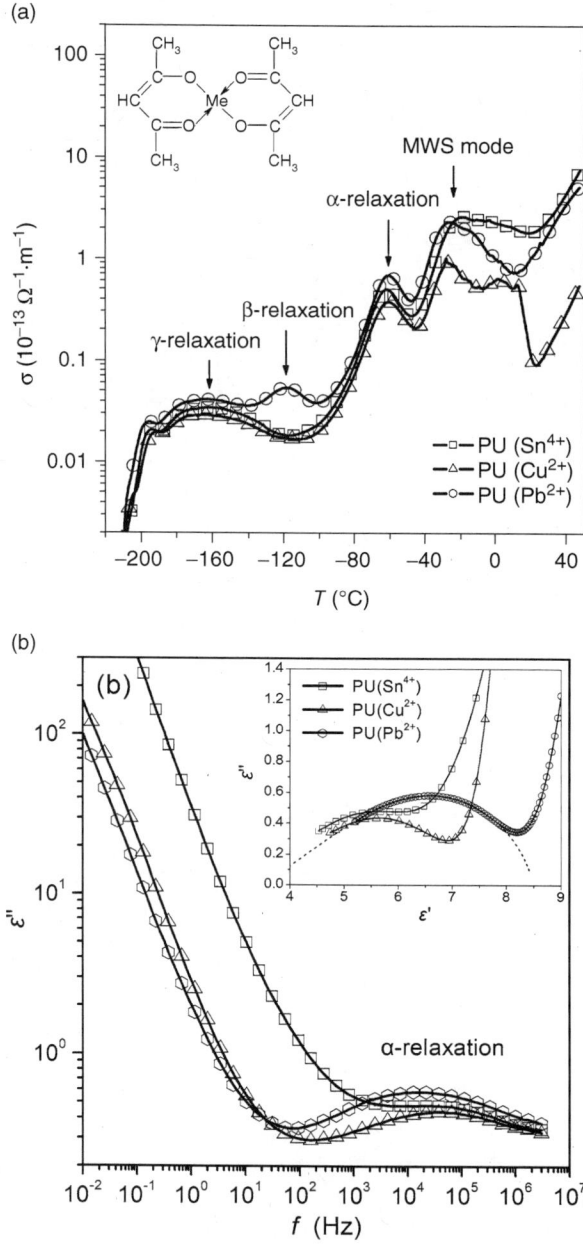

Figure 6.23. (a) TSC spectra for dry 0.5-mm polyurethane films. Soft segments consist of oxytetramethylene glycol (OTMG) units and hard segments consist of diisocyanate units (MDI). The samples were prepared with different transition metal acetyl acetonates, $Me^{x+}(AcAc)_2Cl_{x-2}$, as chain extenders (inset; Me^{x+} = Sn^{4+}, Cu^{2+}, or Pb^{2+}). Experimental conditions: $T_p = 27\,°C$, $E_p \approx 1\,MV/m$, $t_p = 10\,min$, $q = 5\,°C/min$. (b) DEA loss factor $\varepsilon''(f)$ plots recorded at $-10\,°C$ for the abovementioned materials placed in a gold-plated brass electrode capacitor. The inset of plot 6.23(b) shows the Cole–Cole plot (ε'' vs. ε') and a fit of the α signal (dashed curve) to a Havriliak–Negami function [Eq. (6.6)]. [Adapted from Kalogeras, I. M., Roussos, M., Vassilikou-Dova, A., Spanoudaki, A., Pissis, P., Savelyev, Yu. V., Shtompel, V. I., and Robota, L. P. (2005), *Eur. Phys. J. E* **18**, 467. Copyright 2005. With kind permission of Springer Science and Business Media.]

component in polymer blends has considerable impact on the α-relaxation strength (Georgoussis et al. 2000b). Moreover, reported anomalies in the shape of the dielectric α signals were connected to a broadened dispersion of microphase sizes (Kanapitsas et al. 1999). At high temperatures, mesoscopic transport of charge carriers (i.e., where the motion of ions confined to relatively short distances) accompanied by Maxwell–Wagner–Sillars polarizations (i.e., trapping of charges at interfaces arising from the nanophase-separated structure) and DC conductivity effects (i.e., macroscopic motion of charges) become important.

Dynamic dielectric experiments offer information complementary to that obtained from TSC. The variety of data presentations, such as those in the form of the loss factor [$\varepsilon''(f)$] (Fig. 6.23b)] of the real part of AC conductivity [$\sigma(f)$ (Fig. 6.24a)] or the electric loss modulus [$M''(f)$ (Fig. 6.24b)], permits a detailed description of the relaxation response. For example, the Argand complex plane plots of dielectric loss factor versus dielectric constant (ε''–ε' curves, also known as *Cole–Cole plots*; inset of Fig. 6.23b) show a strongly asymmetric shape for the curve related to the segmental mode. This should be contrasted to the perfect semicircle, with its center on the ε' axis, that is to be expected on the basis of the Debye theory of dielectric relaxation [Eq. (6.5)]. The complex structure observed in the frequency-dependent conductivity spectrum is resolved in the electric modulus representation (inset of Fig. 6.24a). The Maxwell–Wagner–Sillars interfacial polarization mechanism, typically hidden under the strong low-frequency conductivity losses, appears as a clear peak between 10^{-1} and 10^3 Hz in the electric loss modulus spectrum and as a "knee" (i.e., a stepwise transition) in the $\sigma(f)$ spectrum (Pissis et al. 1998; Georgoussis et al. 1999, 2000a; Charnetskaya et al. 2003). The latter feature provides a means for distinguishing between "conductivity relaxation" (CR) and MWS effects. The low-frequency divergence in ε'', caused by strong ionic mobility, appears as a clear CR peak in the $M''(f)$ plots (Fig. 6.24b). The shift of this peak with temperature provides information for the conductivity process, such as the conductivity relaxation time (τ_M) and σ_{dc} (which is proportional to the frequency of the peak) (Howell et al. 1974). The overlap of the interfacial and conductivity relaxations depends on the electrode material; for example, compare the spectra obtained with Au-plated brass (inset of Fig. 6.24a) and brass electrodes (Fig. 6.24b). This reveals the role of the sample–electrode interface and the necessity for careful preparation of the capacitor (see Section 6.4).

The Arrhenius plot of these polyurethanes is representative of thermoplastics with phase-separated structure (Fig. 6.25). The data for the DC conductivity show a typical VTFH-type behavior [Eq. (6.24)] (Tuncer et al. 2005), consistent with the coupling of the conductivity mechanism with cooperative segmental motions usually observed in linear polyurethanes and several other thermoplastics. The glass transition temperatures determined by DEA ($T_{g,diel}$), DSC ($T_{g,DSC}$), and thermally stimulated current ($T_{g,TSC}$) show very good agreement. In addition, the majority of published works on polymers [e.g., see

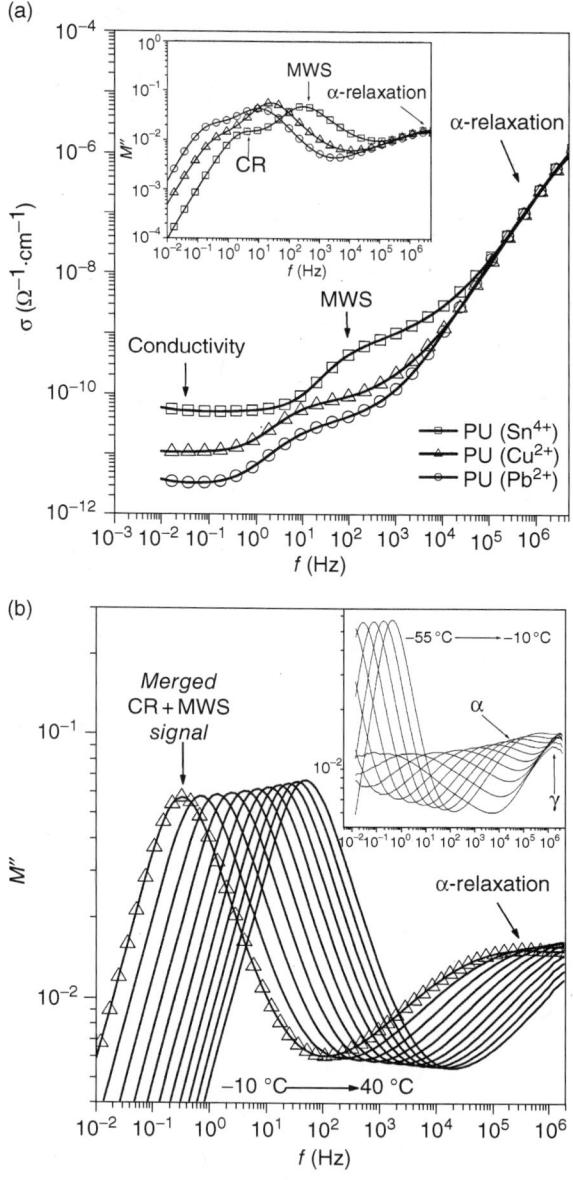

Figure 6.24. DEA results for thermoplastic polyurethane films with various metals in the chain extender: (a) comparison between $\sigma(f)$ and $M''(f)$ spectra recorded at 25 °C; (b) $M''(f)$ plot for a polyurethane sample PU(Cu^{2+}) placed in a brass electrode capacitor. Isothermal scans were performed between −55 °C and 40 °C, in 5° increments. The progressive upshift of various relaxations with increasing temperature of the scan is a typical behavior. [Adapted from Kalogeras, I. M., Roussos, M., Vassilikou-Dova, A., Spanoudaki, A., Pissis, P., Savelyev, Yu. V., Shtompel, V. I., and Robota, L. P. (2005), *Eur. Phys. J. E* **18**, 467. Copyright 2005. With kind permission of Springer Science and Business Media.]

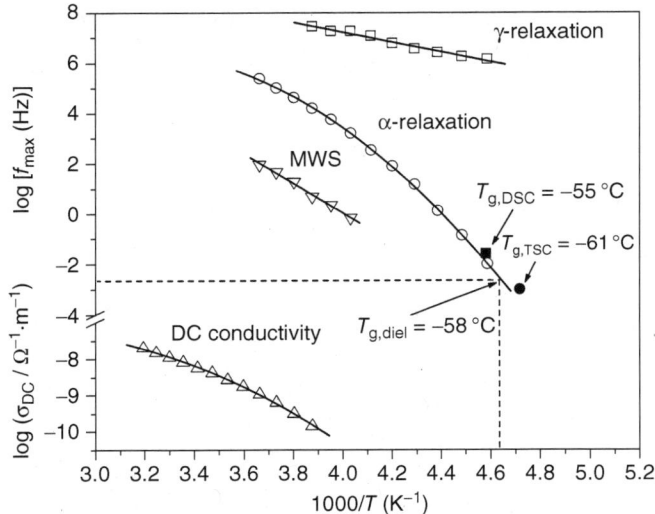

Figure 6.25. Representative Arrhenius diagram for σ_{DC} and relaxations isolated in thermoplastics [PU(Cu^{2+})]. The lines correspond to Arrhenius [γ relaxation: $E = 38$ kJ/mol; Eq. (6.8)] and Vogel–Tammann–Fulcher–Hesse [α relaxation, $T_V = -112\,°C$, Eq. (6.10); DC conductivity, $T_0 = -85\,°C$, Eq. (6.24)] function fittings of the data. The TSC glass transition temperature was obtained from a scan at a heating rate of 5 °C/min, and the DSC T_g is the midpoint of the heat capacity change (second heating at a rate of 20 °C/min). For the method used to determine $T_{g,diel}$ the reader is referred to Section 6.5.2.2. (Kalogeras and Vassilikou-Dova, unpublished data.)

Vatalis et al. (2000) for polyurethane systems] demonstrate excellent agreement between the Arrhenius plots of the α-relaxation mechanism on the basis of dielectric and dynamic mechanical analyses.

It is interesting to observe that not all the cooperative relaxation signals present in a mechanical spectrum can be found in dielectric and DSC traces and vice versa. The comparative DSC, DMA, and TSC data of Hsu et al. (1999) on polyurethanes, based on 2,4-toluene diisocyanate with various NCO/OH ratios, demonstrate the enhanced sensitivity of dielectric analysis in recording the α relaxations in both amorphous regions (see T_g data for NCO/OH = 1.2 in Fig. 6.26). Depolarization current signals clearly occur for the α relaxations in soft- and hard-segment amorphous phases ($T_{g,TSC}$ and $T_{g,global}$). On the contrary, the DSC run reveals only the soft-segment α transition ($T_{g,DSC}$), while the dynamic mechanical loss tangent spectrum shows an appreciable signal only for the global α transition.

Besides the detailed characterization of molecular relaxation mechanisms and space charge effects, several research groups focused on the information extracted from dielectric and other thermal techniques on phase separation that develops in polyurethanes during the stochastic polyaddition reaction [e.g., Pissis et al. (1998), Roussos et al. 2004; Tsonos et al. 2004;

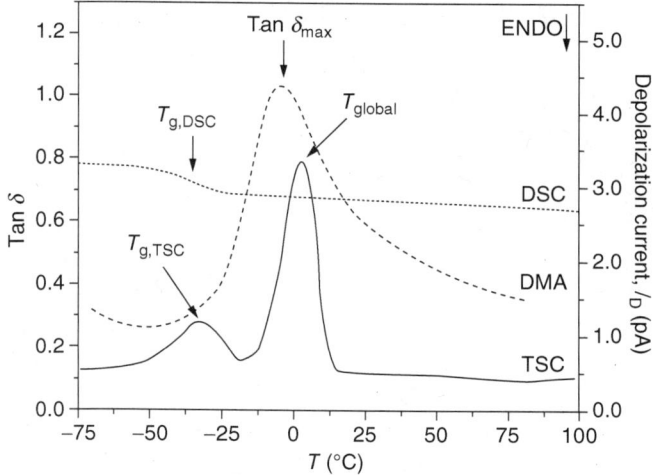

Figure 6.26. Comparative data from DMA (at $f = 1\,\text{Hz}$), TSC ($q = 7\,°\text{C/min}$) and DSC measurements ($q = 10\,°\text{C/min}$) for a polyurethane based on 2,4-toluene diisocyanate (NCO/OH = 1.2). The thermally stimulated currents spectrum is the only one to present the current signatures of both the soft-segment and hard-segment glass transitions. [Adapted from plots presented by Hsu and coworkers (1999), with permission of Elsevier. This article was published in *Thermochimica Acta*, Volume 333, by J.-M. Hsu, D.-L. Yang, and S.-K. Huang, TSC/RMA study on the depolarization transitions of TDI-based polyurethane elastomers with the variation in NCO/OH content. pp. 73–86, Elsevier (1999).]

Kalogeras et al. 2005b]. The degree of nanophase separation can be controlled either through chemical modification (by changing the intrinsic flexibility of hard and soft chain segments or by ionization), or through physical modification (e.g., incorporation of additives). Representative parameters used to study morphological aspects are the glass transition temperature (i.e., lower-temperature amorphous-phase T_g for higher-phase segregation) and the difference between the TSC peak temperatures of the MWS and the α signals (e.g., smaller $\Delta T_M = T_{MWS} - T_\alpha$ suggests greater mixing). In addition, several studies suggest higher "fragility" for materials with lower dissolution of hard segments into soft-segment domains. The fragility index (m) or the "strength" parameter [$D = 590/(m - 16)$ (Crowley and Zografi 2001)] can be obtained from DEA measurements [Eq. (6.23)], and two borderline limits are reached at $m = 16$ and $m \geq 200$ for, respectively, "strong" and "fragile" polymers (Angel 1991); an Arrhenius relationship describes the relaxation phenomena in the former case, while a Vogel–Tammann–Fulcher–Hesse relationship must be used to describe the latter.

6.6.5.2. Polyamides Extensive scientific research in the area of polyamides has been devoted to the exploration of the relaxation dynamics of Nylon 11, owing to its ferroelectric properties (Neagu et al. 2000; Frübing et al. 2006), and of Nylons 6 and 12, with the aim to establish sorption models for the moisture-accessible regions (Patmanathan et al. 1992; Frank et al. 1996). In most cases, dry aliphatic polyamides with variable chain lengths (Scheme 6.3) demonstrate two well-defined relaxations in the temperature range between −170 and +120 °C. The nature of these relaxations, denoted as the γ relaxation for the lowest temperature peak and the α relaxation for the signal associated with the glass transition, has been studied in detail for several representative polymers. The γ relaxation is assigned to local motion of the methylene segments between the amide and the carbonyl groups in the polyamide chain (with participation of the polar amide group in the dielectric relaxation), while the α relaxation is related to the collective motion of CONH groups linked by hydrogen bonds leading to the breaking of these bonds. Between the γ and α signals, a peak labeled as the β relaxation is associated with the motion of water–polymer complexes, such as water molecules forming hydrogen bonds with C=O and N—H groups. At a fixed frequency of 10 Hz, the γ and the β relaxations have been observed in several polyamides, by both mechanical and dielectric spectroscopy, at about −110 °C and −50 °C, respectively. Their dynamics and relaxation strengths have been found to be independent of the crystallinity (McCrum et al. 1967).

The intra and interchain hydrogen bonds between two neighboring amide groups can be modified by moisture sorption, leading to important changes in the mobility of the molecular species that participate in either local or long-range transitions. When comparing dielectric spectra of nylons with various lengths of the methylene sequence, one observes that odd nylons exhibit markedly higher intensities for the α-relaxation mode than the even ones (Yemni and Boyd 1979). The orientation and the crystallinity of the material influence the intensity of the segmental relaxation as has been shown, for example, in Nylon 6,10 (Yemni and Boyd 1979) and Nylon 11 (Frübing et al. 2006). The

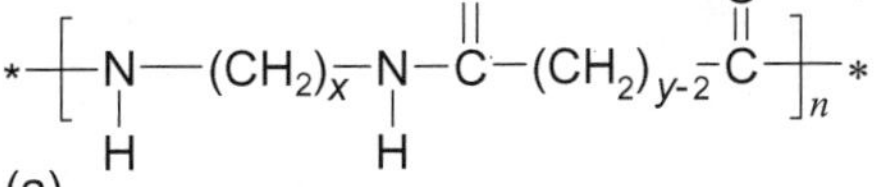

(a) Nylon 6,6 [$x = y = 6$ (Hedvig, 1977)]
Nylon 7,7 [$x = y = 7$ (Yemni and Boyd 1979)]
Nylon 6,10 [$x = 6, y = 10$ (Boyd and Porter 1972)]
Nylon 10,10 [$x = y = 10$ (Lu and Zhang 2006)]

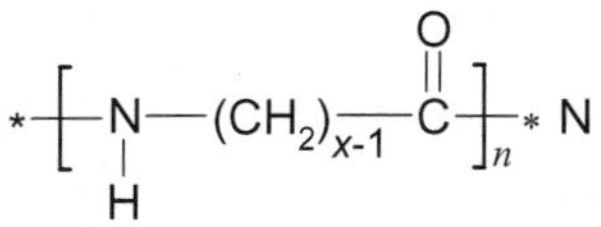

(b) Nylon 6 [$x = 6$ (Laredo et al. 2003; Frank et al. 1996; Laredo and Hernandez, 1997)]
Nylon 11 [$x = 11$ (Neagu et al. 2000; Frubing et al. 2006)]
Nylon 12 [$x = 12$ (Pathmanathan et al., 1992)]

Scheme 6.3. General chemical formulas of two main classes of polyamides. Selected references of DEA/TSC studies of the relaxation dynamics in typical polyamides are given to the right of both structures.

plasticizing effects of moisture sorption are very important and well described for a number of polyamides, such as Nylon 6 (Laredo and Hernández 1997) and Nylon 12 (Pathmanathan et al. 1992). By increasing the hydration level, the α- and β-relaxation signals become predominant, while the relaxation strength of the γ peak decreases, and vice versa. Pathmanathan and Johari (1993) observed an additional "crystalline" $α_c$ relaxation in Nylon 12, appearing at temperatures higher than the glass transition signal. A strong MWS interfacial polarization mode usually appears at the high-temperature/low-frequency regions of the dielectric spectra of semicrystalline nylons, due to the presence of strong DC conductivity and extensive structural heterogeneities (Laredo et al. 2003; Lu and Zhang 2006).

6.6.6. Hybrids and Particulate Systems Incorporating Thermoplastics

The relaxation dynamics of a thermoplastic polymer in intercalated or exfoliated systems was extensively studied by dielectric techniques, since these systems form unique models for studying subtle changes in dynamic properties of polymers under severe nanoconfinement [i.e., with layer spacing << R_g (the radius of gyration) and comparable to the statistical segment length of the polymer]. Interesting results have been reported, for example, for *cis*-polyisoprene, poly(ethylene oxide), poly(butylene oxide), poly(propylene glycol), and PMMA chains intercalated in the interlayers of Li^+ (Vaia et al. 1997) or Na^+–montmorillonite (Elmahdy et al. 2006; Mijović et al. 2006; Page and Adachi 2006; Tran et al. 2005). In addition to the contribution of dielectric analysis in attempts to decipher theoretical problems, a number of studies have remarkable practical implications. As an example, Noda et al. (2005) and Bur et al. (2005) attempted to correlate the aspects of morphology and microstructure of such nanohybrids (e.g., the extent of exfoliation of the nanosize flakes from the larger aggregate particles) with the dielectric relaxation behavior of the polymer.

Mixing of fillers into a polymer containing polar groups alters mechanical, thermal, and electrical properties of the polymer matrix, to a degree determined primarily by the nature and amount of the filler and the interaction between the two components. Several dielectric studies concentrated on monitoring composition-dependent perturbations in the sidechain (noncooperative) and the segmental (cooperative) relaxation dynamics of the thermoplastic component. The relaxational response of the polymeric matrix could sometimes be modified by a competition between several factors. One of them is the looser packing of the polymer chains, due to the presence of nanoparticles and interactions with them. This factor leads to increased free volume and enhanced molecular mobility. Another factor is the formation of a layer of modified polymer around the nanoparticles, which leads to decreased molecular mobility [e.g., see Mohomed et al. (2005)].

6.6.7. Polymer Blends

Blending of polymers is a technological way for providing materials with a full set of desired specific properties (e.g., combination of strength and toughness or strength and solvent resistance). Since the early 1960s or so, many thermal analysis studies of polymer blends (A/B) have been conducted, with particular emphasis on their morphological characteristics and the degree of mixing of their components. The vast majority of polymer blends studied so far (~90%) are multiphase, either partially miscible (i.e., with some limited mixing between the component polymers) or immiscible. The remaining fraction comprises completely miscible polymer blends (single-phase materials). A simple but in some cases unreliable experimental criterion for determining complete miscibility in binary polymer blends is based on the measurement of a single glass transition (Kalogeras and Brastow, 2009), and accordingly a single α-relaxation dielectric signal located in between the glass transition temperatures of the components and consistent with the composition of the blend (Fig. 6.27a). In the case of partially miscible polymer blends with polar components, the α-relaxation signals of both components are recorded in the dielectric spectrum (Fig. 6.27b): convergence of the glass transition temperatures of the two phases is observed, along with a change in the width of the signals. Furthermore, the magnitude of these changes depends on the type and extent of the interactions between the components, as well as on mixing-induced variations in the polymer free volume. Completely immiscible polymer blends also exhibit two α-relaxation signals for all blend compositions, but in this case peak positions are composition-independent (Fig. 6.27c), coinciding with the transition temperature of the corresponding pure component.

Dielectric analysis is frequently applied to address questions related to the strength of molecular interactions and their role in miscibility of components (Genix et al. 2005; Zhang et al. 2004a,b) or to study crystallization processes (Laredo et al. 2005). In cases where complete or partial miscibility is detected by DEA or other thermal techniques, attempts are made to relate it with structural similarity and particularly the presence of various secondary interaction forces (e.g., hydrogen bonding δH, ion–dipole forces and electron donor–acceptor interactions). Dielectric analysis also provides information about the state of the blend on a submicroscopic level. For example, in most miscible polymer blends, the blend α-relaxation loss peak is broader than that of the individual components and this change is frequently accepted as a measure of the molecular environment of the segments (concentration fluctuations and the degree of coupling between various macromolecular chains). Local compositional fluctuations give rise not only to a distribution of relaxation times but also to a dynamic heterogeneity, caused by intrinsic differences in the mobility of the component polymers. For miscible polymer blends, and in particular, polymer blends exhibiting strong hydrogen bonding between the components [e.g., poly(vinyl methyl ether) (PVME)/poly(4-vinylphenol)

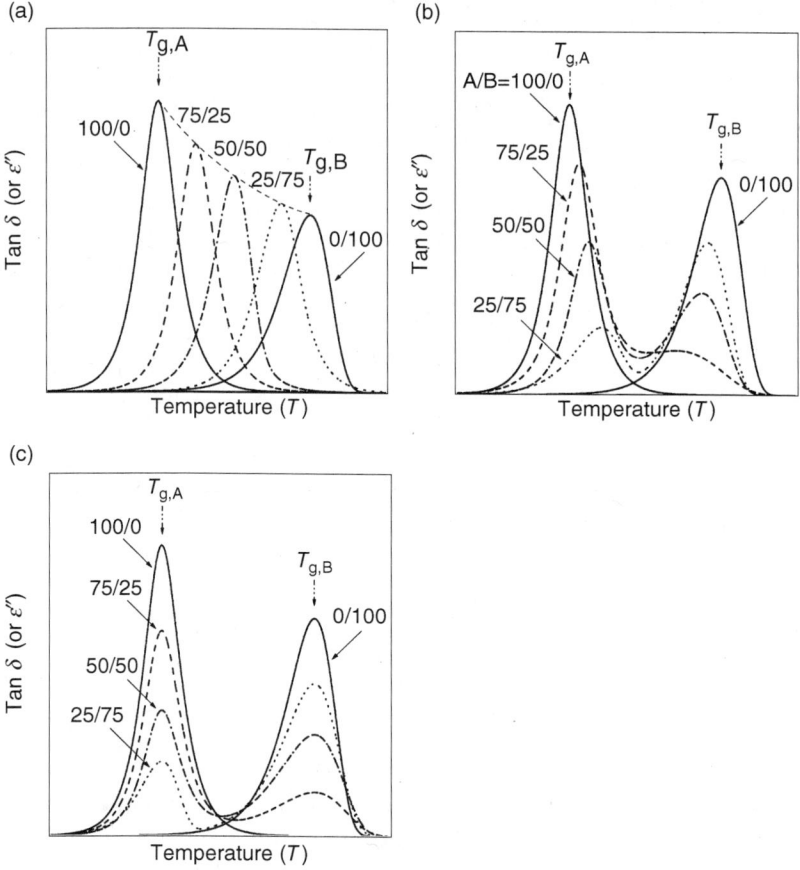

Figure 6.27. Schematic plot of the expected shifts of the dielectric α transitions for polar polymer (A)/polar polymer (B) mixtures, in the case where (a) miscible, (b) partially miscible, or (c) immiscible polymer blends are formed. A behavior intermediate to those shown in plots (a) and (b) frequently appears in miscible binary polymer mixtures with components showing strong dynamic heterogeneity.

(PVPh) (Zhang et al. 2004a)], dielectric studies revealed a reduction of the dynamic heterogeneity. This is more evident for blend compositions with strong nanometer-scale interactions between their components, and is manifested by the coupling of the segmental motions in the two components (single-dielectric glass transition signal). Effects of blending on the secondary relaxation strongly depend on the intermolecular interaction strength. In most cases, the secondary relaxation dynamics of the blend components show little, if any, influence by blending, as has been reported, for instance, by Arbe et al. (1999) for polyisoprene (PI)/poly(vinyl ethylene) (PVE), by Cendoya et al. (1999) for polystyrene (PS)/PVME, and by Urakawa et al. (2001) for PVME/

poly(2-chlorostyrene) (P2CS) blends. However, there are reports indicating that stronger hydrogen bonding in the blend environment, compared to the interaction strength in the neat components, may retard or even suppress the local relaxations of proton-accepting polymers [e.g., see the results obtained by Zhang et al. (2004b) for the β process of poly(2-vinylpyridine) (PVPy) in PVPy/PVPh blends].

Several miscible blends are characterized by the absence of any strong specific intermolecular interaction between their components (e.g., PVME/PS with van der Waals interactions and PVME/P2CS with weak polar–polar interactions). In such cases, a single calorimetric glass transition temperature is usually reported, determined from a single—but anomalously broad—DSC signal, with a number of unresolved glass transitions. Contrary to these findings, the advanced sensitivity of dielectric techniques demonstrated that the very weakly interacting segments [e.g., of PS (weakly polar) or P2CS (strongly polar) and PVME] have very different relaxation rates (Lorthioir et al. 2003), even at temperatures far above the blend T_g. On a theoretical level, the self-concentration model of Lodge and McLeish (2000) for miscible blends has been applied to explain quantitatively the presence of two segmental (dielectric) relaxation processes, and consequently of two effective local T_g values, in the blend environment of PVME/P2CS (Leroy et al. 2003), in which both components are dielectrically active and the dielectric signatures of the segmental motions can be isolated (Urakawa et al. 2001).

Several problems in the applicability of the single glass transition criterion for complete miscibility arise also in the case of melt-miscible crystalline blends, where multiple glass transition–like transitions may appear, although the polymers themselves are miscible (Runt 1997). Furthermore, studies on the miscibility of blends prepared from polymers having close T_g values may provide ambiguous results, depending on the sensitivity of the technique. For example, conventional DSC may fail to differentiate between the presence of complete, partial, or nonexistent miscibility in a polymer blend, if the glass transitions of the initial components do not differ by more than 15–20 °C. This often leads to discrepancies between the DSC results and those from other thermal techniques, like DMA and DEA. The dielectric study of the mainchain segmental motions (α relaxation in amorphous polymers) provides the most common and straightforward criterion for assessing miscibility in amorphous polymer blends, provided that both components are—or can be made—dielectrically active. In semicrystalline miscible polymer blends, dielectric spectra can yield information about the nature of the crystal–amorphous interface within the blends [e.g., see discussion of the DEA results for PEO/PMMA and PVDF/PMMA blends by Runt (1997)]. Finally, in immiscible blends, that is, blends of polymers with usually very weakly interacting or noninteracting components, interfacial polarization strongly influences the dielectric spectra and can lead to a greater understanding of phase morphology.

Dielectric analysis can also be used to study nanoconfinement effects in the polymer chains contained in a polymer blend. Here, the most reasonable physical interpretation of the dielectric results obtained for *strongly asymmetric* blends (i.e., blends with $\Delta T_g = |T_{g,A} - T_{g,B}| > \sim 100$ degrees) studied below their glass transition temperature, considers the motional processes of the mobile chains of the low-T_g component embedded within a frozen matrix formed by the high-T_g component. In this way, little pockets of higher mobility are created where the low-T_g component can move. The confined islands of mobility may reach a size of ~1 nm, as reported for the thermoplastic PMMA/PEO blend by Genix et al. (2005).

6.6.8. Interpenetrating Polymer Networks

As with other multicomponent systems, the extent of miscibility of the components is an important factor in determining, through morphology, the properties of the interpenetrating polymer networks (IPNs, abbreviated A-*i*-B, for a network formed by polymers A and B). The advanced sensitivity of dielectric techniques to nanoheterogeneities has been extensively used to verify the existence of a single glass transition signal (i.e., the common criterion for miscibility) in several fully or semi-IPNs [e.g., polycyanurate-*i*-polyurethane (Georgoussis et al. 2000b) and epoxy resin-*i*-PEO (Kalogeras et al. 2006b)] or of multiple α relaxations in phase-separated systems [e.g., poly(methyl acrylate)-*i*-poly(hydroxyethyl acrylate) (Ribelles et al. 1999) and polyurethane-*i*-poly(vinylpyrrolidone) (Karabanova et al. 2003)]. In most of the latter systems, no differences were found between the glass transition temperatures of the two phases present in the networks and those of the pure components (Ribelles et al. 1999). Either phase-separated or miscible IPNs of selected polymers may develop, depending on the crosslink density achieved by specific amounts of crosslinking agents; in sequential IPNs, forced compatibilization of the components may be achieved by increasing the crosslink density of the first polymer component (Sanchez et al. 2001). There are also dielectric reports where phase separation is reported for highly crosslinked IPNs, but in their case the segmental mobility of each of the two components is significantly modified by the presence of the other [e.g., poly(methyl acrylate)-*i*-polystyrene, (Meseguer-Dueñas and Gómez-Ribelles 2005)]. The compositional variation of dielectric and calorimetric glass transition temperatures has also been used to monitor structural changes [e.g., crystallization (Kalogeras et al. 2006a)]. Several studies indicated that local relaxations exhibit very little, if any, dependence on the composition of the interpenetrating polymer network, with slight acceleration of secondary (β- or γ-) transition rates usually related to a reduced packing density or a weakening of the intermolecular interactions.

6.7. DIELECTRIC ANALYSIS OF THERMOSETS

Dielectric techniques have the particular advantage over other thermal analysis methods of being able to monitor cure-controlled dielectric and electrical properties, *continuously* and *in situ*, as the resin changes from low-viscosity monomers or prepolymer to a crosslinked solid of infinite molecular weight (Senturia and Sheppard 1986; Mijović et al. 1993; Mijović and Bellucci 1996; Pethrick and Hayward 2002). The evolution from the parallel-plate capacitor systems to commercially available wafer-thin planar sensors (e.g., comblike electrodes) allows real-time measurements of parameters—such as dielectric constant, complex permittivity, loss tangent, and conductivity—on various reacting systems with high precision. In this manner one acquires valuable insight into practical aspects, such as the characterization and optimization of the time and temperature required for cure development and the online quality verification of the fabrication process. This section examines dielectric properties of thermosets (e.g., dipolar relaxations, segmental mobilities, and ionic conductivity) and their evolution during cure. Mixtures of a typical epoxy resin [diglycidyl ether of bisphenol A (DGEBA), Epon 828] with various amines (Girardreydet et al. 1995) were used as models. Experimental limitations of dielectric analysis are also discussed. The results of DEA investigations on these model systems were compared with those obtained through studies of dynamic mechanical properties and rheological measurements.

6.7.1. Evolution of Relaxation Dynamics during Cure

The dipole units in the chemical structures of neat epoxies are the source of various relaxations in the dielectric spectra (Senturia and Sheppard 1986). Customarily, these are denoted by the letters γ, β, and α, in order of increasing transition temperature in isochronal plots, or in order of decreasing relaxation frequency in isothermal plots. The molecular origin of the α relaxation in the dielectric spectra of DGEBA prepolymers (Fig. 6.28) lies in segmental motions that cause reorientation of dipole moments associated with the terminal epoxy groups and other less polar groups. The loss factor signal appears at low frequencies [e.g., ε'' peak at $\sim 10^1$–10^2 Hz at $-5\,°C$; (Beiner and Ngai 2005)], but shifts rapidly to higher ranges with increasing temperature. As anticipated, the same signal when present in an isochronal plot will shift to higher temperatures with increasing measuring frequency. The glass transition temperature determined by DSC has a value similar to the temperature of the dielectric α-relaxation peak measured at low frequencies [e.g., ≤ 0.1 Hz (Senturia and Sheppard 1986)], while, the segmental $\varepsilon''(f)$ peak appears near 10^{-2} Hz when the frequency sweep is performed at a constant temperature close to T_g. The β relaxation has shorter relaxation times, much weaker intensity [(100 kHz at $-5\,°C$; (Beiner and Ngai 2005)], and can be resolved only when the experiment is performed at temperatures in the glass transition region and below (Corezzi et al. 2002). For the amine crosslinked epoxy systems, this secondary relaxation is attributed to the crankshaft motion of the hydroxyl-ether groups (glyceryl segments), with an activation energy of $E_\beta = 48$ kJ/mol (Corezzi et al. 2002). Also, it has been reported that the diphenyl propane units contribute to the β relaxation with the "flip-flop" motion of the phenyl rings (Butta et al. 1995). When the Arrhenius dependence of its relaxation time (τ_β) in the glassy state is extrapolated to temperatures above T_g, a tendency of merging with the

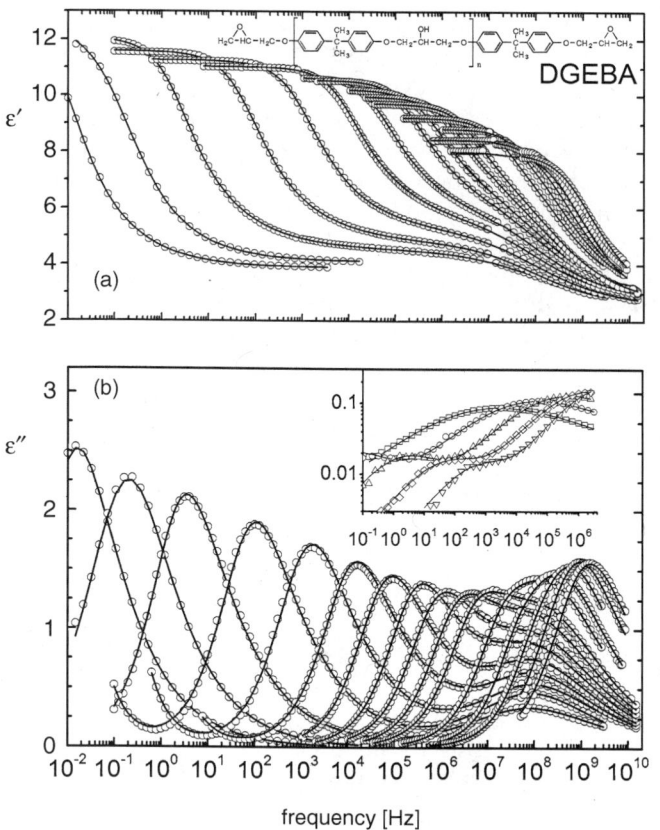

Figure 6.28. (a) Real and (b) imaginary parts of $\varepsilon^*(f)$ for diglycidyl ether of bisphenol A (DGEBA) at several temperatures [80, 70, 60, 50, 40, 30, 25, 20, 15, 10, 5, 0, −5, −10, −14, −17 (°C)] above T_g [note T_g(DSC) ≈ −17 °C] in the range 10^{-2}–10^{10} Hz. In the inset: dielectric loss at some temperatures [−50, −70, −90, −110, −130 (°C)] below T_g, showing the γ process at higher frequencies and the β process at lower frequencies. Solid lines correspond to fits of the data with appropriate functions. [Parts (a) and (b) reused with permission from S. Corezzi, M. Beiner, H. Huth, K. Schröter, S. Capaccioli, R. Casalini, D. Fioretto, and E. Donth, *Journal of Chemical Physics*, **117**, 2435 (2002). Copyright 2002, American Institute of Physics.]

α-relaxation time (τ_α) is found (Fig. 6.29). The γ relaxation (at ~0.1 GHz at −5 °C, see Fig. 6.28b), with the shortest relaxation time, τ_γ (in the nanosecond range near T_g), does not seem to merge with the α relaxation at a finite temperature (Fig. 6.29). This signal is related to localized intramolecular motions involving mainly the epoxide endgroups, with an activation energy of $E_\gamma = 28$ kJ/mol.

Mixing of DGEBA-type epoxies with multifunctional amines [e.g., 4,4′-diaminodiphenyl-methane (DDM), ethylene diamine (EDA), 4,4′-methylenebis 2,6-diethylaniline (MDEA), 4,4′-methylenebis(3-chloro-2,6-

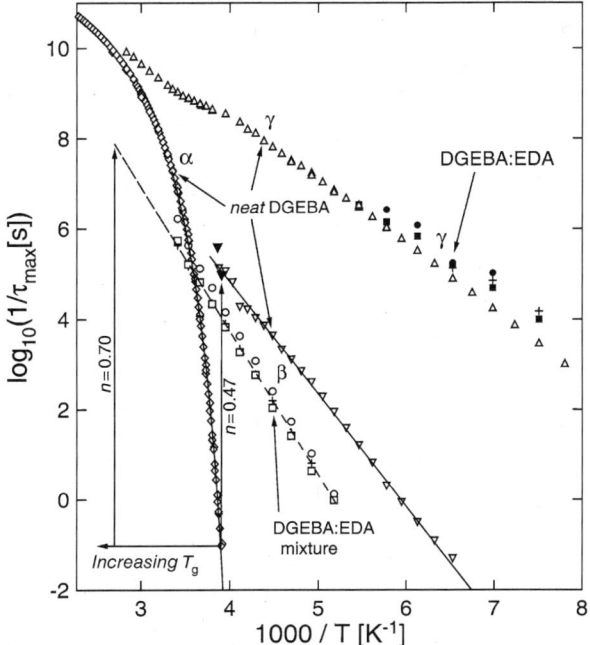

Figure 6.29. Relaxation map of the temperature dependences of the α-, β-, and γ-relaxation angular frequencies for neat DGEBA, unreacted mixture of DGEBA/EDA (ethylene diamine), and DGEBA/EDA cured for 5 h at 40 °C (circles), 55 °C (squares), or 70 °C (crosses). [Reprinted with permission from Beiner and Ngai (2005). Copyright 2005, American Chemical Society.]

diethylaniline) (MCDEA), and 4,4′-diaminodiphenylsulfone (DDS)] initiates cure reactions [see Girardreydet et al. (1995), also Section 2.10 in this book (above)]. These proceed by steps in which secondary and tertiary amine groups are formed and hydroxyl groups appear. After the reaction starts, the covalent bonds formed during the network growth lead to the continuous increase of configurational constraints. This, and intermolecular interactions of DGEBA molecules, produce a more densely packed system. Before significant conversion can be attained, the low-frequency regions of the $\varepsilon'(f)$ and $\varepsilon''(f)$ plots are dominated by blocking electrode effects and DC conductivity losses (Fig. 6.30), which run parallel to the relaxational activity of dipolar units (α- and β-relaxation modes). In view of that, the loss factor signal can be modeled by a superposition of contributions from dipolar effects (usually expressed by a Havriliak–Negami $\varepsilon''_{HN}(\omega)$ function for each relaxation) and ionic displacements (ionic conductivity and electrode polarization; second and third terms, respectively):

$$\varepsilon''(\omega) = \varepsilon''_{dip}(\omega) + \varepsilon''_{ion}(\omega) = \varepsilon''_{HN} + \frac{\sigma_{DC}}{\omega \varepsilon_0} - C_0 Z_0 \cos\left(\frac{n\pi}{2}\right) \omega^{-(w+1)} \left(\frac{\sigma_{DC}}{\varepsilon_0}\right)^2 \quad (6.25)$$

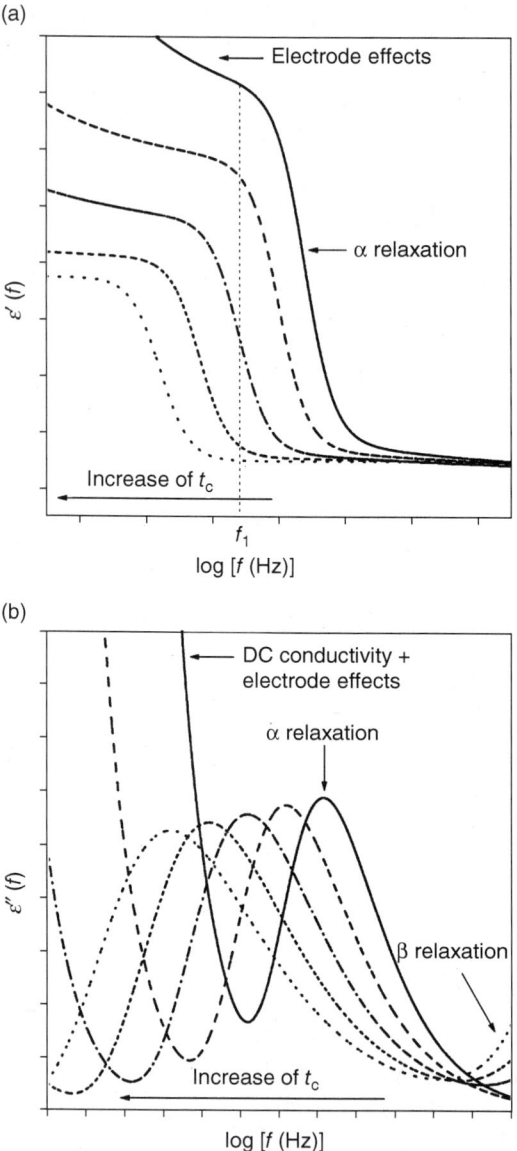

Figure 6.30. Schematic plots of (a) the real part ε' of dielectric permittivity and (b) the loss factor ε'' versus frequency with cure time (t_c) as a parameter (T_{cure} = constant). The higher the frequency f_1, the shorter the time at which the dipolar component (α relaxation) appears as a drop in $\varepsilon'(f)$ and as a maximum in $\varepsilon''(f)$. Experimental evidence of the low-frequency shift of the α relaxation with the increase of t_c is provided by Eloundou et al. (2002).

Here, Z_0 is a constant depending on the electrode–polymer interface and w is a parameter of the Warburg impedance [$0 < w < 1$; see Macdonald (1987)].

Direct-current (DC) conductivity and electrode polarization are dielectric effects controlled by the translational mobility of charged species and the quality of the electrode–polymer interface, respectively. Characteristics of the two mechanisms were discussed in Sections 6.2 and 6.4. In the early dielectric studies of thermosetting systems, such effects were considered to arise merely from impurity ions, typically sodium and chloride ions with concentrations of less than 1 ppm. The presence of Na^+ and Cl^- ions was attributed to the remaining precursors from the synthesis of epoxy resins [extrinsic conductivity; see Senturia and Sheppard (1986)]. The potentially important role of intrinsic charge carriers (e.g., H^+ and OH^-) passed more or less unnoticed, until the mid-1990s, when Bellucci et al. (1995) revealed that even the addition of 100 or 1000 ppm of Na^+ and Cl^- ions had no effect on the measured ionic conductivity of DGEBA/DDM systems. But even after 1995, the research groups active in this area did not properly address this issue; apparently, there exists great difficulty in establishing correlations between the observed phenomena and the presence of specific ionic species. The reason for this becomes clear if one considers the variety of epoxy systems and their precursors, and the unavoidable variations in batch-to-batch characteristics and hydrothermal histories.

Careful inspection of the qualitative plots in Fig. 6.30 reveals that the relaxation frequency of the dipole motions moves to lower values as cure proceeds, indicative of the loss of free space and the increase of the effective T_g of the matrix. Also, ε_r decreases and $\Delta\varepsilon$ gets smaller, somewhat like ΔC_p. The curing time at which the frequency of the α-relaxation loss peak is approximately 0.1 Hz can be taken as indicative of the vitrification point of the matrix.

A schematic of the time variation of ε' and ε'' obtained at a constant frequency f_1, sufficiently high so as to suppress electrode blocking-effects, is given in Fig. 6.31. The sigmoidal shape of the ε' spectrum is a typical behavior. Densification has the effect of slowing down the molecular motion with a concomitant decrease of ionic conductivity and disappearance of electrode effects. During cure the α-relaxation peak shifts continuously towards lower frequencies in isothermal scans (Venkateshan and Johari 2004), or to higher temperatures in isochronal recordings (Mangion et al. 1992). Parallel to the increase in segmental relaxation times τ_α (inset of Fig. 6.31b) with the formation of the polymer network from small epoxy resin molecules [e.g., see Tombari and Johari (1992), Cassettari et al. (1993), Fitz and Mijović (1998)], one can observe an increase of the width of their distribution [i.e., parameter β_{KWW} decreases; see Andjelić et al. (1997)]. The γ relaxation remains at almost the same high frequency, but its intensity is reduced as a result of the continuous consumption of the epoxide groups. The initially weak β-relaxation signal grows in intensity with increasing curing time and shifts to lower frequencies (Fig. 6.31b), eventually merging with the α relaxation at high temperatures.

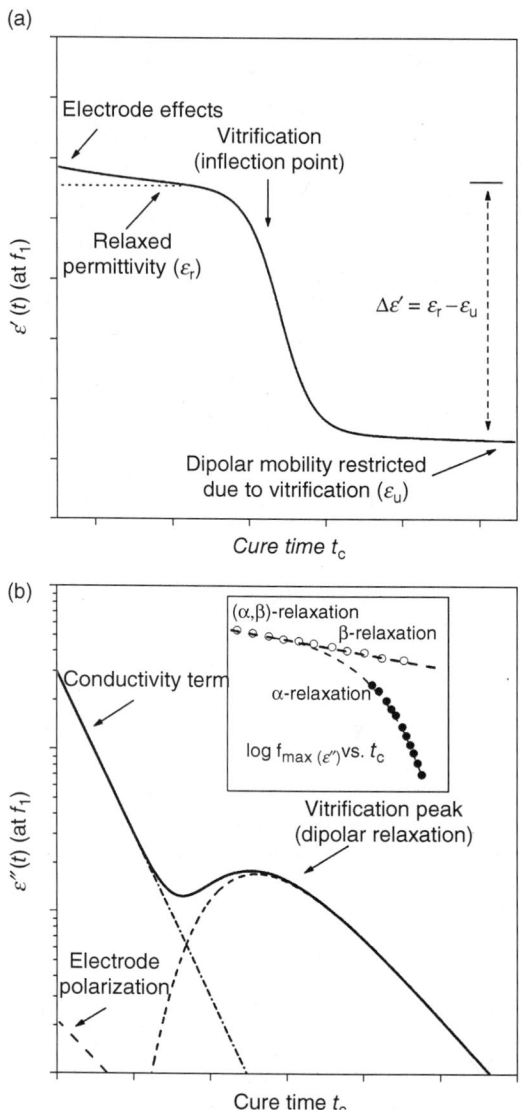

Figure 6.31. Schematic of the time variation of (a) ε' and (b) the loss factor ε'' (plotted in a log scale) obtained at frequency f_1 (see Fig. 6.30a); f_1 is high enough so that blocking electrode effects at low cure times are suppressed. For actual experimental data following the dependences described in this figure, the reader is referred to the works of, for example, Fournier et al. (1996), Eloundou et al. (2002), Montserrat et al. (2003) and Núñez-Regueira et al. (2005). For example, in the case of the DGEBA/DDS system, Eloundou et al. (2002) use T_{cure} values of 140–160 °C, and examine frequencies (f_1) in the range between 1 and 10^4 Hz.

The merging of the α- and β-relaxation signals, recorded in several polymers, into a single (αβ) relaxation process comes into sight when the permittivity–frequency plot is recorded at temperatures well above the glass transition temperature of the material (e.g., see comments in Section 6.2.4.1).

Combined DSC, DMA, and DEA studies of *m*-xylylene diamine–cured DGEBA provided important information regarding the relative sensitivity of these methods in the study of cured epoxy resins (Núñez-Regueira et al. 2005; Núñez et al. 1998a,b). As a result of the low polarity of the dipole units involved in the β relaxation, this secondary transition made a greater contribution to the dynamic mechanical response than it did to the dielectric response. The agreement of the different thermal techniques in the estimation of the glass transition temperature (123 °C by DMA, peak in tan δ; 122 °C by DEA, peak in ε''; and 116 °C by DSC, midpoint of ΔC_p) and the β-relaxation spectral position (−67 °C by DMA and −77 °C by DEA) is striking, especially if one considers the difference in their equivalent frequencies (1, 0.1, and ~0.01 Hz, respectively).

Mechanical, thermal, and electrical properties of the cured products can be controlled in the fabrication process by introducing appropriate amounts of elastomeric modifiers and thermoplastic polymers into the initial epoxy + hardener mixture. Examples include the addition of poly(2,6-dimethyl-1,4-phenylene oxide) (Venderbosch et al. 1995), poly(ether imide) (Li et al. 2004), or carboxy-terminated butadiene acrylonitrile to the blends of DGEBA (Pethrick and Hayward 2002). Dielectric analysis can be used to monitor changes in characteristics of local relaxations and more importantly to study the variation of the glass transition signal and σ_{DC} at different loadings of the modifier (Wang et al. 1992). For instance, for cured DGEBA/DDM + polyethylene oxide (PEO) blends, the shift and strength of the dielectric mechanisms at different blend compositions can be monitored (Fig. 6.32a). Interesting results can be obtained on the extent of phase interactions, structural characteristics of the cured material (e.g., onset of PEO crystallization for $w_{PEO} \geq 50\,wt\%$; Fig. 6.32b), and the compositional dependence of T_g (Fig. 6.32b) (Kalogeras et al. 2005a, 2006b, 2007b).

6.7.2. Dielectric Characterization and Monitoring of Cure Process

Direct-current conduction of ionic impurities in thermosetting resins and the relaxation of dipolar segments of the polymer molecules are the main processes detected by dielectric techniques in thermosetting systems. Several models relating the cure kinetics to either conductivity or relaxation have been proposed for materials cured under isothermal as well as nonisothermal conditions (Nass and Seferis 1989; Day 1989). Kranbuehl (1997) has provided a review of the dielectric characterization and monitoring of polymerization and cure, in which he showed a number of examples of the use of sensors for monitoring the dielectric response during cure in molds, autoclaves, adhesives, films, and coatings. The application of dielectric analysis for monitoring

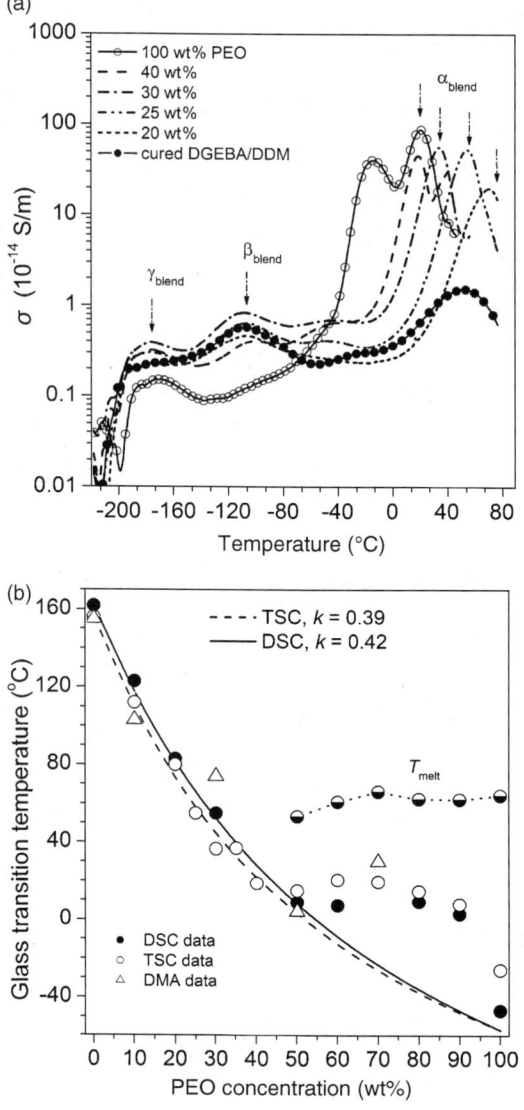

Figure 6.32. (a) Transient conductivity $\sigma(T) = J_D(T)/E_p$ (TSC; see Section 6.3.5) spectra of selected DDM-cured DGEBA + PEO ($M_w \approx 2.0 \cdot 10^4$) blends. [Reproduced from Kalogeras et al. (2006b) with permission of Wiley-VCH.] (b) Compositional dependence of dielectric, DSC, and dynamic mechanical (at 1 Hz) T_g estimates. Fitting lines to the TSC and DSC data obtained at low PEO loadings (completely amorphous PEO in the blend) are based on the Gordon–Taylor (Gordon and Taylor 1952) equation: $T_g^{\text{blend}} = (kw_{\text{ER}}T_g^{\text{ER}} + w_{\text{PEO}}T_g^{\text{PEO}})/(kw_{\text{ER}} + w_{\text{PEO}})$. [Compiled from data published in Kalogeras et al. (2005a, 2006b).]

thermoset cure is based on the influence of the structural transformation on intrinsic dielectric and electrical properties. In brief, the changes in dielectric properties during cure involve (1) a decrease of ionic conductivity induced by an increase of the viscosity, which hinders the translational diffusion of ions in the medium; (2) adjustment of the relaxation characteristics of field-induced orientation and oscillation of permanent dipoles during structural evolution; and (3) changes in the dipole moment per unit volume due to an increase of branching, increase of the crosslink density, and changes in the nature of the dipole population in the course of curing. Early dielectric studies attempted to derive the rate constants and subsequently the activation energy of the reaction (Fava and Horsfie 1968; Acitelli et al. 1971). Since the early 1980s dielectric research has focused on a number of resin cure processing properties, which include the reaction onset, gelation time, vitrification, degree of cure (α), cure time dependence of T_g, reaction completion, viscosity (η), and point of minimum viscosity or maximum flow (Gotro and Prime 2004). With respect to the latter issue, an interesting result, with great importance in processing and cure monitoring, is the correlation observed between the frequency-independent dielectric loss factor peak (maximum mobility) and the viscosity minimum (Ciriscioli and Springer 1989). Gotro and Yandrasits (1989) further suggested that for thermosetting polymers cured at a temperature (T_{cure}) below the ultimate glass transition temperature of the resin ($T_{g,\infty}$), vitrification could be observed as frequency-dependent dipole peaks above the maximum of the loss factor.

The basic strategy in designing a meaningful dielectric experiment calls for the careful selection of experimental conditions, where only one type of dielectric response (dipolar or conductive) plays a major role and as such can be reliably selected, measured, and connected with the fundamental mechanism of network formation. Changes in the high-frequency (dipolar) dielectric response of the reacting material form the basis for the development of microwave monitoring techniques. Research for rapid and accurate microwave band frequency measurements on reactive systems, based on simple and fast apparatuses for narrowband or even monofrequency measurements, with use of disposable sensors (Das et al. 1987), has become popular. The concept of microwave monitoring is based on the hypothesis that in a cure process the dipoles of unreacted species are much more mobile than those of products, so that the dipoles of the unreacted species have a greater effect on the microwave permittivity than do the dipoles of the products. Considering the disappearance of the dipoles of unreacted species with time, the observed linear decrease of the $\varepsilon'(t)$ and $\varepsilon''(t)$ values measured at microwave frequencies should yield a good estimate of the progress of the reaction (Gallone et al. 2001; Zong et al. 2005). The behavior of the normalized loss factors $\varepsilon''(t)/\varepsilon''(0)$ (where $\varepsilon''(0)$ is the dielectric loss value immediately after mixing) obtained isothermally at a microwave frequency (~2 GHz) shows analogy to the $1 - \alpha(t)$ function measured by DSC (Carrozzino et al. 1990). Moreover, the dielectric degree of conversion, defined by

$$\alpha_{\text{diel}}(t) = \frac{\varepsilon''(t) - \varepsilon''(0)}{\varepsilon''(0) - \varepsilon''(\infty)} \qquad (6.26a)$$

or

$$\alpha_{\text{diel}}(t) = \frac{\varepsilon'(t) - \varepsilon'(0)}{\varepsilon'(0) - \varepsilon'(\infty)} \qquad (6.26b)$$

provides estimates that compare well with FTIR results (Gallone et al. 2001). The entire concept of microwave monitoring is embodied in the isolation and measurement of the time-dependent relaxation activity (dielectric strength) of the unreacted dipolar species. Given that this group of dipoles has extremely high relaxation frequencies, the use of microwave frequencies and special measuring systems (e.g., capacitors for coaxial line reflectometry; see Section 6.3) is necessary. Lower frequencies that could be easily attained with a standard dielectric apparatus (such as impedance or frequency response analyzers and parallel-plate capacitors) cannot be used in this case.

The considerable scientific interest for low-frequency dielectric studies and especially the use of ionic conductivity measurements to monitor curing reactions dates back to the pioneering work of Kienle and Race (1934). During cure, the conductivity (σ) is best described as a function of frequency, temperature, and degree of conversion (α)

$$\sigma(f, T, \alpha) = \sigma_{\text{DC}}(T, \alpha) + \sigma_{\text{AC}}(f, T, \alpha) \qquad (6.27)$$

where σ_{DC} is the frequency-independent contribution of purely conductive phenomena and σ_{AC} is the dipolar contribution. Dielectric events, such as electrode polarization and the Maxwell–Wagner–Sillars interfacial polarization, have major impact on the measured dielectric properties and can lead to erroneous estimates of σ_{DC}. However, at a sufficiently high scanning frequency such phenomena can be neglected and the measured ionic conductivity is representative only of the mobility of ionic species. At the same time, there is a limiting low frequency at which the dipolar orientation can be neglected. Since the frequency at which the abovementioned phenomena will not impede conductivity assessments is not known a priori, a trial-and-error approach has to be undertaken by performing experiments at several frequencies. Several methods have been proposed for the calculation of ionic conductivity of thermosetting polymers from dielectric measurements [see e.g., Senturia and Sheppard (1986), Kranbuehl et al. (1986)]: (1) from an experiment conducted at several frequencies ($f_1, f_2, \ldots f_n$), the conductivity curve is built up by superimposing curves of $\varepsilon''\varepsilon_0\omega$ versus time and taking the parts for which $\varepsilon''\varepsilon_0\omega_1 = \varepsilon''\varepsilon_0\omega_2 = \cdots = \varepsilon''\varepsilon_0\omega_n$ (solid line in Fig. 6.33a); or (2) as an alternative approach, it is possible to extract the value of ionic conductivity directly from the low-frequency data (i.e., data at frequencies well below the spectral position of the

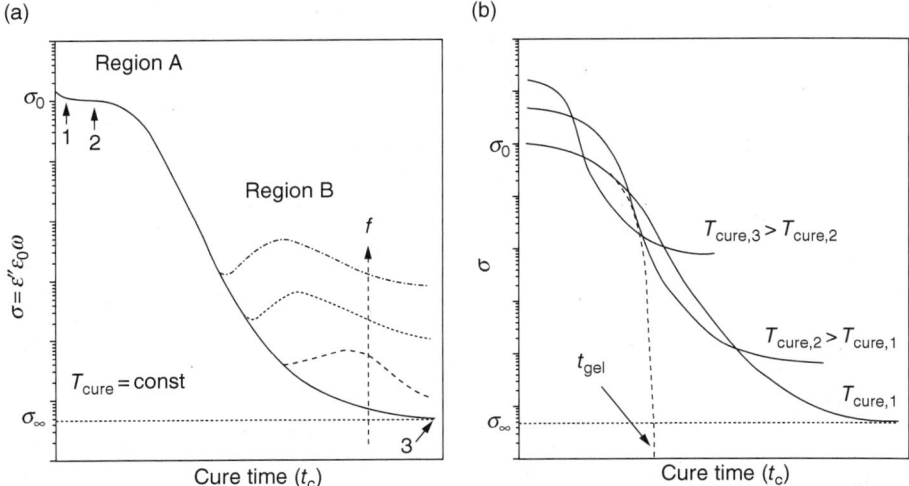

Figure 6.33. Schematic plots of cure time variation of the conductivity measured: (a) at constant cure temperature; (b) at a constant low frequency at which charge migration dominates. An estimate of the time to reach gelation (t_{gel}) for curing at a relatively low temperature ($T_{cure,1}$), based on the percolation model for the $\sigma(t_c)$ function, is shown (see Section 6.7.2.2). For plots with actual experimental data following the dependencies described in this figure the reader is referred to the works of, for example, Eloundou et al. (1998b, 2002) and Núñez-Regueira et al. (2005).

dielectrically active molecular relaxation modes), assuming that the limiting low-frequency response is due entirely to charge migration.

Representative log plots of the time evolution of σ at various frequencies and temperatures, during isothermal cure, are shown in Fig. 6.33. Overlap of σ–t_c curves at different frequencies indicates the time at which translational diffusion becomes the dominant physical process for the loss (region A in Fig. 6.33a). Peaks at higher cure times are indicative of dipole processes (region B in Fig. 6.33a). Early in the reaction, impurity ions move freely in the system and ionic conductivity is high. The point of reaction onset (position 1 in Fig. 6.33a) is reflected by a decrease in the rate of change in the low frequency ionic mobility trace. The point of maximum flow and minimum viscosity is determined by measurement of position 2 (onset of the strong conductivity drop) and is critical for the application of pressure to a fiber-reinforced composite in order to achieve optimum consolidation of the resin matrix. With time, the ionic conductivity decreases further as the curing system offers more resistance to the movement of ions. Eventually, there usually appears an inflection point in the σ_{DC}–t_c curve (Fig. 6.33b) that marks the slowing of the reaction; when there is no further change (i.e., $d\sigma_{DC}/dt \to 0$), the reaction is considered complete (position 3 in Fig. 6.33a). With increase in the temperature of cure, the DC conductivity of the initial mixture (σ_0) and of the fully cured network (σ_∞) increase and the reaction requires less time for completion (Fig. 6.33b).

6.7.2.1. Relationship between Ionic Conductivity and Viscosity Ionic conductivity has evolved as a molecular probe that can be used as a quantitative monitoring tool for the viscosity of the resin during cure. Assuming that the conducting ionic species are rigid spheres moving through the liquid monomer(s), the measured DC conductivity is related to viscosity by Stoke's law

$$\sigma_{DC} = \frac{Ze^2 N}{6\pi R \eta} \propto \eta^{-1} \qquad (6.28)$$

where Z, e, N, and R are the average number of electric charges of ionic species, the electron charge, the concentration of ionic species, and their average radius, respectively. In the crucial processing stage of thermosets before gelation (hydrodynamic regime), ionic conductivity is expected to be inversely proportional to viscosity (Simpson and Bidstrup 1995). After this point, η tends to increase to infinity and σ_{DC} attains a constant value (σ_∞) (Pethrick and Hayward 2002). If the ionic concentration does not change during cure, then σ_{DC} is directly proportional to ion mobility. However, the variation of the ionic charge carrier density during the reaction limits the applicability of the basic assumption of electrical techniques for reaction monitoring, specifically, that the ionic conductivity is related directly to the medium viscosity (Friedrich et al. 2001). Using a combination of ion time-of-flight and dielectric measurements, Krouse et al. (2005) measured separately the two components of the conductivity, ion mobility, and number of mobile charge carriers, in an unreacted DGEBA/MCDEA system and during buildup in the network structure during cure over a similar range of viscosities. They concluded that mobility, not the conductivity, correlates with the changing viscosity both with changes in temperature and network growth. The conductivity changes in such hydrogen-bonding epoxies were thus considered to arise both from the increase of viscosity with cure time and changes in the number of conducting ions. This provides a plausible explanation for a $\sigma_{DC} \propto \eta^{-x}$ dependence with $0.42 \leq x \leq 0.70$, frequently observed in such systems (Eloundou et al. 1998a,b; Bonnet et al. 2000; Krouse et al. 2005) from the beginning of the reaction up to a conversion limit lower than the gel point α_{gel}.

6.7.2.2. Relationship between Ionic Conductivity and Gelation Time (t_{gel}) Molecular gelation takes place at a well-defined and generally calculable step in the course of the chemical reaction (Flory 1988). The reaction mechanism is assumed to be temperature-independent and free of *noncrosslinking* side reactions but depends on stoichiometry, reactivity, and functionality of the reactants. Several attempts have been made to relate dielectric results and the gel point [e.g., see Fournier et al. (1996)]. An inflection point on the logarithm of the ionic conductivity–time curve during isothermal cure of epoxy/amine systems, which correlated with the time to reach the calculated α_{gel} by DSC (Acitelli et al. 1971) and with the point at which the mechanical loss tangent became independent of frequency (Boiteux et al. 1993), has been taken as the time to gel. Applying a percolation model to dielectric measurements, Mangion and Johari (1990, 1991) proposed that ionic conductivity is connected with the time to reach gelation (t_{gel}) by

$$\sigma = \sigma_0 \left(\frac{t_{gel} - t}{t_{gel}} \right)^x \quad (6.29)$$

where the exponent x depends on the curing temperature. Among others, Mijovic et al. (1995) take the value of t_{gel} at fictitious zero conductivity, which is obtained by extrapolating the step of the conductivity drop (Fig. 6.33b). The argument that this fictitious zero corresponds to the inflection point is highly erroneous, as the diffusion of ions continues at the gel point, and has therefore been strongly—and quite logically—criticized by several researchers. For example, Eloundou et al. (1998a) report that when an inflection point exists on log $\sigma(t)$ curves, the times and conversions corresponding to these points are lower than those at the gel point. In spite of this, Eq. (6.29), which is similar to a relation that uses viscosity data at zero shear rate to estimate gel times, has been found to provide results in reasonable agreement with those obtained by rheometry (Núñez-Regueira et al. 2005) and the gelation time determined from measurements in real time of the propagation of ultrasonic shear waves through a polymerizing sample (Zhou and Johari 1997).

6.7.2.3. Relationships between Ionic Conductivity and Degree of Conversion
The establishment of relations between ionic conductivity, on one hand, and the glass transition temperature or the degree of conversion, on the other, has attracted much interest. The degree of conversion at the time when the reactive mixture becomes a glass depends on the temperature of cure. The data in Fig. 6.34 illustrate the variation of the logarithm of conductivity with conversion, which is expected to be dependent on temperature; the absence of a distinctive point on this variation should be noted. On the basis of the initial curvature of the experimental $\sigma(\alpha)$ data, a percolation theory approach to the modeling of the $\sigma(\alpha)$ plot, described by the relation

$$\sigma = \sigma_0 \left(\frac{\alpha_{gel} - \alpha}{\alpha_{gel}} \right)^x \quad (6.30)$$

where α_{gel} is the conversion at gel point, can been considered. However, as with Eq. (6.29), the applicability of Eq. (6.30) to describe the evolution of conductivity with conversion has been questioned (Eloundou et al. 1998a). Considering the frequently observed linear behavior of the ratio log σ/log σ_0 versus T_g (Eloundou et al. 1998a,b), and the relation of DiBenedetto (Nielsen

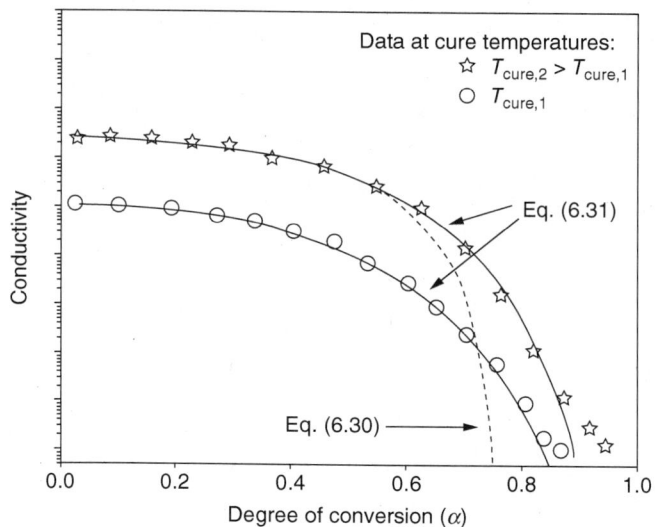

Figure 6.34. Schematic of the variation of conductivity as function of conversion α, corresponding to cure temperatures below the ultimate glass transition temperature of the resin ($T_{g\infty}$) [e.g., see Eloundou et al. (1998a)]. At high conversion levels the hypothetical "experimental" data show positive deviations from the theoretical line [Eq. 6.31)] due to vitrification.

1969), modified by Pascault and Williams (1990), for the evolution of the T_g with conversion, one may predict the behavior of conductivity during isothermal cure using the relation

$$\frac{\log\sigma(t)-\log\sigma_0}{\log\sigma_\infty-\log\sigma_0} = \frac{T_g(t)-T_{g0}}{T_{g\infty}-T_{g0}} = \frac{\lambda\alpha(t)}{1-(1-\lambda)\alpha(t)} \quad (6.31)$$

where T_{g0} and $T_{g\infty}$ are respectively the glass transition temperatures of the uncured mixture (i.e., $\alpha = 0$) and of the fully cured network, and $\lambda = \Delta C_{p,\infty}/\Delta C_{p,0}$ (where $\Delta C_{p,0}$ and $\Delta C_{p,\infty}$ are the heat capacity steps at T_g of the uncured monomer mixture and of the fully cured network, respectively). The first part of Eq. (6.31) is referred to as the *cure index*. The values $T_{g\infty}$, σ_∞, and $\Delta C_{p,\infty}$ are measured at 100% conversion or $\alpha = 1$, and $\alpha(t)$ can be calculated by independent kinetic studies using an autocatalytic cure model. Equation (6.31) is equivalent to the relation proposed by Stephan et al. (1997) for modeling the $\log \sigma/\log \sigma_0$–$\alpha$ dependence. Notice that the applicability of both equations to experimental data obtained at high conversions levels is restricted by vitrification, which develops when $T_{cure} < T_{g\infty}$ (Eloundou et al. 1998a).

6.7.2.4. Conductivity and Coefficient of Thermal Expansion

The change of the coefficient of linear thermal expansion (α_{CLTE}) during the cure reaction is important for industrial applications such as composite processing since it influences the development of initial stresses in the material.[5] The possibility of estimating the change of this coefficient during cure indirectly from non-isothermal real-time dielectric analysis is an attractive option, since direct measurement of α_{CLTE} for a reacting system is difficult. The modified WLF equation can help to model the ionic conductivity of a thermoset. This equation can be written as

$$\log \sigma (a, T) = \log \sigma_0 + \frac{C_1 (T - T_g(\alpha))}{C_2 + (T - T_g(\alpha))} \quad (6.32)$$

and it can be used to estimate α_{CLTE} using the relation $\alpha_{CLTE} \approx 1/C_1 C_2$ (from the WLF theory). Bartolomeo et al. (2001) fitted Eq. (6.32) to ionic conductivity before and during cure reaction for a dicyanate ester thermoset. They observed that the adjusted coefficient C_1 remained constant (Senturia and Sheppard 1986), whereas C_2 strongly varied with progress of the cure reaction. Therefore, the development of α_{CLTE} is determined mainly by the evolution of C_2. Parameters σ_0 and C_1 can be determined by fitting $\sigma(T)$ data obtained at temperatures prior to the reaction to Eq. (6.32) (Leroy et al. 2005). With ln σ_0 and C_1 fixed, the value of C_2 can be adjusted to each conversion level in order to ensure a minimum least-squared difference between experimental σ_{DC} and the prediction of Eq. (6.32). The results obtained by Leroy et al. (2005) showed good agreement between the calculated absolute values of the coefficient α_{CLTE} before and after the complete reaction, and the experimental data. In addition, the evolution of α_{CLTE} with the cure reaction is consistent with the knowledge of the physicochemistry of thermosetting polymers, and in particular the marked changes of their properties and behavior when the system passes through the gel point.

6.7.3. Dielectric Signatures of Water Molecules Confined in Epoxy Resins

The moisture ingress into neat epoxy resins and related systems toughened by thermoplastic or rubber elastomer modifiers has a significant impact on mechanical properties of the matrix, and its presence is important for the stability of the resin–modifier interface. Water penetration in epoxy resins tends to decrease the modulus of the matrix, and if water is allowed to segregate, can cause blistering and environmental stress crazing (Ellyin and Rohrbacher 2000; Chaplin et al. 2000). Epoxy composites may exhibit various degradation mechanisms due to the presence of water. The degree of damage, caused by the aqueous environment and temperature, strongly depends on the temperature (Ellyin and Rohrbacher 2000) and the organization of water within the material. Therefore, it is necessary to clarify the effect of both the rate and the

---

[5] The term α_{CLTE} refers to the liquid or rubbery states (i.e., not vitrified).

extent of water absorption, as well as the nature of sites trapping the water molecules. In a field undoubtedly dominated by gravimetric measurements (see Chapter 3, on TGA), only a few real-time dielectric measurements of water absorption have been reported, and are reviewed by Pethrick and Hayward (2002). These measurements demonstrate the effect of various chemical factors on the water absorption process. Dielectric analysis was also used to evaluate the changes in physical properties due to resin age and exposure to moisture (Pethrick and Hayward 2002). Spectroscopic investigations, including studies of the dielectric relaxation behavior of water molecules in epoxy composites, focused on water molecules filling structural microvoids and other morphological defects in different bonding states. Water in a polymer may be dispersed either as free water or bound water, which, depending on the binding energy, may be classified as loosely bound water or tightly bound water. The molecules of bound water are hydrogen bonded to hydrophilic groups (such as hydroxyl, amine, or carbonyl groups). Proton NMR and near-IR (NIR) spectroscopy are common methods to test the state of the water in polymers, but dielectric analysis can be very helpful, especially through the temperature dependence of the state of water in these polymers. Room temperature permittivity (ε', ε'') plots typically show a dipolar contribution in the kHz region due to bound water molecules, a high-frequency shifting of the molecular relaxation (in the region 1 Hz–1 kHz) associated with the hydroxyl groups of the polymer matrix (plasticization of the β relaxation), and ionic contributions below ~10 Hz [DC protonic conductivity, Maxwell–Wagner–Sillars polarizations, and blocking electrode effects (Boinard et al. 2005)]. It is assumed that the variation of ε' at about 10 Hz reflects the total relaxation spectrum of the absorbed water molecules, while complex permittivity measurements above ~0.1 MHz are considered to deal with the relaxation activity of clustered (free) water molecules. The total water uptake for an amine epoxy resin is of the order of 3–4 wt% (from gravimetric analysis), which is significantly greater than the value that can be calculated from the dielectric increment (1.5–1.7 wt% from DEA). Considering that dielectric estimates assume the value of 80 for the permittivity of free water, the difference exemplifies the role of the topological constraints imposed from the polymeric environment on the rotation of water dipoles.

6.8. INSTRUMENTATION

Dielectric analysis incorporates a number of measuring techniques, spanning from near DC (<10^{-5} Hz) up to microwave frequencies (GHz region). In this section, we describe state-of-the-art instrumentation in the field of dielectric analysis, presenting selected characteristics and specifications of several sophisticated commercial systems for DEA experiments with temperature variation. Special emphasis is placed on turnkey systems and suppliers featuring strong penetration in the current market. The instruments discussed herein are characterized by availability, high accuracy, versatile design and connectivity, and possibility of expansion. Dielectric instruments no longer available on the market are not listed here, although extensive reference to such systems may be found in previous sections of this chapter (e.g., see Section 6.3.2).

6.8.1. Dielectric Spectrometers for Measurements in Frequency Domain

Table 6.2 summarizes features of the most important DEA techniques and dedicated equipment for measuring dielectric permittivity and loss in polymers and related materials. The frequency range between 1 mHz and 10 MHz, in combination with temperature variation in the range between −160 °C and 200 °C, is regarded the most informative range for the dielectric analysis of materials properties and phenomena. Therefore, most commercially available measuring systems and thermal analysis laboratories concentrate on this frequency window.

Novocontrol Instruments (www.novocontrol.com) is the main provider of a variety of general-purpose turnkey dielectric/impedance spectrometers. The specifications of these instruments cover broad ranges. For example, the BDS 40 system is offered for dielectric experiments in the extremely broad frequency range of 3 µHz–20 MHz (12.5 decades). This system is based on the Alpha-A modular measurement system for dielectric, conductivity, electrochemical, and impedance spectroscopy with several test interfaces. With a loss factor accuracy better than 3×10^{-5}, and in combination with the high-precision, rapidly settling temperature control Quatro Cryosystem, this spectrometer is the cornerstone of the currently available commercial instruments with temperature control, as it is highly efficient for studying practically all polymeric materials, including low-loss polymers, like polyethylene. The RF option extends the frequency range up to 8 GHz, providing an excellent turnkey solution for temperature-controlled electric material analysis. The capabilities of the BDS 40 system undoubtedly surpass those of the well-known TA 2970 system (www.tainst.com; no longer available in the marketplace), which uses an Agilent (Hewlett-Packard) 4284A *LCR* bridge (20 Hz–1 MHz) as the impedance analyzer.

Micromet Instruments (*Netzsch*) also offers a range of specialized dielectric spectrometers with certain limitations in temperature or frequency range selection. For example, the DEA 230/2 (MDE Series 20) and DEA 230/10 (Eumetric 100A) are high-performance dielectric measurement systems, both designed for laboratory dielectric analyses of polymers with conductivity values as low as 10^{-14} S/m. Frequency sweeps between 10^{-3} and 10^{5} Hz are possible. These systems are recommended for research and development of resins, composites, coatings, oils, and lubricants, using a wide range of commercially available dielectric sensors, and with the potential to monitor simultaneously two (DEA 230/2) or more (DEA 230/10) sensors. Those interested in specialized applications of DEA with respect to in-process cure monitoring of

TABLE 6.2. Summary of Principal Aspects of Most Common Experimental AC Techniques for Dielectric Analysis of Polymers

Dielectric Technique	Frequency Range (Typical)	Measurement Principle	Characteristics	Representative Systems
Frequency response analysis (FRA; 6.3.2)	100 μHz–10 MHz	A sinusoidal voltage signal is applied and the analyzer measures the frequency dependence of the complex impedance between the electrical ports of a system under test, at a given temperature	Extremely precise results for any form of polymeric materials (liquid, solid, thin film, etc.)	The turnkey BDS concept 10, 20, or 40 systems, available from Novocontrol; the first two are economical versions of BDS 40, but the sample cell has to be ordered separately
Impedance analysis (6.3.2)	100 Hz–10 MHz	A sinusoidal voltage signal is applied; analyzers are based on autobalancing bridges or current/voltage methods	Economical alternative of FRA, for fast and relatively accurate measurements on any form of polymeric materials	The Agilent HP 4284A AC bridge, combined with a typical dielectric sample holder and a compatible temperature control system
Coaxial line reflectometry (6.3.3)	1 MHz–1 GHz	The sample capacitor is placed at the end of a low-loss precision coaxial; impedance is determined from complex reflection factor, measured with a microwave reflectometer at analyzer end of coaxial line	Special sample cells are available for sample thicknesses ranging from 1 μm (thin films) to 20 mm (liquids or solid samples)	The turnkey BDS concept 70 system, available from Novocontrol, combines the Agilent HP E4991A RF impedance analyzer with the BDS 2200 RF sample cell; the complex permeability of polymers can also be measured by using a magnetic sample cell
Network analysis (6.3.3)	100 MHz–10 GHz	Both reflected wave and wave transmitted through sample are analyzed in terms of phase and amplitude using common network analyzers	High precision attained for materials of sufficient conductivity and dielectric strength	The Rhode & Schwarz ZVRE vector network analyzer in combination with a dielectric sample holder and a compatible temperature system
Time domain reflectometry (TDR; 6.3.4)	100 kHz–20 GHz	A rapid voltage pulse is applied at input to sample cell; Fourier transforms of incident and reflected pulse waveforms are used to determine complex permittivity function $\varepsilon^*(\omega)$	Fast method; for materials in a liquid (preferably) or a granular state, including biopolymers	The DSA8200 digital serial analyzer sampling oscilloscope with single-ended (P8018) or differential (P80318) probes, available from Tektronix (www.tek.com)

thermosets should consult the information supplied in the Micromet Website (www.micromet.com).

In addition to the turnkey dielectric systems, which supply good-quality results from the first day of installation, it is also possible to build multipurpose dielectric systems, by simply combining an impedance analyzer with a suitable sample cell (commercial or homemade) and a compatible temperature control system. Most of the commercially available units are modular. This gives researchers options to adapt the system according to their needs, such as including existing equipment, building up a system in several steps, or extending the performance at a later time. The most important module in a dielectric system is the impedance analyzer, the selection of which should be based on the type of experiments and the range of frequencies that are of interest. Table 6.3 lists some of the most common modules from various manufacturers, along with their scanning frequency bandwidths and other interesting characteristics. Most of these analyzers can be combined with several common temperature control systems, while several impedance analysis modules may be combined with one temperature control module.

The sample cell is usually designed and selected to satisfy requirements imposed by the nature and size of the materials under study. The BDS 1200 connection head is the standard sample cell of Novocontrol for the low-frequency range from DC to 10 MHz, including a PT100 temperature sensor, with operational temperature range between −200 °C and 450 °C (optional 500 °C). This cell can be connected by four- or two-wire BNC cables to virtually any kind of the currently available impedance analyzer and is compatible with common temperature control systems. After preparation, the sandwich capacitor is mounted between the electrodes of the BDS 1200 sample cell. Round gold-plated electrodes are available with 10, 20, 30, or 40 mm diameters. Interdigitated electrodes may also be used in combination with the BDS 1200 or the Alpha active sample connection head for temperature-dependent measurements, or can be directly connected to the impedance analyzer for experiments at ambient conditions (for process or part monitoring, e.g., by embedding in manufactured part; it can also track the part through its lifetime). Liquid parallel-plate (BDS 1308) or liquid cylindrical (BDS 1307) sample cells can also be combined with the BDS 1200 holder (see Section 6.4.1). Each of the abovementioned sample holders and measuring cells for dielectric and impedance spectroscopy can be combined with a range of temperature control systems supplied by Novocontrol. For example, the Quatro cryosystem is a high-quality turnkey temperature control system for applications in material research (specifications: temperature range −160–400 °C; temperature ramps from 0.01 to 30 °C/min; temperature stability and accuracy of ±0.01 °C). The Novocool cryosystem is an economical version of the Quatro cryosystem, recommended for the temperature range from −100 °C to 250 °C. The Novotherm system is recommended for the temperature range 20–400 °C, while the Novotherm-HT system extends the accessible temperature window up to 1400 °C.

TABLE 6.3. A Selection of Commercially Available Analyzers[a] Used for Dielectric Analysis of Polymers

Type	Instrument	Frequency Range	Impedance Range (Ω)	tan δ Accuracy (Limit)	Description and Remarks
Broadband frequency analyzers	Novocontrol Alpha-A analyzer	30 µHz–20 MHz	0.001–10^{15}	3×10^{-5}	These systems combine features of general impedance analyzers and dielectric measurement systems, extending tan δ accuracy; the Alpha-A supports several high-performance test interfaces
	Novocontrol Alpha and Beta analyzer series	30 µHz–20 MHz	0.01–10^{14}	3×10^{-5}	
General impedance analyzers	Stanford Research SR850 + BDC	1 mHz–100 kHz	0.01–10^{8}	3×10^{-4}	Lock-in amplifier + broadband dielectric converter
	Agilent HP 4192A	5 Hz–13 MHz	—	10^{-3}	Impedance analyzer (production discontinued)
	Agilent HP 4194A	10 Hz–40 MHz	—	—	Impedance/gain-phase analyzer (production discontinued)
	Quad Tech model 7600	10 Hz–2 MHz	—	—	Precision LCR meter
	Agilent HP E4980A	20 Hz–2 MHz	—	—	Precision LCR meter
	Agilent HP 4284A	20 Hz–1 MHz	—	5×10^{-4}	Precision LCR meter (production discontinued)
	Agilent HP 4294A	40 Hz–110 MHz	—	—	Impedance analyzer (measurement technique: autobalancing bridge)
	Solartron SI 1260A	10 µHz–32 MHz	—	—	Frequency response analyzer
	Solartron SI 1260A + BDC	10 µHz–10 MHz	10–2×10^{14}	—	Frequency response analyzer + active interface (broadband dielectric converter)

	Solartron SI 1260A + CDI	10 μHz–1 MHz	—	—	Frequency response analyzer + active interface (Chelsea dielectric interface)
RF impedance analyzers and LCR meters	Agilent HP E4991A	1 MHz–3 GHz	$0.1–10^5$	3×10^{-3}	RF impedance/material analyzer (Measurement technique: RF I/V)
	Agilent HP 4291B	1 MHz–1.8 GHz	—	—	RF impedance analyzer (production discontinued)
	Agilent HP 4191A	1 MHz–1 GHz	—	10^{-3}	RF impedance analyzer (production discontinued)
	Agilent HP 4287A	1 MHz–3 GHz	$0.2–3 \times 10^3$	—	RF *LCR* meter
Network analyzers	Rohde & Schwarz R&S ZVA	20 kHz–8 GHz	$0.1–10^4$	3×10^{-2} or 10^{-2}	Vector network analyzer
	Agilent HP 4396B	100 kHz–1.8 GHz	$2–5 \times 10^3$	—	Vector network analyzer (measurement technique: RF I/V)
	Agilent HP 8752A	300 kHz–1.3 GHz	$0.1–10^5$	—	Network analyzer
	Rohde & Schwarz R&S ZVRE	300 kHz–4 GHz	—	—	Vector network analyzer (production discontinued)

[a]Analyzers with discontinued production, which—however—currently find extensive use in materials research labs worldwide, are also included. Homepages: www.home.agilent.com (Agilent Technologies), www2.rohde-schwarz.com (Rohde & Schwarz), www.solartronanalytical.com (Solartron Analytical), www.quadtech.com (Quad Tech), www.lambdaphoto.co.uk (Stanford Research Systems) and www.novocontrol.com (Novocontrol Instruments).

6.8.2. Dielectric Spectrometers for DC Measurements

Dielectric spectrometers for TSC (Section 6.3.5) and *slow* time-domain spectroscopy (Section 6.3.1) measurements of liquids and solids (films or pellets), and related software for data acquisition and analysis are commercially available from Novocontrol and Setaram Instruments (www.setaram.com). For instance, the Setaram TSC II system allows temperature scans in the range −170–350 °C while the sample is under high-vacuum conditions (typically ranging from 10^{-4} to 10 mbar). The polarizing electric field can be adjusted between 0 and 500 V/mm, with accuracy on the order of ± 0.15%. The heating rate can be selected at any value between 0.1 and 40 °C/min. Similar characteristics apply for the Novocontrol TSDC system, which can cover the temperature range of −160–400 °C, for example, with the use of the Quatro cryosystem. Both manufacturers provide control and evaluation multitasking software. With it, nearly any kind of TSC experiment can be programmed by flexible experimental setup procedures, which are defined by a series of time intervals. During each time interval, the polarization voltage and temperature are defined as fixed values or changing continuously with time (ramp). The highly informative thermal sampling technique is available on most commercial TSC systems (e.g., using optional software, such as the WinTSC program developed for the Novocontrol TSDC spectrometer). For such experiments, the main function of the software is the evaluation of relaxation maps from a thermal windowing experiment. The sample polarization is calculated by numerical integration from the measured depolarization current, with several procedures available for integration and baseline corrections. After the conversion of the polarization and depolarization current into the relaxation time, the common relaxation maps are created. These show the single (Debye-type) relaxation modes in several representations, such as Arrhenius or a free-energy diagram, and the relaxation time data can be analyzed using built-in curve-fitting model functions (e.g., Arrhenius, Vogel–Tammann–Fulcher–Hesse, or Williams–Landel–Ferry). For representative examples of the application of the abovementioned techniques, the reader is referred to Section 6.5 of the present chapter. In addition to the general experimental setup mode (i.e., the global TSC scan), predefined setups such as, batch modes for a series of TSC sweeps with different heating rates are also supported.

Besides the abovementioned commercial systems, it is also possible to build homemade systems for TSC and *slow* time-domain dielectric experiments by constructing a simple electric circuit that includes the sample capacitor and either a typical voltage source for the polarization step (e.g., a Keithley 450 source) or an electrometer for the depolarization step (e.g., a Keithley 617 or 642 electrometer). A suitable sample cell (e.g., a parallel-plate capacitor placed within a homemade or a commercial cryostat with controlled inert gas flow) and a compatible temperature control system are also required. The creative minds of the technicians working in the laboratory and the use of the technical information and diagrams available in several journal publications, books, or network libraries, can be used in several cases to minimize the cost. However, the problem that usually makes this solution less attractive is the need for developing software that will be used to achieve computer-controlled measurements and subsequently analyze the experimental discharge current (I_D)–temperature (T) data.

APPENDIX

Standards for dielectric analysis under the jurisdiction of ASTM International Committee E37 on Thermal Measurements are listed below (for more information see www.astm.org):

E2038, *Standard Test Method for Temperature Calibration of Dielectric Analyzers*

E2039, *Standard Test Method for Determining and Reporting Dynamic Dielectric Properties*

ABBREVIATIONS

Symbols

$\alpha, \beta, \gamma, \ldots$	relaxation peak classification
α	degree of conversion
α_f	free-volume dilatation coefficient
α_{CLTE}	coefficient of linear thermal expansion
α_{gel}	conversion at gel point
α_{HN}	HN relaxation function power
A, B	symbols designating adjustable parameters in various functions
β_{HN}	fractional power in HN relaxation function
β_{KWW}	parameter of the KWW function
C	capacitance
C_0	geometric capacitance of a capacitor
C_1, C_2	WLF parameters
l	sample thickness, spatial dimension
D	dielectric induction
d	electrode diameter
δT	mean temperature fluctuation
$\Delta \varepsilon$	dielectric strength
ΔC_{p0}	heat capacity of an uncured monomer mixture
$\Delta C_{p\infty}$	heat capacity of a fully cured network
ΔS^*	activation entropy

e	electron charge
E^*	complex modulus
$E, E_\alpha, E_\beta, \ldots$	activation energy of relaxation processes
E_s	static electric field
E_p	polarizing electric field
ε_0	permittivity of vacuum
ε_s	static permittivity (ε' at $f \to 0$)
ε_∞	high-frequency permittivity (ε' at $f \to \infty$)
$\varepsilon^*(\omega)$	dielectric permittivity function (complex)
ε'	dielectric permitivity (real part)
$\varepsilon'(0), \varepsilon'(\infty)$	permittivity values measured at a microwave frequency immediately after components mixing and at end of cure, respectively
ε''	dielectric loss (imaginary part of complex permittivity)
$\varepsilon''(0), \varepsilon''(\infty)$	dielectric losses measured at a microwave frequency immediately after components mixing and at end of cure, respectively
ε_u	unrelaxed permittivity
ε_r	relaxed permittivity
f	frequency (of a DRS scan)
f_0	relaxation–frequency-related parameter of the HN equation
$f_{eq,DSC}$	equivalent frequency of a DSC scan
$f_{eq,TSC}$	equivalent frequency of a TSC scan
f_g	kinetic free volume fraction at T_g
f_{max}	relaxation frequency
$\Phi(t)$	macroscopic relaxation function
$g(\log \tau)$	relaxation time distribution function (normalized)
G	conductance
h	planck constant
i	imaginary number
I	current
I_D	depolarization current
J^*	complex compliance
J_D	depolarization current density
k	Boltzmann constant
L	inductance
m	fragility index
$M^*(\omega)$	electric modulus function (complex)
M_w, M_n	molecular masses (weight and number averages)
μ	dipole moment
N	concentration (of dipoles, ionic species, etc.)
η	viscosity
Ω	ohms
P_d	dielectric polarization
P_0	equilibrium polarization

q	heating rate
q'	cooling rate
Q_{el}	depolarization charge
R	resistance; universal gas constant
R_g	radius of gyration
$\rho^*(\omega)$	resistivity function (complex)
S	surface area; Siemens unit ($1/\Omega$)
σ	conductivity
σ_t	transient conductivity TSC signal
$\sigma^*(\omega)$	conductivity function (complex)
σ_0, σ_∞	conductivities of initial mixture and of fully cured network
σ_{DC}	specific DC conductivity
σ_{AC}	frequency-dependent conductivity
t	time
t'	thickness
t_p	isothermal polarization time
t_c	cure time
t_{gel}	time to reach gelation
$\tan \delta$	dissipation factor (or dielectric loss tangent)
T	temperature (absolute)
T_0	reference temperature for DC conductivity
T_{cl}	isotropization or clearing temperature
T_{comp}	compensation temperature
T_{cure}	cure temperature
T_{cut}	cutoff temperature
T_g	calorimetry-determined glass transition temperature
T_{g0}	glass transition temperature of uncured thermoset
$T_{g\infty}$	glass transition temperature of fully cured thermoset
$T_{g,diel}$	dielectric glass transition temperature
T_{max}	temperature of peak maximum
T_m	melting temperature
T_p	polarization temperature
T_r	reference temperature
T_V	Vogel temperature (VTFH equation)
τ	relaxation time
$<\tau>$	mean relaxation time
τ_{HN}	relaxation-time-related parameter of HN function
τ_{comp}	compensation parameter related to relaxation time
τ_M	relaxation time in electric modulus function
τ_0	preexponential factor of Arrhenius equation
V	voltage, volume
w	parameter of Warburg impedance
ω	angular frequency of applied field
$Z^*(\omega)$	impedance function (complex)
Z_0	a constant depending on electrode–polymer interface

Acronyms

AP	aminophenol
BDC	broadband dielectric converter
BDS	broadband dielectric spectrometer
CR	conductivity relaxation
DDM	4,4'-diaminodiphenylmethane
DDS	4,4'-diaminodiphenylsulfone
DEA	dielectric analysis
DGEBA	diglycidyl ether of bisphenol A
DMA	dynamic mechanical analysis
DRS	dielectric relaxation spectroscopy
DSC	differential scanning calorimetry
EDA	ethylenediamine
FET	field effect transistor
FRA	frequency response analysis
FTIR	Fourier transform infrared
HBA	p-hydroxybenzoic acid
HDPE	high-density polyethylene
HN	Havriliak–Negami
HNA	6-hydroxy-2-naphthoic acid
IPNs	interpenetrating polymer networks
IR	infra red
KWW	Köhlrausch–Williams–Watts
LCPs	liquid crystalline polymers
LCR	inductance–capacitance–resistance
LDPE	low-density polyethylene
LLDPE	linear low-density polyethylene
MCDEA	4,4'-methylenebis(3-chloro-2,6-diethylaniline)
MDEA	4,4'-methylenebis 2,6-diethylaniline
MWS	Maxwell–Wagner-Sillars
NIR	near infrared
NMR	nuclear magnetic resonance
OTMG	oxytetramethylene glycol
P2CS	poly(2-chlorostyrene)
PE	polyethylene
PEEK	poly(ether ether ketone)
PEN	poly(ethylene-2,6-naphthalate)
PEO	polyethylene oxide
PET	poly(ethylene terephthalate)
PI	polyisoprene
PMMA	poly(methyl methacrylate)
POM	polyoxymethylene
PPO	polypropylene oxide
PS	polystyrene
PTFE	poly(tetrafluoroethylene)

PU	polyurethane
PVC	poly(vinyl chloride)
PVDF	poly(vinylidene fluoride)
PVE	poly(vinyl ethylene)
PVF	poly(vinyl fluoride)
PVME	poly(vinyl methyl ether)
PVPh	poly(4-vinylphenol)
PVPy	poly(2-vinylpyridine)
RF	radiofrequency
RHS	right hand side
RMA	relaxation map analysis (TSC variant)
TA	terephthalic acid
TS	thermal sampling (TSC variant)
TSC	thermally stimulated current technique
VTFH	Vogel–Tammann–Fulcher–Hesse
WP	windowing polarization (TSC variant)

REFERENCES

Acitelli, M. A., Prime, R. B., and Sacher, E. (1971), *Polymer* **12**, 335.

Adachi, K. (1997), "Dielectric relaxation in polymer solutions," in *Dielectric Spectroscopy of Polymeric Materials—Fundamental and Applications*, Runt, J. P. and Fitzgerald, J. J., eds., ACS, Washington, DC, pp. 261–282.

Adams, G. and Gibbs, J. H. (1965), *J. Chem. Phys.* **43**, 139.

Alcoutlabi, M. and McKenna, G. B. (2005), *J. Phys. Condens. Matt.* **17**, R461.

Anagnostopoulou-Konsta, A., Apekis, L., Christodoulides, C., Daoukaki, D., and Pissis, P. (1991), "Dielectric study of the hydration process in biological materials," in *Biologically Inspired Physics*, Peliti, L., ed., Plenum, New York, p. 229.

Andjelić, S., Fitz, B., and Mijović, J. (1997), *Macromolecules* **30**, 5239.

Angel, C. A. (1991), *J. Non-Cryst. Solids* **131**–133, 13.

Arbe, A., Alergia, A., Colmenero, J., Hoffmann, S., Willner, L., and Richter, D. (1999), *Macromolecules* **32**, 7572.

Avakian, P., Coburn, J. C., Connolly, M. S., and Sauer, B. B. (1996), *Polymer* **37**, 3843.

Avakian, P., Starkweather, H. W., Jr., Fontanella, J. J., and Wintersgill, M. C. (1997), "Dielectric properties of fluoropolymers," in *Dielectric Spectroscopy of Polymeric Materials—Fundamental and Applications*, Runt, J. P. and Fitzgerald, J. J., ACS, Washington, DC, pp. 379–393.

Bartenev, G. M., Sinitsyna, G. M., and Barteneva, A. G. (2000), *Polym. Sci. Series A* **42**, 1233.

Bartolomeo, P., Chailan, J. F., and Vernet, J. L. (2001), *Polymer* **42**, 4385.

Baskaran, R., Selvasekarapandian, S., Hirankumar, G., and Bhuvaneswari, M. S. (2004), *Ionics* **10**, 129.

Beiner, M. and Ngai, K. L. (2005), *Macromolecules* **38**, 7033.

Bellucci, F., Maio, V., Monetta, T., Nicodemo, L., Kenny, J., Nicolais, L., and Mijovic, J. (1995). *J. Polym. Sci.* **B33**, 433–443.

Berberian, J. G. and King, E. (2002), *J. Non-Cryst. Solids* **305**, 10.

Boersma, A., van Turnhout, J., and Wübbenhorst, M. (1998a), *Macromolecules* **31**, 7453.

Boersma, A., van Turnhout, J., and Wübbenhorst, M. (1998b), *Macromolecules* **31**, 7461.

Böhmer, R., Ngai, K. L., Angell, C. A., and Plazek, D. J. (1993), *J. Chem. Phys.* **99**, 4201.

Boinard, P., Banks, W. M., and Pethrick, R. A. (2005), *Polymer* **46**, 2218.

Boiteux, G., Dublineau, P., Feve, M., Mathieu, C., Seytre, G., and Ulanski, J. (1993), *Polym. Bull.* **30**, 441.

Bone, S. (1988), *Biochim. Biophys. Acta* **967**, 401.

Bonnet, A., Pascault, J. P., Sautereau, H., Rogozinski, J., and Kranbuehl, D. (2000), *Macromolecules* **33**, 3833.

Bottcher, C. J. F. and Bordewijk, P. (1978), *Theory of Electric Polarization*, Elsevier, Amsterdam.

Boyd, R. H. and Liu, F. (1997), "Dielectric spectroscopy of semicrystalline polymers," in *Dielectric Spectroscopy of Polymeric Materials—Fundamental and Applications*, Runt, J. P. and Fitzgerald, J. J., eds., ACS, Washington, DC, pp. 107–136.

Boyd, R. H. and Porter, C. H. (1972), *J. Polym. Sci. B* **10**, 647.

Boyer, R. F. (1973), *Macromolecules* **6**, 288.

Boyer, R. F. (1977), "Transitions and relaxations in amorphous and semicrystalline organic polymers and copolymers," *Encyclopedia of Polymer Science and Technology*, Suppl. 2, pp. 745–839.

Boyer, R. F. (1979a), *J. Macromol. Sci. Phys.* **18**, 461.

Boyer, R. F. (1979b), *Polym. Eng. Sci.* **19**, 10.

Bucci, C., Fieschi, R., and Guidi, G. (1966), *Phys. Rev.* **148**, 816.

Buixaderas, E., Kamba, S., and Petzelt, J. (2004), *Ferroelectrics* **308**, 131.

Bur, A. J., Lee, Y. H., Roth, S. C., and Start, P. R. (2005), *Polymer* **46**, 10908.

Butta, E., Livi, A., Levita, G., and Rolla, P. A. (1995), *J. Polym. Sci. B* **33**, 2253.

Carrozzino, S., Levita, G., Rolla, P., and Tombari, E. (1990), *Polym. Eng. Sci.* **30**, 366.

Casalini, R. and Roland, C. M. (2005), *Macromolecules* **38**, 1779.

Cassettari, M., Salvetti, G., Tombari, E., Veronesi, S., and Johari, G. P. (1993), *J. Mol. Liquids* **56**, 141.

Cendoya, I., Alegria, A., Alberti, J. M., Colmenero, J., Grimm, H., Richter, D., and Frick, B. (1999), *Macromolecules* **32**, 4065.

Chaplin, A., Hamerton, I., Herman, H., Mudhar, A. K., and Shaw, S. J. (2000), *Polymer* **41**, 3945.

Charnetskaya, A. G., Polizos, G., Shtompel, V. I., Privalko, E. G., Kercha, Yu. Yu., and Pissis, P. (2003), *Eur. Polym. J.* **39**, 2167.

Christodoulides, C., Pissis, P., Apekis, L., and Daoukaki-Diamanti, D. (1991), *J. Phys. D, Appl. Phys.* **24**, 2050.

Ciriscioli, P. R. and Springer, G. S. (1989), *SAMPE J.* **25**, 35.

Cohen, M. H. and Turnbull, D. (1959), *J. Chem. Phys.* **31**, 1164.

Cole, R. H. (1975), *J. Phys. Chem.* **79**, 1469.

Cole, R. H., Berberian, J. G., Mashimo, S., Chryssikos, G., and Burns, A. (1989), *J. Appl. Phys.* **66**, 793.

Cole, R. H., Mashimo, S., and Winsor, P., IV (1980), *J. Phys. Chem.* **84**, 786.

Collins, G. and Long, B. (1994), *J. Appl. Polym. Sci.* **53**, 587.

Corezzi, S., Beiner, M., Huth, H., Schröter, K., Capaccioli, S., Casalini, R., Fioretto, D., and Donth, E. (2002), *J. Chem. Phys.* **117**, 2435.

Correia, N. T. and Moura Ramos, J. J. (2001), *Phys. Chem. Chem. Phys.* **2**, 5712.

Craig, D. Q. M. (1995), *Dielectric Analysis of Pharmaceutical Systems*, Taylor & Francis, London.

Creswell, R. A. and Perlman, M. M. (1970), *J. Appl. Phys.* **41**, 2365.

Crowley, K. J. and Zografi, G. (2001), *Thermochim. Acta* **380**, 79.

Dargent, E., Bureau, E., Delbreilh, L., Zumailan, A., and Saiter, J. M. (2005), *Polymer* **46**, 3090.

Das-Gupta, D. K. (1994), *IEEE Electric. Insul. Mag.* **10**, 5.

Das-Gupta, D. K. and Scarpa, P. C. N. (1996), *IEEE Trans. Dielectric Electric. Insul.* **3**, 366.

Das, N. K., Voda, S. M., and Pozar, D. M. (1987), *IEEE Trans. Microwave Theory* **35**, 636.

Day, D. R. (1989), *Polym. Eng. Sci.* **29**, 334.

de Deus, J. F., Souza, G. P., Corradini, W. A., Atvars, T. D. Z., and Akcelrud, L. (2004), *Macromolecules* **37**, 6938.

Debye, P. (1921), *Ann. Physik* **39**, 789.

Debye, P. (1929), *Polar Molecules*, Chemical Catalog Co, New York (reprinted by Dover, New York).

Díaz-Calleja, R. (2000), *Macromolecules* **33**, 8924.

Díaz-Calleja, R. and Riande, E. (1997), "Calculation of dipole moments and correlation parameters," in *Dielectric Spectroscopy of Polymeric Materials—Fundamental and Applications*, Runt, J. P. and Fitzgerald, J. J., eds., ACS, Washington, DC, pp. 139–173.

Donth, E. (1992), *Relaxation and Thermodynamics in Polymers, Glass Transition*. Akademie, Berlin.

Doulout, S., Demont, P., and Lacabanne, C. (2000), *Macromolecules* **33**, 3425.

Dudognon, E., Bernès, A., and Lacabanne, C. (2002a), *J. Phys. D, Appl. Phys.* **35**, 9.

Dudognon, E., Bernès, A., and Lacabanne, C. (2002b), *Macromolecules* **35**, 5927.

Dudognon, E., Bernès, A., and Lacabanne, C. (2003), *J. Macromol. Sci.-Phys.* **B42**, 215.

Dudognon, E., Bernès, A., and Lacabanne, C. (2004), *J. Macromol. Sci.-Phys.* **B43**, 591.

Elicegui, A., Del Val, J. J., Millan, J. L., and Mijangos, C. (1998), *J. Non-Cryst. Solids* **235**, 623.

Ellyin, F. and Rohrbacher, C. (2000), *J. Reinforced Plast. Compos.* **19**, 1405.

Elmahdy, M. M., Chrissopoulou, K., Afratis, A., Floudas, G., and Anastasiadis, S. H. (2006), *Macromolecules* **39**, 5170.

Eloundou, J. P., Gerard, J. F., Pascault, J. P., Boiteux, G., and Seytre, G. (1998a), *Angew. Makromol. Chem.* **263**, 57.

Eloundou, J. P., Ayina, O., Nga, H. N., Gerard, J. F., Pascault, J. P., Boiteux, G., and Seytre, G. (1998b), *J. Polym. Sci. B* **36**, 2911.

Eloundou, J. P., Gerard, J. F., Pascault, J. P., and Kranbuehl, D. (2002), *Macromol. Chem. Phys.* **203**, 1974.

Enns, J. B. and Simha, R. (1977), *J. Macromol. Sci. Phys.* **B13**, 11.

Fava, R. A. and Horsfie, A. E. (1968), *Br. J. Appl. Phys.* **2**, 1, 117.

Fitz, B. D. and Andjelić, S. (2003), *Polymer* **44**, 3031.

Fitz, B. D. and Mijović, J. (1998), *Polym. Adv. Technol.* **9**, 721.

Flory, P. J. (1988), *Statistical Mechanics of Chain Molecules*, Hanser, Munich.

Floudas, G., Mierzwa, M., and Schönhals, A. (2003), *Phys. Rev. E* **67**, 031705. (article number).

Fournier, J., Williams, G., Duch, C., and Aldridge, G. A. (1996), *Macromolecules* **29**, 7097.

Frank, B., Frübing, P., and Pissis, P. (1996), *J. Polym. Sci. B* **34**, 1853.

Friedrich, K., Ulanski, J., Boiteux, G., and Seytre, G. (2001), *IEEE Trans. Dielectric Electric. Insul.* **8**, 572.

Fröhlich, H. (1958), *Theory of Dielectrics*, 2nd ed. (Oxford University Press, Oxford).

Frübing, P., Krüger, H., Goering, H., and Gerhard-Multhaupt, R. (2002), *Polymer* **43**, 2787.

Frübing, P., Kremmer, A., Gerhard-Multhaupt, R., Spanoudaki, A., and Pissis, P. (2006), *J. Chem. Phys.* **125**, 214701. (article number).

Fulcher, G. S. (1925), *J. Am. Ceram. Soc.* **8**, 339.

Gallone, G., Capaccioli, S., Levita, G., Rolla, P. A., and Corezzi, S. (2001), *Polym. Intl.* **50**, 545.

Garlick, G. F. J. and Gibson, A. F. (1948), *Proc. Phys. Soc. Lond.* **A60**, 574.

Garwe, F., Schönhals, A., Lockwenz, H., Beiner, M., Schröter, K., and Donth, E. (1996), *Macromolecules* **29**, 247.

Gaur, U. and Wunderlich, B. (1980), *Macromolecules* **13**, 445.

Genix, A. C., Arbe, A., Alvarez, F., Colmenero, J., Willner, L., and Richter, D. (2005), *Phys. Rev. E* **72**, 031808 (article number).

Georgoussis, G., Kanapitsas, A., Pissis, P., Savelyev, Yu. V., Veselov, V. Ya., and Privalko, E. G. (2000a), *Eur. Polym. J.* **36**, 1113.

Georgoussis, G., Kyritsis, A., Bershtein, V. A., Fainleib, A. M., and Pissis, P. (2000b), *J. Polym. Sci. B* **38**, 3070.

Georgoussis, G., Kyritsis, A., Pissis, P., Savelyev, Yu V., Akhranovick, E. R., Privalko, E. G., and Privalko, V. P. (1999), *Eur. Polym. J.* **35**, 2007.

Girardreydet, E., Riccardi, C. C., Sautereau, H., and Pascault, J. P. (1995), *Macromolecules* **28**, 7599.

Goodwin, A. A. and Hay, J. N. (1998), *J. Polym. Sci. B* **36**, 851.

Gordon, M. and Taylor, J. S. (1952), *J. Appl. Chem.* **2**, 493.

Gotro, J. and Prime, R. B. (2004). "Thermosets," in *Encyclopedia of Polymer Science Technology*, 3rd ed. J. Kroschwitz, ed., Wiley, Hoboken, NJ.

Gotro, J. and Yandrasits, M. (1989), *Polym. Eng. Sci.* **29**, 278.

Gourari, A., Bendaoud, M., Lacabanne, C., and Boyer, R. F. (1985), *J. Polym. Sci. B* **23**, 889.

Graff, M. S. and Boyd, R. H. (1994), *Polymer* **35**, 1797.

Gregorio, R. and Ueno, E. M. (1999), *J. Mater. Sci.* **34**, 4489.

Grimau, M., Laredo, E., Bello, A., and Suarez, N. (1997), *J. Polym. Sci. B* **35**, 2483.

Guarrotxena, N., Del Val, J. J., Elicegui, A., and Millan, J. (2004), *J. Polym. Sci. B* **42**, 2337.

Gun'ko, V. M., Zarko, V. I., Goncharuk, E. V. et al. (2007), *Adv. Colloid Interface* **131**, 1.

Hakme, C., Stevenson, I., David, L., Boiteux, G., Seytre, G., and Schönhals, A. (2005), *J. Non-Cryst. Solids* **351**, 2742.

Hamon, B. V. (1952), *Proc. IEE (Lond.) Pt. IV* **99**, 151.

Hardy, L., Fritz, A., Stevenson, I., Boiteux, G., Seytre, G., and Schönhals, A. (2002), *J. Non-Cryst. Solids* **305**, 174.

Hartmann, L., Gorbatschow, W., Hauwede, J., and Kremer, F. (2002), *Eur. Phys. J. E* **8**, 145.

Havriliak, S. and Negami, S. (1967), *Polymer* **8**, 161.

Hayakawa, T. and Adachi, K. (2001), *Polymer* **42**, 1725.

Hedvig, P. (1977), *Dielectric Spectroscopy of Polymers*, Adam Hilger, Bristol.

Heijboer, J. (1978), in *Molecular Basis of Transitions and Relaxations*, Meier, D. J., ed., Gordon & Breach, New York.

Hensel, A., Dobbertin, J., Schawe, J. E. K., Boller, A., and Schick, C. (1996), *J. Therm. Anal.* **46**, 935.

Higgenbotham-Bertolucci, P. R., Gao, H., and Harmon, J. P. (2001), *Polym. Eng. Sci.* **41**, 873.

Hino, T., Suzuki, K., and Yamashita, K. (1973), *Jpn. J. Appl. Phys.* **12**, 651.

Hirose, Y. and Adachi, K. (2005), *Polymer* **46**, 1913.

Hodge, I. M., Ngai, K. L., and Moynihan, C. T. (2005), *J. Non-Cryst. Solids* **351**, 104.

Howell, F. S., Bose, R. A., Moynihan, C. T., and Macedo, P. B. (1974), *J. Phys. Chem.* **78**, 639.

Hsu, J.-M., Yang, D.-L., and Huang, S. K. (1999), *Thermochim. Acta* **333**, 73.

Huo, P. and Cebe, P. (1992), *Macromolecules* **25**, 902.

Jayathilaka, P. A. R. D., Dissanayake, M. A. K. L., Albinsson, I., and Mellander, B. E. (2003), *Solid State Ionics* **156**, 179.

Johari, G. P. (1973), *J. Chem. Phys.* **58**, 1766.

Johari, G. P. and Goldstein, M. (1970), *J. Chem. Phys.* **53**, 2372.

Jonscher, A. K. (1983), *Dielectric Relaxation in Solids*, Chelsea Dielectrics Press, London.

Kalika, D. S. and Krishnaswamy, R. K. (1993), *Macromolecules* **26**, 4252.

Kalogeras, I. M. (2004), *J. Polym. Sci. B* **42**, 702.

Kalogeras, I. M. (2005), *Acta Mater.* **53**, 1921.

Kalogeras, I. M. (2008), "Contributions of dielectric analysis in the study of nanoscale properties and phenomena in polymers," in *Chapter 10 Progress in Polymer Nanocomposite Research*, Thomas, S. and Zaikov, G., eds., Nova Science.

Kalogeras, I. M. and Brastow, W. (2009). *J. Polym. Sci.* **B47**, 80.

Kalogeras, I. M. and Neagu, E. R. (2004), *Eur. Phys. J. E* **14**, 193.

Kalogeras, I. M., Vassilikou-Dova, A., and Neagu, E. R. (2001), *Mater. Res. Innova.* **4**, 322.

Kalogeras, I. M., Neagu, E. R., and Vassilikou-Dova, A. (2004), *Macromolecules* **37**, 1042.

Kalogeras, I. M., Roussos, M., Christakis, I., Spanoudaki, A., Pietkiewicz, D., Brostow, W., and Vassilikou-Dova, A. (2005a), *J. Non-Cryst. Solids* **351**, 2728.

Kalogeras, I. M., Roussos, M., Vassilikou-Dova, A., Spanoudaki, A., Pissis, P., Savelyev, Yu. V., Shtompel, V. I., and Robota. L. P. (2005b), *Eur. Phys. J. E* **18**, 467.

Kalogeras, I. M., Pallikari, F., Vassilikou-Dova, A., and Neagu, E. R. (2006a), *Appl. Phys. Lett.* **89**, 172905 (article number).

Kalogeras, I. M., Vassilikou-Dova, A., Christakis, I., Pietkiewicz, D., and Brostow, W. (2006b), *Macromol. Chem. Phys.* **207**, 879.

Kalogeras, I. M., Pallikari, F., Vassilikou-Dova, A., and Neagu, E. R. (2007a), *J. Appl. Phys.* **101**, 094108 (article number).

Kalogeras, I. M., Stathopoulos. A., Vassilikou-Dova, A., and Brostow, W. (2007b), *J. Phys. Chem. B* **111**, 2774.

Kanapitsas, A., Pissis, P., Gomez-Ribelles, J. L., Monleon-Pradas, M., Privalko, E. G., and Privalko, V. P. (1999), *J. Appl. Polym. Sci.* **71**, 1209.

Karabanova, L. V., Boiteux, G., Gain, O., Seytre, G., Sergeeva, L. M., Lutsyk, E. D., and Bodarenko, P. A. (2003), *J. Appl. Polym. Sci.* **90**, 1191.

Karlinsey, R. L., Bronstein, L. M., and Zwanziger, J. W. (2004), *J. Phys. Chem. B* **108**, 918.

Kessairi, A., Napolitano, S., Capaccioli, S., Rolla, P., and Wübbenhorst, M. (2007), *Macromolecules* **40**, 1786.

Kienle, R. H. and Race, H. H. (1934), *Trans. Electrochem. Soc.* **65**, 87.

Korbakov, N., Harel, H., Feldman, Y., and Marom, G. (2002), *Macromol. Chem. Phys.* **203**, 2267.

Kranbuehl, D. E. (1997), "Dielectric monitoring of polymerization and cure," in *Dielectric Spectroscopy of Polymeric Materials—Fundamental and Applications*, Runt, J. P. and Fitzgerald, J. J., eds., ACS, Washington, DC, pp. 303–328.

Kranbuehl, D. E., Delos, S. E., and Jue, P. K. (1986), *Polymer* **27**, 11.

Kremer, F. (1997), "Broadband dielectric spectroscopy on collective and molecular dynamics in ferroelectric liquid crystals," in *Dielectric Spectroscopy of Polymeric Materials—Fundamental and Applications*, Runt, J. P. and Fitzgerald, J. J., eds., ACS, Washington, DC, pp. 423–444.

Kremer, F., Hartmann, L., Serghei, A., Pouret, P., and Leger, L. (2003), *Eur. Phys. J. E* **12**, 139.

Kremer, F., and Schönhals, A. (2002), *Broadband Dielectric Spectroscopy*, Springer, Berlin.

Krishnaswamy, R. K. and Kalika, D. S. (1994), *Polymer* **35**, 1157.

Krouse, D., Guo, Z., and Kranbuehl, D. E. (2005), *J. Non-Cryst. Solids* **351**, 2831.

Krygier, E., Lin, G. X., Mendes, J., Mukandela, G., Azar, D., Jones, A. A., Pathak, J. A., Colby, R. H., Kumar, S. K., Floudas, G., Krishnamoorti, R., and Faust, R. (2005), *Macromolecules* **38**, 7721.

Kyritsis, A. and Pissis, P. (1997), *Macromol. Symp.* **119**, 15.

Lacabanne, C., Lamure, A., Teyssedre, G., Bernes, A., and Mourgues, M. (1994), *J. Non-Cryst. Solids* **172–174**, 884.

Laj, C. and Berge, P. (1966), *Comptes Rend. Acad. Sci. Paris* **B263**, 380.

Landau, L. D., Lifshitz, E. M., and Pilaevskii, L. P. (1984), *Electrodynamics of Continuous Media*, 2nd ed., Pergamon, New York.

Langlet, R., Arab, M., Pacaud, F., Devel, M., and Girardet, C. (2004), *J. Chem. Phys.* **121**, 9655.

Laredo, E., Bello, A., Hernández, M. C., and Grimau, M. (2001), *J. Appl. Phys.* **90**, 5721.

Laredo, E., Grimau, M., Bello, A., and Muller, A. J. (2005), *Polymer* **46**, 6532.

Laredo, E., Grimau, M., Sanchez, F., and Bello, A. (2003), *Macromolecules* **36**, 9840.

Laredo, E. and Hernández, M. C. (1997), *J. Polym. Sci. B* **35**, 2879.

Laverge, C. and Lacabanne, C. (1993), *IEEE Electric Insul. Mag.* **9**, 5.

Lefebvre, D. R., Han, J., Lipari, J. M., Long, M. A., McSwain, R. L., and Wells, H. C. (2006), *J. Appl. Polym. Sci.* **101**, 2765.

Leroy, E., Alegria, A., and Colmenero, J. (2003), *Macromolecules* **36**, 7280.

Leroy, E., Dupuy, J., Maazouz, A., and Seytre, G. (2005), *Polymer* **46**, 9919.

Lightfoot, S. and Xu, G. (1993), *Polym. Plast. Technol.* **32**, 21.

Li, H. M., Fouracre, R. A., Given, M. J., Banford, H. M., Wysocki, S., and Karolczak, S. (1999), *IEEE Trans. Dielectr. Electric. Insul.* **6**, 295.

Li, L., Liu, M. J., and Li, S. J. (2004), *Polymer* **45**, 2837.

Lodge, T. P. and McLeish, T. C. B. (2000), *Macromolecules* **33**, 5278.

Lorthioir, C., Alegría, A., and Colmenero, J. (2003), *Phys. Rev. E* **68**, 031805 (article number).

Lu, H. B. and Zhang, X. Y. (2006), *J. Appl. Phys.* **100**, 054104 (article number).

Macdonald, J. R., ed. (1987), *Impedance* Spectroscopy, Wiley.

Macedo, P. B., Moynihan, C. T., and Bose, R. (1972), *Phys. Chem. Glasses* **13**, 171.

Mangion, M. B. M. and Johari, G. P. (1990), *Macromolecules* **23**, 3687.

Mangion, M. B. M. and Johari, G. P. (1991), *J. Polym. Sci. B* **29**, 437.

Mangion, M. B. M., Wang, M., and Johari, G. P. (1992), *J. Polym. Sci. B* **30**, 445.

Mano, J. F. (2003), *J. Macromol. Sci. Phys.* **B42**, 1169.

Mano, J. F. and Gómez Ribelles, J. L. (2003), *Macromolecules* **36**, 2816.

Mashimo, S. (1997), "Application high-frequency dielectric measurements of polymers," in *Dielectric Spectroscopy of Polymeric Materials—Fundamental and Applications*, Runt, J. P. and Fitzgerald, J. J., eds., ACS, Washington, DC, pp. 201–225.

McGrum, N. G. and Morris, E. L. (1964), *Proc. Roy. Soc. (Lond.)* **A281**, 258.

McGrum, N. G., Read, B. E., and Williams, G. (1967), *Anelastic and Dielectric Effects in Polymeric Solids*, Wiley, New York.

Menczel, J. and Wunderlich, B. (1981), *J. Polym. Sci. Lett.* **19**, 261.

Menczel, J. and Wunderlich, B. (1986), *ACS paper Abstr.* **191**, 93.

Meseguer-Dueñas, J. M. and Gómez-Ribelles, J. L. (2005), *J. Non-Cryst. Solids* **351**, 482.

Mijović, J. and Bellucci, F. (1996), Trends Polym. *Sci.* **4**, 74.

Mijović, J., Bellucci, F., and Nicolais, L. (1995), *J. Electrochem. Soc.* **142**, 1176.

Mijović, J., Kenny, J. M., Maffezzoli, A., Tristano, A., Bellucci, F., and Nicolais, L. (1993), *Compos. Sci. Technol.* **49**, 277.

Mijović, J., Lee, H. K., Kenny, J., and Mays, J. (2006), *Macromolecules* **39**, 2172.

Mohomed, K., Gerasimov, T. G., Abourahma, H., Zaworotko, M. J., and Harmon, J. P. (2005), *Mater. Sci. Eng. A Struct* **409**, 227.

Montserrat, S., Roman, F., and Colomer, P. (2003), *Polymer* **44**, 101.

Moscicki, J. K. (1992), in *Liquid Crystal Polymers: From Structures to Applications*, Collyer, A. A., ed., Elsevier Applied Science, London, Chapter 4.

Moura Ramos, J. J., Mano, J. F., and Sauer, B. B. (1997), *Polymer* **38**, 1081.

Murthy, S. S. N. (1993), *J. Polym. Sci. B* **31**, 475.

Muzeau, E., Perez, J., and Johari, G. P. (1991), *Macromolecules* **24**, 4713.

Muzeau, E., Vigier, G., Vassoille, R., and Perez, J. (1995), *Polymer* **36**, 611.

Nad, F., Monceau, P., Carcel, C., and Fabre, J. M. (2000), *Phys. Rev. B* **62**, 1753.

Nakamura, H., Mashimo, S., and Wada, A. (1982), *Jpn. J. Appl. Phys.* **21**, 467.

Nass, K. A. and Seferis, J. C. (1989), *Polym. Eng. Sci* **29**, 315.

Neagu, E. R., Hornsby, J. S., and Das-Gupta, D. K. (2002), *J. Phys. D. Appl. Phys.* **35**, 1229.

Neagu, E. R. and Neagu, R. M. (2000), *Thin Solid Films* **358**, 283.

Neagu, R. M., Neagu, E., Kyritsis, A., and Pissis, P. (2000), *J. Phys. D. Appl. Phys.* **33**, 1921.

Neagu, R. M., Neagu, E. R., Kalogeras, I. M., and Vassilikou-Dova, A. (2001), *Mater. Res. Innovations* **4**, 115.

Ngai, K. L. (2005), *J. Non-Cryst. Solids* **351**, 2635.

Ngai, K. L. and Paluck, M. (2004), *J. Chem. Phys.* **120**, 857.

Nielsen, L. E. (1969), *J. Macromol. Sci. Rev. Makromol. Chem.* **C3**, 69.

Noda, N., Lee, Y. H., Bur, A. J., Prabhu, V. M., Snyder, C. R., Roth, S. C., and McBrearty, M. (2005), *Polymer* **46**, 7201.

Nogales, K., Ezquerra, T. A., Batallan, F., Frick, B., Lopez-Cabarcos, E., and Balta-Calleja, F. J. (1999), *Macromolecules* **32**, 2301.

Núñez-Regueira, L., Gómez-Barreiro, S., and Gracia-Fernández, C. A. (2005), *J. Therm. Anal. Calorim.* **82**, 797.

Núñez, L., Fraga, F., Castro, A., and Fraga, L. (1998a), *J. Appl. Polym. Sci.* **74**, 1013.

Núñez, L., Fraga, F., Núñez, M. R., Castro, A., and Fraga, L. (1998b), *J. Appl. Polym. Sci.* **74**, 2997.

Okrasa, L., Zigon, M., Zagar, E., Czech, P., and Boiteux, G. (2005), *J. Non-Cryst. Solids* **351**, 2753.

Page, K. A. and Adachi, K. (2006), *Polymer* **47**, 6406.
Pascault, J. P. and Williams, R. J. J. (1990), *J. Polym. Sci. B* **28**, 85.
Pathmanathan, K. and Johari, G. P. (1993), *J. Polym. Sci. B* **31**, 265.
Pathmanathan, K., Cavaillé, J.-Y., and Johari, G. P. (1992), *J. Polym. Sci. B* **30**, 341.
Perez, J., Cavaille, J. Y., and David, L. (1999), *J. Mol. Struct.* **479**, 183.
Pethrick, R. and Hayward, D. (2002), *Prog. Polym. Sci.* **27**, 1983.
Pissis, P., Apekis, L., Christodoulides, C., Niaounakis, M., Kyritsis, A., and Nedbal, J. (1996), *J. Polym. Sci. B* **34**, 1529.
Pissis, P., Kanapitsas, A., Savelyev, Y. V., Akhranovich, E. R., Privalko, E. G., and Privalko, V. P. (1998), *Polymer* **39**, 3431.
Poncet, S., Boiteux, G., Pascault, J. P., Sautereau, H., Seytre, G., Rogozinski, J., and Kranbuehl, D. (1999), *Polymer* **40**, 6811.
Pratt, G. J. and Smith, M. J. (2001), *Polym. Intl.* **51**, 21.
Priestley, R. D., Rittigstein, P., Broadbelt, L. J., Fukao, K., and Torkelson, J. M. (2007), *J. Phys. Condens. Matt.* **19**, 205120.
Prime, R. B. and Gotro, J. (2004), "Thermosets," in *Encyclopedia of Polymer Science & Technology*, Kroschwitz, J., ed., Wiley, New York.
Psarras, G. C., Beobide, A. S., Voyiatzis, G. A., Karahaliou, P. K., Georga, S. N., Krontiras, C. A., and Sotiropoulos, J. (2006), *J. Polym. Sci. B* **44**, 3078.
Ribelles, J. L. G., Pradas, M. M., Ferrer, G. G., Torres, N. P., Giménez, V. P., Pissis, P., and Kyritsis, A. (1999), *J. Polym. Sci. B* **37**, 1587.
Roth, C. B. and Dutcher, J. R.. (2005), *J. Electroanal. Chem.* **584**, 13.
Roussos, M., Konstantopoulou, A., Kalogeras, I. M., Kanapitsas, A., Pissis, P., Savelyev, Y., and Vassilikou-Dova, A. (2004), *E-Polymers*, article number 042.
Runt, J. P. (1997), "Dielectric studies of polymer blends," in *Dielectric Spectroscopy of Polymeric Materials—Fundamental and Applications*, Runt, J. P. and Fitzgerald, J. J., eds., ACS, Washington, DC, pp. 283–302.
Saiter, J. M., Dargent, E., Kattan, M., Cabot, C., and Grenet, J. (2003), *Polymer* **44**, 3995.
Sanchez, M. S., Ferrer, G. G., Cabanilles, C. T., Duenas, J. M. M., Pradas, M. M., and Ribelles, J. L. G. (2001), *Polymer* **42**, 10071.
Sauer, B. B., Avakian, P., and Starkweather, H. W., Jr. (1996), *J. Polym. Sci. B* **34**, 517.
Sauer, B. B., Avakian, P., Flexman, E. A., Keating, M., Hsiao, B. S., and Verma, R. K. (1997), *J. Polym. Sci. B* **35**, 2121.
Sauer, B. B., Beckerbauer, R., and Wang, L. X. (1993), *J. Polym. Sci. B* **31**, 1861.
Sawada, K. and Ishida, Y. (1975), *J. Polym. Sci. B* **13**, 2247.
Schäfer, H., Sternin, E., Stannarius, R., and Kremer, F. (1996), *Phys. Rev. Lett.* **76**, 2177.
Schatzki, T. F. (1962), *J. Polym. Sci.* **57**, 496.
Schönhals, A. (1991), *Acta Polym.* **42**, 119.
Schönhals, A., Goering, H., Schick, C., Frick, B., and Zorn, R. (2004), *Colloid Polym. Sci.* **282**, 882.

Sencadas, V., Costa, C. M., Moreira, V., Monteiro, J., Mendiratta, S. K., Mano, J. F., and Lanceros-Mendez, S. (2005), *E-Polymers*, article number 002.
Senturia, S. D. and Sheppard, N. F. (1986), *Adv. Polym. Sci.* **80**, 1.
Shieh, J.-Y., Yang, S.-P., Wu, M.-F., and Wang, C.-S. (2004), *J. Polym. Sci. A* **42**, 2589.
Shinyama, K. and Fujita, S. (2001), *IEEE Trans. Dielectric. Electric. Insul.* **8**, 538.
Simon, G. P. (1997), "Dielectric properties of polymeric liquid crystals," in *Dielectric Spectroscopy of Polymeric Materials—Fundamental and Applications*, Runt, J. P. and Fitzgerald, J. J., eds., ACS, Washington, DC, pp. 329–378.
Simpson, J. O. and Bidstrup, S. A. (1995), *J. Polym. Sci. B* **33**, 55.
Stephan, F., Fit, A., and Duteurtre, X. (1997), *Polym. Eng. Sci.* **37**, 436.
Starkweather, H. W., Jr. (1984), *Macromolecules* **17**, 1178.
Starkweather, H. W., Jr., Avakian, P., Matheson, R. R., Fontanella, J. J., and Wintersgill, M. C. (1992), *Macromolecules* **25**, 1475.
Starkweather, H. W., Jr., Avakian, P., Fontanella, J. J., and Wintersgill, M. C. (1994), *Macromolecules* **27**, 610.
Stockmayer, W. H. (1967), *Pure Appl. Chem.* **15**, 539.
Stracke, A., Wendorff, J. H., Mahler, J., and Rafler, G. (2000), *Macromolecules* **33**, 2605.
Suljovrujic, E. (2002), *Polymer* **43**, 5969.
Suljovrujic, E., Kacarevic-Popovic, Z., Kostoski, D., and Dojcilovic, J. (2001), *Polym. Degrad. Stabil.* **71**, 367.
Sy, J. W. and Mijovic, J. (2000), *Macromolecules* **33**, 933.
Takahara, K., Saito, H., and Inoue, T. (1999), *Polymer* **40**, 3729.
Tammann, G. and Hesse, G. (1926), *Z. Anorg. Allg. Chem.* **56**, 245.
Tombari, E. and Johari, G. P. (1992), *J. Chem. Phys.* **97**, 6677.
Tran, T. A., Said, S., and Grohens, Y. (2005), *Macromolecules* **38**, 3867.
Tsonos, C., Apekis, L., Zois, C., and Tsonos, G. (2004), *Acta Mater.* **52**, 1319.
Tuncer, E., Wegener, M., Frübing, P., and Gerhard-Multhaupt, R. (2005), *J. Chem. Phys.* **122**, 084901.
Urakawa, O., Fuse, Y., Hori, H., Tran-Cong, Q., and Yano, O. (2001), *Polymer* **42**, 765.
Vaia, R. A., Sauer, B. B., Tse, O. K., and Giannelis, E. P. (1997), *J. Polym. Sci. B* **35**, 59.
van den Berg, O., Sengers, W. G. F., Jager, W. F., Pichen, S. J., and Wübbenhorst, M. (2004), *Macromolecules* **37**, 2460.
van Turnhout, J. (1975), *Thermally Stimulated Discharge of Polymer Electrets* (Elsevier, Amsterdam).
van Turnhout, J. (1980), Electrets, in *Electrets*, Topics in *Applied Physics Series*, Vol. 33, Sessler, G. M., ed., Springer, Berlin, pp. 81–215.
van Turnhout, J. and Wübbenhorst, M. (2002), *J. Non-Cryst. Solids* **305**, 50.
Vanderschueren, J. and Gasiot, J. (1979), "Field-induced thermally stimulated currents," in *Thermally Stimulated Currents in Solids*, Topics in Applied Physics Series, Vol. 37, Braunlich, P., ed., Springer, Berlin, pp. 135–223.
Varotsos, P. A. (2005), *The Physics of Seismic Electric Signals*, Terrapub, Tokyo.

Vatalis, A. S., Delides, C. G., Grigoryeva, O. P., Sergeeva, L.M., Brovko, A. A., Zimich, O. N., Shtompel, V. I., Georgoussis, G., and Pissis, P. (2000), *Polym. Eng. Sci.* **40**, 2072.
Venderbosch, R. W., Meijier, H. E. H., and Lemstra, P. J. (1995), *Polymer* **36**, 2903.
Venkateshan, K. and Johari, G. P. (2004), *J. Phys. Chem. B* **108**, 15049.
Vogel, H. (1921), *Phyz. Z.* **22**, 645.
Yemni, T. and Boyd, R. H. (1979), *J. Polym. Sci. B* **17**, 741.
Wang, M., Johari, G. P., and Szabo, J. P. (1992), *Polymer* **33**, 4747.
Wasylyshyn, D. A. (2005), *IEEE Trans. Dielectric Electric. Insul.* **12**, 183.
Williams, G., Landel, R. F., and Ferry, J. D. (1955), *J. Am. Chem. Soc.* **77**, 3701.
Williams, G., Smith, I. K., Holmes, P. A., and Varma, S. (1999), *J. Phys. Condens. Matt.* **11**, A57.
Williams, G. and Watts, D. C. (1971), *Trans. Faraday. Soc.* **67**, 2793.
Wübbenhorst, M., de Rooij, A. L., van Turnhout, J., Tacx, J., and Mathot, V. (2001), *Colloid Polym. Sci.* **279**, 525.
Wübbenhorst, M. and van Turnhout, J. (2002), *J. Non-Cryst. Solids* **305**, 40.
Zhang, S. H., Jin, X., Painter, P. C., and Runt, J. (2004a), *Polymer* **45**, 3933.
Zhang, S. H., Jin, X., Painter, P. C., and Runt, J. (2004b), *Macromolecules* **37**, 2636.
Zhou, J. F. and Johari, G. P. (1997), *Macromolecules* **30**, 8085.
Zielinski, M. and Kryszewski, M. (1977), *Phys. Status Solidi* **A42**, 305.
Zong, L., Kempel, L. C., and Hawley, M. C. (2005), *Polymer* **46**, 2638.

CHAPTER 7

MICRO- AND NANOSCALE LOCAL THERMAL ANALYSIS

VALERIY V. GORBUNOV
Veeco Instruments, Santa Barbara, California

DAVID GRANDY
University of Loughborough, United Kingdom

MIKE READING
University of East Anglia, Norwich, United Kingdom

VLADIMIR V. TSUKRUK
School of Materials Science and Engineering, Georgia Institute of Technology, Atlanta

7.1. INTRODUCTION

The term *microscale* or *nanoscale thermal analysis* (micro/nano-TA) encompasses techniques or combinations of techniques that use a method of highly local material property characterization carried out at the micrometer or nanometer scale, on a sample subjected to a controlled temperature regime. In its broadest sense, micro/nano-TA could therefore include hot-stage optical microscopy and nano-TA hot-stage scanning electron microscopy. It is generally accepted, however, that the term is used to describe methods that exploit a combination of atomic or scanning force microscopy (AFM, SFM) and one or more of the following techniques: thermomechanical analysis (TMA), dynamic mechanical analysis (DMA), differential thermal analysis (DTA), spectroscopy, or analytical pyrolysis. When referring to local thermal analysis, the term *differential scanning calorimetry* (DSC) is not used because a quantitative measure of heat (J/g) is not made. This is because the mass of the sample that melts cannot be determined. In a sophisticated arrangement it is,

---

Thermal Analysis of Polymers: Fundamentals and Applications, Edited by Joseph D. Menczel and R. Bruce Prime
Copyright © 2009 by John Wiley & Sons, Inc.

in principle, possible to combine all these techniques in a single "benchtop" instrument. These are all well-established material characterization methods that, with the introduction of AFM, can now be carried out on preselected parts of a sample typically a few cubic micrometers in volume or smaller. The atomic force microscope is used to acquire topographic and other types of image of a surface and/or to position accurately a near-field probe at a particular location on that surface. Control over the temperature of a sample is provided by the use of a thermally active electrically resistive probe (thermal probe) and/or a variable-temperature microscope stage (temperature stage). If the latter is used, practically any type of probe normally available for AFM may be mounted in the microscope. A thermal probe may function as a thermometer as well as a heat source. This enables a further type of micro/nano-TA to be carried out, in which heat is applied to the sample from an external energy source (e.g., infrared radiation) and the probe is used to sense the resulting change in temperature of the material. This enables spectroscopy to be carried out with a spatial resolution that is, in theory, better than the diffraction limit.

Using the established "macro" forms of these characterization methods, typically carried out on a minimum of a few cubic millimeters or milligrams of material, it is often possible to deduce that a particular sample is heterogeneous, a polymer blend, for example, to assign thermal transitions to different components, and perhaps to identify the constituents and the proportion in which they are present. However, no information on the size, shape, and spatial distribution of phases is acquired. The development of micro/nano-TA technology significantly enhances the utility of these techniques by enabling variations in properties or composition to be mapped at the microscopic level. This will increase their usefulness in the field of materials science and technology, which is becoming increasingly concerned with the control of material structure and hence properties on the *micro*- and, increasingly, the *nano*scale.

7.2. THE ATOMIC FORCE MICROSCOPE

The invention of the atomic or scanning force microscope (AFM, SFM) by Binnig et al. (1986) has been a major factor in enabling the development and commercialization of micro/nano-TA. Prior to its development, the scanning probe microscopes (SPM) available were restricted to use on electrically conductive samples. This is because they utilized the tunneling current between a surface and a conductive near-field probe or, more specifically, the dependence of its magnitude on the probe tip–sample distance. This technique, developed by Binnig and Rohrer (1982), is known as *scanning tunneling microscopy* (STM) and can be used to acquire extremely high-resolution elevation maps or topographic images of a surface down to the atomic scale. The operation of the AFM, on the other hand, depends on the nature of the short-range repulsive or attractive forces that exist between all solids in close proximity. Its use, therefore, is not restricted to conductive materials. The microscope provides the mechanism by which a near-field probe can be positioned and/or raster-scanned, with *nanometer*-level accuracy on or across a surface, while maintaining a controlled contact force. Figure 7.1 is a schematic diagram of a typical AFM with a thermal probe installed. At its heart are two independent but interconnected mechanisms. The first is a piezoelectric scanner, which controls and measures the position of the fixed end of the probe *cantilever* in three dimensions. In certain instruments an additional height measurement is made by a strain gauge bonded to the surface of the z-piezo actuator. The second is an optical lever, in which laser light is focused on the reflective top surface of the cantilever (or, in the case the thermal probe illustrated, a discrete but integral probe mirror) and collected in a four-quadrant photodetector. The vertical position of the laser spot on the detector is a measure of the degree of bending of the cantilever and, therefore, the normal force acting between the tip and the sample. This is calculated from the signal arriving at the top two quadrants minus that detected at the lower two [sometimes referred to as the *top–bottom* (T-B) *signal*]. Similarly, the horizontal position of the laser spot is a measure of cantilever twist and hence frictional forces. In the simplest AFM scanning mode (the so-called contact mode), the average normal force is kept constant by means of a feedback signal from the photodetector to the scanner. The response of the scanner is to raise or lower the fixed end of the cantilever to accommodate the effects of surface topography on the normal force. In this manner, a three-dimensional topographic map of the sample surface is obtained. In certain applications, this may be all that is required. Usually and often more usefully, images can be acquired simultaneously that are constructed from variations in some physical property across the surface of the sample. These may be mechanical, electrical, magnetic, or, when using

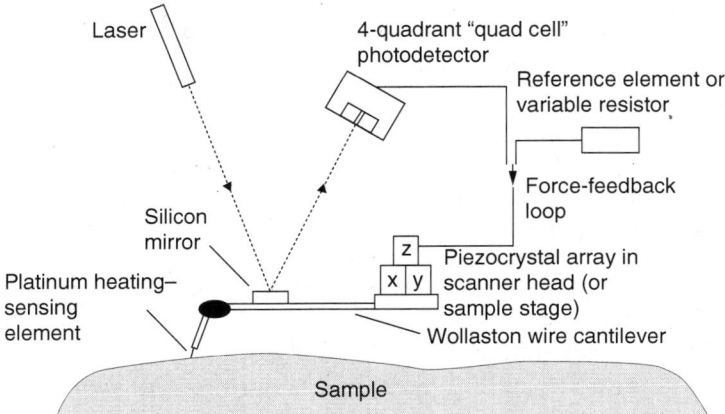

Figure 7.1. Schematic diagram of a typical atomic force microscope.

a thermal probe, thermal properties. Description of the various imaging modes that can be used in micro/nano-TA are described in detail below.

7.3. SCANNING THERMAL MICROSCOPY

Scanning thermal microscopy (SThM) enables the acquisition of images of the surface of a sample constructed from spatial contrast in one or more thermal properties of the material (Pollock and Hammiche 2001). In constant-temperature mode, the thermal probe is held at a fixed temperature by means of a thermal feedback loop as it is raster-scanned across the surface of the sample. The AFM force-feedback mechanism holds it at a constant contact force. The vertical movement of the AFM scanner required to maintain constant force is used to construct a topographic image. The power supplied to the probe to maintain it at the selected temperature is recorded and used to construct the *thermal conductivity image*. The instrument software assigns each measurement to a pixel in the image, colored or shaded according to its magnitude. The ratio of the sampling rate to the image resolution (in this context, meaning the total number of pixels required) determines the number of measurements per pixel. When multiple measurements are used to construct a single pixel, the average value is calculated and the pixel shaded accordingly. Areas of relatively high thermal conductivity will result in more power being supplied to the probe than to neighboring areas with lower conductivity. This leads to the possibility of using SThM measurements to quantify thermal conductivity of unknown materials; this topic is discussed in detail later in the chapter. Here we consider qualitative mapping of differences in thermal conductivity.

By convention, areas with relatively high or low conductivity appear, respectively, as relatively bright or dim areas in the resulting image. In a multiphase material, then, provided there is sufficient contrast in thermal conductivity between phases, the spatial distribution, shape, and size of phases will be mapped. As all materials conduct heat, a significant amount of information in the image will originate from the subsurface structure of the material. Most other AFM modes, such as mechanical property-based imaging, are generally restricted to measuring the response of the material in the immediate vicinity of the surface. The ability of SThM to carry out subsurface imaging can therefore provide a significant advantage. Often the microstructure at the surface of a material is unrepresentative of that of the subsurface region, or the sample may be a multilayer film or contain a buried structure. On the other hand, it may well be that the immediate surface region of a sample is of most interest and the greater effective sample volume provided by SThM is an unnecessary complication.

There is a fundamental difficulty in the interpretation of thermal images acquired by SThM. Specifically, the power required to maintain the probe at constant temperature is strongly affected by the surface topography proximal to the probe at any given point in the scan. Indeed, the *deconvolution* of topographic effects, called *topographic artifacts*, from images constructed from the measurement of a physical property is a problem common to many AFM imaging modes. Conversely, and usually less importantly, variation in mechanical properties can disturb the topographic image. The nature of this effect in SThM is illustrated in Fig. 7.2. When the probe is at or near the summit of a relatively high feature (position A), it is surrounded predominantly by air, whereas at or near to the bottom of a hole or valley (position B), the effective volume of the sample is larger. This is due to (1) the absolute increase in probe–sample *contact area* and (2) if radiative heat transfer is significant (see text below), the greater amount of material in close proximity to the probe.

Variation in contact area is the most common cause of topographic artifacts among AFM imaging techniques. In this case, the relatively high thermal conductivity of the solid means that more heat is transferred from the probe at position B and so more power has to be supplied to the probe to maintain constant temperature. This appears in the resulting thermal image as a region with a shade or color different from (by convention, brighter than) that of the region around position A. Hence, even a thermally homogeneous material, unless it is perfectly smooth, gives rise to thermal image contrast. This means that a careful comparison of the corresponding topographic and thermal images must be made before concluding that features in the thermal image are indeed due to genuine spatial variations in thermal conductivity. On the other hand, the effects of topography may act to mask or considerably reduce the degree of contrast in thermal images from multiphase materials in which exist otherwise detectable variations in thermal conductivity. This is because raised features with relatively high thermal conductivity may cause the probe to consume a similar amount of power as low areas with relatively low conductivity. An extreme case of this would be a binary mixture or blend in which occluded high-conductivity domains were raised above a surrounding higher-conductivity continuous phase. Such samples are rare, and it could be argued that in cases where systematic topographic variation exists between different phases, the thermal conductivity image becomes, to an extent, superfluous. In

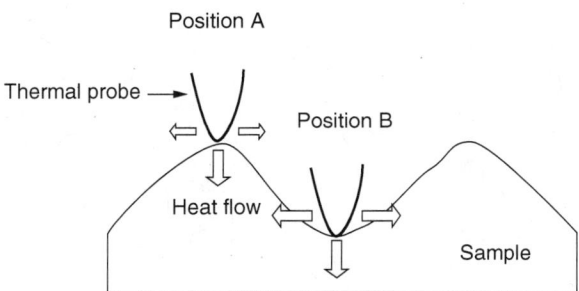

Figure 7.2. An illustration of the effect of sample topography on the heat flux from a thermal probe to its surroundings.

practice, for a given sample, the local topography in some areas usually acts to increase the apparent thermal contrast and in other areas to reduce it. For this reason, a careful visual comparison of the topographic and thermal images is necessary to determine how closely the location, size, and shape of features in one image are reflected in the other. The success of this approach is dependent on the experience and judgment of the experimenter, as well as on the complexity of the sample. This problem is greatly reduced when a flat sample surface can be prepared by sectioning or polishing.

7.4. THERMAL PROBE DESIGN AND SPATIAL RESOLUTION

The form of SThM most relevant to the subject of this discussion is carried out using near-field electrical resistance thermometry, and this method has been adopted in the work reported in this chapter. This is because miniaturized resistive probes have the considerable advantage that they can be used both in passive mode as a thermometer and as an active heat source. This enables local thermal analysis (L-TA; see text below) as well as SThM to be carried out. At present the most common type of resistive probe available is the "Wollaston" or Wollaston Wire probe, developed by Dinwiddie et al. (1994) and first used by Balk et al. (1995) and Hammiche et al. (1996a) The construction details of this probe are illustrated in Fig. 7.3. A loop of 75-μm-diameter coaxial bimetallic Wollaston wire is bent into a sharp V-shaped loop. The wire consists of a central 5-μm-diameter platinum/10% rhodium alloy core surrounded by silver. The loop is stabilized with a small bead of epoxy resin deposited approximately 500 μm from its apex. The probe tip or sensor is made

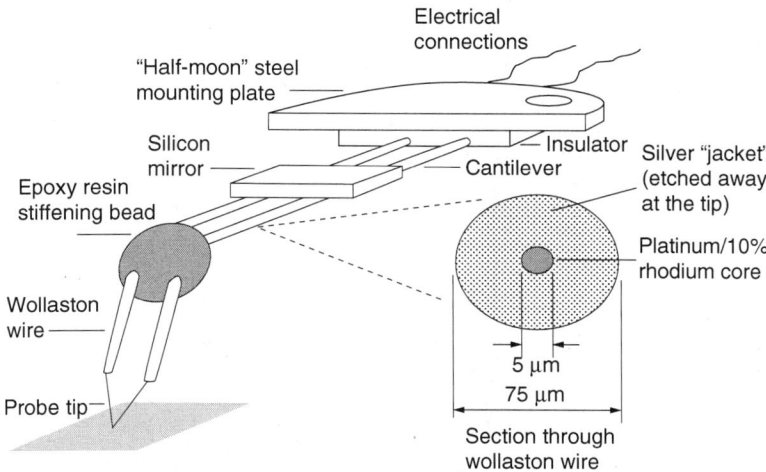

Figure 7.3. A Wollaston resistive thermal probe (not to scale), including a section through Wollaston process wire.

by etching the outer silver wire in the immediate vicinity of the apex of the loop. This exposes the 5-µm platinum alloy over a length of approximately 50 µm. Because the wire at the tip is so much thinner than the rest of the wire in the loop, this is where most of the electrical resistance in the probe circuit is concentrated (typically 2–3 Ω). Hence, when a current is passed through the probe, heating is substantially restricted to this element (usually known as the *probe tip*). The temperature–resistivity dependence for the tip material is well known, and for the temperature range 0–850 °C can be modeled using a second-order polynomial equation (see http://www.w-dhave.inet.co.th/index/RTD/RTD.pdf). The other end of the loop is fixed, via an insulator, to a "half-moon" steel mounting plate compatible with the AFM scanner magnetic chuck. A silicon mirror is bonded to the top side of the loop to allow operation of the laser optics. The loop is bent downward between the epoxy bead and the apex, typically at an angle of 45–70° to what is now the cantilever. The precise angle is determined largely by the shape of the bead, which is somewhat variable. Some variability also exists in the length of the exposed tip and the bend radius of the loop. The minimum bend radius of the apex of the platinum loop is determined by the overall diameter (and formability) of the Wollaston wire, so it is considerably larger than would be possible by bending a 5-µm wire independently. This has a considerable effect on the minimum spatial resolution of the probe (see text below for a fuller discussion on resolution and related issues).

The Wollaston probe is a relatively massive structure compared with most inert probes used in other forms of AFM. These usually incorporate hard ultrasharp tips, made from silicon or silicon nitride, whose contact radius may be as small as 10 nm or so, mounted at the end of a relatively simple stiff elastic cantilever. The spatial resolution of such probes is therefore far superior to that of the Wollaston probe, whose high and variable spring constant (5–20 N/m) and complexity also render it unsuitable for all except the *contact-mode* imaging described above (various alternative AFM imaging modes are described below).

The best results for robust routine local thermal analysis with high spatial resolution comes from probes whose design is based on the approach adopted by King et al. (2006). The advantage that this probe design has over alternatives is that the conductors are produced by doping a thin (100-nm) layer with boron or phosphorus, thus allowing the entire surface of the pyramid to be heated and, therefore, cleaned at will. The spatial resolution of these probes is the same as conventional AFM tips, and so, for imaging topography, nanometer and even subnanometer resolution can be achieved. Photographs of two of these probes are shown in Fig. 7.4. However, these probes are not suitable for high-resolution thermal imaging. The heated area is on the top of the inverted pyramid that is the tip, which means that the resistive element that is sensitive to temperature is relatively large, of the order of 10 µm, and is remote from the surface. The effect of this combination is that the heater serves very well to heat the tip and enable local thermomechanical measure-

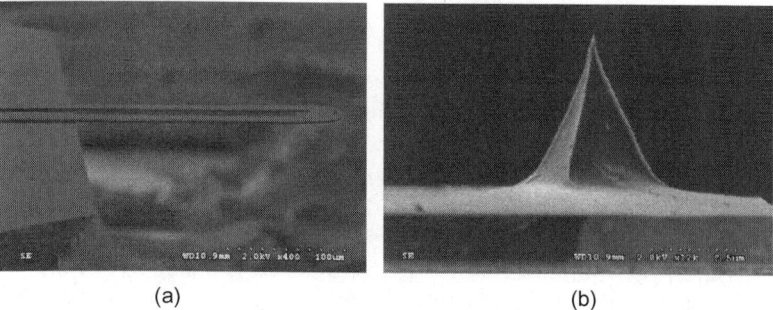

(a) (b)

Figure 7.4. Micromachined silicon nanoprobe: (a) the cantilever; (b) the tip, which has the same sharpness as a conventional AFM probe (reproduced with permission of Anasys Instruments Inc.).

ments with very high spatial resolution; however, it cannot be easily used to map thermal properties such as conductivity.

As discussed above, the spatial resolution of a particular probe design depends largely on the size of the probe tip or, more specifically, the contact area between tip and sample. This will gradually increase with tip wear. When considering resolution, we consider both the smallest topographic feature that can be detected (the detection limit), how accurately its size and shape is measured, and the minimum distance between adjacent features that allows them to be revealed as distinct entities (spatial resolution). With a thermal probe, the spatial resolution of a topographic image may well differ from the *thermospatial* resolution in the corresponding DC and/or AC thermal image. Furthermore, the subsurface resolution of the thermal image will degrade with depth. Besides contact area, other more subtle factors may influence resolution; principal among these is the aspect ratio of the tip (contact area/projected area), as shown in Fig. 7.5. This demonstrates that, although all the probe tips illustrated will *detect* the surface feature shown, it is only the pyramidal tip with the higher aspect ratio [tip (c)] that will map it accurately. Similar tip geometry effects may be readily envisaged on the detection and mapping of narrow and/or steep-sided and/or deep holes or valleys. This may have a powerful effect on the thermal image, particularly if the sides of the tip are in contact with the walls of the valley while the apex remains out of surface contact. An additional effect caused by the shape of the Wollaston probe is that its optimum spatial resolution occurs in the direction perpendicular to the plane of the tip loop. However, if the probe normal force is too high, distortion of the tip can take place most easily in this direction, leading to a deterioration in resolution and definition. Under normal scanner operation, the plane of the loop is parallel to the raster or fast-scan direction. This is itself at right angles to the direction of probe advancement across the surface (slow-scan direction). Moreover, the edge of a particle or valley will be detected most accurately along the aspect of the feature facing the slow-scan direction. The effect of

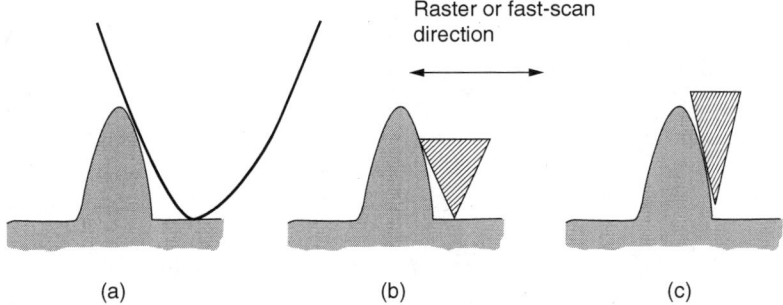

Figure 7.5. Diagrams illustrating effects of shape of probe tip on spatial resolution: (a) Wollaston wire; (b) low-aspect-ratio pyramidal tip; (c) high-aspect–ratio pyramidal tip.

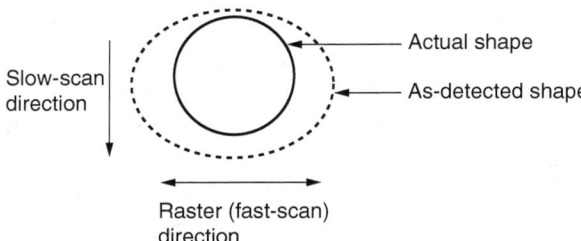

Figure 7.6. The effect of the Wollaston probe geometry on the two-dimensional shape of a circular particle as acquired in the resulting image.

this feature on the as-detected shape of a relatively high circular feature is shown in Fig. 7.6. Such spatial distortion can be counteracted to some extent by varying the raster direction in successive scans of the same area and comparing the resulting images, but this is time-consuming. There may also be differences, caused by in-plane asymmetry of the loop, in the shape of the feature as detected in images acquired in the forward and reverse fast-scan directions, but these can be acquired concurrently.

Various papers have been published that give a good indication of the spatial and thermospatial resolution that may be expected from both Wollaston and microfabricated bowtie probes (these are probes where the metal conducting layer has a bowtie shape at the tip so that electrical resistance is located at the narrow middle area). Submicrometer spatial resolution has been demonstrated on polymer blends using Wollaston probes (Hammiche et al. 1996a, b; Pollock et al. 1998) and Balk et al. (1995) have reported detection of features as small as a few tens of nanometers using AC imaging on a GaAs diode, although for most practical samples, such high resolution is unattainable. Work on a paracetamol (acetaminophen) tablet in the as-supplied condition has been reported by Price et al. (1999a, b, and 2000) in which the contrast in the thermal conductivity image is truly independent of surface topography. Furthermore, the resolution obtained in a thermal image may be superior to that in the corresponding topographic image (Pollock and Hammiche 2001). The resolution of the bowtie probe has been demonstrated on model samples of silica coated with a layer of PMMA (Pollock and Hammiche (2001)). Elongated rectangular discontinuities in the coating are detected in thermal images down to a width of 200 nm, but only when lying parallel to the raster direction. The resolution perpendicular to this is shown to be roughly a factor of 2 poorer. In another direct comparison of the two types of probe, Pollock and Hammiche (2001) have demonstrated the superior edge definition as well as absolute spatial and thermospatial resolution attainable with the micromachined bowtie probe.

The silicon nanoprobes, as used in micro/nano-TA, are a far more recent development, and consequently the spatial resolution that can be achieved with these probes is not yet well established but must be of the same order for topographic imaging as conventional AFM tips because the sharpness of the tip is the same. The capabilities for thermal property imaging remain, at the time of writing, unexplored.

7.5. MEASURING THERMAL CONDUCTIVITY AND THERMAL FORCE–DISTANCE CURVES

A change in the heat dissipation by the thermal tip approaching a surface depends strongly on the ratio between surface thermal conductivity and tip thermal conductivity (Fig. 7.7). Using the Block (1937) and Jaeger (1942) theories, we can present the relationship between the heat dissipation through the contact area during physical contact $[\Delta Q(W)]$, for the quasi-steadystate process in the form (Gorbunov et al. 2000a, b)

$$\Delta Q = \frac{3}{4}\pi \lambda R_C \Delta T \tag{7.1}$$

where ΔT is the initial temperature difference between the probe and the surface, R_C is the effective contact radius of the thermal probe, and λ is the "composite" thermal conductivity in W/(m·K), defined as

$$\frac{2}{\lambda} = \frac{1}{\lambda_S} + \frac{1}{\lambda_P} \tag{7.2}$$

where λ_S and λ_P are thermal conductivities of the surface of the sample and thermal probe and, respectively.

Equation (7.1) shows that the ratio $\Delta Q/\Delta T \sim \lambda \cdot R_C$ should be constant for a given material if the contact radius and the thermal conductivity are

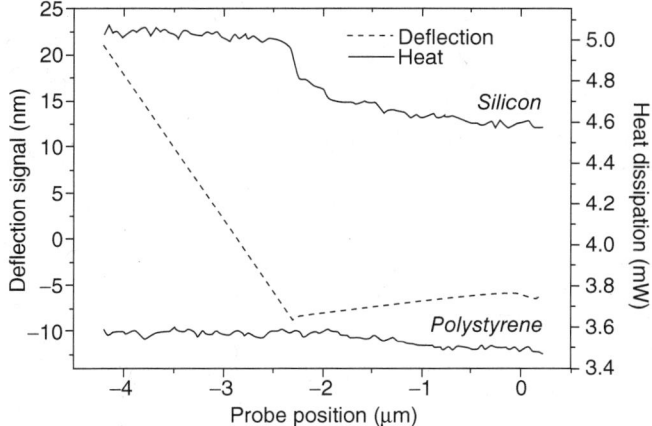

Figure 7.7. Thermal signal versus distance between the thermal probe and silicon and polystyrene surfaces (solid lines) together with deflection data (dashed line) [from Gorbunov (2000a), reprinted with permission of Taylor & Francis Group].

unchanged. The sensitivity of the SThM method highly depends on the mechanisms of heat transfer from the thermal probe to a sample through the physical tip–surface contact. Equation (7.1) can be represented in a different form

$$\frac{\Delta Q}{\Delta T} = \frac{\frac{3}{4}\lambda_p \pi R_c}{1-(\lambda_p/\lambda_s)} \quad (7.3)$$

which allows for direct evaluation of thermal conductivity of the surface probed by conducting a series of experiments with differently preheated SThM tips. The contact radius for materials with different Young's moduli was evaluated using an AFM-adapted Hertzian approximation (Chizhik et al. 1998).

Figure 7.8 presents measurements of the heat dissipation for selected materials with different thermal and mechanical properties presented as a ratio of the thermal conductivities of the surface to the probe. From the simple Eq. (7.2) for the "composite" thermal conductivity coefficient, we can conclude that the method should be more sensitive for materials such as polymers with thermal conductivity less than the conductivity of the tip material than it is for materials with higher thermal conductivity (small variation in ΔQ) (Fig. 7.8).

Equation (7.1) also demonstrates that the amount of heat dissipated to the surface depends on the thermal conductivity of the tested material as $\Delta Q \sim \lambda \cdot R_C \Delta T$. Of interest is an experiment where the tip starts above the surface of the sample and is then moved toward it until it makes contact, in effect a thermal force–distance curve. The results of this measurement on materials with different thermal conductivity are presented in Table 7.1. Conditions of the experiment were chosen in such a way that the contact radius

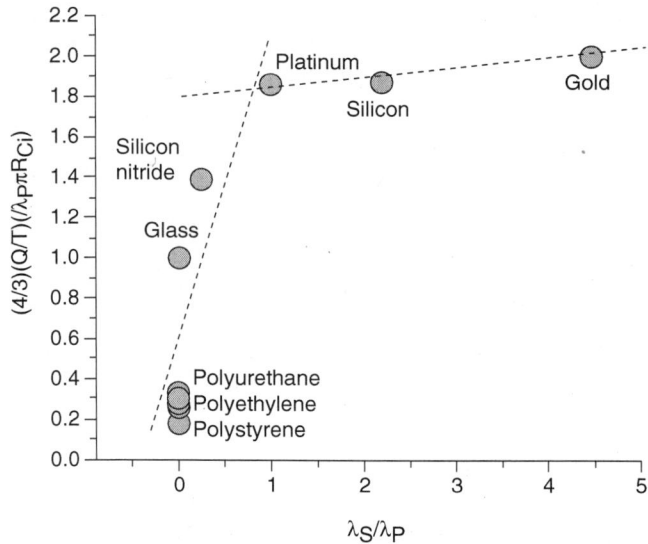

Figure 7.8. Variation of the reduced heat dissipation during physical contact versus the thermal conductivity of various materials reduced to dimensionless units [from Tsukruk et al. (2003), reprinted with permission of Elsevier].

TABLE 7.1. Thermal properties of materials. Thermal conductivity (λ) and thermal diffusivity (a), data taken from MatWeb, MEMS Materials Database, CenBASF/ Materials). (Gorbunov et al. 2000b)

Materials	λ (W/mK)	a ($10^{-8}\,m^2/s$)
polystyrene (PS)	0.142	7.52
polyurethane (PU)	0.147	8.91
polypropylene (PP)	0.18	9.47
poly(methyl methacrylate)(PMMA)	0.19	11.77
poly(vinyl chloride) (PVC)	0.21	14.93
polyethylene (HDPE)	0.37	17.52
glass slide	1.6	67.34
silicon nitride	19.0	863.24
graphite	24.0	1,543.66
platinum	71.0	2,507.59
silicon	156.0	9,394.33
gold	317.0	12,830.60
air	0.024	18.4

for both measurements can be considered equal. A probe temperature of 40 °C was selected to prevent any mechanical property changes for polystyrene and to keep the contact radius constant during the experiment. The results of the measurements show that the expected probe heat loss "jump" is higher for the material with higher thermal conductivity. The contact point can be monitored by the concurrently obtained cantilever deflection data (Fig. 7.7). Note that displacement rather than force is given because the spring constant of these probes is difficult to calibrate.

The behavior of materials with thermal conductivity below that of platinum (first group) is very different from that of high-thermal-conductivity materials (Fig. 7.8). A significant increase in heat dissipation was observed for the first group. In contrast, when the thermal conductivity of the tip material is higher than that of Pt (second group, e.g., silicon or gold), only a small increase in the value of $\Delta Q/\Delta T$ was observed. This difference between first- and second-group behavior can be understood considering the nature of "composite" thermal conductivity (λ), for the tip/surface entity, as presented by Eq. (7.2). This relationship demonstrates that for all surfaces with $\lambda_S \ll \lambda_P$ (polymers, glass, and semiconductors), the "composite" thermal conductivity for the tip/surface entity is determined primarily by the less thermally conductive component of this entity, namely, the surface. Therefore, under these conditions, the "composite" thermal conductivity that is responsible for heat dissipation at the contact point is directly proportional to the surface thermal conductivity of the materials tested, in other words, $\lambda \sim \lambda_S$. In contrast, for high-thermal-conductivity materials (with thermal conductivity higher than that of platinum), the surface becomes a thermal sink and the thermal tip becomes the poorly conductive counterpart. In this case, Eq. (7.2) predicts a virtually constant composite thermal conductivity ($\lambda \cong \lambda_P$) with essentially no cognizance of the actual material tested. Therefore, a virtually constant value of measured heat dissipation should be anticipated for highly conductive materials but not for polymers. Finally, we can conclude that from a practical perspective, for a wide variety of polymeric materials, a linear relationship between the measured heat dissipation and the bulk thermal conductivity holds with reasonable accuracy.

The thermal force–distance curve experiment can be modeled by using heat dissipation theories. The gradual increase in the thermal signal up to the physical contact point is observed to start far away from a surface for both materials in Fig. 7.7. This behavior of the thermal signals indicates that when the tip meets the surface, the area around the contact point is already "preheated" by the approaching thermal probe (Fig. 7.9). The temperature of the surface when contact occurs is determined by the tip temperature, the thermal properties of test material, and the velocity of the heated tip that engages the surface. The temperature distribution of a surface can be analyzed on the basis of heat flow theory. The ratio between the temperature close to the edge of the heating zone (T) and the central temperature (T_c), is calculated to be 0.97 for a polystyrene surface (Fig. 7.9). For a gold surface, this ratio is close to 0.999. Therefore, the temperature within the thermal contact is virtually homogeneous (<3% of variation for all materials), and the temperature in the center of the heated zone (T_c) can be used to estimate the average temperature (Tsukruk et al. 2003). Also, since the thermal probing of a surface can easily satisfy a

628 MICRO- AND NANOSCALE LOCAL THERMAL ANALYSIS

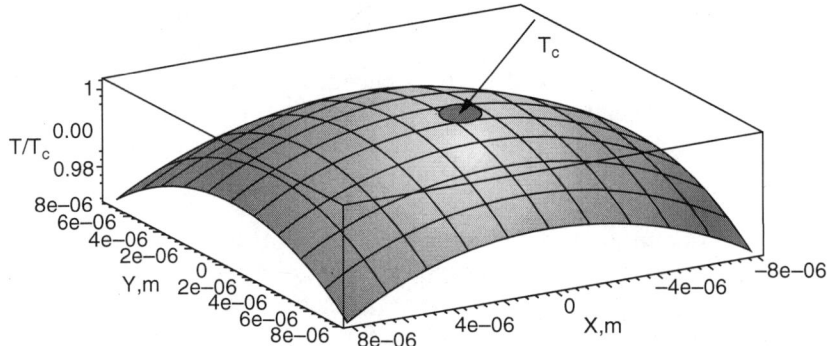

Figure 7.9. Temperature distribution inside heated zone beneath thermal source with radius $R = 4.5\,\mu m$ and measuring time $t = 0.01\,s$. [from Gorbunov (2000b), reprinted with permission of Taylor & Francis Group].

quasistationary case for heat flow from the thermal probe to a surface, the heat transfer can be described by a quasi-stationary equation to analyze the dynamics of the heat dissipation as well (Gorbunov et al. 2000b).

7.6. LOCAL THERMAL ANALYSIS

Local thermal analysis (L-TA) (Lawson et al. 1994; Hammiche et al. 1996a, b) exploits the ability of the AFM, in conjunction with a suitable microscope stage having x–y translation, to place a thermal probe at any point on the surface of a sample. A preliminary SThM experiment, for example, acquisition of thermal conductivity contrast and topography images, may be performed before L-TA, but it is not necessary. If an image (topographic and/or thermal) has been acquired previously, the probe can be directed to any feature of interest within the image (the maximum scan area available is normally of the order of $100 \times 100\,\mu m$). With the probe tip or sensor in position on the surface and exerting a predetermined downward threshold force (i.e., utilizing the signal from the photodetector, which is proportional to cantilever deflection and, hence, force exerted), a temperature ramp can be applied to the sample via the probe. This is usually a linear heating program or linear heating followed by linear cooling. Heating and cooling may be set at different rates. Essentially, two signals are acquired simultaneously. These are (1) the vertical deflection of the probe and (2) the power required by the probe to maintain the temperature ramp. The measurement of probe deflection with temperature is the micro- (or local) analog of thermomechanical analysis used on bulk samples (micro/nano-TMA, L-TMA). Similarly, the measurement of probe power consumption is the microanalog of differential thermal analysis (micro-DTA, L-DTA). Both of these techniques have been used extensively in the study of thermal transitions in polymers and other materials (Gorbunov et al. 1999, 2000c). The information on thermal properties garnered from such studies need not be an end in itself and is indeed used more often as a tool for understanding material structure and processes. The microanalogs of traditional thermal analysis techniques yield results that are similar to those obtained using macroscale techniques, but with the additional benefit that these results are spatially resolved.

Once the probe is in contact with the surface and the temperature program is initiated, the force-feedback mechanism is disabled. This is to prevent the probe from being driven continuously into a soft or molten sample, as the z-axis actuator would move the fixed end of the probe downward in an attempt to maintain constant contact force. Instead, for a given material and heating rate, the depth to which the probe penetrates is controlled by the initial contact force. Hence, the contact force steadily increases as a relatively hard sample expands, or decreases as the probe tip sinks into a soft material, with the fixed end of the cantilever remaining at constant height throughout the experiment. Before a particular probe is used, the relationship between the degree of bending of the cantilever (i.e., contact force) and the vertical displacement of the tip or sensor must be calibrated. This is a rapid procedure included as part of the instrument control software.

For a material that undergoes no thermal transitions over the temperature range of the experiment, the probe deflection with temperature will be essentially linear and upward as the sample beneath the probe heats and expands. Under a fixed set of instrument parameters, the rate of upward deflection will depend on the coefficient of thermal expansion, thermal conductivity, and heat capacity of the material. Heating of the probe element will itself cause some movement of the cantilever, but for the relatively massive Wollaston wire thermal probe, this effect should be minimal. Provided a baseline subtraction procedure is carried out (acquired from a run with the probe in free air), the rate of power consumption of the probe over the duration of the same experiment on a sample should remain constant. When a polymer or other material that may undergo one or more transitions over the temperature range of the experiment (a glass transition, cold crystallization, curing, melting, or degradation) is heated, the response of the micro-TA signals will be very different, as illustrated in Fig. 7.10. This shows typical micro-DTA and micro-TMA results for a crystallizable, but initially mostly amorphous polymer. There is a large indentation at the glass transition temperature (more accurately, this is the "softening" temperature), then a further indentation at the melting temperature. Whether the crystalline material that melts was present at the start of the experiment or formed during the heating cannot be decided from this one experiment; further experiments would be required at other heating rates. For our purposes here, we will not deal with this question. It should be noted that, because the DTA signal is not sensitive to enthalpies like DSC, but is dominated by changes in contact area, we cannot detect the exotherm associated with crystallization and so, it cannot be determined just from these data,

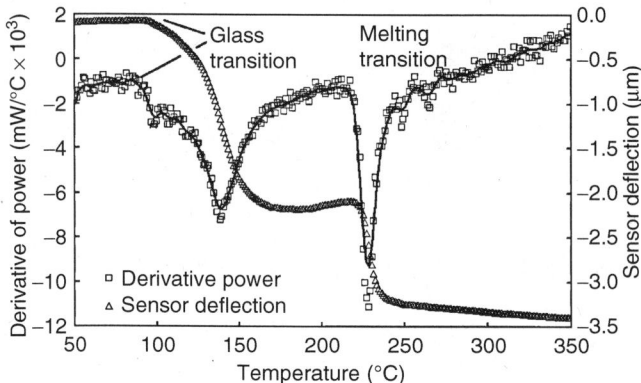

Figure 7.10. Micro-TMA (sensor deflection) and micro-DTA (derivative power) curves for quenched poly(ethylene terephthalate).

whether crystallization has occurred. This is, then, a disadvantage of the micro/nano-TA approach compared to conventional calorimetry. However, the glass transition and melting events are clearly detected.

A beneficial consequence of the small mass of the sample (of the order of a few micrograms compared with several milligrams) is that heating (and cooling) rates can be much higher than in conventional thermal analysis techniques. Currently, the maximum heating rate available in the commercial instrument (Anasys Instruments) is 100 °C/s, compared with several hundred degrees per minute as a typical maximum for conventional experiments. The minimum heating rate can be arbitrarily low. This allows many more experimental runs to be carried out in a given time, which may be of particular benefit to the industrial user. Another benefit is that multiple runs can be carried out on samples that are available in a very limited quantity. In certain cases, samples can be produced only in a form unsuited to conventional TA, but which are readily amenable to L-TA, for example, a thin film on an inert substrate. L-TA can be used to study differences between surface and bulk properties through the use of suitable sectioning techniques (Chui et al. 1998). Indeed, because the volume of material heated during a run is so small, several L-TA experiments can be carried out on the surface of, for example, an individual polymer granule without unduly affecting the results from a conventional TA experiment conducted subsequently on the same sample.

Alternating-current heating may be applied to the thermal probe. This produces a fixed temperature modulation in the range from ±1 °C to ±10 °C, although it is usually confined to ±2 °C to ±5 °C. This may be seen as analogous to the development by Reading (1993) and coworkers (Jones et al. 1997) of modulated temperature differential scanning calorimetry (MTDSC). In micro/nano-TA the response of the sample to the modulated and underlying heat flows can be separated using a deconvolution program. The modulated regime is sensitive to the reversible changes in the heat capacity of the material, associated with molecular vibrations and the latter detects changes due to kinetically controlled processes that are unable to reverse at the temperature and the rate of the modulation. An obvious advantage of this technique is its ability to characterize heterogeneous samples in which different types of transition occur over the same temperature range. In the case of L-TA, a lock-in amplifier is used to extract two signals (AC signals) that are dependent on the response of the sample to the modulated heating. The first is the phase lag between the response of the sample and the input signal to the probe. The second is the amplitude of the response of the sample compared with the input amplitude. Acquisition of both signals normally requires the use of a differential technique in which the response of the sample probe is compared with the response of a near-identical reference probe or suitable reference resistor. Theoretically, the use of AC heating offers similar advantages to those of MTDSC (see Section 2.13) over conventional DSC, and it has been shown that the AC signals may be particularly sensitive to transitions that produce a relatively large change in heat capacity for a small heat input (Hässler and zur Müehlen 2000).

It should be noted that the use of the L-TA power signal was previously termed *micro-DTA* and not *micro-DSC*. This is because, as discussed above, the volume and hence the mass of the heated sample is unknown and its calculation at present does not seem possible. Hence it is argued that micro/nano-TA cannot be described as a form of calorimetry, because it cannot measure the heat capacity of a material or the enthalpies associated with particular transitions. This is a major limitation of thermal probing. However, the ability of this technique to detect transitions occurring in a sample at the micrometer and submicrometer levels and to measure the temperature range over which those transitions occur is a major advantage.

In order to measure the temperature of local transitions, a temperature calibration of the thermal probe must first be carried out. The subject of temperature calibration has been addressed comprehensively by Blaine et al. (1999) and Meyers et al. (2007). A sample calibration graph is given in Fig. 7.11. This procedure currently has some complications; for instance, the L-TA technique acquires two signals (L-TMA and L-DTA), so which one should be selected to use as the method of measuring the transition temperature for the purpose of calibration? Presently it seems better to use the L-TMA signal as this has the higher signal-to-noise ratio and, in any case, this is the only signal available when using the nanothermal probes. Then the temperature calibration can be carried out on two or more substances whose melting temperature (T_m) is well known from the literature. These may be low-molecular-mass organic crystals with well-defined melting temperatures, such as biphenyl, benzoic acid, and 2-chloroanthraquinone (T_m = 69.3 °C, 122.4 °C, and 209.6 °C, respectively). The preparation of relatively smooth samples consisting of large crystals from such compounds can be problematic (however, it can be done; see text below). For this reason, it is often more convenient to use polymer

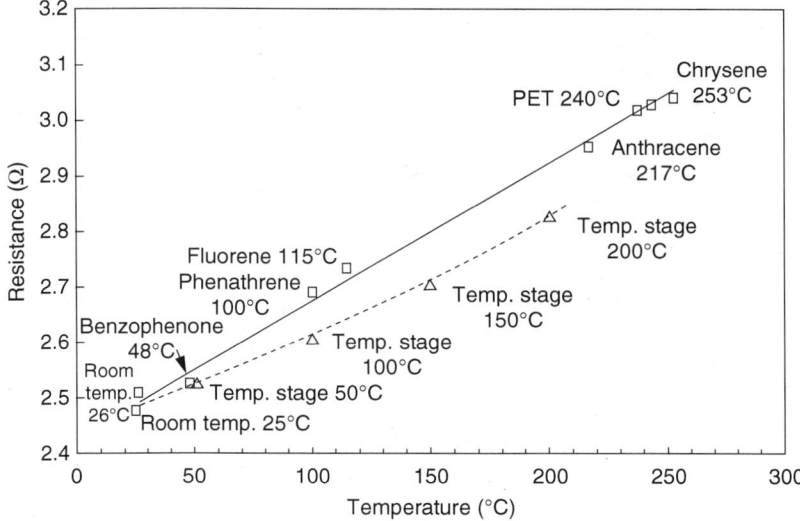

Figure 7.11. Wollaston thermal probe temperature calibration curves. The solid line was constructed by melting organic crystals of known T_m. The broken line was constructed by holding the probe just out of contact with the variable temperature stage at various temperatures.

films whose melting point has previously been measured using DSC or another technique and can be shown to be spatially invariant. This does make the assumption that the enthalpies measured in DSC experiments can be considered to mirror their mechanical analogs (i.e., softening). There is now a reasonable body of data in the literature, based mostly on calibration using polymers that shows that good results can be obtained (good in the sense that transition temperatures have been shown to be capable of identifying materials in many cases); it would, therefore, seem that this assumption is, broadly speaking, appropriate. When using DSC data as the point of comparison, there is the question of which characteristic temperature on the DSC curve should be chosen as the one corresponding to the penetration temperature. Several researchers have suggested using the extrapolated onset temperature of melting, but it is known that the leading edge of melting endotherms of polymers is not straight, so determination of the "extrapolated onset temperature of melting" is subjective. The peak temperature is a function of heating rate because of possible reorganization, superheating effects, and the intrinsic response time of the combination of the sample and measuring system. In addition, the melting transition of polymers can be very broad, adding to the error in the temperature calibration. Thermal probes are typically heated much faster than are conventional DSC experiments (by orders of magnitude), and so there is the question of which DSC heating rate to use when measuring peak temperatures and how this relates to the heating rates used in L-TMA experiments. Blaine et al. (1999) used extrapolated onset, whereas Moon et al. (2000), who calibrated their micro-TA apparatus using the melting of semicrystalline polymers, used the peak temperature of melting (Wunderlich, personal communication, 2007). They both obtained a linear relationship between the polymer melting point and the probe resistance (or applied voltage). Meyers et al. (2007) compared the transition temperatures of a range of polymers as measured by local thermal analysis, onset of the DSC peak, and onset of softening as measured by bulk thermomechanical analysis (TMA). They found a good correlation in all cases with a correlation coefficient of 0.98 or better for DSC and 0.99 or better for TMA. They concluded that transition temperature values for local thermal analysis gave the same value as TMA but were 6 °C higher than the onset of the transition as measured by DSC. In both cases the variation was ±3 °C. This suggests that the best procedure is to use TMA to measure the reference transition temperatures; onset temperature by DSC is a viable alternative, and the peak temperature, where it is only a few degrees different from the onset, would also be acceptable. The reader is advised to consult the literature as this is a topic that requires more work and so will evolve over time.

7.7. PERFORMING A MICRO/NANOSCALE THERMAL ANALYSIS EXPERIMENT

In principle, it can be envisaged that a thermal probe could be used in conjunction with an optical microscope that would image the sample surface to facilitate selection of the locations for analysis; in other words, an atomic force microscope is not strictly necessary. In practice, all micro/nano-TA to date has been performed using an AFM, so we will consider this option only in this section. In its modern form micro/nano-TA can be interfaced with a wide range of AFMs, and so exact details of the procedure will vary from instrument to instrument, here we will discuss generic requirements.

7.7.1. Sample Preparation

Atomic force microscopes fall into two categories: (1) instruments where the probe is moved and the sample remains stationary during imaging and (2) instruments where the sample is moved and the probe remains stationary in the x–y plane (although it can translate in the z axis because the probe must be lowered onto the sample at the start of an experiment and removed at the end). When the probe is moved, then the AFM can often be placed on any surface, and so, in this case, little or no sample preparation is required. Where a stage is used for large-scale movements of the sample (i.e., the probe is moved for imaging but a stage is used to move the sample so that an area can be selected prior to imaging), then the sample size must be commensurate with the size of the stage. When imaging is achieved by moving the sample, then the sample has to be small enough to be accurately and quickly moved by the actuators; typically this will mean that the sample is approximately 1 cm or less in diameter and only millimeters thick. Besides these restrictions on the total size of the sample, little sample preparation is required in many cases. The other restriction is that the z-direction travel of many AFMs is ~10 μm, and so the sample surface cannot be too rough. Where this is the case, the sample can be sectioned or polished to provide a sufficiently flat surface.

7.7.2. Probe Selection

At the time of writing, there are two major types of probe, both of which are described above: the Wollaston and the nanoprobes. The Wollaston probe is very robust, can be used at temperatures of ≤700 °C, and can provide the calorimetric measurement required for the thermal force–distance curves described above as well as AC and DC thermal imaging. Its disadvantages are that the spatial resolution is of the order of a micro-meter and can be used only in contact mode.

The nanoprobes are less robust, but as robust as most conventional silicon AFM probes, and can be used in all of the conventional AFM imaging modes, including contact, tapping, and pulsed-force mode. When used in this way, the spatial resolution is similar to that in conventional AFM microscopy, (i.e., of the order of 1 nm). For L-TA the spatial resolution is of the order of 100 nm or better. Their disadvantages are that they cannot, at the time of writing, be used for calorimetric measurements or thermal imaging. This is because the heater is located at the top of the pyramid that forms the tip rather than immediately adjacent to the surface. However, they perform well for local TMA. The maximum temperature is ~250 °C; undoubtedly more versions of these probes will become available in the future with a wider range of capabilities.

7.7.3. Calibration

The process of temperature calibration has been discussed above and is also mentioned below in reference to the use of nanoprobes (see Section 7.8.3). In essence, a series of standard materials is used to construct a calibration curve; this is the procedure routinely adopted in many thermal analysis techniques. It is preferable that the surfaces of the calibrants are smooth. Polymers are often available in films or coupons that readily lend themselves to use as calibrants. However, as discussed previously, polymers have relatively broad melting transitions (sometimes very broad), and so polymer samples must be selected with as narrow a melting transition as possible. Simple organic materials such as benzoic acid can be obtained in high-purity form with very sharp melting transitions but are usually presented in a granular form not very suitable for use in L-TA. Consequently, such materials are more effectively melted onto a very flat surface and then allowed to crystallize before being removed.

In this way a sample can be obtained with at least one flat surface, that formed against the substrate. The major transitions of these calibration standards should match the temperature range of the materials being studied, and the melting temperature should normally be taken as the onset. Users are advised to seek recommendations from the instrument manufacturer and/or make measurements on materials that they are familiar with and can check against TMA or DSC results. The instrument software will enable a function to be fitted to the measured transition temperatures to construct a calibration curve that will then be used for the analysis of unknown materials.

It is worth making some comments on temperature calibration for the thermal nanoprobes. The first task after installing a new probe is to calibrate its heating voltage–tip temperature response. It might be believed that, at such a small level of scrutiny, it might not be possible to find samples that have spatially invariant melting temperature. Typical cantilever deflection–heating voltage curves for PCL and PET are shown in Fig. 7.12 that demonstrate excellent repeatability, and so we can make the empirical observation that such samples do exist and that temperature calibration is possible. This was confirmed by Meyers et al. (2007). In the examples given below of application of the thermal nanoprobes, a three-point calibration curve was obtained from samples of polycaprolactone (PCL), high-density polyethylene (HDPE), and poly(ethylene terephthalate) (PET), whose onset melting temperatures were determined previously by differential scanning calorimetry.

Two important considerations must be kept in mind. An L-TA experiment involves very large temperature gradients that can affect the quality of the temperature calibration. The calibrants and the sample must be thermal insulators, that is, materials such as polymers and most organic materials [i.e., with

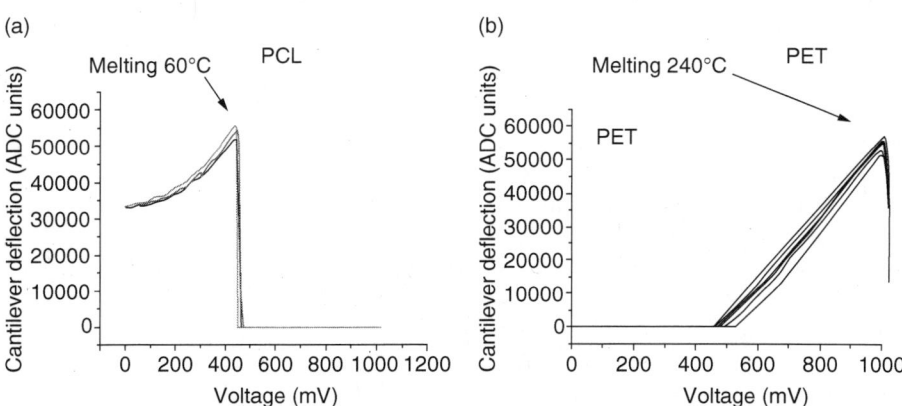

Figure 7.12. Typical graphs of thermal nanoprobe cantilever deflection versus heating voltage used to construct a cantilever deflection–tip temperature calibration curve: (a) polycaprolactone; (b) poly(ethylene terephthalate) (reproduced with permission of Anasys Instruments Inc.).

thermal conductivity on the order of ~0.1 W/(m·K)]; otherwise the thermal gradient in the region of the tip will be so steep that the temperature measurement will be compromised. This includes thin insulating films on good thermal conductors; for example, a 100-nm polymer film on a copper block cannot be measured accurately. The second point that must be borne in mind is that this issue and others such as the contact area of the tip mean that L-TA is not an ideal method for high-accuracy measurement of transition temperatures; rather, it is a method of monitoring changes in these temperatures with reasonable precision across a surface. The best accuracy that can be expected is several degrees. However, this is sufficient for a wide range of applications such as those illustrated below.

The other measurement that must be calibrated is the depth of penetration of the probe. To do this, the probe must be placed on a hard surface that will not yield; then it must be bent further by moving the z-axis actuator downward by a known amount. In this way the degree of displacement of the laser beam reflected from the back of the probe cantilever can be calibrated (e.g., in micrometers). Details of how this is done will vary from AFM to AFM. In general terms, there are two methods. One is to carry out a deflection curve and monitor the photodetector signal during this process. Another is to manually move the setpoint of the feedback control while the probe is in feedback and monitor z displacement.

7.7.4. The L-TA Experiment

After selecting a suitable probe and calibrating it, the LTA experiment will typically consist of

1. Imaging a surface in a selected AFM imaging mode
2. Selecting a point for analysis
3. Moving the probe to the selected point
4. Turning off the z-axis feedback
5. Programming the temperature

Steps 1–4 depend on the type of AFM that is being used, and so the manufacturer's instruction manuals should be consulted. It is necessary to turn off the z-axis feedback because otherwise, as the sample softens, the feedback will drive the probe downward to try to keep the degree of bending of the cantilever constant. This is generally undesirable. The temperature programming will be performed by the micro/nano-TA add-on unit. Heating rates are typically of the order of tens of degrees per second; the overwhelming majority of work to date uses rates of this magnitude, and so the new user is safest when following this practice. However, rates of hundreds of degrees per second are undergoing experimentation, and many thousands of degrees per second can certainly be achieved. This area will likely be addressed in the near future.

7.8. EXAMPLES OF MICRO/NANOSCALE THERMAL ANALYSIS APPLICATIONS

7.8.1. Glass Transitions for Ultrathin Polymer Films

Analysis of ultrathin polymeric films on a high-conductivity substrate such as silicon is a significant challenge. The huge difference in the thermal conductivity of a polymeric film and a substrate results in heat dissipation mostly to the substrate through the tested film, and only a very minor part of the heat dissipates into the polymer. In this case, L-TMA and L-DTA measurement procedures should be significantly modified (Gorbunov et al. 2000c). To balance heat dissipation between thermal and reference probes, the reference probe should be engaged on an identical bare substrate using a microscopic manipulator on a separate microstage under a stereo microscope. The authors of the present chapter used two thermal probes with similar thermal characteristics that were independently tested prior to their selection. Without this modification, a large imbalance between heat dissipation of main and reference probes would have prevented any meaningful microthermal measurements of ultrathin polymer films. With the modified measurement setup, the thermal sensitivity of the thermal probe increased dramatically, thus allowing detection of minute heat dissipation variations ($<1\,\mu W$) associated with the polymer film itself. Obviously, the high thermal conductivity of the silicon substrate under the given experimental conditions prevents the detection of very minor modulations related to nanometer-thick PS films.

Analysis of the experimental data shows that the glass transition temperature decreases when the film thickness is less than 400 nm (compare these data with results for bulk PS film (1 µm thick) in Fig. 7.13). Note that the x axis is not continuous in this figure. For the thinnest film presented in this plot, the glass transition temperature decreases by 20 °C from its bulk value. These results follow general trends observed for ultrathin polymeric films deposited on solid substrates with weak film–substrate interactions (Tsukruk et al. 2003). These data demonstrate the sensitivity of the present micro/nano-TA design to probe nanometer-thick polymer films.

7.8.2. Photodegradation of Polycarbonate

Many polymers undergo degradation under the influence of light. This is a subject of particular interest in the case of polycarbonate, which is a widely used polymer for car windshields. Since the windshield should be transparent, and thus cannot be protected by adding pigments, it is important to understand the mechanism of photodegradation as a function of depth. Photodegradation can be studied by subjecting a sample to intense UV radiation in both the presence and absence of oxygen, as this may play a crucial role in the photochemistry of polycarbonate. Generally speaking, photodegradation in the absence of oxygen leads to chain scission, whereas in the presence of oxygen

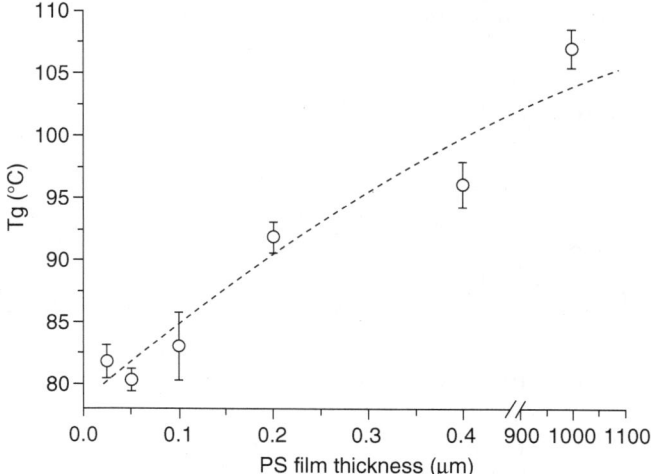

Figure 7.13. Change of the glass transition temperature of polystyrene (PS) film with the film thickness. Datum at 1 μm represents the glass transition temperature for bulk PS; notice that the x axis is discontinuous. [From Gorbunov (2000c), reprinted with permission of Sage Publications.]

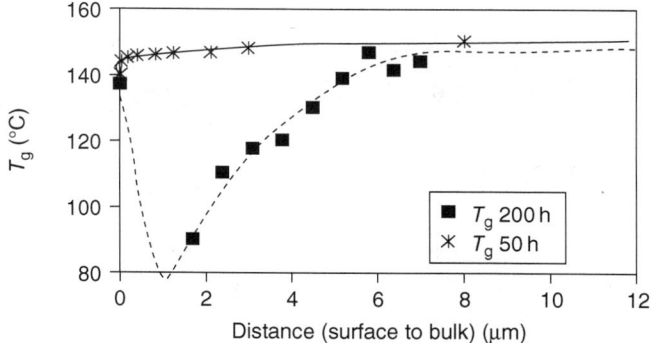

Figure 7.14. Change of glass transition temperature of polycarbonate as a function of depth with different exposure times to UV radiation and oxygen. After 100 h in an inert environment oxygen was introduced. (From Gonnon, personal communication, 2007).

crosslinking occurs. Both of these processes can be studied by measuring changes in the glass transition temperature both on the surface and as a function of depth, which can be accomplished by sectioning the sample. In one study, for the first 100 h the sample was in an oxygen-free environment, during which time the T_g at the surface, and to some extent beneath the surface, decreased as expected. After the oxygen was introduced at 100 h, the T_g started to rise. Figure 7.14 shows the glass transition temperature as a function of depth for two exposure times, one before the introduction of oxygen (50 h) and one afterward (200 h). At the shorter exposure time there is a decrease in T_g at the surface relative to the bulk, where photodegradation is less as the radiation is absorbed. After 200 h (thus 100 h without oxygen and then 100 h with oxygen), the T_g at the surface, having decreased and then increased, was the same as at 50 h of exposure without oxygen. However, as a function of depth, T_g drops because the oxygen has not penetrated sufficiently to reverse the decline in the glass transition temperature caused during the 100 h of UV irradiation without oxygen. The T_g tends back to the bulk value at greater depths. In this way mirco/nano-TA can be used to study the kinetics of the photodegradation process.

7.8.3. Glass Transition and Melting Behavior of Polymeric Materials Analyzed by Nanoprobes

Melting does not show a significant dependence on heating rate or force applied by the tips, whereas, as would be expected, the glass transition temperature shows a marked dependence on both the heating rate and the force. The dependence of the melting point and the glass transition temperature on force is illustrated in Fig. 7.15. While melting and the glass transition both lead to a softening of the sample, this difference in behavior enables them to be differentiated.

7.8.4. Spatial Resolution of Thermal Nanoprobes

Figure 7.16 shows an array of L-TA holes made with nanoprobes in a poly(ethylene terephthalate) (PET) film sample. The craters range in diameter

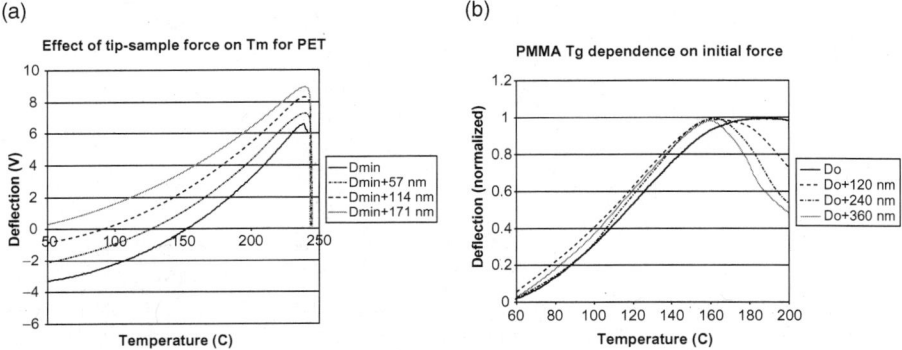

Figure 7.15. Melting (a) of PET with different applied probe forces. Note that force is proportional to probe deflection. Glass transition (b) of PMMA with different applied probe forces. The melting temperature is not influenced by changes in applied force, but the glass transition temperature shows a significant decrease as the applied force is increased, as expected. (reproduced with permission of Anasys Instruments Inc.).

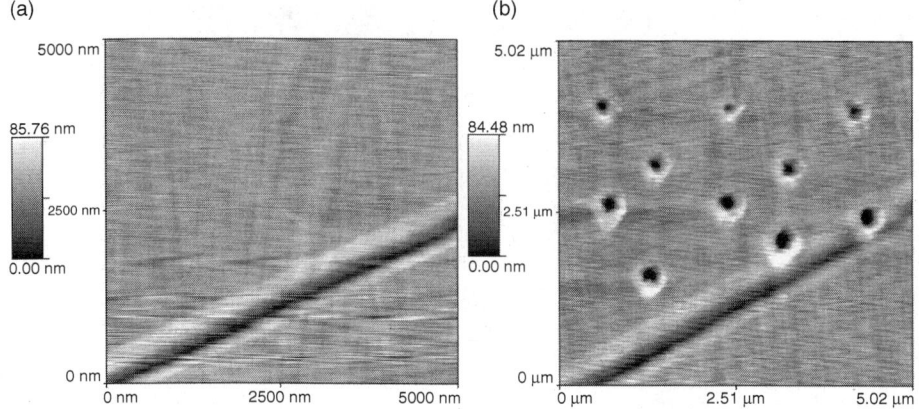

Figure 7.16. Craters made in a PET film by local thermal analysis with nanoprobes: (a) before L-TA and (b) right after L-TA (reproduced with permission of Anasys Instruments Inc.).

from approximately 100 to 200 nm, and their size depends on the maximum probe temperature and initial cantilever force. This crater size is at least an order of magnitude smaller than that produced by a Wollaston wire. This enables L-TA to be usefully applied to many more materials, such as binary polyolefin blends and ethylene–vinyl alcohol (EVOH) copolymer films, which are the subject of the following practical examples.

7.8.5. Identification of Matrix and Occluded Phases in a Polymer Blend

The topographic and tip deflection images in Fig. 7.17 are those of a sectioned 50/50 polyolefin blend. They were acquired with a thermal nanoprobe and are comparable in lateral spatial resolution to those obtainable with a conventional AFM probe. The 20 μm × 20 μm images clearly reveal a phase-separated microstructure, with micrometer-scale isolated or occluded phases distributed in a continuous matrix. However, as the material is a 50/50 blend, it is not possible from imaging alone to ascertain which of the parent polymers form the matrix and which forms the occluded domains. The ability to direct the thermal probe to any location in the acquired image and measure the melting temperature of the surface allows us to identify unequivocally which material is which (provided, of course, that the constituents have different melting temperatures). Probe targeting is best achieved by using the "zoom" imaging facility available in any AFM control software. Carrying out low-resolution scans (e.g., of 50 or 100 lines) of the target area at increasingly higher magnification is a relatively rapid operation and ensures that factors such as sample drift can be readily counteracted. Once the probe is scanning within the

EXAMPLES OF MICRO/NANOSCALE THERMAL ANALYSIS APPLICATIONS   641

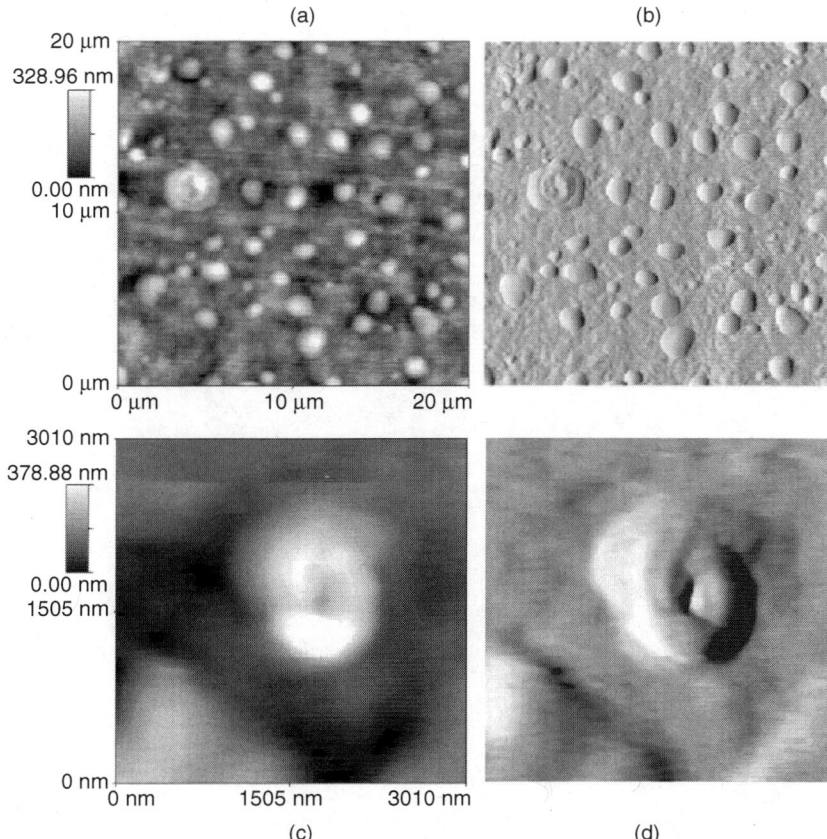

Figure 7.17. (a) Topography and (b) error signal (difference between setpoint and actual force, useful for delineating the edges of features) images of a polymer blend suggesting a structure with continuous and occluded phases; (c) topography and (d) error signal images showing zoom-in on a crater caused by local thermal analysis with a nanoprobe (reproduced with permission of Anasys Instruments Inc.).

selected area, it is moved to the point of interest and immobilized, then the force feedback is deactivated (details of how this is achieved vary with the specific AFM hardware and software). An L-TA run is then carried out (in the examples below, at a heating rate of 20 °C/s) and the probe is disengaged from the surface (while it is above the melting or softening temperature of the sample). It is good practice to image the sample surface after each L-TA measurement to confirm that it was acquired from the desired area.

The $3 \times 3\,\mu m$ images in Fig. 7.17 show an L-TA crater on the surface of one of the isolated domains and confirm that the measurement is confined to this phase. L-TA results from several occluded domains and locations in the matrix are shown in Fig. 7.18. They show that, at approximately 115 °C, the matrix has

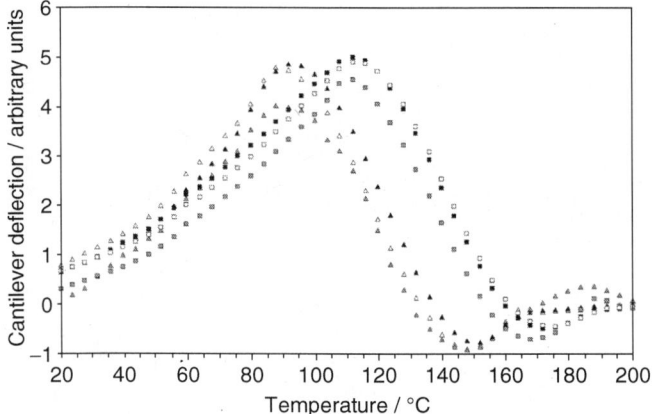

Figure 7.18. Results of local thermal analysis on the polymer blend shown in Fig. 7.17: occluded phase (diamonds); continuous phase (squares) (reproduced with permission of Anasys Instruments Inc.).

an onset melting temperature approximately 20 °C above that of the isolated phase. This information can be readily used to identify each phase. Wherever possible, it is desirable to carry out L-TA on the individual constituents of a heterogeneous sample. In this case, however, such samples were not available.

7.8.6. Identification of a Particulate Contaminant

The next example is a forensic application. It had been identified, by SEM, that a particulate contaminant was entering an ethylene vinyl alcohol (EVOH) copolymer film. There were, however, several candidate materials that could have entered the feedstock stream, and this result was insufficient to isolate the contaminant and enable corrective measures to be taken. These materials included polypropylene (PP), nylon, and an adhesive. Figure 7.19 shows images of a cryofractured sample of the contaminated material before and after local thermal analysis. The resulting L-TA curves are shown in Fig. 7.20. They clearly show a reproducible difference in onset melting temperature between particles (155 °C) and matrix (180 °C). The next step was to carry out L-TA on the raw feedstock materials. By comparing the results of each candidate feedstock material with those from the film, it was shown, unambiguously, that the material forming the particulate contaminant was polypropylene.

7.8.7. Analysis of Biaxially Oriented Polypropoylene Films

Biaxially oriented polypropylene film is extensively used in the packaging industry. These films are single- or multilayered with a typical total thickness

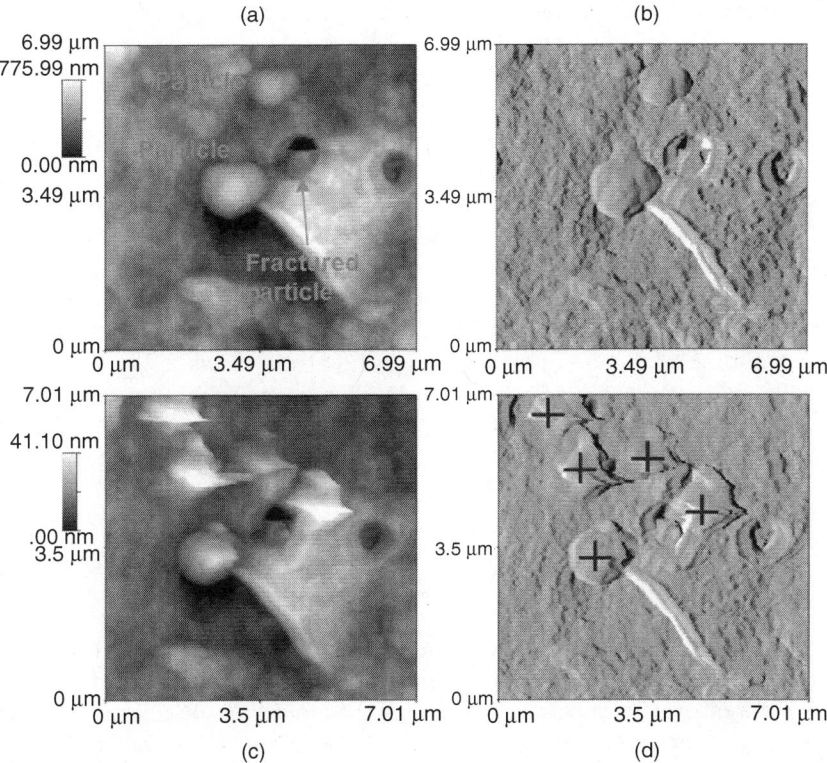

Figure 7.19. (a) Topography and (b) error signal (difference between setpoint and actual force, useful for delineating the edges of features) images of an EVOH sample with a particulate contaminant; (c) topography and (d) error signal images showing the sample after several local thermal analysis experiments (reproduced with permission of Anasys Instruments Inc.).

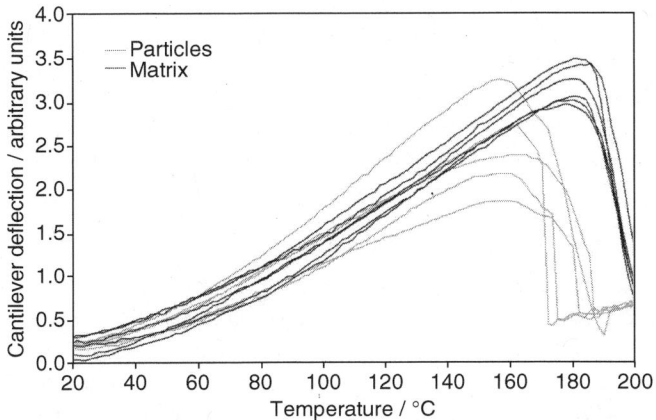

Figure 7.20. Results of local thermal analysis on the sample shown in Fig. 7.19; note that onset melting of particles is 155 °C and of the matrix, 180 °C (reproduced with permission of Anasys Instruments Inc.).

of 15–20 μm. Here we consider a three-layer structure: a thick layer of polypropylene homopolymer sandwiched between two thin layers of a polypropylene copolymer that are usually of the order of 1 μm. The core layer provides the rigidity of the film, while the skin layers provide sealing and/or surface properties. Analysis was performed in cross section by embedding the film in epoxy resin and microtoming the embedded structure. Figure 7.21a shows a topographic image, obtained with a thermal nanoprobe in contact mode, in which the 1-μm skin layer can be clearly seen. In Fig. 7.21b we see the results of local thermal analysis, which can clearly distinguish the different materials. This type of experiment can help elucidate the effects of processing; for example, evidence of a glass transition implies lower crystallinity than if only a melting transition is detected. Lower melting temperatures imply less perfect crystals. Consequently, L-TA can be used to do more than simply identify materials.

7.9. OVERVIEW OF LOCAL THERMAL ANALYSIS

The results discussed above show examples of generic problems in the characterization of polymeric materials. Glass transition temperatures and melting temperatures are fundamental properties that significantly influence polymer performance. As more complex composites are used, the need to determine the properties of individual components becomes more and more important. While these can sometimes be inferred from bulk measurements, it is clearly desirable to be able to make measurements directly on each phase, and micro/nanoscale thermal methods provide this capability. Whether it is the effects of degradation, contamination, processing, or simply reducing the scale of a structure down to a nanolevel, micro/nanoscale thermal analysis can provide unique information. L-TA is always used with some form of visualization of the sample surface so that regions of interest can be identified. In some cases the visualization will simply be optical microscopy; in some, scanning thermal microscopy; and in others, the various intermittent contact modes of AFM. Increasingly, as smaller scales of scrutiny are needed, nano-TA will be used as an adjunct to AFM.

7.9.1. Local Chemical Analysis

The techniques mentioned so far provide no direct spatially resolved information on the chemical composition of materials. Various established analytical techniques, such as secondary-ion mass spectrometry (SIMS), X-ray photoelectron spectrometry (XPS), and infrared (IR) and Raman microspectrometry, can be used to carry out local compositional studies. In the case of SIMS and XPS, the sample needs to be held in high vacuum, while in IR and Raman microspectrometry, the resolution is limited by the relatively large wavelength

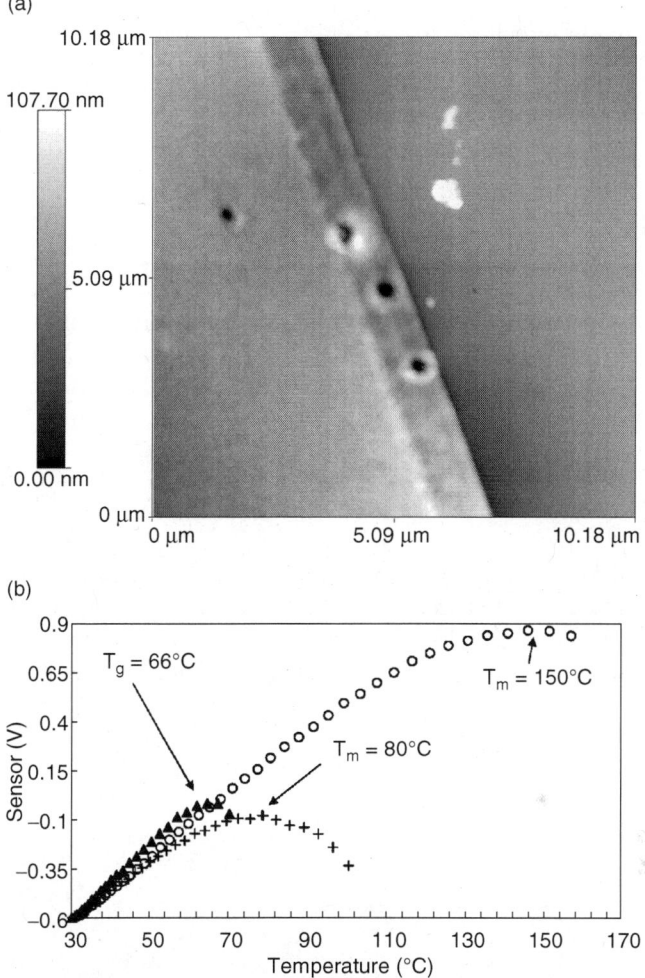

Figure 7.21. (a) Topographic image of packaging film showing the 1 μm skin layer with holes from the local thermal analysis; (b) local thermal analysis results showing transition temperatures of skin layer ($T_m = 80\,°C$), core layer ($T_m = 150\,°C$), and epoxy matrix ($T_g = 66\,°C$). [from Gotzen and Van Assche (2007)].

of the excitation radiation. Two developments of microthermal analysis have attempted to overcome these limitations. The first of these is local evolved gas analysis (L-EGA) coupled with gas chromatography–mass spectrometry (GC-MS) (Price et al. 1999a, b, 2000) or directly with mass spectrometry (Reading et al. 2001). This may be regarded as the microanalog of the established hybrid technique of thermogravimetry (TGA)-GC-MS (McClennen et al. 1993), in which the gases evolved from a sample heated in a thermobalance are analyzed. The second technique is near-field photothermal infrared spectroscopy (Pollock and Hammiche (2001)). It is anticipated that both of these techniques will be further developed as high-spatial-resolution analytical techniques in the future.

7.9.2. Micro/Nano-TA Equipment

At the time of writing there is only one manufacturer of micro/nano-TA equipment; Anasys Instruments, (www.anasysinstruments.com). They supply the hardware and software for local thermal analysis and thermal imaging that can be interfaced with the most popular types of atomic force microscope. More recently they have launched an instrument based on an optical microscope, the Vesta system, which is simpler to use than an atomic force microscope but the spatial resolution is limited to approximately 1.5 micrometers, see Fig. 7.22. The Wollaston probes are supplied by Veeco (www.veeco.com) and are, therefore, compatible only with Veeco AFMs. The nanoprobes are supplied by Anasys Instruments and can be used with most popular makes of AFM. In addition, Anasys Instruments supply calibration kits containing temperature standards. More recently they have launched an instrument based on an optical microscope, the Vesta system, which is simpler to use than an atomic force microscope, but the spatial resolution is limited to approximately 1.5 µm.

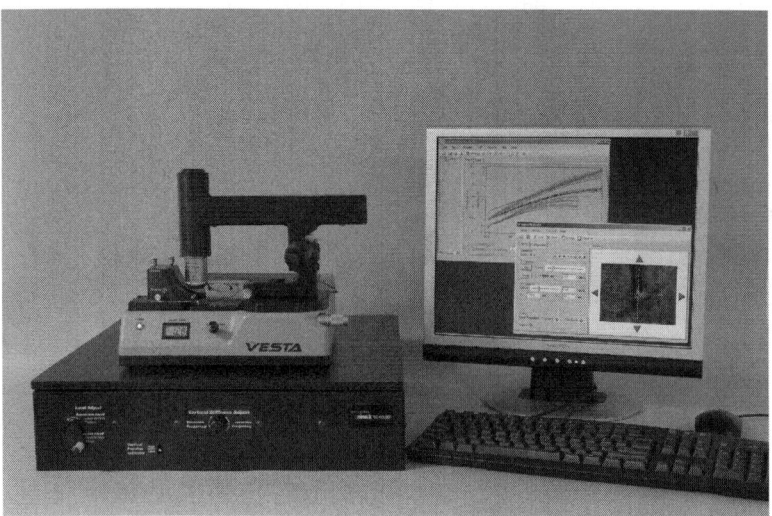

Figure 7.22. Vesta system for micro-thermal analysis. Courtesy of Anasys Instruments.

ABBREVIATIONS

Symbols

a	thermal diffusivity
Ω	ohms
Q	heat
q	heating rate
R_c	effective thermal probe contact radius
T	temperature, temperature close to heating zone
t	time
T_c	central temperature
T_g	glass transition temperature
T_m	melting point
λ	composite thermal conductivity
λ_p	thermal conductivity of thermal probe
λ_s	thermal conductivity of sample surface
ΔQ	heat dissipation
π	3.1416

Acronyms

AC	alternating current
ADC	analog to digital converter
AFM	atomic force microscopy
DC	direct current
DMA	dynamic mechanical analysis
DSC	differential scanning calorimetry
DTA	differential thermal analysis
EVOH	ethylene–vinyl alcohol
GC	gas chromatography
GC-MS	gas chromatography–mass spectroscoscopy
HDPE	high-density polyethylene
IR	infrared
L-DTA	local differential thermal analysis
L-EGA	local evolved gas analysis
L-TA	local thermal analysis
L-TMA	local thermomechanical analysis
MS	mass spectroscopy
MTDSC	modulated temperature differential scanning calorimetry
PCL	polycaprolactone
PET	poly(ethylene terephthalate)
PMMA	poly(methyl methacrylate)
PP	polypropylene
PS	polystyrene

SFM   scanning force microscopy
SIMS   secondary-ion mass spectroscopy
SPM   scanning probe microscope
SThM   scanning thermal microscopy
STM   scanning tunneling microscope
TA   thermal analysis
TGA   thermogravimetric analysis
TGA-GC-MS thermogravimetric analysis–gas chromatography–mass spectroscopy
TMA   thermomechanical analysis
XPS   X-ray photoelectron spectrometry

REFERENCES

Balk, L. J., Maywald, M., and Pylkki, R. J. (1995), *Proc. 9th Conf. Microscopy of Semi-conducting Materials* (Inst. Phys. Conf. Series 146), Oxford, UK, pp. 655–658.

Binnig, G., Quate, C. F., and Gerber, C. (1986), *Phys. Rev. Lett.* **12**, 930.

Binnig, G. and Rohrer, H. (1982), *Helv. Phys. Acta* **55**, 355.

Blaine, R. L., Slough, C. G., and Price, D. M. (1999), in Williams, K. R. and Kociba, K., eds., *Proc. 27th NATAS Conf.*, Savannah, GA, North American Thermal Analysis Society, Omnipress, pp. 691–696.

Block H. (1937), *Proc. General Discussion Lubrication and Lubricants*, Vol. 2, Institute of Mechanical Engineering, p. 222.

Bozec, L., Hammiche, A., Pollock, H. M., Conroy, M., Chalmers, J. M., Everall, N. J., and Turin, L. (2001), *J. Appl. Phys.* **90**, 5159.

Chizhik, S. A., Huang, Z., Gorbunov, V. V., Myshkin N. K., and Tsukruk, V. V. (1998), *Langmuir* **14**, 2606–2609.

Chui, B. W., Stowe, T. D., Ju, Y. S., Goodson, K. E., Kenny, T. W., Mamin, H. J., Terris, B. D., Ried, R. P., and Rugar, D. (1998), *J. Microelectromech. Syst.* **7**, 69.

Dinwiddie, R. B., Pylkki, R. J., and West, P. E. (1994), *Thermal Conductivity*, Vol. 22, Tong T. W., ed., Technomics, Lancaster, PA, p. 668.

Gorbunov, V. V., Fuchigami, N., Hazel, J. L., and Tsukruk., V. V. (1999), *Langmuir* **15**, 8340–8343.

Gorbunov, V. V., Fuchigami, N., and Tsukruk, V. V. (2000a), *Probe Microsc.* **2**, 53.

Gorbunov, V. V., Fuchigami, N., and Tsukruk, V. V. (2000b), *Probe Microsc.* **2**, 65.

Gorbunov, V. V., Fuchigami, N., and Tsukruk, V. V. (2000c), *High Perform. Polym.*, **12**, 603–610.

Gotzen, N. A. and Van Assche, G. (2007), *Proc. 35th NATAS Conf.*

Hammiche, A. and Pollock, H. M. (2001), *J. Phys. D: Appl. Phys.* **34**, R23–R53.

Hammiche, A., Hourston, D. J., Pollock, H. M., Reading, M. and Song, M., (1996a), *J. Vac. Sci. Technol. B* **14**, 1486.

Hammiche, A., Pollock, H. M., Song, M., and Hourston D. J. (1996b), *Meas. Sci. Technol.* **7**, 142.

Hammiche, A., Pollock, H. M., Reading, M., Claybourn, M., Turner, P. H., and Jewkes, K. (1999), *Appl. Spectrosc.* **53**, 810.

Hässler, R. and zur Müehlen, E. (2000), *Thermochim. Acta* **361**, 113 (available at http://www.w-dhave.inet.co.th/index/RTD/RTD.pdf).

Jaeger, J. C. (1942), *Proc. Roy. Soc.* **76**, 22.

Jones, K. J., Kinshott, I., Reading, M., Lacey, A. A., Nikolopoulos, C., and Pollock, H. M. (1997), *Thermochim. Acta* **305**, 187.

King, W. P., Saxena, S., Nelson, B. A., Weeks, B. L., and Pitchimani, R. (2006), *Nano Lett* **6**, 2145–2149.

Lawson, N.S., Ion, R. H., Pollock, H. M., Hourston, D. J., and Reading, M. (1994), *Phys. Scripta* **T55**, 199–205.

McClennen, W. C., Buchanan, R. M., Arnold, N. S., Dworzanski, J. P., and Meuzelaar, H. L. C. (1993), *Anal. Chem.* **65**, 2819.

Meyers, G, Pastzor, A., and Kjoller, K. (2007), *Am. Lab*. (Nov./Dec. 2007).

Moon, I., androsch, R., Chen, W., and Wunderlich, B. (2000), *J. Therm. Anal. Calorim.* **59**, 187–203.

Pollock, H. M. and Hammiche, A. (2001), *J. Phys. D: Appl. Phys.* **34**, R23–R53.

Pollock, H. M., Hammiche, A., Song, M., Hourston, D. J., and Reading, M. (1998), *J. Adhesion* **67**, 217.

Price, D. M., Reading, M., and Trevor, T. J. (1999a), in *Proc. 27th NATAS Conf.*, Williams, K. R. and Kociba, K., eds., Omnipress, pp. 420–425.

Price, D. M., Reading, M., Hammiche, A. and Pollock, H. M. (1999b). *Int. J. Pharm.* **192**, 85.

Price, D. M., Reading, M., Hammiche, A. and Pollock, H. M. (2000), *J. Therm. Anal. Cal.* **60**, 723.

Reading, M. (1993), *Trends Polym. Sci.* **1**, 248.

Reading, M., Price, D. M., Grandy, D. B., Smith, R. M., Pollock, H. M., Hammiche, A., Bozec, L., Mills, G., and Weaver, J. M. R. (2001), in *Recent Advances in Scanning Probe Microscopy of Polymers; Macromolecular Symp.*, Tsukruk, V. V. and Spencer, N. D., eds., Vol 167, p. 45.

Tsukruk, V., Gorbunov, V., and Fuchigami, N. (2003), *Thermochim. Acta* **1**, 7043.

INDEX

Abbreviations
   DEA-related, 599–601
   DMA-related, 489–491
   DSC-related, 225–228
   Micro/nano-TA-related, 647–648
   TGA-related, 312
   TMA-related, 380
Absolute temperature scale, 11
Accelerating models, 279
AC dielectric experiments, measuring conditions in, 533–535. *See also* Alternating current entries
ACR formulation, 368, 369
Acronyms
   DEA-related, 602–603
   DMA-related, 491
   DSC-related, 228–229
   related to micro-scale local thermal analysis, 647–648
   TGA-related, 313
   TMA-related, 380–381
Acrylate polymers, 556
   DMA data for, 462
Acrylic, highly crosslinked, 439–440
Acrylonitrile–butadiene (ACN/BD) copolymer, commercial blend with phenolic resin, 306–308
Acrylonitrile–butadiene–styrene (ABS) terpolymers, 296
Activation energies, 142, 148–149, 543, 550
   calculating, 151

   estimating, 152
   evaluating, 153
   measuring, 147
   of the relaxation process, 510
Additives
   low-mass, 269
   low-molecular-mass, 77
Adhesion, in dual-coated optical fibers, 371–374
Adhesive blend, analysis of, 306–308
Adiabatic calorimetry data, 165
Adjacent re-entry, 87
Advanced Thermal Analysis System (ATHAS) Research Group, x
AFM control software, 640. *See also* Atomic force microscopy (AFM)
AFM scanning mode, 617
AFM thermal probe, 628–629
Agilent HP (Hewlett-Packard) 4284A bridge, 524
Aging, of glasses, 73
Aging process, analysis of, 185
Akinay, Ali E., 319
Aliphatic polyamides, relaxations of, 571
$\alpha\beta$ relaxation process, 583
α_{CLTE} relation, 591. *See also* Coefficient of linear thermal expansion (CLTE)
α_c transition, 433
α dispersion, 18
α-relaxation curve, 514
α-relaxation peak, 581

Thermal Analysis of Polymers: Fundamentals and Applications, Edited by Joseph D. Menczel and R. Bruce Prime
Copyright © 2009 by John Wiley & Sons, Inc.

652 INDEX

α relaxations, 415, 512. *See also*
  Segmental α-relaxation process
  amorphous, 433
  in polyamides, 571
  in polyurethanes, 565–567
α-relaxation signals, 573
Alternating-current heating, 630–631. *See
  also* AC dielectric experiments
Alternating-current impedance bridges,
  523
Alternating DSC (ADSC), 168. *See also*
  Differential scanning calorimeters
  (DSCs)
Amine crosslinked epoxy systems, 577
Amine–epoxy reaction, 136–137, 148
Amine epoxy resin, total water uptake
  for, 592
Amorphous α relaxation, 433
Amorphous polycarbonate, thermal
  properties of, 165
Amorphous polymers
  first heating–cooling–second heating
    program for, 213–214
  glass transition in, 60, 184–187
  miscibility of, 185–186
  secondary relaxation processes in,
    512–514
  secondary transitions in, 423–424
  segmental α-relaxation process in,
    514–515
  viscoelastic transitions in, 410–411
Amorphous poly(methyl methacrylate),
  538
Amorphous thermoplastics, 388
Analyzers, commercially available,
  596–597
Anasys Instruments, 646
Annealing
  in as-received fibers, 121
  of crystalline polymers, 109–110
  of glasses, 74–76
Annealing peaks, 110
Antiplasticization effect, 425
Antiplasticized aromatic polymer
  systems, mechanical property data
  for, 431
Antiplasticizers, 429–430
Apolar mainchains, thermoplastics with,
  556–559

ARES rheometers, 523
Arrhenius diagrams, analysis of, 551
Arrhenius equation, 145, 147, 149, 279
  fitting shift factors and, 408
Arrhenius plots, 291, 510, 511
  for α and β transitions, 416
  construction and analysis of, 549–553
  of polyurethanes, 567–569
Arrhenius rate constant, 154
AR Series rheometers, 486
As-received fibers, 121
As-spun fibers, 120, 432
  dynamic mechanical curves for, 434,
    436
A stage of cure, 134
ASTM E698 kinetics software, 142, 145,
  151
ASTM International Committee E37 on
  Thermal Measurements, 225, 378
ASTM method, 151–152, 287, 293–295
ASTM standards,
  DEA-related, 599
  DMA-related, 488–489
  DSC-related, 255
  TMA-related, 378–379
  TGA-related, 311–312
Atactic nonpolar thermoplastics, 555
Atactic PMMA, permittivity and
  dielectric loss spectra of, 548. *See
  also* Poly(methyl methacrylate)
  (PMMA)
ATHAS databank, 53
Athermal nucleation, 85
Atmosphere, in TMA experiments, 336
Atmospheric pressure chemical
  ionization (APCI), 250
Atomic force microscopes, 616–618,
  633–634
Atomic force microscopy (AFM), 615.
  See also AFM entries
Atomic polarization, 512
Autocatalytic processes, 279, 280
Autocatalytic rate equation, 147
Autosamplers, 3, 36
Autostepwise method/technique, 247,
  265, 306
Autostepwise TGA technique, 276. *See
  also* Thermogravimetric analysis
  (TGA)

Autotensioning, 466–467, 469
AutoZero accessory, 27
Avrami equation, 88
Avrami–Erofeev models, 280
Avrami evaluation, of nonisothermal crystallization of polymers, 92
Avrami exponent, 89–90

Baffling, in thermobalances, 245
Bähr Thermoanalyse, 308
Bair, Harvey E., 7, 241, 319
Baseline, 54
  in differential scanning calorimetry, 36
Baseline method, 95, 96
  heat of fusion and, 100–102
BCH-52 liquid crystal standard, 50
BDS 40 system, 593
BDS 1200 connection head, 530
BDS 1200 sample cell, 595
BDS 1308 parallel-plate shield liquid sample cell, 530
BDS 2200/2100 RF sample cells, 531
BDS concept 70 system, 524–525
Bending modes, 470
Benzocyclobutene, 296–297
β* relaxation, 561
β relaxations, 415, 424, 512–514, 560–561, 565
  dielectric spectra and relaxation frequencies of, 547
  effect of moisture on, 425–426, 427
  in polyamides, 571
Biaxially oriented polypropylene films, analysis of, 642–644
Biaxially oriented polypropylene, crystallization of, 86–87
Biaxial orientation, 124
Biopolymers, application of time-domain reflectometry in, 526
Bisphenol A polycarbonate, 303
  TMA curves of, 347
"Blank" experiments, 244–245
Blank runs, 54
Blank TSC, 537–538. *See also* Thermally stimulated current (TSC) technique
Blends
  DSC of, 61, 77–80, 186, 202
  DMA of, 388, 409
  TGA of, 304–306

Block, 21
Block temperature, 43
Boltzmann distribution, 10
Bose Corporation, 483
Bose ElectroForce (ELF) DMA Instruments, 483–484
Bound water, 592
Bowing, in integrated circuit packages, 374–375
Bowtie probes, 623
  resolution of, 624
Branched polymers, 110
Brittle fracture, 429
Broadband dielectric converter (BDC), 522
B stage of cure, 134
Bulk chemical reactions, 130
Bulk properties, in confining environments, 499–500
"Bulk" sample temperature, 4
Bulk thermomechanical analysis, 633
Buoyancy phenomenon, 271
  in thermogravimetric analysis, 243–244

Cable jacket, orientation recovery in, 370–371
Cahn Instruments, 225
Cahn VersaTherm bottom-loading thermogravimetric analyzers, 308–309
Calculation software products, 215
Calibrant run, 54
Calibrants, 465
Calibration
  of differential scanning calorimeters, 41–52
  of dynamic mechanical analyzers, 414, 460–465
  enthalpy, 50
  heating, 41–46
  instrument, 460–465
  in a micro/nanoscale local thermal analysis experiment, 634–636
  of thermal lag, 50–52
  thermocouple, 462–463
  of thermomechanical analyzers, 332–335
  of thermogravimetric analyzers, 251–256

Calibration (*cont'd*)
  thermogravimetric analysis and, 252–256
Calibration curve, 634
Calibration samples, 464–465
Calibration standards, 634–635
  for fiber measurements, 334–335
  frequently used, 43–44
Calorimeter time constants, 25
Calorimetric purity method, 37
Calvet DSC, 221. *See also* Differential scanning calorimeters (DSCs)
Cantilever mode, 466
Carbon black (CB) analysis, 275–276
Cationically cured systems, 161
Cationic photopolymerization, 161–162
Celcon® copolymer, 111
Cell measuring, in dielectric experiments, 528–531
Chain-folding principle, 87
Chain mobility, 556
Chain relaxation capability (CRC), 323
Charge migration, long-range, 515–516
Charge transport, evaluation of, 500
Charging/discharging currents, isothermal, 520–521
Chartoff, Richard P., 387
Chelsea dielectric interface (CDI), 522
Chemical reactions
  online monitoring of, 500
  time scale for, 154
Chiral nematic crystals, 47
Cholesteric liquid crystals, 47
Cholesteric phase, 48
Clausius, R., 14
Cleaning procedures, in TGA experiments, 258–259
Clearing point, 47, 48
Closed loop, 28, 30
CLTE measurements, 362, 336. *See also* Coefficient of linear thermal expansion (CLTE)
Coatings, polymeric, 371–374
Coating thickness, determination of, 244–245
Coaxial-line reflectometry, 524–525

Coefficient of linear thermal expansion (CLTE), 320, 321, 326, 335, 338, 365. *See also* α_{CLTE} relation; CLTE measurements
  techniques for measuring, 345
Coefficient of thermal expansion, conductivity and, 590–591
Coefficient of volumetric expansion, 321
Cold crystallization, 84, 188–189, 420, 433
  isothermal, 92
  nonisothermal, 92–94
Cole–Cole plots, 567
Comb electrodes, 530–531
Commercial DMA instrumentation, 477–488
Commercial frequency response analyzers, 522
Commercial instruments
  special calibration procedures for, 52
  temperature scale of, 44
Commercially available analyzers, 596–597
Commercial photocalorimetry accessories, 156
Commercial thermal analysis instrumentation, 2–3
Comonomer units, 114
"Compensation point," 543
Compensation rule, 546–547
Completely amorphous PEEK, 562
Completely miscible polymer blends, 573
Complex modulus, 397–399
Complex permittivity, 504, 520–521
Complex permittivity data, analysis of, 549
Complex reflection factor, 524
Complex signals, experimental means for analyzing, 528
Complex TSC peaks, experimental analysis of, 543
Complex viscosity, 400
Compliance calibration, 460
Composites, 469–472
  thermoplastic, 375–376
  TMA of, 363–364
"Composite" thermal conductivity, 624, 625, 627
Compositional analyses, thermogravimetric, 274–277

INDEX

Compression geometry, 459
Computerized nomograph, 475
Conditioning a sample, in TMA experiments, 339
Conductivity
   coefficient of thermal expansion and, 590–591
   cure time variation of, 587
   defined, 586
   extrinsic, 515
   variation of, 590
Conductivity relaxation (CR), 567
Confining environments, information on surface and bulk properties in, 499–500
"Consistency" criterion, 473
Constant-force mode, 480
Constant-frequency experiments, relaxations for, 410
Constrained-state DSC measurements, on oriented films, 124. *See also* Differential scanning calorimetry (DSC)
Constrained-state measurement, 115–117
Contact area, 619
Contact mode, 617
Contact-mode imaging, 621
Contact problems, in dielectric experiments, 532
Contact radius, 624–626
Continuous-fiber composites, 470
Contraction curves, for primary and secondary films, 373, 374
Controlled rate thermogravimetric analysis (CRTGA), 246–248, 305. *See also* Thermogravimetric analysis (TGA)
Conversion, 137, 138
   in isothermal TGA experiments, 268
Conversion-time curves, 148–149
Conversion-time data, 139
Conversion-time plots, 280–282
Cooling
   in DSC measurements, 213
   glass transition on, 64
Cooling calibration, 46–50
Cooling curves, 72, 74
Cooling media, 335–336

Cooling rates, 32, 537
   in a commercial DSC, 163
   importance in DSC measurements, 214
   in Perkin-Elmer units, 219–220
   in TMA experiments, 337–338
   transforming from DSC into frequencies, 552
Cooperative (long-range) processes, dielectric monitoring of, 497
Cooperative relaxation signals, 569
Copolymers
   ethylene–vinyl acetate, 304–306
   glass transition in, 77
   melting of, 110–115
Cox–Merz rule, 401
Creep, 325–326, 329
   measuring, 358–359
Creep compliance curves, 361
Creep compliance master curve, 359
Creep curves, 405
Creep–recovery curve, 358
Creep testing, 394–395
Crimper, 34
Crimping, 210
Crivello salts, 161
Crosslink density, 326
Crosslinked polymers, DMA characterization of, 438–456
Crosslinking, 344–345
   effect on glass transition, 77
   in Nylon 6 fibers, 121
Cryostats, 525
Cryosystems
   Novocool, 595
   Quatro, 593, 595, 598
Crystal → crystal transitions, 107–108, 434
Crystal growth process, 85–94
"Crystalline" α_c relaxation, 572
Crystalline macromolecules, melting of, 96–98
Crystalline melting point, 517
Crystalline polymers, 388
   annealing of, 109–110
   equilibrium melting point of, 17
   heat of fusion of, 99
   reporting characteristic data of, 103–104
   secondary transitions in, 423–424

Crystalline relaxation, 434
Crystallinity, 70–71
　β-relaxation signal and, 561
　determination of, 190
　effect on glass transition temperature, 344, 433–434
　in measuring T_g, 417–421
Crystallinity calculation software, 100
Crystallizable polymers, 420–421
Crystallization, 81–94
　induction time of, 89
　of low-molecular-mass substances, 82–83
　measured by modulated temperature differential scanning calorimetry, 187–191
　of polymers, 83–84
　recording heat effects of, 88
　stepwise isothermal, 110
　two-step process of, 94
Crystallization exotherm, 182
Crystallization processes, modified method for studying, 188
Crystallization temperature, 17
Crystal → nematic transition, 224
Crystal perfection, 90
Crystal phase relaxations, 517
Crystals
　extended-chain, 85–87
　irradiation of, 162–163
　reorganization during melting, 104–105
C stage of cure, 134
Cure(s). *See also* Curing entries; Time–temperature transformation cure diagram
　changes in dielectric properties during, 585
　chemical reactions of, 131–132
　DSC measurements of, 134–136
　infrared, 274
　isothermal, 445–450
　nonisothermal, 450–456
　stages of, 134
　T_g-conversion relationship and, 142
　thermoset, 440–441
　volatile products from, 271–274
Cure data, obtaining at low temperatures, 139

Cured thermosets, transitions in, 450
Cure exotherm, deconvolution of, 195
Cure index, 590
Cure kinetics, 144–154
　defined, 145
Cure parameters, 133
　DMA techniques for developing, 445–456
Cure processes
　dielectric characterization and monitoring of, 583–591
　optimization of, 144
Cure reaction factors, evaluation by TMA, 357
Cure reactions
　for applications of TTS kinetics, 148
　diffusion control of, 153–154
Cure stages, chemically controlled and diffusion-controlled, 154
Cure temperature, 442, 450
　reference materials for, 253
Curing, thermomechanical analysis of, 356–357
Curing experiments, modulation amplitudes in, 180
Curing process, characteristics of, 441–445
Curing system monitoring, MTDSC in, 192–194
Curved Arrhenius diagrams, 510
Curve-fitting model functions, 598
Curve-fitting procedure, 406
Curve-fitting techniques, 528
Cutoff temperature, 543
Cyclic processes, 14
Cyclic strain, 387
Cycloaliphatic epoxy, 161–162
Cytop
　stability of, 300
　thermal stability of, 298–299
Cytop compounds, mass loss behavior of, 271

$d\alpha/dT$ values, 283–284
Damping peak, 443
Data acquisition, computer-controlled automated, 534

INDEX

DC measurements, dielectric spectrometers for, 598–599. *See also* Direct-current (DC) conductivity
DEA techniques, 520. *See also* Dielectric analysis (DEA)
Debye functions, 504, 505
Debye-type mechanisms, 507–508
Decelerating models, 279–280
Decomposition behavior, 305
Deconvolution procedure, 170, 172–173, 618–619
Deconvolution program, 630
Deflection-heating voltage curves, 635
Degradation, mechanisms in epoxy composites, 591
Degree of conversion, relationship with ionic conductivity, 589–590
Degree of cure, 137, 139
  monitoring with penetration measurements, 357
Degree of swelling, 326, 354–355
Dehydration, 5
δ relaxation, 563–564
ΔT correction factor, 49
Depolarization current density plot, 527
Depolarization current signals, 569
DETA accessory, 523–524
Devitrification, 453, 454, 456
  during nonisothermal cure of epoxy–amine system, 198–199
DGEBA/DDM + polyethylene oxide (PEO) blends, 583. *See also* Diglycidyl ether of bisphenol A (DGEBA)
DGEBA epoxy, moisture in, 270
DGEBA-type epoxies, mixing with multifunctional amines, 578–579
Diaminodiphenylsulfone (DDS), 135
Diamond DSC, 219. *See also* Differential scanning calorimeters (DSCs)
DiBenedetto equation, 142–143
DICLAD® 880 composite, 375–376
Dielectric, electrical properties of, 503
Dielectric α-relaxation band, 519
Dielectric α transition shifts, 574
Dielectric analysis (DEA), ix, 1, 3, 497. *See also* DEA techniques
  advantages of, 498

based on distributed-circuit methods, 524–525
based on lumped circuit methods, 521–524
experimental AC techniques for, 594
how to perform an experiment, 528–538
of liquid crystalline polymers, 562–565
for monitoring thermoset cure, 583–585
of polyamides, 570–572
of polyesters, 560–562
of polymer blends, 573–576
of polyoxides, 559–560
of polyurethanes, 565–570
runs with thick PMMA films, 547–553
standards for, 599
theory and background of, 502–518
as a thermal analysis approach, 498–499
of thermoplastic polymers, 572
for thermoplastic polyurethane films, 568
of thermoplastics, 553–576
of thermosets, 576–592
Dielectric analyzers, commercially available, 596–597
Dielectric characterization, of polymers, 501
Dielectric data, representations of, 505–509
Dielectric δ-relaxation signal, 563–564
Dielectric depolarization spectroscopy, 526–528
Dielectric events, impacts of, 586
Dielectric experiments
  AC, 533
  designing, 585–586
  involving relaxation function, 504
  performing, 528–538
Dielectric glass transition temperature, 551
Dielectric increment, 503
Dielectric induction, 504
Dielectric instrumentation, 592–599
Dielectric losses, 506
Dielectric measurement, 522
Dielectric measurement systems, high-performance, 593

Dielectric permittivity, 506
Dielectric permittivity function, 503
Dielectric plot structure, inspection of, 518
Dielectric polymer thick films, 273
Dielectric probes (labels), 554
Dielectric relaxation spectroscopy (DRS), 501. *See also* DRS entries
Dielectric relaxation time, 510–511
Dielectric reports, 502
Dielectric research, focus of, 585
Dielectric response, 505
  of polymeric materials, 512–518
Dielectric results, complex intensive parameters and, 505–506
Dielectrics, frequency-dependent response of, 502–505
Dielectric signatures, of water molecules confined in epoxy resins, 591–592
Dielectric spectra, comparisons with published literature, 519
Dielectric spectrometers, 534
  for DC measurements, 598–599
  for measurements in frequency domain, 593–595
Dielectric spectroscopy, 500
  time-domain, 520–521
Dielectric spectrum
  decomposition of, 498
  peak assignments in, 518–520
Dielectric strength, 503
Dielectric systems, high-precision, 520
Dielectric techniques, 520–528
Dielectrometers, 520
Differential photocalorimetry (DPC), 134, 154–162
  background and principles of, 155–159
  major objective of, 158
Differential photocalorimetry experiments, 162
Differential scanning calorimeters (DSCs). *See also* Power compensation DSCs
  calibration of, 41–52
  enthalpy calibration of, 50
  heat capacity calibration of, 50
  temperature calibration of, 41–50
  types of, 36

Differential scanning calorimetry (DSC), ix, 1, 3, 7–239, 615. *See also* DSC entries; Heat-hold-cool DSC measurements; Modulated temperature DSC (MTDSC); Photo-DSC; Power compensation DSCs; Step scan DSC (SSDSC)
  basics of, 18–36
  calibration on cooling, 48–49
  defined, 18
  elements of thermodynamics in, 9–18
  enthalpy change in, 14
  fast-scan, 162–168
  of fibers, 115–123
  of films, 123–130
  first-order phase transitions in, 16
  heat flux, 21–22
  major applications of, 8
  meanings of, 8
  Mettler Toledo, 25–27
  power compensation, 27–31
  purity determination of low-molecular-mass compounds by, 37–41
  standards for, 225
  usefulness of, 8
Differential thermal analysis (DTA), 18, 615, 628–629
  thermal resistance in, 19
Diffusion, time scale of, 154
Diffusion control, 153–154
Diffusion-controlled reactions, 198
Diffusion factor (DF), 193–194
Diffusion of species, 5
Diglycidyl ether of bisphenol A (DGEBA), prepolymers, 577. *See also* DGEBA entries
Dilatometry, 321
Dillman, Steven H., 387
Diluents
  effect on T_g values, 422–423
  sensitivity to, 424
Dipolar orientation, 586
Dipolar polymer molecule segments, relaxation of, 583
Dipole motions, relaxation frequency of, 581
Dipole processes, 587
Dipoles, relaxation mechanisms of, 510

Direct-current (DC) conductivity, 505, 515, 581. *See also* DC measurements
　Arrhenius plot of, 552
Dispersion, 503
Dissipation factor, 506
Distributed-circuit methods, dielectric analysis based on, 524–525
DMA characterization. *See also* Dynamic mechanical analysis (DMA)
　calibration, 460–465
　of crosslinked polymers, 438–456
　for thermoplastics, 424–432
　of thermoset cure, 440–441
DMA/DEA sample configuration, 524. *See also* Dielectric analysis (DEA)
DMA experiments
　modes of operation of, 456–458
　practical aspects of, 456–477
DMA instrumentation, commercial, 477–488
DMA instruments, operating limits of, 461
DMA measurements, on fibers and thin films, 467–469
DMA operation, 388
DMA scan, 402
DMA sinusoidal stress–strain response curves, 396, 399
DMA solid-mode experiments, 448–450
DMA standards, 488–489
DMA systems, categories of, 477–478
DMA tests, polymer state during, 389–390
DMA test samples, 465–472
DMA T_g values, 476–477
dm/dt values, 246, 247
Doolittle viscosity equation, 323
Double crystallization, 87
Double endothermic peak, 191
Double glass transition, 554
DPC accessories, 159. *See also* Differential photocalorimetry (DPC)
Drawn fibers, 120, 432
　dynamic mechanical curves for, 434, 436
　melting point of, 116, 117
Drawn Nylon 6 films, 126, 129

Draw ratio, 115, 432
DRS data, 547. *See also* Dielectric relaxation spectroscopy (DRS)
DRS dielectric analysis data, 552
Drying, of coatings and paints, 272
DSC-1, 7. *See also* Differential scanning calorimeters (DSCs); Differential scanning calorimetry (DSC)
DSC131 line, 220
DSC 200 F3 Maia, 218
DSC 204 F1 Phoenix, performance characteristics of, 218
$DSC823^e$ unit, performance characteristics of, 217, 218
DSC analysis, of thermoplastic composite, 375–376
DSC ASTM E698 methodology, 142
DSC calibration, in a DSC purity determination, 39
DSC capsules, high-pressure, 35–36, 211
DSC cell
　covering, 217
　maintaining, 215–216
DSC curves, 28, 62, 63, 66–67, 69
　annealing and, 74–75
　of Vectran fiber, 123
DSC data, using, 632
DSC exotherm, 148
DSC experiments
　glass transition temperature in, 60–61
　purpose of, 212
DSC glass transition temperature, 551
DSC heat flux cell, 21
DSC heating curves, 72, 70
DSC heating rate, 165–166
DSC instrumentation, 9, 217–225
　resolution and precision of, 223–224
DSC measurements, 208–217
　carrying out, 212–215
　experimental parameters for, 214–215
　hardware considerations related to, 215–217
　weighing samples for, 224–225
DSC pans, 34–35, 210–211
　loading samples into, 211–212
DSC purity determination
　parameters during, 38–39
　simplicity of, 39
DSC purity technique, examples of, 41

DSC runs, 155, 156
DSC sensor assembly, 23
DSC systems, suppliers of, 223
DSC temperature scan series, 141
DSC T_g values, 476–477
DSC traces, evaluating, 215
DTG curve, 260, 272
Dual-coated optical fibers, adhesion in, 371–374
Dual-cure mechanism, of thermosetting materials, 131
Dual/single cantilever geometry, 459
Dutch Society for Thermal Analysis (TAWN), 223–224
Dynalyser, 487
Dynamic dielectric experiments, 567
Dynamic DMA scans, 452
Dynamic glass transition process, 514
Dynamic heterogeneity, in miscible polymer blends, 573–574
Dynamic Hi-Res software, 265, 276, 305. *See also* Dynamic rate Hi-Res™ TGA
Dynamic mechanical analysis (DMA), ix, 1, 3, 134, 387–495, 498, 615. *See also* DMA entries
  applications of, 388, 410–424
  in characterizing viscoelastic materials, 396–399
  determining shrinkage using, 367–369
  determining T_g by, 411–414
  in developing cure parameters, 445–456
  finding time to reach gelation using, 443
  rheology and, 399–401
  rheology mode in, 445–446
  in the solids mode, 446–448
  viscoelastic analysis for, 457–458
Dynamic mechanical analyzer, compliance calibration of, 460
Dynamic-mode DMA measurements, 367–368
Dynamic rate Hi-Res™ TGA, 247. *See also* Dynamic Hi-Res software; Thermogravimetric analysis (TGA)
Dynamic viscosity data, 401
Dynamic viscosity profiles, 445

Eastern Analytical Symposium, 2
Edge capacitance, 535
Educational programs, in thermal analysis, 1–2
Effective crosslink density, 355–356
Effective temperature of polarization, 537
Eigendeformation calibration, 332
Elastic/elasticity modulus, 322, 356, 390
Elastic response region, 395
Elastic solids, ideal, 390–393
Elastomeric modifiers, 583
Elastomers, DMA instruments and, 467
Electrical effects, nonlinear, 500–501
Electrical loss modulus, 506
Electric modulus, 506
Electrode diameters, measurements of, 535
Electrode material, in dielectric experiments, 532–533
Electrode polarization, 532, 581
Electrode system, parallel-plate, 529–530
Electromagnetic fields, undesired, 251
Electronic microbalances, 224–225
Electronic polarization, 512
Electronic thermocouple calibration, errors in, 462–463
Electron impact (EI), 249–250
Electrostatic disturbances, 251
Empirical evaluation, 180–181
Endothermic hysteresis peaks, 62, 66–67, 71
Endothermic reaction, temperature and, 252
Endotherms, 18
  multiple melting, 110
Energy barrier, of α- and β-relaxation processes in PMMA phases, 550
Engineering applications, dielectric characterization of polymers for, 501
Engineering strain, 390
Enthalpies, 13–14, 94
  glassy and melt, 63–64
  of relaxation, 185
Enthalpy calibration, of DSCs, 50
Entropy, 14
Entropy jump, 94

Environment temperature, 56
Epoxies, 131
  high-performance, 135
Epoxy–amine reaction, 137, 147
Epoxy–amine systems, 136
  conversion versus time data for, 140
  isothermal cure of, 195–197
  T_g-conversion relationship for, 143
  vitrification and devitrification during nonisothermal cure of, 198–199
Epoxy–aromatic diamine systems, 135
Epoxy resins, dielectric signatures of water molecules confined in, 591–592
Epoxy systems, multicomponent, 199
Epoxy thermoset, 147
Equilibrium heat of fusion, 17, 97, 98–100
Equilibrium melting, of polymers, 96–97
Equilibrium melting point, 82, 97–98, 107
Equilibrium region, 395
Equilibrium systems, 10
Equilibrium thermodynamics, 9
Equipment, micro/nano-TA, 646
Ethylene copolymers, 111
Ethylene–propylene copolymers, 111
Ethylene–vinyl acetate (EVA), two-step decomposition of, 288
Ethylene–vinyl acetate copolymers, quantitative analysis of, 304–306
Evolved gas analysis (EGA), 249
Exothermic heat of reaction, 136–137
Exothermic hysteresis peaks, 62, 66–67, 76
Exothermic reaction, temperature and, 252
Exotherms, 18
  asymmetric or multiple-reaction, 152
  shape of, 136–137
Expansion behavior
  of coupled UV coatings, 371–373
  of polymers, 363
Expansion measurements, checking performance in, 335
Expansion-mode TMA, 354–356. *See also* Thermomechanical analysis (TMA)
Expansion probes, 333

Experimental AC techniques, for dielectric analysis of polymers, 594. *See also* AC dielectric experiments
Experimental data
  extrapolation of, 474
  mathematical analysis of, 519
Experimental heating rates, 292
Experimental parameters
  for DSC measurements, 214–215
  effects on thermally stimulated current peaks of polymers, 542
EXSTAR 6000 DSC line, performance characteristics of, 220. *See also* Differential scanning calorimeters (DSCs)
EXSTAR 6000 Dynamic Mechanical Spectrometer, 483
Extended-chain (equilibrium) crystals, 85–87, 97
Extended-chain PE crystals, 125. *See also* Polyethylene (PE)
Extent of conversion, 278
External thermal resistances, 19
Extrapolated ending-point of melting, 104
Extrapolated onset temperature, 104
  of melting, 632
Extrinsic charges, migration of, 515

Fahrenheit scale, 11
Fast DSC rates, 163. *See also* Differential scanning calorimeters (DSCs)
Fast heating rates, 32
Fast-reacting thermoset systems, 135
Fast-scan DSC, 8, 162–168. *See also* Differential scanning calorimeters (DSCs)
Fast-scan heat flux DSC, 163
Fast scan mode, 456
Fast-scan power compensation method, 163
Fast scans, improvement in sensitivity with, 166
Fast time-domain dielectric spectroscopy, 525–526
Fiber, drawing conditions of, 434. *See also* Fibers

Fiber crystallinity/orientation, effect on glass transition temperature, 433–434
Fiber macromolecules, structure of, 115
Fiber measurements, calibration standards for, 334–335
Fibers
  characteristics of, 432–438
  differential scanning calorimetry of, 115–123
  DMA measurements on, 433, 467–469
  gel-spun polyethylene, 117–119
  isotactic polypropylene, 119
  liquid crystalline polymeric, 121–123
  polyacrylonitrile, 121
  TMA of, 363–364
Fiber samples, running, 215
Fiber shrinkage, 350
Field-effect transistors (FETs), 531
Fillers, in, polymer containing polar groups-572
Film(s)
  blowing, 433
  differential scanning calorimetry of, 123–130
  DMA measurements on, 433
  liquid crystalline polymer, 128
  Nylon 6, 126–127
  oriented, 352, 353
  polyethylene, 124–126
  poly(ethylene terephthalate), 126
  polypropylene, 642–644
  poly(vinylidene fluoride), 127
  TMA of, 363–364
  ultrathin polymer, 637
Film thickness, 337
Fingerprinting, 115. See also Sample "fingerprint"; TGA "fingerprint"
First heating–cooling–second heating runs, 212–213
First law of thermodynamics, 9–10, 67
First-order phase transitions, 81, 94, 464
First-order transition, 16
Fixed-length measurement, 115–117
Flexible polar sidegroups, 556–557
Flexural modulus, 391–392
Flexural probes, 333
Flexural properties, of polymeric material, 354

Flexural stress, 326
Flory–Huggins solvent–polymer interaction parameter (χ), 356
Flow rate, of purge gas, 34
Flynn–Wall method, 268, 285, 293, 299
Folded-chain lamellae, 87, 107
Force calibration, 332
Force curves, 351
Force motors, computerized, 328
Fourier transform (FT), generating, 174
Fourier transform infrared (FTIR) spectrometer, 256. See also FTIR gas analysis
Fourier transform infrared spectroscopy, 249
Fox equation, 78
Fractional polarization, 528
Fracture behavior, 428
Fragility index, 552, 570
Free enthalpy, 15
Free-radical systems, UV cure of, 159–160
Free-to-shrink DSC melting curves, of Nylon 6 films, 129. See also Differential scanning calorimetry (DSC)
Free-to-shrink measurements, 115–117
Free volume (v_f)
  estimating, 362
  measuring, 321–322
Free-volume approach, 344
Free-volume dilatation coefficient, 552
Free-volume theory, 322–324
Freezing point, 17
Frequencies, range of, 534–535
Frequency domain, dielectric spectrometers for measurements in, 593–595
Frequency effects, 184, 459–460
Frequency measurements, multiple, 473
Frequency response analysis (FRA), 521–523
Fringed micelles, 87
FRS5 sensor, 27
FTIR gas analysis, 250. See also Fourier transform infrared entries
Furnace temperatures, 335
Fusible-link technique, 254–255

Gallagher, Patrick K., 1, 241
Gallium, as a melting point standard, 45
γ relaxation, 513, 555, 578, 581
 in oxide polymers, 559–560
 in polyamides, 571
Gas chromatography–mass spectrometry (GC-MS), 645
Gases, polymer-evolved, 250
Gas-switching accessories, 259
Gaussian functions, 543
Gelation, 356
 mixing of reactants and, 445
Gelation curve, 441
Gelation time, relationship with ionic conductivity, 588–589
Gel point, 131, 356, 442, 443, 445
Gel point conversion, 443
 degree of, 444
Gel-processing technology, for PE films, 126
Gel-spun fibers, 432
 polyethylene, 117–119
$_{gel}T_g$ temperature, 133, 441, 455. See also Glass transition temperature (T_g)
Gel time, 133, 326
Geometries, correction factors for, 470
G/E ratio, 470
Gibbs free energy, 15
Gibbs free energy curves, 81–82
Glasses, physical aging and annealing of, 73–76
Glass → melt transition, 61
Glass-reinforced thermoplastic, 377
Glass–rubber transition, 424
 in linear or high-density polyethylene, 428
Glass transition, 58–81
 of amorphous polymers, 184–187
 characteristic features of, 415–417
 DSC curve of, 184–185
 effect of crosslink density on, 438–439
 effect of crosslinking on, 77
 effect of low-molecular-mass additives on, 77
 effect of orientation on, 76
 effect of pressure on, 76
 identifying, 411
 measurement techniques for, 65–66
 of nanoprobe-analyzed polymeric materials, 639
 polyblend miscibility and, 77–80
 of a polymer, 69–70
 in polymer blends and copolymers, 77
 in semicrystalline polymers, 68–73
 step scan DSC monitoring of, 205–207
 temperatures necessary to define, 62
 for ultrathin polymer films, 637
 viscoelastic response and, 389
Glass transition regions
 broad, 564
 overlapping, 195
 stress effects in, 186
Glass transition signal, 562
Glass transition temperature (T_g), 17–18, 59, 61–68, 133–134, 138, 326, 514. See also $_{gel}T_g$ temperature; T_g entries
 assigning for highly crystalline polymers, 418–419
 broadening of, 417, 418
 compositional variation of, 576
 criterion for specifying, 413–414, 439–440
 determining by DMA, 411–414
 factors influencing measurement of, 414–423
 frequency dependence of, 417
 importance of, 59–60
 measurement with modulated temperature DSC, 80–81
 of miscible polymer blends, 78
 molecular mass and, 65, 66
 suppression of relaxation at, 417–418, 419
 in TGA experiments, 263
 in TMA experiments, 338
Global TSC experiments, measuring conditions in, 536–538. See also Thermally stimulated current (TSC) technique
Global TSC spectrum, features of, 538–540
Goldstone mode, 565
Gorbunov, Valeriy V., 615
Gordon–Taylor equation, 78
Graft ratios, 80
Grandy, David, 615

Haake MARS rheometer, 487–488
Hamon procedure, 521
Handbook of Chemistry and Physics, 335
Hard-core volume, 321, 322–323, 326
Hartman equation, 362–363
Havriliak–Negami function, 514, 543
Havriliak–Negami model, 505
Havriliak–Negami parameter estimates, 549
HDPE fibers, shrinkage of, 352. *See also* High-density polyethylene (HDPE)
Heat, 11–13
  defined, 9
  latent, 13
Heat capacity, 15–16
  in the two-phase model of semicrystalline polymers, 71–72
Heat capacity at constant pressure, 53
  DSC runs for, 57
Heat capacity at constant volume, 53
Heat capacity calibration, of DSCs, 50
Heat capacity change, evolution in, 200
Heat capacity determination
  accuracy of, 56
  using MTDSC, 192
Heat capacity increase, 71
Heat capacity measurement(s), 52–58, 203–204
  on cooling, 57
  with modulated temperature DSC, 58
  software for, 55
  with traditional DSC, 54–57
Heat capacity signal, 41, 177, 196, 198
  excess contribution in, 202–203
Heat conduction, 12–13
Heat convection, 13
Heat–cool–heat protocol, 329
Heat deflection temperature (HDT), 345, 377
Heat dissipation, 624, 625, 626, 627
  theories of, 627
Heat distortion temperature (HDT), 377. *See also* Heat deflection temperature (HDT)
Heat flow, 12–13. *See also* Heat flux entries
Heat flow cycle, 181
Heat flow equation, four-term, 24–25
Heat flow modulation, 181

Heat flow phase angle signal, 195–196
Heat flow rate, 26, 30
Heat flow signal, 41
  terms of, 173
Heat flux, effect of sample topography on, 619
Heat flux DSC, 8, 18, 21–22. *See also* Differential scanning calorimetry (DSC); Heat flux DSCs
Heat flux DSCs, 216. *See also* Differential scanning calorimeters (DSCs)
  advantage of, 36
  glass transition in, 59
Heat-hold-cool DSC measurements, 149. *See also* Differential scanning calorimetry (DSC)
Heating
  in DSC measurements, 212–213, 214
  glass transition on, 64–65
Heating calibration, 41–46
Heating experiments, temperature calibration methods in, 334
Heating program, for thermogravimetric analysis, 260
Heating rate(s), 27–28, 31–32, 46, 137, 152
  in a commercial DSC, 163
  in a DSC purity determination, 38
  effect of, 461–464
  effect on indium melting peak height, 164, 165
  effect on ordinate displacement of DSC signal, 163–165
  fast, 253, 282, 292
  influence of, 252
  in kinetic analyses, 287
  loss modulus peaks and, 451–452
  in TMA experiments, 337–338
  during TSC experiments, 537
Heating rate methods, multiple, 140–142, 289–290
Heat leak, 20
Heat of crystallization, determination of, 91–92
Heat of fusion, 17, 104, 212
  determination of, 96
  of PET film samples, 128
Heat of fusion calibration, 42, 50

Heat of reaction, 136–137
Heat of transition, 13, 61
Heat-only mode, 177
Heat-with-some cooling mode, 177
Height:width ratio, for a high-purity metal standard, 224
Helium, as a purge gas, 33, 157, 167, 172, 219, 220, 241, 252, 289, 328, 336, 364–366, 371–373
Helmholtz free energy, 15
Hermetically sealed DSC pans, 210–211. *See also* Differential scanning calorimeters (DSCs)
Hermetically sealed pans, 35, 58
Heterogeneous nucleation, 85, 89
High-density polyethylene (HDPE), glass–rubber transition in, 428. *See also* HDPE fibers
Higher force tabletop systems, 477
High-frequency dielectric data, 535
High-frequency (dipolar) dielectric response, changes in, 585
High-frequency permittivity, measuring, 525
Highly crystalline polymers, 418–420
High-pressure dielectric analyzer, 530
High-pressure DSC capsules, 35–36. *See also* Differential scanning calorimeters (DSCs)
High-purity metals, melting points of, 254
High-temperature interfacial polarization modes, analysis of, 499
High-T_g phase, vitrification of, 200–201
High-thermal-conductivity materials, 627
High-thermal-conductivity purge gas, 289
Hildebrand's rule, 94
Hoffman–Weeks method, 98, 99, 107
Homogeneous nucleation, 85, 89
Homopolymers, 113–114
Hooke's law, 390, 391, 396
HP-53 liquid crystal standard, 50
HP DSC827^e module, 217–218. *See also* Differential scanning calorimeters (DSCs)
HSS7 sensor, 27
Hybrids, incorporating thermoplastics, 572
Hydrogel polymers, 354

Hydrogels, mechanical measurements on, 355
Hydrogen bonds, modification by moisture sorption, 571
Hydrophilic polymers, 269
Hygroscopic expansion, 355
HyperDSC^{TM}, 163. *See also* Differential scanning calorimeters (DSCs)
Hysteresis peaks, 62, 66–67, 71, 75–76, 194
  absence of, 72k
  intensity of, 184

ICTAC Certified Curie Temperature Reference Materials, 253. *See also* International Confederation of Thermal Analysis and Calorimetry (ICTAC)
Ideal elastic solids, 390–393
Ideal liquid behavior, 393–394
Ideal sample capacitance, 535
IDEX series sensors, 531
Immiscible blends, interfacial polarization in, 575
Impact behavior, 428–432
Impedance analysis, 521
Impedance analyzers, 523, 595
Impedance bridges, alternating-current, 523
Incompressible volume, 322–323
Indium
  crystallization of, 83
  as a melting point standard, 44–45
Indium films, 254
Industrial applications, of thermomechanical analysis, 363–379
Infrared (IR) analysis, 308
Infrared cure, 274
Infrared microspectrometry, 644. *See also* Near-IR (NIR) spectroscopy; Real-time infrared (RT-IR) spectroscopy
Initial rise method, 543
Injection-molded GFPPS components, thermomechanical tests on, 378
Inorganic compounds, in calibration experiments, 44
Instrumental baseline, 36

Instrumental factors, in measuring T_g, 414
Instrumentation. *See also* Instruments
 DEA, 592–599
 DMA, 477–488
 DSC, 217–225
 Micro/nano-TA, 646
 thermal analysis, 1, 2–3
 TGA, 308–311
 TMA, 326–332
Instrument baseline, checking, 216
Instrument calibration, 460–465
Instrument constant, 50, 56
Instrument manufacturers, 3
Instruments. *See also* Instrumentation
 calibration of, 41
 drive systems used for, 478
 turning off, 216
Instrument Specialists I Series DSC, 217. *See also* Differential scanning calorimeters (DSCs)
Instrument Specialists iSeries TGA, 309. *See also* Thermogravimetric analysis (TGA)
Integral isoconversional methods, 288–289
Integrated circuit packages, bowing in, 374–375
"Interchain" effect, 539–540
Interdigitated electrodes, 530–531
Interfacial polarization, 515–516, 518
Interfacial polarization mode, 558
Intermolecular interaction strength, 574–575
Internal plasticization effect, 556
Internal thermal resistance, 19
International Confederation of Thermal Analysis and Calorimetry (ICTAC), 6, 284. *See also* ICTAC Certified Curie Temperature Reference Materials
International Temperature Scale, 4
International Union of Pure and Applied Chemistry (IUPAC), 6
Interpenetrating polymer networks (IPNs), 576
Intersection of tangents method, 445
"Intrachain" effect, 539
Intrinsic charge carriers, 581

Intrinsic charges, migration of, 515
"Introduction to Thermosets," 131
Ionic conductivity
 relationship with degree of conversion, 589–590
 relationship with gelation time, 588–589
 relationship with viscosity, 588
 of thermosetting polymers, 586–587
Ionic impurities, direct-current conduction of, 583
Ion mobility, evaluation of, 500
Ion transport, dynamics of, 521
Irradiation, baseline shift caused by, 155
Irradiation curing, 161–162
Irradiation cycles, repeating, 156
Irreversible dimensional change, 329
Irreversible processes, 14
Irreversible shrinkage, 350
Isochronal DEA spectra, features of, 547–549. *See also* Dielectric analysis (DEA)
Isochronal DMA scan, 450. *See also* Dynamic mechanical analysis (DMA)
Isochronal experiments, 502
 on amorphous polymers, 423
 relaxations for, 410
Isochronal plots, 509, 533
Isochrone, 475
Isoconversional analysis, 145
Isoconversional gelation, 132
Isoconversional kinetic methods, 140
Isoconversional methods, 285
 advances in, 288–289
 situations encountered when applying, 287–288
Isoconversional temperature-heating rate data, 152
Isostructural state, 535
Isotactic nonpolar thermoplastics, 555
Isotactic polypropylene fibers, 119
Isothermal baselines, starting or ending, 214
Isothermal charging/discharging currents, 520–521
Isothermal cold crystallization, 92

Isothermal crystallization, 84
  DSC recording of, 90
  measurements of, 216
Isothermal crystallization curve, 90
Isothermal cures, 445–450
  conductivity behavior during, 590
  of epoxy–amine systems, 195–197
Isothermal DEA spectra, features of, 547–549. *See also* Dielectric analysis (DEA)
Isothermal dielectric frequency scans, 516
Isothermal kinetic analysis, 283
Isothermal kinetic measurements, 138–140, 146–147
Isothermal kinetics, traditional, 145, 146–147
Isothermal measurements, 4
  baseline quality in, 216
Isothermal melt crystallization, 88–90
Isothermal plots, 502, 509
Isothermal polarization temperature, 539
Isothermal runs, 155
  kinetic analyses based on, 280–283
Isothermal scans, 502, 547
Isothermal stepwise crystallization, 126
Isothermal temperatures, in TGA experiments, 262
Isothermal TGA, of a phenolic-containing system, 272–273. *See also* Thermogravimetric analysis (TGA)
Isothermal TGA experiments, 266–269
Isothermal TGA mass loss curves, 267
Isothermal time–temperature transformation (TTT) cure diagrams, 441, 444
Isotropic film, 124
Isotropic phase, 48

Jade DSC, 219. *See also* Differential scanning calorimeters (DSCs)
Jaffe, Michael, 319
Johari–Goldstein mode, 513
Judovits, Lawrence, 7

Kalogeras, Ioannis M., 497
Kevlar®, 121, 123

Kinetic analyses
  applications of, 290–295
  based on a single nonisothermal run, 283–284
  based on isothermal runs, 280–283
  based on multiple nonisothermal runs, 284–290
  lifetime projection and, 299–300
  objective of, 280
Kinetic curves, 280
Kinetic data, from isothermal measurements, 236
Kinetic free volume fraction, 552
Kinetic information
  continuous, 248
  importance of, 277
Kinetic predictions, 292–295
Kinetic rate equation, 193
Kinetics, in thermogravimetric analysis, 277–295
"Kinetic triplet," 279, 287, 288
  evaluating, 283
Kinetic viscoelasticity, 451
Kissinger method, 284

Lab environment temperature, 45
Lamellar crystals, 107
Large sample mass, 259
Latent heat, 13
Laws of thermodynamics, 9–10
L-DTA signal, 631. *See also* Differential thermal analysis (DTA)
"Leading edge," 43, 95, 96
Lexan polycarbonate, water permeability of, 303
Lifetime estimates, for polymers, 295–301
Lifetime predictions, for polymeric materials, 360–361
Light guide, 159
Light meter, 159
Linear amorphous polymers, 388
  transition region for, 413
Linear block copolymers, 565
Linear growth rate, of crystals, 85
Linear hydrogen-bonded mainchains, polymers with, 565–572
Linear isobaric expansivity, 321

Linear low-density polyethylene (LLDPE), 111
Linear macromolecules, equilibrium melting point of, 98
Linear polyethylene, glass–rubber transition in, 428
Linear stress–strain behavior, 392
Linear viscoelastic range, 398
Linear voltage differential transformer (LVDT), 327
Linseis thermogravimetric analyzers, 309
Liquid behavior, ideal, 393–394
Liquid crystalline copolyesters, rigid aromatic, 564
Liquid crystalline polymer films, 128
Liquid crystalline polymeric fibers, 121–123
Liquid crystalline polymers (LCPs)
  dielectric analysis of, 562–565
  ferroelectric properties of, 565
  thermal expansion of, 353–354
Liquid crystalline standards, 46, 48, 49–50
Liquid crystals, classes of, 47
Liquid cylindrical sample cells, 595
Liquid–liquid transition, in polymers, 557
Liquid nitrogen, quenching with, 93–94
Liquid-nitrogen-based cooling system, 223
Liquid parallel-plate sample cells, 595
Liquids, dynamic mechanical analysis of, 399–401
Liquid thermoset samples, 466
Lissajous figures, 182–183
"Living" polymerization, 161
Load-bearing applications, 350
Local chemical analysis, 644–646
Local evolved gas analysis (L-EGA), 645
Local thermal analysis (L-TA), 620, 621, 628–633. *See also* L-TA entries
  overview of, 644–646
  use for, 630
Lock-in amplifier, 631
Log plots, 587
Long-range charge migration, mechanisms related to, 515–516
Loss factor, 582
Loss factor signal, 577, 579

Loss modulus, 398, 401
  compliance limits and, 461
Loss tangent (tan δ), 398, 399, 476, 506. *See also* Tan δ entries
Low-density polyethylene (LDPE), 102–103, 110, 111
  crystallinity of, 126
  isothermal stepwise crystallization of, 126
  melting curves of, 109
  in TGA experiments, 263–264
Low-force tabletop systems, 477
Low-frequency dielectric data, 535
Low-frequency dielectric studies, 586
Low-molecular-mass additives, effect of glass transition, 77
Low-molecular-mass components, 307
  in polymers, 269–271
  purity determination of, 37–41
Low-molecular-mass substances
  Avrami equation for, 88
  crystallization of, 82–83
  melting of, 94–96
Low-temperature heat capacity measurements, 55
Low-temperature measurements, 217
L-TA crater, 641. *See also* Local thermal analysis (L-TA)
L-TA experiments
  components of, 636
  temperature gradients in, 635–636
L-TA power signal, 631
L-TMA signal, 631. *See also* Thermomechanical analysis (TMA)
Lumped circuit methods, dielectric analysis based on, 521–524

M24 liquid crystal standard, 50
Machine compliance, errors due to, 460–461
Macromolecular systems, impurity concentrations in, 37
Macromolecules, crystallization of, 81
Macroscopic gelation, versus molecular gelation, 131–132
Magnetic standard, 255
Mainchain LCPs, 564. *See also* Liquid crystalline polymers (LCPs)

INDEX   669

Mainchain segmental motions, dielectric study of, 575
Mass, 6
　in thermogravimetric analysis, 251
Mass fraction crystallinity, 99
Mass loss, 241–242
　in TGA experiments, 262–263
Mass loss curves, 305–306
　decomposition kinetic analysis of, 299
Mass loss processes, isothermal experiments for, 266, 267, 269, 271–274
Mass measurements
　accuracy of, 256
　routine for, 245–246
Mass spectrometer (MS), 256
Mass spectroscopy (MS), 249
Mass standard, 45–46
Mass-to-charge ratio (m/z), 249
Master cure curves, 145, 149, 150
Master curves, 361, 404–406
　shifting, 407
Material characteristics, in measuring T_g, 417–423
Material loss factor, 398
Material properties
　determining key aspects of, 529
　measuring, 339
Materials
　thermal properties of, 626
　selection of, 428
　viscoelastic, 394
Mathematical analysis, of experimental data, 519
Matrix, in polymer blends, 640–642
Maximum heat-only mode, 178, 179
Maximum loss modulus, 412
Maxwell–Wagner–Sillars (MWS) polarization, 515–516, 518, 519, 567, 592
MDSC2920 flow diagram, 175. *See also* Differential scanning calorimeters (DSCs); Modulated DSC (MDSC®)
MDSC diagram, 176
Measured capacitance, 535
Measurement times, high scan rates and, 167
Measuring systems, with AC bridges, 523

Mechanical energy input, influence on glass transition, 415
Mechanical properties, studying using TMA, 355
Melt crystallization, 58–59, 84
　isothermal, 88–90
　nonisothermal, 91–92
Melt → crystal transition, 108
Melt → glass transition, 61
Melting
　of copolymers, 110–115
　of crystalline macromolecules, 96–98
　in differential scanning calorimetry, 94–115
　of low-molecular-mass substances, 95–96
　recrystallization during, 105
　reorganization and recrystallization during, 190–191
Melting behavior
　of nanoprobe-analyzed polymeric materials, 639
　of Nylon 6 fibers, 121
Melting curves, 45, 102, 104–105
　comparative, 40
　of gel-spun polyethylene fibers, 117, 118
　of low-density polyethylene thin film, 127
　of polyethylene films, 125
Melting peaks, 38, 105, 109–110
　for gel-spun polyethylene fibers, 117–119
　of a low-molecular-mass crystalline substance, 42–43
Melting point(s), 16–17, 43, 103
　determination of, 340, 343
　for tau lag adjustment, 52
Melting point lowering, of a low-molecular-mass material, 37
Melting point standard, 255
Melting processes, step scan DSC for studying, 207–208
Melting transition, analysis by MTDSC, 189–190
Menczel, Joseph D., 1, 7, 319, 387
Metallized polymers, TSC measurements of, 537
Metal powders, 44

Metal standards, 44–45
Metastable crystals, heating of, 106–107
Metastable glasses, 73
Metastable polymeric crystallites, 190–191
Methacrylate polymers, 556
Methylenedianiline (MDA), 135
Mettler Toledo, Inc., 478
Mettler Toledo DMA SDTA 861, 478–479. *See also* Dynamic mechanical analysis (DMA)
Mettler Toledo DSC, 25–27, 50–52. *See also* Differential scanning calorimeters (DSCs)
Mettler Toledo DSC1 unit, 217–218
Mettler Toledo TGA/DSG1, 309. *See also* Thermogravimetric analysis (TGA)
MHTC DSC, 221. *See also* Differential scanning calorimeters (DSCs)
Microanalogs, of traditional thermal analysis techniques, 628–629
Microbalances, 224–225
Microdielectrometer sensor, 531
Micro DSC III/VII, 221. *See also* Differential scanning calorimeters (DSCs)
Micro-DTA results, 629, 630. *See also* Differential thermal analysis (DTA)
Micromet Instruments, 531, 593–595
Micro/nanoscale local thermal analysis, applications of, 637–644. *See also* Micro/nano-thermal analysis (μ/n-TA); Microscale thermal analysis; Nanoscale local thermal analysis; Thermal analysis (TA)
Micro/nanoscale local thermal analysis experiment, 633–636
Micro/nano-TA equipment, 646. *See also* Micro/nano-thermal analysis (μ/n-TA)
Micro/nano-thermal analysis (μ/n-TA), ix, 1, 3, 630. *See also* Micro/nanoscale local thermal analysis entries
Microscale thermal analysis, 615–649. *See also* Micro/nanoscale local thermal analysis; Micro/nano-thermal analysis (μ/n-TA)

Micro-TMA results, 629, 630. *See also* Thermomechanical analysis (TMA)
Microwave monitoring techniques, 585–586
Microwave techniques, 524
Miniaturized resistive probes, 620
Miscibility
 in polyblends, 77–80
 relationship with structural similarity and secondary interaction forces, 573–574
Miscibility gaps, 202
Miscible blends, intermolecular interaction in, 575
Miscible interpenetrating polymer networks, 576
Mixed-mode deformation, 470
Mobility factor (MF), 154, 193–194
Model-free kinetics (MFK), 140, 145, 151, 153
Model-free kinetics software, 142, 151, 153
Model-free predictions, 294–295
Modeling, 627
Modified WLF equation, 591. *See also* Williams–Landel–Ferry (WLF) equation
Modulated DSC (MDSC®), 168, 171. *See also* Differential scanning calorimetry (DSC); MDSC entries; Modulated temperature DSC (MTDSC)
Modulated heat flow, amplitude of, 58
Modulated techniques, 5
Modulated temperature DSC (MTDSC), 8, 134, 133, 161, 168–208, 630. *See also* Differential scanning calorimetry (DSC); MTDSC entries; Temperature-modulated differential scanning calorimetry (TMDSC)
 advantages of, 177, 184
 background and principles of, 168–170
 in characterizing reacting polymer systems, 203
 crystallization measured by, 187–191
 disadvantages associated with, 183
 as an extension of DSC, 183
 heat capacity measurement with, 58

INDEX 671

measurement of glass transition temperature with, 80–81
Modulated temperature TMA (MTTMA), 329–332. *See also* MTTMA signals; Thermomechanical analysis (TMA)
Modulated thermogravimetric analysis (MTGA), 248–249. *See also* Thermogravimetric analysis (TGA)
Modulation amplitude, 178
   change in, 180
Modulation cycles, 181
Modulation frequencies, range of, 180
Modulation parameters, verification of, 181
Modulus accuracy, 460
Modulus–log time curve, 401–402, 404
Modulus of elasticity, 390–393
Modulus–temperature curve, 404
Modulus–temperature data, 434–436, 437
Modulus values, 472
Moisture
   effect on T_g values, 422–423
   effects of, 424–428
   as a plasticizer, 426
Moisture absorption, effects on polymer relaxation activity, 517–518
Moisture content, of samples, 339
Moisture sorption, effect on hydrogen bonds, 571
Molds, casting, 465–466
Molecular dipole vectors, orientation of, 512
Molecular gelation, versus macroscopic gelation, 131–132
Molecular/ionic species, movements of, 498
Molecular mass, effect of, 343
Molecular mobility, dynamics of, 521
Molecular motions
   relaxation characteristics of, 556
   thermal-analysis-related spectroscopic studies of, 563
Molecular relaxation processes, 387
Monofilaments, 432
MTDSC experiments, 105, 169–170, 214–215. *See also* Modulated temperature DSC (MTDSC)

modulation selections for, 178
sample mass in, 179–180
MTDSC parameter selection, for thermosets, 180
MTDSC runs, good, 180–183
MTDSC terminology, 176–177
MTTMA signals, 331. *See also* Modulated temperature TMA (MTTMA)
Multifrequency DEA data, 549
Multifrequency DMA data, 449, 451
Multifrequency measurements, 509
Multiphase polymer blends, 573
Multiple heating rate, 284
Multiple heating rate kinetics, 145, 151–153
Multiple heating rate methods, 140–142, 289–290, 293–295
Multiple melting endotherms, 110
Multiple nonisothermal runs, kinetic analyses based on, 284–290
Multiple-peak melting, of polymers, 107–110
Multiple reaction exotherms, 152
Multiple temperature programs, 284
Multipurpose dielectric systems, 595
Multistage degradation, 305

Nanoconfinement effects, dielectric analysis of, 575–576
Nanoconfinement-induced effects, 499–500
Nanometer-level accuracy, 617
Nanophase separation, 570
Nanoprobe-analyzed polymeric materials, glass transition and melting behavior of, 639
Nanoprobes, 634
Nanoscale local thermal analysis, 615. *See also* Microscale thermal analysis
Near-IR (NIR) spectroscopy, 592. *See also* Infrared entries
Nematic → isotropic transition, 47, 224
Nematic liquid crystals, 47
Nematic phase, 48
Nematic → smectic transition, 563
Neoprene rubber, TMA information obtained from, 346
Network analysis, 524, 525

Network-forming materials, 130, 131
Network-forming polymers, 366
Netzsch DMA 242 C, 479–480
Netzsch instrumentation, 218
Netzsch Instruments, Inc., 479, 531, 593–595
Netzsch thermogravimetric analyzers, 309–310
Newton's law of cooling, 22
Newton's law of viscosity, 393, 397
Nitrogen, as a purge gas, 33–34
Nomograph
  computerized, 475
  time–temperature superposition, 473–476
Noncrosslinking side reactions, 588
Nonequilibrium melting, of polymers, 100–104
Nonequilibrium polymer crystals, free energy of, 82
Nonequilibrium thermodynamics, 9
Nonisothermal cold crystallization, 92–94
Nonisothermal conditions, kinetic measurements performed under, 283–284
Nonisothermal cure, 450–456
  to characterize a thermoset, 194
  vitrification and devitrification during, 198–199
Nonisothermal cure experiments, 180
Nonisothermal curves, 283
Nonisothermal melt crystallization, 84, 91–92
Nonlinear electrical effects, analysis of, 500–501
Nonlinear optical effects, analysis of, 500–501
Nonmiscible polyblends, 77–78
Nonpolar polymers, simple, 554–556
Nonpolar thermoplastics, 555
  dielectric analysis of, 554
Nonreversing heat flow, 172, 197
Nonreversing signal, 329
Nonwoven polymer matrices, 334
Normal-mode relaxation process, 515
North American Thermal Analysis Society (NATAS), 2
North American Thermal Analysis Society Conference, 163

Novocontrol Instruments, 524, 530, 531, 593
Novocontrol spectrometers, 520
Novocontrol TSDC system, 598
Novocool cryosystem, 595
Novolac, 269, 270
Novolac–anhydride system, dynamic mechanical behavior of, 454–456
Novotherm system, 595
nth-order kinetics, 138, 146
Nucleation, 84–85
Nucleation types, Avrami exponent for, 89
Numerical differentiation, 284, 286
Numerical integration methods, 153
Nylon(s), 121
  dielectric spectra of, 571
  melting peak of, 122
  sorption of water by, 424–425
  water as an antiplasticizer for, 430–432
Nylon 6, 571
Nylon 6–6, major relaxations in, 424
Nylon 6 films, 126–127
  free-to-shrink DSC melting curves of, 129
Nylon 11, relaxation dynamics of, 570–571
Nylon 12, 571, 572
Nylon M5T, melting curves of, 108

Occluded phases, in polymer blends, 640–642
Offset, 52
Ohm's law, 22
One-part thermoset systems, 135
Open loop, 28, 30
Optical effects, nonlinear, 500–501
Optical fibers, dual-coated, 371–374
Optical microscopy, 85
Optical pyrometry, 4
Order–disorder transition, 517
Organic crystalline compounds, low-molecular-mass, 37
Organic crystals, charge transport and ion mobility in, 500
Organic liquids, dielectric measurements of, 530
Orientation
  effect on glass transition, 76

effect on glass transition temperature, 433–434
Orientation recovery, in a cable jacket, 370–371
O-ring leak rate, 302–303
Oscillating DSC (ODSC), 168. *See also* Differential scanning calorimetry (DSC)
Oscillator strength, 503
Oscillatory shear for fluids geometry, 459
Oven temperature, variations in, 463
Overlapping phenomena, separating, 195
Oxidation/oxidative induction time (OIT), 39, 297
  thermogravimetric analysis for, 265
Oxidative stability, in TGA experiments, 260–265
Oxygen inhibition, 157
Ozawa method, 92

Pan crimping, in a DSC purity determination, 38–39
Pan masses, 55–56
Pan type, in a DSC purity determination, 38
Parallel-plate compression mode, 359
Parallel-plate dielectric test fixtures, 530
Parallel-plate electrode system, 529–530
Partial heating technique, 543, 545
Particulate contaminants, identification of, 642
Particulate fillers, polymers containing, 421–422
Particulate systems, incorporating thermoplastics, 572
Peak assignments, in dielectric spectrum, 518–520
Peak cleaning techniques, 540–543
Peak exotherm temperature, 152
Peak intensity, dependence on rate of cooling, 537
Peak temperature, 632
  of melting, 103, 100, 633
Pendant groups, effect of, 343
Penetration-mode TMA, 358. *See also* Thermomechanical analysis (TMA)
Penetration probes, 333

PEN materials, completely amorphous, 561. *See also* Poly(ethylene-2,6-naphthalate) (PEN); Poly(ethylene naphthalate) (PEN)
Percolation theory approach, 589
Perkin-Elmer (PerkinElmer), 225
Perkin-Elmer DMA7e, 368
Perkin-Elmer DMA 8000, 480–481
Perkin-Elmer DPA7 double-beam photocalorimeter accessory, 157
Perkin-Elmer DSC1 power compensation instrument, 27. *See also* Differential scanning calorimeters (DSCs)
Perkin-Elmer Hyper® DSC, 32. *See also* Differential scanning calorimeters (DSCs)
Perkin-Elmer Life and Analytical Sciences, 480
PerkinElmer Life and Analytical Sciences instrumentation, 218–220
Perkin-Elmer power compensation DSCs, 29, 51. *See also* Differential scanning calorimeters (DSCs)
  glass transition in, 59
Perkin-Elmer thermogravimetric analyzers, 310
Permanent set, 359, 395
Permeability, calculation of, 303
Permittivity, 503
  frequency dependence of, 547
Permittivity components, temperature dependence of, 509
Permittivity data, constructing Arrhenius plots from, 550
Permittivity plots, 505–506
PET film, 126. *See also* Poly(ethylene terephthalate) (PET)
  expansion and penetration measurements on, 347
PET materials, completely amorphous, 561
PET yarn, storage modulus of, 435
Phase lag, 179
  changes in, 176
  correction for, 177
Phase-separated interpenetrating polymer networks, 576

Phase separation, 200–201
  indication of, 202–203
  information on, 569–570
  reaction-induced, 199–203
Phase transitions, 16
  in amorphous and crystalline polymers, 58–115
Phenolic bonding compound, mass loss from, 272
Phenolic resin systems, cure of, 278
Photocalorimetry, differential, 154–162
Photocalorimetry accessories, 156
Photochemical reactions, 155
  correction technique for tracking, 155
Photodegradation, of polycarbonate, 637–639
Photo-DSC, 155. See also Differential scanning calorimeters (DSCs)
Photoinitiated free-radical systems, 160
Photoinitiated polymerization shrinkage, 367–369
Photoinitiated systems, crosslinking of, 154
Photooxidation, of polycarbonate, 369–370
Photopolymerization, cationic, 161–162
Physical aging process, 406
  analysis of, 185
  of glasses, 73
Physical processes, online monitoring of, 500
Pigment volume concentration (PVC), 276–277
Planar sensors, 577
Plasticization effects, 565
Plasticizer efficiency, 359–360
Plasticizers, 77
  effect on T_g values, 422–423
  effects of, 424–428
Plastic materials, CLTE and T_g of, 379
Platelet fillers, 422
Platinum resistance thermometers (Pt sensors), 27
PMMA films, thick, 547. See also Poly(methyl methacrylate) (PMMA); Thick PMMA films
Polarization
  of a dielectric, 512
  interfacial, 515–516

Polarization optical microscopy, 85, 98, 100
Polar mainchain, polymers with, 559–565
Polar polymers, 427
Polar sidegroups, 557–559
Poly(2-hydroxyethyl acrylate) (PHEA), 356
Poly(2-pentamethylene terephthalamide), equilibrium melting point for, 99
Poly(3-hydroxybutyrate-co-hydroxyvalerate), melting behavior of, 208
Polyacrylonitrile (PAN), 110–111
Polyacrylonitrile fibers, 121
Polyamides, 121
  dielectric analysis of, 570–572
Polyamid–imide polymer, DMA data for, 426–427
Polyblend grafts, 80
Polyblends, miscibility in, 77–80
Polybutadiene–polyisoprene blends, DSC curves of, 78–79
Poly(butylene terephthalate), nonisothermal crystallization curve of, 91, 92
Polycarbonate (PC)
  change of glass transition temperature of, 638
  photodegradation of, 637–639
  thermal properties of, 165
  UV-stabilized, 369–370
Polyesters, 120–121, 515
  dielectric analysis of, 560–562
  flexibility of the mainchains of, 561–562
  thermotropic mainchain liquid crystalline, 564
Poly(ether ether ketone) (PEEK), molecular dynamics of, 562
Polyethylene (PE). See also Polyethylenes
  heat of fusion of, 99
  hydrophilic sites in, 269
  LDPE, HDPE, and LLDPE of, 554
  predicting stability of, 297–298
Poly(ethylene-2,6-naphthalate) (PEN), 560. See also PEN materials

Poly(ethylene-2,6-naphthalene dicarboxylate) (PEN), 69, 70
Polyethylene fibers, gel-spun, 117–119
Polyethylene films, 124–126
  gel-processing technology for, 126
Poly(ethylene naphthalate) (PEN), 120. *See also* PEN materials
  degradation of, 281–282
Poly(ethylene oxide) (PEO), 559
  reversing component of, 209
Polyethylenes, crystallization and melting curves of, 112–113
Poly(ethylene terephthalate) (PET), 97–98, 120, 560. *See also* PET entries
  glass transition of, 72, 75
  glass transition parameters of, 69
  heating curve of, 93
  plasticization of, 186
  preventing crystallization of, 166–167
  quenched, 108
  reversible melting of, 190
Polyimide precursors, 272
Polyisobutylene, 555
Polyisoprene (PI)/poly(vinyl ethylene) (PVE) blends, 574
Poly(L-lactide) (PLLA), 207
Polymer blends
  dielectric analysis of, 573–576
  glass transition in, 77
  identifying matrix and occluded phases in, 640–642
Polymer chain, random scission of, 290
Polymer characterization, glass transition temperature and, 64
Polymer composites, charge transport and ion mobility in, 500
Polymer electrolytes, charge transport and ion mobility in, 500
Polymer-evolved gases, 250
Polymer films
  liquid crystalline, 128
  ultrathin, 637
Polymer Handbook, 335, 362
Polymeric coatings, behavior of, 371–374
Polymeric crystallites, 104
  types of, 85–88
Polymeric fibers, liquid crystalline, 121–123
Polymeric films, 123–130

Polymeric material(s)
  absorption and desorption of water by, 270
  compositional analyses of, 274–277
  flexural properties of, 354
  key processes in, 323
  nanoprobe-analyzed, 639
  origin of dielectric response of, 512–518
  service performance of, 360–361
  thermal expansion of, 345
Polymeric stabilizer systems, studying, 297
Polymerization, free-radical, 159–160
Polymerization shrinkage, photoinitiated, 367–369
Polymer Laboratories, 520
Polymer melting
  characteristic temperatures of, 103
  time-dependent processes during, 104–106
Polymer networks, interpenetrating, 576
Polymer oxidation products, 293
Polymer–polymer interactions, 356
Polymer–polymer miscibility, 186
  determining, 78–80
Polymer properties, free volume and, 322
Polymer relaxation activity, effects of moisture absorption on, 517–518
Polymers. *See also* Amorphous polymers; Semicrystalline polymers
  analyzing dielectric spectra of, 538–553
  characteristic parameters of, 362–363
  compositional heterogeneity of, 369–370
  configuration/dynamic properties/applications relationships in, 499
  containing particulate fillers, 421–422
  cooling rate for, 32
  crystallization of, 83–84
  degradation mechanisms of, 261
  dielectric analysis of, 594
  dielectric characterization of, 501
  equilibrium melting of, 96–97
  equilibrium melting points of, 97–98, 111
  fabricated, 319
  with flexible polar sidegroups attached to an apolar mainchain, 556–557

Polymers (cont'd)
    fracture behavior of, 428
    glass transition temperatures for, 68
    heat capacity of, 53–54
    investigating the stability of, 296–297
    lifetime estimates for, 295–301
    with linear hydrogen-bonded mainchains, 565–572
    liquid–liquid transition in, 557
    low-molecular-mass components in, 269–271
    mass loss of, 241–242
    measuring mechanical properties of, 387
    melting points of, 114
    melting transition of, 632
    miscibility of, 185–186
    multiple-peak melting of, 107–110
    network-forming, 131
    nonequilibrium melting of, 100–104
    overlapping transitions in, 170–171
    PAN, 110–111
    peak assignments in dielectric spectrum of, 518–520
    with polar mainchain, 559–565
    with polar sidegroups, rigidly attached to an apolar mainchain, 557–559
    propensity for change of, 162
    structural weaknesses in, 296
    thermal analysis of, ix, 1
    thermal conductivities of, 463
    thermal degradation of, 290
    two-phase model of, 69
    unzipping of, 261
    use of, 65
    viscosity of, 393
Polymer structure–property characterization, 388
Polymer structures, thermoplastic, 560–562
Polymer thermal stability, in TGA experiments, 260–265
Polymer thick films (PTF), 272
    composition and cure of, 273
    cure of, 259
Poly(methyl methacrylate) (PMMA). See also Atactic PMMA; PMMA films; Thick PMMA films
    DRS and TSC data of, 549
    measurements on, 538–553
    partial heating technique in, 545
Poly(n-alkyl methacrylates), 557
Polyoxides, dielectric analysis of, 559–560
Polyoxymethylene (POM), 559
Polyphenylene sulfide, glass-filled, 377–378. See also Poly(p-phenylene sulfide)
Poly(phenylene terephthalamide) (PPT), 121. See also Kevlar®
Poly(p-phenylene sulfide), melting curves for, 188
Polypropylene, 555
    crystallization of, 86–87
Polypropylene fibers, isotactic, 119
Polypropylene films, biaxially oriented, 642–644
Poly(propylene oxide) (PPO), 559
Polystyrene (PS), change of glass transition temperature of, 638
Polystyrene (PS)/PVME blends, 574
Poly(tetrafluoroethylene) (PTFE), 97, 130, 375, 376
    intensity of glass transition in, 418
    relaxation activity in, 555
Polyurethanes, dielectric analysis of, 565–570
Poly(vinyl acetate) (PVAc), 269
Poly(vinyl alcohol) (PVA) fibers, 434
    storage modulus of, 437
Poly(vinyl chloride) (PVC), 77
    degradation of, 261
    plasticized, 359, 360
    relaxations of, 558
    stabilizing additives for, 263
Poly(vinylidene fluoride) (PVDF)
    melting of, 191
    relaxations of, 558
Poly(vinylidene fluoride) crystallites, reorganization of, 192
Poly(vinylidene fluoride) films, 127
Poly(vinyl methyl ether) (PVME)
    degradation of, 297
    oxidative thermal degradation of, 263
    thermal and thermooxidative stability of, 267–268
    thermooxidative degradation of, 259, 269, 292

Position calibration, 335
   in thermomechanical analysis, 332
Postcure, 444
Postmelting baseline, 36, 100, 102
Power compensation DSCs, 8, 27–31, 159, 163, 204–205. *See also* Differential scanning calorimeters (DSCs); Differential scanning calorimetry (DSC)
   advantage of, 36
   burn off procedure for, 216–215
   operation schematic of, 29
Power compensation sample holder, 28
Power-law models, 279
Precision difference method, 526
Predictions
   for long time periods, 405–406
   model-free, 294–295
   TTS-based, 409
Preload force, 466
Premelting baseline, 36, 100, 102
Prepolyimide resin, 366
Pressure
   crystallization control and, 530
   effect on glass transition, 76
Pressure DSC, 8, 217–218. *See also* Differential scanning calorimetry (DSC)
Pressure-sensitive tape, 130
Pretension, 349, 468–469
Pretensioning, 467
Prigogine, I., 9
Primary nucleation, 84
Primary nuclei, 84
Prime, R. Bruce, 1, 7, 241, 319
Principle of energy conservation, 9–10
Probe heat loss "jump," 626–627
Probe normal force, 622
Probe power consumption, measurement of, 628–629
Probes
   design of, 621
   penetration depth of, 636
   spatial resolution of, 622
   types of, 329
   weight of, 335
Probe selection, in a micro/nanoscale local thermal analysis experiment, 634

Probe tip, 621
   geometry of, 622
Program temperature, 43
Projected results, confidence in, 300
Proton NMR spectroscopy, 592
Published literature, comparisons of dielectric spectra with, 519
Pultrusion processes, 149
   time-temperature profile for, 150
Pultrusion systems, 135
Purge gas(es), 4, 22, 33–34, 336
   dry, 212
   flow rate of, 57
   temperature calibration and, 46
   thermal conductivity of, 252
   in TMA measurements, 328
Purge gas options, in TGA experiments, 259–260
Purity determination, of low-molecular-mass compounds, 37–41
Purity software, 38
PVME/poly(2-chlorostyrene) (P2CS) blends, 574–575

Q Series DSCs, 221–223. *See also* Differential scanning calorimeters (DSCs)
   controlled cooling rates for, 223
Q Series modules, 23–25, 176
Quasi-isothermal analysis, 190
Quasi-isothermal experiments, 178–179
Quasi-isothermal mode, heat capacity measurement with modulated temperature DSC in, 58
Quasi-isothermal modulation, 177, 182
Quatro cryosystem, 593, 595, 598
Quenching methods, 93–94
Quick Freeze Deep Etch (QFDE), 94

Rabinowitch model, 154
Radiofrequency (RF) dielectric measurements, 531
Radiofrequency impedance analyzers, 524, 525
Raman microspectrometry, 644
Ramp rate(s), 4, 6, 204, 207–208
   modulation of, 181
Random chain scission, 261

Random copolymers, 111–114
  melting equation of, 114
Rate constant, 282
Rate equation, in the single-step process, 278–279
Rate of conversion, 137, 138–139
  measuring, 146
Rate of photoinitiation, 160
Reaction heat capacity, 196
Reaction-induced phase separation (RIPS), 199–203
Reaction-induced vitrification, 198–199
Reaction order models, 279–280, 281
Reaction rate, conversion dependence of, 279
Reaction temperature, in cationically cured systems, 161
Reactive systems, MTDSC for monitoring, 192–194
Reading, Mike, 7, 615
Real-time infrared (RT-IR) spectroscopy, 159. *See also* Infrared entries
Recorded dielectric plot structure, inspection of, 518
Recording, of well-resolved thermal transitions, 528
Recovery region, 395
Recrystallization, 181
  during melting, 105, 190–191
Reduced-frequency nomograph, 474–475
Reference calorimeter, 30
Reference disk thermocouple, 24
Reference holder, 31
Reference materials, for thermogravimetric analysis, 251–256
Relaxation, normal-mode, 515. *See also* α-relaxation entries; β relaxations; δ relaxation; γ relaxation; Relaxation behavior; Relaxations; Secondary relaxation entries; Stress relaxation entries
Relaxation activity, effects of moisture absorption on, 517–518
Relaxation behavior
  generalizations about types of, 423
  subglassy, 562
Relaxation dynamics, 497–498
  of thermoplastic polymers, 572
Relaxation frequencies, 512, 551

Relaxation function, dielectric experiments involving, 504
Relaxation map(s)
  evaluation of, 598
  of temperature dependencies, 579–578
Relaxation map analysis (RMA), 528
Relaxation mechanisms, dielectric strength of, 540
Relaxation processes, in semicrystalline polymers, 423–424
Relaxation rates, in miscible blends, 575
Relaxations, 410–411
  crystal phase, 517
  dielectric strength of, 544
  of polyethylene, 554–555
  of polyoxides, 559–560
  in semicrystalline polymers, 516–517
  of the TSC spectrum, 539
Relaxation spectrum, damping mechanisms affecting, 422
Relaxation temperatures, dependence on absorbed water level, 425
Relaxation time, 510–511
"Relaxed" permittivity, 503
REOLOGICA Instruments, Inc. US, 486
Reorganization, during melting, 190–191
Reorientational dynamics, in LCPs, 563
Resolution, increasing, 171
"Reversible" expansion/contraction, 355
Reversible melting, 95
Reversible processes, 5, 9, 14
Reversing heat capacity, 171–172
Reversing heat flow, 170, 171–172
Reversing signal, 329
Rheological properties, at gelation, 443
Rheology
  dynamic mechanical analysis and, 399–401
  measurements of, 387
Rheology mode, in dynamic mechanical analysis, 445–446
Rheometers, rotational, 486–488
Richard's rule, 94
Riga, Alan, 241
Rigaku TG/DTA 8120 system, 310. *See also* Differential thermal analysis (DTA)
Rigid amorphous fraction, 71
Rigid amorphous phase, 346, 516

Rigid pendant groups, effect of, 344
Roe–air–Gieniewski (RBG) method, 297–298
Room temperature permittivity plots, 592
Rotational diffusion, 547
Rotational rheometers, 477–478, 486–488
RSA-III rheometers, 523
RT Instruments, Inc., 310, 482
Rubber products, compositional analysis of, 276
"Rubbery flow," 389
Run parameters, 4–5
  for thermoplastics, 177–179

Sample(s)
  applying initial load to, 339–340
  encapsulating, 34–36
  interaction with atmosphere, 5
  irregularly shaped, 337
  size and geometry of, 259
  thermal history of, 339
Sample calorimeter, 30
Sample cell, 595
Sample cycling technique, 364
Sample dimensions, control of, 529
Sample environment, in dielectric experiments, 532
Sample "fingerprint," 212. *See also* Fingerprinting
Sample geometry modes, clamping errors and, 469–470
Sample holder, 31
  coolant of, 46
  maintaining, 215–216
Sample impedance, 523, 524
Sample mass, 4, 179–180
  in a DSC experiment, 33
  in DSC measurements, 211
  large, 259
  in TGA experiments, 257–259
Sample pans, TGA, 257, 258
Sample platform, 326, 328
Sample preparation
  in dielectric experiments, 528–531
  for differential scanning calorimetry, 210–211
  in a micro/nanoscale local thermal analysis experiment, 633–634

thermoset data and, 134–135
  in TMA experiments, 338–339
Sample probes, 326–327, 328
  TMA, 332
Sample requirements, in TMA experiments, 336–337
Sample run, 54
Sample size, in a DSC purity determination, 38–39
Sample strain, 461
Sample temperature, 82, 463–464
  deviations in, 289
Sample test geometries, 458–459
Sample thermocouple, 24
  calibration of, 464–465
Sample thickness, measurements of, 535
Santonox R, 39
Sapphire run, 54
Sartorius, 225
Scanning force microscopy (SFM), 615
Scanning probe microscopes, 616
Scanning rates, in TMA experiments, 337–338
Scanning thermal microscopy (SThM), 618–620
Scanning tunneling microscopy (STM), 616
Scientific instrumentation, evolution of, 3
"Seating" process, 372–373
Secondary crystallization, 90
Secondary-ion mass spectrometry (SIMS), 644
Secondary nucleation, 84
Secondary relaxation(s), 577. *See also* Relaxation entries
  importance of, 429
Secondary relaxation dynamics, 574
Secondary relaxation loss peak, 429
Secondary relaxation processes, in amorphous polymers, 512–514
Secondary transitions, 410
  in amorphous and crystalline polymers, 423–424
Second law of thermodynamics, 10, 11
Second-order transitions, 16, 433
Segmental α-relaxation process, in amorphous polymers, 514–515
Segmental relaxation mode, 516
Segmented polyurethanes, 565

Seiko EXSTAR6000 TGA/DTA series, 310. *See also* Differential thermal analysis (DTA); Thermogravimetric analysis (TGA)
Seiko Instruments, 220, 482
Self-nucleation, 85
Semiconductors, charge transport and ion mobility in, 500
Semicrystalline miscible polymer blends, 575
Semicrystalline PEEK, 562
Semicrystalline polycarbonate, thermal properties of, 165. *See also* Polycarbonate (PC)
Semicrystalline polymers
  α relaxation in, 514
  glass transition in, 68–73
  melting range in, 16
  relaxation processes for, 410, 423–424, 516–517
  softening point for, 346
  transitions in, 60
  melting of, 100
Sensors
  crosstalk between, 20
  implantable, 531
Sensys DSC line, 220–221. *See also* Differential scanning calorimeters (DSCs)
Service performance, of polymeric materials, 360–361
Setaram DSC instrumentation, 220–221
Setaram Instruments, 598
Setaram thermogravimetric analyzers, 310
Shear deformation, 470–471
Shear modulus, 326
"Shear sandwich" experiment, 450
Shear sandwich geometry, 459
Shear strain, 391
Shear stress, 391, 393
Shift factor, 407
Shimadzu instrumentation, 221
Shimadzu TGA-50/50H, 310–311. *See also* Thermogravimetric analysis (TGA)
Shrinkage, 349–350
  of fibers and thin films, 469
  photoinitiated, 367–369

predicting, 369
thermal cure, 366–367
Shrinkage force, 350–351
Sidechain LCPs, 563. *See also* Liquid crystalline polymers (LCPs)
Sidegroups
  elimination of, 261
  flexible polar, 556–557
  polar, 557–559
Sigma-Aldrich, 334
Sigmoidal baseline, 100–102
Silicone gel, heating in nitrogen, 300–301
Silicon nanoprobes, 622, 624
Simple nonpolar polymers, 554–556
Single glass transition criterion, applicability of, 575
Single heating rate methods, 284, 293
Single isotherm mode, 456
Single-relaxation-time mechanism, 507
Single-relaxation-time model, 504
Sinusoidal strain, 396, 397
Slow heating rates, 32
Slow time-domain dielectric experiments, 598
Slow time-domain dielectric spectroscopy, 520–521
Slow time-domain measurements, 536
Small-molecule reactions, Rabinowitch model for, 154
Small sample mass, 630
Smectic $\rightarrow$ isotropic transition, 563
Smectic liquid crystals, 47
Smectic phase, 48
Softening point, 412
  determining, 345–346
  and heat deflection temperature, 326
"Softening" temperature, 629
"Soft" mode, 565
Software issues, 3
Software products, 8–9
Sol fraction, 442
Solid-mode experiments, 448–450
Solids, ideal elastic, 390–393
Solid $\rightarrow$ solid transition, 130
Solvent removal, 5
Spatial resolution, 620–624
  of thermal nanoprobes, 639–640
Specific heat capacity, 15, 53

Spectrometers, 506
  dielectric, 593–595
Spherulites, 88
Standard DSC pans, 34–35. *See also* Differential scanning calorimeters (DSCs)
Standards, DMA-related, 488–489
Starting point of melting, 103
Static force, 466
Static-mode technique, 367
Static stress, 468
Static stress–strain measurements, 469
Stationarity, 180
Steady state, 180, 182–183
Step heating technique, 543
Step isothermal experiments, 509
Stepped isotherm mode, 456, 457
Step scan DSC (SSDSC), 168–169, 204–205. *See also* Differential scanning calorimeters (DSCs)
Stepwise isothermal approach, 247
Stepwise isothermal crystallization, 110
Stereochemistry, influence on glass transition temperature, 344–345
Stoichiometrically balanced reaction, 147
Storage kinetics, 144
Storage modulus, 398, 401
  compliance limits and, 461
  frequency-dependent increase in, 450
  during gelation, 443
Storage modulus–cure time plot, 447
Straight (linear) baseline, 100–102
Strain rate, 393, 397
"Strength" parameter, 570
Stress effects, in the glass transition region, 186
Stress relaxation, 326, 395
Stress-relaxation behavior, 359
Stress relaxation curves, 405
Stress–strain behavior, 392
Stress–strain curves, 348–349
Strongly asymmetric blends, 576
Structural fingerprinting, 115. *See also* Fingerprinting
Structure formation, online monitoring of, 500
Styrene–acrylonitrile copolymer, DSC heating curves of, 79–80

Subambient sample thermocouple calibrations, 464–465
Submicrometer spatial resolution, 623–624
Submilligram samples, 158
Substrate effect, 336
Subsurface imaging, scanning thermal microscopy for, 618
Sub-T_g relaxations, 423. *See also* Glass transition temperature (T_g)
Superheating, 98, 102, 106, 117
Supermolecular structures, 88
Surface properties, in confining environments, 499–500
Surfaces, thermal probing of, 627–628
Swelling
  kinetics of, 354
  measuring using expansion-mode TMA, 354–356
Swelling ratio, 354
Swier, Steven, 7
Swollen systems, parameters for, 355–356
Symbols
  DEA-related, 559–601
  DMA-related, 489–491
  DSC-related, 225–228
  Micro/nano-TA-related, 647
  TGA-related, 312
  TMA-related, 380
Système International de Unités (SI), 9

T^* (magnitude of intermolecular interactions), estimating, 362
TA 2970 system, 593
TAI 2900 DSC modules, 176. *See also* Differential scanning calorimeters (DSCs); TA Instruments entries
  temperature modulation for, 174
TAI modules, 22–25
  910, 2910, and 2920 modules, 22–23
  MDSC phase-corrected signal comparison for, 177
  Q Series modules, 23–25
TA Instruments, Inc. (TAI), 221–223, 311, 481, 486, 520, 523
TA Instruments Q800 DMA, 481–482
Tan δ function, 506. *See also* Loss tangent (tan δ)
Tan δ peak, 412, 413

Tau lag, 50, 52
TAWN. *See* Dutch Society for Thermal Analysis (TAWN)
Temperature
  of the α relaxation, 18
  crystallization, 17
  in differential scanning calorimetry, 10–11
  of half-unfreezing, 60
  molecular fractionation by, 274–275
  in thermal analysis, 4
  in thermally activated processes, 277–278
  in thermogravimetric analysis, 251–256
Temperature accuracy, kinetic data and, 256–257
Temperature axis calibration approaches, 253–255
Temperature calibration
  of differential scanning calorimeters, 41–50
  of dynamic mechanical analyzers, 461–464
  in a micro/nanoscale local thermal analysis experiment, 634–636
  of thermogravimetric analyzers, 251–256
  in thermomechanical analysis, 332
  in a TMA instrument, 334–335
Temperature calibration graph, 632
Temperature Calibration of Thermomechanical Analyzers, 334
Temperature dependence, in step reaction thermosets versus photoinitiated free-radical polymerization, 160
Temperature gradients, 19
  in L-TA experiments, 635–636
Temperature-induced phase separation, 202
Temperature integral approximation, 284
Temperature measurements, accuracy of, 256, 461–464
Temperature-modulated differential scanning calorimetry (TMDSC), 169. *See also* Differential scanning calorimetry (DSC); Modulated temperature DSC (MTDSC)
Temperature ramp, 628

Temperature scales, 11
Temperature scans, 402–403, 509
Temperature sensor, 252
Temperature shift factor, 407
Temperature step increments, 457
Temperature–time/frequency analysis, 473–477
Temperature values, 11
  display of, 104
Tenacity, 432
Tensile mode, problems with, 466
Tensile-mode TMA study, 370–371. *See also* Thermomechanical analysis (TMA)
Tensile modulus, 322, 390
Tensile properties, measuring by thermomechanical analysis, 348–354
Tensile storage modulus, of polymeric films, 436–438
Tensile storage modulus curves, 437
Tensile stress–strain plot, 392
Tension geometry, 459
Tension mode, thermomechanical analysis in, 349, 350
Tension-mode experiments, 340
Tension probes, 333
Tertiary nucleation, 84
Test frequency, in measuring T_g, 415–417
Testing material quality, in dielectric experiments, 533
Testing modes, for DMA test samples, 466
Test samples, DMA, 465–472
Tetraglycidylmethylene dianaline (TGMDA), 135
T_g. *See* Glass transition temperature (T_g)
T_{g0} temperature, 133, 441–442
$T_{g\infty}$ temperature, 133, 441, 442
TGA curve, 260. *See also* Thermogravimetric analysis (TGA)
TGA/DSC, 4, 250, 254. *See also* Differential scanning calorimetry (DSC)
TGA/DTA, 4, 250, 254, 255. *See also* Differential thermal analysis (DTA)
TGA experiments
  choices associated with, 257
  designing and performing, 256–260

extent of reaction during, 266
isothermal, 266–269
objectives of, 256–257
pros and cons of, 301
TGA "fingerprint," 308. *See also* Fingerprinting
TGA/FTIR, 4. *See also* Fourier transform infrared (FTIR) spectroscopy
TGA instruments
heating rates of, 289
temperature calibration of, 255
TGA kinetic analysis, confidence in projected results of, 300
TGA kinetics software, 285
TGA/MS, 4. *See also* Mass spectroscopy (MS)
TGA multiple heating rate methodology, 305
TGA runs, heating rates used for, 292
TGA sample pans, 257, 258
TGA vaporization studies, 369–370
TGA/WVT capsule technique, 302–304. *See also* Water vapor transport (WVT)
T_g-conversion relationship, 138, 197, 198, 514
vitrification and, 142–144
T_g–degree of cure relationship, 440–441
T_g measurement, thermomechanical analysis in, 340–343
T_g relaxations, measuring in crystallizable polymers, 420–421
T_g-time data, 144
T_g values, DMA and DSC, 476–477
Thermal analysis (TA)
in combinations, 4
educational programs for, 1–2
of fibers, 115–123
general methods and techniques of, 2
micro- and nanoscale local, 615–649
studies in, 2
uses for, 1
Thermal analysis experiment, micro/nanoscale local, 633–636
Thermal analysis measurements, 1
goal of, 115
Thermal analysis techniques
for measuring, 1

underlying principles of, ix
Thermal analysis vendors,
DEA instruments, 592–599
DMA instruments, 477–488
DSC instruments, 217–225
Micro/nano-TA instruments, 646
TGA instruments, 308–311
TMA instruments, 347
Thermal Characterization of Polymeric Materials (Turi), x
Thermal conductivity, 20, 30, 203–204, 624–628
behavior of materials with, 627
images via, 618, 619–620
of purge gases, 252
Thermal contact, 289
Thermal cure shrinkage, 366–367
Thermal decomposition, 247
Thermal degradation, 293
initiation at weak link sites, 290
Thermal degradation kinetics, predicting, 287
Thermal diffusivity, 19, 203
Thermal environment, rapidly changing, 5
Thermal equilibrium, 12
Thermal events, timescale of, 180
Thermal expansion
determining the coefficient of, 340
of polymeric materials, 345
measurements of, 320
Thermal force–distance curves, 624–628
Thermal gradients, 464
Thermal history, reproducible, 32
Thermal lag, 19, 20, 21, 337
calibration of, 50–52
Thermal lag constant, 51–52
Thermally activated processes, rate of, 277–278
Thermally stimulated current (TSC), 517. *See also* TSC entries
Thermally stimulated current analysis, of thick PMMA films, 538–547
Thermally stimulated current peaks of polymers, effects of experimental parameters on, 542
Thermally stimulated current technique, 501, 526–528, 540
advantages of, 528

684   INDEX

Thermal nanoprobes
  spatial resolution of, 639–640
  temperature calibration for, 635
Thermal nucleation, 85
Thermal probes
  design of, 620–624
  heating of, 632–633
Thermal probe temperature, calibration of, 631–633
Thermal radiation, 13
Thermal resistance, 19–20, 25
Thermal sampling (TS), 528, 543, 546
Thermal scan mode, 456–457
Thermal scanning rates, 465
Thermal scans, 162
Thermal shrinkage, 350
Thermal stability
  estimating, 292
  in inert and oxidizing atmospheres, 296–297
  lifetime estimates and, 295–301
Thermal transitions, sensitive recording of, 528
Thermobalance(s)
  commercial, 244
  in thermogravimetric analysis, 242–243
Thermobalance components, thermal expansion of, 245
Thermocouple calibration, errors in, 462–463
Thermocouples, 4, 18, 22, 289, 328
  in DMA instruments, 461–462
Thermodilatometry (TD), 320. See also Thermomechanical analysis (TMA)
Thermodynamic C_p, 205
Thermodynamic parameters, change of, 94
Thermodynamics
  goal of, 9
  laws of, 9–10
Thermodynamic temperature scale, 11
ThermoFisher Scientific Materials Characterization, 487
Thermogravimetric analysis (TGA), ix, 1, 3, 241–317. See also TGA entries; Thermogravimetry (TG)
  applications of, 295–308

  background principles and measurement modes of, 242–250
  benefits of, 276–277
  calibration and reference materials for, 251–256
  controlled rate, 246–248
  of copolymers, 304–306
  factors affecting, 243–246
  information from, 250
  instrumentation for, 308–311
  kinetics in, 277–295
  measurements and analyses in, 256–277
  for measuring percent solids and pigment volume concentration, 276–277
  modulated, 248–249
  temperature axis calibration approaches in, 253–255
Thermogravimetric compositional analyses, 274–277
Thermogravimetry (TG), 241. See also Thermogravimetric analysis (TGA)
  standards for, 311–312
Thermomagnetometry (TM)
  temperature axis calibration method, 253
Thermomechanical analysis (TMA), ix, 1, 3, 319–386, 615. See also TMA entries
  expansion-mode, 354–356
  of films, fibers, and composites, 363–364
  of glass-reinforced thermoplastic, 377–379
  industrial applications of, 363–379
  instrumentation for, 326–332
  key applications of, 340–363
  in measuring softening point, 345–346
  measuring tensile properties by, 348–354
  modulated-temperature, 329–332
  physical quantities measured by, 325–326
  principles and theory of, 320–326
  shrinkage information from, 366–367
  standards for, 378–379
  of thermoplastic composite, 376

for thin bonded film measurements, 364–365
Thermometers, 11
  platinum resistance, 4
Thermooxidative degradation, 263
Thermooxidative stability, 264–265
Thermopiles, 26
Thermoplastic composite, 375–376
Thermoplastic polymers, 583
  structures of, 560–562
Thermoplastics
  with apolar mainchains and polar sidegroups, 556–559
  dielectric analysis of, 553–576
  DMA characterization for, 424–432
  glass-reinforced, 377–379
  hybrids and particulate systems incorporating, 572
  run parameters for, 177–179
Thermoplastic samples, 466
Thermoplastic/thermoset adhesive, yield from, 266
Thermoset coatings, ambient cure conditions of, 194
Thermoset cure
  chemical reactions of, 131–132
  extent of, 440–441
  model-free kinetics for, 153
Thermoset cure processes, 193
  temperatures in, 149
Thermosets, 130–154, 438
  dielectric analysis of, 576–592
  handling and processing of, 134
  MTDSC parameter selection for, 180
  nonisothermal cure to characterize, 194
  partially cured, 140
  samples of, 472
  uncured, 131
  use of MDSC for, 194–203
  versus thermoplastic polymers, 131
Thermosetting systems, 135
  heat of reaction of, 136
Thermospatial resolution, 622
Thermotropic LCPs, 563. *See also* Liquid crystalline polymers (LCPs)
Theta Industries, 311
Thick-film screen process, 26

Thick PMMA films. *See also* Poly(methyl methacrylate) (PMMA)
  DEA runs with, 547–553
  thermally stimulated current analysis of, 538–547
Thin bonded films, measurements on, 364–365
Thin-film (chip) calorimetry, 163
Thin films
  characteristics of, 432–438
  DMA measurements on, 467–469
  dynamic mechanical analysis data for, 436–438
4,4′-Thiobis(3-methyl-6-tert-butylphenol) determination of purity of, 39
Third law of thermodynamics, 10
Thomson–Gibbs equation, 98, 99, 107, 114–115
Three-point bending, 391
Three-point bending geometry, 459
Three-point bending mode, 466
Time, for describing dielectric phenomena, 503–504
Time constants, calorimeter, 25
Time-dependent discharge current, 536
Time-dependent phenomena/processes, 5–6
  during polymer melting, 104–106
Time-dependent stress relaxation modulus, 395–396
Time-domain dielectric spectroscopy
  fast, 525–526
  slow, 520–521
Time-domain reflectometry (TDR), 525–526
Timescale, in TSC experiments, 527
Time–temperature equivalence principle, 402, 404, 409
Time–temperature–frequency relationship, 401–409
Time–temperature superposition (TTS), 409
  modified, 406–407
Time–temperature superposition kinetics, 145, 148–150
Time–temperature–superposition master curves, 275, 362, 404
Time–temperature superposition nomograph, 473–476

Time–temperature superposition principle, 359, 361, 398, 403–409, 421, 511
   significance of, 409
Time–temperature superposition software, 473
Time–temperature transformation cure diagram, 441–456
Time-temperature-transformation (TTT) diagram, 197
Time to steady state, 21
TMA curves, 321, 322. *See also* Thermomechanical analysis (TMA)
   interpretation of, 364
   in machine and transverse directions, 352–353
TMA experiments
   applying initial load to sample in, 339–340
   performing, 335–340
   preparations for, 335–336
TMA instruments, 330
   calibration of, 332–335
   expansion response of, 335
   vertical-design, 327–328
TMA penetration measurements, 360
TMA probes/applications, 333
TMA sample cell, 336
TMA shrinkage measurements, 369
TMA specifications, 330
Top-bottom (T-B) signal, 617
TOPEM®, 168
   software, 27
   technique, 174–175
Topographic artifacts, 619
Torsional (shear) mode, 466
Torsional shear geometry, 459
Torsion pendulum technique, 417
Total heat flow, 170
Toughness, 348–349
Traditional amorphous fraction, 71
Traditional amorphous phase, glass transition of, 71
Traditional DSC, measurement of heat capacity with, 54–57. *See also* Differential scanning calorimetry (DSC)
Training, thermal analysis, 2

Transition region, 395
   for linear amorphous polymers, 413
Transition temperature, 253
Triclinic → hexagonal transition, 130
Trouton's rule, 94
True differential heat flow, 31
True reference temperature, 30
TSC experiments, global, 536–538. *See also* Thermally stimulated current entries
TSC peaks, complex, 543–547
TSC spectra, 565, 566
   analysis of, 538
   global, 538–540
TS scans, 543
Tsukruk, Vladimir V., 615
Turi, Edith, x
Turnkey dielectric/impedance spectrometers, 593
Twisted nematic crystals, 47
Two-part thermoset systems, 135
Type A/B/C polymers, 512
Tzero calibration, 52
Tzero cell design, 222
Tzero modes, 192
Tzero sensor temperature, 43
Tzero technology, 23, 24
Tzero thermocouple, 24

Ultrathin films, dielectric methods for, 500
Ultrathin polymer films, glass transitions for, 637
Ultraviolet (UV) radiation. *See* UV entries
Unconstrained measurement, 115–117
Uncured thermosets, 131
Underlying heating rates, 178–179
Underlying rate, 168
Undrawn Nylon 6 films, 126, 129
Uniaxial orientation, 124
Unidirectional composites, 470, 471
Uniform temperature distribution, 20
Universal values, 361
"Universal" WLF equation, 407
Unoriented films, DSC experiments on, 124
"Unrelaxed" permittivity, 503

UV (ultraviolet)-curable materials, 154–155
UV cure, of free-radical systems, 159–160
UV exposure cycles, multiple, 162
UV irradiation, 157
UV-stabilized polycarbonate, 369–370

van't Hoff calorimetric purity method, 39
van't Hoff equation, 37
van't Hoff plot, 38, 39
Variable-amplitude phase generator, 523
Variable-heating-rate experiments, 123
Vassilikou-Dova, Aglaia, 497
Vectra®, 121–123
Vectra group, relaxations of, 564
Vectran®, 123
Vectran film, 128
  as-extruded, 129
  stress–strain curves of, 353
Vectran liquid crystalline film, tensile storage modulus of, 436, 438
Veeco, 646
Vendors
  commercial DMA instrumentation, 477–488
  DMA instruments from, 483–485
  thermal analysis, 478–483
Vesta system, 646
VICAT softening point, 345
Vinyl monomer-containing polymer (VACR), 368, 369
Viscoelastic behavior, characterization of, 394–401
Viscoelastic characterization, 390
Viscoelastic damping devices, 476
Viscoelastic data, time–temperature superposition of, 473
Viscoelasticity, 324, 357, 360
  of polymers, 319–320
Viscoelastic materials, 394
Viscoelastic transitions, 410–411
Viscosity
  measuring, 444
  Newton's law of, 393, 397
  relationship with ionic conductivity, 588

Vitrification, 132–134, 149, 153, 442, 453, 454
  absence of, 455
  of a high-T_g phase, 200–201
  during nonisothermal cure of epoxy–amine system, 198–199
  T_g-conversion relationship and, 144
Vitrification curve, S-shaped, 441
Vitrification time, 448
Vogel–Tammann–Fulcher–Hesse (VTFH) equation, 511, 551
Vogel–Tammann–Fulcher–Hesse relaxation time function, 519, 533
Vogel temperature, 511
Volatile cure products, 266
Volatile products. *See also* Volatiles
  condensation of, 245
  in TGA experiments, 261–262
Volatile reaction products, loss of, 271–274
Volatiles
  containment of, 135–136
  loss of, 158
Volume change, on melting, 94
Vyazovkin, Sergey, 241

"Waiting time," 534
Warping, in integrated circuit packages, 374–375
Water
  as an antiplasticizer for nylon, 430–432
  as a melting point standard, 44
  permeability of, 302–304
Water absorption, real-time dielectric measurements of, 592
Water molecules, dielectric signatures of, 591–592
Water–polymer interactions, 356
Water vapor transport (WVT), 301–304
Water vapor transport capsule, 302
Well-resolved thermal transitions, sensitive recording of, 528
Wide-angle X-ray diffraction (WAXD), 99
Williams–Landel–Ferry (WLF) equation, 323–324, 407–408, 510–511. *See also* Modified WLF equation; WLF entries

Windowing polarization (WP), 528
Winter method, 443
WinTSC program, 598
Wire mesh method/technique, 443–444, 449
WLF parameters, 510–511. *See also* Williams–Landel–Ferry (WLF) equation
WLF relationships, nonlinear, 417
WLF shift factor curve, 408
Wollaston probes, 646
  geometry of, 623
Wollaston Wire probe, 620–621, 622
  thermospatial resolution expected from, 623
Wunderlich, Bernhard, x
Wunderlich's rule, 61

X-ray photoelectron spectrometry (XPS), 644

m-Xylylene diamine-cured DGEBA, 583

Yarns, 432
Young's modulus, 322, 326, 390
  measuring by thermomechanical analysis, 348

z-axis feedback, 636
Zero-entropy-production melting, 82, 106–107
Zero-entropy-production melting points, 98, 102, 107
Zero heating rate, 41
01dB-Metravib Instruments, 484–485
Zeroth law of thermodynamics, 9, 10–11
Zone drawing–zone annealing method, 433
"Zoom" imaging facility, 640